P9-CKU-646

THE LARGE-SCALE CHARACTERISTICS OF THE GALAXY

INTERNATIONAL ASTRONOMICAL UNION
UNION ASTRONOMIQUE INTERNATIONALE

SYMPOSIUM No. 84

HELD IN COLLEGE PARK, MARYLAND, U.S.A.
12–17 JUNE, 1978

THE LARGE-SCALE CHARACTERISTICS OF THE GALAXY

EDITED BY

W. B. BURTON
Department of Astronomy
University of Minnesota

D. REIDEL PUBLISHING COMPANY
DORDRECHT : HOLLAND / BOSTON : U.S.A. / LONDON : ENGLAND

Library of Congress Cataloging in Publication Data

Main entry under title:

The large-scale characteristics of the galaxy.

(International Astronomical Union Symposium; no. 84)
Symposium jointly sponsored by IAU Commission 33, Structure and dynamics of the galactic system, and by Commission 34, Interstellar matter and planetary nebulae.
Includes index.
1. Milkyway–congresses. I. Burton, William Butler, 1940–
II. International Astronomical Union. Commission 33. III. International Astronomical Union. Commission 34. IV. Series; International Astronomical Union. Symposium; no. 84.
QB857.7.L37 523′.12 79-19483
ISBN 90-277-1029-5
ISBN 90-277-1030-9 pbk.

Published on behalf of
the International Astronomical Union
by
D. Reidel Publishing Company, P.O. Box 17, Dordrecht, Holland

Sold and distributed in the U.S.A., Canada, and Mexico
by D. Reidel Publishing Company, Inc.
Lincoln Building, 160 Old Derby Street, Hingham,
Mass. 02043, U.S.A.

TABLE OF CONTENTS

IV. GALACTIC KINEMATICS AND DISTANCES

V. PHYSICAL PROPERTIES OF THE INTERSTELLAR MEDIUM

VI. THE GALACTIC NUCLEUS

VII. COMPARISONS OF OUR GALAXY WITH OTHER GALAXIES

VIII. THE SPHEROIDAL COMPONENT

IX. THE GALACTIC WARP

FOREWORD

The International Astronomical Union has encouraged the study of our galaxy through a series of symposia. This volume contains the proceedings of IAU Symposium No. 84 on the Large-Scale Characteristics of the Galaxy, held in College Park, Maryland, from June 12 to 17, 1978.

Symposium No. 84 was jointly sponsored by IAU Commission 33, Structure and Dynamics of the Galactic System, and by Commission 34, Interstellar Matter and Planetary Nebulae. The Scientific Organizing Committee consisted of F. J. Kerr (chairman), B. J. Bok, W. B. Burton, J. Einasto, K. C. Freeman, P. O. Lindblad, D. Lynden-Bell, R. Sancisi, S. E. Strom, H. van Woerden, and R. Wielen. The topics and speakers were chosen in order to emphasize current observational material and theoretical results pertaining to various morphological aspects of our galaxy. In preparing the program particular care was taken to relate recent work on other galaxies to the situation in our own galaxy.

The meetings were held in the Center for Adult Education on the campus of the University of Maryland. The Local Organizing Committee consisted of B. M. Zuckerman (chairman), A. P. Henderson, P. D. Jackson, T. A. Matthews, B. F. Perry, V. C. Rubin, P. R. Schwartz, F. W. Stecker, J. D. Trasco, and G. Westerhout. Joan Ball ably assisted this committee. The National Science Foundation made a financial contribution to the general support of the meeting. The National Radio Astronomy Observatory assisted with some of the participants' travel expenses.

During the editing of these proceedings for the press I have been helped considerably by several people. At the meeting, Joan Ball initially organized the preparation of the discussion remarks; subsequently at the NRAO Phyllis Jackson retyped the discussion and a number of the contributions, and M. B. Weems and Patricia Smiley prepared the typescript format; at the University of Minnesota, Iona Quesnell helped with the editing of the material into its present form. I am very grateful to each of these people for their help.

W. B. Burton

Minneapolis
March, 1979

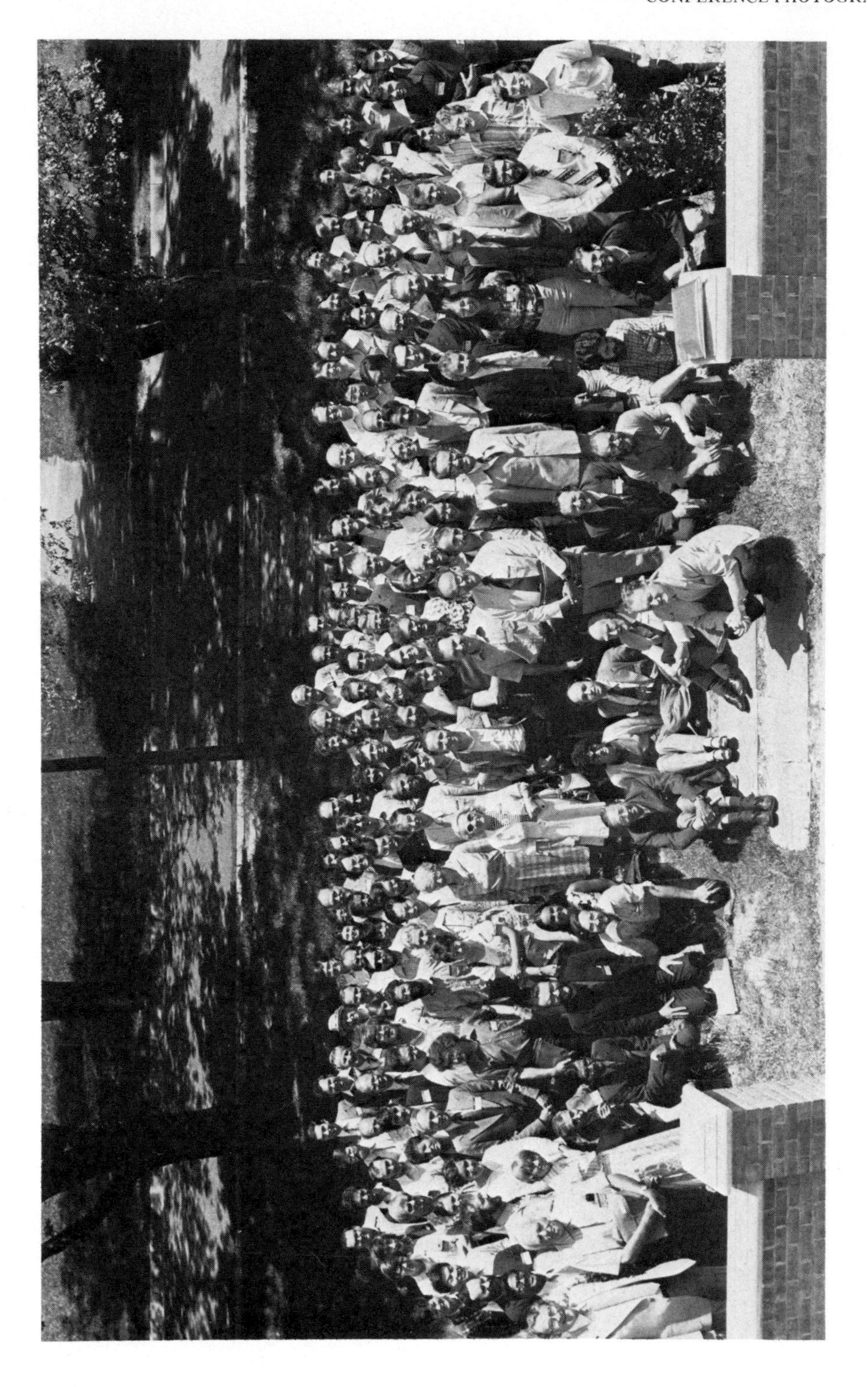

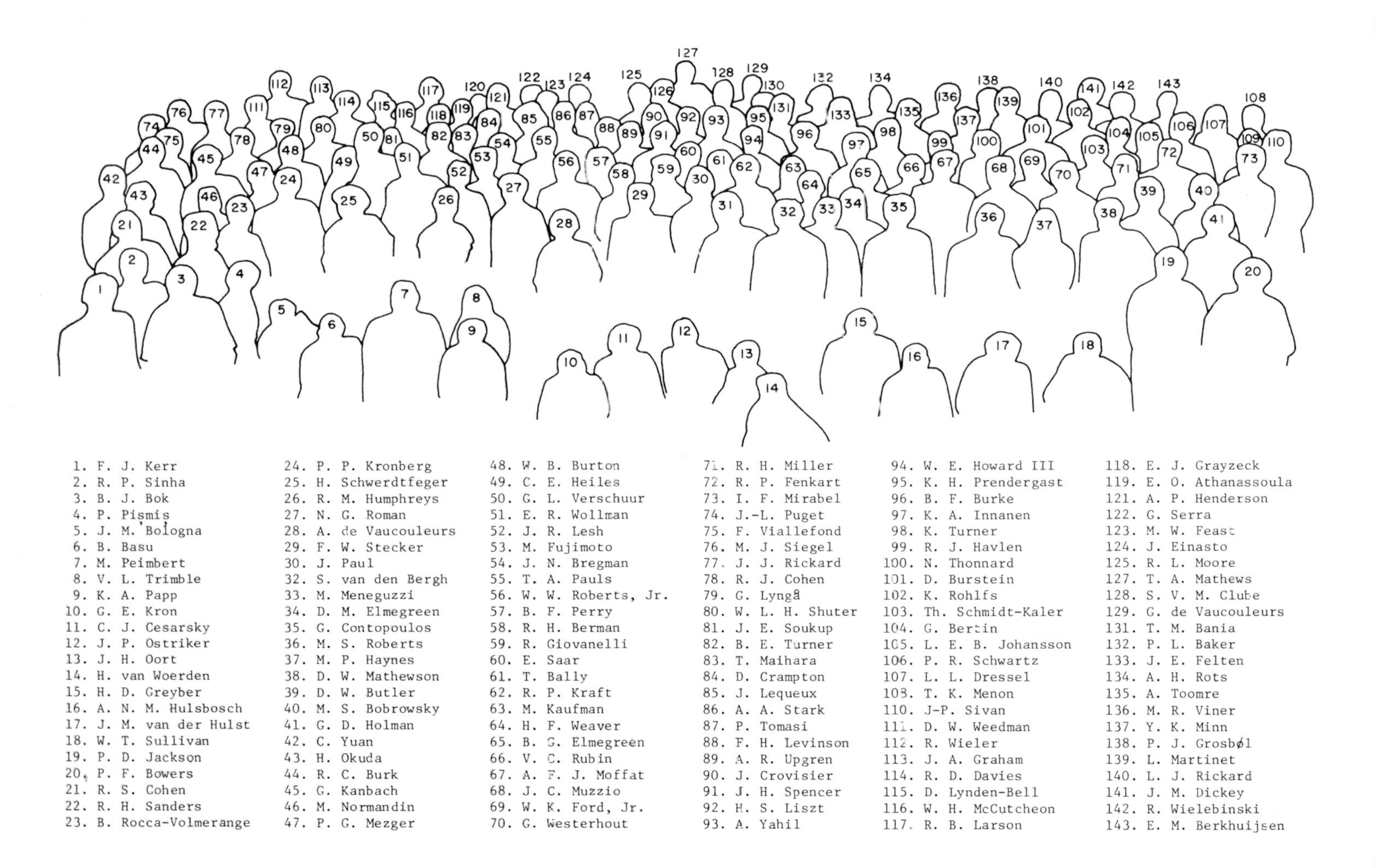

1. F. J. Kerr
2. R. P. Sinha
3. B. J. Bok
4. P. Pişmiş
5. J. M. Bologna
6. B. Basu
7. M. Peimbert
8. V. L. Trimble
9. K. A. Papp
10. G. E. Kron
11. C. J. Cesarsky
12. J. P. Ostriker
13. J. H. Oort
14. H. van Woerden
15. H. D. Greyber
16. A. N. M. Hulsbosch
17. J. M. van der Hulst
18. W. T. Sullivan
19. P. D. Jackson
20. P. F. Bowers
21. R. S. Cohen
22. R. H. Sanders
23. B. Rocca-Volmerange
24. P. P. Kronberg
25. H. Schwerdtfeger
26. R. M. Humphreys
27. N. G. Roman
28. A. de Vaucouleurs
29. F. W. Stecker
30. J. Paul
32. S. van den Bergh
33. M. Meneguzzi
34. D. M. Elmegreen
35. G. Contopoulos
36. M. S. Roberts
37. M. P. Haynes
38. D. W. Mathewson
39. D. W. Butler
40. M. S. Bobrowsky
41. G. D. Holman
42. C. Yuan
43. H. Okuda
44. R. C. Burk
45. G. Kanbach
46. M. Normandin
47. P. G. Mezger
48. W. B. Burton
49. C. E. Heiles
50. G. L. Verschuur
51. E. R. Wollman
52. J. R. Lesh
53. M. Fujimoto
54. J. N. Bregman
55. T. A. Pauls
56. W. W. Roberts, Jr.
57. B. F. Perry
58. R. H. Berman
59. R. Giovanelli
60. E. Saar
61. T. Bally
62. R. P. Kraft
63. M. Kaufman
64. H. F. Weaver
65. B. G. Elmegreen
66. V. C. Rubin
67. A. F. J. Moffat
68. J. C. Muzzio
69. W. K. Ford, Jr.
70. G. Westerhout
71. R. H. Miller
72. R. P. Fenkart
73. I. F. Mirabel
74. J.-L. Puget
75. F. Viallefond
76. M. J. Siegel
77. J. J. Rickard
78. R. J. Cohen
79. G. Lyngå
80. W. L. H. Shuter
81. J. E. Soukup
82. B. E. Turner
83. T. Maihara
84. D. Crampton
85. J. Lequeux
86. A. A. Stark
87. P. Tomasi
88. F. H. Levinson
89. A. R. Upgren
90. J. Crovisier
91. J. H. Spencer
92. H. S. Liszt
93. A. Yahil
94. W. E. Howard III
95. K. H. Prendergast
96. B. F. Burke
97. K. A. Innanen
98. K. Turner
99. R. J. Havlen
100. N. Thonnard
101. D. Burstein
102. K. Rohlfs
103. Th. Schmidt-Kaler
104. G. Bertin
105. L. E. B. Johansson
106. P. R. Schwartz
107. L. L. Dressel
108. T. K. Menon
110. J-P. Sivan
111. D. W. Weedman
112. R. Wieler
113. J. A. Graham
114. R. D. Davies
115. D. Lynden-Bell
116. W. H. McCutcheon
117. R. B. Larson
118. E. J. Grayzeck
119. E. O. Athanassoula
121. A. P. Henderson
122. G. Serra
123. M. W. Feast
124. J. Einasto
125. R. L. Moore
127. T. A. Mathews
128. S. V. M. Clube
129. G. de Vaucouleurs
131. T. M. Bania
132. P. L. Baker
133. J. E. Felten
134. A. H. Rots
135. A. Toomre
136. M. R. Viner
137. Y. K. Minn
138. P. J. Grosbøl
139. L. Martinet
140. L. J. Rickard
141. J. M. Dickey
142. R. Wielebinski
143. E. M. Berkhuijsen

LIST OF PARTICIPANTS

G. E. Assousa, Carnegie Institution of Washington, Washington, DC, USA
E. O. Athanassoula, Observatoire de Besançon, Besançon, France
A. F. Aveni, Colgate University, New York, NY, USA
P. L. Baker, Max-Planck-Institut für Radioastronomie, Bonn, FRG
D. L. Ball, University of Maryland, College Park, MD, USA
T. Bally, University of Massachusetts, Amherst, MA, USA
T. M. Bania, National Astronomy and Ionosphere Center, Arecibo, PR
F. N. Bash, University of Texas, Austin, TX, USA
B. Basu, University of Calcutta, Calcutta, India
E. M. Berkhuijsen, Max-Planck-Institut für Radioastronomie, Bonn, FRG
R. H. Berman, University of Reading, Reading, UK
G. Bertin, Scuola Normale Superiore, Pisa, Italy
B. J. Bok, Steward Obs., Tucson, AZ, USA
J. M. Bologna, U.S. Naval Research Laboratory, Washington, DC, USA
P. F. Bowers, National Radio Astronomy Obs., Charlottesville, VA, USA
J. N. Bregman, Columbia University, New York, NY, USA
B. F. Burke, Massachusetts Institute of Technology, Cambridge, MA, USA
D. Burstein, Carnegie Institution of Washington, Washington, DC, USA
W. B. Burton, National Radio Astronomy Obs., Charlottesville, VA, USA
D. S. Butler, Yale University, New Haven, CT, USA
C. J. Cesarsky, Centre d'Etudes Nucléaires de Saclay, Gif-sur-Yvette, France
S. V. M. Clube, Royal Observatory, Edinburgh, UK
R. J. Cohen, Nuffield Radio Astronomy Laboratories, Jodrell Bank, UK
R. S. Cohen, Institute for Space Studies, New York, NY, USA
G. Contopoulos, University of Athens, Athens, Greece
D. Crampton, Dominion Astrophysical Obs., Victoria, BC, Canada
J. Crovisier, Observatoire de Paris, Meudon, France
T. M. Dame, Columbia University, New York, NY, USA
R. D. Davies, Nuffield Radio Astronomy Laboratories, Jodrell Bank, UK
A. de Vaucouleurs, University of Texas, Austin, TX, USA
G. de Vaucouleurs, University of Texas, Austin, TX, USA
J. M. Dickey, University of Massachusetts, Amherst, MA, USA
D. Downes, Max-Planck-Institut für Radioastronomie, Bonn, FRG
A. J. Dragt, University of Maryland, College Park, MD, USA
L. Dressel, University of Virginia, Charlottesville, VA, USA
J. Einasto, W. Struve Astrophysical Obs., Tôravere, Estonia, USSR
B. G. Elmegreen, Harvard College Obs., Cambridge, MA, USA
D. M. Elmegreen, Harvard College Obs., Cambridge, MA, USA
M. W. Feast, South African Astronomical Obs., Cape, South Africa
J. E. Felten, NASA Goddard Space Flight Center, Greenbelt, MD, and University of Maryland, College Park, MD, USA

R. P. Fenkart, University of Basel, Binningen, Switzerland
C. E. Fichtel, NASA Goddard Space Flight Center, Greenbelt, MD, USA
G. B. Field, Harvard College Obs., Cambridge, MA, USA
W. K. Ford, Jr., Carnegie Institution of Washington, Washington, DC, USA
M. Fujimoto, Nagoya University, Nagoya, Japan
R. Giovanelli, National Astronomy and Ionosphere Center, Arecibo, PR
J. A. Graham, Cerro Tololo Interamerican Obs., La Serena, Chile
E. J. Grayzeck, University of Nevada, Las Vegas, NV, USA
H. D. Greyber, Potomac, MD, USA
P. J. Grosbøl, Copenhagen University Obs., Copenhagen, Denmark
J. P. Harrington, University of Maryland, College Park, MD, USA
R. J. Havlen, University of Virginia, Charlottesville, VA, USA
M. P. Haynes, National Radio Astronomy Obs., Charlottesville, VA, USA
C. E. Heiles, University of California, Berkeley, CA, USA
A. P. Henderson, University of Maryland, College Park, MD, and
Manhattan College, New York, NY, USA
W. Herbst, Carnegie Institution of Washington, Washington, DC, USA
R. W. Hobbs, NASA Goddard Space Flight Center, Greenbelt, MD, USA
G. D. Holman, University of Maryland, College Park, MD, USA
W. E. Howard, III, National Science Foundation, Washington, DC, USA
A. N. M. Hulsbosch, Catholic University, Nijmegen, The Netherlands
R. M. Humphreys, University of Minnesota, Minneapolis, MN, USA
K. A. Innanen, York University, Toronto, Ont., Canada
P. D. Jackson, University of Maryland, College Park, MD, USA
L. E. B. Johansson, Onsala Space Obs., Onsala, Sweden
D. H. P. Jones, Royal Greenwich Obs., Hailsham, Sussex, UK
F. C. Jones, NASA Goddard Space Flight Center, Greenbelt, MD, USA
G. Kanbach, Max-Planck-Institut für Physik und Astrophysik, München, FRG
M. Kaufman, Ohio State University, Columbus, OH, USA
F. J. Kerr, University of Maryland, College Park, MD, USA
M. Kesteven, Queen's University, Kingston, Ont., Canada
G. R. Knapp, California Institute of Technology, Pasadena, CA, USA
D. A. Kniffen, NASA Goddard Space Flight Center, Greenbelt, MD, USA
S. H. Knowles, U.S. Naval Research Laboratory, Washington, DC, USA
R. P. Kraft, University of California, Santa Cruz, CA, USA
G. E. Kron, Pinecrest Obs., Flagstaff, AZ, USA
P. P. Kronberg, University of Toronto, West Hill, Ont., Canada
R. B. Larson, Yale University, New Haven, CT, USA
J. Lequeux, Observatoire de Paris, Meudon, France
J. R. Lesh, NASA Goddard Space Flight Center, Greenbelt, MD, USA
F. H. Levinson, University of Virginia, Charlottesville, VA, USA
C. C. Lin, Massachusetts Institute of Technology, Cambridge, MA, USA
H. S. Liszt, National Radio Astronomy Obs., Charlottesville, VA, USA
F. J. Lockman, Carnegie Institution of Washington, Washington, DC, USA
R. V. E. Lovelace, Cornell University, Ithaca, NY, USA
D. Lynden-Bell, The Observatories, Cambridge, UK
G. Lyngå, Lund Obs., Lund, Sweden
T. Maihara, Kyoto University, Kyoto, Japan
J. W-K. Mark, Massachusetts Institute of Technology, Cambridge, MA, USA
L. Martinet, Observatoire de Genève, Sauverny, Switzerland
G. M. Mason, University of Maryland, College Park, MD, USA

D. S. Mathewson, Mount Stromlo Obs., Canberra, ACT, Australia
T. A. Matthews, University of Maryland, College Park, MD, USA
M. Maucherat-Joubert, Laboratoire d'Astronomie Spatiale, Marseille, France
W. H. McCutcheon, University of British Columbia, Vancouver, BC, Canada
M. Meneguzzi, Princeton University Obs., Princeton, NJ, USA
T. K. Menon, Tata Institute of Fundamental Research, Bombay, India
P. G. Mezger, Max-Planck-Institut für Radioastronomie, Bonn, FRG
R. H. Miller, University of Chicago, Chicago, IL, USA
Y. K. Minn, National Astronomical Obs., Seoul, Korea
I. F. Mirabel, Nuffield Radio Astronomy Laboratories, Jodrell Bank, UK
A. F. J. Moffat, Université de Montréal, Montréal, PQ, Canada
T. Montmerle, Centre d'Etudes Nucléaires de Saclay, Gif-sur-Yvette, France
R. L. Moore, Steward Obs., Tucson, AZ, USA
D. C. Morton, Anglo-Australian Obs., Epping, NSW, Australia
J. C. Muzzio, Observatorio Astronómico, La Plata, Argentina
P. C. Myers, Massachusetts Institute of Technology, Cambridge, MA, USA
M. Normandin, University of Toronto, West Hill, Ont., Canada
H. Okuda, Kyoto University, Kyoto, Japan
J. H. Oort, Sterrewacht, Leiden, The Netherlands
J. P. Ostriker, Princeton University Obs., Princeton, NJ, USA
J. Paul, Centre d'Etudes Nucléaires de Saclay, Gif-sur-Yvette, France
T. A. Pauls, Max-Planck-Institut für Radioastronomie, Bonn, FRG
B. F. Peery, Howard University, Washington, DC, USA
M. Peimbert, Universidad Nacional Autonoma de Mexico, Ciudad Universitaria, Mexico
P. Pişmiş, Universidad Nacional Autonoma de Mexico, Ciudad Universitaria, Mexico
K. H. Prendergast, Columbia University, New York, NY, USA
J.-L. Puget, Centre National de la Recherche Scientifique, Paris, France
J. J. Rickard, University of Iowa, Iowa City, IA, USA
L. J. Rickard, National Radio Astronomy Obs., Charlottesville, VA, USA
M. S. Roberts, National Radio Astronomy Obs., Charlottesville, VA, USA
W. W. Roberts, University of Virginia, Charlottesville, VA, USA
B. Rocca-Volmerange, Centre National de la Recherche Scientifique, Paris, France
K. Rohlfs, Ruhr-Universität Bochum, Bochum, FRG
N. G. Roman, NASA Astrophysics Division, Washington, DC, USA
A. H. Rots, Radiosterrenwacht, Dwingeloo, The Netherlands
V. C. Rubin, Carnegie Institution of Washington, Washington, DC, USA
E. Saar, W. Struve Astrophysical Obs., Tôravere, Estonia, USSR
E. E. Salpeter, Cornell University, Ithaca, NY, USA
D. B. Sanders, State University of New York, Stony Brook, NY, USA
R. H. Sanders, Kapteyn Astronomical Laboratory, Groningen, The Netherlands
Th. Schmidt-Kaler, Ruhr-Universität Bochum, Bochum, FRG
P. R. Schwartz, U.S. Naval Research Laboratory, Washington, DC, USA
H. Schwerdtfeger, Astronomisches Rechen-Institut, Heidelberg, FRG
N. Z. Scoville, University of Massachusetts, Amherst, MA, USA
P. O. Seitzer, University of Virginia, Charlottesville, VA, USA
G. Serra, Centre d'Etudes Spatiales et des Rayonnements, Toulouse, France
W. L. Shuter, University of British Columbia, Vancouver, BC, Canada
S. C. Simonson, III, Greenbelt, MD, USA

R. P. Sinha, University of Maryland, College Park, MD, USA
J.-P. Sivan, Laboratoire d'Astronomie Spatiale, Marseille, France
E. v. P. Smith, University of Maryland, College Park, MD, USA
P. M. Solomon, State University of New York, Stony Brook, NY, USA
J. E. Soukup, The City College of New York, New York, NY, USA
J. H. Spencer, U.S. Naval Research Laboratory, Washington, DC, USA
A. A. Stark, Princeton University Obs., Princeton, NJ, USA
F. W. Stecker, NASA Goddard Space Flight Center, Greenbelt, MD, USA
S. E. Strom, Kitt Peak National Obs., Tucson, AZ, USA
G. Strongylis, NASA Goddard Space Flight Center, Greenbelt, MD, USA
W. T. Sullivan, III, University of Washington, Seattle, WA, USA
J. H. Taylor, Jr., University of Massachusetts, Amherst, MA, USA
Y. Terzian, Cornell University, Ithaca, NY, USA
D. J. Thompson, NASA Goddard Space Flight Center, Greenbelt, MD, USA
N. Thonnard, Carnegie Institution of Washington, Washington, DC, USA
B. M. Tinsley, Yale University, New Haven, CT, USA
P. Tomasi, Instituto di Fisica, Bologna, Italy
A. Toomre, Massachusetts Institute of Technology, Cambridge, MA, USA
J. D. Trasco, University of Maryland, College Park, MD, USA
V. L. Trimble, University of California, Irvine, CA, and
University of Maryland, College Park, MD, USA
B. E. Turner, National Radio Astronomy Obs., Charlottesville, VA, USA
K. C. Turner, Carnegie Institution of Washington, Washington, DC, USA
A. R. Upgren, Van Vleck Obs., Middletown, CT, USA
S. van den Bergh, Dominion Astrophysical Obs., Victoria, BC, Canada
J. M. van der Hulst, National Radio Astronomy Obs., Charlottesville,
VA, USA
H. van Woerden, Kapteyn Astronomical Laboratory, Groningen,
The Netherlands
G. L. Verschuur, University of Colorado, Boulder, CO, USA
F. Viallefond, Kapteyn Astronomical Laboratory, Groningen,
The Netherlands
M. R. Viner, Queen's University, Kingston, Ont., Canada
J. A. Waak, U.S. Naval Research Laboratory, Washington, DC, USA
H. F. Weaver, University of California, Berkeley, CA, USA
D. W. Weedman, Vanderbilt University, Nashville, TN, USA
G. Westerhout, U.S. Naval Observatory, Washington, DC, USA
R. Wielebinski, Max-Planck-Institut für Radioastronomie, Bonn, FRG
R. Wielen, Astronomisches Rechen-Institut, Heidelberg, FRG
A. S. Wilson, University of Maryland, College Park, MD, USA
E. R. Wollman, Kitt Peak National Obs., Tucson, AZ, USA
P. R. Woodward, Lawrence Livermore Laboratory, Livermore, CA, USA
J. P. Wright, National Science Foundation, Washington, DC, USA
A. Yahil, State University of New York, Stony Brook, NY, USA
C. Yuan, City College of New York, New York, NY, USA
B. M. Zuckerman, University of Maryland, College Park, MD, USA

I. PERSPECTIVE

LARGE-SCALE STRUCTURE OF SPIRAL GALAXIES: PROBLEMS OLD AND NEW

Morton S. Roberts
National Radio Astronomy Observatory,* Green Bank, W.Va. U.S.A.

Introductory remarks for IAU Symposium No. 84 are made. These include a brief history of previous symposia dealing with galactic structure, basic research problems in this area, and parallels regarding our galaxy which can be drawn from extragalactic research.

Twenty-five years ago, almost to the day, Symposium No. 1 of the International Astronomical Union convened at Groningen in the Netherlands. The subject matter was coordination of galactic research; the bulk of the discussion was devoted to optical studies. Topics such as radio spectral line work, so prominent in today's work, were barely touched upon. This particular subject was in its infancy; the 21-cm line of neutral hydrogen had first been observed only two years previous to that Symposium. The discovery of the OH radical lay 10 years in the future. Radio spectroscopy is but one example of the remarkable increase in the scope of galactic studies and of all areas of astronomical research that has occurred in the intervening quarter of a century.

At this Symposium we shall hear of results derived from essentially the entire electromagnetic spectrum; from γ-rays to decametric radio waves, covering a span of fifteen orders of magnitude in wavelength and in photon energy. The emphasis will be the opposite of the meeting 25 years ago, for surprisingly little will be presented on optical studies of our own galaxy. The reasons for this change are obvious; the greater wavelength coverage yields extensive new information on different constituents of our galaxy both thermal and nonthermal. Of equal import is our ability to see much further into our galaxy. The peculiar circumstance (or is it just selection?) of a galaxy filled with dust particles whose size is comparable to the wavelength of visible light restricts our view to a distance of just a few kiloparsecs in the galactic plane. The corresponding area that can be surveyed is a few percent of this plane and the information so gathered, though interesting and detailed, is only of the local structure. The prominent spiral arms,

* Operated by Associated Universities, Inc., under contract with the National Science Foundation.

W. B. Burton (ed.), The Large-Scale Characteristics of the Galaxy, 3–7.

located some 5 kpc from the center, as well as the nuclear region, features that define other galaxies, are not visible to our narrow band eyes; but they are prominent at other wavelengths. Out of the plane we are far less restricted by dust but are hindered by a fainter population of stars so that detailed studies are again limited to comparable distances.

It is traditional in an introductory talk such as this to list the significant problems of research peculiar to the topic of the Symposium. It is a somewhat strange custom, in that the participants have already assembled and either are prepared to talk about recent, often puzzling observations posing even newer problems, or else they wish to present solutions to problems and questions they themselves have posed. I do not mean to imply that these observations and solutions are not significant but rather that the two lists may be uncoupled.

Instead of presenting my own list, I prefer again to look back, this time 15 years to the Canberra Symposium on "The Galaxy and the Magellanic Clouds". At that time Professor Oort summarized, under the heading Structure of the Galaxy, the relevant areas of research and the associated problems. Here is his list, somewhat recast into a series of questions:

1. What is the rotation curve for our galaxy?
2. Why is the HI distribution warped in the outer regions?
3. How may we explain the general problem of the origin and conservation of spiral structure?
4. How might we understand the large scale HI distribution in our galaxy?
5. What conditions give rise to the complex observations of the central region of our galaxy?
6. How do we explain the high-velocity gas that is observed?
7. What is the strength and distribution of the galactic magnetic field?
8. What is the distribution and origin of the thermal and non-thermal components of the general galactic background?
9. What is our distance from the galactic center, and what are the components of our motion with respect to the center?
10. What is the mass density near the sun and how is this mass distributed among the various constituent components?
11. What is the density and velocity distribution of old disk objects and of halo objects of different metal abundances?
12. What can we say of the origin and evolution of our galaxy?

I think we have made progress in most, if not all, of these areas. If the questions are not firmly answered we at least have a better understanding of them with, in a number of instances, viable and competing theories. How well any special problem, or all of them, are understood and answerable today, I leave to you to decide, preferably at the end of this Symposium.

Let me instead turn to galaxies other than our own, to systems which we are able to study from outside their confines. By finding common, large-scale properties of other spiral systems we can be guided in describing and understanding our own galaxy. Turning to other spiral systems for such guidance dates back to 1852 when Stephen Alexander suggested that our Milky Way had spiral arms. This approach was further elaborated on at the turn of the century by Cornelis Easton. We now know that the features they describe were local--that an overall view was limited by interstellar extinction. An inauspicious beginning whose lesson is caution. But there are similarities that we do recognize and with the proper caution we may well draw other parallels.

In spiral galaxies, the distribution of light is centrally concentrated, in marked contrast with that of the neutral hydrogen which lacks such a concentration and often shows a pronounced central minimum. This feature in the HI distribution in our galaxy was first described by Hugo van Woerden. There are a number of explanations, including efficient star formation in this region of a galaxy. The parallel between the nuclear bulge and elliptical galaxies, the latter also being poor in neutral hydrogen, is often made. That the interstellar gas may be in some form other than HI is also a viable suggestion, one prompted by the abundance of observed molecules and inferred presence of significant amounts of molecular hydrogen in the galactic nuclear region.

In the outer part of a galaxy the generally planar HI distribution is often found to be warped or bent. So often, in fact, that it appears that an undistorted plane is the exception to the rule. In our galaxy such a warp was recognized by Frank Kerr and independently by Bernard Burke over 20 years ago. Several suggestions as to the mechanism responsible for these warps have been put forth: tidal effects, galactic winds, and precession; the first has received the most attention. However, the large sample of warped galaxies now available places severe constraints on the tidal model. Renzo Sancisi in a recent review of this problem concludes that the origin and survival of warps is puzzling--that frequently only distant dwarf companions are visible so that a tidal explanation becomes difficult. Although tidal effects may be important in some instances, there is a need to re-examing other explanations.

A third aspect of the HI distribution concerns the correlation between HI and optical spiral arms. The Westerbork data for such galaxies as M81 and M101 show what at first sight appears to be excellent agreement between these two. But a more detailed examination shows significant and extensive regions of anticorrelation. There are bright, optical arm features with only relatively weak HI emission and the converse, high HI surface density in regions of faint spiral features. We can understand the first of these anticorrelations as possibly due to more extensive ionization by the high-luminosity stars which define the arms. But the second anticorrelation indicates regions in which the star formation rate is low although at least one constituent for such processes, HI, is plentiful. Perhaps we are witnessing two different star formation mechanisms: In the inner regions, a large-scale efficient one such as shocks caused by density waves while in the outer regions only a local mechanism is operative in triggering star formation.

The measurement of the distribution of light and of radio continuum radiation out of the plane of the galaxy is an observationally difficult experiment. We turn to other galaxies for at least an indication of what might be expected. The observations here are difficult too, but tractable; modern data are few. Westerbork continuum data for two edge-on systems, NGC 891 and NGC 4631, are available. They both show an extension of several kiloparsecs perpendicular to the plane. These results imply a thick disk of continuum radiation rather than a spherical halo. Extension of such a model to our Galaxy is tempting but premature.

Information on the optical luminosity at large distances from the plane is also scant. What little is available is consistent with a spherical halo of low luminosity light out to several tens of kiloparsecs. This leads us to the next aspect of extragalactic studies which have an important bearing on our understanding of galactic structure--rotation curves.

The modern data, both 21-cm and optical, show that most rotation curves, when they extend far enough, become flat and remain so out to the last measured point. Here I ignore the important small-scale oscillations in the rotation curve which are probably indicative of streaming motions. As with warps so with flat rotation curves; they are common, so common that a decreasing rotation curve at a large radius becomes an interesting exception. In at least two instances, M81 and NGC 4361, a decrease is seen and can be attributed to observations of material that lies in a tidal link to a neighboring galaxy.

These kinematic data, together with the pertinent photometry, demand an increasing M/L ratio with distance from the center; the mix of stellar types clearly changes with distance. A not too surprising result for the high-luminosity stars which are responsible for much of the light and little of the mass in the inner regions of a spiral galaxy have become very rare at the large radii to which the more extensive measurements refer.

It must also be stressed that rotation curves contain no information on the 3-dimensional distribution of matter. One must assume a model and various geometries, such as a spherical halo or a thin disk, can equally well fit the data. The latter model has the awkward feature that, unlike a sphere, the gravitational attraction at a given point is determined by _both_ the mass interior and exterior to that point. In essence you have to know the answer, or to anticipate it, as with a Brandt curve, in order to obtain the answer.

As a final aspect of extragalactic studies, let me turn to the immediate environs of galaxies. I mentioned earlier that the bending of galaxies might be difficult to explain in terms of a tidal mechanism, at least for some of the warped systems now recognized. Such a mechanism has had much greater success in explaining the tails and plumes extending from galaxies and the bridges connecting galaxies, features seen both optically and in 21-cm. The Magellanic Stream, so prominent in

21-cm Southern Hemisphere observations, is far from unique. There are about a dozen other instances of such HI features now recognized. Martha Haynes, who has found a number of new examples, will summarize the status of these data later in the Symposium.

That such perturbations are frequent and ongoing, even for our own galaxy, will add a new complication to the previous picture of isolated systems. Matter is being pulled out; what happens to it and how much will fall back in? Yet another preview of later discussion.

The period dating from the first IAU Symposium has been exciting and productive. The continuing development of new instrumentation for both space and ground-based observations promises an even more fruitful period ahead. But it is more than instrumentation that is needed; Democritus over two millennia ago eloquently described another vital ingredient:

> "Who seeks will find the good only with labor and pains; the bad, however, is found by everybody without seeking".

DISK AND SPHEROIDAL COMPONENTS OF EXTERNAL GALAXIES: AN OVERVIEW

S. E. Strom and K. M. Strom
Kitt Peak National Observatory *

A brief review of current theoretical views on how disk and spheroidal galaxies form and evolve provides the background for a summary of recent optical observations of external galaxies. Primary emphasis is placed on a discussion of the large-scale distribution and chemical composition of the stellar and gaseous constituents of relatively isolated galaxies. New studies of halo and disk surface-brightness distributions in spiral and S0 galaxies are summarized. The "missing mass" in galactic halos, the relationship between disk size and luminosity, the nonexponential character of disk light distributions and very low-surface-brightness disk systems are highlighted in this section. Next discussed are observations which may provide insight into the factors which regulate the star-forming history of galactic disks and the post-astration appearance of spiral galaxies. Finally, the observed properties of relatively isolated disk galaxies are compared with those located in dense groups. It appears from this comparison that environment plays a significant role in governing the evolutionary history of a galaxy.

1. INTRODUCTION

The Milky Way Galaxy is a relatively luminous, star-forming disk system of intermediate Hubble type located in a region of comparatively low galaxy density. From our vantage point in the solar system, we have, over the past half century, built up considerable understanding of the kinematics, chemical compositions, and ages of the Galaxy's stellar constituents. The past two decades have witnessed a rapid growth in our knowledge of the varied physical conditions and dynamical interactions in the plasmas pervading the disk and halo. From these investigations, astronomers have attempted syntheses aimed at developing plausible models for the formation of the Galaxy, and the evolution, with time, of its stellar and interstellar constituents. Our views of galactic structure and evolution are to a large extent based on an

* Operated by the Association of Universities for Research in Astronomy, Inc., under contract with the National Science Foundation.

W. B. Burton (ed.), The Large-Scale Characteristics of the Galaxy, 9–26.

appreciation of the range of phenomena accessible to observation near our "local swimming hole." However, the development of more cosmopolitan outlooks necessitates studies of the large-scale characteristics of external galaxies. Determination of the scales and masses of disk and halo components, the properties of near-nuclear regions, the nature of wave phenomena in galactic disks, the factors influencing the rates of star formation and chemical-element production are among those problems which depend to a great extent on observations of other galactic systems.

Furthermore, it has become increasingly clear in recent years that a proper understanding of the evolutionary history of any particular galaxy cannot be achieved until we learn whether and how the galactic environment affects the gaseous and stellar components of that system. Only by examining galaxies located in regions characterized by significantly different physical properties can we hope to evaluate the effects on galactic evolution of close gravitational encounters among galaxies and interactions between galaxies and the intergalactic medium. A careful synthesis of "local" and "cosmopolitan" experiences appears to offer the best hope for refining our knowledge of the main factors which determine the structure and evolution of both the Milky Way and external galaxies.

We would like to take this opportunity to summarize recent, primarily optical work on the disk and spheroidal components in external galaxies located in a variety of environmental settings. In the course of this review we will attempt to identify a series of problems which appear, to our prejudiced eyes, to be critical and ripe for solution.

2. THEORETICAL OVERVIEW

At present, we believe that the bulge and disk components of a typical disk galaxy form from a rotating protogalactic cloud in which star formation accompanies gravitational collapse. Current models (cf. Gott and Thuan 1976; Larson 1976) suggest that the relative prominence of the bulge and disk is determined by the efficiency of star formation during collapse. Systems in which star formation is highly efficient at early epochs form large spheroidal components (since the stellar system undergoes an essentially dissipationless collapse). When the early star-forming efficiency is low, the collisions between gas clouds in the protogalaxy lead to the dissipation of energy in the gas and the eventual formation of a thin disk; presumably, stars begin to form in this disk when the average gas density is sufficiently large. The factors which determine the initial star-forming efficiency in a protogalactic cloud are currently unknown. Speculation centers on the initial protogalactic density or the distribution of angular momentum in the protogalactic cloud as the agents responsible for differences in early star-forming history. Although existing models appear to rule out differences in initial angular momentum as the overriding factor which determines whether or not a galaxy forms a dominant bulge or disk, recent observational work (Bertola and Capaccioli 1978) suggests that disk galaxies have larger values of angular momentum per unit mass than do elliptical galaxies. The significance of this observation in the context of current galaxy-formation theory is not at all well understood.

The star-forming history of a protogalactic cloud also determines the degree of element production and the distribution of heavy elements in the collapsing cloud. Clouds in which star formation is rapid, relative to the time scale for collapse of the gas, produce significantly different "abundance profiles" than do slow star-forming clouds (Larson 1974).

The star-forming history of a galaxy immediately following the appearance of a gaseous disk may be critical to determining the current epoch morphological appearance of the system. Sandage et al. (1970) argue that the fraction of gas remaining after the formation of the first generation of disk stars is essentially determined at the time of formation. The efficiency with which star formation consumes disk gas in these initial phases determines whether a galaxy becomes an S0 system, or a spiral or irregular galaxy. Whether this view — the "genetic" outlook — is correct is as yet not clear; moreover, it is difficult to test.

Advocates of the "evolutionary outlook" appeal to a continuous range in the rates of disk astration activity as the prime determinant of galaxy appearance. As our understanding of the factors influencing star-forming efficiency improves, the evolutionary outlook can be put to more and more severe tests. A derivative of the density-wave theory (DWT) of spiral structure, the "spiral-shock" model (Roberts et al. 1975) appears to offer a particularly promising starting point for understanding the evolution of galactic disks.

A large number of disk galaxies (possibly all) appear capable of sustaining density-wave patterns in their disks during a significant fraction of their evolutionary history. The manner in which the wave pattern is initially induced remains controversial. However, the susceptibility of systems of a given mass distribution to the growth of particular wave modes has been studied extensively in recent years (Mark 1976), and, as a result, many theoretical objections to the DWT appear to be diminishing. Compression of disk gas in spiral shocks induced by acceleration of gas in the vicinity of the density-wave crests (the spiral arms) is believed capable of inducing star formation (see Roberts et al. 1975 for a review). The frequency of star formation and the degree of post-shock gas compression are related to the angular speed of the wave pattern in the disk and to the circular velocity of the disk gas at a given galactocentric distance; in turn, these quantities are related to the distribution of mass in the galaxy. Galaxies with a relatively high degree of central concentration tend to form stars more frequently and possibly more efficiently. Consequently, one expects that an Sb galaxy with a prominent nucleus will form stars and deplete its disk gas more rapidly than will an Sc-type system. The chemical-enrichment patterns should also reflect differing rates of star formation across the disk. It is not yet clear to what degree the differences in star-forming activity along the Hubble sequence can be explained in the context of the spiral-shock picture. For example, do S0 and irregular galaxies represent extrema in the range of disk star-forming histories? Are the former, systems in which astration has proceeded so rapidly that no disk gas remains? If so, what has happened to the density-wave pattern in their disks? Are irregulars galaxies in which star formation has not been triggered "efficiently" by spiral shocks but rather has proceeded chaotically (perhaps because these systems, for some reason, cannot support the growth of a dominant wave pattern or because the gas-to-stellar-density ratio is insufficiently small)?

The roles of injection and removal of gas from the disks of evolving galaxies may also be essential factors in determining the current epoch appearance of galaxies and the chemical histories of their disks. Not nearly enough data are yet available regarding the distribution and velocity fields of gas in galactic halos surrounding disk systems. Yet, in our own Galaxy, we believe that the amount of gas injected from halo-gas clouds is sufficient to account for a large fraction of the current disk-gas content and possibly much of the star formation occurring at present. It will be extremely important to our understanding of galactic evolution to pursue high-sensitivity searches (Sargent and Knapp, private communication) for neutral hydrogen located in galactic halos and in the intergalactic medium.

Removal of gas by the action of galactic winds emanating from nuclear-bulge regions has been studied by Mathews and Baker (1971) and later by Faber and Gallagher (1976). The combined effects of supernova explosions and collisional heating of ejecta from dying stars raise the bulge-gas temperature to values sufficiently high to drive a wind of outflow velocity sufficient to escape the bulge region. [However, Chevalier and Oegerle (1978) argue that, at least in our own Galaxy, conditions appear to favor the establishment of a hot, bound corona rather than an outflowing wind.] If the density of disk gas is sufficiently low and if the density and velocity of the outflowing material is high enough, a galactic disk may be swept free of interstellar gas. Furthermore, the wind continually purges the disk of gas ejected by evolving low-mass stars. Because winds should be "stronger" in bulges of high mass, systems with large bulges (Sa or Sb spirals) may be more susceptible to early removal of disk gas than are galaxies that have small bulges (late-type spirals and irregulars).

External factors, such as collisions with other galaxies and with intergalactic clouds of hydrogen, as well as interactions of galaxies with intergalactic gas, may also affect the course of post-disk-formation evolution. Particular attention, in recent years, has been paid to the role of "ram-pressure stripping" in removing disk gas from galaxies. As a spiral galaxy traverses a rich cluster of galaxies, it encounters the hot gas (10^8 $^{\circ}$K gas, 10^{-4} atoms cm^{-3}) believed to pervade such clusters at the current epoch. When the velocity of the galaxy, with respect to the intracluster gas and the density of the gas, is high enough, compared with the density of gas in a galactic disk, gas may be "stripped" from the system. Unless gas from dying stars is produced at a rate sufficient to replenish the disk before the system undergoes further stripping (either through repeat passage through the cluster medium or by action of galactic winds), further star formation in the disk cannot be sustained.

The above represents a brief summary of "mainstream" wisdom regarding disk-system evolution. Recent work has emphasized the possible significance of massive, unseen halos in affecting disk galaxy evolution. Stability arguments (Ostriker and Peebles 1973) suggest that cold disks are unstable to the growth of bar-like instabilities. An attractive mechanism for stabilizing the disk against the formation of a bar is to surround the system with an extensive, "hot" halo component of mass comparable to, or even much greater than, the disk mass. At present, optical and radio searches are aimed at determining whether halos have a mass (and mass distribution) great enough, first, to stabilize the disk and, second, to explain the mass discrepancy between galactic masses and the virial masses of rich clusters. If optical counterparts of halos cannot be observed, even after careful searches at all wavelengths, we must entertain the hypothesis that invisible constituents (e.g., neutral stars, black holes) contribute most of the halo mass (and indeed most of the mass in the universe!). If so, when did these currently invisible objects form, and what effect did they have on the early star-forming and heavy-element-producing history of a galaxy? Some very recent theoretical work (White 1978) has explored the possibility that luminous components of galaxies formed significantly after the aggregation of "invisible" material. It is becoming increasingly clear that the formation of galaxies may not be well understood without first understanding much more about the distribution of nonluminous material in galaxies.

The next sections review the current highlights of optical studies of disk and halo components of galaxies. We have tried, except insofar as continuity demands, to avoid duplication of the material presented in our recent review of disk galaxy evolution presented last summer in Bonn (Strom and Strom 1978d). The reader should refer to the more extensive Bonn review for additional details and reference material.

3. SPHEROIDAL COMPONENTS OF DISK GALAXIES

Kormendy's (1977) study of S0 galaxies suggests that the surface-brightness distribution for the spheroidal component (SC) of disk systems can be well represented by the de Vaucouleurs (1959) law. In this respect the SC's resemble elliptical galaxies. Recent work on the edge-on spiral NGC 4565 suggests that, at least in some cases, the SC light distribution can fall off much more rapidly than a de Vaucouleurs law. However, because this latter result is also true for E galaxies (Strom and Strom 1978a,b), its significance cannot be evaluated as yet.

Little is known about the distribution of mean ellipticities for SC's. Kormendy and Bruzual (1978) note that not all SC's appear spheroidal in form; "box-like" nuclear bulges, similar to that of NGC 128, are not uncommon. This observation suggests that SC's may not be analogous to E galaxies, either morphologically or dynamically.

Strom et al. (1976b) find that the nuclear bulge regions of the edge-on S0 galaxies NGC 3115 and NGC 4762 show large abundance gradients extending over scales of many kpc, in contradiction to the small halo gradients observed for a large majority of E galaxies (Strom and Strom 1978a,b). NGC 7332 (Strom and Strom 1978c) and our Galaxy also appear to show halo abundance gradients. If the frequent occurrence of large-scale (extending over many kpc's) halo composition gradients is confirmed for a more significant sample of disk-system SC's, it would suggest a star-forming and chemical-element-producing history significantly different from that characterizing most E galaxies. Interpreted in the context of Larson's (1974) gas-dynamical models, this result would imply that spiral bulges might have been formed from lower-density protogalactic clouds.

No evidence is yet available to permit a test of the similarity between the color(composition)-luminosity relation for E's and disk galaxy SC's. T. Boroson of the Steward Observatory is just beginning such a study.

A number of recent studies have attempted to extend observations of disk system SC's to extremely low light levels. The desire to better understand galactic halos is motivated in large measure by the hope of discovering the objects responsible for the missing mass in galaxies. Most notable among recent studies have been those of Hegyi and Gerber (1977; NGC 4565), Spinrad et al. (1978; NGC 4594, NGC 4565, and NGC 253), and Kormendy and Bruzual (1978; NGC 4565). Their studies reveal the presence of halos out to galactocentric distances well beyond 50 kpc. Assuming a mass-to-light ratio characteristic of the inner regions of either disk galaxy SC's or E galaxies ($M/L_B \sim 10$), the mass contained in the ratio of halo-to-disk mass for NGC 4565 is estimated to be comparable to or greater than the disk mass. Similar results were found by Strom et al. (1977) for NGC 3115 and Rubin et al. (1978) for NGC 4378. In all cases the observed halo surface brightness falls off more rapidly than r^{-2} at large galactocentric distances. We conclude that the halo mass in visible objects may be sufficiently large to preclude the development of bar-like instabilities in the disk. However, there is no evidence indicating the presence of an optically detectable massive ($M_H/M_D >> 10$) component for which the surface brightness falls off less rapidly than r^{-2}. Hegyi and Gerber report that the V - I color at the faintest isophotal level observed in their study is suggestive of an increase in the population of faint, red stars (possibly dwarfs later in type than K7 V). However, in our opinion their result requires careful reexamination before it can be accepted. Spinrad et al. find their (B - R) photometry to be consistent with either no color changes or with a slight "bluing" at large galactocentric distances. Strom et al. (1977) find that the halo colors for NGC 3115 become blue [in the (U - R) color system] out to galactocentric distances r $\sim$ 10 kpc. Moreover, an infrared ($\lambda \sim 2.2\ \mu$) study of this galaxy (Strom et al. 1978) suggests that the (V - K) color index also decreases with increasing galactocentric distance; however, their infrared results extend only to r $\sim$ 2 kpc.

We conclude this section with a list of questions, the answers to which may be of some importance to furthering our understanding of disk galaxy SC's.

1. How do the light distributions measured along the minor axes of the nuclear bulges in edge-on disk galaxies compare with those of E galaxies? At a fixed luminosity, are their characteristic sizes similar?

2. What is the distribution of ellipticities among the nuclear-bulge components of edge-on disk systems? How frequent are box-like nuclei?

3. Is the relationship between velocity dispersion and luminosity the same in disk galaxy SC's as in E galaxies?

4. Is the color-luminosity relation the same for E galaxies and disk galaxy SC's?

5. Do all disk galaxy SC's exhibit extensive composition gradients? If so, does this property imply a fundamental difference in the protogalactic clouds which gave birth to spiral and E galaxies?

4. THE DISK COMPONENT

4.1 Distribution of the Stellar Light

Both de Vaucouleurs (1959) and Freeman (1970) argue that the disk light distribution for spirals and S0 galaxies can be well fitted by an exponential law. Characteristic distances for a 1/e decrease in surface brightness range between 2 and 10 kpc. Freeman also observes that the extrapolated central surface brightnesses (obtained by evaluating the best-fit exponential law at a galactocentric distance of zero) for most disk galaxies are identical to within 30 percent [B(0) = 21.65 mag arcsec^{-2}]. An extensive discussion of disk-system light distributions by Kormendy (1977) leads to significantly different conclusions. Kormendy states that it is essential to subtract the contribution of the bulge component prior to fitting an exponential law to the combined disk-bulge light distribution. By following a prescription in which a best-fit de Vaucouleurs profile to the bulge light distribution is subtracted from the observed surface-brightness profile, he finds that very few disk components can be represented by an exponential law. In many cases the disk components contribute insignificantly to the system light near the galactic nucleus; indeed, it is not clear to what extent a disk component exists in the near-nuclear regions of some of Kormendy's sample galaxies. There appears to be little evidence of a universal central or even a "mean" disk surface brightness.

Further evidence, which contradicts the concept of a universal B(0), has been presented recently by Romanishin et al. (1977). They discuss the properties of a group of "low-surface-brightness" (LSB) spiral galaxies in which the projected central surface brightness (adopting Freeman's definition) ranges between a factor of 2 and 6 smaller than the "canonical" value of B(0).

Recent work by B. Peterson of the Steward Observatory has led to an increased understanding of the extent of galactic disks. From measures of the light distribution in the disks of 36 Virgo and 30 Hercules cluster spiral galaxies, he finds that, to an isophotal level, $\mu_B = 26.6$ mag arcsec^{-2} (the level used to define the Holmberg radius), $\log r(\mathrm{kpc}) = -0.166\ M_B - 2.150$ (if $H_0 = 50$ km s^{-1} kpc^{-1}); M_B is the blue absolute magnitude of a galaxy. From a study of the Virgo systems, he finds that $r(\mu_B{=}28.6)/r(\mu_B{=}26.6) = 1.4$. Thus the most luminous spiral galaxies of disks with observable light extend to radii of more than 100 kpc.

More detailed studies of the light distribution in the underlying disk component of spiral galaxies have been carried out recently by Schweizer (1976) and by Jensen (1977). Both authors report the presence of spiral wave patterns which appear to be surface density crests in the "old-disk" population. The study of "smooth-arm" spiral galaxies (Strom et al. 1976a; Wilkerson et al. 1977) also supports the notion that spiral arms not only manifest themselves in the distribution of newly formed stars, but also in the underlying disk population. Hence, at least to these reviewers, the case for a spiral wave pattern rooted in the overall mass distribution of spiral galaxies is overwhelming.

The following observations appear essential to furthering our understanding of the distribution of stars in galactic disks:

1) obtain bulge light distributions from photometry along the minor axes of edge-on disk systems. This will reduce uncertainties in "modeling" the bulge component implicit in Kormendy's deconvolution of bulge and disk light.

2) compare the light distributions of a much larger sample of disk systems of spiral, S0, and irregular types. Although Freeman's study suggests that the ranges of disk parameters overlap among these types, a more careful treatment seems in order. In addition to contributing to our empirical understanding of the differences among disk galaxies of differing type, these data might well shed some light on the question of why some disk systems (spirals) can support stable wave patterns and why others (S0's and irregulars) cannot.

3) carry out harmonic analyses of the wave patterns in spiral

galaxy disks in order to learn which modes can grow (and at what rate) in galaxies characterized by differing mass distributions. Efforts should concentrate on those galaxies for which accurate rotation curves and surface photometry can be obtained. Such analyses could be quite valuable in testing the newly developed wave-mode approach of the MIT group.

4) obtain disk photometry carried out at wavelengths ($\lambda \geq 6000$ Å) which primarily map the underlying old stellar population rather than the combined new and old populations. Such observations should provide a better measurement of the true mass distribution, independent of changes in the ratio of young-to-old stars across the disk.

4.2 Mass Distributions in Galactic Disks

It is commonly assumed that, to a first approximation, the observed surface-brightness profile in an appropriately chosen bandpass provides an indication of the true mass distribution in the system. Based on an analysis of rotation curves and surface photometry of galaxies, Nordsieck (1973) concludes that the mass-to-light ratio within individual disk galaxies is indeed constant. More recent optical studies, however, appear to contradict this result. For example, both Schweizer's (1978) analysis of the Sombrero galaxy (NGC 4594; Sa) and Rubin et al.'s (1978) study of NGC 4378 (Sa) suggest that the ratio of mass-to-photographic-luminosity increases with increasing galactocentric distance, at least in galaxies of early Hubble type. Rotation curves derived from neutral hydrogen studies (which extend to greater galactocentric distances) by Krumm and Salpeter (1978) and Sancisi (1978) also point to larger values of M/L in the outer parts of disk galaxies. The increase in M/L is most commonly attributed to the effects on the rotation curve of the inner regions of a massive halo characterized by M/L. However, the possibility of changes in the population mix in the disk stars has not been thoroughly investigated.

Gas motions can be used as a probe of the gravitational field of the underlying disk stars. Recent 21-cm maps of M81 made at Westerbork (Visser 1978) provide strong evidence that gas motions in the vicinity of spiral arms can be well understood if one adopts as a potential field that inferred from the combined effects of the observed, smooth, axisymmetric disk light distribution and of the spiral wave crests observed by Schweizer (1976). Optical work by Rubin and Ford (private communication), who used the very high spectral and spatial resolution available with the KPNO and CTIO Cassegrain spectrographs, also suggests that the gas flows near the arms are consistent with the presence of a spiral disturbance in an otherwise axisymmetric gravitational field; more quantitative results may be available within the next year or two. Surface photometry directed at the determination of the variation of wave amplitude with position, in a bandpass ($\lambda > 6000$ Å) sensitive primarily to the distribution of the old-disk population, will be a necessary complement to the Rubin-Ford study.

4.3 Bulge-to-Disk Ratios in Spiral Galaxies

In Sec. 1 we reviewed current speculation regarding the factors which influence the star-forming histories of spiral galaxies. If star formation is driven by spiral shocks, then systems with a high degree of central concentration (such as Sa- and Sb-type galaxies) should exhibit higher star-formation and gas-depletion rates. Moreover, systems that have massive nuclear bulges should be most susceptible to disk-gas removal by galactic winds emanating from the bulge. These two predictions lead to the following working hypothesis: If there is an evolutionary progression from actively star-forming spiral galaxies to "inactive" S0 galaxies, then the systems most likely to have undergone such transmutation are those with the highest ratio of bulge-to-disk mass. A recent study by Burstein (1977) of bulge-to-disk ratios inferred from photometry of spirals and S0's (located in regions of relatively low galaxy density) reveals that the frequency of large bulge-to-disk ratios is highest among galaxies of the S0 type. Relatively few S0's have small bulges, while few actively star-forming spirals have large bulge-to-disk ratios. While Burstein's result is entirely consistent with the above working hypothesis, we cannot rule out the possibility that genetic factors influence both the bulge-to-disk ratio and the initial consumption of gas by star-forming events in the disk.

If spiral galaxies can become S0's following the depletion of disk gas, what is the fate of the spiral wave pattern in the disk stars? For a while, the wave pattern may persist. Strom et al. (1976a) and Wilkerson et al. (1977) find spiral wave patterns in the disks of galaxies in which active star formation has ceased. The colors of the spiral arms and of the disks in their sample of "smooth-arm spiral galaxies" are identical and tend to be more typical of S0 and E galaxies than of actively star-forming spiral galaxies. These authors believe these systems to be the immediate descendents of spiral galaxies in which disk gas has been removed, either through astration, galactic winds, or by stripping in rich clusters. However, since S0 galaxies show no evidence of spiral structure, it must be presumed that the wave pattern cannot persist for very long after gas has been removed. Lin (private communication) suggests that the spiral wave amplitude at first increases, since the damping provided by galactic shocks no longer curtails the growth of the dominant modes (see also Dekkar 1974). When the wave amplitude becomes sufficiently large, the motions of the disk stars may become significantly perturbed. An increase in the random velocities of the disk stars ensues which, in turn, eventually damps the wave.

Further theoretical work on the growth and damping of spiral waves in gas-free systems will be required in order to check these speculations. Such work would represent a critical step in our understanding of the Hubble sequence, since at present the hypothesis that spirals can be transmuted to S0's rests directly on the belief that the spiral density waves can be damped subsequent to the removal of disk gas.

4.4 Star Formation in Disk Galaxies

Understanding the factors which control star-formation efficiency as a function of time is fundamental to synthesizing plausible models of galactic evolution. Because spiral galaxies are systems which (a) are forming stars at the current epoch and (b) contain within their disks regions of significantly differing chemical and physical conditions, they appear to be attractive "laboratories" in which we can "test" various proposed mechanisms for triggering star formation.

Sargent et al. (1973) attempt to derive estimates of the star-forming history of disk galaxies from analysis of available wide-band photometry of such systems. They assume that the star-forming history of a galaxy can be represented by two separable terms: (a) an exponential birth-rate function characterized by a decay time, β^{-1}, and (b) an initial mass function (IMF) of the form $\phi(m) = Cm^{-\alpha}$. Except for irregular galaxies where the IMF may be weighted more toward the formation of massive stars, Sargent et al. find α lies close to the estimated solar-neighborhood value of $\alpha \sim 2.45$. Values of β were found to range widely, small values ($\beta^{-1} \sim 10^9$ yr) are characteristic of early-type spirals, while large values ($\beta^{-1} \gtrsim 10^{10}$ yr) are characteristic of late-type spirals and irregulars.

While these results may be correct for the galaxies as a whole (disk and bulge), the Sargent et al. analysis cannot properly describe the <u>disk</u> star-forming histories, since the aperture photometry available to them is affected to varying degrees by contributions from bulge light; for early-type spirals, the bulge contribution to the total system light may be very large. It is believed almost universally that star formation in the bulge is completed on a time scale $\lesssim 10^9$ years after "formation" and probably well prior to the bulk of disk star formation. Hence it is perhaps not surprising that small values of β^{-1} characterize early-type systems, since light from old bulge stars dominates. To learn more about <u>disk</u> star-forming histories, it would seem best to follow the spirit of the Sargent et al. analysis of integrated colors but to measure these colors as a function of disk position. Obtaining the required broad-band surface photometry of a large sample of disk galaxies is well within the capabilities of modern techniques.

Aaronson (1978) summarizes the potential of UVK photometry for making more precise estimates of plausible star-forming histories than is possible with the UBV system (see also Larson and Tinsley 1978). Measurements in this system are particularly helpful in sorting the relative contributions of newly formed stars and old-disk population. In fact, for galaxies in which the majority of star formation has taken place at recent epochs, measurement of a color similar to (V - K) may provide the only means of detecting early generations of stars.

Recent work has centered on studies of proposed mechanisms for initiating star formation in galactic disks. Models in which star formation is induced by compression behind shocks propagating in the interstellar medium have been studied extensively (see, for example, Roberts et al. 1975; Woodward 1976; Elmegreen and Lada 1977). Perhaps because of its "global" predictive powers, the spiral-shock model has received considerable attention in the last several years. First, the rate of star formation in this picture depends upon the rate at which disk gas encounters the density-wave pattern. Second, the efficiency of star formation is assumed to be related to the degree of compression suffered by disk gas in the post-shock region. Both the rate and efficiency of star formation across the disk are ultimately linked to the mass and size of a galaxy and the distribution of matter within the system. Hence, for a galaxy in which the above quantities can be estimated, it is possible to predict (qualitatively) relative star-formation rates across the disk.

Based on such reasoning, Oort (1970) suggests that the "holes" in the distribution of neutral hydrogen observed in the central regions of several nearby galaxies resulted from rapid depletion of disk gas in regions in which the encounters between the gas and the wave pattern are very frequent. However, recent observations (Burton and Gordon 1978, and references therein) of the CO distribution in our Galaxy suggest that the hydrogen holes might not represent minima in galactic gas density distribution but result instead from the predominence of H_2 compared with neutral hydrogen in the inner disk region. If so, then we must also explain why the inner disk regions in external galaxies do not appear to give birth to massive stars capable of producing observable H II regions. Can only low-mass stars form the inner disks of Sa and Sb galaxies at the current epoch, or are the massive stars and H II regions obscured by optically opaque dust clouds?

An indirect test of the spiral shock model was attempted by Jensen et al. (1976). These authors reason that in regions of high expected-star-formation rate, the rate of gas depletion and heavy element enrichment would be highest. They therefore selected a group of 14 galaxies which, on the spiral-shock picture, should exhibit a wide range of star-forming rates and efficiencies. From a comparison of emission-line strength ratios believed to be indicative of metal-to-hydrogen ratios in the gas, Jensen et al. find that the indicated heavy element abundances were indeed highest in regions expected to suffer high rates of star formation at high efficiency.

Kaufman (1978) attempts a more sophisticated test. She models the star-forming history in our Galaxy by assuming that two primary mechanisms operate: (a) supernova (or H II region) shock and (b) spiral shock-induced star formation. From optical observations of the distribution of H II regions in our Galaxy, she concludes that most stars did not originate in star-forming events triggered by galactic shocks; rather, over two-thirds of the stars were formed as a consequence of compression behind supernovae or H II region induced shocks. No attempts have been made to apply Kaufman's analysis to other galaxies.

Jensen et al. (1978; private communication with E.B.J.) have embarked on an ambitious project to map the current-epoch star-forming efficiencies across the disks of a number of spiral galaxies. Their results will be based on analysis of U, B, V, and R maps of galactic disks. From such maps, they believe that the number and ages of newly formed stars can be accurately charted. As yet, results are available in preliminary form only for the southern, barred spiral galaxy M83 (Jensen et al. 1977). They find that while most newly formed stars are found near spiral arms, there is evidence for significant star formation outside arm regions. Hence spiral shock-induced star formation cannot be the sole agent for triggering new generations of stars. Thus far, Jensen et al. find no strong evidence from their age maps for a "drift" of newly formed stars relative to the spiral wave pattern [see Schweizer (1976) for a description of a similar attempt]. However, it appears possible that the spread in formation times subsequent to compression by the spiral shock is sufficient to obscure the theoretically expected variation of age with angular position relative to the spiral arm.

Until very recently, it has been assumed that star formation in disk galaxies has proceeded continuously from the epoch when the disk gas was first assembled until the present. However, several authors (Sargent et al. 1973; Huchra 1977) argue that the star-forming histories of some disk galaxies may be extremely chaotic and characterized by bursts of star formation followed by long quiescent periods.

Even more surprising is the recent work on the ages of disk stars in our own Galaxy (Demarque and McClure 1977) and in the Large Magellanic Cloud (Butcher 1977). This latter work suggests that most stars did not start to form in the Milky Way until nearly 5 x 10^9 years following the formation of the SC, while in the LMC most star formation may have been delayed until only 3 to 5 billion years ago. Some of the LSB spiral galaxies discussed by Romanishin et al. (1977) appear to have formed the majority of their disk stars within the last 5 billion years.

5. EFFECTS OF ENVIRONMENT ON DISK GALAXY EVOLUTION

Astronomers have gradually begun to realize the importance of environment on the evolution of disk galaxies. The theoretical and observational works described in Secs. 1 through 4 have implicitly assumed that galactic evolution proceeds in isolation. However, conditions in the central regions of rich clusters of galaxies may greatly alter the course of galactic evolution. Because galaxies are moving through a hot plasma, disk and halo gas can, in principle, be removed by the effects of ram-pressure stripping (Gunn and Gott 1972) or by evaporation (Cowie and Songaila 1977). If halo or disk gas extends to galactocentric distances in excess of $\sim$100 kpc. collisions between galaxies will remove some of this gas. Tidal encounters in the dense central regions may be important in (a) "heating" the stellar subsystems in disk galaxies (Richstone 1976; Marchant and Shapiro 1977), (b) stripping stars from the outer regions of their disks and halos, and (c) inducing star formation. How efficacious these processes are depends upon the density and temperature of intracluster gas, the galaxian density, the cluster velocity dispersion, and the time since the cluster was first assembled. Evaluating the interplay between these factors and the natural evolutionary processes in galaxies presents a significant challenge. However, observations of galaxies in differing environmental settings appear to offer the possibility of both testing many pictures of disk galaxy evolution and deriving a sounder understanding of cluster evolution.

The possible effects of ram-pressure stripping on disk galaxy evolution have received the most observational attention in the last few years. In this picture it is implicitly assumed that stripping of gas from a spiral or irregular galaxy will result in a transmutation to a system of the S0 type. Recent work (Melnick and Sargent 1977; Bahcall 1977) regarding the distribution of spiral and S0 types in rich clusters demonstrates that (1) S0's dominate those clusters in which the 2-10 kev X-ray luminosity and the cluster velocity dispersion are highest, and (2) spirals are most frequently found in the outskirts of rich clusters where the density of intracluster material is expected to be lowest. These observations are consistent with the stripping hypothesis, since removal of disk gas by this process takes place most effectively when the velocity of a galaxy, relative to the intracluster medium, and the density of the medium are highest. Other optical studies (cf. Strom and Strom 1978d) also support the stripping hypothesis.

Sullivan (1978) reports the results of a survey of the neutral hydrogen content of disk galaxies in two X-ray clusters (Coma and Abell 1367). He concludes that the hydrogen mass-to-luminosity ratios for spirals in these clusters are significantly below those expected for field galaxies of similar morphology, and suggests that ram-pressure stripping has removed some of the disk gas in these systems.

Butcher and Oemler (1977) discuss observations of galaxy colors in two distant ($Z \sim 0.4$), rich clusters of galaxies similar in morphological appearance to the Coma cluster. They find a large excess of blue galaxies compared with Coma, and conclude that in these clusters stripping has not yet removed sufficient gas to preclude active star formation. W. Rice of Harvard is presently compiling data on the galaxian color distribution in ten nearby galaxy clusters of differing morphological types. When more information on distant clusters becomes available, his data should form the basis for a sound comparison of colors of nearby and distant clusters.

Gisler (1978) at Kitt Peak suggests that the rate at which gas is expelled from dying disk stars is sufficiently high during much of a typical galaxy's lifetime to preclude stripping in a cluster of properties similar to Coma until relatively recent epochs. Until the gas-expulsion rate from dying stars decreases sufficiently, Gisler finds that ram-pressure stripping cannot remove all the gas from a galaxy. Hence, in the context of his computations, the Butcher-Oemler inference of a large number of "unstripped" galaxies at look-back times of $\sim 5 \times 10^9$ years is not terribly surprising. Gisler adds an important caveat:

the IMF cannot be biased too heavily in favor of the production of low-mass stars else stripping should take place relatively soon after the formation of a rich cluster.

Wilkerson et al. (1977) claim to have detected a class of galaxies, smooth-arm spirals, which appears to be a good candidate for spiral systems stripped in the relatively recent past. Hence, not only do we find the presumed ultimate descendents of stripped spirals — S0 galaxies — but the transition cases as well.

Sandage and Visvanathan (1978), however, argue that S0 galaxies are not formed by stripping. From a study of the integrated colors of a sample of over 300 S0 galaxies, they find no evidence for systematic differences between S0 colors in the field or in clusters. If stripping were preferentially operative in clusters, one might expect cluster galaxies to show less evidence of recent star formation in their integrated colors; they do not. It is difficult, however, to evaluate the Sandage and Visvanathan results. First, some of their "cluster" samples are not located in regions thought to be pervaded by a sufficiently dense intracluster medium. Second, as with the Sargent et al. study of galaxy-integrated colors, it is not clear what fraction of the observed light arises from a bulge component (in which star formation has presumably been long inactive) as opposed to the disk. Systems in which the observed light is dominated by the bulge component will tend to have similar colors (to the extent that their mean metallicity is similar) despite the star-forming activity, recent or otherwise, in the disk. The Sandage-Visvanathan result seems to these reviewers neither damning nor supportive of the stripping hypothesis.

The effects of stripping in rich clusters can be used to test the efficacy of this proposed mechanism for S0 galaxy formation. Burstein (1977) finds that for field and low-density groups and clusters the distribution of bulge-to-disk ratios (B/D) for S0 galaxies tends to be weighted toward higher values than that found for spirals. If stripping "truncates" star formation by removing disk gas and if stripped spirals eventually become S0 galaxies, then the distribution of B/D for a sample of cluster S0 galaxies should include more small B/D ratio systems than the Burstein field S0 sample. Moreover, the distribution of S0 B/D ratios should vary both with position with the cluster (cf. Melnick and Sargent 1977) and from cluster to cluster (cf. Bahcall 1977).

Not only do the distribution of morphological types, integrated colors, and hydrogen mass-to-light ratios differ between disk galaxies located in rich clusters and in the field, but disk sizes appear to be smaller as well. Strom and Strom (1978d) report that S0 galaxies located in the central region of the Coma cluster appear smaller than those found in the cluster outskirts or in lower-density regions. It is not clear whether this results from the effects of tidal stripping of stars in the outer disk, or from the indirect effects on disk-system star-forming histories of ram-pressure stripping or galaxy collisions.

It is also possible that close encounters between galaxies or with intergalactic clouds can stimulate star-forming events (Toomre and Toomre 1972; Larson and Tinsley 1978; Lynds and Toomre 1976). As yet, it is not clear to what extent these events may affect the star-forming history in systems located in groups of differing density and velocity dispersion.

We should not ignore the possibility that the difference in galaxy-type distributions between rich clusters and lower-density galaxy regions might be an artifact of formation conditions rather than evolution. However, this view presupposes that galaxies "know" they will be cluster members from the time of formation. We feel that considerable effort should be invested in delineating not only morphological but also detailed structural differences between cluster and field galaxies. For example, are the bulge components of cluster disk galaxies different from those characteristic of the field? Can these differences plausibly be ascribed to differences in initial conditions? Do composition gradients in disk-system, spherical components differ significantly from those in the field? From answers to these questions, we may hope to begin to understand and separate genetic and environmental effects.

REFERENCES

Aaronson, M.: 1978, *Astrophys. J.* (in press).
Bahcall, N. A.: 1977, *Astrophys. J. Letters* 218, pp. 93-95.
Bertola, F. and Capaccioli, M.: 1978, *Astrophys. J. Letters* 219, pp. 95-96.
Burstein, D.: 1977, Univ. of California, Santa Cruz, (Ph.D. Thesis).
Burton, W. B. and Gordon, M. A.: 1978, *Astron. Astrophys.* 63, pp. 7-27.
Butcher, H.: 1977, *Astrophys. J.* 216, pp. 372-380.
Butcher, H. and Oemler, A.: 1977, *Astrophys. J.* 219, pp. 18-30.
Chevalier, R. A. and Oegerle, W. R.: 1978, *Astrophys. J.* (in press).
Cowie, L. L. and Songaila, A.: 1977, *Nature* 266, pp. 501-503.
Dekkar, E.: 1974, in Proc. CNRS Int. Colloquium, *La Dynamique des Galaxies Spirales*, Bures-Sur-Yvette, France.
Demarque, P. and McClure, R. D.: 1977, in B. M. Tinsley and R. B. Larson (eds.) *The Evolution of Galaxies and Stellar Populations*, Yale Univ. Obs., New Haven, pp. 199-217.
de Vaucouleurs, G.: 1959, *Handbuch der Physik* 53, pp. 311-372.
Elmegreen, B. G. and Lada, C. J.: 1977, *Astrophys. J.* 214, pp. 725-741.
Faber, S. M. and Gallagher, J. S.: 1976, *Astrophys. J.* 204, pp. 365-378.
Freeman, K. C.: 1970, *Astrophys. J.* 160, pp. 811-830.
Gisler, G.: 1978, *Astrophys. J.* (in press).
Gott, J. R. and Thuan, T. X.: 1976, *Astrophys. J.* 204, pp. 649-667.
Gunn, J. E. and Gott, J. R.: 1972, *Astrophys. J.* 176, pp. 1-19.
Hegyi, D. and Gerber, G. L.: 1977, *Astrophys. J. Letters* 218, pp. 7-11.
Huchra, J. P.: 1977, *Astrophys. J.* 217, pp. 928-939.
Jensen, E. B.: 1977, Univ. of Arizona (Ph.D. Thesis).
Jensen, E. B., Strom, K. M., and Strom, S. E.: 1976, *Astrophys. J.* 209, pp. 748-769.

Jensen, E. B., Talbot, R. J., Davis, J. C., and Dufour, R. J.: 1977, *Bull. Am. Astron. Soc.* 9, 648.
Kaufman, M.: 1978, *Astrophys. J.* (in press).
Kormendy, J.: 1977, *Astrophys. J.* 217, pp. 406-419.
Kormendy, J. and Bruzual, G.: 1978, *Astrophys. J. Letters* (in press).
Krumm, N. and Salpeter, E.: 1978, in R. Wielibinski (ed.), Proc. IAU Sym. 77, *Structure and Properties of Nearby Galaxies*, Reidel, Dordrecht.
Larson, R. B.: 1974, *Monthly Notices Roy. Astron. Soc.* 166, pp. 585-616.
Larson, R. B.: 1976, *Monthly Notices Roy. Astron. Soc.* 176, pp. 31-53.
Larson, R. B. and Tinsley, B. M.: 1978, *Astrophys. J.* (in press).
Lynds, C. R. and Toomre, A. R.: 1976, *Astrophys. J.* 209, pp. 382-388.
Marchant, A. B. and Shapiro, S. L.: 1977, *Astrophys. J.* 215, pp. 1-10.
Mark, J. W-K.: 1976, *Astrophys. J.* 205, pp. 363-378.
Mathews, W. G. and Baker, J. C.: 1971, *Astrophys. J.* 170, pp. 241-259.
Melnick, K. and Sargent, W. L. W.: 1977, *Astrophys. J.* 215, pp. 401-407.
Nordsieck, K. H.: 1973, *Astrophys. J.* 184, pp. 735-751.
Oort, J. H.: 1970, *Astron. Astrophys.* 7, pp. 381-404.
Ostriker, J. H. and Peebles, P. J. E.: 1973, *Astrophys. J.* 186, pp. 467-480.
Richstone, D. O.: 1976, *Astrophys. J.* 204, pp. 642-648.
Roberts, W. W., Roberts, M. S., and Shu, F. H.: 1975, *Astrophys. J.* 196, pp. 381-405.
Romanishin, W., Strom, K. M., and Strom, S. E.: 1977, *Bull. Am. Astron. Soc.* 8, 538.
Rubin, V. C., Ford, W. K., Strom, K. M., Strom, S. E., and Romanishin, W.: 1978, *Astrophys. J.* (in press).
Sancisi, R.: 1978, in R. Wielibinski (ed.), Proc. IAU Sym. 77, *Structure and Properties of Nearby Galaxies*, Reidel, Dordrecht.
Sandage, A. R., Freeman, K. C., and Stokes, N. R.: 1970, *Astrophys. J.* 160, pp. 831-844.
Sandage, A. R. and Visvanathan, N.: 1978, *Astrophys. J.* (in press).
Sargent, W. L. W., Searle, L., and Bagnulo, W. G.: 1973, *Astrophys. J.* 179, pp. 427-438.
Schweizer, F.: 1976, *Astrophys. J. Suppl.* 31, pp. 313-332.
Schweizer, F.: 1978, *Astrophys. J.* 220, pp. 98-106.
Spinrad, H., Ostriker, J. P., Stone, R. P. S., Chiu, L. G., and Bruzual, G.: 1978, *Astrophys. J.* (in press).
Strom, K. M. and Strom, S. E.: 1978a, *Astron. J.* 83, pp. 73-134.
Strom, K. M. and Strom, S. E.: 1978b, *Astron. J.* (in press).
Strom, K. M. and Strom, S. E.: 1978c (unpublished).
Strom, K. M., Strom, S. E., Jensen, E. B., Moller, J., Thompson, L. A., and Thuan, T. X.: 1977, *Astrophys. J.* 212, pp. 335-346.
Strom, K. M., Strom, S. E., Wells, D. C., and Romanishin, W.: 1978, *Astrophys. J.* 220, pp. 62-74.
Strom, S. E., Jensen, E. B., and Strom, K. M.: 1976a, *Astrophys. J. Letters* 206, pp. 11-14.
Strom, S. E. and Strom, K. M.: 1978d, in R. Wielibinski (ed.) Proc. IAU Sym. 77, *Structure and Properties of Nearby Galaxies*, Reidel, Dordrecht.
Strom, S. E., Strom, K. M., Goad, J. W., Vrba, F. J., and Rice, W.: 1976b, *Astrophys. J.* 204, pp. 684-693.

Sullivan, W. B.: 1978, *Astrophys. J.* (in press).
Toomre, A. and Toomre, J.: 1972, *Astrophys. J.* 178, pp. 623-666.
Visser, H. C. D.: 1978, Rijksuniversiteit te Groningen (Ph.D. Thesis).
White, S. D. M.: 1978, *Comments Astron. Astrophys.* (in press).
Wilkerson, M. S., Strom, S. E., and Strom, K. M.: 1977, *Bull. Am. Astron. Soc.* 9, 649.
Woodward, P. R.: 1976, *Astrophys. J.* 207, pp. 484-501.

DISCUSSION

Ostriker: You quoted a paper by Spinrad et al. with regard to the mass ratio M_{Disk}/M_{Halo}. In that paper, which was based on two dimensional photometry, we derived ratios of the light L_{Disk}/L_{Halo} for several spiral galaxies. No information about mass ratios is obtainable from photometry without unwarranted assumptions about mass to light ratios. There are dynamical arguments that may be made concerning the ratio M_D/M_H and these, which will be noted in my talk, indicate a ratio $\lesssim$ 1.

Strom: Agreed. I should emphasize that M_H/M_D was deduced from the observation of visible light only and involved the assumption that $(M/L)_H = (M/L)_D$. However it is of some interest that M_H/M_D deduced from optical photometry under these assumptions suggests that the halo mass may be sufficient to stabilize the disk.

II. THE DISK COMPONENT

INTRODUCTION TO THE SESSION

James Lequeux, Observatoire de Meudon

The main purpose of this session is to gather observational data which should ultimately yield a better understanding of the relationships between the interstellar matter and the stars in our Galaxy. Ideally we would like to know how the Galaxy evolves in this respect and in particular to understand quantitatively the laws of star formation: What is the relation between star formation and the density of gas? (for a study of this kind see Guibert, Lequeux, and Viallefond, Astron. Astrophys. 68, 1, where Schmidt's law is rediscussed); Does the Initial Mass Function (IMF) vary within the Galaxy? (for recent studies see Burki, Astron. Astrophys. 57, 135, and Puget, Serra, and Ryter, this Symposium). We are obviously still far from being able to give definitive answers to these questions, but I hope that we will obtain at least a better understanding of the basic ingredients: the distribution of interstellar matter and of young stars.

There are many ways of obtaining the galactic distribution of interstellar matter, most of which will be covered today:

- The 21-cm line gives the distribution of atomic hydrogen(Kerr).

- The distribution of the molecular component is indirectly derived from the distribution of the CO millimeter line emission and from the distribution of other molecules such as OH, H_2CO, and CH (Solomon, Cohen, Scoville, et al.) A big problem remains here in the conversion of CO intensities to H_2 masses, a conversion which is probably not more accurate than a factor 3 or more.

- Interstellar extinction and far-infrared radiation give (provided a grain model is adopted) the distribution of the dust and indirectly of the gas (Lyngå, Puget).

W. B. Burton (ed.), The Large-Scale Characteristics of the Galaxy, 27–28.

- The distribution of interstellar matter in all forms can also be inferred from the γ-ray surveys (Paul) although hypotheses must be made on the flux of cosmic rays in various parts of the Galaxy.

- The distribution of heavy elements (mostly O) in all forms can be derived from low-energy X-ray absorption studies. Few new data have been obtained, and this will not be discussed at this Symposium.

Turning now to young stellar populations, data on their local distribution, formation rate and initial mass function are obtained from optical studies of young stars (Humphreys) and of HII regions (Sivan). Only radio and IR observations can yield data for the entire Galaxy. Surveys of OH-IR stars will be described by Oort at this meeting; Puget will discuss far-IR and near-IR surveys of the galactic plane. These surveys also provide very interesting new insights into star formation as a function of galactocentric distance. Other information comes from radio observations of HII regions and from the thermal continuum (Lockman; additional data will be presented by Downes and by Mezger), and also from radio studies of pulsars, supernova remnants and the non-thermal continuum (Wielebinski, Taylor).

The most interesting result of all these studies is that there is a very large rate of star formation at 5 kpc from the galactic center, in the region where there is also a concentration of gas. The rate is so high that the lifetime of the gas against astration is only a few 10^8 years, so that the gas has to be replenished by some mechanism; there are also some indications that the IMF could be different in this region, with relatively less very massive stars formed. These results obviously have a very strong bearing on galactic evolution.

THE GALACTIC DISTRIBUTION OF OH/IR STARS

B. Baud, H. J. Habing, and J. H. Oort
Sterrewacht
Leiden, Netherlands

Through systematic surveys (Johansson et al., Caswell and Haynes, Bowers, Baud, Caswell et al.) some 200 OH masers have been detected that are presumably associated with long period variables of very late spectral type (>M5). Tentatively these stars will be called "OH/IR stars". They are characterized by their strong emission in the 1612 MHz line which shows a double peak due to the expansion of the shell surrounding these stars. The velocity difference ΔV between the peaks is generally between 10 and 50 km s^{-1}.

The distribution of these objects can be studied through a large fraction of the Galaxy. They show a strong concentration to the galactic plane, and a strong concentration to lower longitudes (cf. Figures 1 and 2). However, their longitudinal distribution reaches a maximum between $\ell = 30^o$ and $\ell = 15^o$. The derivation of the (true) galactic distribution (Bowers, Baud) shows an increase in spatial density as R^{-4} for $R > 4.5$ kpc, but a decrease in spatial density as R^{+7} for $R < 4.5$ kpc. (Here R is the distance from the Galactic Centre). This decrease cannot be explained by observational selection. Its reality is supported by a discussion of the radial velocity distribution and by a more sensitive pilot survey made with the Effelsberg telescope. The derived density distribution in our Galaxy is similar to that of CO.

However, not all OH/IR stars belong to the *extreme* population I; about half of them have velocity dispersions, and presumably ages, comparable to early A-type stars. It is possible to separate them roughly into age groups by the velocity difference ΔV between the two peaks in the spectra. Figure 3 indicates how the random motions decrease with increasing ΔV. While among stars with $\Delta V > 31$ km s^{-1} there are only 2 with negative radial velocity and only 3 with V beyond the dashed curve the diagram for $\Delta V \leq 26$ km s^{-1} shows a great fraction outside these limits.

The variation in the velocity dispersion is also reflected in the latitude distribution: stars with small ΔV show a wider distribution in z than those with large ΔV, as indicated in the table:

W. B. Burton (ed.), The Large-Scale Characteristics of the Galaxy, 29–34.

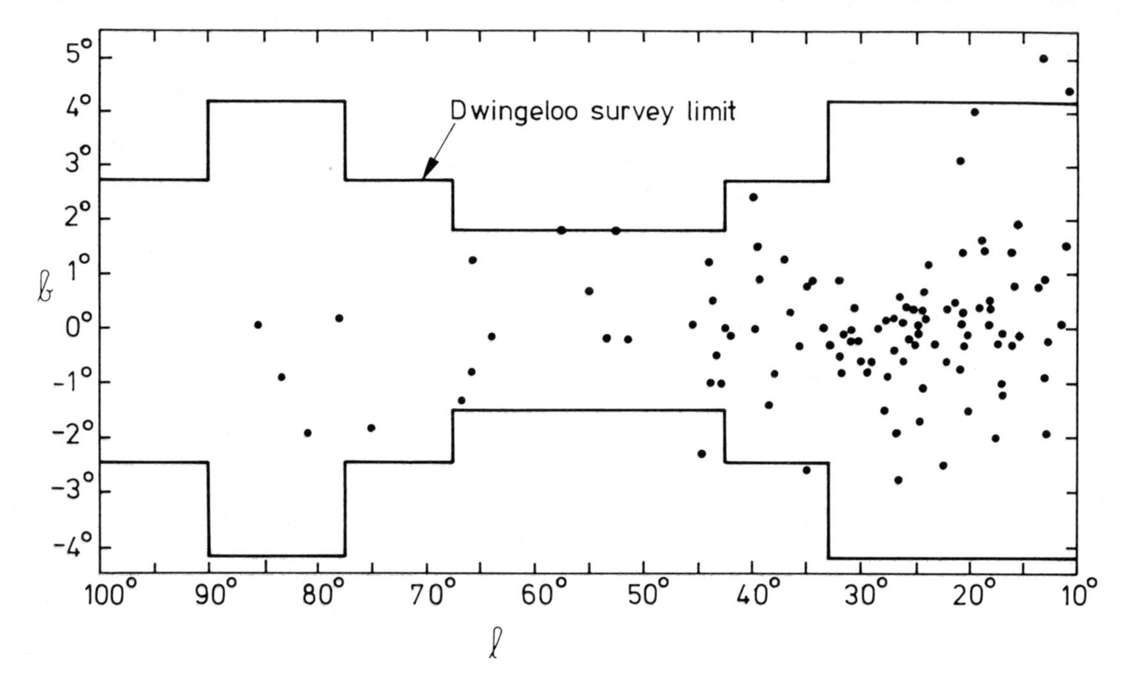

Figure 1. Distribution of representative sample of OH/IR sources with $\ell > 10°$ in galactic co-ordinates (Baud 1978).

ΔV km s^{-1}	$<\|b\|>$ o	$<\|z\|>$ pc	disp Z km s^{-1}	disp Π km s^{-1}	Age y
$\gtrsim 29$	0.5	~ 60	~ 8	~ 10	$\sim 10^7$
<29	1.1	~ 140	10-15	~ 35	$(1\text{-}2)\times 10^9$

For the small-ΔV stars the ratio of the velocity dispersions in the Z- and Π-directions appear to be less than 0.5 at R $\sim$ 5 kpc where most of the stars are situated. The quantity ΔV appears to be correlated with the period (Dickinson *et al.* 1975).

The kinematic behaviour of the OH/IR stars appears to be similar to the behaviour of Mira-type stars, as analyzed by Feast (1963) but extended to much longer periods and sampled over a larger part of the Galaxy. Direct identification of a few OH/IR stars with oxygen-rich long period variables supports the view that OH/IR stars are a mixture of late type giants and supergiants - well evolved stars of masses $\gtrsim 1.5\ M_\odot$, and ages $\lesssim 2 \times 10^9$ years.

The number of OH/IR stars increases strongly with decreasing luminosity L: $N(L)dL \propto L^{-\alpha}dL$, where α is estimated to lie between 1.1 and 1.6, down to a cut-off at perhaps 1 Jy.

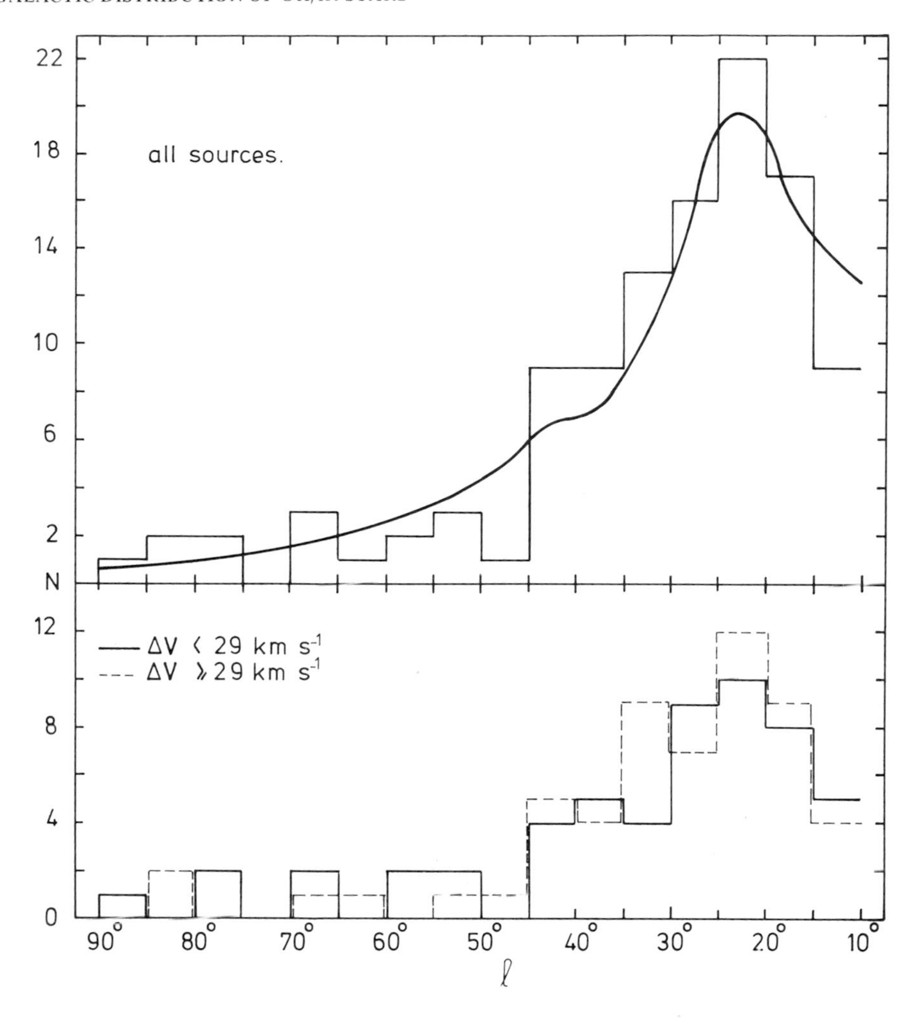

Figure 2. Longitude distribution between $\ell = 10°$ and $90°$ for all OH/IR sources (top) and for the sources with large and small ΔV separately (bottom). The curve is the distribution predicted from a density model (Baud 1978).

The distribution of OH/IR stars near the galactic center is known very incompletely. A recent survey by Baud and others between 358° and 14° longitude has yielded 43 sources, mostly new. A new deep survey within 1° of the center is being made at Effelsberg. The density in the central region is generally low, but shows an increase within $0\overset{\circ}{.}5$ from the center. For $|\ell| < 5^o$ the radial velocities show a symmetric distribution around V = 0 km s^{-1}, with an increasing dispersion for decreasing $|\ell|$. At $\ell = 0^o$ $\langle V^2\rangle^{\frac{1}{2}} \approx$ 130 km s^{-1}. One object was found to have an exceptionally high velocity, of −343 km s^{-1}; it is situated at $\ell = 0\overset{\circ}{.}3$, $b = -0\overset{\circ}{.}2$ (Baud et al.). In the center region the kinematic

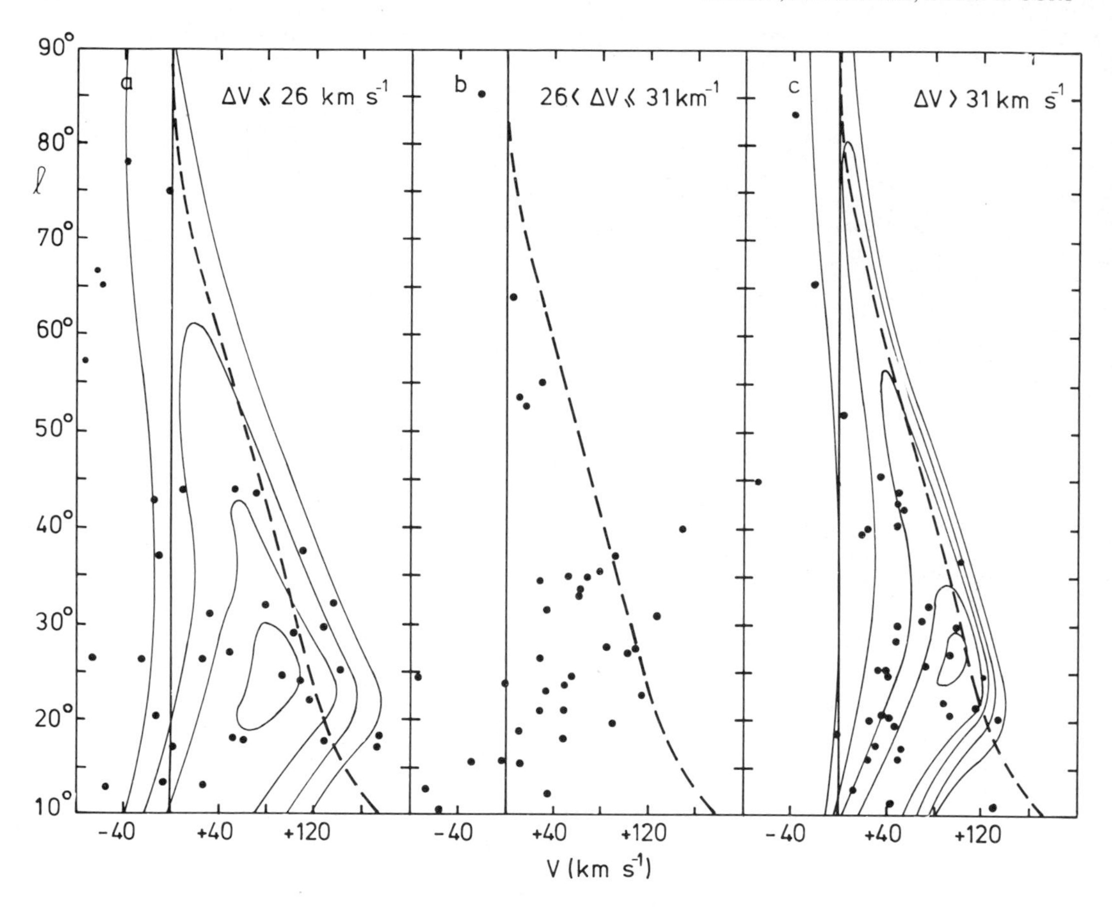

Figure 3. Radial-velocity distribution of OH/IR sources for three intervals of ΔV, containing approximately equal numbers of sources. Thin drawn lines indicate model velocity distributions derived with dispersions of 35 km s^{-1} (a) and 10 km s^{-1} (c) of the velocity components Π in the direction towards and away from the galactic center. The dashed curve (Burton 1974) corresponds to the maximum radial velocities corresponding to pure rotation (Baud 1978).

properties of OH/IR stars of short and long periods are the same; differences are marginal.

REFERENCES

Baud, B., Habing, H.J., Matthews, H.E., O'Sullivan, J.D., Winnberg, A. 1975, Nature 258, 406.

Baud, B. 1978, Thesis, Leiden University.

Bowers, P.F. 1978, Astron.Astrophys. Suppl. 31, 127.

Bowers, P.F. 1978, Astron.Astrophys. 64, 307.
Burton, W.B. 1974, Galactic and Extra-Galactic Radio Astronomy. Eds. G.L. Verschuur, K.I. Kellermann, Springer Verlag, Heidelberg, p. 82.
Caswell, J.L., Haynes, R.F. 1975, M.N.R.A.S. 173, 649.
Caswell, J.L., Goss, W.M., Haynes, R.F., Mebold, U. 1978 (in preparation).
Dickinson, D.F., Kollberg, E., Yngvesson, S. 1975. Astrophys.J. 199, 131.
Feast, M.W. 1963, M.N.R.A.S. 125, 367.
Johansson, L.E.B., Andersson, C., Goss, W.M., Winnberg, A.1977, Astron. Astrophys. Suppl. 28, 199.
Johansson, L.E.B., Andersson, C., Goss, W.M., Winnberg, A. 1977, Astron. Astrophys. 54, 323.

DISCUSSION

Bowers: First, although I agree that the unidentified OH/IR stars with larger values of ΔV may tend to be younger, this does not imply that stars with larger values of ΔV (> 29 km s^{-1}) are necessarily supergiants. For example, there are identified OH Miras with ΔV > 29 km s^{-1}. From comparisons of the 1612-MHz profile structures of identified and unidentified sources, I have suggested that possibly $\lesssim$ 10% of the unidentified sources may actually be supergiant stars (Bowers, Astr. Ap. 64, 307; Bowers and Kerr, Astr. J. 83, 487). The good correlation in ℓ,v space between the OH/IR stars with ΔV > 29 km s^{-1} and CO also does not imply that these stars are only ∿ 10^7 years old because it has been suggested that the massive molecular clouds are distributed in interarm regions and age estimates of a few times 10^7 to 10^8 years have been suggested for these clouds.

Second, if the period and ΔV are correlated for the unidentified OH/IR stars, then the age of 1-2 x 10^9 years, given for the stars with smaller ΔV(∿ 20-29 km s^{-1}) is inconsistent with periods > 500 days. The age-velocity dispersion relation used to estimate ages $\gtrsim$ 10^9 year is only well known in the solar neighborhood, not at R $\gtrsim$ 5 kpc where most of these unidentified sources are located. If, indeed, the periods of these stars are > 500 days, it seems more likely that their ages do not exceed 5 x 10^8 years, as indicated by Bowers (1978) based on a calibration of the period-age relation originally derived by Feast (MNRAS 125, 367). For this case I have suggested that most of the unidentified sources are a rare class ($\lesssim$ 1%) of Mira variables which are not only relatively young but also relatively massive ($\gtrsim$ 2.5 $M_\odot$), with extensive mass loss. Because of their rarity, it is no surprise that we find few of them in the solar neighborhood.

Third, I would like to make the general comment that frequency distributions as a function of galactic coordinates are strongly influenced by the sensitivity of the particular survey. Thus such diagrams must be considered cautiously when comparing results of different surveys. For example, both my survey and the Onsala survey showed that these sources are systematically below the galactic plane at 1 $\simeq$ 25° to 30°. I interpreted this to imply that the sources are systematically below the plane near the maximum in their radial density distribution. Although Baud's results do not show this deviation, his survey is of higher sensitivity and thus is sampling a larger volume of the Galaxy in these directions. The deviation from the galactic plane for the mean z-height of these sources strengthens the result that they are concentrated in the 5-kpc annulus because such deviations are known for numerous other "young" species in this part of the Galaxy.

Finally, I would like to suggest that because these stars are likely progenitors of planetary nebulae, it seems probable that there is a subsystem of planetaries evolved from these stars with a density concentration at about 5 kpc from the galactic center. As with the unidentified OH/IR stars, however, such a subsystem should represent a very small percentage of planetary nebulae.

Oort: First point. The small average $|z|$ for the stars indicates that they are young, not older than about 10^8 years. I agree that there is no reason to assume them to be still younger.

Second point. We do not know whether ΔV is strongly correlated with period; wasn't this only a suggestion? The only criterion for estimating an age is the z-direction round R = 5 kpc. It is true that Wielen's relation between age and $\langle |z| \rangle$ is based on data at $R \approx 8.5$; but Baud has estimated how the relation would change when going to $R \approx 5$ kpc.

Third point. This was fully recognized by Baud, who made a special effort to take account of the differences in sensitivity and resolution of the different surveys. I do not see why the higher sensitivity of this survey could cause a systematic error at R = 5 kpc.

As regard the planetary nebulae, I do not know of any sign of a concentration near R = 5. They are undoubtedly much older. Is there a good reason to think that the OH/IR stars would be their progenitors?

Burton: How do you measure from the data the quantity disp z?

Oort: It was obtained firstly from a combination of the dispersion in b and z at the longitude of maximum density with a value of K_z estimated from a mass model of the galaxy, like Schmidt's model. It was independently inferred from the value of disp π, which may be estimated from ℓ,v diagrams, in combination with an assumption concerning the ratio disp z/disp π.

Heiles: How does the large velocity dispersion in the galactic interior (R < few kpc) arise?

Oort: It may well be of the same nature as the high velocities which are observed in the CO clouds; these reflect apparent expansion velocities in a tilted central disk.

GIANT MOLECULAR CLOUDS IN THE GALAXY: DISTRIBUTION, MASS, SIZE and AGE

P. M. Solomon†, D. B. Sanders†, and N. Z. Scoville††

†Astronomy Program, State University of New York, Stony Brook, N.Y.
††Astronomy Dept., University of Massachusetts, Amherst, Mass.

1. INTRODUCTION

Millimeter wave observations of emission from the CO molecule have become, over the past eight years, the dominant method for determining the physical properties of dense interstellar clouds, composed primarily of molecular hydrogen and for exploring the structure and kinematics of the galactic disk. In this paper we briefly review the CO survey results in the literature (Section 2) and then present new results (Section 3-7) of an extensive ^{13}CO and ^{12}CO survey of the galactic distribution, size, mass and age of molecular clouds. The interpretation of this survey leads to a new picture of the interstellar medium dominated by very massive stable long-lived clouds which we refer to as Giant Molecular Clouds. We find that Giant Molecular Clouds (GMC's) with $M = 10^5 - 3 \times 10^6 M_\odot$ are a major constituent of the galactic disk, the dominant component of the interstellar medium in the galaxy interior to the sun and the most massive objects in the galaxy. We find that the interstellar medium and star formation are dominated by massive gravitationally bound clouds in which stars and associations are forming but at a very low rate in comparison to the free fall time. The galactic distribution of the molecules as traced by CO emission is interpreted as the distribution of GMC's. As the most massive objects in the galaxy they are also basic to the dynamics of the disk.

2. SUMMARY OF PREVIOUS RESULTS ON RADIAL DISTRIBUTION OF CO EMISSION

The most striking large scale characteristic of molecular clouds in the galaxy is their concentration in the galactic center region and in a ring at a galactocentric distance (R) extending from 4-8kpc. Both of these features are in sharp contrast to the flat atomic hydrogen distribution. The shape (see Figure 1) and magnitude of the effect have remained unchanged since the original CO survey by Scoville and Solomon (1975) of 90 points in the galactic plane, carried out with the NRAO 36-foot antenna (beamwidth of 1!1 at the CO wavelength of 2.6mm). The observations included a sampling every 1° in longitude from $\ell = 0^o$ to $\ell = 90^o$.

W. B. Burton (ed.), The Large-Scale Characteristics of the Galaxy, 35–52.

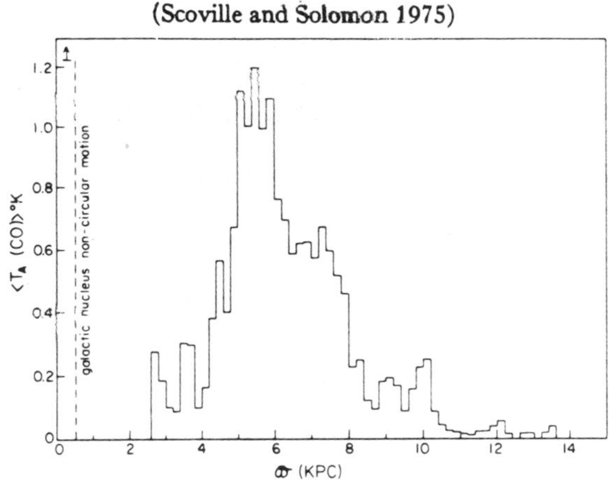

Figure 1.
Distribution of CO emission as a function of galactocentric distance showing the 5.5kpc peak and ring like structure between 4-8kpc (Reproduced from Scoville and Solomon, 1975, Ap. J. 199 L105)

The use of the small beam combined with the large cloud sizes made it possible to truly sample the cloud distribution rather than just the distribution of emission from confused sources. The interpretation of the well defined features in the line profiles was in terms of emission from discrete clouds. Estimates of the total mass were based on counting clouds to obtain the number of clouds per unit length, using 3 observations of ^{13}CO emission in the plane combined with radiative transfer calculations to estimate the H_2 density within each cloud and adopting a standard path length of 20pc through each cloud. The necessity for more ^{13}CO observations to obtain better quantitative densities was stressed; however the ratio of the total number of clouds at the 5kpc peak compared to 10kpc, the essential feature of the distribution, can be deduced just from the counting of emission lines and the galactic rotation law. The main results are summarized in Table 1.

A subsequent ^{12}CO survey was carried out by Burton and Gordon (see Table 1) with higher sensitivity and containing about 5 times as many points, all at $b = 0^o$. Gordon and Burton quoted a mass and H_2 density based on an assumed ^{13}CO intensity ratio of 1/3 for all emission, but ^{13}CO was not observed. Burton (1976) has reviewed this work. Cohen and Thaddeus have presented ^{12}CO survey results, including data taken out of the plane, obtained with a 4 foot antenna. The scale height was observed to increase between 5 and 8kpc. The general shape of the radial distribution was similar to that found by the earlier surveys except for a total absence of emission between 2 and 4kpc. These results however are not confirmed by the new observations reported here.

3. OUTLINE OF FOUR PART SURVEY

In the following we present the principal results of a new four part survey (Solomon, Sanders and Scoville, 1979) designed to determine the distribution and total mass of molecular clouds and consequently star formation regions in the galaxy as well as the properties of the

TABLE 1.

SUMMARY OF GALACTIC CO SURVEY RESULTS (Prior to 1978)

AUTHORS	RADIAL DISTRIBUTION	H_2 DENSITY	Z(FWHM)	COMMENTS
SCOVILLE & SOLOMON ApJ 199 L105, 1975	"RING" AT 4-7kpc PEAK AT 5.5kpc AND AT <400pc MIN. AT 2-4kpc	1-5cm^{-3} 5cm^{-3} AT PEAK	130pc (FWHM) 2 ℓ's	TOTAL MASS 1-3x10^{9}M$_\odot$ MASS PER CLOUD ~10^{5}M$_\odot$,MOLECULES IN CLOUDS STRONG CLUMPING MORE H_2 THAN H
BURTON et al. ApJ 202, 30 1975	4-8kpc	NOT GIVEN	70pc	CONSIDERABLE STRUCTURE
GORDON & BURTON ApJ 208,346 1976	4-8kpc RING MAX. at 5.7kpc MIN. at 2-4kpc	2cm^{-3} AT PEAK	117pc 1 ℓ	CLOUD DIAMETER 5-17pc 10^6 CLOUDS TOTAL STOCHASTIC MODELS TOTAL GAS INCREASING TO CENTER
COHEN & THADDEUS ApJ 217 L155	4-8kpc RING NO EMISSION AT <4kpc	NOT GIVEN	90-170 pc	SCALE HEIGHT INCREASES BETWEEN 5-8kpc CENTER BELOW b=0°

individual clouds in terms of size, mass, temperature, stability and age. The first three parts containing about 1500 separate observations have been carried out at the NRAO[1] 36-foot antenna with some additional data from the FCRAO 45-foot antenna. The purpose and type of observations are:

I. H_2 Densities from ^{13}CO Observations.

II. Size and Mass of Molecular Clouds from High Spatial Resolution ^{12}CO Strips in ℓ and b.

III. Galactic Distribution in Radius, R and Height Above the Plane Z, Total Mass and Possible Spiral Structure (if any) from ^{12}CO Observations every 1° in ℓ, every 12' in b.

IV. Two Dimensional Maps of Selected Clouds to determine individual cloud Size, Shape and Mass.

4. ^{13}CO OBSERVATIONS

The use of CO as a tracer of molecular hydrogen within molecular clouds and of molecular clouds throughout the galaxy, is based on two related premises. First the CO to H_2 abundance ratio is assumed to be reasonably constant and therefore a measure of the CO column density will

[1]Operated by Associated Universities, Inc., under contract with the NSF.

yield a measure of the hydrogen column density, and secondly the photons emitted by CO, and consequently the line intensity, result from collisions between H_2 and CO.

However, the high intensity and ubiquity of the ^{12}CO emission, which allows extensive observations of otherwise unknown molecular clouds, is due to a high opacity and therefore is not easily interpreted in quantitative terms. At moderately high densities the ^{12}CO intensity reflects primarily the cloud temperature. In order to quantitatively interpret the CO survey of the galaxy and test the validity of the molecular distribution derived from ^{12}CO, we have observed the much less abundant isotope ^{13}CO at even galactic longitudes with $b = 0^o$ and $10^o \leq \ell \leq 44^o$ and $\ell = 0^o$.

4.1 Radial distribution of ^{13}CO

A sample spectra of ^{13}CO and ^{12}CO is presented in figure 2; the close correspondence between them is evident but the intensity ratio T^{13}/T^{12} is different in each feature. In our total sample the observed ratio varies from less than 1/20 to 1/3 with an average value of 1/5.5. In figure 3 we compare the ^{13}CO and ^{12}CO emissivity, $J = T\ dv/dr$, from the same data set, as a function of distance from the galactic center. The extremely good agreement of the radial distribution between ^{13}CO and ^{12}CO establishes the reality of the high concentration of molecules between 4-8kpc as deduced from ^{12}CO. Thus ^{12}CO emission is a good indication of the relative mass distribution in the galaxy even though it is saturated. This can be explained if the total emission depends on the number of clouds rather than the intensity of each cloud.

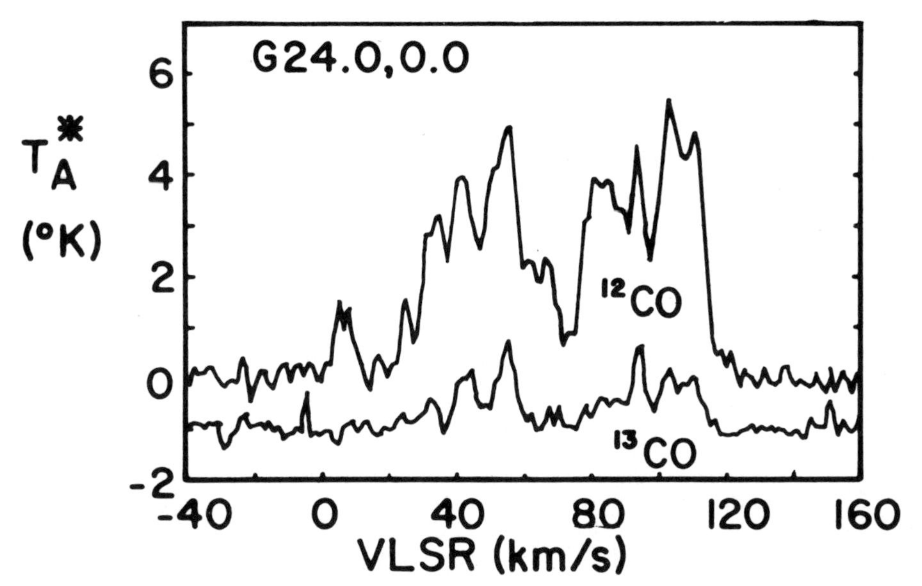

Figure 2. Sample spectra of ^{12}CO and ^{13}CO. The intensity is in units of T_A^*. $T_A^*/\eta = 1.6\ T_A^*$.

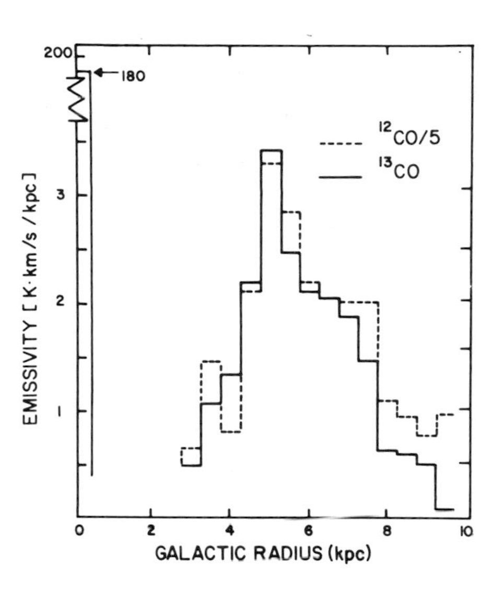

Figure 3. The ^{13}CO and ^{12}CO emissivity, J, as a function of galactic radius.

4.2 Calibration of H_2 densities

We have analyzed the ^{13}CO data and relative intensities T^{13}/T^{12} using the escape probability approximation for radiative transfer (Scoville and Solomon 1974). Assuming $^{12}CO/^{13}CO = 40$, we find that a good value for $^{13}CO/H_2$ is 5×10^{-7} corresponding to $CO/H_2 = 2 \times 10^{-5}$ or about 3% of C in CO. As can be seen from figure 3, there is no systematic variation of T^{13}/T^{12} with galactic radius and we therefore assume that $^{13}CO/H_2$ and $^{13}CO/^{12}CO$ are constant, in interpreting the data. There is insufficient sampling at $R \geq$ 9kpc to judge the reality of the apparent decrease in T^{13}/T^{12}. The one observation of the galactic center shows $T^{13}/T^{12} = 1/7$, very similar to the rest of the galaxy. Using the above values and a typical observed ^{12}CO intensity ($T_A^* = 9$-13K) which determine the stimulated emission factor and partition function, the H_2 column density can be shown to be

$$N(H_2) = 3.6 \times 10^{21} \int T\,(^{13}CO)\,dv\ \mathrm{cm}^{-2} \quad ,$$

or

$$\frac{N(H_2)}{\Delta r} = \bar{n}_{H_2} = 1.2\ J\,(^{13}CO)\ \mathrm{cm}^{-3} \quad ,$$

where J [K·km/s/kpc] is the ^{13}CO emissivity. Since the observed mean value at $3 < R < 9$ kpc of J^{12}/J^{13} is 5-6, we adopt $\bar{n}_{H_2} = 0.2\ J\,(^{12}CO)$ in order to derive molecular hydrogen densities and total mass from the large scale ^{12}CO survey which samples emission from a full range of b and ℓ (see section 6). The mean H_2 density at 5.5kpc is therefore $n_{H_2} \sim 5\ \mathrm{cm}^{-3}$. This result is however sensitive to the true ratio $^{13}CO/H_2$.

5. CLOUD SIZE AND MASS DISTRIBUTION FROM HIGH SPATIAL RESOLUTION DATA

The goal of these observations is to determine the nature of the objects containing interstellar molecules. This project is designed specifically to obtain the cloud properties from a random sample.

Two strips in galactic longitude have been observed at $b = 0^\circ.0$, $23^\circ.0 \leq \ell \leq 30^\circ.5$ with 2' spacing and at $b = -0^\circ.6$, $11^\circ.5 \leq \ell \leq 16^\circ.0$ with 4' spacing. In addition a few strips perpendicular to the plane have been observed at $\ell = 12^\circ$, 24° and 30°. The complete data can be found in Solomon et al. (1979) and Solomon and Sanders (1979). The emission in the velocity-longitude plane (Figure 4), summarizing 215 separate observations, shows a very clear breakup into clouds. The approximate galactocentric distance of the features is indicated on the top axis; the concentration in the 4-7kpc region is very striking. The large number of features near the maximum permitted velocity is due to the small radial velocity gradient near the tangential point.

5.1 Cloud size distribution

In order to display the observations in a form convenient for measuring cloud sizes, they have been divided into sections of about 2 1/2 degrees in ℓ; the distance to a feature is calculated assuming it is on the near side of the tangential point, and the displacement in longitude is then converted to apparent displacement parallel to the galactic plane. The result is a drawing showing the projected size versus velocity as in figure 5.

Typical spacing between observations ($R \sim 5.5$kpc, $\ell \sim 27^\circ$, $\Delta\ell = 2'$) corresponds to about 3pc and the determination of cloud dimensions has a lower limit of about 10pc. The contour levels in Figures 4 and 5 are in units of $T_A^* = 1$K or $T_A^*/\eta \sim 1.6$K. Cloud sizes at the level of $T_A^* = 3$, 4 and 5K were measured for all high resolution observations along the galactic plane. Clouds were counted only if the size > 10pc at the 4K contour level with at least one contour at $T_A^* = 5$K. Measurement of size was obtained at a constant velocity; the velocity of maximum emission was restricted to a full range of only 3km/s in order for a feature to be designated as a single cloud. As can be seen from the figures, there is significant overlap at $T_A^* \leq 2$K both in velocity and space. The observed sizes are lower limits to the true sizes since the near distance has been assumed, the emission is traced only to the level $T_A^* = 3$K and the size represents a random slice through the cloud. Figure 6 shows the distribution of cloud sizes at the $T_A^* = 3$K ($T_A^*/\eta = 5$K) level. The average cloud size in the sample is 40 pc, a factor of 4 above the minimum detectable size. There are no emission holes inside these regions, although there may be several local maxima. These objects can best be described as Giant Molecular Clouds or possibly GMC Complexes and represent a class of objects which are the most massive in the galaxy.

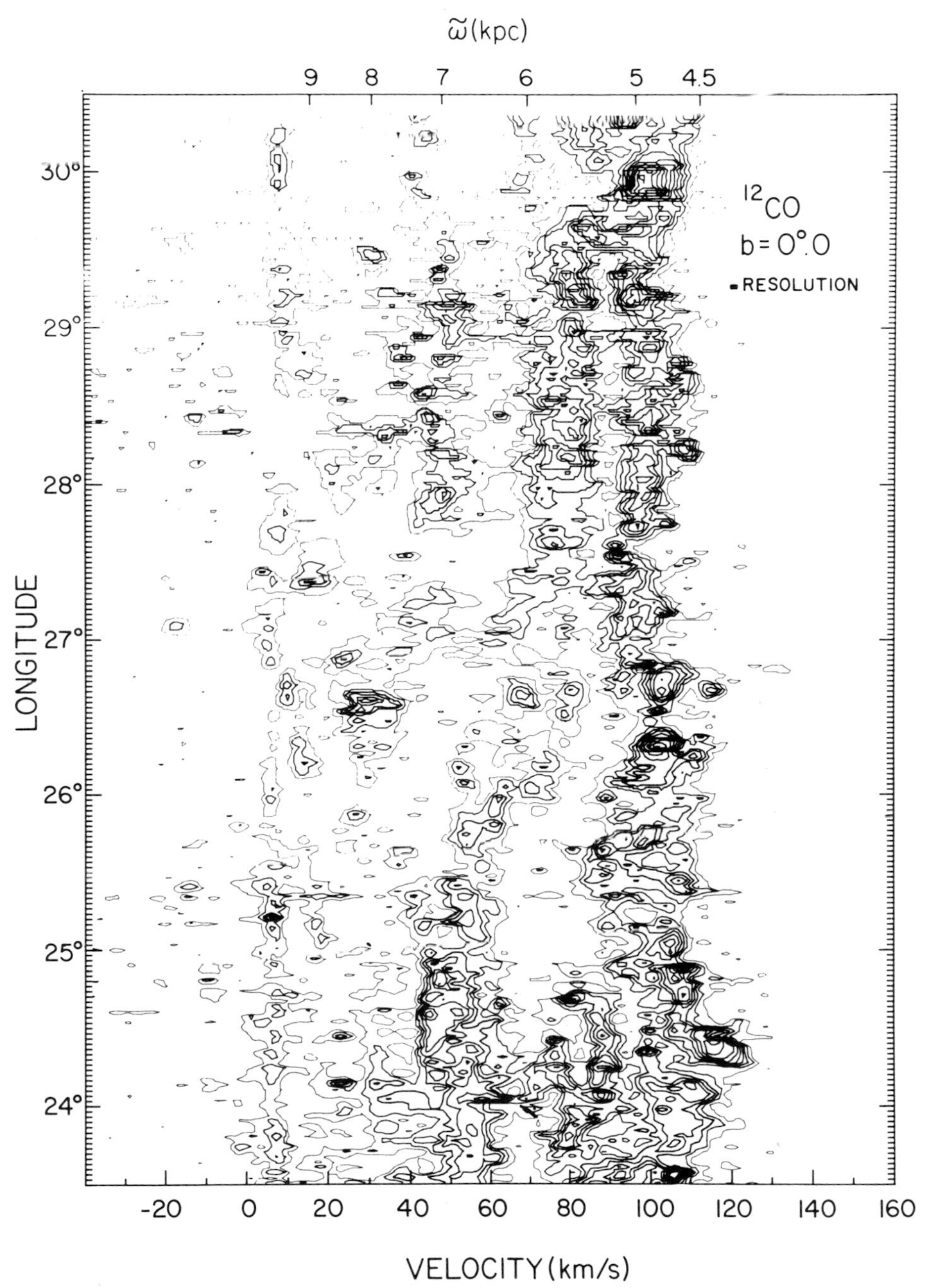

Figure 4. CO emission from data taken every 2' displayed in the longitude-velocity plane. Contour intervals are in steps of $T_A^* = 1K(T_A^*/\eta = 1.65)$. The diagram is a synthesis of 215 separate observations. The resolution is indicated by the black rectangle. The breakup into Giant Molecular Clouds is very striking above the third contour.

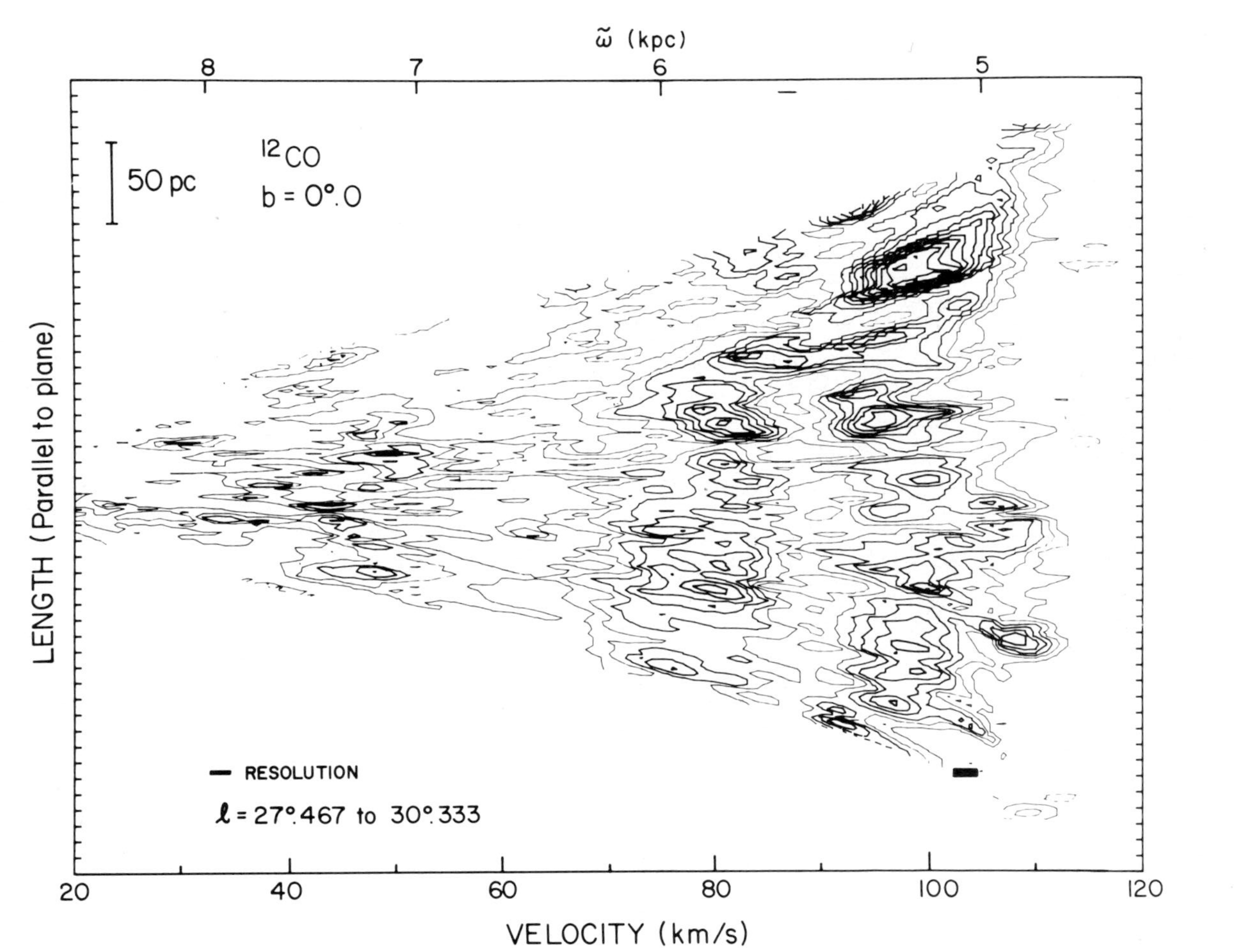

Figure 5. A sample of CO emission showing the sizes of the clouds along the galactic plane. In deriving the sizes, the near distance has been assumed. The resolution is indicated by the black rectangles. Lowest contour unit is $T_A^* = 1K$ ($T_A^*/\eta = 1.6K$) with intervals of 1K.

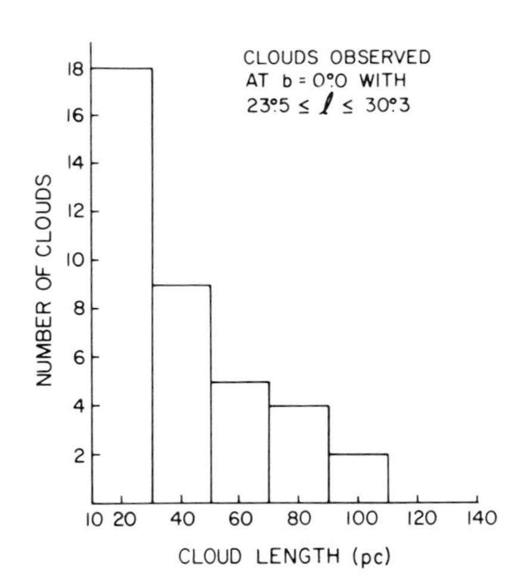

Fig. 6. The distribution of length parallel to the galactic plane for a typical chord intersecting a GMC, (size) determined from 2' resolution mapping parallel to the galactic plane between $\ell=23^{\circ}\!.5$ and $30^{\circ}\!.5$ at $b=0^{\circ}$. See text for a description of criteria used for measurement.

With a typical dimension of 40pc and a mean H_2 density of 300 cm^{-3} the mass per GMC in the sample is about $5 \times 10^5 M_{\odot}$. The density of 300 cm^{-3} is consistent with the observed ^{12}CO and ^{13}CO intensities. Radiative transfer calculations show that $T^{13}/T^{12} = 1/6$ for these conditions. A small fraction of the mass is in condensed cores within the clouds, which are the actual sites of star formation.

The observed distribution of cloud length (Fig. 6) has been obtained from a sample along a line parallel to the galactic plane. This represents a fair sampling of cloud area. In order to obtain the true distribution of cloud lengths a correction must be employed allowing for the fraction of clouds of length Λ in the sample which is $\propto 1/\Lambda$. The total number of clouds in the 4-8kpc ring meeting the selection criteria described above is

$$N(\Lambda)d\Lambda = \frac{N'(\Lambda)}{\varepsilon} \; \frac{2.55\, Z_{\frac{1}{2}}}{\Lambda} \; d\Lambda \qquad (1)$$

where N' is the number of clouds counted in the sample (Fig. 6), ε is the fraction of the galactic ring in ℓ and r sampled and $2.55\, Z_{\frac{1}{2}}/\Lambda$ is the ratio of the total number in the disk to the number sampled at $b=0^{\circ}$. $Z_{\frac{1}{2}}$ is the HWHP height of the cloud distribution from Section 6. We estimate that $\varepsilon \approx 0.05$ for the data at $23^{\circ}\!.5 < \ell < 30^{\circ}\!.5$, using the radial distribution of Section 6. (See Solomon et al. 1979 for further discussion.

The sample between $\ell = 23^{\circ}\!.5$ and $30^{\circ}\!.5$ at $b = 0^{\circ}\!.0$ contains 38 GMC's. Using the scale height and radial distribution from Section 6, we estimate that this represents approximately 0.75 percent of all clouds, leading to a total of 4000 Giant Molecular Clouds in the Galaxy. The lower limit for counting in this estimate is clouds with a chord larger than 10pc with some CO emission at $T_A^*/\eta > 8K$, ($T_{kinetic} > 11K$) in a random slice through the region.

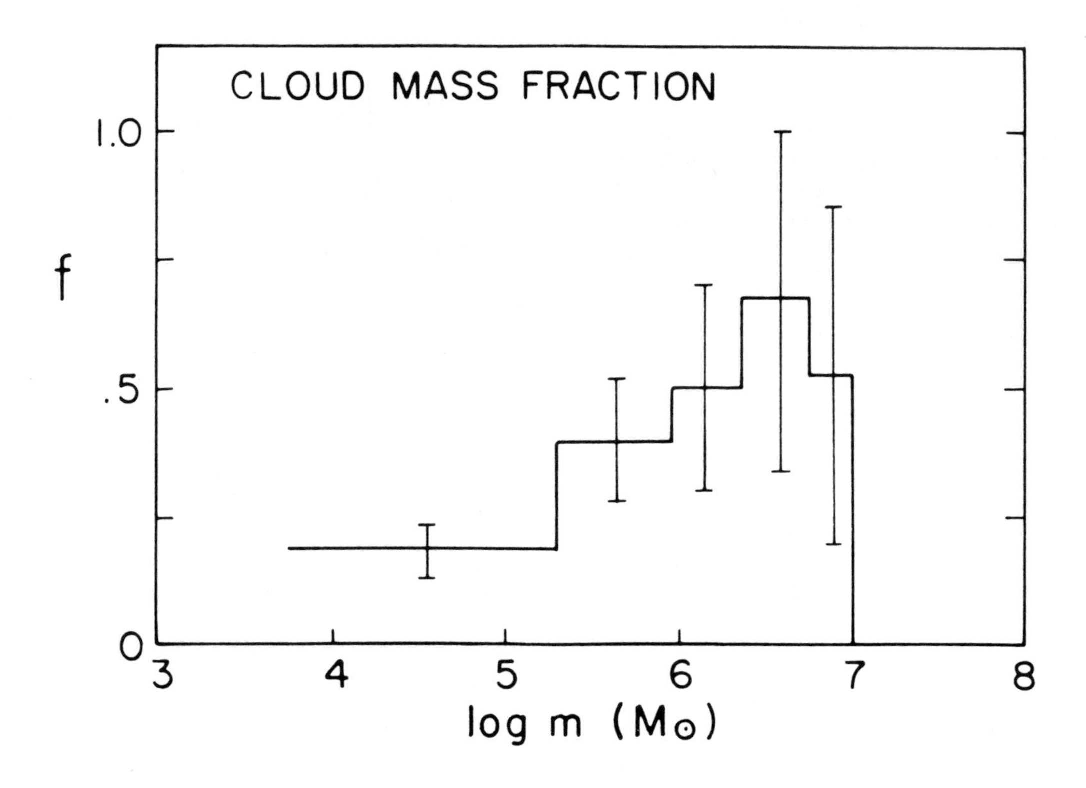

Fig. 7. The fraction of total mass contained in Giant Molecular Cloud Complexes of mass m, per logarithmic interval. Observational selection of Fig. 6 has been corrected by equation (1). The error bars are determined by the sample size. The greatest uncertainty in m stems from the uncertainty in $^{13}CO/H_2$.

5.2 Cloud mass function

The most important parameter in determining the origin and evolution of the clouds is the mass distribution. From our sample of 38 clouds corrected by equation (1) we can derive the fraction of total mass per logarithmic cloud mass interval. In order to proceed from linear size measurements, we assume that the measured length represents a random slice through an irregular cloud and that the true cloud mass can be approximated by a sphere of this dimension with a density $n(H_2) \sim 300\ cm^{-3}$. The effect of elongation along the galactic plane introduces a small error; however the greatest uncertainty comes from the very high mass end of the spectrum where the sampling becomes insufficient since a few GMC's may contain a large fraction of the mass. We have observed no GMC's in this sample with a length (as defined above) > 100pc, and the statistics give large uncertainties in the mass fraction in GMC's > 60pc with $m > 10^6 M_\odot$. Figure 7 shows the observed mass fraction per logarithmic mass interval

$$f = \frac{1}{M}\frac{dM}{d\log m}$$

where M is the total mass in clouds and m is the GMC mass. The total mass of the 38 GMC's is about $4 \times 10^7\ M_\odot$. The 4000 GMC's in the galactic ring have a total mass $\simeq 2 \times 10^9 M_\odot$. J. Kwan (private communication) has considered theoretical mass distributions for colliding molecular clouds which give reasonable agreement with Fig. 7. From our sample corrected by equation (1) we find $f \propto m^{0.2}$.

The formal gravitational collapse time (free fall time) for these objects is $t_c \simeq 3 \times 10^6$ yrs and yet most of the mass in clouds is contained in them. They are clearly not systematically collapsing to form stars since the galactic star formation rate of about $3\ M_\odot$/yr is only a small fraction of the total formal collapse rate. The efficiency of star formation can be defined as the ratio of the observed rate to the formal mass collapse rate in GMC's,

$$\varepsilon = \frac{\left(\frac{dM}{dt}\right)_{observed}}{\left(\frac{dM}{dt}\right)_{GMC}}$$

Using (dM/dt)GMC $\approx 3 \times 10^9/3 \times 10^6 = 10^3\ M_\odot$/yr yields $\varepsilon \approx .003$. We thus find that star formation is inhibited by a factor of $\simeq 300$ compared with the formal free fall time.

6. GALACTIC DISTRIBUTION IN R AND Z.

In the section of the galaxy $-4^\circ < \ell < 70^\circ$ we have obtained about 730 observations of ^{12}CO emission with full velocity coverage, spaced every 1° in longitude and every 12' in latitude out to positions where the total integrated intensity has dropped at least to 1/4 of the maximum ($|b| \sim 2\overset{\circ}{.}0$) and no features other than local clouds are apparent.

The large number of line profiles, showing emission from molecular clouds along the entire line of sight, can be viewed more comprehensively in the form of latitude-velocity contour diagrams; each latitude velocity drawing is a composite of all observations at a single galactic longitude. In figure 8 we present a sample b-v diagram which demonstrates some of the salient features of the distribution. The emission is discrete in both the velocity and spatial dimensions indicating that only a small fraction of the volume is occupied by molecular clouds. The clouds are often centered below $b = 0^\circ$ and there are irregular bumps in the location of the center. Observations confined only to $b = 0^\circ$ will therefore miss significant emission.

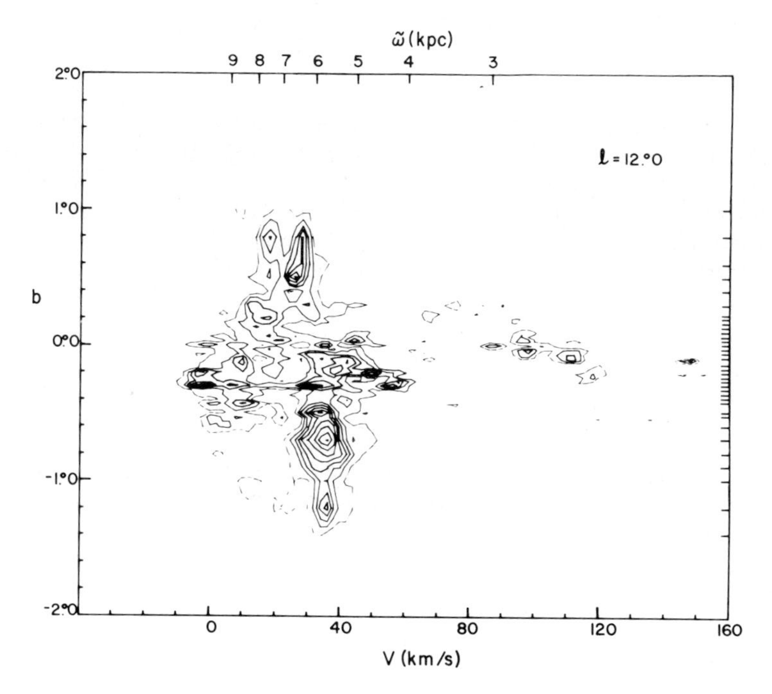

Figure 8. A sample latitude velocity diagram of CO emission from the survey at $\ell = 12^{o}$. The righthand vertical axis tick marks indicate points observed. Each contour interval is $T_A^* = 1K$ ($T_A^*/\eta = 1.6K$)

The intensity is typically $T_A^* = 3\text{-}5K$ or $T_A^*/\eta = 5\text{-}8K$ which translates to Planck brightness temperatures and therefore kinetic temperatures $T_{kinetic} = 8\text{-}11K$. This represents a typical kinetic temperature in a random region of a molecular cloud. Cloud cores with $T_k >> 10K$ are sampled only a small fraction of the time.

The survey data has been analyzed on the assumption of cylindrical symmetry to derive the radial, R, and height, Z, dependence of molecular clouds in the galaxy, the center of the distribution, Z_c, and the total surface emission of CO perpendicular to the galactic plane. The mean molecular hydrogen density and mass has been determined using the ^{13}CO calibration of Section 4.

For the purposes of our analysis the inner galaxy, $R < R_{\odot}$, has been divided into annuli $\Delta R = 0.5$kpc. The emissivity $J = T\ dv/dr$ was computed for all line segments passing through a given annulus determined from our samping in longitude $0^{o} < \ell < 70^{o}$. The appropriate fitting function (assuming cylindrical symmetry) is a double Gaussian which allows for contributions from the "near" and "far" segments along the line of sight with identical galactocentric distance R. Such points at distances r_1 and r_2 from the sun are equidistant from and on opposite sides of the subcentral point. The three variable parameters in the expression for J are the emissivity at the center of the distribution J_o, displacement of the center (from $b = 0^o$) Z_c, and scale height (half width at half maximum) $Z_{1/2}$.

$$J(R,b) = J_o \exp - \left[\frac{(Z_1 - Z_c)}{1.2\ Z_{1/2}}\right]^2 + J_o \exp - \left[\frac{(Z_2 - Z_c)}{1.2\ Z_{1/2}}\right]^2 \quad \text{K·km/s/kpc}$$

The distance perpendicular to the plane for near and far points contributing to $J(R,b)$ is given simply by $Z_1 = r_1 b$ and $Z_2 = r_2 b$.

The scale height of the molecular cloud distribution $Z_{1/2}$ and the center of the "plane" Z_c are presented in Figure 9. A displacement of Z_c below the $b = 0^o$ plane is found for the entire region $2.5 < R < 8.5$kpc with $Z_c = -26$pc in the ring. The region $4.5 \leq R < 8.5$kpc shows a relatively constant value of $Z_{1/2} = 60 \pm 9$pc. The HI scale height in this region determined primarily from emission near the loci of subcentral points has an average value of 130pc, about twice that for molecular clouds. (Jackson and Kellman, 1974).

The radial distribution in emissivity $J_o(R)$ over the range 2.5 - 9.5kpc and at the galactic center is presented in Figure 10. We have not included regions exterior to the sun which are inadequately sampled over this longitude range. At the peak of the distribution where $\bar{n}(H_2) \sim 5\ cm^{-3}$, there is a factor of 30 more H nucleons in molecular hydrogen than atomic hydrogen. Even if we assume that the calibration of section 4 has overestimated by a factor of three the ratio of emissivity to $n(H_2)$, by using too low a value of $^{13}CO/H_2$, we still find 10 times more nucleons in H_2 than H.

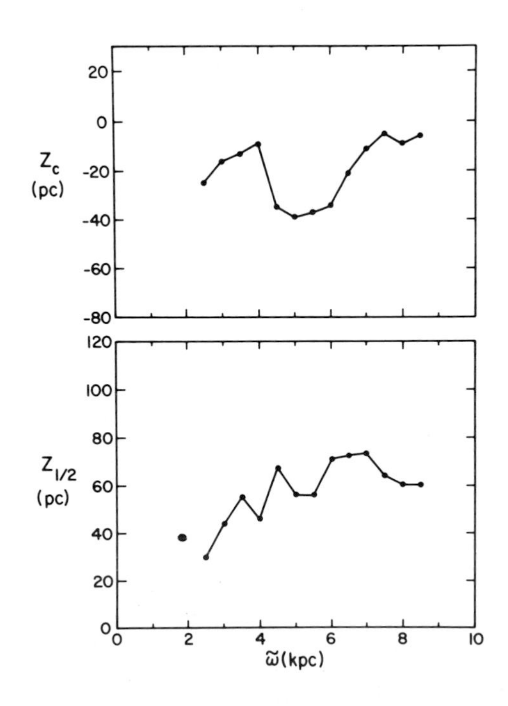

Figure 9. The scale height of molecular clouds in the galaxy $Z_{1/2}$ and the displacement from $b = 0^o$ represented by Z_c as a function of galactic radius.

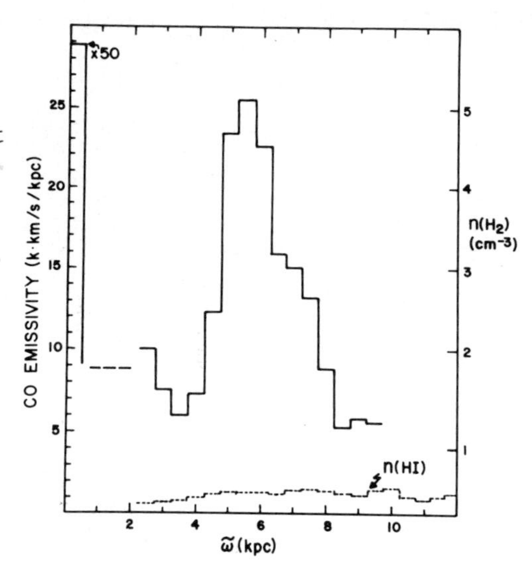

Figure 10. The CO emissivity J_o at the center of the distribution and the mean H_2 density. The transfer from emissivity to H_2 density has employed the ^{13}CO data. (see text). The $^{13}CO/H_2$ ratio is the largest uncertainty in the calibration.

The ring has half intensity at b = - 0°.8 and b = + 0°.3. The total surface density of molecular clouds perpendicular to the galactic plane is presented in figure 11. Integrating between 2 and 9.5kpc, we find a total mass in molecular clouds of 4 x $10^9 M_\odot$. Only approximately 1/15 of the total surface density in the interstellar medium is in atomic hydrogen at the peak of the ring. Again allowing for as much as 10% of available ^{13}C in ^{13}CO the ratio is still only 1/5.

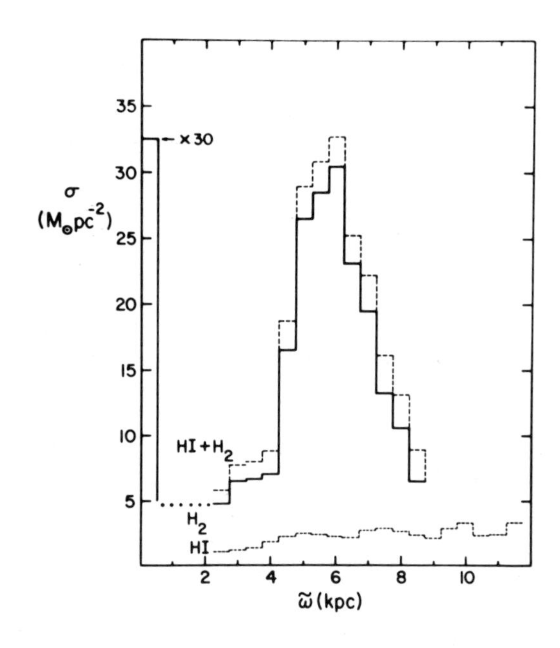

Figure 11. The mass surface density of molecular clouds in the galaxy. The total mass is about $4x10^9 M_\odot$.

7. TWO DIMENSIONAL ^{12}CO MAPS OF SELECTED CLOUDS

In order to explore the structure of individual clouds in greater detail, a set of 15 cloud features was chosen at random from the initial large scale ^{12}CO survey. The objective was to determine cloud sizes in ℓ and b, orientation with respect to the galactic plane, peak temperatures and the presence of core regions as well as kinematic information on the internal structure of the clouds.

Data for seven objects has been reduced and is summarized in Table 2. All clouds were mapped at 3' resolution. Sizes have been measured to a level $T_A^* = 4K$ and, where sufficient data exists, to $T_A^* = 3K$. The ℓ/b ratio in Table 2 indicates that there is some tendency for elongation along the galactic plane. The average size in ℓ is 34pc at $T_A^* = 4K$ using all 7 clouds and 45pc at $T_A^* = 3K$, using 4 of 7 clouds where sufficient data exists. The ℓ sizes are in good agreement with the high resolution longitude strip results of Section 5. ^{13}CO observations of each cloud have allowed the determination of column densities. Using the measured cloud cross sections and column densities, mass estimates for each cloud were also determined. The average mass is $6.2 \times 10^5 M_\odot$ similar to the mass found for a mean cloud size in the high resolution survey.

TABLE 2.

AVERAGE CLOUD SIZES† AT INTENSITY $T_A^*(^{12}CO) = 4K$

ℓ	34pc
b	26pc
maximum	49pc
ℓ/b	1.3
angle††	30°
mass	$6.2 \times 10^5 M_\odot$

†Sample of 7 clouds in the region $10^o \leq \ell \leq 40^o$, $4 < R < 8$kpc

††Angle of maximum chord from galactic plane

8. SUMMARY: AGE AND ORIGIN OF GMC'S

Although interstellar clouds are generally regarded as extremely young objects ($\leq 10^7$yrs) we show below that the large mass of individual clouds and the large fraction of the ISM in molecular clouds necessitates an age of at least 3×10^8yrs. We first summarize the cloud properties, based on the high resolution data (Section 5), in Table 3. The clouds have several small cores with much higher densities and in this sense should be regarded as complexes.

TABLE 3

AVERAGE PARAMETERS FOR GIANT MOLECULAR CLOUDS

Length $\langle Area \rangle^{1/2}$	40 pc
Kinetic Temperature	10K
Density, n_{H_2}	300 cm^{-3}
Mass	$5 \times 10^5 M_\odot$
Number in the Galaxy $4<R<8$kpc	4000

On a galactic scale we have deduced the smoothed out properties for the molecules including all CO emission without considering individual cloud masses. These are summarized from Sections 4 and 6 in Table 4. The largest uncertainty in the mean densities is the $^{13}CO/H_2$ ratio.

TABLE 4

MOLECULE DENSITIES AND SCALE HEIGHT AT 4-8kpc

Mean Density at 4-8kpc, $\bar{n}_{H_2}$	3 cm^{-3}
Scale Height, $Z_{1/2}$	60pc
Center, Z_c	- 26pc
Mass Surface Density, σ	20 $M_\odot/pc^2$
$2\ n(H_2)/n_{HI}$	20
Mass Ratio $\sigma(H_2)/\sigma_{HI}$	10

The age of GMC's, τ(GMC) can be simply determined from the observations by considering the relationship between the total mass in a form available for their formation, M_i, the time scale for formation by a particular mechanism τ_i and the total mass in GMC's M(GMC). In a steady state the total mass in a form is proportional to the lifetime. If form i is the dominant source for GMC's, then

$$\frac{M_i}{\tau_i} = \frac{M(GMC)}{\tau(GMC)} \quad \text{or ,} \qquad (2)$$

$$\tau(GMC) = \tau_i \frac{M(GMC)}{M_i}$$

In the most general case we have

$$\frac{1}{\tau(GMC)} = \frac{1}{M(GMC)} \sum_i \frac{M_i}{\tau_i} \qquad (3)$$

The generally accepted mechanism for formation of dense clouds is the compression of the diffuse interstellar medium or low density clouds by a spiral density wave. The time scale for this mechanism is the phase rotation period for the spiral density wave $\tau_R \approx 2 \cdot 10^8$yrs. If all of the interstellar medium represented by HI is available to be compressed, then

$$\tau_{GMC} \geq 2 \times 10^8 \text{yrs} \frac{M_{GMC}}{M_{HI}}$$

From the survey data we have $M_{GMC}/M_{HI} = 2\ \bar{n}_{HI} = 30$ if 3% of C is in the form of CO, and $2\ \bar{n}(H_2)/\bar{n}_{HI} = 10$ if 10% of C is in CO, giving $2 \times 10^9 < \tau_{GMC} < 6 \times 10^9$ yrs a significant fraction of the age of the galaxy. If the compression is viewed as working on the intercloud medium, the time scale is greater than the age of the universe. We conclude that GMC's do not form by compression of diffuse clouds in a spiral density wave.

The largest pool of mass available to form GMC are molecular clouds themselves. Cloud collisions, a mechanism considered for much smaller masses by Field and Saslaw (1965) are therefore an obvious candidate. In this case M_i, from (2) is the total mass in clouds available for collision with the GMC. What is clear from the mass distribution of clouds is that at least half of the total mass is in GMC's. We therefore take $M(GMC)/M_i = 1$ for this case and τ_i becomes τ_c, the growth time for a GMC by collisions. We have for a GMC of mass m ,

$$\tau_c \sum_j n_j\ \sigma\ v_j\ m_j = m(GMC) \tag{4}$$

with n_j clouds of mass m_j; σ is the GMC crossection πr^2. Substituting $\sum n_j m_j = \bar{n}_{H_2}\ m_{H_2}$ and $m = 4/3\ \pi\ r^3 n_{H_2}$ gives

$$\tau(GMC) = \frac{n_{H_2}}{\bar{n}_{H_2}}\ \frac{4/3\ r}{v} \tag{5}$$

when v is the weighted velocity of the clouds represented by $\bar{n}_{H_2}$. For our current estimate we merely adopted $\bar{n}_{H_2}$ from Table 4 using all available mass, n_{H_2} and r from Table 3 and set v = 10 km/sec giving

$$\tau(GMC) = 3 \times 10^8 \text{ yrs.}$$

This age which is 100 times the collapse time may be a lower limit, since we have assumed an equal mass in small cold clouds to that observed in GMC's. The problem of star formation in GMC is then one of an overabundance of opportunities and the question is not how do stars form but rather, why is star formation so inefficient in the current stage of galactic evolution.

ACKNOWLEDGEMENT

This work has been supported in part by NSF Grant AST 77-23419 at the State University of New York at Stony Brook

REFERENCES

Burton, W. B. 1976, Ann. Rev. Astron. Astrophys., 14, 275
Burton, W. B., Gordon, M. A., Bania, T. M. and Lockman, F. J. 1975, Astrophys J., 202, 30.
Cohen, R. S. and Thaddeus, P. 1977, Astrophys. J., 217, L155.
Field, G. B. and Saslaw, W. C. 1965, Astrophys. J., 142, 568.
Gordon, M. A. and Burton, W. B. 1976, Astrophys. J., 208, 346.

Jackson, P. D. and Kellman, S. A. 1974, Astrophys. J., 190, 53.
Scoville, N. Z. and Solomon, P. M. 1974, Astrophys. J., 187, L67.
Scoville, N. Z. and Solomon, P. M. 1975, Astrophys. J., 199, L105.
Solomon, P. M. and Sanders, D. B. 1979, "Proceedings of Gregynog Conference on Giant Molecular Clouds", eds. M. Edmunds and P. M. Solomon, Pergamon Press, Oxford, England.
Solomon, P. M., Sanders, D. B. and Scoville, N. Z. 1979, Astrophys. J. to be published.

DISCUSSION

Stark: In the bulk of your data, what is the antenna temperature of the noise and of the baseline ripple?

Cohen: The 1σ noise level is always less than 0.3 K in a 500 kHz bandwidth spectrum. The baseline curvature is generally about 0.1 K and is always less than the 1σ noise.

Solomon: $\Delta T_{rms} \simeq 0.3$ K. The baselines in our earliest data have some ripples of order 0.5 K but the recent data have negligible ripples. Some of our data (including ^{13}CO) has $\Delta T_{rms} \sim 0.1$ K.

Strom: Optical photographs of M33 reveal dust clouds with $\tau_V \geq 1$ of sizes $\geq$ 10 arc sec (linear size $\gtrsim$ 30 pc). Many of the clouds lie outside the spiral arms, in the disk of M33. Hence, optical observations of M33 suggest the presence of clouds, comparable in dimension with those Solomon reports for our own Galaxy, outside the arms.

Stecker: Concerning the important question of the absolute amount of H_2 in the 5-kpc annulus, have you made any assumptions to account for possible large-scale radial abundance gradients of [C]/[H] and [O]/[H] between R = 5 and 10 kpc?

Solomon: No. Our observations of the $^{13}CO/^{12}CO$ intensity ratio as a function of galactic radius do not show an increase with decreasing R. If $^{13}CO/H_2$ and $^{12}CO/H_2$ were both increasing inward we would expect to observe increasing saturation of ^{13}CO. The stronger emissivity at 5 kpc results primarily from a larger number of clouds and not from differences in the emissivity of individual clouds.

AN OUT-OF-PLANE CO SURVEY OF THE FIRST GALACTIC QUADRANT

R.S. Cohen, G.R. Tomasevich, and P. Thaddeus
NASA/Goddard Institute for Space Studies, New York City

We are undertaking at Columbia University an out-of-plane survey of 2.6 mm CO emission from the first quadrant of the galactic disk with a 1.2 meter telescope. With the 8' beamwidth of this instrument at 2.6 mm (about that of the largest existing steerable instruments at 21 cm) it is possible to completely map the first quadrant in only a few hundred days of observation.

The first systematic out-of-plane survey (Cohen and Thaddeus 1977; Cohen 1978), consisting of 179 spectra taken with the 1.2 m telescope, confirmed the findings of two previous in-plane surveys (Gordon and Burton 1978; Scoville and Solomon 1975) that galactic CO emission is concentrated in a ring 6 kpc in radius. A fit of a cylindrically symmetric galactic model to this data provided the first systematic determination of the thickness of this molecular ring as a function of distance from the galactic center, and showed that the ring lies roughly 40 pc below the b=0° plane.

Since this first survey was made, several significant improvements have been made in the 1.2 m telescope. A fast drive now allows rapid position switching over large angles. The spectral resolution has been improved from 2.6 km s^{-1} to 0.65 km s^{-1}, and the spectral range of the backend has been increased from 104 km s^{-1} to 166 km s^{-1}. The single sideband system noise temperature of the uncooled front end has been lowered from 1200°K to 860°K.

A new galactic survey was undertaken with this improved system during the past winter (1977-8). To date, about 1600 spectra have been obtained of the CO disk at longitudes from 12° to 60°, and latitudes from -1.25° to +1.25°. Specifically, b=0° has been sampled every beamwidth (0.125°) in longitude, b= ±0.25° and ±0.50° have been sampled every 0.25°, b= ±0.75°, ±1.00°, and -1.25° have been sampled every 0.50°, and some observations have been made at b=+1.25°. The LSR velocity range covered was from -13 to 153 km s^{-1} for l<55°, and from -55 to 112 km s^{-1} for $l\geq$55°. Integration times were adjusted to give a 3σ noise level of 0.8°K per 0.5 MHz spectral channel. Base-

W. B. Burton (ed.), The Large-Scale Characteristics of the Galaxy, 53–56.

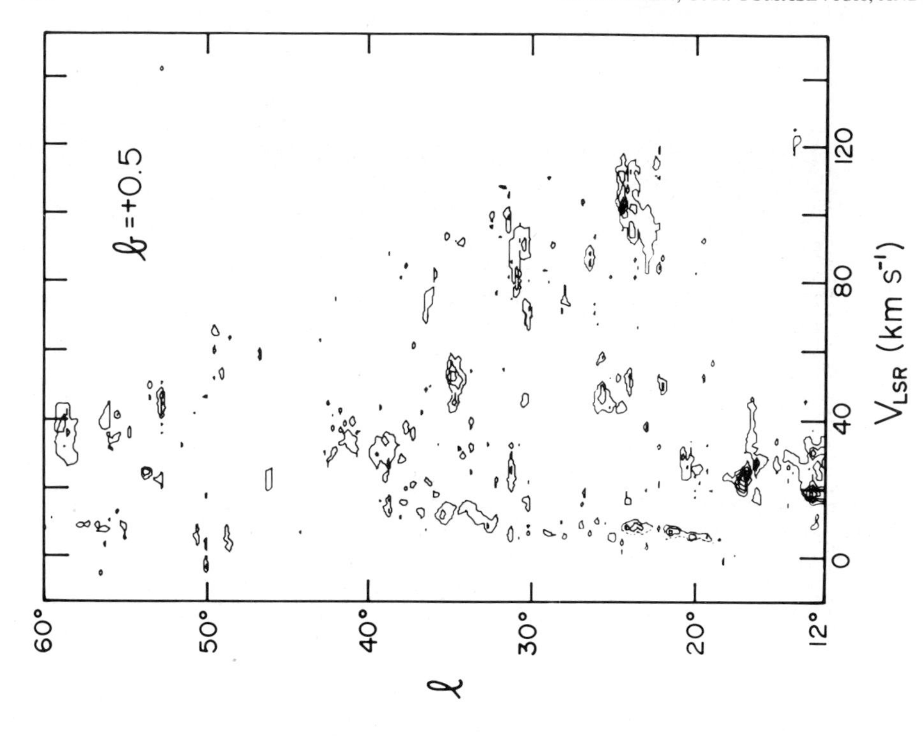

Figure 2. Longitude-velocity diagram of CO emission at b = +0°5.

Figure 1. Longitude-velocity diagram of CO emission at b = -0°5. The contour interval is 1 K of antenna temperature corrected for beam efficiency and atmospheric absorption.

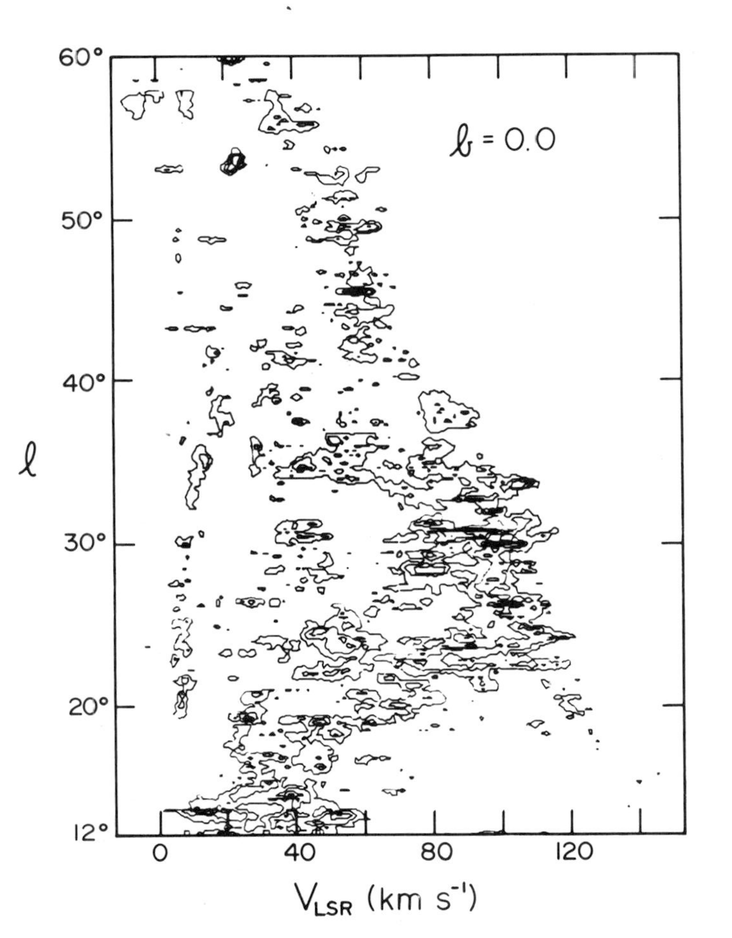

Figure 3. Longitude-velocity diagram of CO emission at b = 0°.

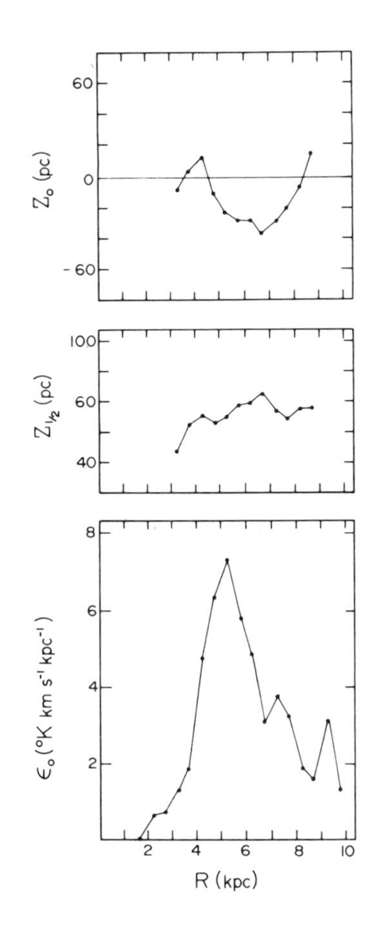

Figure 4. Displacement (top), half-thickness at half-maximum (middle), and centroid emissivity (bottom) of the CO disk as functions of galactic radius.

lines were linear to within the 1σ noise level.

Longitude-velocity diagrams for the galactic plane and for $b=\pm0.5°$ are shown in figures 1, 2, and 3. (Figure 3 is preliminary and represents only half the existing in-plane data.) These figures corroborate the main features of the galactic distribution of CO obtained in our previous survey. The most intense emission comes from the roughly triangular area with corners at $l=12°$, v=20 km s^{-1}; $l=22°$, v=110 km s^{-1}; and $l=33°$, v=100 km s^{-1}. When transformed to galactocentric coordinates using a circular rotation model of the Galaxy, this area corresponds to the region of the molecular ring. The displacement of CO emission with respect to the $b=0°$ plane is immediately apparent on comparing the $b=+0.5°$ and $b=-0.5°$ diagrams.

The new survey has been analysed quantitatively in terms of the cylindrically symmetric galactic model described previously (Cohen 1978; Cohen and Thaddeus 1977). The half-thickness of the CO plane, its displacement, and the emissivity of CO, all as functions of galactic radius, are shown in figure 4. These new results are generally consistent with those reported before; an apparent but statistically uncertain rise in the thickness of the CO disk for R between 7 and 8 kpc in the first survey is now absent. Near the peak of the molecular ring at R=5.2 kpc, the ring is 50 pc thick, and is 23 pc below the $b=0°$ plane.

With the large amount of CO data now available, large scale features of the Galaxy in addition to the molecular ring are starting to appear. For example, on the $b=0°$ map, there appear to be two roughly parallel ridges, one running from $l=12°$, v=10 km s^{-1}, to $l=30°$, v=80 km s^{-1}; and the other from $l=12°$, v=20 km s^{-1} to $l=25°$, v=90 km s^{-1}. From $l=24°$ to $l=27°$, these ridges are separated by a large hole. There is another ridge extending almost vertically at v$\approx$60 km s^{-1} from $l=34°$ to 53°. There is an approximate correspondence between these features and some of the features identified with spiral arms by Burton and Shane (1970) and Shane (1972).

REFERENCES

Burton, W.B., and Shane, W.W.: 1970, in Becker and Contopoulos (eds.), The Spiral Structure of Our Galaxy, IAU Symposium No. 38, p.397, D. Reidel, Dordrecht.
Burton, W.B., and Gordon, M.A.: 1978, Astron. Astrophys. 63, p.7.
Cohen, R.S.: 1978, Ph.D. Thesis, Columbia University, reprinted as NASA Technical Memorandum 78071, Goddard Space Flight Center, Greenbelt, MD.
Cohen, R.S., and Thaddeus, P.: 1977, Ap. J. Lett. 217, p.L155.
Scoville, N.Z., and Solomon, P.M.: 1975, Ap. J. Lett. 199, p.L105.
Shane, W.W.: 1972, Astron. Astrophys. 16, p.118.

RADIO OBSERVATIONS OF THE GALACTIC PLANE CH DISTRIBUTION

L.E.B. Johansson, Å. Hjalmarson and O.E.H. Rydbeck
Onsala Space Observatory
S-430 34 Onsala, Sweden

ABSTRACT

The CH distribution in the galactic plane has been determined through observations of the ground state main line transition $^2\Pi_{1/2}$, J=1/2,F=1-1 at 3335 MHz. The longitude ranges $10^o \leq \ell \leq 60^o$ and $60^o < \ell \leq 230$ have been covered with the spacings 2°.5 and 5°, respectively. The derived radial distribution is similar to that of CO, although differencies may exist. From observations at $\ell=30^o$ and 43°.26 the CH layer halfwidth is estimated to be 50 pc, in close agreement with that of CO. A CH displacement below the standard plane is observed in these directions.

INTRODUCTION

Radio observations of the three hyperfine transitions in the $^2\Pi_{1/2}$,J=1/2 ground state of CH have shown that this radical is widespread in the Galaxy and generally behaves like a weak maser(see e.g.Rydbeck et al., 1976; Hjalmarson et al.,1977 and references therein).The negative excitation temperature observed is believed to be a fundamental property of interstellar CH; collisions with H and H_2 is the likely net inversion mechanism of the Λ-doublet. No significant difference in the main line excitation temperature is observed in clouds of varying density.Furthermore, since CH shows weak maser characteristics the column density determination is quite insensitive to variations in the excitation temperature in the case of moderately strong continuum background radiation(Hjalmarson, et al.,1977). These properties of CH make it a useful tool to investigate the interstellar matter of the Galaxy and a complement to the CO surveys which reveal very cool and dense regions. CH surveys presumably trace the more diffuse constituents of the Galaxy, since this reactive radical most likely is tied up in heavier molecules in the densest regions.

EQUIPMENT, OBSERVATIONS AND DATA REDUCTION

The survey was carried out at 3335 MHz(the CH main line)with the 25.6 m radio telescope of Onsala Space Observatory during spring-fall 1977 and spring 1978. The telescope was equipped with a TW-maser giving a system

W. B. Burton (ed.), The Large-Scale Characteristics of the Galaxy, 57–60.

noise temperature in the range 30-35 K. The HPBW is 15 arcmin and the beam efficiency is 0.6. The backend used consists of a 100 channel filterbank with a frequency resolution of 10 kHz(the total velocity coverage is 90 km s^{-1} at 3335 MHz).

The galactic plane was covered in the longitude ranges $10^o \leq \ell \leq 60^o$ and $60^o \leq \ell \leq 230^o$ with a spacing of 2°.5 and 5°, respectively. The typical integration time of each spectrum is 5 hours. In addition, the local spiral arm was traced from $\ell=69^o$ to $\ell=87^o$ with an interval of 1°. Observations out of the galactic plane have been performed at three longitudes so far: $\ell=30^o$, ℓ=43°.26 (W49 region) and $\ell=71^o$(the local spiral arm). The observations in the W49 area are an extension of the earlier observations by Sume and Irvine (1977).

Low order (not greater than two) polynomial baselines have been removed from the raw spectra. Since each spectrum just covers 90 km s^{-1} in velocity, the polynomial fitting process introduces some baseline uncertainties in spectra of broad, weak features. Thus we can not rule out the possibility that features of this kind are lost.

THE LONGITUDE VELOCITY DISTRIBUTION

The observed CH antenna temperatures are displayed in a longitude velocity diagram in Fig.1. Due to the insufficient velocity coverage of the filterbank, the spectra were essentially centered to cover allowed velocities according to the Schmidt rotational model. For ℓ>37°.5 the velocity coverage is 90 km s^{-1}, for $\ell \leq$37°.5 it is about 160 km s^{-1}.

The CH longitude-velocity diagram is similar to those obtained for CO (Burton and Gordon,1978) and diffuse ionized hydrogen(Lockman,1976) in that most emissions originate from regions inside the solar circle and lacks significant high velocity components for $\ell<25^o$.

THE RADIAL DISTRIBUTION

The radial distribution of CH has been determined with the aid of the rotational curve obtained by Burton and Gordon (1978). In deriving the CH density, equation (7) in the paper by Rydbeck et al.(1976) was used. The excitation temperature was assumed to be -10 K and effects of clumping were neglected. The relative positions in the line of sight of the CH clouds and the discrete continuum sources are in most cases impossible to determine. Therefore, the radial distribution of CH(Fig.2) is derived for the two extreme cases: the discrete sources are supposed to be located in front of (1) and behind (2) the CH gas, respectively. The continuum contribution has been estimated from the 2.695 GHz map of Altenhoff et al.(1970). As seen from Fig. 2, the CH and CO maxima coincide. However, there is a lack of CH inside the 4 kpc ring and the fall off towards larger distances is not as steep as in the CO case. Whether these differencies are real or not is an open question. The apparent lack of CH at small distances may simply be due to the previously mentioned baseline uncertainties. The CH excess at larger distances could possibly

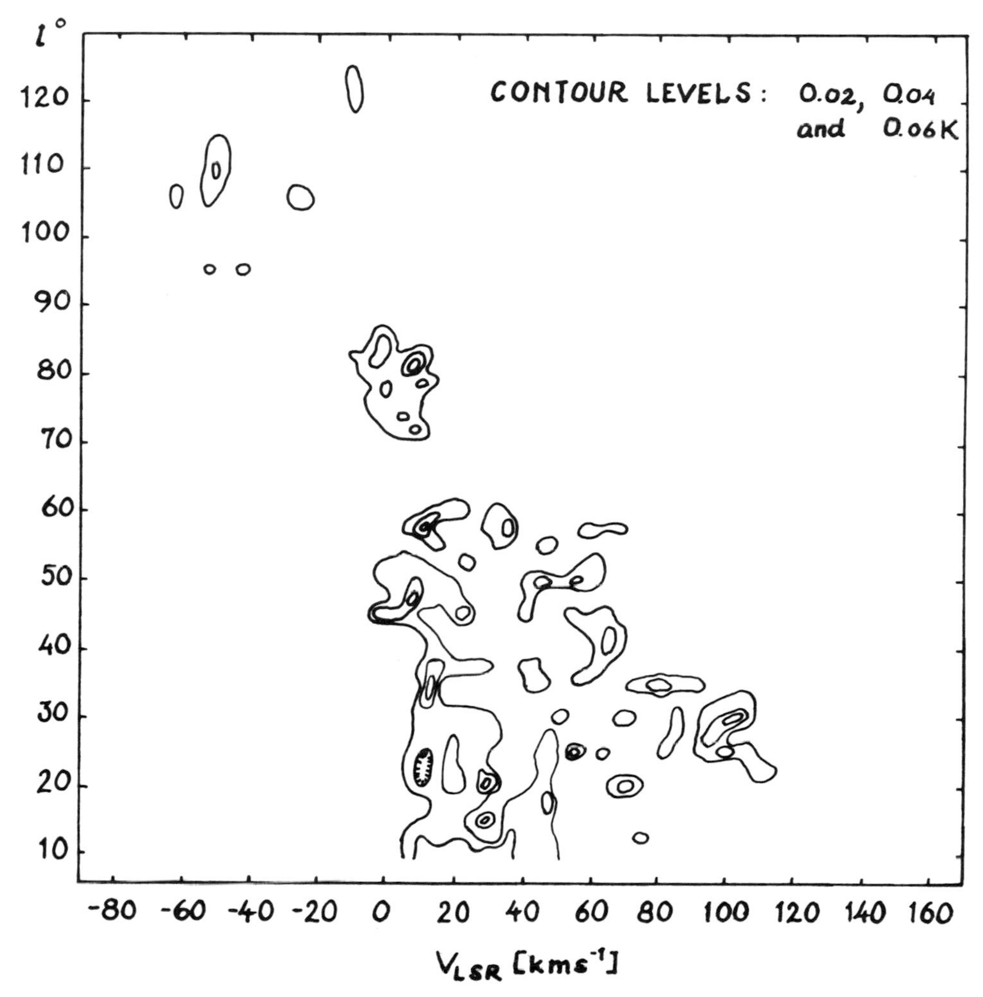

Fig. 1. Longitude velocity diagram of the CH antenna temperatures.

indicate a tendency for CH to follow the large scale HI distribution.

THE DISTRIBUTION PERPENDICULAR TO THE GALACTIC PLANE

The latitude velocity diagrams of the antenna temperatures for $\ell=30^o$ and 43°26 are displayed in Fig.3. Using the high velocity features we have estimated the CH layer halfwidth $z_{1/2}$ and the displacement z from the standard plane to be $z_{1/2}=50\pm10$ pc, $z=-35\pm10$ pc at $\ell=30^o$ and $z_{1/2}=45\pm15$ pc, $z=-65\pm25$ pc at $\ell=43^\circ.26$. These numbers are corrected for beam broadening; the errors include uncertainties due to noise fluctuations, continuum background and in the case of $\ell=43^\circ.26$, the kinematic distance ambiguity. From Fig. 2 in the paper by Cohen and Thaddeus (1977) we get the corresponding numbers for CO: $z_{1/2}=50$ pc, $z\approx0$ pc for $\ell=30^o$ and $z_{1/2}=50$ pc, $z=-40$ pc at the subcentral point of $\ell=43^\circ.26$. We thus may conclude

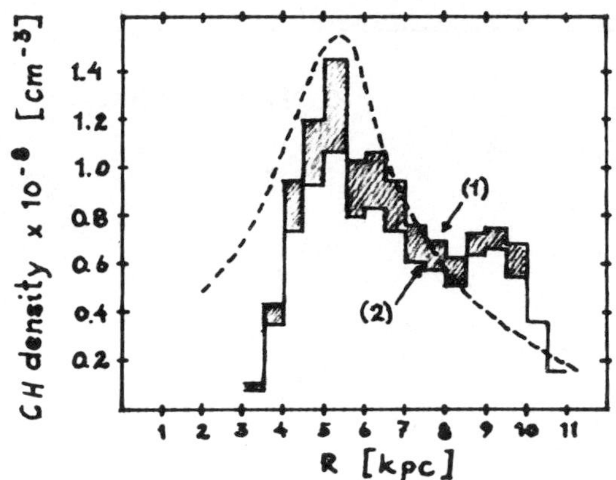

Fig. 2. The CH radial distribution as determined for the two extreme cases (1) and (2)(see text). The relative CO density distribution is shown by the dashed curve(Burton and Gordon,1978).

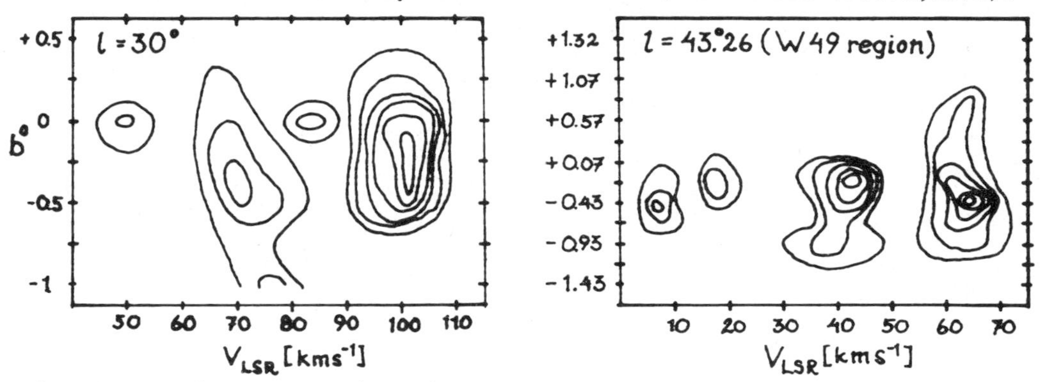

Fig. 3. Latitude velocity diagrams of the CH antenna temperatures for $\ell=30^o$ and $43^{\circ}\!.26$. Contour levels are 0.01(0.01)0.06 K.

that the CH distribution perpendicular to the galactic plane resembles that of CO but differs significantly from the HI distribution in this respect.

A more detailed analysis of the CH data will appear in a forthcoming scientific report from the Onsala Space Observatory.

REFERENCES

Burton,W.B.,and Gordon,M.A.:1978,Astron.Astrophys.63,pp.7-27.
Cohen,R.S.,and Thaddeus,P.:1977,Ap.J.(Letters)217,pp.L155-159.
Hjalmarson,Å.,Sume,A.,Elldér,J.,Rydbeck,O.E.H.,Moore,E.L.,Huguenin,G.R., Sandqvist,A.,Lindblad,P.O.,and Lindroos,P.:1977,Ap.J.Suppl.35,pp.263-280.
Lockman,F.J.:1976,Ap.J.209,pp.429-444.
Rydbeck,O.E.H.,Kollberg,E.,Hjalmarson,Å.,Sume,A.,Elldér,J.,and Irvine,W.M.: 1976,Ap.J.Suppl.31,pp.333-415.
Sume,A.,and Irvine,W.M.:1977,Astron.Astrophys.60,pp.337-343.

HI IN THE GALACTIC DISK

F. J. Kerr
University of Maryland, College Park, Md. USA

This review of 21-cm studies of the galactic disk will be rather short, because several aspects of such studies are being covered by other speakers. These include the galactic rotation curve, the warp, the outer limits of the hydrogen layer, high-velocity clouds, large supernova remnants and comparisons between observations and theory with respect to density waves.

Large-scale 21-cm survey work continues. The table gives the surveys which have been carried out since the publication of earlier comprehensive lists of surveys by Kerr (1968) and Burton (1974):

Surveys of Hydrogen Emission from Near the Galactic Equator, Part III

Author	Publication Date	Beam (arcmin)	Longitude Range (degrees)
Garzoli	1972	30	270-310
Wrixon and Sanders	1973	21	357- 3
Lindblad	1974	13,21	339- 72
Weaver and Williams	1974	36	10-250
Cohen	1975	31x35 13x13	355- 10
Kerr, Harten and Ball	1976	14	236-345
Jackson	1976	30	305-309
Mirabel	1977	28	348- 36
Bystrova and Rakhimov	1977	7'x5°	δ = -29° to +40°
Burton, Gallagher and McGrath	1977	21	349- 12
Braunsforth and Rohlfs	1978	9	20- 42
Sinha	1978	21	339- 11
Kerr, Bowers and Henderson	1978	14	230-350

Our group has recently carried out two 21-cm surveys of the Southern Milky Way at the Australian National Radio Astronomy Observatory at

W. B. Burton (ed.), The Large-Scale Characteristics of the Galaxy, 61–63.

Parkes, NSW, Australia. A new longitude-velocity diagram for the galactic equator from ℓ = 230° to 350°, obtained with the 64-meter telescope (Kerr, Bowers and Henderson 1978), was presented and discussed. Observations were obtained every 3 arcminutes with a beamwidth of 14 arcmin and an effective channel bandwidth of 2 km/sec. This and other contour maps will be published shortly.

This type of diagram tends to emphasize steep gradients, such as the slope at extreme negative velocities over the whole longitude range. The sharp step at zero velocity, first noted by Burton (1973) in the first quadrant, is also very clear; this is the place on the profile where we move from having two distances along the line of sight at the same velocity to only a single distance.

The spiral structure is patchy and not very clear, but we can see for example an arm bending around, with a tangential point at about 308°-310°. There is also a long outer ridge near the highest positive velocities; this feature is strongest well below the plane, in the region of the warp.

The diagram shows many holes and troughs, for example a deep hole at $\ell \sim 276°$, which is right beside a very high peak of brightness temperature. This hole also shows up well in scans across the plane at constant longitude.

The second survey was taken with the Parkes 18-meter telescope (Kerr, Bowers and Jackson 1978). This was a fully-sampled survey of the region ℓ = 240° to 350°, b = -10° to +10°, with a beamwidth of 48 arcmin and a channel width of 2 km/sec. It is a southern counterpart of the northern survey by Weaver and Williams (1974). Reduction of the 45,000-line profiles is well along; some of the results for the region of the warp are being shown by Henderson.

Various attempts have been made in recent years to derive the global spiral structure, but no generally-agreed solution has yet been obtained. The most recently published diagrams are those by Simonson (1976) and Henderson (1977). Simonson's model, based on density-wave kinematics, has a two-armed pattern with a pitch angle of 6°-8°, extending out to near the solar circle, where two additional major arms originate; in the outer parts the pattern is multi-armed, with a pitch angle of 16°. Henderson finds that a four-armed spiral with a 13° inclination can fit two optically-observed and four radio-observed regions quite well.

There is renewed interest in the degree of flatness of the hydrogen layer. For example, Lockman (1977) discusses quite systematic deviations which place the center of the layer at 35-50 pc below the galactic plane in the inner Galaxy, perhaps including the fourth, as well as the first, quadrant. There is also evidence of corrugations in the layer, with adjacent spiral arms being alternately above and below the plane in some regions. A new study of the warp in the outer parts has been carried out by Henderson, who finds that the warp on the southern side extends to smaller z-distances than that on the north, if similar velocity fields are used on the two sides.

REFERENCES

Braunsfurth, E., and Rohlfs, K.: 1978, Astron. Astrophys. Suppl. Series, 32, 177-204.
Burton, W. B.: 1973, Publ. Astron. Soc. Pacific, 85, 679-703.
Burton, W. B.: 1974, "Galactic and extra-galactic radio astronomy", Springer-Verlag, New York, pp. 82-117.
Burton, W. B., Gallagher, J. S., and McGrath, M. A.: 1977, Astron. Astrophys. Suppl. Series, 29, 123-138.
Bystrova, N. V., and Rakhimov, I. A.: 1977, "Pulkovo sky survey in the interstellar neutral hydrogen radio line. III". Akademiya Nauk SSSR, Leningrad, 62 pp.
Cohen, R. J.: 1975, Mon. Not. Roy. Astron. Soc., 171, 659-696.
Garzoli, S. L.: 1972, Carnegie Instn. Wash. Publ. No. 629.
Henderson, A. P.: 1977, Astron. Astrophys., 58, 189-196.
Jackson, P. D.: 1976, Astron. Astrophys. Supp. Series, 25, 433-447.
Kerr, F. J.: 1968, "Nebulae and Interstellar Matter", eds. B. M. Middlehurst and L. H. Aller, U. Chicago Press, 575-622.
Kerr, F. J., Bowers, P. F., and Henderson, A. P. 1978, in preparation.
Kerr, F. J., Bowers, P. F., and Jackson, P. D. 1978, in preparation.
Kerr, F. J., Harten, R. H., and Ball, D. L. 1976, Astron. Astrophys. Suppl. Series 25, 391-432.
Lindblad, P. O.: 1974, Astron. Astrophys. Suppl. Series., 16, 207-236.
Lockman, F. J.: 1977, Astron. J., 82, 408-413.
Mirabel, F.: 1977, Astron. Astrophys. Suppl. Series, 28, 327-331.
Simonson, S. C., III: 1976, Astron. Astrophys., 46, 261-268.
Sinha, R. P.: 1978, Ph.D. thesis, University of Maryland.
Weaver, H. F., and Williams, D. R. W.: 1974, Astron. Astrophys. Suppl. Series, 17, 1-249.
Wrixon, G. T., and Sanders, R. H.: 1973, Astron. Astrophys. Suppl. Series, 11, 339-345.

OPTICAL HII REGIONS

Y. M. Georgelin, Y. P. Georgelin, and J.-P. Sivan
Observatoire de Marseille and Laboratorie d'Astronomie Spatiale, Marseille, France

1. INTRODUCTION: GALACTIC AND EXTRAGALACTIC HII REGIONS

The ionized hydrogen regions seen on monochromatic Hα photographs of nearby spiral galaxies (e.g. Fig. 1) can be divided into four classes (Monnet, 1971):(1) bright, condensed, classical HII regions; (2) diffuse emission in the arms surrounding and connecting the classical regions (emission measure (EM) $\sim$ 150 cm^{-6} pc, for a gas electron temperature of 6000 K); (3) much fainter, diffuse emission extending over the entire disk (EM $\sim$ 50 cm^{-6} pc) in most Sc and Sd galaxies; (4) diffuse emission in the nuclear region.

In our Galaxy class 4 cannot be detected, class 3 is not yet observed, class 2 (i.e., the general Hα emission produced by a uniformly distributed component of the ionized interstellar medium) has been studied in several optical and radio recombination lines (see Courtès et al., 1978), class 1 is of course well known. The first section of this paper presents the apparent distribution of the galactic HII regions as seen from the Sun. We also point out some large-scale characteristics of the ionized interstellar medium. The second section deals with the distribution of the HII regions in the galactic plane.

2. THE APPARENT DISTRIBUTION: MORPHOLOGICAL CHARACTERISTICS

Since the first photographic surveys of the Milky Way, it has been clear that the classical HII regions (class 1) are distributed along the galactic equator (except for some nearby nebulae). Only recently has the actual appearance of the diffuse emission out of the condensed regions (class 2) been shown by the two following studies: (1) Reynolds et al. (1974) have constructed a new contour map of this emission, from Fabry-Perot Hα observations of the northern Milky Way, with a limiting EM of 4 cm^{-6} pc and a spatial resolution of 5°. (2) Sivan (1974) has carried out a very-wide-field photographic survey of the entire Milky Way which sums up, as completely as possible, the Hα emission features brighter than 20 cm^{-6} pc, with a spatial resolution of 10 arc min.

This survey reveals a number of new large-angular-diameter HII regions and a very extended HII complex in Orion, Eridanus and Cetus

W. B. Burton (ed.), The Large-Scale Characteristics of the Galaxy, 65–72.

(Sivan, 1977). Furthermore, it gives a large-scale view of the diffuse Hα emission in excellent agreement with the results of Reynolds et al., and shows the large-scale distribution of the classical HII regions throughout the galactic arms. Owing to the low angular resolution of the wide-field photographs, these arms (Fig. 2) are seen under strictly similar conditions to those of spiral galaxies recorded on large telescope Hα plates. The chaotic and filamentary structure of the diffuse component and the typical ring-like structure of a number of classical regions should be pointed out.

Ring-like HII regions were previously noted in the Galaxy and in other galaxies by Gum and de Vaucouleurs (1953). Owing to improved photographic techniques, numerous extragalactic HII rings have recently been

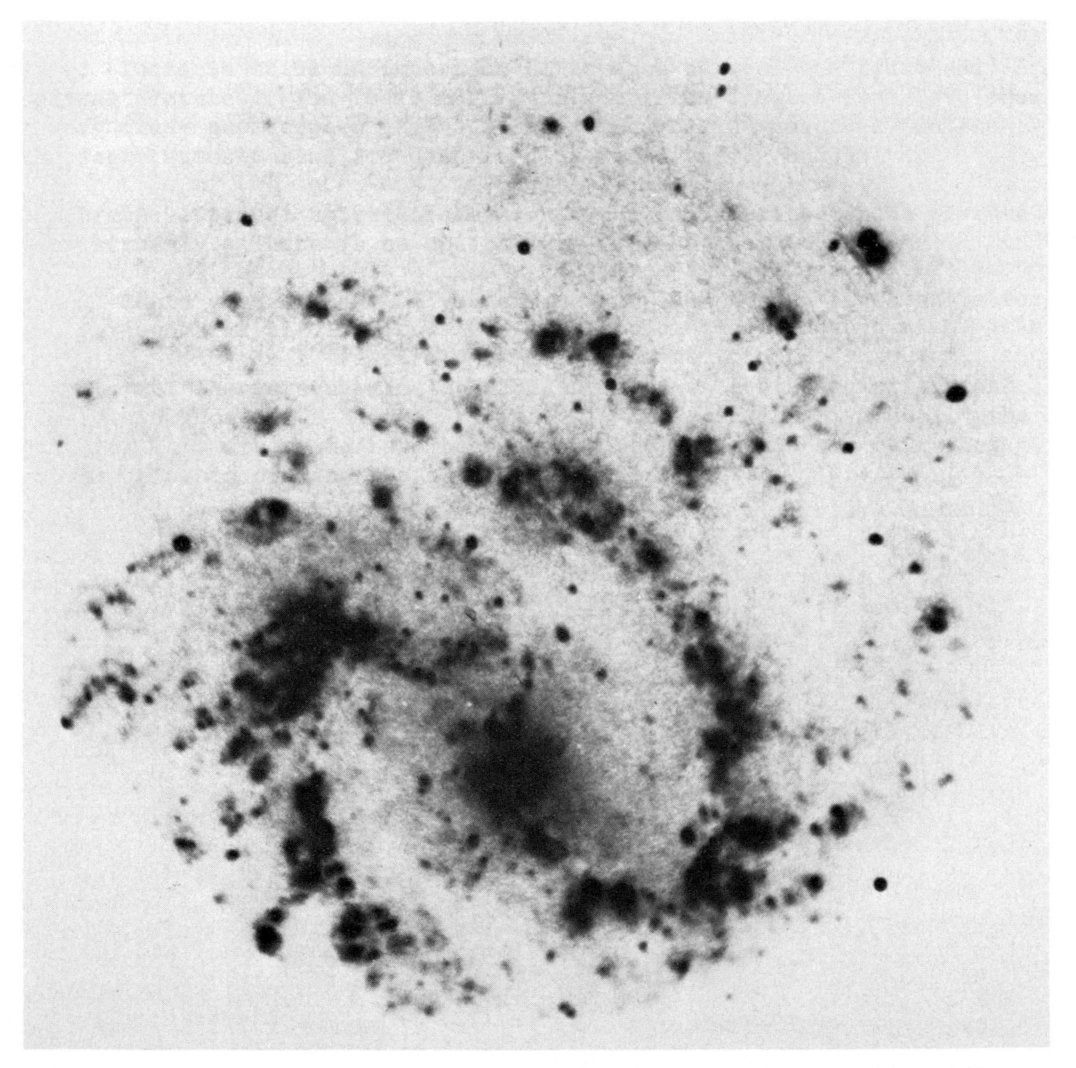

Fig. 1. Hα image tube photograph of M 83 (Comte and Georgelin, f/2 focal reducer of Courtès on the ESO 3.6 m telescope).

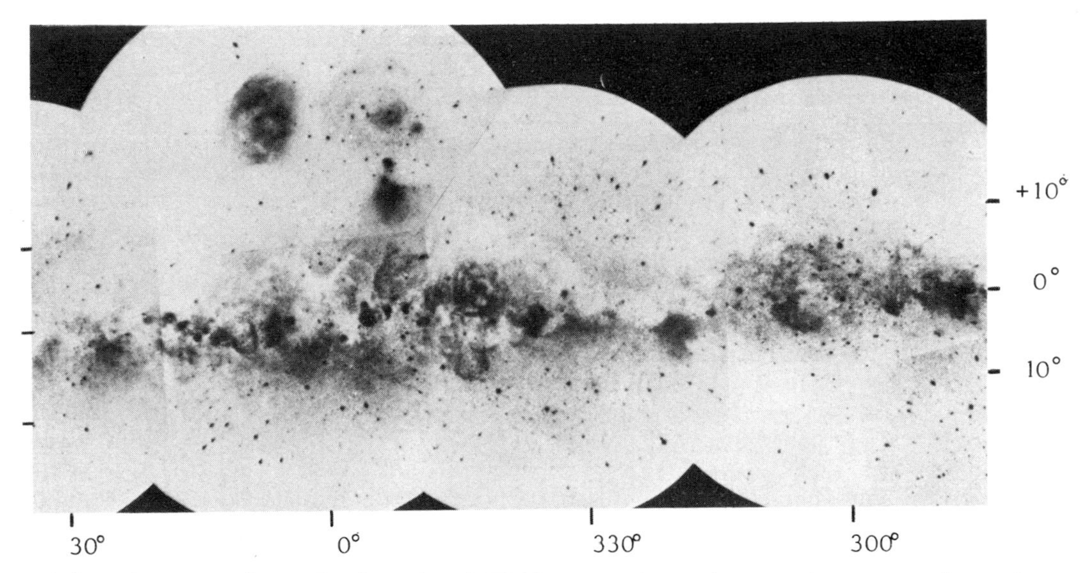

Fig. 2. Section of Sivan's (1974) Hα atlas of the Milky Way (mosaic of 60° field monochromatic plates) showing the entire Sagittarius-Carina arm, and high latitude nearby nebulae.

detected (e.g. Fig. 1; see de Vaucouleurs, 1978). These are good secondary extragalactic distance indicators (de Vaucouleurs, 1978). The following table lists the largest rings visible on the plates of the wide-field Hα survey. The first five regions are single well-defined rings; for the others the ring-shaped structure is more complex (several interlaced rings), or uncertain.

Region	Ring diameter (pc)	Expansion velocity (km s^{-1})	
Cetus Loop	70	23	(HI)
NGC 2237	20	15	(HI)
Barnard Loop	115	9	(HI)
RCW 59	80		
[328.5 - 1.0]	85		
S34 + S35	80		
A0 Cas	90		
IC 1805	35		
(Per OB2)	50	5	(HI)
S264 (λ Ori)	45	8	(HI)
Gum Nebula	∿220	20	(Hα)
[303.5 + 4.0]			
IC4628+RCW113+...	∿200	8	(Hα)
RCW 114			
(Sco OB2)	35	5	(HI)
[354.0 + 5.5]			

For half of these regions there is strong evidence for an expanding shell structure: either an expansion velocity is measured in Hα, or the Hα ring coincides with a 21 cm expanding shell. One million year old supernova explosions may be at the origin of the shells (c.f. Sivan, 1977). However, mechanisms of wind-driven circumstellar bubbles cannot be excluded. In both galactic and extragalactic rings, exciting stars are frequently observed at the edge of the ring. These regions are probably

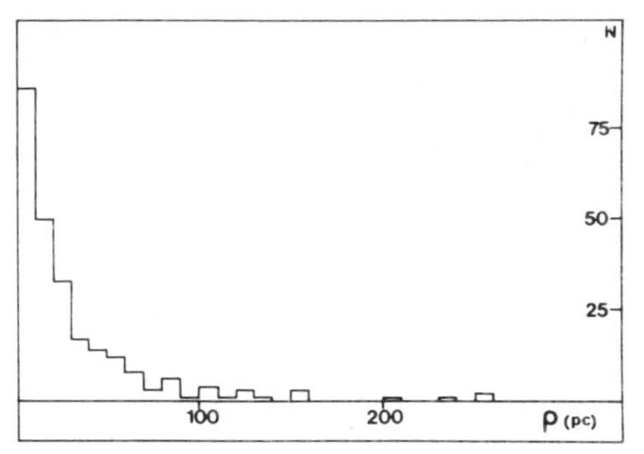

Fig. 3. Histogram of diameters of (galactic) optical HII regions.

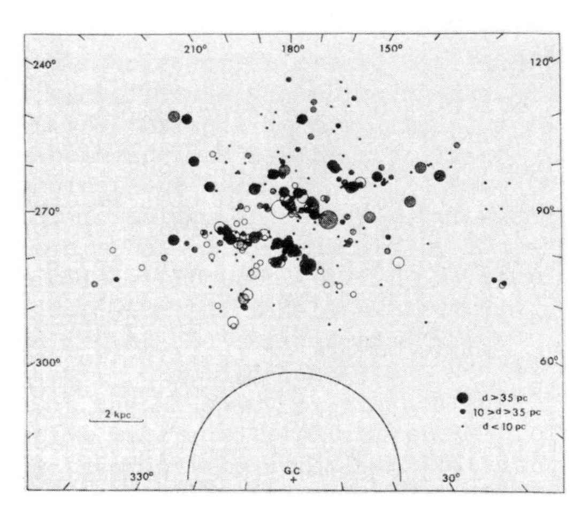

Fig. 4. The distribution of optical HII regions projected on the galactic plane. The points are weighted according to the size and brightness of the regions.

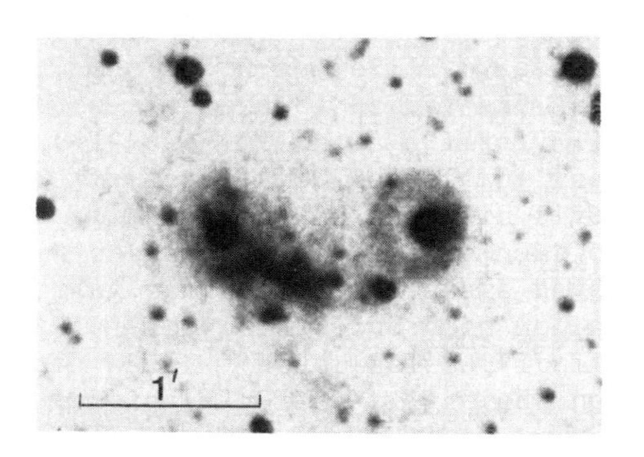

Fig. 5. Hα photograph of G 298.2 - 0.3 (Comte and Georgelin, ESO 3.6 m telescope) distance 11.7 kpc.

a late stage in the life of an expanding ionized region and a second generation of stellar formation may occur (Boulesteix et al., 1974).

Two HII rings (Barnard Loop and Cetus Loop) belong to the Orion-Eridanus-Cetus complex. Spectrophotometric measurements on the whole complex (Sivan, 1977) indicate normal HII region line ratios and a low excitation - i.e., results similar to those obtained by Reynolds (1976) on the Gum Nebula, in agreement with the above interpretation. Also, this complex may be considered as a nearby sample of the uniformly distributed component of the ionized interstellar medium (class 2). The measured low excitation is in agreement with an ionization by *in situ* OB stars (Torres-Peimbert et al., 1974; Comte and Monnet, 1974).

Determination of the distance to the Sun of most of the observed HII regions (§ 3) has allowed us to construct a new histogram of the frequency distribution of the intrinsic diameters for 246 regions (Fig.3): 50% of the regions have linear diameters smaller than 15 pc and 7 regions are larger than 150 pc.

3. THE SPIRAL STRUCTURE OF OUR GALAXY

The distance determination of the optical regions has allowed us to plot them on the galactic plane (Crampton and Georgelin, 1975). The revised diagram in Fig. 4 has been constructed according to more recent data (Crampton et al., 1978). 80% of the regions are plotted using the distances of the exciting stars. 20% are kinematic distances, based upon the Hα radial velocities and the rotation model of the Galaxy of Georgelin and Georgelin (1976). Fig. 4 shows the same distribution as that obtained from other spiral arm tracers (cf. Humphrey's paper at this symposium). Nevertheless, greater distances are reached in some directions: 5 regions in Carina beyond 4.5 kpc, S 99 - 100 at 9 kpc, etc. Recently, optical HII regions have been detected in Hα, coinciding with very distant radio sources: W 51 (Crampton et al., 1978), G 298.2 - 0.3 (Fig. 5), etc.

Such very distant optical regions are not numerous enough to provide

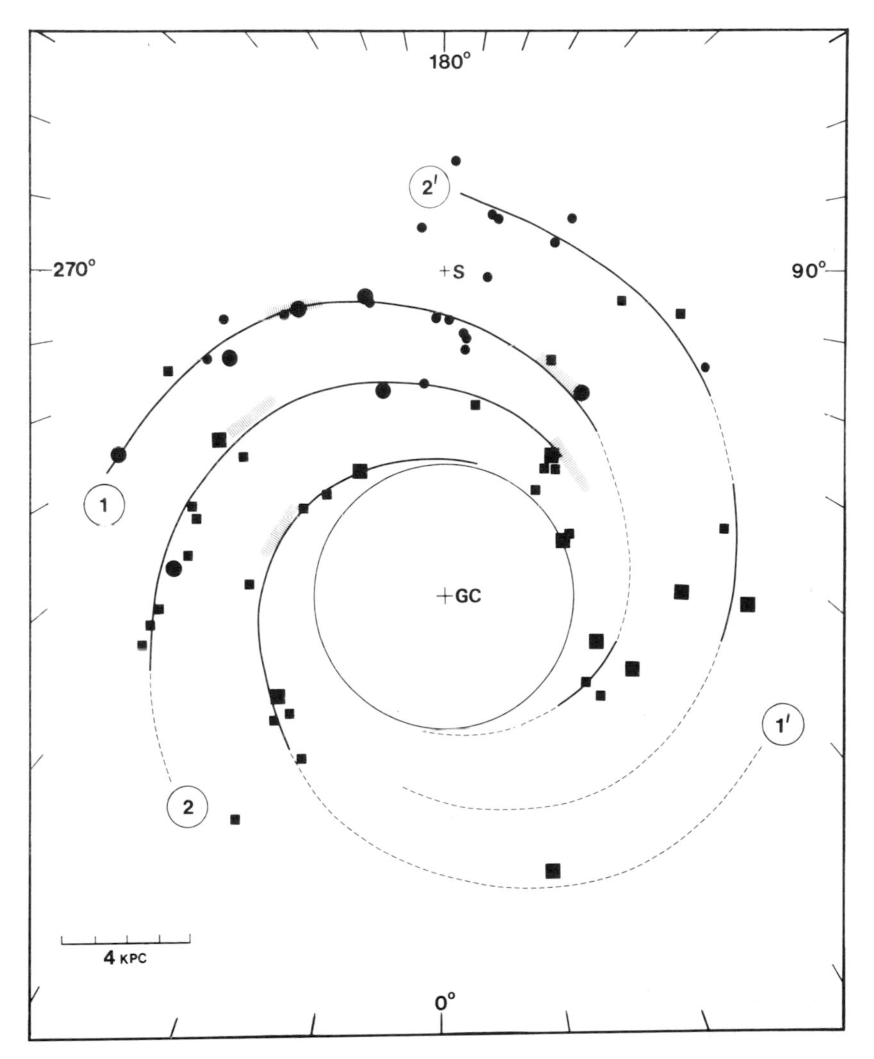

Fig. 6. Spiral model of the Galaxy obtained from high excitation parameter HII regions. Circles = optical; squares = radio regions. Larger symbols : U > 200 pc cm^{-2}; smaller ones : 200 > U > 100 pc cm^{-2}.
1 : Sagittarius-Carina major arm; 2 : Scutum-Crux intermediate arm;
1': Norma internal arm; 2': Perseus external arm. Hatched areas correspond to intensity maxima in the radio continuum and in neutral hydrogen (the arms are seen tangentially at the same longitude as in radio).

useful information on the spiral structure beyond 4 kpc. To compensate for this lack of optical data and in order to establish a coherent picture of the spiral structure of our Galaxy, Georgelin and Georgelin (1976) have combined optical and radio data, making extensive use of the H 109α radial velocities observed principaly by Reifenstein et al. (1970) and Wilson et al. (1970). This was quite possible because (1) there is excellent agreement between Hα and H 109α velocities, (2) H 109α sources must be considered not as independant regions but as parts of extended optical HII regions . This study uses a homogeneous population (HII regions + exciting stars). It is based upon a method of analysis which has been fully described elsewhere (Georgelin and Georgelin, 1976).

Fig. 6 shows the spiral model obtained from 60 high excitation parameter ($U > 100$ cm^{-2} pc) HII regions - i.e., giant regions as defined by Mezger. These are class 1 regions (§ 1) similar to those observed in external galaxies and which are known to define spiral arms (Fig. 1). The diagram in Fig. 6 differs from the previously published one only by the higher limit put on the U's. As a consequence of the selection of the HII regions according to their U's, our local region is practically insignificant. Two symmetrical pairs of arms are found - i.e., 4 arms altogether - with a pitch angle of 12°. The continuous curves are the best fits to the plotted regions; the dashed curves are their symmetric images. No arm is observed to go all the way around the Galaxy, in agreement with observation of spiral structure in external galaxies.

REFERENCES

Boulesteix, J., Courtès, G., Laval, A., Monnet, G., Petit, H.: 1974, Astron. Astrophys. 37, 33.
Comte, G., Monnet, G.: 1974, Astron. Astrophys. 33, 161.
Courtès, G., Saïsse, M., Sivan, J-P.: 1978 (in preparation)
Crampton, D., Georgelin, Y.M.: 1975, Astron. Astrophys. 40, 317.
Crampton, D., Georgelin, Y.M., Georgelin, Y.P.: 1978, Astron. Astrophys. 66, 1.
de Vaucouleurs, G.: 1978, Astrophys. J. (in press)
Georgelin, Y.M., Georgelin, Y.P.: 1976, Astron. Astrophys. 49, 57.
Gum, C.S., de Vaucouleurs, G.: 1953, Observatory 73, 152.
Monnet, G.: 1971, Astron. Astrophys. 12, 379.
Reynolds, R.J.: 1976, Astrophys. J. 203, 151.
Reynolds, R.J., Roesler, F.L., Scherb, F.: 1974, Astrophys. J. Letters 192, L 53.
Reifenstein, E.C., Wilson, T.L., Burke, B.F., Mezger, P.G., Altenhoff, W.F.: 1970, Astron. Astrophys. 4, 357.
Sivan, J-P.: 1974, Astron. Astrophys. Suppl. 16, 163.
Sivan, J-P.: 1977, Thesis, University of Provence
Torres-Peimbert, S., Lazcano-Araujo, A., Peimbert, M.: 1974, Astrophys. J. 191, 401.
Wilson, T.L., Mezger, P.G., Gardner, F.F., Milne, D.K.: 1970, Astron. Astrophys. 6, 364.

DISCUSSION

de Vaucouleurs: On two slides you showed the diameter of the Gum Nebula to be 220 and 250 pc; is this due to different definitions or different distance scales?

Sivan: This is due to different definitions. For the study of the frequency distribution of intrinsic diameters of the galactic HII regions, I use the diameter of the outer limits of the Gum Nebula (250 pc). But in the tables listing the ring-like regions, the diameter is measured at the maximum intensity of the ring (about 220 pc).

Crampton: It is obvious that several spiral features or arms must continue in the longitude range $25° < 1 < 90°$ but they are not seen either in the optical or radio observations of HII regions. There must be more HII regions there, and I urge radio astronomers to try to fill in this longitude range with deeper surveys. Optically, the region is quite obscured but there are some holes, and because this direction is so important to our understanding of spiral structure, a major effort is warranted.

Heiles: I was pleased to see that many of the HII regions are rings. A great many HI rings exist in the galactic plane which have similar properties.

Terzian: Have you made a comparison of the distribution of HII region sizes in our Galaxy with those of other galaxies like M33 and M31?

Sivan: I have compared my galactic results with those obtained using the same observational techniques, by Pellet et al. (A.&A. Suppl. 31, No. 3) for M31, and by Boulesteix et al. (1974, A.&A. 37, 33) for M33. The Galaxy is closer to M31 than to M33. Half of the galactic regions of M31 have diameters of 40 pc or less. We do not observe in the Galaxy a most probable value, as is observed in M33. The diameters of the largest regions are 220 pc for M31, 250 pc for the Galaxy, and 350 pc for M33.

Rubin: Do you see HII regions in the anticenter direction, beyond the Perseus arm?

Sivan: The outer Perseus arm appears to be more tenuous than the others. We do not see HII regions beyond it. The most distant region in the Georgelins' diagram in the anticenter direction, is IC 410. It is situated on the outer part of the Perseus arm at 3.4 kpc from the Sun (spectrophotometric distance).

Bok: Congratulations to Dr. Sivan and Dr. Georgelin for the beautiful spiral diagram of our Galaxy that we have just seen. My colleagues Roberta Humphreys, Juan Carlos Muzzio, and Ellis Miller, and I, have long tried to obtain an overall spiral diagram and found this most difficult. Dr. Georgelin drew a diagram three years ago and while it was a good one the spiral structure was ambiguous. Now we have a diagram, based mostly on optical data but in agreement with radio data, which shows the spiral structure of our Galaxy in its full scope.

RADIO STUDIES OF THE DISTRIBUTION OF IONIZED GAS

Felix J. Lockman
Department of Terrestrial Magnetism,
Carnegie Institution of Washington

INTRODUCTION

Although radio recombination line and continuum observations are very useful for investigating galactic structure, it is well to remember their limitations. First, they only provide measurements of coordinates and velocities; a kinematic model is needed to derive the distance and thus the actual location of every nebula. In some directions, particularly all longitudes $\gtrsim 50°$ from $\ell=0°$, kinematic distance estimates are prone to systematic errors arising from velocity crowding or uncertainties in the rotation curve, and are of little use in quantitative studies. I will only discuss the distribution of radio nebulae in the inner $\sim 100°$ of the galactic plane, since in this area kinematic analyses can give reasonable results and, in any case, here we must rely on radio observations for most of our information.

The second limitation is that it is not easy to quantify the properties of the gas being observed at radio wavelengths. Surveys in different recombination line transitions are selectively sensitive to different components of the ISM. In addition, radiative transfer, antenna beam convolution and confusion effects can be so severe that meaningful estimates of the mass of ionized gas or even of the number of HII regions in the inner Galaxy are difficult to make. However, it is possible to distinguish between two types of nebulae that have been studied at radio wavelengths. The first are those observed in high frequency discrete-source surveys. I will call these "dense" nebulae since they were observed because they appeared as bright features against the galactic continuum; they are generally small, high emission measure objects. Dense HII regions are one of the few species which has been observed with complete latitude coverage and comparable sensitivity in both hemispheres. Secondly, there are lower frequency surveys made with relatively large antenna beamwidths which aim to cover a portion of the Galaxy in the manner of HI and CO surveys. These observations are most sensitive to large, moderate density nebulae, although to some extent they detect recombination line emission from the entire hierarchy of ionized structures along the line of sight.

W. B. Burton (ed.), The Large-Scale Characteristics of the Galaxy, 73–79.

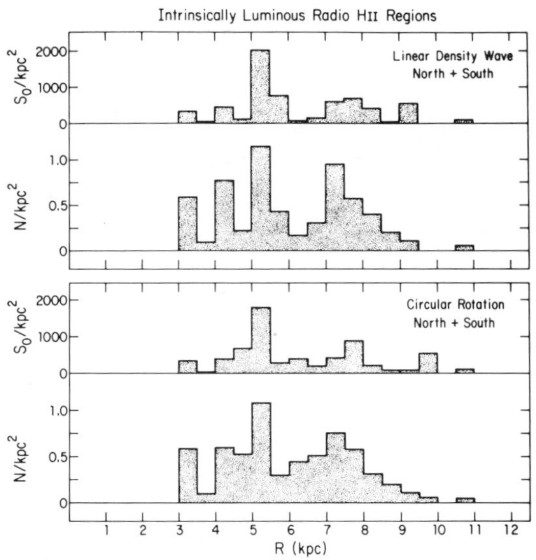

Figure 1: The surface density of dense radio HII regions vs. distance from the galactic center.

The basic radio data on ionized gas in the inner Galaxy come from the discrete source (H109α) surveys of Reifenstein et al. (1970) and Wilson et al. (1970) (supplemented by a few other observations, e.g. Dickel and Milne, 1972; Caswell, 1972), and from the uniformly-spaced lower frequency recombination line surveys of Hart and Pedlar (1976) and Lockman (1976). Our task is to convert these measurements into a function $N(R,\theta,z)$ which describes the amount of ionized gas at a given distance from the galactic center, azimuthal angle, and distance from the plane.

THE DISTRIBUTION OF DENSE NEBULAE

I have selected a set of 110 dense nebulae which 1) have longitudes in the range $5°\leq \ell \leq 55°$ or $305°\leq \ell \leq 355°$ (to be called the northern and southern intervals, respectively); 2) have accurate 5 GHz recombination line and continuum measurements; 3) have been observed in enough absorption species to allow resolution of distance ambiguities and 4) are intrinsically luminous in that each must be ionized by at least one O star. This set was analyzed using models based on circular motion and density-wave theory; full results are given elsewhere (Lockman, 1979). All of the nebulae lie within 15 kpc of the sun, so the following remarks refer mostly to the HII regions in the half of the Galaxy nearest the sun. Because of the longitudinal restrictions, the galactic nucleus is not included and the sample is incomplete at R>8.2 kpc.

Figure 1 shows radial surface density functions (not corrected for distance selection effects) for dense nebulae. In many ways

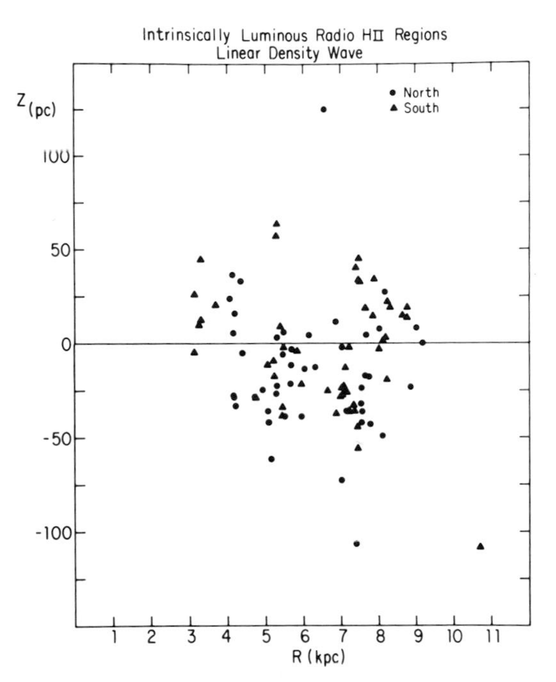

Figure 2: The distance of dense nebulae from the galactic plane, z, vs. their distance from the galactic center.

these are similar to previously published figures (Mezger, 1970). Both the number and absolute flux density (defined as the observed continuum flux density times the square of the source distance) per kpc^2 are shown; the latter quantity is a measure of the number of UV photons maintaining the ionization. Note the secondary peak in the surface densities near 7.5 kpc. This is not a feature in the N(R) functions of other population I-type species like CO and SNRs (Burton, 1976), whose radial abundance has fallen to quite low values at R=8 kpc. While the secondary peak may be partly an artifact of distance selection effects, it is unlikely that it is entirely so.

The inner boundary of dense nebulae is at R=4 kpc in the North and at R=3 kpc in the South (neglecting those at the galactic center). There is no compelling evidence that the 6 southern nebulae with R<4 kpc are in any way connected with the 3-kpc arm.

The distance of nebulae from the galactic plane is plotted vs. their distance from the galactic center in Figure 2. Although these results were derived from a linear density-wave analysis, the assumption of pure circular rotation produces little change in the distribution. Dense nebulae tend to lie in a well defined z(R) pattern. I have discussed this phenomenon elsewhere (Lockman, 1977), and it is seen in HI, CO, SNRs and OH/IR stars (Quiroga, 1974; Cohen and Thaddeus, 1977; Lockman, 1977; Bowers, 1978). The similarity of the pattern for northern and southern objects indicates

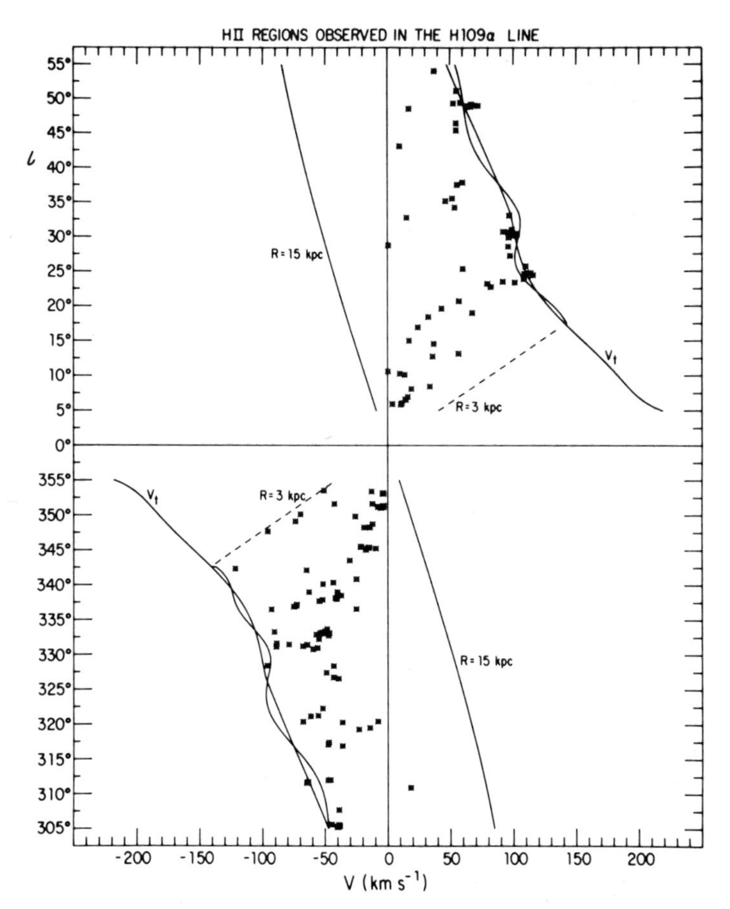

Figure 3: The observed velocities and longitudes of dense radio nebulae. This figure includes not only the intrinsically luminous nebulae, but fainter objects as well.

that the displacement from planarity is a large-scale characteristic of the Galaxy. The layer of dense HII regions has the smallest z-extent of any species: its mean z is -10 pc with a dispersion of only 30 pc; half the dense nebulae are located within 20 pc of the mean z.

The azimuthal distribution of dense HII regions cannot be derived from a simple plot of the nebulae in polar (R,θ) coordinates, for kinematic distance estimates are so uncertain that the appearance of a polar diagram is determined by kinematic assumptions as much or more than by the actual distribution of nebulae. Instead, the azimuthal distribution must be derived by comparing model distributions directly with the observations.

Figure 3 shows the sample of dense nebulae in the observed coordinates. Their v-ℓ distribution is singular among galactic species. Note the absence of nebulae near the terminal velocity (V_t) at $\sim$35° to 45° from ℓ=0° in both the North and South. These "gaps" along V_t are not seen in HI or CO and are unlikely to arise by chance, for projection of the galactic velocity field causes a rather large area in the Galaxy to have LSR velocities near the terminal velocity. (This effect, as pointed out by Burton, 1971, is the reason why HI profiles show a peak at V_t). The gaps along V_t are a distinctive feature of dense HII regions, and indicate that they have a distribution quite different from that of HI or molecular clouds.

Numerous calculations, using circular rotation, linear density-wave and nonlinear density-wave kinematics, show that the gaps along V_t are not produced in models with N(R,θ)=N(R). However, the gaps are reproduced if nebulae are located in a spiral pattern. The northern observations are matched by models with all nebulae confined to the vicinity of the two-armed spiral pattern derived by Burton (1971) from HI streaming motions. The Burton pattern was not intended to fit southern HI, and although use of it in models gives qualitative agreement with the southern nebulae, the "gaps" it produces are not at the correct longitudes. Models with the nebulae confined to rings also produce gaps along V_t, but these models fit the data better if the ring radius increases from North to South, i.e. in the sense of trailing spiral arms. These results are essentially independent of the adopted kinematic model. In sum, observations of dense nebulae are consistent with a confinement of nebulae to a spiral pattern, although there are too few data to allow accurate specification of the pattern. There is no compelling evidence for the existence of any interarm dense nebulae.

MODERATE DENSITY IONIZED GAS

Moderate density ionized gas is seen in very weak H166α recombination line emission at all positions in the plane with $\ell \lesssim 45°$. Because there have been no southern hemisphere observations and the northern surveys have been made almost entirely at b=0°, relatively little is known about this medium. It is possible that the H166α emission arises in large structures with emission measures $\sim 10^3$ cm^{-6} pc. Although this moderate density gas is abundant in the inner Galaxy, it has a smaller radial and longitudinal extent than the dense HII regions. It's N(R) function has no secondary peak near 7.5 kpc (Lockman, 1976; Hart, 1977), and its abundance has fallen below the detection level $\sim$8 kpc from the galactic center. Estimates of its scale height give dispersions ranging from 30 to 70 pc about z=0; I have made recent scans in H127α which generally support the smaller value, but which indicate that the scale height may vary.

Figures 2 and 3 in the review by Hart (1977) show an absence of H166α emission along V_t at the longitude of the northern "gap" in dense nebulae, indicating that moderate density ionized gas may be confined to the spiral pattern suggested by dense HII regions. However, there may well be some weak H166α emission in areas of the v-ℓ diagram lacking dense HII regions; the question of weak "interarm" HII regions cannot be decided without more sensitive observations. Except in the direction of the galactic center, northern data show no H166α emission at the velocity of the 3-kpc arm. It is reasonable to assume that, except for their more restricted radial abundance, low density nebulae are distributed like their dense counterparts.

CONCLUDING REMARKS

Dense galactic HII regions have a unique distribution. They have the smallest known scale height; they are found to within 3 rather than 4 kpc from the galactic center; they are the only known species which may be totally confined to a spiral pattern. Lower density nebulae are potentially more important than their dense counterparts as tracers of star formation, but their radio emission is so weak that, despite a substantial commitment of telescope time, relatively little is known about them. Existing data indicate that, except in their radial extent, they are distributed like the dense nebulae.

A model consistent with all recombination line observations from the inner Galaxy places all dense HII regions in trailing spiral arms (which, in the North, are identical to those derived by Burton, 1971), of extremely thin z-extent, located with or embedded in larger clouds of lower density ionized gas. It is likely that dense HII regions have a broader radial distribution than other species.

REFERENCES

Bowers, P.F.: 1978, Astron. Astrophys. 64, pp. 307-318.
Burton, W.B.: 1971, Astron. Astrophys. 10, pp. 76-96.
Burton, W.B.: 1976, Ann. Rev. Astron. Astrophys. 14, pp. 275-306.
Caswell, J.L.: 1972, Aust. J. Phys. 25, pp. 443-450.
Cohen, R.S. and Thaddeus, P.: Astrophys. J. 217, pp. L155-L159.
Dickel, J.R. and Milne, D.K.: 1972, Aust. J. Phys. 25, pp. 539-544.
Hart, L. and Pedlar, A.: 1976, Mon. Not. Roy. Astron. Soc. 176, pp. 547-559.
Hart, L.: 1977, in H van Woerden (ed.), "Topics in Interstellar Matter", Reidel, Dordrecht.
Lockman, F.J.: 1976, Astrophys. J. 209, pp. 429-444.
Lockman, F.J.: 1977, Astron. J. 82, 408-413.
Lockman, F.J.: 1979, Astrophys. J. (in the press).
Mezger, P.G.: 1970, in W. Becker and G. Contopoulos (eds.), The Spiral Structure of Our Galaxy", Reidel, Dordrecht.
Quiroga, R.: 1974, Astrophys. Space Sci. 27, pp. 323-342.
Reifenstein, E.C., Wilson, T.L., Burke, B.F., Mezger, P.G. and Alterhoff, W.J.: 1970, Astron. Astrophys. 4, pp. 257-377.
Wilson, T.L., Mezger, P.G., Gardner, F.F. and Milne, D.K.: 1970, Astron. Astrophys. 6, pp. 364-384.

DISCUSSION

Cohen: The b = 0° CO ℓ,v map is very similar to the HII distribution you have shown. Also, the radial distribution of CO shows a small peak at 7 kpc from the galactic center.

Lockman: Although CO ℓ,v diagrams may show "gaps" along the terminal velocity at certain latitudes, the surface density of CO is rather uniformly distributed along v_t. This is not the case for the dense HII regions.

Roman: What distance did you use for the Sun's distance to the galactic center?

Lockman: R_o = 10 kpc.

LINE AND CONTINUUM SURVEYS OF THE GALACTIC PLANE WITH THE 100-M TELESCOPE

D. Downes
Max-Planck-Institut fur Radioastronomie, Bonn, West Germany

A new continuum survey of the galactic plane has been made with the Effelsberg 100-m telescope at a frequency of 4.875 GHz with a beamwidth of 2!6. The data are available in the form of radio contour maps covering ℓ = 357°5 to 60°, b = $\pm$1°, together with a list of 1186 radio sources. (Altenhoff, Downes, Pauls and Schraml, 1978, Astron. Astrophys. Suppl. 35, 1).

All of the sources with antenna temperatures greater than 1 K in the continuum survey have been observed in the hydrogen 110α recombination line at 4.874 GHz and the formaldehyde line at 4.830 GHz. The line data provide a wealth of new information on the kinematic distances of the HII regions and their associated molecular clouds in this part of the Galaxy. (Downes, Bieging, Wilson and Wink, in preparation).

We have also searched for H_2O emission at 22 GHz near the peaks of 476 compact sources from the continuum survey, including all continuum sources in the survey with antenna temperatures > 0.3 K. In this longitude range, there are 59 H_2O masers, with concentrations at ℓ = 32° and 47° as in the distributions of CO line intensity and radio continuum sources. (Genzel and Downes, Astron. Astrophys. in press).

W. B. Burton (ed.), The Large-Scale Characteristics of the Galaxy, 80.

THE DISTRIBUTION OF FORMALDEHYDE IN THE GALAXY

R. D. Davies and R. W. Few
Nuffield Radio Astronomy Laboratories

A deep survey of formaldehyde absorption along the galactic plane has been made using the MkII telescope (beamwidth 9 arcmin). Observations were made with 1° spacing between ℓ = 14° and 36° and 2° spacing elsewhere in the longitude range 8 to 60 . Integration times were typically 7 hr; the rms noise was 4 mK.

A comparison of the H_2CO longitude-velocity distribution with that of neutral hydrogen shows that H_2CO has a more clumpy distribution, although not as clumpy as CO (Burton and Gordon 1978). The major concentrations of H_2CO lie within HI concentrations (spiral arms) and most are coincident with ionized hydrogen features seen in the 166α recombination line ℓ-v diagram of Hart and Pedlar (1976). The radial distribution of H_2CO in the Galaxy was derived using the rotation curve of Burton and Gordon (1978). The H_2CO distribution is similar to that of CO and HII, showing a broad peak between R = 4 and 6.5 kpc. An estimate of the molecular hydrogen content of the Galaxy was made by assuming $N(H_2CO) = 1.25 \times 10^{-9}\ N(H_2)$ as given by Scoville and Solomon (1973). This leads to a maximum density (at R = 5 kpc) of 1 cm^{-3}, a value about half that suggested by Gordon and Burton (1976). The density at the solar radius is 0.15 cm^{-3}, which compares closely with 0.143 cm^{-3} derived from UV observations.

It should be emphasized that radio observations of different molecules give probes of different density regimes of interstellar molecular hydrogen. CO samples the highest densities (say 10^2 to 10^3 cm^{-3}), H_2CO intermediate densities (say 1 to 100 cm^{-3}), and CH lower densities (say 1-10 cm^{-3}). Observations of a range of molecular species are required to give the full picture.

REFERENCES

Burton, W. B., and Gordon, M. A.:1978, Astr. and Astrophys. 63, 7.
Gordon, M. A., and Burton, W. B.: 1976, Astrophys. J. 208, 346.
Hart, L., and Pedlar, A.: 1976, M.N.R.A.S. 176, 547.
Scoville, N. Z., and Solomon, P. M.: 1973, Astrophys. J. 180, 31.

W. B. Burton (ed.), The Large-Scale Characteristics of the Galaxy, 81–83.

DISCUSSION

Davies: We consider that the main contribution to the background radiation field is the 2.7 K cosmic background. Evidence for this is that the H_2CO absorption is not limited to the regions of highest galactic continuum emission.

Burton: Uncertainties in the galactic rotation curve may result in discrepancies among the radial abundance distributions derived from the various tracers which we have heard discussed today. Although there is not much doubt about the rotation curve at galactocentric distances greater than 4 kpc, at $R < 4$ kpc there are several different curves available in the literature. Numerical experiments show that the differences are reflected in radial abundance derivations.

Davies: We have looked at the H_2CO radial distribution as a function of the assumed rotation curve and found that there is no significant effect for H_2CO. This is possibly because the H_2CO is mainly concentrated at $R < 4$ kpc where the galactic rotation curve is well-determined. Differences in the various model rotation (and expansion) curves occur at $R < 4$ kpc; these affect the derived HI distribution particularly.

Field: You quoted "H_2 densities" even though you noted that H_2CO lines are formed in low-density regions, which are largely HI. Will you clarify this point?

Davies: We observe a formaldehyde density which we convert to a total gas density (atomic plus molecular) by multiplying by 8×10^8. This relation is supported by observations of HI and H_2CO absorption in the direction of strong galactic non-thermal sources. At higher densities (> 100 cm^{-3}, say) the gas may be predominantly molecular.

Cohen: Again, it seems to me that the H_2CO and CO (from the Cohen and Thaddeus survey presented earlier) ℓ,v diagrams are completely consistent.

Davies: I agree.

Solomon: The radial distribution in the galaxy of H_2CO is identical (from your measurements) with the CO distribution that we measure and very different from HI. Therefore, the H_2CO must be sampling regions which are molecular.

Your analysis of abundance and H_2 density assumes absorption only of the 2.7 K background. The continuum from HII regions will add significantly in some locations, leading to an underestimate of τ. The excitation temperature of formaldehyde may also vary.

Davies: I agree that the H_2CO is sampling the H_2 distribution; my point is that although the H_2CO and HI have different radial distributions, the H_2CO concentrations still lie within the main HI spiral features as delineated in the ℓ,v diagram. Our analysis of the H_2CO abundance included the effect of the galactic continuum as well as the 2.7 K background. An excitation temperature of 1.7 K was assumed.

Stecker: First, as a point of clarification, you stated that your derived H_2 abundance in the 5-kpc annulus was about 1/2 that derived by Gordon and Burton. Would that then be about 1/4 of that given by Dr. Solomon this morning? Second, because your survey samples clouds which are less dense than those sampled in the CO surveys, could most of the H_2 be tied up in the CO clouds in the annulus rather than in the less dense H_2CO clouds and could that help account for the apparent disrepancy?

Davies: The answer to the first question is yes, and to the second is possibly yes: there will be some extra contribution from the CO, but I have not calculated how much extra.

Heiles: Instead of absorbing the 3 K background, the H_2CO you observe might just be absorbing only the extra 0.6 K (average) contributed by the HII regions. This would bias your observations because you would see H_2CO only on the front side of HII regions.

SPECTRAL-LINE SKY SURVEYS FROM THE NASA-JPL SETI PROJECT

G. R. Knapp
Owens Valley Radio Observatory, Caltech
T.B.H. Kuiper
Jet Propulsion Laboratory

The Jet Propulsion Laboratory (California Institute of Technology) has recently begun preparations for a modest SETI (Search for Extra-Terrestrial Intelligence) project at radio wavelengths. The proposed project is two-fold: (1) to search the entire sky visible from Goldstone, California (site of one of the NASA Deep Space Network Stations), i.e. that north of $\delta \sim -30^{o}$, at all frequencies between 1 and 24 GHz using a horn receiver and (2) to carry out a more sensitive sky search at a few selected frequency bands, using dedicated 26 and 10m antennas. The search will be for signals of the type unlikely to be produced by natural causes, e.g. narrow spikes in frequency space. The basic rationale behind this search is discussed by Murray, Gulkis and Edelson (1978).

The interest of this project to Galactic astronomy is, of course, that a possible offshoot of the search at selected frequency bands is the acquisition of sensitive surveys of the Galactic emission at several useful frequencies (Cuzzi and Gulkis 1977). The proposed instrumentation is: the 26 m dish for wavelengths $\lambda \gtrsim 3$ cm and the 10m dish for shorter wavelengths: (2) a dual-polarization maser receiver, with a series of parametric down-converters, operable between 26 and 1 GHz; the target system temperature is 12K on the telescope: (3) a 10^{6} channel fourier-transform spectrometer of bandwidth 320 MHz.

The frequency bands of primary interest to SETI are (1) the 'water hole', containing the four OH ground-state 18 cm-lines and the HI 21-cm line, and (2) the regions near the 22.3 GHz H_2O line and (possibly) the 24-GHz NH_3 line. The possible observing strategy to optimize the radio astronomical output would be to devote one year to the H_2CO and H_2O lines and two to the water-hole band.

In Table 1 we list the following system parameters at each frequency: telescope beamwidth, and system sensitivity to flux and to brightness temperature. This list has been calculated assuming a 'reasonable' (astronomical) value of the width of the observed line,

W. B. Burton (ed.), The Large-Scale Characteristics of the Galaxy, 85–86.

which is also given in the table, that each point is observed for the same amount of time and that half of the one- or two-year period is spent observing. The limit has been taken as five times the rms noise. It can easily be seen by consulting Table 1 that very useful astronomical surveys can be extracted from this project.

The purpose of this presentation is to solicit advice from the astronomical community regarding priorities and, particularly from colleagues experienced in spectral line sky surveys, techniques. The surveys will be conducted with SETI as the primary goal, but the desire to obtain useful astronomical data will play a major role in the planning. Because the SETI rationale tends to emphasize the search for very narrow spectral features, the motivation and techniques for obtaining flat baselines and well-calibrated spectra will need to come from the radio astronomical community. Colleagues who feel that they are able to contribute towards the planning in this area are particularly urged to communicate with us.

Table 1 Survey Sensitivity

Wave-length	Line	Beamwidth arcmin	Flux Limit (Jy)	Ass.line width (km s^{-1})	Brightness temp.limit	Ass.line width
					(K)	
21 cm	HI 1420.4	35	0.2	10	0.20	0.2
18 cm	OH 1665.7	30	0.7	1	0.10	0.5
6 cm	H_2CO	11	2	1	0.20	0.5
2 cm	H_2CO	10	9	1	0.20	0.5
1 cm	H_2O 22.3GHz	4	10	1	0.25	0.5
1 cm	NH_3 24GHz	4	10	1	0.25	0.5

REFERENCES

J.N. Cuzzi, S. Gulkis, "Summary of Possible Uses of an Interstellar Search System for Radio Astronomy", in The Search for Extraterrestrial Intelligence. NASA SP-419, U.S. Govt. Printing Office, Washington, D.C. (1977), pp. 147.

B. Murray, S. Gulkis, R.E. Edelson, 1978, "Extraterrestrial Intelligence: An Observational Approach", Science, 199, pp. 485 (1978).

GALACTIC DUST AND EXTINCTION

Gösta Lyngå
Lund Observatory, Sweden

The ratio R between visual extinction and colour excess, is slightly larger than 3 and does not vary much throughout our part of the Galaxy. The distribution of dust in the galactic plane shows, on the large scale, a gradient with higher colour excesses towards $l=50^\circ$ than towards $l=230^\circ$. On the smaller scale, much of the dust responsible for extinction is situated in clouds which tend to group together. The correlation between positions of interstellar dust clouds and positions of spiral tracers seems rather poor in our Galaxy. However, concentrated dark clouds as well as extended regions of dust show an inclined distribution similar to the Gould belt of bright stars.

1. THE VALUE OF $R=A_V/E_{B-V}$

The extinction of light A by galactic dust has a reddening effect described by the colour excess E. For the UBV system, in which most photometric studies of distant objects are made, one finds a value $R=A_V/E_{B-V}$ slightly larger than 3 when observing early type stars. The variation of R with intrinsic colour due to the wide passbands has been studied by Olson (1975).

Several studies of the value of R have been made, generally involving early type stars. Schalén (1975) demonstrated that several different methods give R=3.1 without much variation from field to field. For some regions much higher values have earlier been proposed but have now been refuted. Penston et al. (1975) have shown that the very high value earlier determined for the Orion cluster (cf. Johnson, 1968) is due to a misinterpretation of infrared excesses for the stars. Moffat and Schmidt-Kaler (1976) have explained the high R value for some associations with reflection nebulae as due to erroneous membership designation. Crézé (1972) derived different R values for spiral arm regions and for interarm regions. However, Sparke (1977) pointed out that these effects can be caused by the use of kinematic distances if, in fact, non-circular motions are present. Infrared photometry (Smyth and Nandy, 1978) for early type stars give a value $R=3.12\pm0.05$. It seems as if R is reasonably constant in the interstellar medium, but there may yet be slight variations. Whittet (1977) finds a systematic change between R=2.91 at longitude 85° and

W. B. Burton (ed.), The Large-Scale Characteristics of the Galaxy, 87–92.

R=3.25 at longitude 265°. This systematic trend is very similar to the variation of λmax, the wavelength of maximum linear polarisation, along the Milky Way. Turner (1976) has used the variable extinction method for open cluster reddenings and finds R=3.08 showing a slight variation with galactic longitude in a similar sense.

A high degree of homogeneity is present in the shortwave extinction as shown by Nandy et al. (1976). Even in that case we deal with the near-by dust (within the nearest few kpc).

2. CHARACTERISTICS OF THE DUST DISTRIBUTION

The most obvious clouds are globules (cf. Bok, 1977) but from the viewpoint of interstellar extinction we are mainly interested in the larger clouds, from a few parsecs upwards and the homogenous dust layer. Scheffler (1967) made an analysis of extinction values for distant O and B stars and found that most dust is situated in clouds and that there is a continuous frequency distribution of these. The more frequent are typically 3 pc in diameter, have extinction measures of $0^m.26$ and occur at a rate of 5 per kpc. Corresponding values for the larger clouds are 70 pc, $1^m.6$ and 0.5 per kpc. In a study of the interstellar extinction at intermediate distances in longitudes 280° - 320° Egret et al. (1978) find, from angular autocorrelation of colour excesses, that typical cloud diameters are less than 20 pc. The cloud structures discussed by Lucke (1978) are considerably larger, several hundred parsecs and presumably they can be resolved into smaller clouds. It seems that dust clouds tend to group in larger scale structures, as is also observed in external galaxies. Kron (1977) finds evidence for a uniform dust layer 400-600 pc thick in addition to the clouds.

While the sizes, masses and other properties of globules are becoming increasingly well known, there is still much to be learnt about the characteristics of the extended dust concentrations.

3. LATITUDE VARIATIONS OF EXTINCTION

At high latitudes the cloud structure of the dust will allow some directions to have much lower extinction than average. However, I shall for reasons of convenience first discuss conditions assuming a uniform dust layer.

The basis for determining high latitude extinction from galaxy counts has been re-examined by Heiles (1976). For the extinction in blue light at the North Galactic Pole (NGP) he finds $A_{90}=0^m.25$ in good agreement with Holmberg (1974). The surface magnitudes of galaxies at different galactic latitudes (Holmberg, 1958) give $A_{90}=0^m.22$. These values are considerably lower than earlier determinations which may not have taken due regard to the cloud structure causing small scale variations in the extinction. The principle of the galaxy count method has been criticized by Knapp and Kerr (1974) on the ground that galaxies as surface objects have surface brightnesses independent of distances. Considering the mentioned agreement with the surface magnitude effects on external galaxies we shall, however, accept $A_{90}=0^m.25$ for blue light.

The colour excess corresponding to this value has been determined from a variety of sources. Holmberg (1974) has collected some of the most reliable results and gives $E_{B-V}=0\overset{m}{.}054$ as a mean value. This would give R=3.6, approximately equal to the standard value.

Heiles (1976) shows that there is a latitude variation of the relation between lg (N) and E_{B-V}. At higher latitudes the galaxy counts decrease relatively slowly with increased colour excesses, which would point to larger R values and possibly indicate larger grain sizes than in the disk (Serkowski et al., 1975).

The colour excess towards the SGP has been determined by Knude (1977) who found $E_{B-V}=0\overset{m}{.}057 \pm 0\overset{m}{.}004$ and by Eriksson (1978) who found $E_{B-V}=0\overset{m}{.}04$ from an extensive material involving stars later than A. As is the case towards the NGP several authors get lower excesses from studies of A stars and other more luminous objects. To some extent this is due to the cloud structure of the dust, but it may also be due to a population related difference between intrinsic colours of stars with the same spectral class.

Comparing the observed extinctions towards the galactic poles it seems that the sun is not significantly displaced from the plane of symmetry of the galactic dust. However, Lucke (1978) confirms earlier views that the dust has an inclined galactic distribution similar to that of Gould's belt of bright stars. The distribution of discrete dark clouds is similar to this, as has been discussed by several investigators since Hubble (1922) and as is clearly shown from the catalogues by Lynds (1962) and by Sandqvist (1977). Turon and Mennessier (1975) have discussed the inclined distribution of clouds in terms of a model with elongated dust clouds aligned at an angle with the galactic plane.

4. RELATION OF DUST TO GALACTIC STRUCTURE

It was shown by Lynds (1970) that the lanes of dark nebulae are situated on the insides of arms in Sc galaxies. A particularly striking case is M 51 as studied by Mathewson et al. (1972) where dust lanes are shown to coincide with radio continuum radiation distribution on the inside of the bright arm structure. Trying to examine whether a similar situation obtains in our galaxy one must first get an impression of the distribution of clouds on a larger scale; the dark patches which show dust distribution in other galaxies correspond to dimensions of 50 pc or more.

4.1. Extinction studies in particular longitudes

Stellar statistical investigations (reviewed by McCuskey, 1976) are useful for comparing the amount of dust at intermediate distances in different galactic directions. The geometrical sizes of the dust clouds are, however, difficult to determine because of the dispersion in absolute magnitude for stars of a certain spectral or colour class as simple numerical experiments will show (Lyngå, 1976). The main use of extinction studies by stellar statistics is to give a sound basis for determinations of luminosity functions.

4.2. Colour excesses observed for distant stars

The investigations by Neckel (1967) and by FitzGerald (1968) give an overall view of the extinction situation. Although some early type stars are observed at large distances, there are two selection effects that become serious for distances of more than 2 kpc: the dispersion in absolute magnitude (cf. Malmquist, 1920) and the cloud structure of the interstellar dust.

Lucke (1978) has made a study which takes the z distribution into account. From a material of 4000 O and B stars he determined the overall structure of interstellar dust clouds within 2 kpc from the sun. Comparing Lucke's fig. 10 with a plot of spiral tracers such as fig. 2 in the paper by Humphreys (1976) one finds no positional correlation or anti-correlation except a slight similarity between the angles made by the different structures and the direction to the galactic centre. The lack of detailed agreement between well studied features makes me doubtful of analogies with a high luminosity galaxy like M 51. The situation reminds rather of less luminous galaxies where dust concentrations do not well correlate with other features and where also in general structural features have dimensions of one kiloparsec or smaller.

4.3. Extinction determined for open clusters

The most reliable determinations of interstellar extinction come from cluster studies. Well determined distances and colour excesses exist for more than 300 open clusters. To interpret the colour excesses in

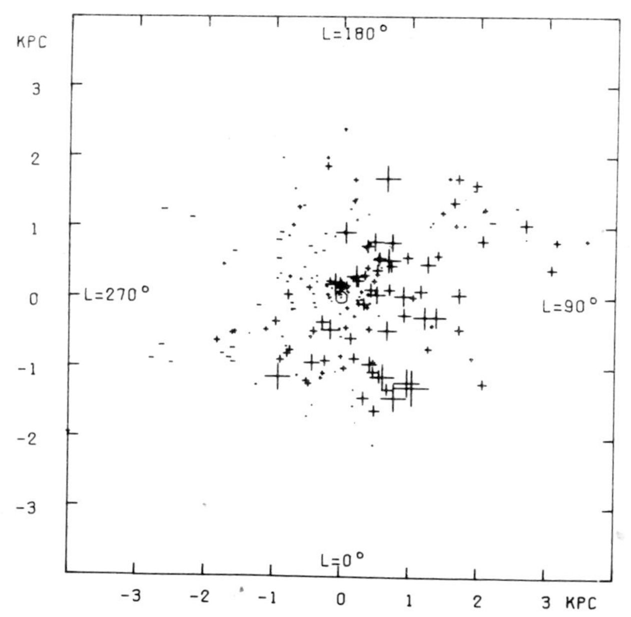

Figure 1. Extinction excesses Δ for open clusters. Scale of marks is 1.5 mm per magnitude.

terms of dust distribution I have assumed a uniform dust layer of 100 pc thickness. Inside the layer there is an extinction rate $a_v = 0^m.7$ per kpc so that a cluster at galactic latitude b and distance r is expected to suffer an extinction $0^m.7$ r if it is inside the dust layer and an extinction $0^m.035$ r cosec b if it is outside that layer. The observed extinction value exceeds the expected by an amount Δ, the sign of which has been displayed in figure 1. The size of each mark corresponds to the numerical value of the excess and the position corresponds either to the position of the cluster or the position where the line of sight enters the dust layer. Figure 1 shows clearly that towards $l=50^o$ there is a preponderance of + signs and in the opposite direction mostly - signs. A least square solution shows that a gradient of about $0^m.1$ per kpc near the sun would describe the asymmetry.

4.4 Galactic windows

In some directions the galactic disk is more transparent than normal. A well established such galactic window is situated in Vela at longitude 245°- 255° (FitzGerald, 1974; Dodd and Brand, 1975) and another is found in Circinus at a longitude of 311° (Lyngå, 1977). In a number of other directions one finds a relatively small extinction out to about 4 kpc from the sun (cf. FitzGerald and Moffat, 1976). Such features are not easily assimilated by a large scale model of spiral structure.

Added to other evidence given above the presence of galactic windows will support the view that the distribution of interstellar dust has a large scale structure which is not well correlated with the distribution of spiral arm tracers.

ACKNOWLEDGEMENTS

I thank Prof. C. Schalén and Dr. S. Wramdemark for useful comments on the manuscript.

REFERENCES

Bok, B.J.: 1977 Publ. Astron.Soc.Pacific 89, p. 597
Crézé, M.: 1972 Astron.Astrophys. 21, p. 85
Dodd, R.J., Brand, P.W.J.L.: 1975 Astron.Astrophys. Suppl. 25, p. 519
Egret, D., Lyngå, G., Ochsenbein, F.: 1978 In preparation
Eriksson, P.-I.: 1978 Uppsala Astron.Obs. Rept. No. 11
FitzGerald, M.P.: 1968 Astron.J. 73, p. 983
FitzGerald, M.P.: 1974 Astron.Astrophys. 31, p. 467
FitzGerald, M.P., Moffat, A.F.J.: 1976 Sky Telesc. 52, p. 104
Heiles, C.: 1976 Astrophys.J. 204, p. 379
Holmberg, E.B.: 1958 Meddelanden Lunds Astron.Obs., Ser. II, No. 136
Holmberg, E.B.: 1974 Astron.Astrophys. 35, p. 121
Hubble, E.: 1922 Astrophys.J. 56, p. 162
Humphreys, R.M.: 1976 Publ. Astron.Soc. Pacific 88, p. 647
Johnson, H.L.: 1968 In Kuiper, Stars and Stellar Systems Vol VII, p. 167
Knapp, G.R., Kerr, F.J.: 1974 Astron.Astrophys. 35, p. 361
Knude, J.K.: 1977 Astrophys. Letters 18, p. 115

Kron, G.: 1977 Bull.Am.Astron.Soc. 9, p. 581
Lucke, P.B.: 1978 Astron.Astrophys. 64, p. 367
Lynds, B.T.: 1962 Astrophys.J. Suppl. 7, p. 1
Lynds, B.T.: 1970 IAU Symp. No. 38, p. 26
Lyngå, G.: 1976 Astron.Astrophys. 46, p. 369
Lyngå, G.: 1977 Astron.Astrophys. 54, p. 71
Malmquist, K.G.: 1920 Meddelanden Lunds Astron.Obs. II, No. 22
Mathewson, D.S., van der Kruit, P.C., Brouw, W.H.: 1972 Astron.Astrophys. 17, p. 468
McCuskey, S.W.: 1976 IAU Trans. XVIA, part 3, p. 44
Moffat, A.F.J., Schmidt-Kaler, Th.: 1976 Astron Astrophys. 48, p. 115
Nandy, K., Thompson, G.I., Jamar, C., Monfils, A., Wilson, R.: Astron. Astrophys. 51, p. 63
Neckel, Th.: 1967 Veröffentl. Heidelberg-Königstuhl 19
Olson, B.I.: 1975 Publ.Astron.Soc.Pacific 87, p. 349
Penston, M.V., Hunter, J.K., O'Neill, A.: 1975 Monthly Notices Roy. Astron.Soc. 171, p. 219
Sandqvist, Aa.: 1977 Astron Astrophys. 57, p. 467
Schalén, C.: 1975 Astron.Astrophys. 42, p. 251
Scheffler, H.: 1967 Z. Astrophys. 65, p. 60
Serkowski, K., Mathewson, D.S., Ford, V.L.: 1975 Astrophys.J. 196, p. 261
Smyth, M.J.,Nandy, K.: 1978 Monthly Notices Roy.Astron.Soc. 183, p. 215
Sparke, L.S.: 1977 Astron.Astrophys. 56, p. 307
Turner, D.G.: 1976 Astron.J. 81, p. 1125
Turon, P., Mennessier, M.O.: 1975 Astron.Astrophys. 44, p. 209
Whittet, D.C.B.: 1977 Monthly Notices Roy. Astron.Soc. 180, p. 29

DISCUSSION

Yahil: Could you comment on the clumpiness of absorption within, say, 60° of the Galactic poles, and the appropriate absorption correction for extragalactic objects.

Lyngå: According to colors for globular clusters as discussed by Sandage a few years ago, one finds directions with practically no extinction. This would be due to clumpiness in the cloud distribution. I believe that Dr. Kron has more details about the situation.

Kron: My work on reddening pertains only to an average minimum covering a considerable volume of the solar neighborhood. No direct measurement of reddening at either pole is implied; however an extrapolated value of $0^m.02$ to $0^m.03$ can be deduced from an application of the cosecant b "law".

THE DISTRIBUTION OF YOUNG STARS, CLUSTERS AND CEPHEIDS IN THE MILKY WAY AND M33 - A COMPARISON

Roberta M. Humphreys
University of Minnesota

Ever since the pioneering work by Morgan and his collaborators (1952, 1953), it has been well known that the distribution of the associations of young stars, HII regions, and young clusters defines the optical spiral features. Although considerable progress has been made in spiral structure studies during these past 25 years, the basic picture of optical spiral structure has not been significantly altered. The three spiral features first described by Morgan are still recognized. Modern work on the various optical spiral features has strengthened and improved the definition of the optical features, especially to larger distances. Most of the improvements and any additions to the basic three-arm pattern have resulted primarily from observations of the spiral tracers in the Southern Milky Way. Specifically, the Sagittarius feature is now generally recognized as the Sagittarius-Carina arm which may indeed be a major arm of the Galaxy. It can now be traced optically to very large distances, up to 6 kpc or more in the direction $\ell = 290°$. The Local arm (Cygnus-Orion) probably extends to 4 kpc in the direction of Puppis ($\ell \cong 240°$), and most astronomers would probably agree that our local spiral feature is not a major arm, but an inter-arm feature.

In this paper, I will briefly review the evidence for optical spiral structure as revealed by the distribution of the assocations of young stars, the youngest open clusters and the long-period Cepheids. I will also discuss the distribution of the young stars, associations and Cepheids in our neighboring spiral galaxy M33. A comparison allows some perspective on the problems of spiral structure in the Milky Way.

The first figure shows the distribution of the recognized OB associations and youngest clusters (B2-B3). Considerable work has recently been done on the clusters in the Southern sky by Moffat, Vogt, and FitzGerald (references given at the end) and their new data has been combined with the young clusters from the compilation by Becker and Fenkart (1971). Although considerable recent work has also been done by a number of astronomers (Humphreys 1973, 1975, Walborn 1973, Miller 1972, Jackson 1976, Muzzio and Orsatti 1977a, b, and Garrison,

W. B. Burton (ed.), The Large-Scale Characteristics of the Galaxy, 93–98.

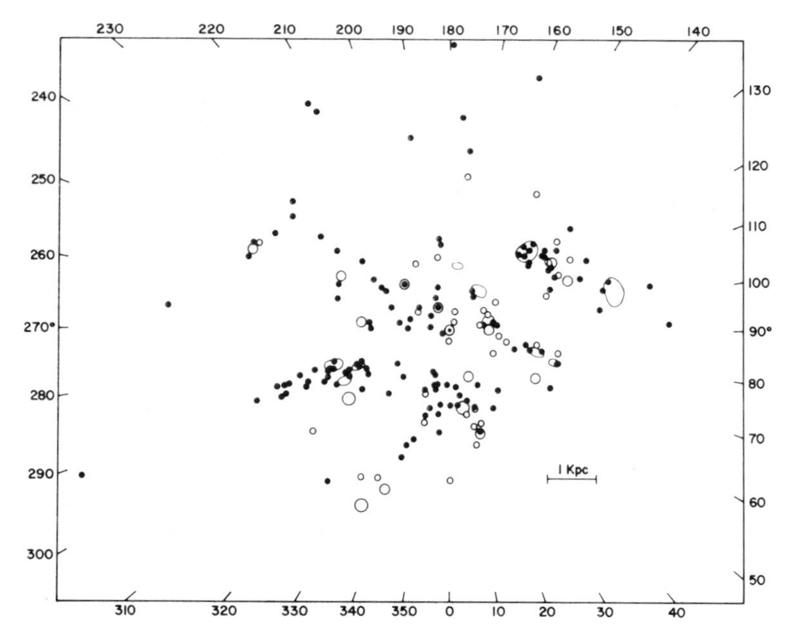

Figure 1 - The space distribution of the associations of young stars, O, and young clusters (B2-B3), ●. The sun's position is indicated by ⊖.

Hiltner and Schild 1977) on the supergiants and OB stars in the Southern sky, I have chosen to show the distribution of the associations of young stars instead of the individual stars to eliminate some of the scatter due to distance errors. The large open circles for the associations are shown in various sizes in the figure in an attempt to weight the associations, first according to their sizes based on their published boundaries (the angular size or extent in figure 1) and secondly by the number of known member stars (radial size in figure 1). Thus in this figure greater weight is given to the largest and most heavily populated associations.

The three major spiral features, Sagittarius-Carina, the Local feature, and the Perseus feature, are clearly defined, and there are significant gaps between the three features. The outer edge to the Sgr-Car arm occurs at $\ell = 30°-40°$ in the North and $\ell = 280°-285°$ in the South. With the addition of supergiants and HII regions this arm can be traced optically to more than 6 kpc from the sun. In addition there is evidence for an inner spur in the direction $\ell = 305°$. A spiral feature, Norma-Scutum, interior to Sgr-Car is also present at 3 to 4 kpc from the sun.

The Local feature can now be traced to large distances towards $\ell = 240°$ and it is possible it may be a spur or bifurcation of the Perseus arm. The Perseus feature or outer arm is not very well-defined at larger distances, although it may extend through Cam OB3 and Aur OB2 towards $\ell = 210°$.

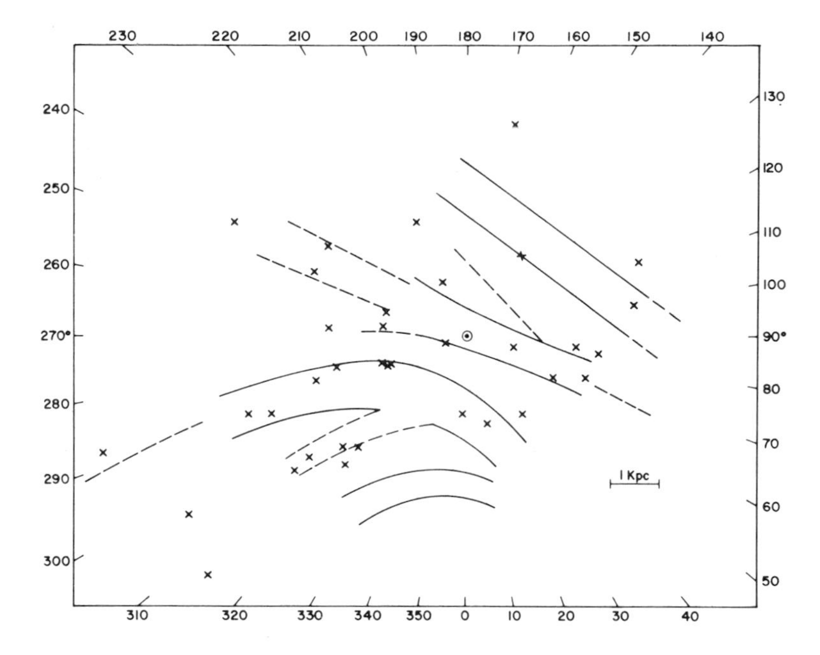

Figure 2 - The space distribution of the long-period Cepheids, $P \geq 15^d$, with the outlines of the spiral features as shown by the objects in Figure 1.

Figure 2 shows the distribution of the longest period Cepheids ($P \geq 15^d$), selected from Tammann (1970) and Grayzeck (1974). This limit was selected because a period of 15^d corresponds to a visual luminosity of $M_V \simeq -5^m$ and most of the supergiants and OB stars are even brighter than this. It was felt that shorter period and consequently less luminous Cepheids might belong to a somewhat older population. It is clear from this figure that there are too few Cepheids of longest period to be used alone as spiral tracers. However the outlines of the spiral features as revealed by the associations and young clusters are also included with the Cepheids, and it is apparent that most of the longest period Cepheids lie in or near the spiral features. From this comparison it is reasonable to conclude that in future spiral structure studies the most luminous Cepheids should serve as beacons or signposts to places which should be studied for other more abundant tracers. However we shall see in M33 that the presence of a Cepheid does not always mean that a spiral arm is present.

The following discussion of M33 is based on the first results of a study of the population I stars in that galaxy. This work is in collaboration with Allan Sandage. The brightest blue and red stars have been identified in M33 by blinking blue, visual and red (IV-N) plates of that galaxy, and the associations of the blue stars have been mapped based on deep blue photographs. A photograph of M33 shows the outlines of the associations of blue stars. 141 associations have been identified and it is clear from this picture that the associations follow the basic spiral pattern of M33. The spiral structure of M33 is rather complex as is obvious from any photograph of that galaxy. Although one has the impression of an underlying two-arm pattern it is complicated by the extensive fragmentation of the arms which begins quite near the nucleus. It was hoped that the distribution of the brightest blue and red stars and particularly the associations would make the basic pattern clearer, and they do to some extent.

The distribution of the brightest blue stars reflects the basic spiral structure of M33, as well as all of the smaller spiral features and branches. Their distribution clearly illustrates the problems one would encounter in mapping the structure from inside M33. There are not as many red stars, so that the spiral structure is not so apparent and contamination from foreground stars is serious for the red stars. When most of the field stars have been removed, the spiral pattern may be more obvious. Those red stars in the associations and the spiral arms are presumably the M supergiants.

The 35 Cepheids from Hubble's (1926) original paper are shown on an outline of the spiral structure in M33. The Cepheids appear to be concentrated towards the nucleus of M33, however this is probably a selection effect since Hubble's plates were of good quality only over the central regions of M33. These Cepheids have periods between 13^d and 70^d. Approximately one third of these long-period Cepheids are in or near the associations and another third in spiral features. The remaining Cepheids including those that appear to be in the nuclear region are clearly _not_ in spiral arms. This result supports the earlier suggestion that Cepheids in the Milky Way should be used in combination with other spiral tracers and as beacons of where possible spiral features might exist.

A comparison of the distribution of the sizes of the associations in the Milky Way and in M33 shows that what we call associations in our galaxy are smaller than those in M33. This is probably not real, but instead due to observational selection. In our galaxy we recognize smaller groupings of stars than in a distant galaxy. It is important to consider what we would call a spiral feature or recognize as a spiral tracer, that is an association of young stars, if the Milky Way were at the distance of M33. Looking at Figure 1 again it is obvious that there are large clumpings of associations, young stars, and clusters - in Carina, Perseus, Cygnus and Sagittarius. These clumps are what we would call spiral tracers in more distant galaxies.

The linear sizes or extent of the associations in our galaxy were determined from their published boundaries and their distances, and the mean linear size of 57 associations is 125 parsecs. In M33 the tangential size refers to the linear size parallel to the spiral arm and the radial size refers to the perpendicular direction. There is a tendency for the associations to be larger in the tangential direction than in the radial. The mean tangential size is 251 parsecs while the mean radial size is 203 parsecs. Thus the associations in M33 appear to be pulled out or stretched parallel to the arm.

In conclusion the optical spiral tracers in our region of the Galaxy reveal wide spiral arms with considerable evidence for branching and fragmentation. Sgr-Car may be a major arm of our galaxy particularly since it may be traced to such large distances towards $\ell = 290°$. It would be very worthwhile to study the spiral structure along the northern edge of this arm $\ell = 30°-40°$ to test this hypothesis. It is possible that the local feature is an inner arm spur branching perhaps from the Perseus arm, or that these two features may be due to a bifurcation of the other major arm. We know that considerable branching of the spiral arms occurs in other galaxies particularly in the outer parts. We see such a small part of our galaxy that it is difficult to extrapolate, but if our region is any indication, the spiral structure of the Milky Way may be quite complex.

REFERENCES

Becker, W. and Fenkart, R.: 1971, Astron. and Astrophys. Suppl. 4, 241.
FitzGerald, M.P., Hurkens, R. and Moffat, A.F.J.: 1976, Astron. and Astrophys., 46, 287.
Garrison, R.E., Hiltner, W.A. and Schild, R.E.: 1977, Ap.J. Suppl., 35, 111.
Grayzeck, E.J.: 1974, Doctoral Dissertation, Univ. of Maryland.
Hubble, E.: 1926, Ap.J., 63, 236.
Humphreys, R.M.: 1973, Astron. and Astrophys. Suppl., 9, 85.
Humphreys, R.M.: 1975, Astron. and Astrophys. Suppl., 19, 243.
Jackson, P.D.: 1976, Doctoral Dissertation, Univ. of Maryland.
Miller, E.W.: 1972, A.J., 77, 216.
Moffat, A.F.J.: 1972, Astron. and Astrophys. Suppl., 7, 355.
Moffat, A.F.J. and FitzGerald, M.P.: 1974, Astron. and Astrophys. Suppl. 16, 25.
Moffat, A.F.J. and Vogt, N.: 1973, Astron. and Astrophys. Suppl., 10, 135.
Moffat, A.F.J. and Vogt, N.: 1975, Astron. and Astrophys. Suppl., 20, 85.
Moffat, A.F.J. and Vogt, N.: 1975, Astron. and Astrophys. Suppl., 20, 125.
Moffat, A.F.J. and Vogt, N.: 1975, Astron. and Astrophys. Suppl., 20, 155.
Morgan, W.W., Sharpless, S., and Osterbrock, D.E.: 1952, A.J., 53, 3.
Morgan, W.W., Whitford, A.E., and Code, A.D.: 1953, Ap.J., 118, 318.
Muzzio, J.C. and Orsatti, A.M.: 1977a, A.J., 82, 345.
Muzzio, J.C. and Orsatti, A.M.: 1977b, A.J., 82, 474.
Tammann, G.A.: 1970, in I.A.U. Symposium No. 38, p. 236.
Vogt, N. and Moffat, A.F.J.: 1972, Astron. and Astrophys. Suppl., 7, 133.
Vogt, N. and Moffat, A.F.J.: 1973, Astron. and Astrophys, Suppl., 9, 97.
Walborn, N.R.: 1973, A.J., 78, 1067.

DISCUSSION

Bok: 1. How do you intend to eliminate the galactic red stars from your statistics for M33? 2. Have you corrected the discussions of your M33 HII regions for the most probable tilt of M33?

Humphreys: 1. Sandage is measuring true colors and magnitudes for the brightest blue and red stars in M33. A color-magnitude diagram will be produced; we will then be able to eliminate most of the foreground stars. 2. We will be determining the tilt of M33 from the spiral structure diagram; we will then be able to correct the dimensions for tilt.

Kerr: There is still an important difference between the radio and optical pictures. The radio continuum and recombination-line data show a very clear gap between the Sagittarius and Carina features, unlike the Sagittarius-Carina arm described by Sivan and by Humphreys.

Humphreys: Optical spiral tracers show the Sagittarius-Carina feature very strongly. There is no optical gap between the two regions.

Yuan: Could you explain how you arrived at such a high pitch angle for the Perseus arm?

Humphreys: The Perseus arm location is based on the associations and clusters with $r > 2 - 2.5$ kpc between $\ell \simeq 100°$ and $180°$. Actually the pitch of this arm could have been even higher. I tried to be rather conservative in drawing the outline of this feature.

FAINT OPTICAL SPIRAL TRACERS

Juan C. Muzzio[1]
Observatorio Astronómico de la Universidad Nacional de La Plata, and Consejo Nacional de Investigaciones Científicas y Técnicas de la República Argentina

The use of modern photographic emulsions and techniques allows for the discovery of optical spiral tracers much fainter than was possible in the past. Our searches for such faint objects deal with the Puppis, Vela, Crux, Circinus, Norma and Ara regions of the Milky Way using Kodak plates (baked in dry nitrogen) obtained with the Curtis Schmidt-telescope at Cerro Tololo Inter-American Observatory. We have discovered OB stars as faint as 15 mag on IIIa-J plates obtained with the thin prism, and Hα emission-line objects (many of which are Be stars) as faint as 16 mag on 127-04 plates obtained with the 4° prism.

The study of those faint objects offers the opportunity to improve the optical picture of the spiral structure of our Galaxy by allowing: a) to reach larger distances from the Sun, and b) to compensate the bias caused by heavy obscuration in some regions. In fact, distances larger than 4 kpc and total visual absorptions exceeding 4 mag are not uncommon among the objects we found.

Unfortunately, the derivation of the distances of the optical spiral tracers to the Sun has not progressed accordingly. It is still difficult to obtain an accuracy better than about 15% for stellar groups or 30% for single stars, and the situation is worst for faint objects. Besides, a warning should be made concerning the use of Hβ photometry to derive distances of early type stars. It is not unfrequent that the wide and narrow filters have different effective wavelengths and, as a result, the measured β index depends on the color excess of the star. We found that this effect is much more important than was previously thought, particularly for the faint optical spiral tracers which are usually highly reddened. The effect is present even in the filters used to define the standard system and the current calibrations may thus need some revision.

[1]Visiting Astronomer, Cerro Tololo Inter-American Observatory, supported by the National Science Foundation under contract No. NSF-C866.

W. B. Burton (ed.), The Large-Scale Characteristics of the Galaxy, 99.

THE GALACTIC DISTRIBUTION OF 60 YOUNG OPEN CLUSTERS

Rolf P. Fenkart
Astronomical Institute of the University of Basle

The photometric distances of 60 young galactic clusters (with sp < b3), all observed in UBV or RGU, have been calculated or recalulated according to "method A" (Becker, 1963). The galactic distribution of these clusters, shown in Figure 1, confirms their role of being good spiral tracers.

20 years have elapsed since Becker and Stock (1958) published their catalogue of 40 three-colour photometrically observed galactic clusters. This was the first one in a series containing increasing numbers of clusters (i.e. 82, 156, 216) and published by the Basle Observatory (Becker, 1961; Becker, 1963; Becker and Fenkart, 1971).

The main aim of these compilations was to provide the given clusters with the best, homogeneously determined distances available. Covering distances up to 5 to 7 kpc from the sun with a mean error of not more than $\pm$ 10%, the three-colour photometric method turns out to be ideal for this purpose, yielding, together with the distance, reddening, absorption, probable physical members and earliest spectral type for the cluster in question. The last parameter is a good indicator of the relative age of the cluster and therefore it is important for the identification of those clusters which are young enough to be spiral tracers. Indeed, all clusters with an earliest spectral type not later than about b2 have a strong tendency to concentrate their positions along relatively well defined fragments of the local spiral arms, representing therefore the most reliable spiral tracers next to HII regions.

For these reasons, it seemed useful to extend the material contained in the previous catalogues by collecting all available three-colour photometric data of those galactic clusters which have been observed since the forth Basle Catalogue (BC IV) (Becker and Fenkart, 1971), either in Johnson's UBV- or in Becker's RGU-system.

In view of their importance as potential spiral tracers, we have reconsidered all those young clusters, i.e. with sp $\leq$ b2, whose distances had not been determined according to "method A" applied at the

W. B. Burton (ed.), The Large-Scale Characteristics of the Galaxy, 101–104.

Basle Observatory and consisting of the evaluation of <u>both</u> colour-magnitude diagrams, because it provides a better identification of probable physical members and a more precise determination of interstellar absorption and distance than "method B" which uses <u>one</u> colour-magnitude diagram only, together with the distance-independent two-colour diagram (Becker, 1963). Among the young clusters collected since BV IV and up to the deadline of the compilation in question (March 18th, 1978), there were 17 whose distances had to be redetermined according to method A. The resulting values differ, in some cases, quite considerably from those obtained by the corresponding authors according to method B.

Figure 1 shows all 60 young clusters of this last Basle Catalogue (BV C; Fenkart and Binggeli, 1978) containing totally 191 three-colour photometrically observed galactic clusters. They are projected upon the galactic plane, where the hatched regions give an idea of the approximate course of Becker's spiral arms -II, -I, 0 and +I which represent the local spiral structure as it is defined by previous plottings of reliable spiral tracers, as HII regions and young open clusters (<u>cf</u>. Becker and Fenkart, 1970).

The original distances (according to method B) of the 17 reconsidered clusters are indicated by the points of the radial arrows starting at the inverted triangles (▼) which represent their recalculated positions according to method A. None of these revised positions is outside a hatched region or a plausible extension of one of them.

Among the 43 clusters (●) whose distances had been determined according to method A by the corresponding authors themselves, only five lie clearly in the inner arm regions between the hatched areas: 95 and 99 between arms 0 and -I, 120, 149 and 183 between arms -I and -II. 95, 120, 149 and 183, however, lie so close to neighbouring arms that even during their short lifetimes they could have withdrawn from there, provided they had been born sufficiently close to the corresponding edges of these arms and assuming sufficiently steep directions of their space motions with respect to them. In the cases of 149 and 183, even a distance correction by less than 10% (the mean error of three-colour photometrically determined cluster distances) would shift them into neighbouring arms. For 99, however, only an unusually large component of its space motion, perpendicular to arm 0, (13 km sec^{-1}) would reconciliate the low age of this cluster with its extreme inner arm position.

The remaining 38 clusters lie either within the hatched arm regions or so far from them that it cannot be decided whether they would fall into their extensions, into inner arm regions or, in the cases of 32, 38, 44, and 45, even into +arms lying farther out from the sun.

The addition of 60 further spiral tracers overwhelmingly confirms the familiar local spiral structure defined by Becker's arms. The clusters whose positions fall outside this system are not sufficiently crowded to give reliable hints for plausible extensions.

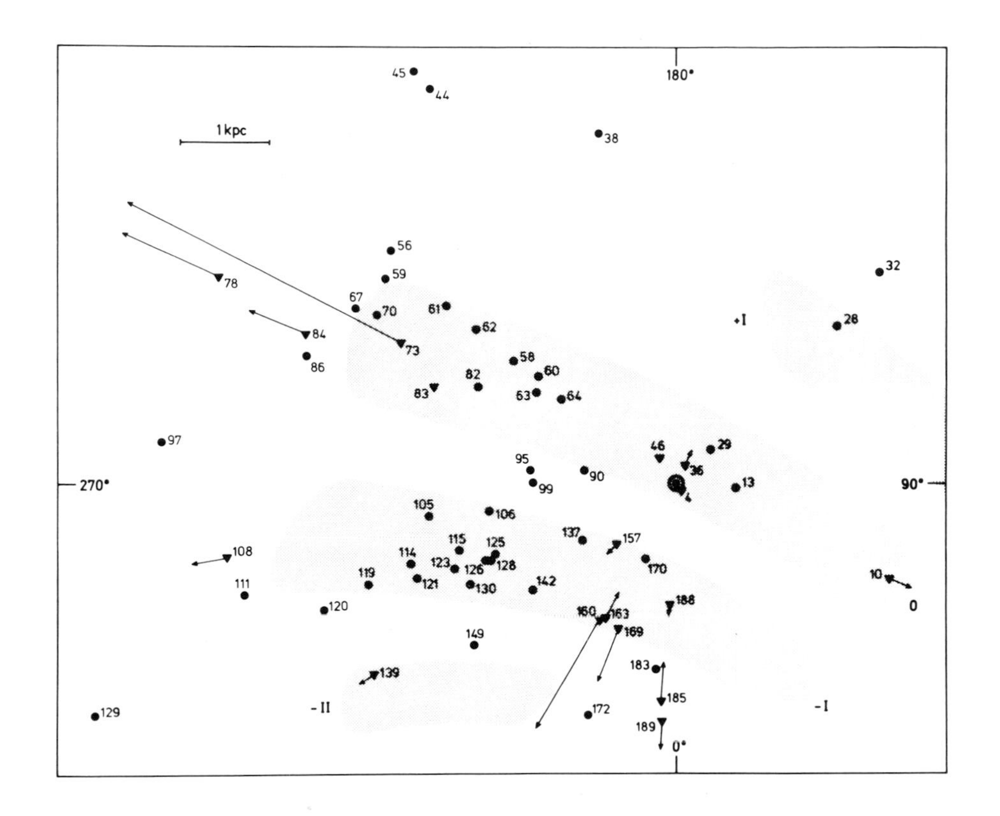

Figure 1. Galactic distribution of 60 young galactic clusters (sp $\leq$ b2) contained in the Fifth Basle Catalogue (BC V).

Hatched regions: Approximate course of Becker's spiral arms -II, -I, 0 and +I.

- ⊙ : Sun
- ▼ : Redetermined positions according to method A (by F)
- ↓ : Original positions according to method B (determined by corresponding authors)
- ● : Positions according to method A(determined by corresponding authors)

Numbers corresponding to those in column 1 of table 1 (BC V).

REFERENCES

Becker, W., Stock, J.: 1958, Z. Astrophys. 45, 269.
Becker, W.: 1961, Z. Astrophys. 51, 151.
Becker, W.: 1963, Z. Astrophys. 57, 117.
Becker, W., and Fenkart, R. P.: 1970, The Spiral Structure of Our Galaxy, IAU Symposium No. 38, Reidel, Dordrecht.
Becker, W., and Fenkart, R. P.: 1971, Astron. Astrophys. Suppl. 4, 241.

DISCUSSION

Kaufman: What width did you adopt for the spiral arms? Whether you assign a cluster to arm or interarm depends on the assumed arm width.

Fenkart: The width of the hatched arm regions is given by the approximate range of spiral tracer positions given in earlier plots (see, e.g., Becker and Fenkart, 1970).

Jackson: In what you call the Arm-I and others call the Sagittarius-Carina arm, you are sure that you are seeing through it to the other side, and not just running into dust at different distances in different directions?

Fenkart: Yes.

de Vaucouleurs: Spiral arm "tracers" have been much discussed at this symposium. At the same time the confused picture of spiral structure in our solar neighborhood given by these "tracers" is a disappointment. Let me remind you that a similar situation is observed in many galaxies whose spiral structure almost disappears if one plots only the "tracers" (HII regions, etc.). In some galaxies, such as M101, the "tracers" are in many parts much displaced from the smooth, old spiral arms.

Bok: 1. What is the Carina Arm in your picture? 2. It seems very desirable always to print side-by-side with one diagram (like the one shown by Dr. Fenkart) showing the suggested spiral arms, a second diagram showing only the raw data. By having these diagrams side-by-side, it is easy for the readers to judge for themselves whether or not it makes sense to draw the spiral features as shown.

Fenkart: 1. The Carina fragment has not been fully hatched, because only very few new objects are found there. 2. I agree with your suggestion.

GALACTIC STRUCTURE FROM INFRARED STUDIES

J. L. Puget
Institut d'Astrophysique, Paris
G. Serra
Centre d'Etudes Spatiales des Rayonnements, Toulouse
C. Ryter
Centre d'Etudes Nulcéaires de Saclay

Star densities on a galactic scale are traced by far infrared emission of dust heated by young stars and by the 2.4 μm radiation of stars in the red giant phase. Coherent results are obtained, pointing to a very strong star formation rate during the last ∿200 My in a ring 5 kpc from the galactic center. A steepening of the initial mass function compared to that observed in the solar vicinity is also suggested.

Interstellar extinction decreased quickly as the wavelength of the observed radiation increases from visible to infrared. At 2.4 μm it is already possible to observe throughout the galaxy at b = 0'. At this wavelength, emission from stars in the red giant phase are expected to be the dominant component. At longer wavelengths, thermal emission from dust becomes more important. Two very important physical properties of the interstellar dust can be obtained from studies in the infrared: its line-of-sight column density and its average temperature, which is a measure of the energy density of the stellar radiation heating it.

1. REVIEW OF THE EXISTING DATA

Recently partial surveys of the galactic plane at 2.4 μm have been made by several groups (Hayakawa *et al*. 1976, Ito *et al*. 1976, Hoffman *et al*. 1977, Ito *et al*. 1977, Okuda *et al*. 1977, and Matsumoto *et al*. 1977). In Figure 1, a compilation of the data is shown. No significant discrepancy can be seen. Furthermore, a rocket observation has been made by Hayakawa *et al*. (1978) near ℓ = 182°, b = -9°. In the far infrared, the observations of the diffuse emission are less complete. There is basically no information on the latitude variation and the longitude profile is still incomplete even at ℓ < 30°. Nevertheless, some basic features can be seen on the existing data shown on Figure 2. They include one rocket observation by Pipher (1973), balloon measurements by Serra *et al*. (1977), Low *et al*. (1977), and Serra *et al*. (1978a,b). Data on the structure of this emission have been obtained with an airborne telescope by Rouan *et al*. (1977) and by Viallefond *et al*. (1978). On the basis of a simple model in which the stellar

W. B. Burton (ed.), The Large-Scale Characteristics of the Galaxy 105–111.

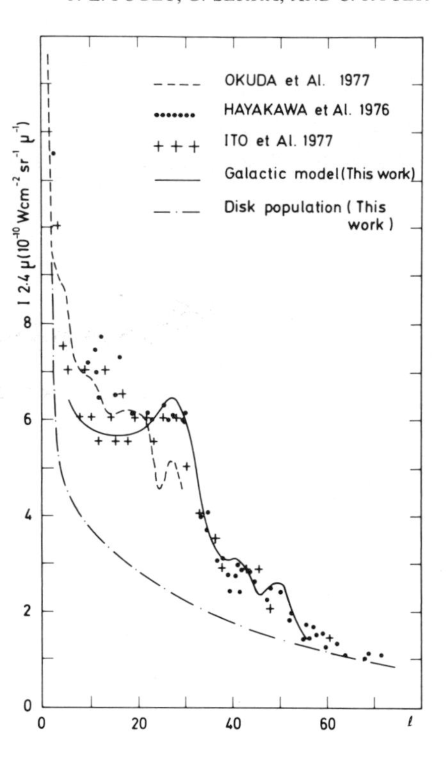

Figure 1. Summary of observations of the λ = 2.4 µm galactic radiation, and values predicted for a conventional mass-to-light ratio and mass model of Galaxy (disc population), and with additional star formation at 5 kpc from the galactic center (see text).

radiation field is uniform in the Galaxy, Stein (1967) and Fazio and Stecker (1976) have predicted values about an order of magnitude lower than what is observed. Much higher values had been predicted by Ryter and Puget (1977) and by Drapatz and Michel (1977), in basic agreement with the data. The significantly higher temperature of the dust predicted in the two later papers is also confirmed by the color temperature found in the balloon observations (T $\gtrsim$ 25 K).

2. IMPLICATIONS OF THE 2.4 µm DATA

The values of the brightness observed at $\ell \geq 55°$ in the galactic plane in the anticenter can be compared to that expected from M and K star counts in the solar vicinity, and from the mass distribution in the disc. The good agreement between the computed and the observed fluxes indicates that no unexpected component is detected in these directions (Maihara et al. 1978, Hayakawa et al. 1977, Serra and Puget (1977), Hayakawa et al. 1978). On the other hand, extensive studies of the galactic center have been made in the near infrared; good agreement between the infrared data and the dynamical mass is found (Becklin et al. 1968, Sanders and Lowinger 1972). In the whole range $5° < \ell < 55°$, models based on a standard mass distribution in the disc yields a predicted 2.4 µm emission much below the observed values, as can be seen in Figure 1 (Serra et al. 1978c). The excess is shown in Figure 3 and is quite similar to the longitude profiles of extreme population I tracers (see for example Burton, 1976). We attribute this excess to massive red giants associated with regions of rapid star

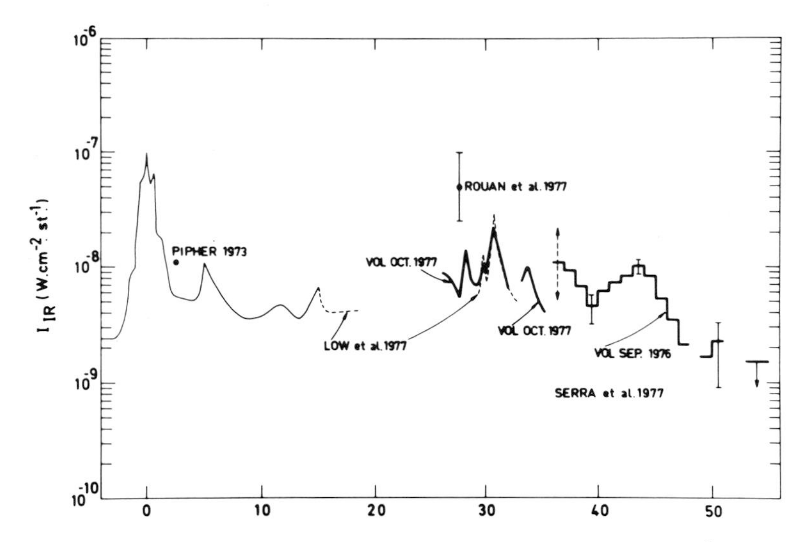

Fig. 2: Summary of observations of the far infrared ($\lambda \approx 50$ to $200\ \mu m$) galactic background.

formation. The ratio

$$g = \frac{\int F(t, 2.4\mu)\ dt\Delta\lambda}{\int dt\ \int F(t,\lambda) d\lambda}$$

where F_λ is the spectral flux density emitted by a star, integrated over the lifetime of the star. It has been computed as a function of star mass. Models of the radial distribution of the 2.4 μm source function have been produced and compared to the data. A high production rate is required in a ring $\sim$5 kpc from the galactic center, and we deduce that the total luminosity of the main sequence stars associated with the red giants in this region is $L_{tot} \approx 300\ L_\odot\ pc^{-2}$. This very high luminosity implies a star formation rate much higher than that in the solar neighborhood.

3. IMPLICATIONS OF THE FAR INFRARED DATA

A fraction f_1 of the stellar radiation is absorbed by the molecular cloud in which the star was formed, a fraction f_2 is absorbed by dust in the diffuse interstellar medium, and a fraction f_3 is absorbed by all other molecular clouds. An evaluation of these factors as a function of the mass of the star has been made by Serra et al. (1978), who found $f = f_i + f_2 + f_3 \approx 0.3$.

A high infrared luminosity $L \approx 100\ L_\odot\ pc^{-2}$ is needed in a ring at $\sim$5 kpc from the galactic center to account for the observed flux. Considering the typical value of f, the luminosity is well accounted for by the total luminosity computed from the 2.4 μm excess.

A summary of data and of the results for the solar neighborhood, the 5-kpc ring, and the galactic center, is given in Table 1 and shows clearly that the formation rate of stars with $M < 1\ M_\odot$ per unit mass of

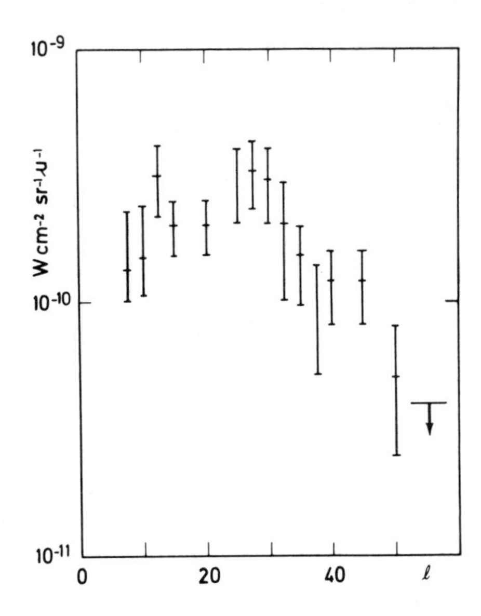

Figure 3. The 2.4 µm intensity in excess of that produced by the disc population, required to account for the observations, and attributed to massive stars in red giant phase.

interstellar gas is much higher in the inner galaxy than in the solar vicinity. If the ratio between the total mass of stars formed below and above one solar mass is everywhere the same as close to the sun (most of the mass goes into stars with $M < 1\ M_\odot$), the gas will be exhausted very quickly in the inner Galaxy (< 100 My). If some stationary state exists, it implies that the initial mass function is not uniform in the Galaxy, or that there is a strong gas infall on the galactic plane or that the gas disc is in radial contraction.

It has also to be noted that a comparison of the total luminosity of the stars recently formed and the Lyman continuum photons produced in the same regions suggest a deficiency of the HII regions in the

TABLE 1 SUMMARY OF DATA AND RESULTS ON DENSITIES AND SOURCE FUNCTIONS IN THE GALAXY

	Mass in Stars	Mass in Gas	$L_{2.4\mu}$ Observed	$L_{2.4\mu}$ Disc	$L_{2.4\mu}$ Excess	$L_{100\mu}$	L_{Lyc}	$\frac{L_{tot,PopI}}{M_{gas}}$	$\frac{L_{Lyc}}{M_{gas}}$
Solar Vicinity	0.145 $M_\odot pc^{-3}$	2.4×10^{-2} $M_\odot pc^{-3}$	1.6×10^{24} $wpc^{-3}\mu^{-1}$	1.6×10^{24} $wpc^{-3}\mu^{-1}$	0		1.1×10^{42} $s^{-1}pc^{-3}$	2 - 4	
	112 $M_\odot pc^{-2}$	5.4 $M_\odot pc^{-2}$	1.2×10^{27} $wpc^{-2}\mu^{-1}$	1.2×10^{27} $wpc^{-2}\mu^{-1}$	0		2.2×10^{44} $s^{-1}pc^{-3}$	$\frac{L_\odot}{M_\odot}$	4×10^{43} $s^{-1}M_\odot^{-1}$
4.5-5.5 kpc ring	0.68 $M_\odot pc^{-3}$	$M_\odot pc^{-3}$	4×10^{25} $wpc^{-3}\mu^{-1}$	8×10^{24} $wpc^{-3}\mu^{-1}$	3.2×10^{25} $wpc^{-3}\mu^{-1}$	2.5×10^{26} $s^{-1}pc^{-3}$	6×10^{42}	15 - 30	
	422 $M_\odot pc^{-2}$	14 $M_\odot pc^{-2}$	1.8×10^{28} $wpc^{-2}\mu^{-1}$	5×10^{27} $wpc^{-2}\mu^{-1}$	1.3×10^{28} $wpc^{-2}\mu^{-1}$	4×10^{28} wpc^{-2}	1.35×10^{45} $s^{-1}pc^{-2}$	$\frac{L_\odot}{M_\odot}$	9.6×10^{43} $s^{-1}M_\odot^{-1}$
Galactic Center R<300pc	1.5×10^{10} $M_\odot$	5×10^{7} $M_\odot$	2×10^{10} $L_\odot$			5×10^{8} $L_\odot$	3.5×10^{52} s^{-1}	20 $\frac{L_\odot}{M_\odot}$	7×10^{44} $s^{-1}M_\odot^{-1}$

µ stands for µm .

inner Galaxy, consistent with a steepening of the initial mass function.

REFERENCES

Becklin, E. E., and Neugebauer, G.: 1968, Astrophys. J. 151, pp. 145-161.
Burton, W. B.: 1976, Ann. Rev. Astron. Astrophys. 14, pp. 275-306.
Drapatz, S., and Michel, K. W.: 1976, Mitt. Astron. Ges. 40, pp. 187-192.
Fazio, G. G., and Stecker, F. W.: 1976, Astrophys. J. Letters 207, pp. L49.
Hayakawa, S., Ito, K., Matsumoto, T., Ono, T., and Uyama, K.: 1976, Nature 261, pp. 29-31.
Hayakawa, S., Ito, K., Matsumoto, T., and Uyama, K.: 1977, Astron. Astrophys. 58, pp. 325-330.
Hayakawa, S., Ito, K., Matsumoto, T., Murakmi, H., and Uyama, K.: 1978 (preprint).
Hofmann, W., Lemke, D., and Thum, C.: 1977, Astron. Astrophys. 57, pp. 11-114.
Ito, K., Matsumoto, T., and Uyama, K.: 1976, Astron. Soc. Pub., Japan 28, pp. 427-436.
Ito, K., Matsumoto, T., and Uyama, K.: 1977, Nature 265, pp. 517-518.
Low, F. J., Kurtz, R. F., Poteet, W. M., and Nishimura, T.: 1977, Astrophys. J. Letters 214, pp. L115-118.
Maihara, T., Oda, M., Sugiyama, T., and Okuda, H.: 1978, Pub. Astron. Soc. Japan 30, p. 1.
Matsumoto, T., Murakami, H., and Hamajima, K.: 1977, Pub. Astron. Soc. Japan 29, pp. 583-591.
Okuda, H., Maihara, T., Oda, N., and Sugiyama, T.: 1977, Nature 265, pp. 515-516.
Pipher, J.: 1973, I.A.U. Symp. 52, pp. 559-566, eds. J. M. Greenberg and H. C. van de Hulst, Reidel Pub. Co.
Rouan, D., Léna, P. J., Puget, J. L., de Boer, K. S., and Wijnbergen, J. J.: 1977, Astrophys. J. Letters 213, pp. L35-39.
Ryter, C., and Puget, J. L.: 1977, Astrophys. J. 215, pp. 775-780.
Sanders, R. H., and Lowinger, T.: 1972, Astron. J. 77, pp. 292-297.
Serra, G., Puget, J. L., and Ryter, C.: 1977, "Symp. on Recent Results in Infrared Astronomy" NASA Tech. Mem. TMX-73, 190, pp. 71-73.
Serra, G., and Puget, J. L.: 1977, Meeting of the French Physical Soc., Poitiers, June 27-July 1 (in press).
Serra, G., Puget, J. L., Ryter, C., and Wijnbergen, J. J.: 1978a, Astrophys. J. Letters 222, pp. L21-25.
Serra, G., Boissé, P., Gispert, R., Wijnbergen, J. J., Ryter, C., and Puget, J. L.: 1978b, Astron. Astrophys. (submitted).
Serra, G., Puget, J. L., and Ryter, C.: 1978 (preprint).
Stein, W. A.: 1967, "Interstellar Grains", eds. J. M. Greenberg and T. Roark (Washington, D. C.: NASA doc. 67-60065).
Viallefond, F., Léna, P. de Muizon, M., Nicollier, C., Rouan, D., and Wijnbergen, J. J.: 1978, Astron. Astrophys. (submitted).

DISCUSSION

Maihara: In connection with Dr. Puget's review, we would like to present a recent far infrared result. We observed an extended area from 345° to 30° in longitude with about 0°.7 resolution at $\lambda_{eff} \simeq 150\ \mu m$. The figure shows the preliminary longitudinal distribution along the ridge of the galactic plane. Our observations have detected a number of discrete sources associated with HII regions. The lower curve is the H166α intensity of Lockman, which may be a good representation of diffuse HII regions. So, we presume that the far infrared emission is concentrated in galactic HII regions. We have tried to detect the diffuse far infrared emission of dust as indicative of the general distribution of interstellar matter.

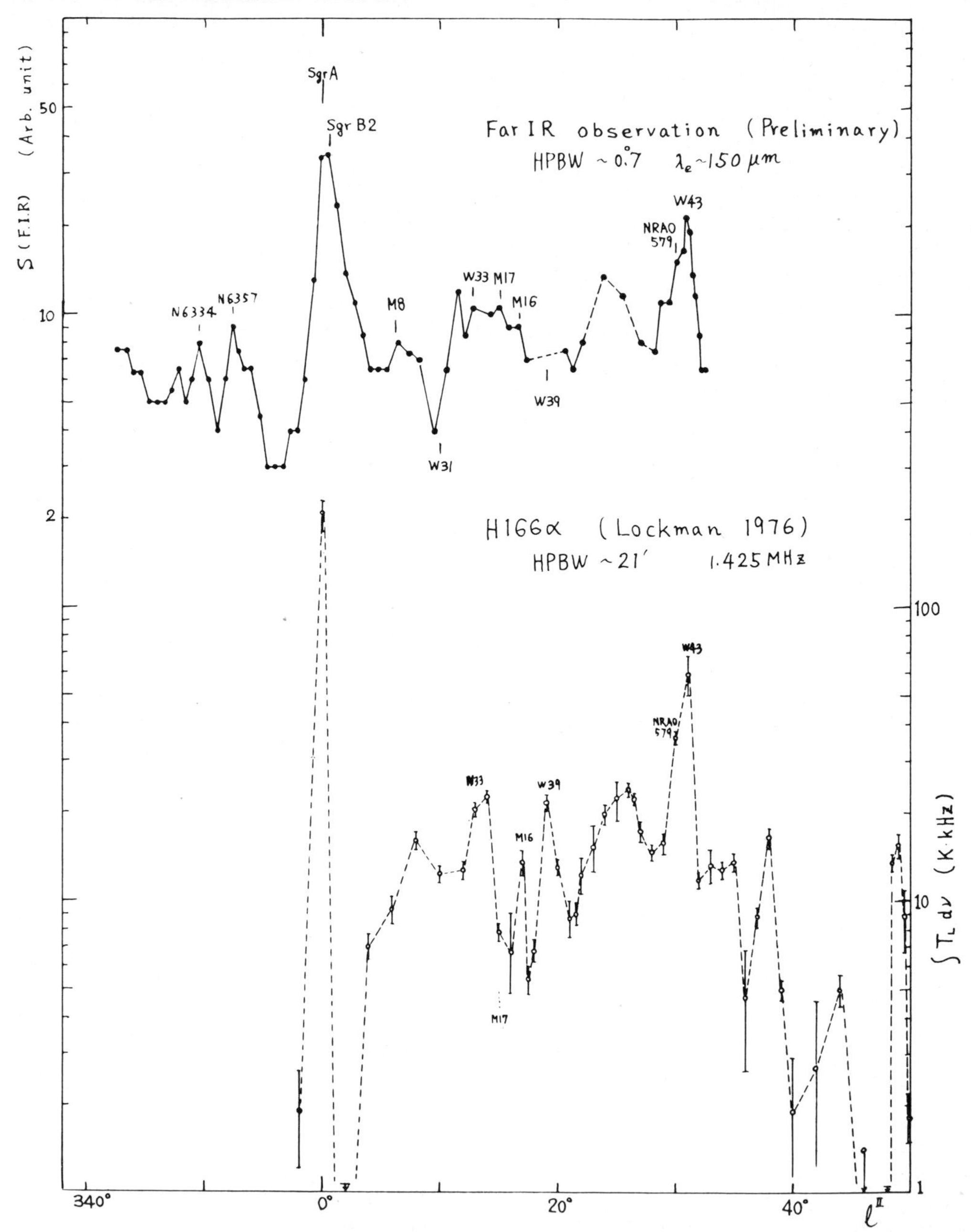

Mezger: I, and independently Drapatz (both papers to appear in Astron. and Astrophys.), also analyzed the diffuse galactic IR emission and investigated its origin. We agree with Puget that OB stars are the main contributors to the heating of the dust. However, contrary to Puget, we find that the O stars (as observed through their Lyman continuum photon production rate) are sufficient to account for the heating of the dust which is responsible for the diffuse FIR emission. Also, and in contradiction to Puget, both I and Drapatz conclude that the bulk of the FIR emission does not come from the dense molecular clouds which dominate the CO emission but comes from (i) radio HII regions, (ii) extended low-density (ELD) HII regions, and (iii) neutral intercloud gas. To observe dust emission from molecular clouds, one must indeed observe in the wavelength range 0.5-1 mm, as pointed out correctly by Puget.

Puget: Although Dr. Drapatz agrees with you about the fact that there is no need for a different initial mass function, he nevertheless finds that most of the flux comes from molecular clouds and that the heating due to stars other than O stars is more important than the heating due to O stars. We disagree with the statement that the O stars can alone explain the 5-kpc infrared brightness. Using the solar-vicinity ratio of Lyman continuum photon to integrated population I luminosity, we find a luminosity too low by a factor 3 to 6 to account for either the far infrared or the near infrared data. The point is even stronger for the 2.4 μ excess at 5 kpc which cannot, by far, be accounted for by O stars.

Solomon: If you use the mass density of H_2 (35 $M_\odot$ pc^{-2}) at 5.5 kpc from our results, the lifetime of the gas for star formation is only a factor of 2 less at 5.5 kpc than at the Sun.

Puget: In that case the star formation rate per unit mass of gas will only be 2.5 times the value in the solar vicinity.

Scoville: The dust temperatures which you have deduced from 2-color measurements are more typical of what is expected for dust at the boundary of the HII region, not for dust distributed randomly through molecular clouds well away from HII regions. Typical gas temperatures (and presumably dust temperatures if there is equilibrium) in these removed regions are only 10 K.

Puget: In the envelopes of molecular clouds, the densities are not high enough for gas and dust to be in thermal equilibrium. The factor of 2 found in the temperature ratio is about what one expects from the calculations of Goldreich and Kwan.

Lynden-Bell: I wonder if there is equally intense infrared emission at ℓ = 330°? If our Galaxy has a bar it should have a large HII region at the bar end at 30°, but the other end of the bar should be further away and produce less flux and an asymmetry in the infrared.

Puget: So far no data have been obtained in the southern hemisphere. It certainly is very important to get a longitude profile in the southern hemisphere.

GALACTIC NONTHERMAL RADIATION. DISTRIBUTION OF SNRs AND PULSARS

R. Wielebinski
Max-Planck-Institut für Radioastronomie, Bonn, F.R.G.

All sky surveys of the radio continuum emission give us the basic information on the distribution of the nonthermal emission in the Galaxy. At metre wavelengths, where nonthermal emission is dominant, good angular resolution is difficult to attain. For many years the best surveys near 2 m wavelength gave us a picture of the galaxy with $\sim 2^o$ resolution. At centimetre wavelengths, where arc min resolution is available, the intense HII regions dominate the radio sky. Supernova remnants have a distribution somewhat similar to that of the discrete HII regions and must be delineated by various methods in high resolution galactic plane surveys in the decimetre wavelength range.

Surveys which cover a substantial part of the sky are available currently for frequencies as low as 1 MHz and as high as 820 MHz. A tabulation of the major all sky surveys is made in Table 1. Below 10 MHz, with a few exceptions, satellites (e.g. Novaco and Brown, 1978) have been used. It should be noted that in the surveys with good angular resolution below 30 MHz parts of the galactic plane are seen in absorption. The only combination survey which covers all the sky with a resolution of $\sim 2\overset{\circ}{.}2$ is still the 150 MHz map of Landecker and Wielebinski (1970). A series of observations made in Jodrell Bank, Effelsberg and Parkes by Haslam et al. (1970, 1974, 1975) have covered nearly all the sky at 408 MHz with a resolution better than 1°. To complete the survey the north polar cap will be surveyed in Jodrell Bank soon. These surveys should in future be the basic starting point for interpretations of the galactic nonthermal emission. There is a sad lack of good quality sky surveys at frequencies above 1 GHz. This is due to the fact that nonthermal emission falls to average values below 1 K at 1420 MHz and to less than 0.1 K at 2695 MHz in 'cold' regions. A survey of the northern sky at 1420 MHz with r.m.s. noise of $\sim$ 15 mK is at present in final stages of reduction (Reich, 1977).

Any survey with good temperature and angular resolution shows a great amount of detailed structure. As an example a section of the 408 MHz survey is shown in Figure 1. At high galactic latitudes extragalactic point sources stand out. Nearer to the disc numerous spurs

W. B. Burton (ed.), The Large-Scale Characteristics of the Galaxy, 113–118.

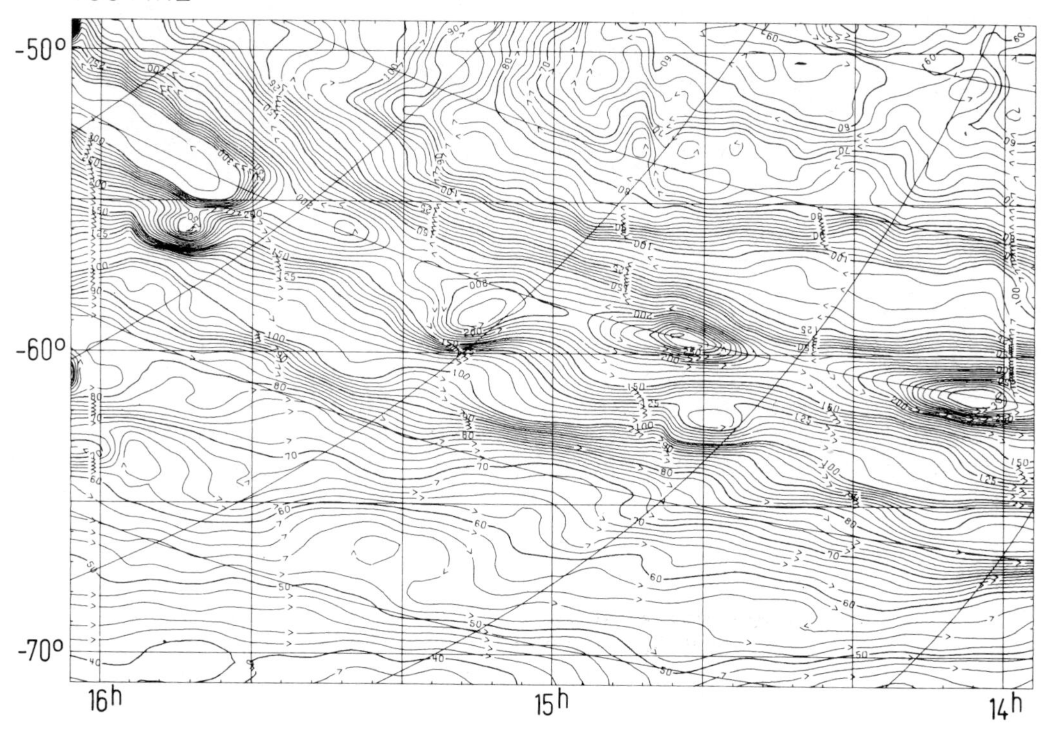

are seen. These have been interpreted to be nearby supernova remnants (e.g. Berkhuijsen, 1973). Recently an alternative explanation, in which the North Polar Spur is interpreted to be due to magnetohydrodynamic waves from the galactic centre, has been put forward by Sofue (1977). Still nearer the galactic plane the intensity of emission rises rapidly. Three dominant features are Cygnus X, the galactic centre, and Vela X. From the symmetry of Cygnus X and Vela X complexes their interpretation as tangential directions of our spiral arm (or interarm) seems to be definitive. Other 'steps' due to the inner spiral arms can also be seen, particularly in the southern maps, and especially somewhat away from $b = 0$. Inside the area $60^{\circ} < \ell < 300^{\circ}$ and $b = \pm 5^{\circ}$ a complicated mixture of thermal and nonthermal emission is observed. A tabulation of the major galactic plane surveys is given in Table 2.

The need for good surveys of a broad strip of the galactic plane at numerous frequencies is dictated by the wish of separation of the thermal component in the total radiation. There is some evidence that in addition to the discrete HII regions diffuse thermal emission is present in the Galaxy. This may be a superposition of separate HII regions or diffuse clouds which escaped from the ionised shells. The separation of thermal emission from the nonthermal component, although quite simple in theory, is hampered by numerous instrumental problems. The determination of the base levels and the adjustment of a survey to

TABLE 1
SELECTED SURVEYS OF ALL SKY

FREQUENCY (MHZ)	BEAM (DEGREES)	COVERAGE	REFERENCE
1.3, 2.2, 3.9, 4.7, 6.6, 9.2	120-45	NEARLY WHOLE SKY	NOVACO, J.C., BROWN, L.W. 1978, AP. J. 221, 114
2.1	7.1	$0^H < \alpha < 24^H$, $\delta < 0^0$	REBER, G. 1968, J. FRANKL INST. 285, 1
10.0	2.0	α INCOMPLETE, $\delta > -5^0$	CASWELL, J.L. 1976, MNRAS 177, 601
150	2.2	ALL SKY	LANDECKER, T.L., WIELEBINSKI, R. 1970, AUST. J. PHYS. AP. SUPPL. 16, 1
408	0.75	$4^H < \alpha < 12^H$, $60^0 > \delta > -20^0$ $0^H < \alpha < 04^H$, $60^0 > \delta > 20^0$	HASLAM ET AL. 1970. MNRAS 147, 405
408	0.6	$12^H < \alpha < 04^H$, $48^0 > \delta > -8^0$	HASLAM ET AL. 1974, A & A SUPPL. 13, 359
408	0.85	$0^H < \alpha < 24^H$, $0^0 > \delta > -90^0$	HASLAM ET AL. 1978 (IN PREPARATION)
820	1.2	$0^H < \alpha < 24^H$, $85^0 > \delta > -7^0$	BERKHUIJSEN, E.M. 1972, A & A SUPPL. 5, 263
1420	0.5	$0^H < \alpha < 24^H$, $90^0 > \delta > -19^0$	REICH ET AL. 1978 (IN PREPARATION)

TABLE 2
SELECTED GALACTIC PLANE SURVEYS

FREQUENCY (MHZ)	BEAM (MIN)	COVERAGE	REFERENCE
29.9	48	$30^0 < \ell < 225^0$ $b \pm 10^0 (15^0)$	JONES, B.B., FINLAY, E.A. 1974, AUST. J. PHYS. 27, 687
408	3	$55^0 < \ell < 195^0$ $b \pm 3^0$	GREEN, A.J. 1974, A & A SUPPL. 18, 267
1410	14	$356^0 < \ell < 281^0$ $b \pm 5^0 (6^0)$	HILL, E.R. 1968, AUST. J. PHYS. 21, 735
1414	10	$55^0 < \ell < 12^0$ $b \pm 4^0$	ALTENHOFF ET AL. 1970, A & A SUPPL. 1, 319
2695	8	$47^0 < \ell < 286^0$ $b \pm 2^0$	DAY ET AL. 1969-70. AUST J. PHYS. SUPPL. 11. 13
2695	11	$75^0 < \ell < 345^0$ $b \pm 2^0$	ALTENHOFF ET AL. 1970, A & A SUPPL. 1, 319
4875	2.6	$60^0 < \ell < 357^0.5$ $b \pm 1^0$	ALTENHOFF ET AL. 1978, A & A SUPPL. (IN PRESS)
5000	4.1	$40^0 < \ell < 190^0$ $b \pm 2^0$	HAYNES, R.F. 1978, AUST. J. PHYS. (IN PRESS)

an absolute temperature scale are only now reaching a stage of accuracy where a meaningful separation may be possible. Furthermore, the degree of linear polarisation of the nonthermal component above 1 GHz is high, possibly up to 75% in some regions (e.g. Brouw and Spoelstra, 1976). The linearly polarised component is pure nonthermal emission. It seems that in order to obtain a definitive separation surveys which contain complete polarisation information at a number of high radio frequencies may be necessary. Although difficult in execution, this is a very important venture, which should be carried out in the next years to further our understanding of the emission processes in the Galaxy.

The intensity contours determined in a sky survey are a superposition of the following components:

1. Extragalactic background
2. Galactic disc emission (HII regions, SNRs, spiral arms)
3. Local emission (Spurs, Cygnus X, Vela X)
4. Remaining "halo" emission.

Components 1-3 may be determined each with a certain degree of reliability. In particular it is difficult to separate the numerous spurs. The subtraction of these three components from a reliable sky survey gives finally the intensity of the remaining emission, of a halo. The halo is of great importance in high energy astrophysics, since confinement (or otherwise) of cosmic rays in the Galaxy can best be tested by observing the nature of any radio halo surrounding our Galaxy (e.g. Ginzburg and Ptuskin, 1976; Ginzburg, 1977 and references therein). Original sky survey results indeed supported the existence of an electron halo. Refined measuring techniques, particularly sidelobe determinations, have gradually reduced the intensity which could be attributed to the halo (e.g. Baldwin, 1967, 1976). Alternative experiments to test the halo hypothesis have been conducted by numerous observers (e.g. Wielebinski and Peterson, 1968) and placed low upper limits on a halo component. Finally a detailed analysis by Webster (1975), using different observing techniques, implied the existence of a weak, nonconfining halo. All the evidence available from direct interpretation of sky surveys points to a weak halo. An alternative approach to this problem is to compute three-dimensional models of the radio emissivity. A number of papers (Brindle et al., 1978 and references therein) use all the available evidence on the magnetic fields, the density wave theory and energy distribution of cosmic rays to produce 'maps' which can then be compared with existing surveys. The best fit is obtained for a model with a thick disc. The increase in spectral index away from the plane of a galaxy which seems to be now definitively documented has not as yet been considered in any such models. A good discussion of the problems associated with modelling is found in Baldwin (1976). Additional evidence which supports the above conclusion comes from the analysis of γ-ray observations. Stecker and Jones (1977) show that the data can best be explained by an electron halo with a half thickness $L_e = 2 \pm 2$ kpc. Observations of nearby edge-on galaxies also support such a model. The galaxy NGC 4631 has the most distinctive ellipsoidal halo at 610 MHz (Ekers and Sancisi, 1977). The edge-on galaxy NGC 891 also shows a rather weak halo which has a steep spectrum away from the plane of this galaxy (Allen et al., 1978). The increase in the spectral index away from the plane appears by now to be a definite feature of galaxies. This allows us to speculate on the reason for this, namely either the fall off in magnetic field strength or deficiency in high energy electrons at high z-distances. Possibly both of these causes are responsible. A new result is available in the form of a recent 8.6 GHz map of NGC 253 by Beck et al. (1978). In this map a halo of nearly spherical symmetry is seen apparently originating in the nucleus. The thick disc of this galaxy, seen at lower frequencies, presumably is supported by electrons originating in SNRs and diffusing from the plane of the galaxy, while the nuclear activity seems to power the young halo.

Supernova remnants are found predominantly in a narrow strip of sky along the galactic plane. Some 125 remnants are known with more suspected remnants awaiting confirmation. Recent studies by Clark and Caswell (1976) in fact showed that up to 30% of SNRs in older catalogues were false identifications. At present good observational data of a homogeneous sample is available for $\delta < 18^o$ only. Recently interest in large low surface brightness objects has been awakened by the ability of modern instruments to map these weak objects. An analysis by Henning and Wendker (1975) of the 408 MHz survey indicated that there were no extended SNRs having $\Sigma_{408} > 1.5 \times 10^{-22}$ W m^{-2} ster^{-1} Hz^{-1}. A recent analysis of the source G65.2+5.7 in Cygnus by Reich et al. (1978) disproves this. Further new objects should be added to the lists soon. Evidence for a class of SNRs resembling the Crab nebula has been put forward by Weiler and Wilson (1977). The origin of cosmic rays is still intimately tied up with the rate of supernova remnants. The figures vary from 30 years to $\sim$ 150 years between events, but the energy production appears to be sufficient to power the nonthermal radio emission.

Over 300 pulsars are now known. The recent Molonglo survey, announcing the discovery of 155 new pulsars (Manchester et al., 1978) has given us a very homogeneous sample. The period distribution for the new pulsars is similar to that for previously known pulsars. None of the new pulsars has a very short period, but several have periods in excess of 2 sec. Some pulsars have been found with dispersion measure DM > 500 cm^{-3} pc but none with extremely high DM. It appears that all the known facts about distribution are only confirmed. We have the clustering of the pulsars along the galactic plane and a local more isotropically distributed pulsar population. The connection between SNR and pulsar, so definite in Crab nebula is not so clear in all other cases.

REFERENCES

Allen, R.J., Baldwin, J.E., Sancisi, R.: 1978, Astron. Astrophys. 62, 397.
Baldwin, J.E.: 1967, IAU Symp. 31, 337 (D. Reidel Publ. Co., Dordrecht).
Baldwin, J.E.: 1976, Proc. Int. Symp. on Structure and Content of the

Galaxy and Galactic Gamma Rays, NASA, p. 286.
Beck, R., Biermann, P., Emerson, D.T., Wielebinski, R.: 1978 (in prep.).
Berkhuijsen, E.M.: 1973, Astron. Astrophys. 24, 143.
Brindle, C., French, D.K., Osborne, J.L.: 1978, Monthly Notices Roy. Astron. Soc. (in press).
Brouw, W.N., Spoelstra, T.A.Th.: 1976, Astron. Astrophys. Suppl. 26, 129.
Clark, D.J., Caswell, J.L.: 1976, Monthly Notices Roy. Astron. Soc. 174, 267.
Ekers, R.D., Sancisi, R.: 1977, Astron. Astrophys. 54, 973.
Haslam, C.G.T., Quigley, M.J.S., Salter, C.J.: 1970, Monthly Notices Roy. Astron. Soc. 147, 405.
Haslam, C.G.T., Wilson, W.E., Graham, D.A., Hunt, G.C.: 1974, Astron. Astrophys. Suppl. 13, 359.
Haslam, C.G.T., Wilson, W.E., Cooke, D.J., Cleary, M.N., Graham, D.A., Wielebinski, R., Day, G.A.: 1975, Proc. Astron. Soc. Australia 2, 330.
Henning, K., Wendker, H.J.: 1975, Astron. Astrophys. 44, 91.
Landecker, T.L., Wielebinski, R.: 1970, Australian J. Phys. Astrophys. Suppl. 16, 1.
Manchester, R.N., Lyne, A.G., Taylor, J.H., Durdin, J.M., Large, M.I., Little, A.G.: 1978 (in press).
Novaco, J.C., Brown, L.W.: 1978, Astrophys. J. 221, 114.
Reich, W.: 1977, Ph.D. Thesis, Bonn University.
Reich, W., Berkhuijsen, E.M., Sofue, Y.: 1978, Astron. Astrophys. (submitted).
Sofue, Y.: 1977, Astron. Astrophys. 60, 327.
Stecker, F.W., Jones, F.C.: 1977, Astrophys. J. 217, 843.
Webster, A.: 1975, Monthly Notices Roy. Astron. Soc. 171, 243.
Weiler, K.W., Wilson, A.S.: 1977, in "Supernovae", Ed. D.N. Schramm, Reidel Co., p. 67.
Wielebinski, R., Peterson, C.E.: 1968, Observatory 88, 219.

DISCUSSION

Tinsley: Dr. Wielebinski commented that there are "too many pulsars and too few supernova remnants". Would he expand on this remark?

Wielebinski: The rates of SNR formation given by various authors lie in the 30-100 years range. Considerations of pulsar beaming leads us to conclude that pulsar occurrences are one every 5-7 years. This is the discrepancy.

PULSARS AS A GALACTIC POPULATION

Joseph H. Taylor
University of Massachusetts, Amherst, Massachusetts

Recent pulsar surveys have increased the number of known pulsars to well over 300, and many of them lie at distances of several kpc or more from the sun. The distribution of pulsars with respect to distance from the galactic center is similar to other population I material such as HII regions, supernova remnants, and carbon monoxide gas, but the disk thickness of the pulsar distribution is rather greater, with $<|z|>\simeq 350$ pc. Statistical analysis suggests that the total number of active pulsars in the Galaxy is a half million or more, and because kinematic arguments require the active lifetimes of pulsars to be $\lesssim 5\times 10^6$ years, it follows that the birthrate required to maintain the observed population is one pulsar every ~10 years (or less) in the Galaxy.

This contribution is intended to serve as an update on what is known about the galactic distribution of pulsars. This subject has received much attention in the past year or so (see for example, Lande & Stephens 1977; Davies, Lyne & Seiradakis 1977; Taylor & Manchester 1977; Manchester & Taylor 1977). All of these studies have concluded that the number of pulsars in the Galaxy, considered together with the estimated active lifetime of pulsars, requires a pulsar birthrate of one every ~6 years in the Galaxy. This computed mean interval between births is somewhat less than (but not grossly inconsistent with) recent estimates of the rate of occurrence of galactic supernova outbursts (Tammann 1977).

In the past 18 months, the number of known pulsars has increased from 148 to 320, thanks largely to 155 pulsars discovered in a sensitive, systematic survey of the entire sky between declinations -85° and +20°, carried out with the Mills Cross antenna at Molonglo, Australia (Manchester, Lyne, Taylor, Durdin, Large & Little 1978). A complementary survey of the northern sky is now underway, and has detected 17 new pulsars (Damashek, Taylor & Hulse 1978). The distribution of all presently known pulsars is shown in equal area galactic coordinates in Figure 1. This picture is unlikely to change markedly in the foreseeable future, because of sensitivity limitations of existing radio telescopes, and it is interesting to note in passing that spiral structure in the pulsar

W. B. Burton (ed.), The Large-Scale Characteristics of the Galaxy, 119–123.

distribution is now suggested by a clumping of pulsars near the "tangential" longitudes 280°, 310°, and 330°.

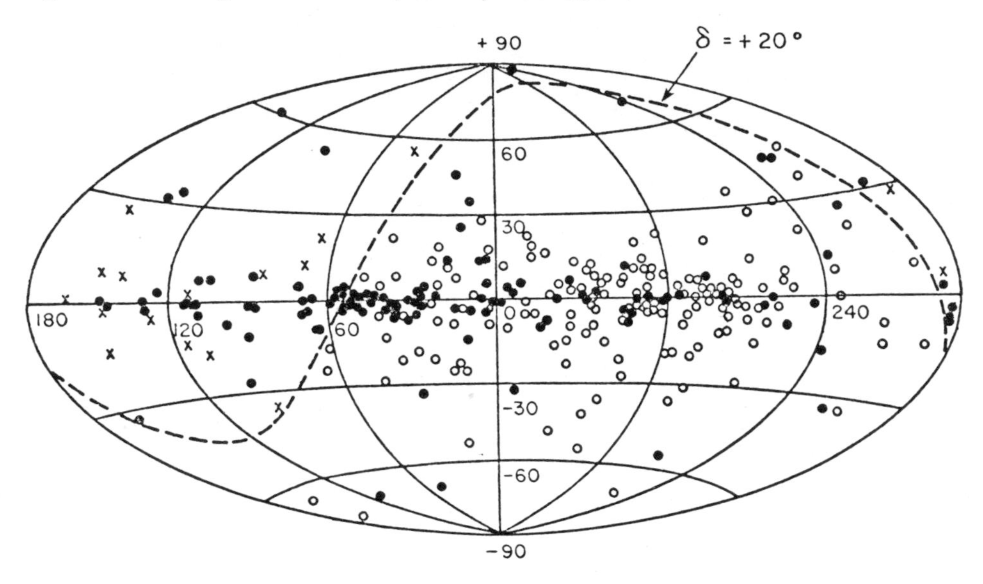

Figure 1. The distribution in galactic coordinates of all pulsars known as of June 1978. Sources newly discovered in the second Molonglo survey (Manchester et al 1978) are denoted by open circles, and those discovered by Damashek, Taylor and Hulse (1978) by crosses. In crowded regions, a few symbols have been omitted for clarity.

A three-dimensional description of the galactic pulsar distribution requires knowledge of pulsar distances. Dispersion measure (the integral of electron density n_e along the line of sight) is for pulsars a directly measurable quantity that is approximately proportional to distance, because n_e has been shown to be reasonably constant throughout much of the galactic disk (see, for example, Ables and Manchester 1976). Therefore pulsar distances can be estimated directly from the observed dispersion measures; such distances are thought to be statistically reliable to within 20 or 30 percent, although individual distances may be in error by a factor of 2 or more because of excess dispersion caused by intervening HII regions.

Comparison of observed pulsar flux densities with the estimated distances shows that pulsar luminosities (crudely defined for this purpose as $L=Sd^2$, where S represents flux density and d represents distance) range over at least 6 orders of magnitude. The observed distribution of luminosities is such that most observed pulsars lie in the medium-to-high range, but consideration of the volumes of space effectively surveyed for different luminosities shows that far more pulsars exist with low L than high L. In fact, the pulsar luminosity function is approximately a power law, with the number of pulsars per logarithmic luminosity interval inversely proportional to luminosity.

Analysis of the pulsar distribution in terms of distance R from the galactic center shows a curve that increases sharply inside the solar circle, reaching a peak at 4 to 6 kpc from the galactic center. The radial distribution curve is very similar to those obtained for the distributions of ionized hydrogen, supernova remnants, carbon monoxide gas, and γ-radiation, and all of them differ markedly from the neutral hydrogen distribution (see Burton 1976 for a review). On the other hand, the pulsar distribution in height above the galactic plane, z, is rather different from the other population I constituents. Whereas the scale height of the other types of material mentioned above is $\lesssim$100 pc, we find the pulsar scale height to be approximately 350 pc. This difference is a direct result of the high velocities that pulsars acquire at birth: proper motion studies show that pulsar peculiar velocities are typically $\gtrsim$200 km s^{-1} (see for example, Taylor & Manchester 1977), and thus,

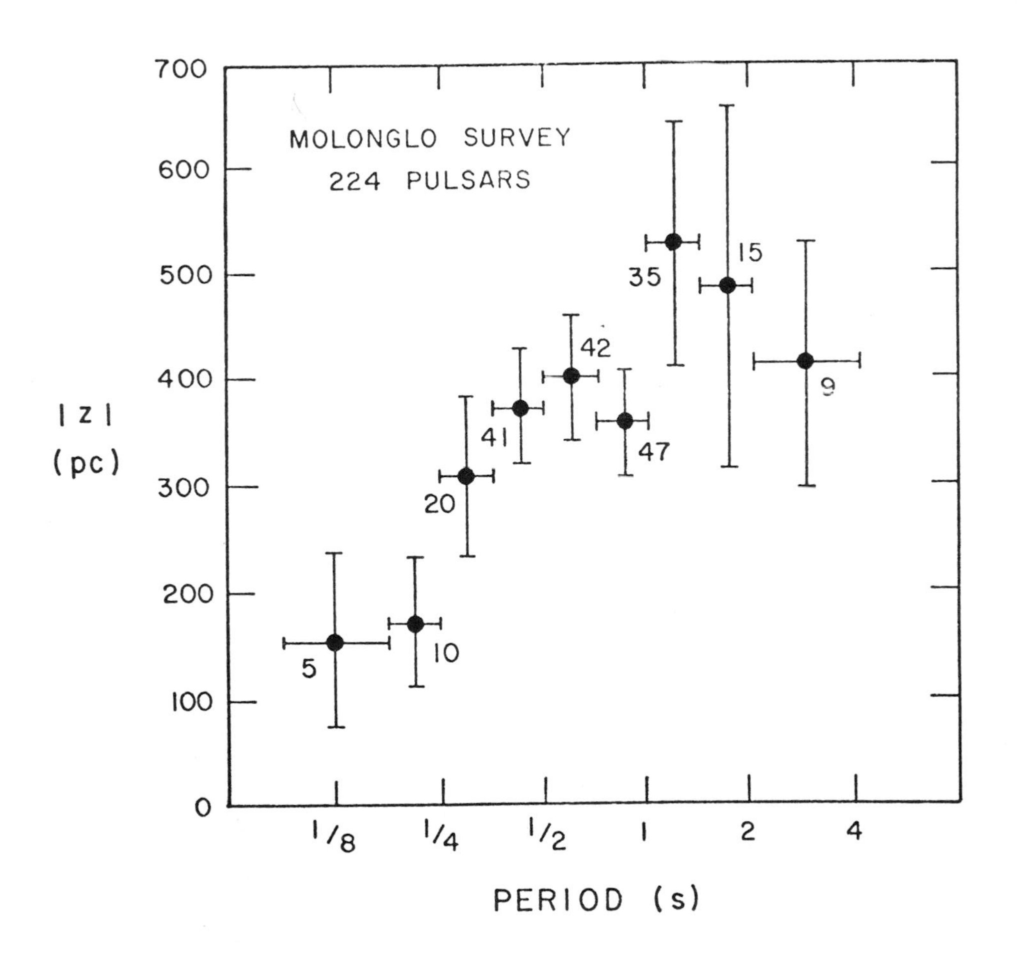

Figure 2. Average distance of pulsars from the galactic plane, plotted as a function of period, for nine period ranges.

although most pulsars are no doubt born at heights $|z| \lesssim 150$ pc, they disperse to considerably greater distances from the plane within a few million years. Some new direct evidence of this effect is illustrated in Figure 2, which is based on the uniform sample of 224 pulsars detected in the second Molonglo pulsar survey. The short-period pulsars, which are known to be the younger ones, have a markedly narrower z-distribution than their older cousins.

The thickness of the pulsar disk, together with the observed pulsar velocities, yields an estimate of the mean pulsar lifetime, $<|z|>/<|v_Z|>$ ≈ 350 pc/100 km $s^{-1} \approx 3 \times 10^6$ years. Integration of the density distributions with respect to R and z suggests a total of at least 10^5 potentially observable pulsars in the Galaxy at the present time. To maintain such a steady-state population would require a pulsar birth at least every ~30 years, and if beaming effects make some ~80% of pulsars unobservable from Earth, the required birthrate would increase to one every ~6 years or so. The well calibrated nature of the recent pulsar surveys will allow the uncertainties on these estimates to be reduced significantly, and work is now underway on a full statistical analysis of the new data.

REFERENCES

Lande, K., and Stephens, W.E.: 1977, Astrophys. Space. Sci., 49, pp. 169-177.

Davies, J.G., Lyne, A.G., and Seiradakis, J.H.: 1977, Mon. Not. Roy. Astron. Soc. 179, pp. 635-650.

Taylor, J.H., and Manchester, R.N.: 1977, Astrophys. J. 215, pp. 885-896.

Manchester, R.N., and Taylor, J.H.: 1977, "Pulsars," W.H. Freeman & Co., San Francisco.

Tammann, G.A.: 1977, Annals N.Y. Acad. Sci. 302, pp. 61-80.

Manchester, R.N., Lyne, A.G., Taylor, J.H., Durdin, J.M., Large, M.I., and Little, A.G.: 1978, Mon. Not. Roy. Astron. Soc. (in press).

Damashek, M., Taylor, J.H., and Hulse, R.A.: 1978, Astrophys. J. (Letters) (in press).

Ables, J.G., and Manchester, R.N.: 1976, Astron. Astrophys. 50, pp. 177-184.

Burton, W.B.: 1976, Ann. Rev. Astron. Astrophys. 14, pp. 279-306.

DISCUSSION

Tinsley: The expected relative numbers of SNR and pulsars depend of course on their relative lifetimes. What do you find for the mean pulsar lifetime? And how does the pulsar birthrate then compare with the formation rate of SNR?

Taylor: The most assumption-free estimate of the average pulsar lifetime is based on kinematics: the observed z-distribution of pulsars is consistent with the observed proper motion velocities of ~ 100 km s^{-1} only if the mean age is not more than a few million years. The corresponding pulsar birthrate (if only $\sim$20% of pulsars are potentially observable because of beaming effects) is one every five years or less. Only a small fraction of pulsar-producing events can result in long-lived supernova remnants.

Innanen: Is the decline in pulsar density with galactocentric distance exponential or a power law?

Taylor: The experimental uncertainties are not small enough to warrant any conclusions beyond the simple statement that the density distribution falls off rather sharply over the range 5 to 10 kpc from the galactic center.

Stecker: Some workers, in analyzing the radial distribution of SN remnants in the Galaxy, have found a monotonic increase all the way to the galactic center; other analyses have yielded a radial distribution peaking in the 5-kpc annulus similar to the pulsar distribution shown by Dr. Taylor. Do you have any comments on this apparent discrepancy?

Wielebinski: It is difficult to see SNR in the direction of the center. Maybe we should recall Berkhuijsen's work on M31: the SNR distribution in M31 does not follow the distribution of the ratio continuum radiation.

Mathewson: The apparent discrepancy between the number of pulsars and the number of supernova remnants in the Galaxy would be removed if there was a region of low electron density in the local neighborhood.

Taylor: I agree completely. The distance scale for nearby pulsars is crucial in this regard; some direct parallax measurements would be nice.

Terzian: We should remind ourselves that $\langle n_e \rangle$ is not really constant in all directions in the Galaxy, consequently distances of pulsars derived on the assumption of constant $\langle n_e \rangle$, say $\sim$0.03, can have errors of factors of 3 or more. Similarly the derived luminosities and $|z|$-distribution will have large errors.

Taylor: I agree that individual pulsar distances estimated from dispersion measures may sometimes be in error by a factor of 2 or 3. Statistically, however, I believe that the distance scale is much better than this, and that the composite distribution functions are reasonably well determined.

GAMMA-RAY OBSERVATIONS AND THE LARGE SCALE STRUCTURE OF THE GALAXY

J. A. Paul
DPh/EP/ES, Centre d'Etudes Nucléaires de Saclay (France)

Within the last few years, γ-ray astronomy has shifted from the discovery phase to the exploratory phase, thanks to the SAS-2 and COS-B satellites. The strongest feature of the γ-ray sky is the overwhelming emission of the galactic disc; even the radiation observed away from the galactic plane appears to be predominantly galactic, on the basis of its latitude dependence (Fichtel et al., 1978). Nevertheless, extragalactic γ-ray astronomy is not hopeless: the γ-radiation of the nearby quasar 3C273 has been very recently detected (Swanenburg et al., 1978). A brief summary of the present status of the galactic γ-ray astronomy follows.

1. LATITUDE AND LONGITUDE DISTRIBUTIONS

In their analysis of the COS-B data obtained during its first year of operation, Bennett et al. (1977) investigated in detail the latitude distribution of the γ-radiation. Figure 1 shows the observed width of the γ-ray disc as a function of the galactic longitude. The central regions are not resolved; this sets an upper limit of $\sim 2°$ on the width of the emitting region in the inner part of the Galaxy. At large angular distances from the galactic centre, the disc is resolved and appears quite broad, suggesting that a large fraction of the γ radiation is emitted close by. Furthmore, the asymmetry with respect to $b = 0$ points towards a local nature of the radiation from the outer disc, in agreement with the conclusion of Strong et al. (1977) that the γ-ray disc does not extend beyond the Perseus arm. Turning now to the inner regions of the Galaxy, Bennett et al. (1977) have derived more precise latitude profiles using only γ rays with $E \geq 300$ MeV, taking advantage of the better angular resolution of the COS-B instrument in this energy range. The latitude profile in the longitude range $10° \leq \ell \leq 40°$ reveals the presence of an additional component at low positive latitudes, possibly related to local dense clouds. This profile also exhibits a small displacement of the maximum emission towards negative latitudes, reminiscent of a similar asymmetry in the CO emission of the same region observed by Cohen and Thaddeus (1977).

W. B. Burton (ed.), The Large-Scale Characteristics of the Galaxy, 125–130.

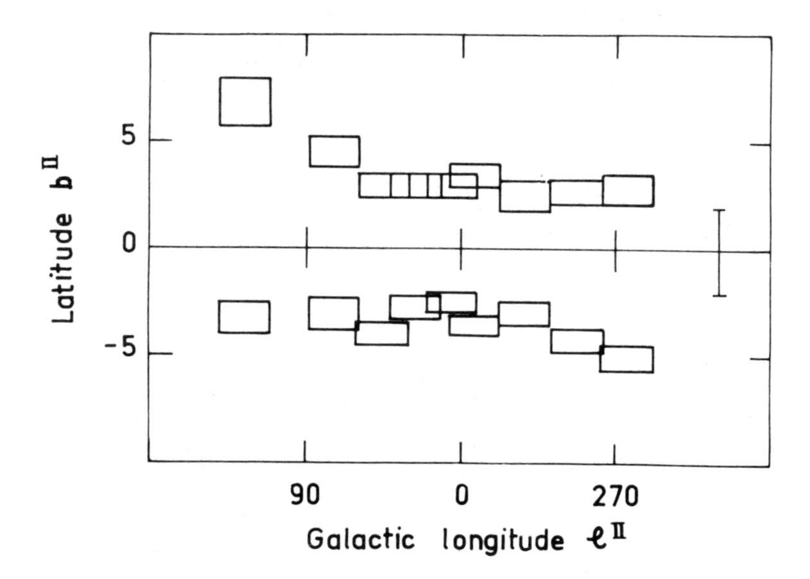

Figure 1. Variation with longitude of the latitude at which the -ray intensity (E $\geq$ 100 MeV) is half the maximum value. The error bar represents the F.W.H.M. for an ideal line source (it reflects the effect of the finite angular resolution of the instrument).

Figure 2 shows the distribution along the galactic equator of the γ radiation, obtained from the SAS-2 (Fichtel et al., 1977) and COS-B (Bennett et al., 1977) data. Note that the COS-B distribution is an average over successive observations not corrected for possible systematic effects. Both distributions are in remarkable agreement within uncertainties (10% to 30% of the quoted intensities). Small scale structures are discussed in detail by Hermsen et al.(1977), including 11 so far unidentified γ-ray sources, from the study of 1/3 of the galactic plane. Three sources of the COS-B catalogue (CG185-5=PSR0531+21, CG263-2=PSR0833-45, and CG195+4), were previously reported by the SAS-2 group (Kniffen et al., 1977a) as well as evidence for γ-ray emission from PSR1747-46, PSR1818-04 and Cygnus X-3, not yet confirmed by COS-B.

2. SPECTRAL CHARACTERISTICS OF THE GALACTIC GAMMA RADIATION

Paul et al. (1978) have presented the γ-ray spectrum relative to different regions of the Galaxy, in which sources may be present to a greater or lesser extent. It is premature to draw astrophysical conclusions before carefully subtracting the source contribution. However, in some regions the contribution of sources seen by COS-B appears quite small (less than 10% of the intensity above 100 MeV for 116° $\leq \ell \leq$ 136°).

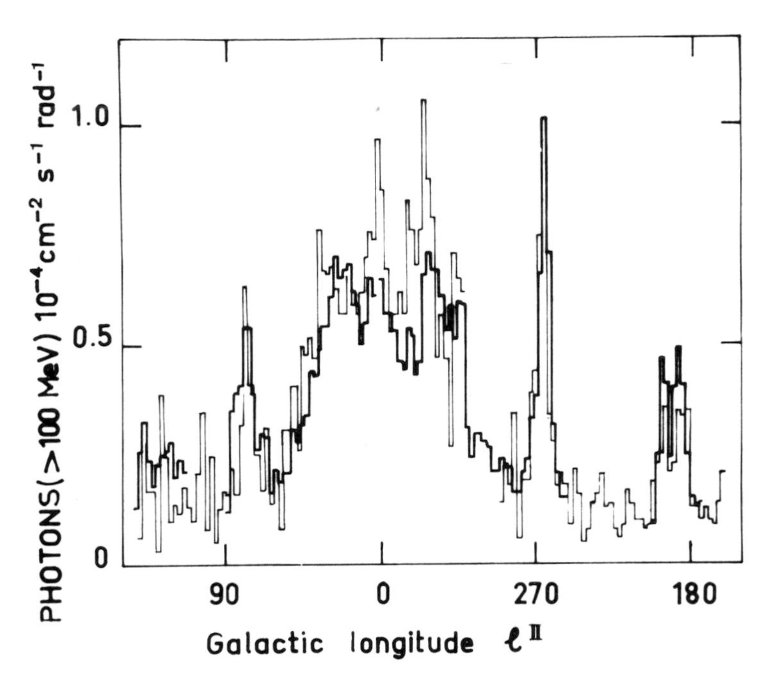

Figure 2. Longitude profile of the γ-ray intensity (E $\geq$ 100 MeV). Thin line: SAS-2 data integrated over the latitude range $|b| \leq 5°$. Thick line: COS-B data integrated over the latitude range $|b| \leq 6°$.

The source contribution seems also insignificant in the high latitude observations performed by the SAS-2 group (Fichtel et al., 1978). In both cases, the observed galactic γ radiation is presumably local, then its spectral characteristics provide some information on the diffuse emission in the local interstellar medium. The main emission mechanisms involve high energy nucleons ($\pi°$ induced emission) and electrons (Bremsstrahlung). Cesarsky et al. (1978) have recently evaluated the local production rate of very high energy photons. Fitting the spectrum of the local diffuse radiation from the galactic region $116° \leq \ell \leq 136°$ (Paul et al., 1978) and from regions away for the galactic plane (Fichtel et al., 1978) they conclude that the flux of cosmic-ray electrons (in the range 50 - 500 MeV) is much higher than that predicted by demodulation theories.

3. LARGE SCALE STRUCTURE OF THE GAMMA-RAY DISC

On the basis of the SAS-2 γ-ray observations, several authors (Kniffen et al., 1977b; Cesarsky et al., 1977; and references therein) have developed models in which most of the high energy γ radiation is of diffuse origin and should be a tracer of the galactic structure. However, the COS-B observations stress the importance of sources which

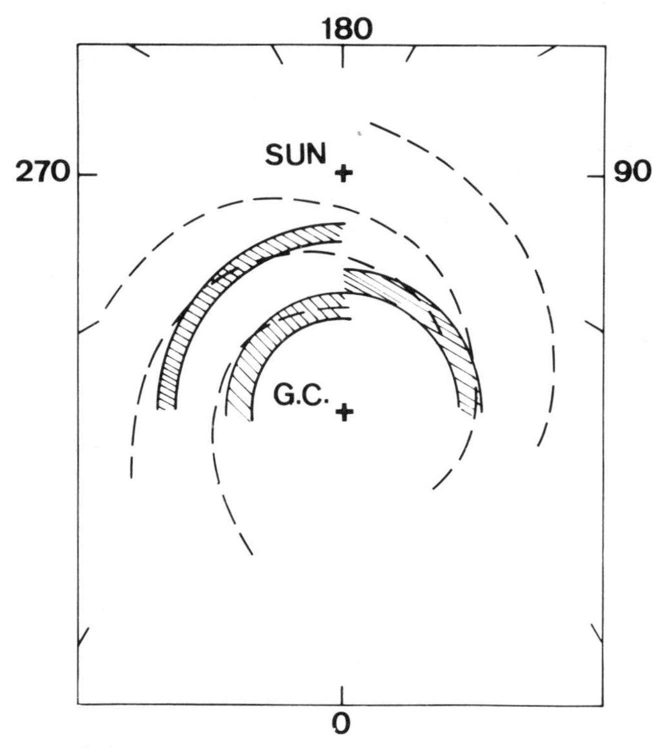

Figure 3. Repartition of high γ-ray emissivity regions across the galactic plane (shaded area) unfolded from the SAS-2 longitude profile, superimposed on the spiral pattern derived by Georgelin and Georgelin (1976). The regions of the galactic centre and beyond are excluded.

may contribute significantly to the overall γ-ray luminosity. In the anticentre region, for instance, they account for most of the emission. It seems now necessary to analyze the large scale structure of the γ-ray disc without any assumption regarding the physical origin of the galactic γ radiation.

This approach is attempted by Caraveo and Paul (1978) using the unfolding technique described by Strong (1975) which only implies the assumption of cylindrical symmetry (at least in one half of the Galaxy). This technique leads to distinct radial distributions of the γ-ray emissivity for "positive" longitudes ($0° \leq \ell \leq 180°$) and "negative" longitudes ($180° \leq \ell \leq 360°$). At positive longitudes, the broad peak at a distance R = 5 to 6 kpc from the galactic centre, is reminiscent of the distribution of interstellar matter and young objects as e.g. supernova remnants and pulsars (Stecker, 1977). At negative longitudes, two peaks clearly appear at R = 4 to 5 kpc and R = 7 to 8 kpc. The γ-ray Galaxy is sketched on Figure 3 after Caraveo and Paul (1978).

4. CONCLUSIONS

High energy γ-rays provide a new view of the Galaxy. The γ-ray disc seems as thin as the gaseous disc but it may not extend well beyond the solar circle; structural features are clearly apparent. The physical origin of the γ radiation from remote galactic regions remain questionable. If the contribution of sources is dominant, leaving a small room to diffuse processes, one would be forced to assume that either the interstellar density is overestimated (Cesarsky et al., 1977) or that the cosmic-ray density is strongly reduced in dense clouds. The latter hypothesis, discussed by Cesarsky and Volk (1977) is not supported by Lebrun and Paul (1978) on the basis of COS-B observations.

I thank Michel Cassé for useful discussions.

REFERENCES

Bennett, K., Bignami, G. F., Buccheri, R., Hermsen, W., Kanbach, G., Lebrun, F., Mayer-Hasselwander, H. A., Paul, J. A., Piccinoti, G., Scarsi, L., Soroka, F., Swanenburg, B. N., and Wills, R. D.: 1977, Proc. of the 12th ESLAB Symp., ESA SP-124, 84.
Caraveo, P. A., and Paul, J. A.: 1978, submitted to Astron. and Astrophys.
Cesarsky, C. J., Cassé, M., and Paul, J. A.: 1977, Astron. and Astrophys. 60, 139.
Cesarsky, C. J., Paul, J. A., and Shukla, P. G.: 1978, submitted to Astrophys. and Sp. Sci.
Cohen, R. S., and Thaddeus, P.: 1977, Ap. J.(Letters) 217, L155.
Fichtel, C. E., Kniffen, D. A., and Thompson, D. J.: 1977, Proc. of the 12th ESLAB Symps., ESA SP-124, 95.
Fichtel, C. E., Simpson, G. A., and Thompson, D. J.: 1978, Ap. J. (in press).
Georgelin, Y. M., and Georgelin, Y. P.: 1976, Astron. and Astrophys. 49, 57.
Hermsen, W., Bennett, K., Bignami, G. F., Boella, G., Buccheri, R., Higdon, J. C., Kanbach, G., Lichti, G. C., Masnou, J. L., Mayer-Hasselwander, H. A., Paul, J. A., Scarsi, L., Swanenburg, B. N., Taylor, B. G., and Wills, R. D.: 1977 Nature 269, 494.
Kniffen, D. A., Fichtel, C. E., Hartman, R. C., Lamb, R. C., and Thompson, D. J.: 1977a, Proc. of the 12th ESLAB Symp., ESA SP-124,
Kniffen, D. A., Fichtel, C. E., and Thompson, D. J.: 1977, Ap. J. 215, 765.
Lebrun, F., and Paul, J. A.: 1978, Astron. and Astrophys. 65, 187.
Paul, J. A., Bennett, K., Bignami, G. F., Buccheri, R., Caraveo, P., Hermsen, W., Kanbach, G., Mayer-Hasselwander, H. A., Scarsi, L., Swanenburg, B. N., and Wills, R. D.: 1978, Astron. and Astrophys. 63, L31.
Stecker, F. W.: 1977, Ap. J. 212, 60.
Strong, A. W.: 1975, J. Phys. A.: Math. Gen. 8, 617.
Strong, A. W., Bennett, K., Wills, R. D., and Wolfendale, A. W.: 1977, Proc. of the 12th ESLAB Symp., ESA SP-124, 167.
Swanenburg, B. N., Bennett, K., Bignami, G. F., Caraveo, P., Hermsen, W., Kanbach, G., Masnou, J. L., Mayer-Hasselwander, H. A., Paul, J. A., Sacco, B., Scarsi, L., and Wills, R. D.: 1978, submitted to Nature.

DISCUSSION

Stecker: Could the outer "ring" or "arm" in your Figure 3 be accounted for by a point source in the southern hemisphere longitude distribution (GC 312-1)?

Paul: The observed longitude profile of the galactic emission in the region $300° < \ell < 360°$ can, in principle, be represented by a few sources (see Hermsen et al., 1977). Due to the limited angular resolution of the γ-ray detector, however, even a spiral arm segment can appear as a source.

Okuda: Is the observed latitudinal distribution intrinsic?

Paul: The 2° angular resolution of the present γ-ray detector is not sufficiently small to resolve the latitude distribution at $\ell = 0°$.

III. SPIRAL STRUCTURE

THE DENSITY-WAVE THEORY CONFRONTED BY OBSERVATIONS

Roland Wielen
Astronomisches Rechen-Institut, Heidelberg, Germany

1. INTRODUCTION

Density waves are probably the most general phenomenon producing spiral structure in disk galaxies. The density-wave theory is able to give a rather successful interpretation of the observed spiral structure and of the related kinematics in external galaxies and in our Galaxy. We assume that the reader is familiar with the basic concepts of density-wave theory; for recent reviews see e.g. Kalnajs (1978), Lin (1975), Lindblad (1974), Marochnik and Suchkov (1974), Roberts (1977a,b), Rohlfs (1977, 1978), Shu (1973), Toomre (1977), Wielen (1974a). In Section 2, we discuss the proper theoretical devices which should be used for a meaningful comparison between observations and density-wave theory. The other sections are devoted to a comparison of density-wave theory with some relevant observations, mainly in our Galaxy.

2. HOW TO APPLY DENSITY-WAVE THEORY

Although the basic concepts of density-wave theory are rather simple, the predicted behaviour of well-observable objects such as HI, HII or young stars, turn out to be quite complicated. Unfortunately, some authors ignore these complications and try to fit the observations with oversimplyfied theoretical models. Obviously, such comparisons are of doubtful significance for an appraisal of the density-wave theory, although they may be sometimes justified for getting a first rough insight into the general capabilities of the theory.

2.1. Stationary density waves

It is generally assumed that the basic spiral structure of galaxies is of a rather permanent nature. Therefore, the density- wave theory assumes primarily a <u>stationary</u> wave in the gravitational potential of a galaxy. In the 'response problem', it is then asked how various populations of objects react on that potential wave. It is of primary importance to keep in mind that the conventional response formulae of the

W. B. Burton (ed.), The Large-Scale Characteristics of the Galaxy, 133–144.

density-wave theory are only valid if the objects under consideration have already reached a stationary dynamical equilibrium in the galaxy. For reaching such a stationary state, typical objects require a few orbital revolutions in the galaxy. Hence only older stars, of an age of more than a few 10^8 years, and the long-lived components of the interstellar gas can be described by the usual response formalism of the density-wave theory. In contrast to the older stars and to HI, no stationary density wave exists among younger stars or HII regions, because these objects have not had enough time to reach a stationary equilibrium state. Therefore, the usual formulae of the density-wave theory unfortunately do not apply at all to these well-observed young objects. The distribution and kinematics of young objects do not represent a wave phenomenon but can mainly be characterized as a migration out of the original spiral arms in which they were born. In order to predict the behaviour of young objects, we have essentially to solve an initial value problem for their orbits (Section 2.3.).

2.2. Linear theory and shock waves

At first, the linear version of the density-wave theory was developed by Lin and Shu. Later, it was shown by Roberts (1969), Shu et al. (1973) and Woodward (1975) that under rather general circumstances the interstellar gas reacts in a very non-linear way, including shocks, on even a small wave in the potential. In Fig. 1, we compare the surface density σ and the velocities U and V of HI according to the shock version with the linear theory. The shock version is calculated for a 'standard' density wave in our Galaxy (e.g. Wielen 1973) at $R \sim 10$ kpc. While the phase and the relative amplitudes of the linear waves are provided by the theory, the absolute amplitude has been chosen freely to match as far as possible the velocities of the shock version. From Fig. 1 it is very obvious that the linear theory is not able to describe the behaviour of the gas properly, not even in a first approximation. Hence today, only the shock version should be used for a realistic discussion of the distribution and kinematics of HI. Additional uncertainties enter here, however, because of magnetic fields, different phases of the gas, etc. .

2.3. Motions of young objects

Star formation is probably very effectively triggered by the sudden increase in gas pressure at the shock front. Hence the birthplaces of most young objects should be close to the shock front. It is more difficult to derive the initial velocities of newly born stars. We may distinguish two plausible alternatives: In the post-shock case, the stars reflect the motion of the gas immediately after the shock. In the pre-shock case, the average initial velocity of a star is the gas velocity before the shock. If dense interstellar clouds do exist all the time, the pre-shock case would be adequate, because the motion of such clouds is not immediately affected by the shock front. If dense clouds are only formed after the shock front by phase transition, then the post-shock case would be appropriate. From the birthplaces and initial velocities, we can obtain the orbits of the stars in the Ω_p-system which corotates with the

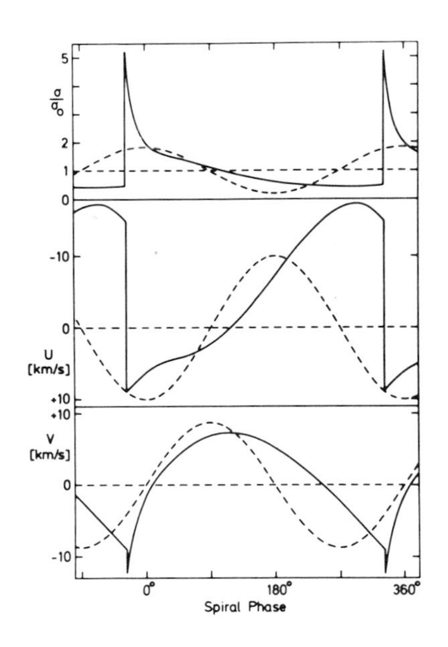

Figure 1. The surface density σ and the velocity components U and V of HI along a streamline. σ is normalized to its mean value σ_0. The U-axis points towards the galactic center, the V-axis in the direction of galactic rotation. The velocities are measured relative to the circular velocity at the actual position of the particle. Solid curve: shock version. Dashed curve: linear theory. The spiral phase of the density wave is zero at the minimum of the potential and increases by 360° from one spiral arm to the next one.

density wave (Fig.2) and the U- and V-velocities along the orbits (Fig.3) for a typical pre- and post-shock case (Schwerdtfeger 1977; Wielen 1975a, b, 1978). In Fig.3, the spiral phase indicates the position relative to the spiral potential. From the orbits of many test stars, we obtain in Figs.4 and 5 the drift of ageing spiral arms (defined by the locations of stars of a common age τ). The broadening of ageing spiral arms due to an initial velocity dispersion (10 km/s in the figures) is indicated by a cloud of stars. Figs.4 and 5 show that the drift and broadening are neither linear nor monotonic with age τ. From Figs.2-5, we must conclude that the motions of young stars are rather complicated, both in position and velocity space. They depend also on additional assumptions about star formation (e.g. pre- or post-shock case) which do not form an integral part of density-wave theory. All this hampers severely any conclusive comparison between theory and observations for younger stars and HII regions.

3. EXTERNAL GALAXIES

The general questions of existence, origin and maintenance of density waves should be studied mainly in external galaxies, where one can distinguish large-scale structure from local perturbations much better than in our Galaxy. In fact, external galaxies now provide the best evidence for the existence of density waves. Visser (1978) gives a very convincing interpretation of the Westerbork HI observations of M81 in terms of density-wave theory. The radio continuum observations of M51 (Mathewson et al. 1972) represent still the most suggestive evidence for the existence of spiral shock fronts (see also van der Kruit and Allen 1976). Optical observations of spiral galaxies (Schweizer 1976) can be interpreted

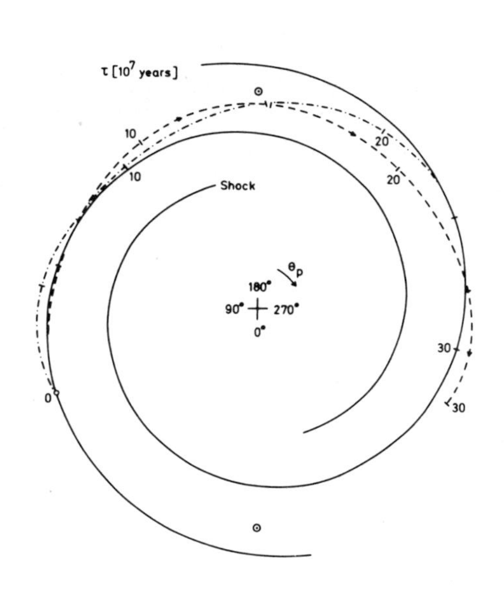

Figure 2. Typical pre-shock (dash-dotted) and post-shock (dashed) orbits of stars in the Ω_p-system. The shock fronts are given by the solid spirals.

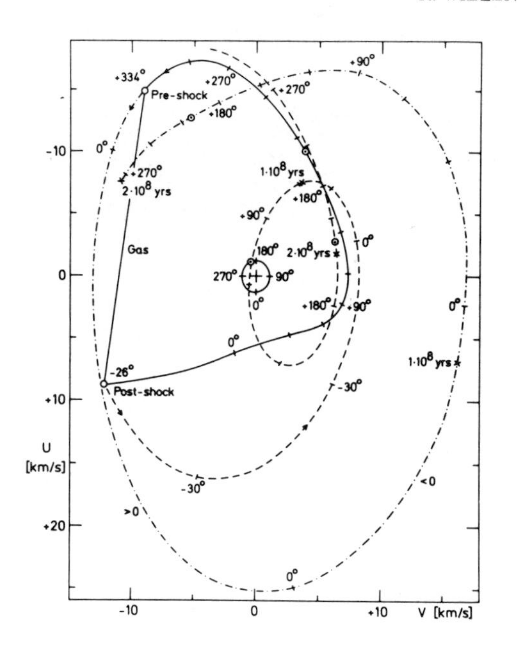

Figure 3. Velocities along the streamline of the gas (large solid curve), along the pre-shock orbit (dash-dotted) and along the post-shock orbit (dashed).

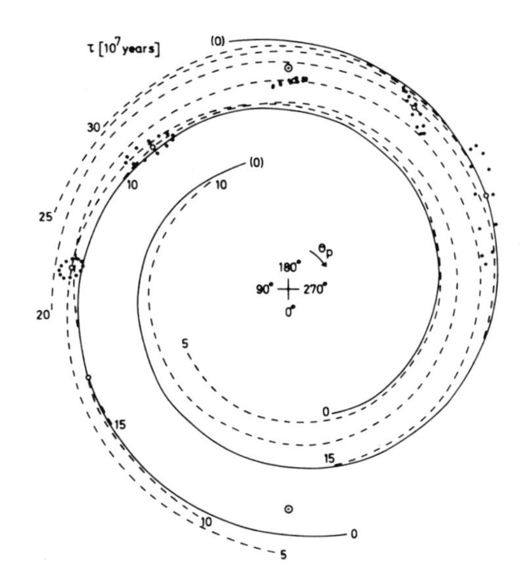

Figure 4. Drift of young stars of age τ for the pre-shock case.

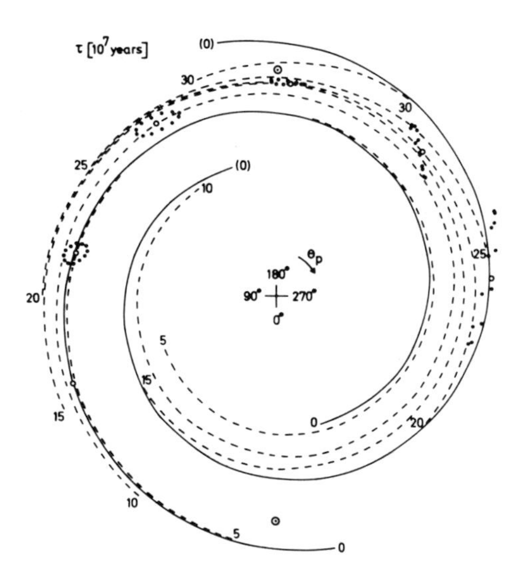

Figure 5. Drift of young stars of age τ for the post-shock case.

as showing directly the basic density wave among the older disk stars, although the implied amplitudes of some of the waves are disturbingly high. It should be emphasized that the well-observed galaxy M33 has probably only a weak density wave without a significant shock front (Roberts et al. 1975), and is therefore less suited for a confrontation with theory.

4. OBSERVATIONS IN OUR GALAXY

4.1. HI

In our Galaxy, the best evidence for a density wave are still the wavy irregularities in the extreme radial velocities of HI at different galactic longitudes (Fig.6). The waves correspond to amplitudes in the tangential streaming velocity of HI of about 10 km/s. No conclusive quantitative confrontation of the observations with the shock version has been presented so far. Either the inappropriate linear theory has been used, or it has been incorrectly assumed that the V-velocity (Fig.1) at the tangential point is equal to the terminal velocity along the line of sight. For a correct procedure see Sawa (1978). It is rather disturbing that the maxima of the column density of HI (Fig.6) seem to be correlated with maxima of the terminal velocity, while the shock-wave theory predicts nearly the opposite behaviour (see Fig.1, where maxima of σ correspond to minima of V).

The detailed profiles of HI at different longitudes can be well interpreted by density-wave models (e.g. Simonson 1976). Due to the many free parameters usually allowed in the models, the obtained agreement is, while certainly encouraging for density-wave theory, not finally decisive.

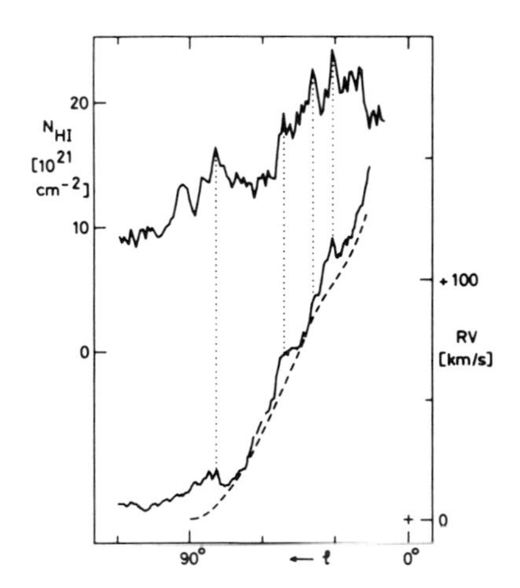

Figure 6. The lower part shows the terminal radial velocities of HI along the line of sight as a function of galactic longitude ℓ (data from Burton 1970). The upper part gives the integrated column density N(HI) as a function of longitude (data from Burton 1976).

4.2. HII

In many external galaxies, giant HII regions are strongly confined to narrow spiral arms. If this is also true in our Galaxy, then the giant HII regions should be located on thin curves in the directly observable diagram of radial velocity RV versus galactic longitude ℓ. Density-wave motions can only deform but not destroy the loops which correspond to the spiral arms in position space. Contrary to HI, velocity crowding would be unimportant for the HII features. The Figs.8, 9, 10 show the loops for a two-armed spiral structure in the cases of circular, pre-shock and post-shock velocities of HII regions. The observed diagram for HII regions (Fig.7) do not show the expected loops clearly. As long as it is rather impossible to deliniate the loops in the RV-ℓ-diagram, there is no reason to expect a better definition of spiral arms in position space, no matter how accurate the velocity-distance relation may be. Fig.9 shows that the loops are more open and overlapping for the pre-shock case. Together with peculiar velocities of HII regions slightly higher than usually assumed, the pre-shock version may explain the blurring of the loops in the observed diagram for galactic HII regions. Star formation with pre-shock velocities is also favoured by observations of external galaxies (Wielen 1978).

The abundance of HII relative to HI as a function of galactocentric distance R can be explained by density-wave theory as due to a higher star formation rate in the inner part of the Galaxy, caused by the higher compression at the shock front and by the higher frequency of passages of the gas through the shocks. Both effects may also explain the high abundance of molecules (H_2, CO, etc.) in the inner region.

4.3. Studies of individual spiral arms

Although density-wave theory is mainly concerned with the grand design of spiral structure, this theory can also help one to understand the detailed structure and kinematics of nearby spiral arms on a smaller scale. Especially, non-circular motions and differences in radial velocities of HI, HII and young objects at the same location can be attributed to the density wave (e.g. Humphreys 1972, Roberts 1972, Minn and Greenberg 1973, Burton and Bania 1974). However, due to the complicated behaviour of young objects discussed in Section 2.3., it is difficult to make other than general qualitative statements of agreement.

4.4. Places of formation

Birthplaces of stars or clusters can be obtained by calculating their orbits backwards in time. The resulting pattern of birthplaces usually agrees well with the predictions of the density-wave theory (e.g. Yuan 1969, Wielen 1973, Grosbøl 1976). Unfortunately, the birthplaces of present nearby stars are strongly biased by a kinematic selection effect: Young objects ($\tau \leq 1\text{-}2 \cdot 10^8$ years) born in the main spiral arms, can have reached the Sun only if they had a rather large peculiar velocity at birth (see Figs.4 or 5). Furthermore, the diffusion of stellar orbits due

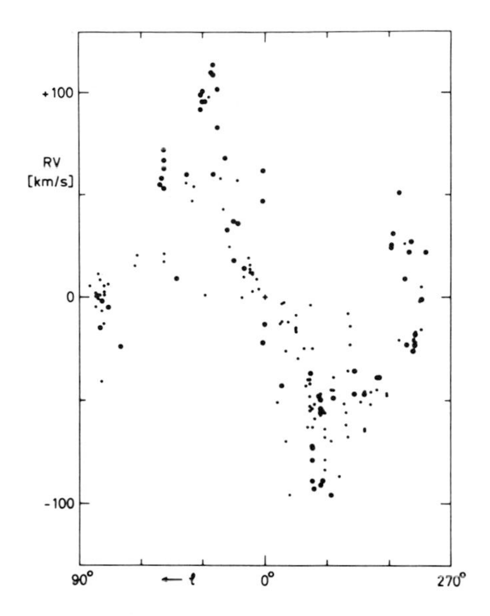

Figure 7. RV-ℓ-diagram for HII regions (giant regions: larger dots), based on H109α recombination line data of Reifenstein et al. (1970) and Wilson et al. (1970).

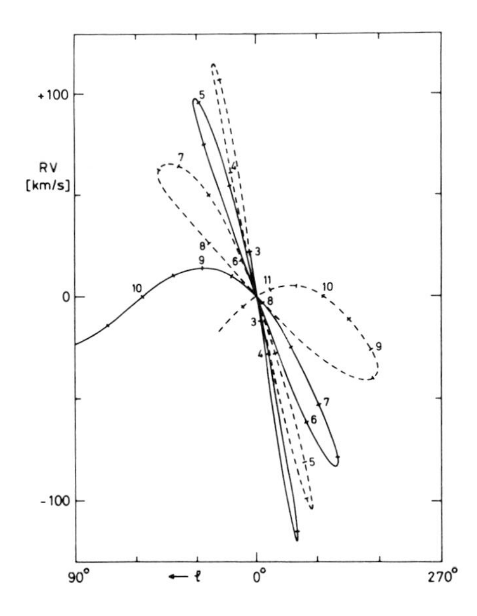

Figure 8. RV-ℓ-diagram for objects with circular velocities. The two spiral arms are distinguished by solid and dashed lines. The numbers indicate the distance from the galactic center.

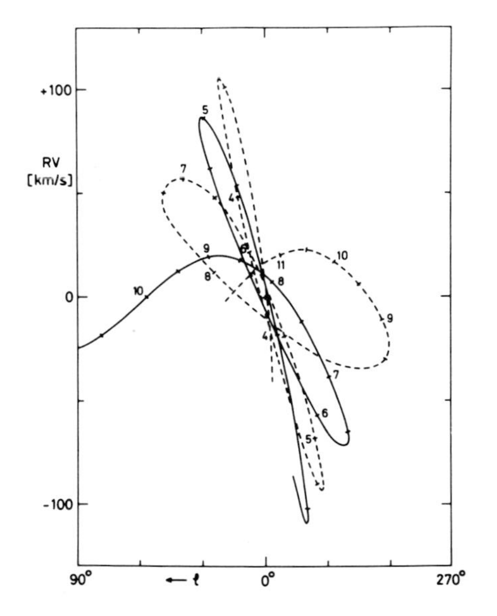

Figure 9. RV-ℓ-diagram of objects with pre-shock velocities at the shock front.

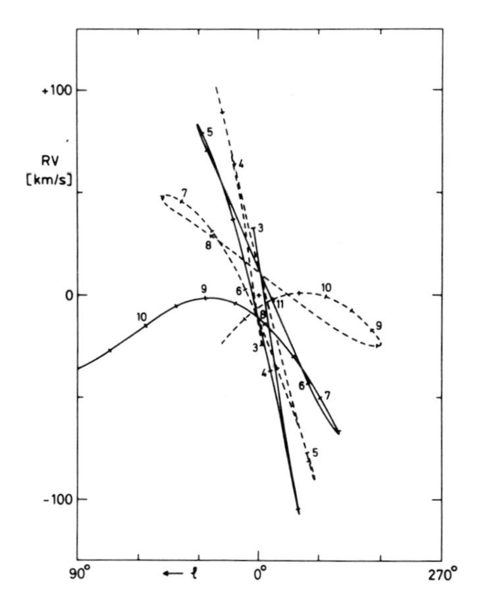

Figure 10. RV-ℓ-diagram of objects with post-shock velocities at the shock front.

to local irregularities of the galactic gravitational field probably causes severe uncertainties in the derived birthplaces of stars older than 10^8 years (Wielen 1977).

4.5. Motions in the solar neighbourhood

Density-wave theory predicts significant non-circular motions for the gas and stars (Fig.3). These motions should be partially detectable in the solar neighbourhood, although local effects (e.g. the Orion arm) may severely disturb the global flow pattern. Lin et al. (1974) have studied the effect of shock waves in HI on the determination of the local galactic differential rotation. Probably the most easily detectable effect should be a positive K-term in HI: $K = 0.5 \operatorname{div} \vec{v} = -0.5\, \dot{\sigma}/\sigma$. Along a streamline, $\dot{\sigma}$ is mostly negative (Fig.1). K should be typically of the order of +6 km/s/kpc near the Sun. The available observations are however inconclusive.

In the presence of a density wave, the local mean velocity of objects of different age and different velocity dispersion should differ, and this can severely affect the determination of the local standard of rest (Blaauw 1970, Lin and Yuan 1975). If the Sun is located near the middle of an interarm region (e.g. at a spiral phase of 202°), the mean HI velocity, relative to the circular velocity, should be about $U = -\dot{R} = -10$ km/s and $V = +4$ km/s. While the predicted mean motion of young stars is rather uncertain (Section 2.3., Fig.3), the mean motion of old stars with a high velocity dispersion can be described by the linear theory. The amplitudes of the non-circular motions of older stars, $\hat{U}_*$ and $\hat{V}_*$, are correlated with the relative amplitude of their density variation, $(\hat{\sigma}_1/\sigma_0)_*$, by $\hat{U}_* = R(\Omega - \Omega_p) \tan i\, (\hat{\sigma}_1/\sigma_0)_*$ and $\hat{V}_* = R(\kappa^2/4\Omega) \tan i\, (\hat{\sigma}_1/\sigma_0)_*$. For $(\hat{\sigma}_1/\sigma_0)_* = 10\%$ and a pitch angle $i = 6\overset{\circ}{.}2$, we expect $\hat{U}_* = 1.2$ km/s and $\hat{V}_* = 1.1$ km/s (central oval in Fig.3).

In Fig.11, we present observational data on the mean velocities of various classes of objects with different velocity dispersions σ_U. Directly observable are not the absolute mean motions, but only the motions of the objects relative to the Sun, i.e. differences of mean velocities. Since young objects and gas probably deviate by about 10 km/s from circular motion, we must use older stars to define the local standard of rest, inspite of the larger mean errors. The most suitable objects seem to be the McCormick K+M dwarfs in Gliese's Catalogue of nearby stars (Wielen 1974c). The velocities of these dwarfs are not biased by selection effects. Since these K+M dwarfs are typical disk stars, their density amplitude $(\hat{\sigma}_1/\sigma_0)$ is probably less than 10%; hence their outward radial motion is about 1 km/s. The observed motion of the Sun relative to these K+M dwarfs, $U_\odot = +5\ (\pm 3)$ km/s, would then lead to a solar motion of $U_\odot \sim +4$ km/s relative to the local circular velocity. This value differs from the standard solar motion by about 6 km/s, and would explain perfectly the observed difference between the northern and southern galactic rotation curves (Kerr 1962) as due to an outward motion of the hitherto used local standard of rest. The V-component of the solar motion with respect to the circular velocity cannot be so easily derived, because

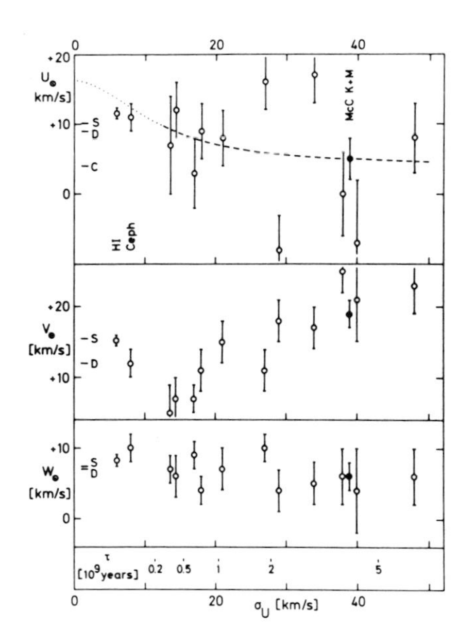

Figure 11. The solar motion for objects with various radial velocity dispersions σ_U (and corresponding ages τ). Data taken from Crovisier (1978), Jahreiß (1974), and Wielen (1974b, c). Mean errors are indicated. S = standard solar motion; D = peculiar solar motion after Delhaye (1965). A possible variation of $U_\odot$ with σ_U according to the density-wave theory is schematically indicated by the dashed curve. C indicates the corresponding circular velocity. The dotted part of the curve is invalid because of the non-stationarity and/or the non-linearity of the motions of the objects with small σ_U.

the V velocities of the K + M dwarfs and of other old objects are affected by the classical 'asymmetric drift', which is difficult to eliminate quantitatively.

5. CONCLUSIONS FOR OUR GALAXY

Many observations in our Galaxy can be well explained by the presence of a density wave. This strongly suggests the actual existence of this wave, although most of the observations are not well suited as decisive tests at present. The observations have not provided a clue for the origin and maintenance of such a density wave. It is even unclear whether the observationally implied density wave in the gas is primarily caused by a tightly-wound spiral wave in the potential (mainly due to disk stars) or by a bar-like distortion of the inner parts of our Galaxy.

REFERENCES

Blaauw,A. : 1970, IAU Symposium No.38, p.199.
Burton,W.B. : 1970, Astron.Astrophys.Suppl. 2, p.261.
Burton,W.B. : 1976, Ann.Rev.Astron.Astrophys. 14, p.275.
Burton,W.B., Bania,T.M. : 1974, Astron.Astrophys. 33, p.425.
Crovisier,J. : 1978, Astron.Astrophys. (in press).
Delhaye,J. : 1965, Stars and Stellar Systems 5, p.61.
Grosbøl,P. : 1976, Dissertation, Univ. Copenhagen.
Humphreys,R.M. : 1972, Astron.Astrophys. 20, p.29.
Jahreiß,H. : 1974, Dissertation, Univ. Heidelberg.
Kalnajs,A.J. : 1978, IAU Symposium No.77 (in press).

Kerr,F.J. : 1962, Monthly Notices Roy.Astron.Soc. 123, p.327.
Lin,C.C. : 1975, in 'Structure and Evolution of Galaxies', Ed. G. Setti, NATO ASI C21, Reidel Publ.Co., Dordrecht, p.119.
Lin,C.C., Yuan,C., Roberts,W.W. : 1974, IAU Highlights of Astronomy 3, p.441, and Preprint.
Lin,C.C., Yuan,C. : 1975, Bull.American Astron.Soc. 7, p.344.
Lindblad,P.O. : 1974, IAU Symposium No.58, p.399.
Marochnik,L.S., Suchkov,A.A. : 1974, Usp.Fiz.Nauk 112, p.275 = Sov.Phys.Usp. 17, p.85.
Mathewson,D.S., van der Kruit,P.C., Brouw,W.N. : 1972, Astron. Astrophys. 17, p.468.
Minn,Y.K., Greenberg,J.M. : 1973, Astron.Astrophys. 24, p.393.
Reifenstein,E.C., Wilson,T.L., Burke,B.F., Mezger,P.G., Altenhoff,W.J. : 1970, Astron.Astrophys. 4, p.357.
Roberts,W.W. : 1969, Astrophys.J. 158, p.123.
Roberts,W.W. : 1972, Astrophys.J. 173, p.259.
Roberts,W.W. : 1977a, Vistas in Astronomy 19, p.91.
Roberts,W.W. : 1977b, in 'The Structure and Content of the Galaxy and Galactic Gamma Rays', Eds. C.E. Fichtel and F.W. Stecker, NASA CP-002, Washington, p.119.
Roberts,W.W., Roberts,M.S., Shu,F.H. : 1975, Astrophys.J. 196, p.381.
Rohlfs,K. : 1977, 'Lectures on Density Wave Theory', Lecture Notes in Physics 69, Eds. J. Ehlers et al., Springer-Verlag, Berlin.
Rohlfs,K. : 1978, Mitt.Astron.Ges. No.43 (in press).
Sawa,T. : 1978, Astrophys.Space Sci. 53, p.467.
Schweizer,F. : 1976, Astrophys.J.Suppl. 31, p.313.
Schwerdtfeger,H. : 1977, Dissertation, Univ. Heidelberg.
Shu,F.H. : 1973, American Scientist 61, p.524.
Shu,F.H., Milione,V., Roberts,W.W. : 1973, Astrophys.J. 183, p.819.
Simonson,S.C. : 1976, Astron.Astrophys. 46, p.261.
Toomre,A. : 1977, Ann.Rev.Astron.Astrophys. 15, p.437.
van der Kruit,P.C., Allen,R.J. : 1976, Ann.Rev.Astron.Astrophys. 14, p.417.
Visser,H.C.D. : 1978, IAU Symp. No.77 (in press); and Dissertation, Univ. Groningen.
Wielen,R. : 1973, Astron.Astrophys. 25, p.285.
Wielen,R. : 1974a, Publ.Astron.Soc.Pacific 86, p.341.
Wielen,R. : 1974b, Astron.Astrophys.Suppl. 15, p.1.
Wielen,R. : 1974c, IAU Highlights of Astronomy 3, p.395.
Wielen,R. : 1975a, in 'La dynamique des galaxies spirales', CNRS Colloquium No.241, Ed. L. Weliachew, Paris, p.357.
Wielen,R. : 1975b, in 'Optische Beobachtungsprogramme zur galaktischen Struktur und Dynamik', Ed. Th. Schmidt-Kaler, Bochum, p.59 = Mitt.Astron.Rechen-Inst. Heidelberg Ser. B No.49.
Wielen,R. : 1977, Astron.Astrophys. 60, p.263.
Wielen,R. : 1978, IAU Symposium No.77 (in press).
Wilson,T.L., Mezger,P.G., Gardner,F.F., Milne,D.K. : 1970, Astron.Astrophys. 6, p.364.
Woodward,P.R. : 1975, Astrophys.J. 195, p.61.
Yuan,C. : 1969, Astrophys.J. 158, p.889.

DISCUSSION

Pişmiş: Dr. Wielen stated that the strongest argument supporting the density wave theory is the existence of the waves observed in the rotation curves of galaxies, the maximum velocity corresponding to the spiral arm and the minimum to the interarm region. The explanation for these waves is not unique. A possible, or even a more logical, explanation is afforded by the hydrodynamical equations of motion in a galaxy in steady state and with rotational symmetry. According to these equations, at a given distance from the galactic center the rotational velocity can have values from the circular velocity as a maximum down to values very small or even zero. It is easy to see that between the spiral arms the velocity of rotation must be low because we are observing there essentially an older population with a lower velocity of rotation with respect to the neighboring arms, where the rotation velocity is high.

Wielen: The wavy irregularities in the rotation curves have been observed mainly in HI, and partly in HII or young stars. All these objects have rather small velocity dispersions. Hence I do not think that the observed wavy irregularities can be explained in the way you propose.

Burton: Regarding the worry that the terminal velocity excursions are associated with local maxima in the HI apparent column densities: optical depth effects might be quite important. If typically $\tau \gtrsim 1$, an increase of some 5 km s^{-1} of the total velocity extent will cause an increase of the integral $T_B(v)dv$ which might be sufficient to account for the observed increase in N_{HI}. It would be straightforward to test if this effect would dominate details in a model density distribution. It would be difficult, however, to correct the observations for these optical depth effects, because the true density distribution is too poorly known.

Wielen: Your data for the HI column density, which I used, were--at least partially--corrected for optical depth. If your explanation is correct, the optical depth effects have really to be extremely dominant for the HI profiles.

Bok: In the southern hemisphere we observe some spiral features (e.g., the Carina arm) over a large range of distances. Can we predict average radial velocities for the young stars at the inside of the arm, in the middle, and at the outside of the arm? Would this allow us to differentiate between pre-shock and post-shock star formation? We also observe the Sagittarius arm cross-wise. Can we predict average velocity differences for stars between the Sun and the Sagittarius arm, those in the arm, and those lying slightly beyond the arm as viewed from the Sun?

Wielen: No detailed prediction of the kind you ask for has been made up to now, although this can be done in a quite straightforward way from our calculations. The velocities of the stars depend sensitively, however, on the age, as shown in Figure 3 of my paper. A meaningful comparison between theory and observations can probably be carried out if the different ages of the observed objects are properly taken into account.

Meneguzzi: Do we see spiral features in old-population stars?

Wielen: We should not expect to see the linear density wave among the older disk stars in our Galaxy, because the amplitude of the density variation is probably of the order of 5 or 10%. The amplitudes of the non-circular motions may be about 1 km s^{-1}. According to Schweizer (1976), the density wave among older stars is visible in some external galaxies. The implied amplitudes, however, are sometimes disturbingly high. Perhaps there is still a significant contribution of younger stars to the observed smooth spiral arms.

Rickard: I got the impression from what you said that we should expect a rather poor agreement between the velocities of the young stars and the gas. But we know that there is good general agreement between the two--certainly not the large differences of $\sim$ 25 km s^{-1} you discussed. I can recall only one case where the gas and stellar velocities in an HII region are as different as 12 km s^{-1}. In the Perseus arm there is a major HI feature well-correlated in velocity with the stellar velocities of the HII regions and young clusters. How do you explain this?

Wielen: The HII gas is probably co-moving with the young stars, on the average. So we would expect the two velocities to coincide except for some random velocity dispersion. The high-velocity differences between HI and young stars for the pre-shock case are maintained only for a short time (see Fig. 3 of my paper). Because the young stars remain for a rather long time in the neighborhood of the spiral shock front, the average difference in velocity between HI and young stars may be small. In the Perseus arm, the shock strength is probably small, because the gas motion is only slightly supersonic ($W_{\perp} \sim a$). Therefore, the difference between the pre- and post-shock velocities should be smaller than those shown in Figure 3, which illustrates the situation at R $\sim$ 10 kpc.

A MECHANISM FOR THE ORIGIN AND DEVELOPMENT OF SPIRAL ARMS IN A GALAXY

Paris Pişmiş
Instituto de Astronomía
Universidad Nacional Autónoma de México

It is generally believed that the constituents of a galaxy like ours are formed from an isolated gaseous assembly at successive stages, starting from a spheroidal sybsystem down to the flattest subsystem.

We envision such a galaxy to have evolved to the stage where star formation is completed everywhere except in the flat gaseous subsystem. We postulate that this subsystem is permeated by a magnetic field of bipolar configuration, say a dipole, of which the axis is nearly perpendicular to the rotation axis of the galaxy. Circumstances leading to the formulation of this hypothesis and its consequences are discussed in the following papers: Pişmiş (1960, Huang and Pişmiş (1960, and Pişmiş (1961, 1963, 1965 and 1969). We now renew emphasis on the hypothesis by giving a sketch of it below.

It is expected that the gaseous subsystem will undergo contraction. The frozen-in magnetic lines of force will carry the gas along, except at the two polar regions where the plasma will leak out. The track of the material left behind the shrinking gaseous subsystem will delineate spiral arms through differential rotation. An application of this mechanism to M31 has rendered a spiral pattern with 1.5 turns in 3×10^9 years.

A number of observed properties in the Galaxy, seemingly unrelated to one another, find an explanation in a unified manner as a direct consequence of this mechanism. These properties are:

a) The existence of a spiral pattern superposed on a "smooth" axisymmetric stellar system.
b) The bi-symmetry of the spiral pattern.
c) Magnetic field lines along the spiral arms. Recent observations of M51, presented by Fujimoto at this Symposium, support this statement.
d) The reversal of the magnetic field where crossing the plane of the Galaxy.
e) The warps observed in our and in other galaxies.

W. B. Burton (ed.), The Large-Scale Characteristics of the Galaxy, 145–146.

f) The alternately up and down position of consecutive spiral arms (Henderson 1967).
g) The gradual thickening of the neutral hydrogen layer towards the edges of the Galaxy.

The leakage of the gas from the magnetic poles may also occur as ejection; in this case, a fortiori, a spiral pattern ensues. We wish to emphasize that the magnetic field is used here only as a funneling agent and not to sustain spiral formations.

REFERENCES

Henderson, A. P.: 1967, Ph.D. Thesis, University of Maryland.
Huang, S. S., and Pişmiş, P.: 1960, Bol. Obs. Tonantzintla Tacubaya 2, 7.
Pişmiş, P.: 1960, Bol. Obs. Tonantzintla Tacubaya 2, 3.
Pişmiş, P.: 1961, Bol. Obs. Tonantzintla Tacubaya 3, 3.
Pişmiş, P.: 1963, Bol. Obs. Tonantzintla Tacubaya 3, 127.
Pişmiş, P.: 1968, Bol. Obs. Tonantzintla Tacubaya 4, 229.
Pişmiş, P.: 1970, in IAU Symposium No. 38, ed. W. Becker and G. Contopoulos (Dordrecht: Reidel) pp. 452-454.

AN EJECTION THEORY OF SPIRAL STRUCTURE

Th. Schmidt-Kaler
Astronomisches Institut der Ruhr-Universität

A number of questions was posed at this Symposium: why is the density of molecules highest near R = 5 kpc? Why is star formation most active near 5 kpc? The life-time of the molecular clouds near 5 kpc is about $2 \cdot 10^8$ years - where does the gas come from? Why is almost no gas observed between R = 1 and 3 kpc?
It seems that an ejection theory can answer these questions.

The density wave theory of spiral structure is apparently a viable description of the physical state in our Galaxy between $6 < R < 16$ kpc (adopting the standard $R_\odot$ = 10 kpc). In this region the velocity variations due to the d.w. exceed those due to radial expansion, $\Delta v(\text{d.w.}) > \Delta v(\text{ej.})$. There are, however, two basic problems the d.w. theory has to cope with: the importance of the d.w. in the presence of the very hot phase of the interstellar gas, and the excitation and maintenance of the d.w. in the presence of strong dissipative forces.

1. THE DENSITY WAVE THEORY AND THE VERY HOT INTERSTELLAR GAS COMPONENT

OVI lines detected by the Copernicus satellite and the soft X-ray background strongly indicate the existence of an extended component of the interstellar gas with temperatures in the range $3\text{-}10 \cdot 10^5$K and densities in the range $5 \cdot 10^{-4}$ to 10^{-2} cm^{-3} (McKee and Ostriker 1977). This very hot gas is believed to result from supernova shock heating (Cox and Smith 1974, Burnstein et al. 1976) and seems to pervade in the form of a network of tunnels and bubbles the whole space around the sun (within distances up to at least 300 pc). The filling factor is at least 0.5, it may be as high as 0.8. This means that the bulk of the volume is filled with a medium in which the sound velocity and hence the group velocity of shock waves is of the order of c_{II} = 120 km/s. The streaming velocities induced in the interstellar gas by perturbations due to the spiral d.w. are of the order $w_\perp$ = 10-25 km/s. Consequently strong shocks would never develop, and the d.w. theory would lose its most attractive feature, namely the spiral galactic shock which gave a theoretical basis for so many observational facts. Furthermore, the kinematic life-time of

W. B. Burton (ed.), The Large-Scale Characteristics of the Galaxy, 147–150.

the d.w. becomes $\tau = \Delta R/v_g \cong 10$ kpc/120 km $s^{-1} = 8 \cdot 10^7$ yrs. This would deprive the whole wave concept of its sense since it was devised to overcome the winding dilemma implying just that time-scale.

A closer inspection, however, reveals that the interstellar medium is not in a steady state. The supernova shock heating is restricted to the spiral arms where supernovae light up, which are massive stars resulting from spiral shock-triggered star formation.
When the gas moves out into the interarm region the supernova activity subsides and the gas is efficiently cooling down to T = $1\text{-}3 \cdot 10^4$K before it enters the next sweep of the d.w. pattern. The cooling time of the very hot gas is $t_c \cong 10^5/n_e \cong 10^8$ years (Shapiro and Moore 1976). The cycle time of the d.w. in a two-armed Galaxy is $t_{dw} = \pi/(\Omega_\odot - \Omega_p) = 2.7 \cdot 10^8$ years. The supernova heating phase takes roughly 5-10 %, the inter-arm cooling most of the rest of this time. When the gas now enters the next sweep of the d.w. pattern the temperature is low enough to produce a strong shock with subsequent compression cooling, formation of low-temperature intercloud gas (T=$1\text{-}8 \cdot 10^3$K), and cool gas and dust clouds part of which may later on collapse to form massive stars. After a few million years the most massive stars become supernovae and start the whole cycle again. It thus appears that the d.w. pumps non-gravitationally bound interstellar matter through a whole phase cycle which is schematically depicted in fig. 1. The Sun is in the local spiral arm located where maximal supernova heating is expected.
The overall sound velocity is about 17 km/s.
A detailed paper will appear in Astron. Astrophys. (Reinhardt and Schmidt-Kaler 1978).

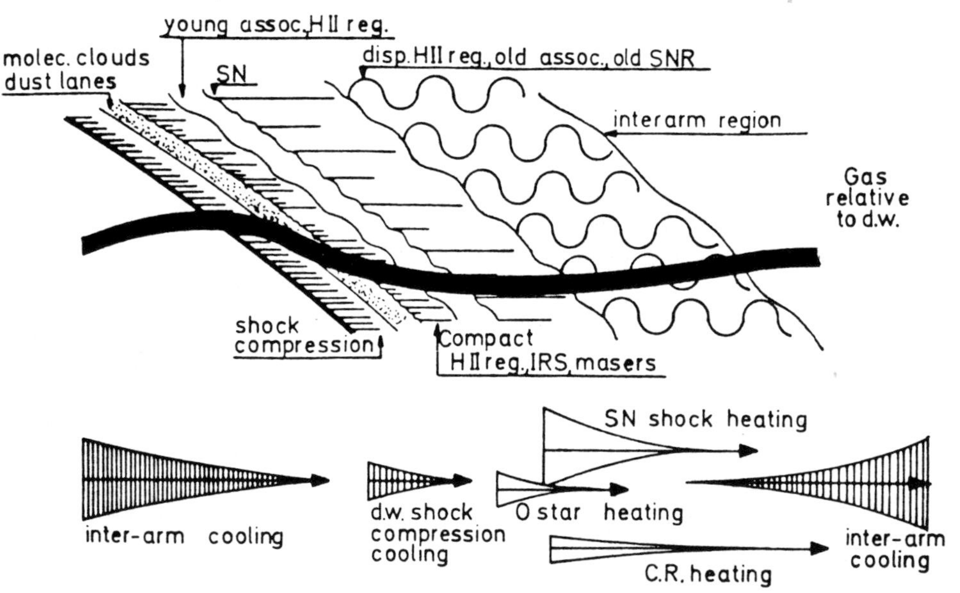

Fig. 1. The phase cycle of the interstellar matter due to the density wave half cycle. The most important cooling and heating processes in the successive phases are schematically indicated.

2. THE ORIGIN AND ENERGY SUPPLY OF THE DENSITY WAVE BY EJECTION OF GAS FROM THE CENTRAL REGION OF THE GALAXY

The total energy density of the d.w. in the corotating system is, according to Shu (1974) $\varepsilon_{dw} = 1\cdot 10^9$ erg cm^{-2}, or within $3 \leqq R \leqq 15.7$ kpc the total energy is $\mathcal{E} = 7.1\cdot 10^{54}$ erg. The kinematic life-time of the d.w. in our Galaxy is $\tau = \Delta R/v_g = 12.7$ kpc/17 km s^{-1} $= 7.5\cdot 10^8$ years; due to the energy used up by the galactic shock the dynamic life-time of the d.w. is still shorter (Feitzinger and Schmidt-Kaler 1978). In order to maintain the d.w. stationary it is therefore necessary to supply energy and angular momentum to the relevant regions of the Galaxy at least equal to

$$\dot{E} = \mathcal{E}/\tau = 3\cdot 10^{38} \text{ erg s}^{-1}$$

$$\dot{I} = J/\tau = \frac{\mathcal{E}}{\Omega - \Omega_p}/\tau = 7\cdot 10^{53} \text{ erg}$$

since the angular momentum in the corotating system becomes

$$J = \frac{m}{\Omega}E = \frac{2}{\Omega}E \cong \frac{E}{\Omega_p} \longrightarrow \frac{\mathcal{E}}{\Omega - \Omega_p}.$$

Lin's short wave mode propagates from the corotation radius to the Lindblad resonance circles. But since the d.w. energy is negative for $R < R_{co}$ and positive for $R > R_{co}$ the SWM d.w. transports effectively positive energy and momentum from the inner to the outer Lindblad resonance.

The necessary energy supply cannot come from the outer regions. A central mass asymmetry, e.g. an oval distortion of the central spheroid or a bar could drive the d.w. (Schmidt-Kaler 1975). I was much pleased by the papers of Roberts and others who take up this idea which I put forward at Sydney 1973. But meanwhile our calculations showed a fundamental difficulty which is also observed in the work of Sanders, Roberts and others: the gas inevitably after the shock is braked and by the continuity equation gets a large component directed radially inward. Thus, the low-density gas shows expanding and the high-density gas ingoing motions which is just contrary to the observations. We do not invent particular scenarios to give explanations for each of the expanding features separately but take the observed expansion motions seriously in their entirety (Rohlfs and Schmidt-Kaler 1978). Also, the two other spiral galaxies whose inner regions have been investigated (M31 and M81) show strong expansion motions. In fact, the state of motion of M31 very much resembles that of our Galaxy (see fig. 2 for our Galaxy, and the work of Rubin et al., summarized by Schmidt-Kaler (1975) for M31). Mass outflow with typically $v_e = 100$ km/s, $\dot{M} = 0.25\, M_\odot$/year near $R_N = 2$ kpc will supply $\dot{E}_{kin} = \frac{1}{2}v_e^2\dot{M} = 8\cdot 10^{38}$ erg s^{-1}, $\dot{I} = \frac{1}{2}\dot{M}R_N^2(\Omega_N - \Omega) = 8\cdot 10^{53}$ erg.

The mass and energy balance of gas shells lost by aging stars in the central spheroid is amply sufficient for continuous replenishment over the last $5\cdot 10^9$ years (life-time of the disk population). Further: the mass loss of stars of the disk implies a net loss of kinetic energy and an increase of gravitational energy which leads to cooling of the disk and stabilizing of the spiral wave.

The oval distortion of the central spheroid may be described by the Riemann instability (Schmidt-Kaler 1978): Model calculations show that explosions in the nucleus with energies of the order of 10^{54} erg on a time-scale of 10^6 yrs can produce the observed situation.

Summarizing, the ejection theory
(1) relates spiral development to activity of galactic nuclei
(2) predicts two, diametral, trailing spiral arms (inside $R = R_{iL}$) which in turn excite trailing d.w. spirals for $R > R_{iL}$
(3) explains by the continuity equation the gas densities (and ensuing processes) inside R = 6 kpc.

REFERENCES

McKee, C.F., Ostriker, J.P., 1977: Astrophys.J. 218, 148
Cox, D.P., Smith, B.W., 1974: Astrophysic.J. 189, L105
Burnstein, P. et al. 1976: Astrophysic.J. 213, 405
Shapiro, P.R., Moore, R.T., 1976: Astrophysic.J. 207, 460
Shu, F.H., 1974: The Interstellar Medium (ed. K.Pinkau) Reidel, p. 219
Feitzinger, J., Schmidt-Kaler, Th., 1978: this Symposium, and Astron. Astrophys. subm.
Schmidt-Kaler, Th., 1975: Vistas in Astronomy 19, 69; Proceeed.Conf.Obs. Progr.Gal.Struct. and Dyn. Bochum, p. 75, 105
Rohlfs, K., Schmidt-Kaler, Th., 1978: this Symposium
Schmidt-Kaler, Th., 1978: in IAU Colloquium No.45, Chemical and Dynamical Evolution of Our Galaxy (ed. Basinska-Grzesik,Mayor), Geneva

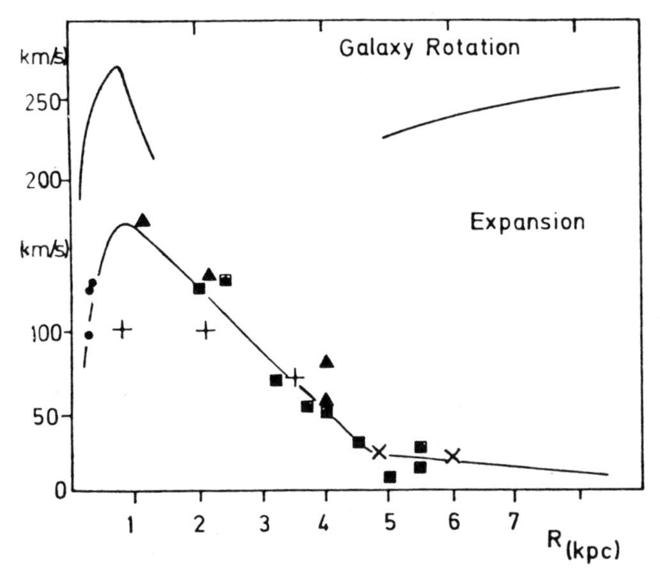

Fig. 2. The velocity curve of our Galaxy (circular component according to Sanders and Lowinger 1972, radial component according to Schmidt-Kaler 1978). The features of Cohen and Davies are partly at $|z| > 150$ pc so that apparently too small expansion components may result

ON THE MAINTENANCE OF SPIRAL STRUCTURE

James W-K. Mark and Linda Sugiyama
Massachusetts Institute of Technology, Cambridge, MA, USA

Robert H. Berman
University of Reading, Reading, UK

Giuseppe Bertin
Scuola Normale Superiore, Pisa, ITALY

A concentrated nuclear bulge with about 30% of the galaxy mass is sufficient (Lin, 1975; Berman and Mark, 1978) to eliminate strong bar-forming instabilities which dominate the dynamics of the stellar disk. Weak bar-like or oval distortions might remain depending on the model. In such systems self-excited discrete modes give rise to global spiral patterns which are maintained in the presence of differential rotation and dissipation (cf. especially the spiral patterns in Bertin _et al._, 1977, 1978). These spiral modes are standing waves that are physically analyzable (Mark, 1977) into a superposition of two travelling waves propagating in opposite directions back and forth between galactic central regions and corotation (a resonator). Only a few discrete pattern frequencies are allowed. An interpretation is that the central regions and corotation radius must be sufficiently far apart so that a Bohr-Sommerfeld type of phase-integral condition is satisfied for the wave system of each mode. The temporal growth of these modes is mostly due to an effect of Wave Amplification by Stimulated Emission (of Rotating Spirals, abbrev. WASERS, _cf._ Mark 1976) which occurs in the vicinity of corotation. In some galaxies one mode might be predominent while other galaxies could exhibit more complicated spiral structure because several modes are present. Weak bar-like or oval distortions hardly interfere with the structure of these modes. But they might nevertheless contribute partially towards strengthening the growth of one mode relative to another, as well as affecting the kinematics of the gaseous component.

Berman, R.H. and Mark, J.W-K.: 1978, manuscript in preparation.
Bertin, G., Lau, Y.Y., Lin, C.C., Mark, J.W-K. and Sugiyama, L.: 1977, Proc. Natl. Acad. Sci. USA, 74, 4726.
Bertin, G. and Mark, J.W-K.: 1978, Astron. Astrophys. 64, 389.
Lin, C.C.: 1975, in Structure and Evolution of Galaxies, ed. G. Setti, p. 119.
Mark, J.W-K.: 1976, Astrophys. J., 205, 363.
Mark, J.W-K.: 1977, Astrophys. J., 212, 645.

W. B. Burton (ed.), The Large-Scale Characteristics of the Galaxy, 151–153.

DISCUSSION

Miller: Your computer simulations look a good deal like the pictures Hohl showed at Besancon (I.A.U. Symposium 69) as an example of a long-lived bar. Your axis ratios look nearly the same as his. Yet you claim this simulation did not show a bar. Could you explain this difference of interpretation?

Mark: When we have a strong bar in the stellar disk, as in the cases run with no spheroidal matter, then our bar really is a thin bar or an open spiral object, and not to be confused with the bar-shaped shock waves of some gas-dynamical calculations. But in the simulation with a 30% bulge, we actually have an open spiral which terminates before reaching the galactic center. If you wish, you may talk of the inner open spiral or part of a bar or oval distortion, but this "bar" does not continue through the center.

Sanders: How do you model the bulge component?

Mark: In order to limit the spheroidal component to a bulge, and not allow spillage into a halo, we have taken a bulge density which decays as R^{-5} rather than the slower decay of observed spheroidal components. Eighty percent of the "bulge" mass lies within 4 or 5 kpc of the galactic center. Note that we have removed disk matter from the central parts of the Kuzmin disk and replaced it by spheroidal matter.

Toomre: In view of the all-trailing nature of your feedback cycle, what assurance can you offer that your relatively slowly growing modes will not be swamped or otherwise overwhelmed by some more general and much more rapidly growing global instabilities possibly admitted by the same model galaxies?

Lin: First, the transient growth rates, obtained for shearing models by you and Julian and by Lynden-Bell and Goldreich, are indeed very impressive; but they are not necessarily indicative of the growth rates for modes. This remark is based on the 1907 result of Orr for infinite flow with uniform shear. It exhibits rapid transient growth of periodic disturbances, but all investigators agree that the flow has no unstable modes. Besides these transient effects, I am not aware of any extraordinarily large growth rates for the kind of mass distribution described. Of course there may be other additional excitation mechanisms, such as that due to a stationary halo, as discussed by Mark and by Marochnik. But our primary aim is to discuss a mechanism of maintenance with a calculated pattern speed. The growth rate is only approximately obtained in our calculations.

Contopoulos: May I address a question to Dr. Toomre? Dr. Lin and his associates have found solutions that give unstable modes: Did you find any different solution that is more strongly unstable.

Toomre: Yes, too often! Of course, in challenging Lin and Mark just now, I was thinking mostly of the severely unstable spiral modes found by Kalnajs, Erickson, and Bardeen starting almost a decade ago, plus the various N-body experiments of Miller, Hohl, and others. Yet I do caution also from some recent personal experience: During the past several years, Tom Zang and I have been studying numerically the global modes of a class of stellar disks with flat rotation curves--and, incidentally, also with artificial inside "holes" or immobile central regions very similar to those favored by Lau, Lin, and Mark. What worries both of us is that these disks again exhibit strongly growing two-armed modes whenever we carve out those holes sharply enough, and even if we also lock as much as half of the remaining disk density into a rigid halo. Worse still, just as in the earlier global-mode work or the N-body experiments, some of these instabilities persist if the local stability parameter, Q, is assigned values as large as 1.5 or 2.0. Mind you, I am not suggesting that such rapid instabilities could still be present in any of the real galaxies. More likely, they are just nuisances to be avoided, or at least to be reduced greatly in severity. But I do suggest that theorists proposing any gentler and more desirable spiral modes should nowadays be asked to guarantee that those nuisances have in fact been avoided.

Lin: Let me re-emphasize that we only claim that our solutions apply to galaxies with the type of mass distribution described.

Mark: In fact we obtained comparable spiral structures and growth rates for models of the type which was studied by Zang; in some cases, there may be differences of factors of two in the growth rate.

Lin and Mark: The models involving large growth rates and large values of Q are not of interest to us. Neither are these resultant modes expected to occur with mass distributions of the type we consider; nor are they expected to lead to quasi-stationary spiral structure. For example, the N-body experiments mentioned by Dr. Toomre all have hot stellar disks because their structure is dominated by a strong bar. After all, is it not the original point of Ostriker and Peebles that these N-body experiments do not give a realistic simulation of normal spiral galaxies? We now know that the difference between these earlier N-body experiments and our results is due to the presence of a sufficiently massive spheroidal component. For the latter galaxy models, we have now shown that there are relatively slower growing modes which may be expected to lead to quasi-stationary spiral structures; we do not find any evidence for strongly growing modes.

UNSTABLE SPIRAL MODES IN THE MILKY WAY SYSTEM

Y. Y. Lau and Jon Haass
Massachusetts Institute of Technology
Cambridge, Massachusetts 02139, USA

Discrete spiral modes of density waves have been calculated for models of the Milky Way system. The calculation is based on a fluid dynamical formulation. (See Lau and Bertin, 1978.) The equilibrium assumes a rotation curve given by Schmidt in 1965. The surface density σ_o of the active disk is less than the projected value implied by the Schmidt model, because of the presence of the spheroidal component and of the finite thickness of the disk. One such model (left figure) supports a spiral mode with a pattern speed of 13.6 km/sec/kpc. The spiral pattern (right figure) is very similar to that calculated earlier by Lin and Shu (1967) who used only the short wave. The perturbation density on the contours shown equals 1/5 of the maxima. There are other unstable modes. The modes are sensitive to the assumed Q profile. The superposition of these modes opens the way to the explanation of the complicated features observed but not well explained by the single spiral pattern computed earlier.

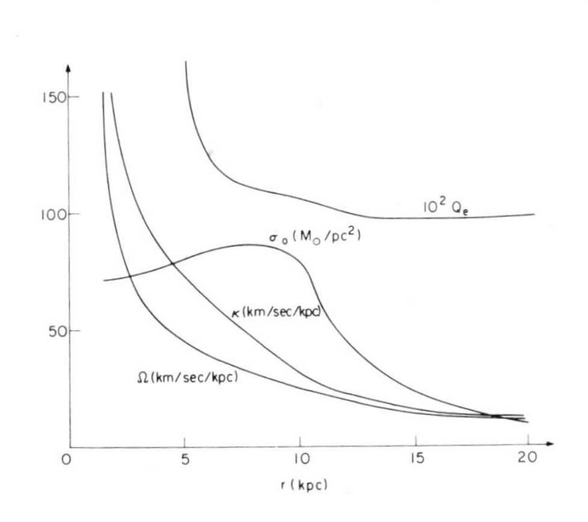

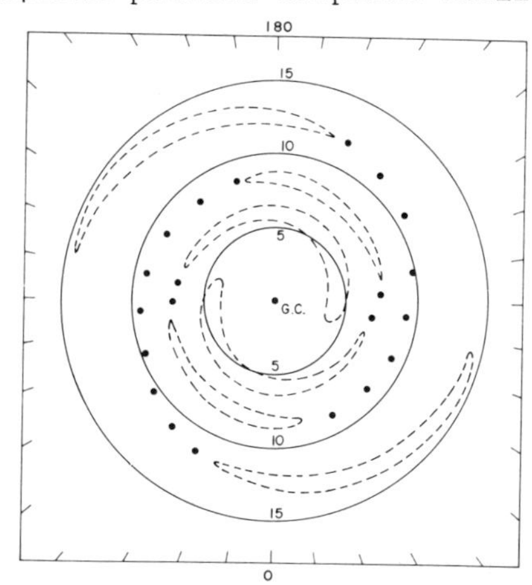

Lau, Y.Y. and Bertin, G.: 1978, Astrophys. J., in press.
Lin, C.C. and Shu, F.H.: 1967, Proc. I.A.U. Symp. No. 31, (Noordwijk), p. 313.

W. B. Burton (ed.), The Large-Scale Characteristics of the Galaxy, 154.

LIFETIME OF DENSITY WAVES AND CLASSIFICATION OF SPIRALS

J. V. Feitzinger and Th. Schmidt-Kaler
Astronomisches Institut der Ruhr-Universität Bochum

Checking the density-wave theory against observations of our own Galaxy has proven very difficult, as witnessed also at this Symposium. Less ambiguous results, however, are obtained for other galaxies. These results involve a) calculating convincing models for a sample of 25 fairly well observed spirals (Roberts et al. 1975) and b) locating the compression zones on the inner edges of the spiral arms.

We have studied the physical parameters of the density wave in more detail. Using Brandt-Belton mass models, and assuming the sound velocity $a = 10$ km s^{-1} and the stability parameter $Q = 1$, we calculated the propagation velocity v_g, the wave energy density E, and the energy loss per shock (assumed isothermal), and from these quantities the kinematic life-time τ and the dynamic life-time T of the density wave. We found $\tau \sim 3.5$ times the rotation period at corotation, therefore continuous or iterating excitation of the density wave is necessary. The galactic shock reduces this time-scale drastically by a further factor of 3. Therefore, a density wave cannot penetrate from one Lindblad resonance to the other, to be amplified, without continuous energy supply. The spiral type T_{sp} of the galaxy (after de Vaucouleurs) is related to the energy density of the spiral wave which in turn depends on the underlying mass distribution:

$$T_{sp} = 6.86 - 3.6 \log E_9$$
$$\pm 0.19 \pm 0.27$$

(correlation coefficient -0.94; rms scatter ± 0.65, mostly due to errors of T_{sp} and E_9 denoting the power 10^9 erg cm^{-2}).

The heavy element abundance (resulting from previous formation of massive stars) is also related to the total energy density of the density wave which continuously triggers stars formation. This is true for the whole sample of galaxies and within individual galaxies. The energy loss agrees well with the measured compression strength (van der Kruit 1973).

REFERENCES

Roberts, W. W., Roberts, M.S., and Shu, F.H.:1975, Astrophys.J. 196,391.
van der Kruit, P. C.: 1973, Astrn. Astrophys. 29, 263.

W. B. Burton (ed.), The Large-Scale Characteristics of the Galaxy, 155–156.

DISCUSSION

Greyber: With your theoretical considerations added to the density wave theory of spiral structure, can one explain Baade's observed dust spiral arms in M31 seen extending inward to much less than 1 kpc from the tiny M31 galactic nucleus, i.e., in a region presumably far inside the Inner Lindblad Resonance? Also, can one explain Arp's observations of the M31 spiral arms extending outward beyond 30 kpc, i.e., presumably far outside the Outer Lindblad Resonance?

Schmidt-Kaler: The mechanism described can work beyond the Lindblad Resonances. The steep onset of the dust arms in the nuclear region is just what is expected as a result of the further development of the Riemann instability.

ON A MECHANISM THAT STRUCTURES GALAXIES

D. Lynden-Bell
Institute of Astronomy, The Observatories, Cambridge CB3 OHA

By considering the interaction of a single stellar orbit with a weak cos 2ϕ potential it is shown that in the central regions of galaxies with slowly rising rotation curves, the elongations of the orbits will align along any potential valley and oscillate about it. This effect is more pronounced for elongated orbits. In such regions any pair of orbits will naturally align under their mutual gravity and so a bar will form. The gravity of this bar will drive a spiral structure in the outer parts of the galaxy where differential rotation is too strong to allow the orbits to be caught by the bar. The spiral structure carries a torque which slowly drains angular momentum from the bar, gradually making its outline more eccentric and slowing its pattern speed. In the outer parts of the bar only the more eccentric orbits align with the potential valley; the rounder ones form a ring or lens about the bar. As the pattern speed slows down, the co-rotation resonance and outer Lindblad resonance, which receive the angular momentun, move outwards. The evolution of the system is eventually slowed down by the weakness of these outer resonances where the material is rather sparse.

DISCUSSION

Miller: In a detailed study of the dynamics of a stellar bar we find that orbits in the bars of numerical experiments form figure-8's (as seen from a frame that rotates with the bar pattern), quite different from the elliptical loops you described. We find no ellipse-like orbits inside the bar region, out of a study of 1500 orbits. The elliptical orbits cannot have been important in earlier stages of the formation of the bar, because there is no way for a self-consistent system to switch over from domination by elliptical orbits to domination by figure-8 orbits. We find the bar to be slowly rotating; elliptical orbits are appropriate to rapidly rotating configurations.

Lynden-Bell: Your bars and my bars are different, although mine secularly slow down and would eventually trap orbits of your type.

W. B. Burton (ed.), The Large-Scale Characteristics of the Galaxy, 157–158.

Bok: The term "bar" is being used to denote all sorts of objects. Can we have some summary about the astrophysical and kinematical properties of observed bars in barred spirals before we refer to bar-like features obtained from numerical calculations?

No response.

Sanders: Taking the Rougoor-Oort rotation curve for the inner region of the Galaxy, the quantity $\Omega - \kappa/2$ increases to high values inside 2 or 3 kpc. Does this mean that particle orbits in the inner Galaxy are "donkeys"; that there is an anti-bar-forming tendency in the inner part of the Galaxy?

Lynden-Bell: Yes, our Galaxy is anti-barring in the middle but may be barring a little further out where the velocity curve rises again.

Contopoulos: I have made some calculations similar to Dr. Lynden-Bell's and would like to give a counterexample. The periodic orbits inside the inner Inner Lindblad Resonance (dashed circle) are elongated along the bar. So at first glance they should enhance the bar, as pointed out by Dr. Lynden-Bell. However, the orbits are more elongated near the resonance than closer to the center, where they are more circular. Thus the orbits are congested near the line PP, perpendicular to the bar. This congestion may overcome the enhancement along the bar and produce a maximum of density perpendicular to the bar. In fact, I found several models where this is the case. Thus one should be careful to take into account all the important effects that contribute to the stellar density. On the other hand, the congestion of gas should be smaller because of pressure effects. Therefore, the response along the bar probably should be larger. This is consistent with the calculations of the gaseous response made by Sanders.

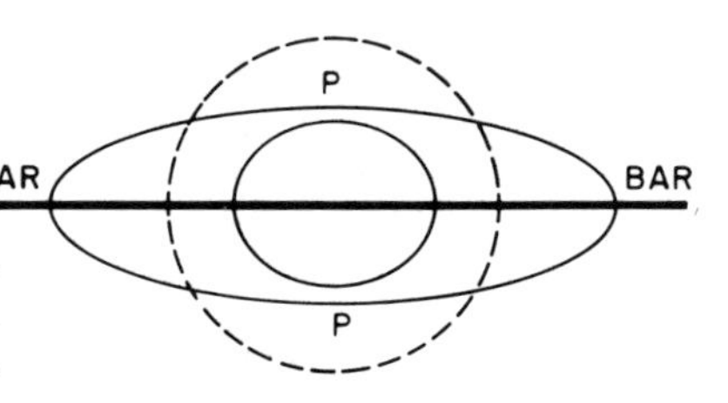

COMPRESSION OF INTERSTELLAR CLOUDS IN SPIRAL DENSITY-WAVE SHOCKS

Paul R. Woodward
Leiden Observatory, The Netherlands, and
Lawrence Livermore Laboratory, Livermore, Calif.

ABSTRACT
A mechanism of triggering star formation by galactic shocks is discussed.

The possibility that shocks may form along spiral arms in the gaseous component of a galactic disk is by now a familiar feature of spiral wave theory. It was suggested by Roberts (1969) that these shocks could trigger star formation in narrow bands forming a coherent spiral pattern over most of the disk of a galaxy. In this paper I will report some results of computer simulations of such a triggering process for star formation.

I begin the simulations with a spherical "standard" interstellar cloud in an interarm region. The cloud has a mass of 524 $M_{\odot}$, radius of 15 pc, and density of 1.5 m_H cm^{-3}. Reasons for this choice of the initial cloud are discussed in an earlier article (Woodward 1976). The results for this cloud can be scaled to different cloud masses by multiplying all lengths and times by a common factor. The cloud was followed numerically using the BBC code (cf Sutcliffe 1973, Noh and Woodward 1976) as it passed through a spiral arm shock in the surrounding isothermal gas. This intercloud gas, with a sound speed of 8.6 km/sec, increases in density from about 0.02 to 0.11 m_H cm^{-3} across the spiral arm shock.

In figure 1 the cloud is shown near the beginning of the computation as it is entering the spiral arm. The cloud is viewed and velocities are displayed in a frame of reference in which the interarm gas moves to the right at 1.5 km/sec. In this frame the intercloud gas of the spiral arm rushes into the cloud at about 15.5 km/sec. In figure 1 the spiral arm shock is delineated by the closely spaced density contours at the bottom left. A bow shock ahead of the cloud has formed because the cloud moves supersonically with respect to the intercloud gas of the spiral arm. The ram pressure from this motion causes the shock driven into the cloud to be strongest at the front. The resulting flattening of the cloud is already apparent in figure 1, about 3×10^6 years after the cloud entered

W. B. Burton (ed.), The Large-Scale Characteristics of the Galaxy, 159–164.

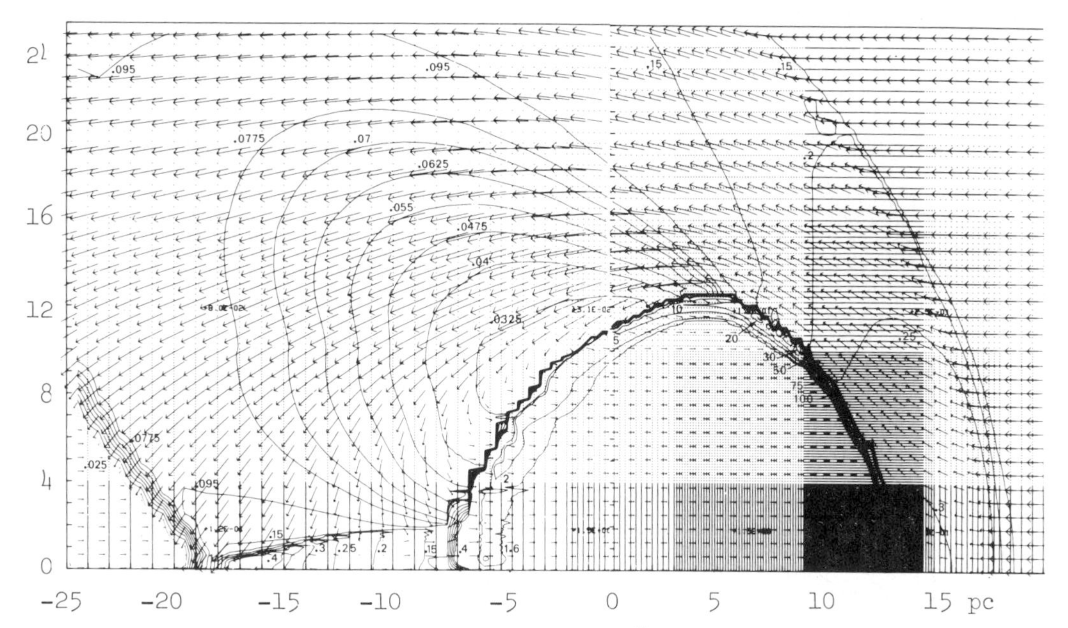

Figure 1. An interstellar cloud shown 3×10^6 yrs after encountering a spiral arm shock. See text for details of display format.

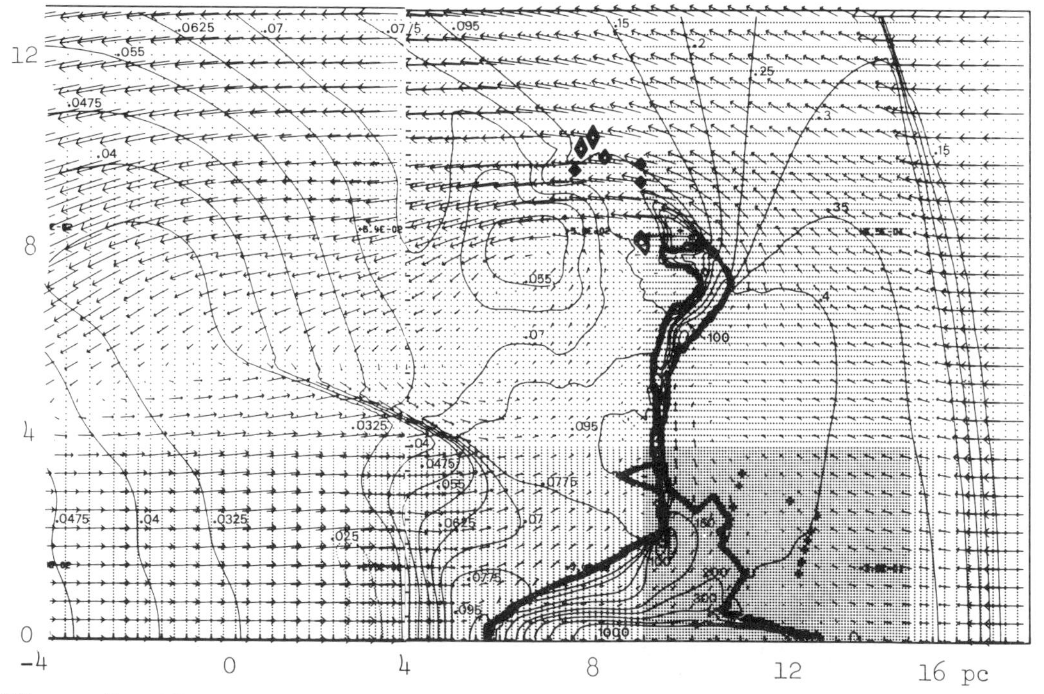

Figure 2. The interstellar cloud of figure 1 shown 7×10^6 yrs after encountering the spiral arm shock.

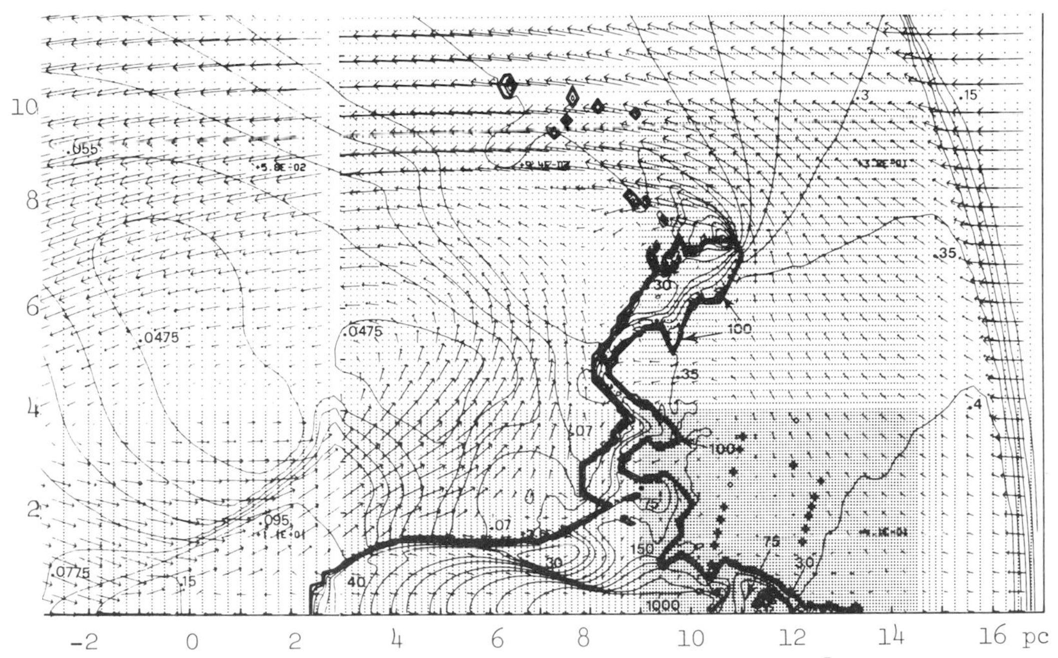

Figure 3. The interstellar cloud of figure 1 shown 8×10^6 yrs after encountering the spiral arm shock.

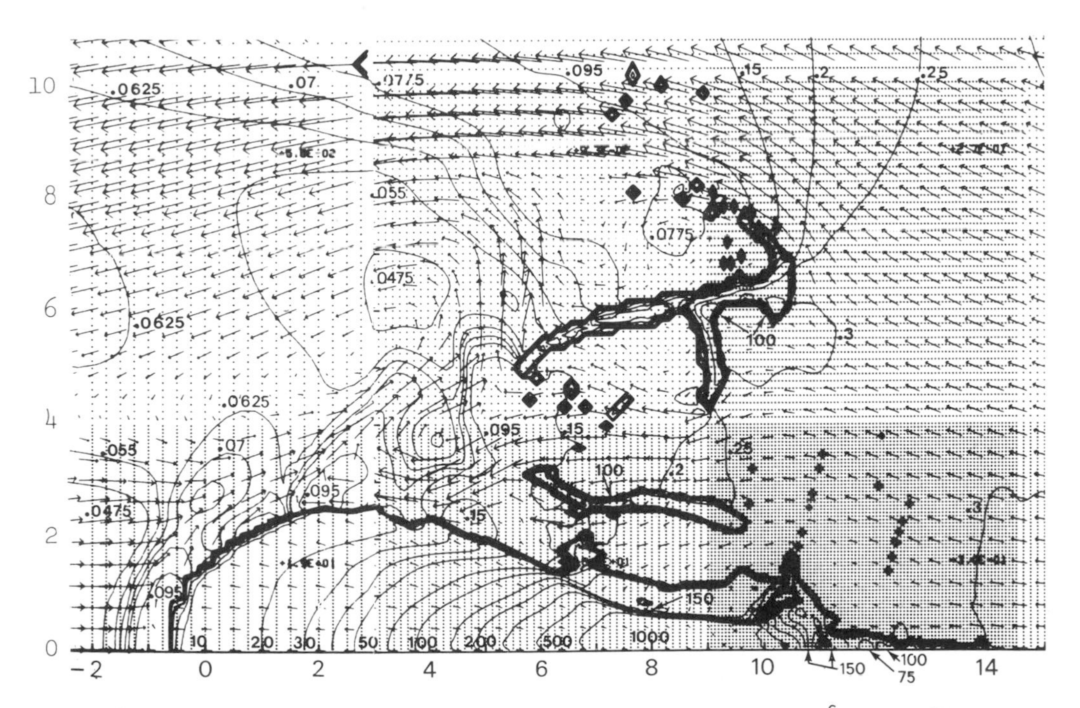

Figure 4. The interstellar cloud of figure 1 shown 9×10^6 yrs after encountering the spiral arm shock.

the spiral arm. The density contours tracing the cloud boundary in the figure give it a step-like appearance. This is due to the plotting routine; the actual data is smooth. The computational grid extends beyond the figure and contains about 34,000 zones. Zones are indicated by dots, which are so close together near the front of the cloud that the region appears completely black in figure 1.

The Kelvin-Helmholtz, or "water-wave," instability which was seen in earlier computations (Woodward 1976) is not seen in figure 1. The isothermal equation of state used here for the intercloud gas results in a supersonic flow of this gas around the cloud. This, together with the effective numerical viscosity of the boundary treatment, causes the Kelvin-Helmholtz instability to be suppressed (cf Gerwin 1968). The Rayleight-Taylor instability of the front cloud surface is apparent from the distortions of that surface in figure 2, which shows the cloud 7×10^6 years after it entered the spiral arm. Here the cloud has been entirely flattened. Focussed flow of shocked cloud gas toward the symmetry axis has given rise to a maximum density there, about 10^4 m_H cm^{-3}. Now that propagation of the shock through the cloud has communicated the high pressure associated with this dense gas on the symmetry axis to the back of the cloud, an expansion into the low-pressure intercloud region behind the cloud has begun. This causes a dense, expanding tail to form along the symmetry axis behind the densest part of the cloud. This tail is more clearly visible in figure 3, where the cloud is shown 8×10^6 years after entering the spiral arm. As the flattened sheet of cloud gas shown in figure 2 expands into the low pressure region behind it, its great acceleration causes rapid growth of the Rayleigh-Taylor instability. This gives rise to the distortion of the sheet in figure 3 and its eventual break up in figure 4, 9×10^6 years after entry into the spiral arm.

The cloud in figure 4 has several features characteristic of this sort of shock-driven implosion. It is elongated in the direction of motion relative to the surrounding intercloud gas. At the front end the density is highest, and this is the preferred location for star formation in the cloud. This model therefore leads to a picture where stars form preferentially at the edges of dense clouds, rather than at their centers. The model therefore ties in naturally with shock-driven star formation mechanisms which demand the presence of a first generation of massive stars in order to initiate a chain reaction of star formation. Thus young stars formed at the head of the cloud in figure 4 could compress the cloud further by means of strong stellar winds, expanding H II regions or supernova explosions. These mechanisms, which all involve a shock compression from one side of the cloud, could drive a wave of star formation down the length of the cloud.

The ionization-driven mechanism is illustrated by the dense clouds near the H II region W3. These are shown in figure 5. The integrated CO brightness temperature is shown within the dashed box, sampled at single-beam intervals. These observations were made with the Columbia 4-ft telescope, with a resolution of 8', by Cong and Thaddeus (1977, private communication). Longitude values are shown along the galactic

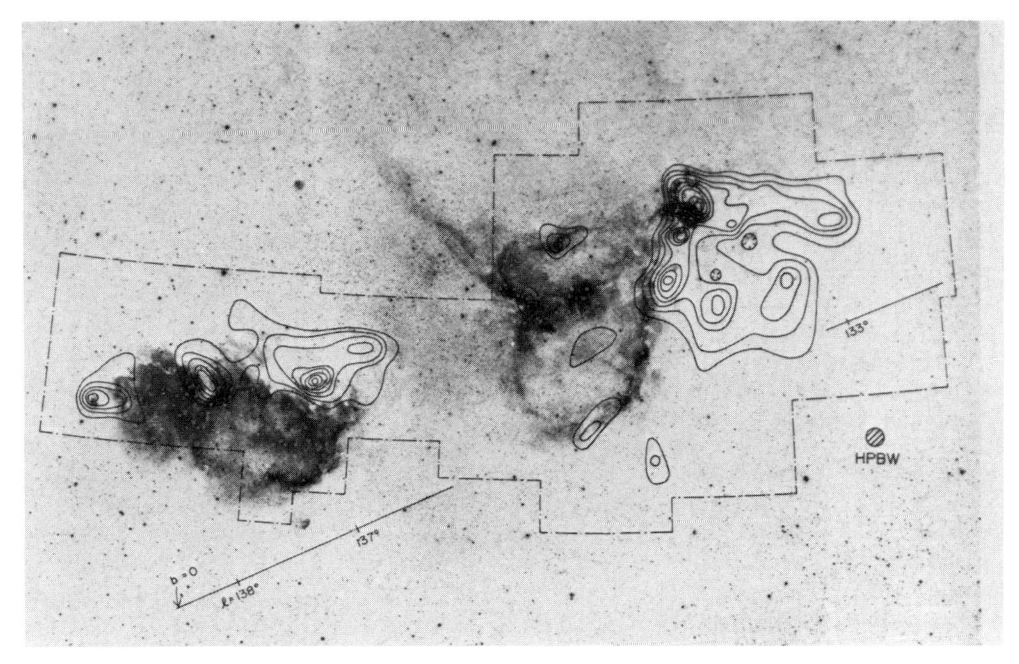

Figure 5. Dense clouds near the H II region W3. Contours of CO brightness temperature from observations of Cong and Thaddeus.

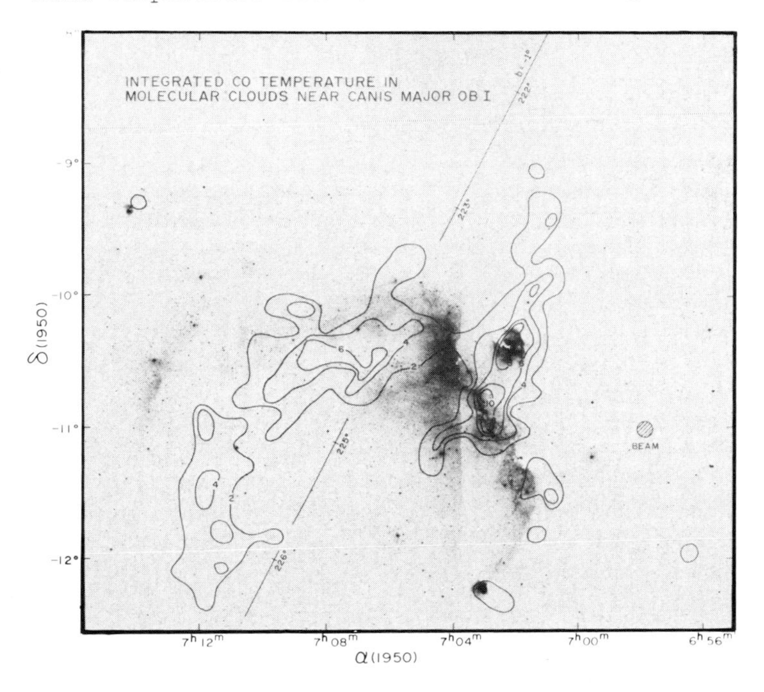

Figure 6. CO map of the Canis Major OB I region (distance 1.15 kpc) from observations of Leo Blitz.

plane in the figure. The clouds are in the Perseus spiral arm at a distance of about 2 kpc. Their orientation is consistent with the idea that the star formation at their edges was initiated by their entry into the Perseus spiral arm. In any case, it is apparent that star formation in these clouds is now being driven by the effects of the H II regions which are pressing against the edges of the clouds. A one-dimensional model of a wave of star formation in such a cloud driven by the ionizing radiation of massive stars has been presented by Elmegreen and Lada (1977). The implosion of a cloud upon entering a spiral arm provides a means of obtaining a first generation of stars at one end of the cloud, which may then cause sequential star formation of this sort to take place.

Once the first generation of massive stars is formed at one end of a cloud, sequential star formation may be driven by ionizing radiation, as mentioned above, or by supernova shocks. This latter alternative has been stressed by Herbst and Assousa (1977), who present the Canis Major OB I association as a likely illustration of supernova-induced star formation. A CO map of the region obtained by Leo Blitz (1977, private communication) is shown in figure 6. Beside the shell-like feature which appears to be a remnant of a supernova, an elongated cloud is being compressed and stars have recently formed. The orientation of this elongated cloud does not suggest an original implosion by a spiral arm shock. Nevertheless, it illustrates how a star formation chain reaction might be triggered and maintained in a cloud which is imploded upon entering a spiral arm.

The above reference to observations in connection with the calculations presented here is made with the following words of caution. CO observers estimate that the masses of clouds such as those near W3 are 10^5 $M_\odot$ or more. However, scaling up the cloud in figures 1-4 beyond a mass of about 2×10^4 $M_\odot$ would make it gravitationally unstable before entering the spiral arm. Either a new model for stable massive clouds in interarm regions or a mistake in the observational mass estimate is needed to make a completely consistent matching of the computer simulations to the observations.

REFERENCES

Elmegreen, B. G., and Lada, C. J. 1977. Ap. J. 214, pp. 725-41.
Gerwin, R. A. 1968. Rev. Mod. Phys. 40, pp. 652-8.
Herbst, W., and Assousa, G. E. 1977. Ap. J. 217, pp. 473-87.
Noh, W. F., and Woodward, P. R. 1976. Proc. Int. Conf. Numer. Methods Fluid Dyn., 5th, Enschede, Netherlands, 1976, pp. 330-40.
Roberts, W. W. 1969. Ap. J. 158, pp. 123-43.
Sutcliffe, W. G. 1973. Tech. Rpt. UCID-17013, Lawrence Livermore Lab, Livermore, Calif.
Woodward, P. R. 1976. Ap. J. 207, pp. 484-501.

THE GALACTIC DENSITY WAVE, MOLECULAR CLOUDS AND STAR FORMATION

Frank N. Bash
Astronomy Department, University of Texas
Austin, Texas 78712, USA

ABSTRACT

A model has been devised for the orbits of molecular clouds in the Galaxy. The molecular clouds are assumed to be launched from the two-armed spiral-shock wave, to orbit in the Galaxy like ballistic particles with gravitational perturbations due to the density-wave spiral-potential and each cloud is assumed to produce an identical cluster of stars. A comparison of the model with observations suggests that each cloud radiates detectable $^{12}C^{16}O$ (J = 1 → 0) spectral line radiation from birth to an age of 30 million years and that stars are seen in the cloud 15 million years after its birth. The model has been tested by comparing its predicted velocity-longitude diagram for CO against the observed one for the Galaxy and by comparing the model's predicted surface brightness in the UBV photometric bands against observed surface photometry for Sb and Sc galaxies.

1. INTRODUCTION

Work on star formation mechanisms is proceeding at a rapid rate. Herbst and Assousa (1977), Ögelman and Moran (1976) and Cameron and Truran (1977) cite evidence connecting supernovae with the formation of stars. Loren (1977) and Elmegreen and Lada (1977) cite evidence that star formation is triggered by an expanding HII region as it interacts with a dense cloud. Wielen (1973) and in several other papers has discussed questions related to density wave star formation. Bash, Green and Peters (1977) discuss evidence that the spiral density wave is causally related to star formation in the Galaxy. Seiden and Gerola (1978) have devised a stocastic model for star formation in galaxies which is capable of giving, at least, spiral arcs and possibly even 2-arm spirals of newly formed stars without invoking the density wave theory.

I believe that these various modes of star formation are not mutually exclusive. The obvious, basic 2-arm spiral symmetry seen in most Sa, Sb and Sc spiral galaxies must result from the galactic density wave, but the fraction of stars which it produces directly is unclear. It may produce

W. B. Burton (ed.), The Large-Scale Characteristics of the Galaxy, 165–172.

only a relatively small number of "primary" stars which, in turn, produce additional "secondary" stars through the supernova or expanding ionization front mechanisms. Since only massive, short-lived stars are believed to be effective in producing supernovae or large HII regions, these secondary stars will also be in a spiral pattern. The mechanism suggested by Seiden and Gerola (1978) may describe star formation in, for example, Sd galaxies; however, it seems incapable of producing the two smooth, wide spiral arms seen in the old disk stars as revealed in very red photographs of, e.g., M51 (Zwicky, 1955). These very red photographs seem to show the response of the old disk stars to the linear density wave.

2. DENSITY WAVE INDUCED STAR FORMATION

Bash and Peters (1976), Bash, Green and Peters (1977), and Bash (1978) have attempted to find observational evidence which connects the spiral density wave, molecular clouds, and star formation in the Galaxy. We have assumed that a spiral density wave exists in the Galaxy, that its current position, gravitational potential and pattern speed are the values fitted to HI 21-cm observations and to the positions and velocities of a group of 25 stars by Yuan (1969a, 1969b). We have also used the values of the gas velocities in the two-armed spiral-shock (TASS) wave given by Shu, et al. (1972) and Shu, Milione and Roberts (1973). We have assumed that molecular clouds are launched from the TASS wave with the predicted shock-wave velocities, and that they ballistically orbit in the Galaxy as perturbed by the spiral-arm potential. By integrating their orbits we can predict their positions and velocities as a function of time since they left the TASS wave. This time, called their dynamical age, is the independent variable in the integration of their orbits.

Bash and Peters (1976) concluded that the model ballistic-particle molecular-clouds whose dynamical ages are no more than thirty million years have the same values of predicted radial velocities as those observed for galactic CO. We suggested that some process must cause the CO-emitting molecular clouds to be no longer observable at ages greater than thirty million years, since the model predicts that older clouds have radial velocities which exceed any observed velocities.

Bash, Green and Peters (1977) tested the above result by looking for CO-emitting molecular clouds associated with young clusters of stars. We found that 90% of young clusters containing O-stars have associated CO emission, while less than 10% of the clusters whose earliest star lies between B0 and B4 have associated CO-emission. This result caused us to assume that molecular clouds launched from the TASS wave form star clusters or associations and that the thirty million year CO "cut-off" corresponds to the time in the cluster's life when the last O-star completes its evolution. Wheeler and Bash (1977) have suggested that perhaps a B0 star is the most massive star capable of becoming a supernova and that the first supernova explosion rids the clusters of its associated CO. Since we estimate that the main sequence lifetime of an B0 star is 15 million years, we concluded that about 15 million years elapse from the time a molecular cloud leaves the TASS wave until the stars begin to form. That star formation time is very close to the values inferred from observations of M51 (Mathewson, van der Kruit and Brouw, 1972) and of M81 (Rots and Shane, 1975 and Rots 1975) and computed by Woodward (1976).

3. STAR FORMATION AND UBV SURFACE BRIGHTNESS

Bash (1978) also computes the UBV surface brightness predicted by the model and compares it with the observations of Schweizer (1976). We assumed that each model-ballistic-particle is a molecular cloud which becomes an open star cluster 15 million years after leaving the TASS wave. For simplicity, this was assumed to be the only mechanism for star formation. The star cluster is assumed to continue in ballistic orbit around the Galaxy and, as it ages, the stars evolve. Each cluster is initially bright and blue due to the light of the early-type stars but, as the cluster ages, its light becomes dimmer and redder as the massive stars die. The distribution of stellar masses in each cluster was taken from observations (Taff, 1974 and Scalo, 1978) and the total mass in each cluster was adjusted to agree with the observed average surface brightness of M81. The orbit of each molecular cloud→star cluster was integrated for 100 million years. To represent the light of the Galactic disk, a smooth stellar disk with a radial brightness gradient was added and the disk colors, brightness and radial brightness gradient were taken directly from observations of external galaxies.

The model resembles the observations quite well except that the model's (U-B) and (B-V) colors were too blue by about $0^{m}.5$. However, it was pointed out that the evolution of the model cluster stars had been simplified by considering only their main-sequence phases and that was at least a part of the cause for the color disagreement.

4. RECENT WORK

We shall now report some new work completed after Bash (1978). The model, described above, has been improved by, a) including the giant and supergiant phases in the evolution of the model cluster stars and b) adding a linear density wave to the underlying disk. The computed and observed (U-B) and (B-V) colors now agree.

Bash (1978) describes a "cluster model 1". We now wish to allow the stars in cluster model 1 to evolve through the giant and supergiant phases. Only stars more massive than 3 $M_\odot$ leave the main-sequence in 100 million years which is the limit for our integration of the cluster orbits. Evolutionary tracks for stars of mass 15 $M_\odot$, 9 $M_\odot$ and 5 $M_\odot$ were taken from Iben (1967). Tracks for stars of mass 63 $M_\odot$, 40 $M_\odot$ and 25 $M_\odot$ come from Stothers (1963, 1964, 1965, 1966a, 1966b, 1968). Bolometric corrections were obtained from Panagia (1973), Morton and Adams (1972) and Johnson (1966), giving values of Mv. Spectral types were obtained from the same references as used for the bolometric corrections and (U-B) and (B-V) colors were obtained from Davis (1977) for stars earlier than G0 and Johnson (1966) for stars later than G0. The total mass of the stars in each model cluster was adjusted so that the average surface brightness of the model galaxy agreed with the values observed by Schweizer (1976) at distances 4.79 kpc and 6.67 kpc from the center of M81. The cluster mass which gives best agreement for cluster model 1 is 405 $M_{\odot}$/cluster. The absolute magnitude, Mv, and the (U-B) and (B-V) colors for cluster model 1 and for 100 million years after the stars turn on are shown in Figure 1.

Schweizer (1976) reports surface photometry measurements on six Sb and Sc galaxies. The surface photometry is measured in three color bands U, B_3 and O and, according to Schweizer, $B_3 = B - 0.3\ (B-V)$. His surface photometry is displayed for annuli, centered on the center of the galaxy, and with a variety of radii. The surface brightness around each annulus is displayed as a function of ϕ, the galactocentric azimuth, which increases in the direction of rotation. He defines the disk component as the surface brightness of a level line passing through the two dimmest points, separated by at least 90° in ϕ. The spiral arm component is defined as the average amount of light above the disk.

For the disk of our model galaxy we have adopted the surface brightness, colors, and radial gradient in brightness found by Schweizer (1976) for M81. To the disk we have added a linear density wave of the form

$$D\ (\text{magnitudes}) = -A\ (1 + \cos\ [-2\phi + \Phi(R)\ ]\ ).$$

At the spiral arm minima, D = 0 and we see the smooth disk. At the spiral arm maxima, the disk brightness increases to the smooth disk value minus 2A magnitudes. The disk color is assumed to be everywhere the same, only its brightness changes. Cluster model 1 with its evolved stars (as described above) plus the value $A = 0^m\!.2$, allows us to fit Schweizer's observations of the surface brightness and color of M81.

Figure 2 shows the results of the model calculations. The data are displayed in the same way as Schweizer (1976) displays his observational data. We have chosen to compute the surface brightness for three annuli at R = 4.79, 6.67 and 8.24 kpc from the center of the model of the Galaxy. The ordinate is absolute magnitude per square kiloparsec, M kpc^{-2}. (M kpc^{-2} + 36.57 = magnitude per square arcsecond.) The shapes of the surface brightness cuts across the arms, and the decrease in the apparent "noise" from U to V all resemble observed data in Schweizer's (1976) Figure 5c for M81. The average surface brightness in the model annuli computed in the U and B_3 filters are within 0.2 magnitudes/square arcsec. of Schweizer's observed values. The average peak height of the computed profiles above the disk brightness lie within 0.3 magnitudes/square arcsec. of Schweizer's observed values in the U and B_3 filters. The average width of the spiral arms, using Schweizer's measure, $\Delta\phi_{1/2}$, is 27°, very close to the average value he measured for M81. The arm width of the model arms does not drop rapidly to very small values at large radii, as Schweizer observes, but unlike the predictions of the models which he quotes.

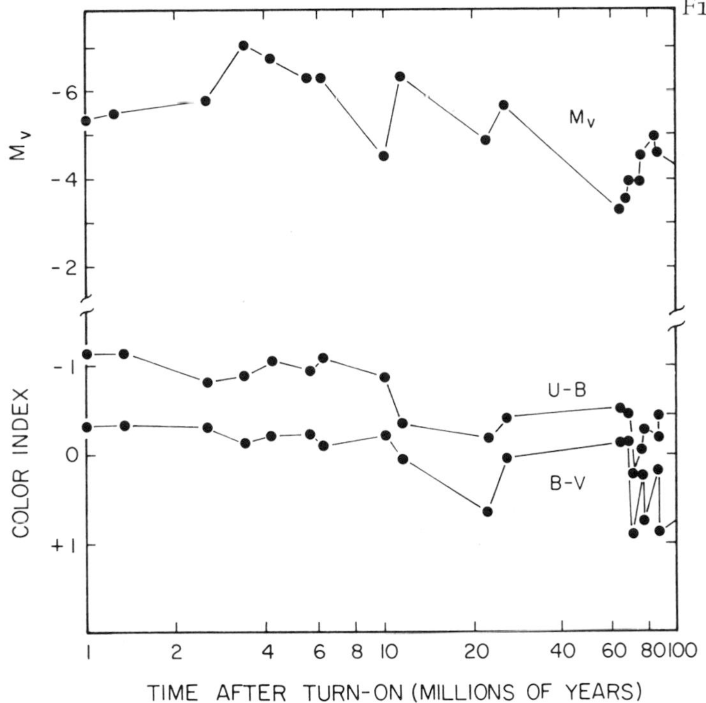

Figure 1. The absolute magnitude, Mv, and (U-B) and (B-V) colors of one model 1 cluster as a function of time after the stars turn on. Cluster model 1 contains 405 $M_\odot$ of stars.

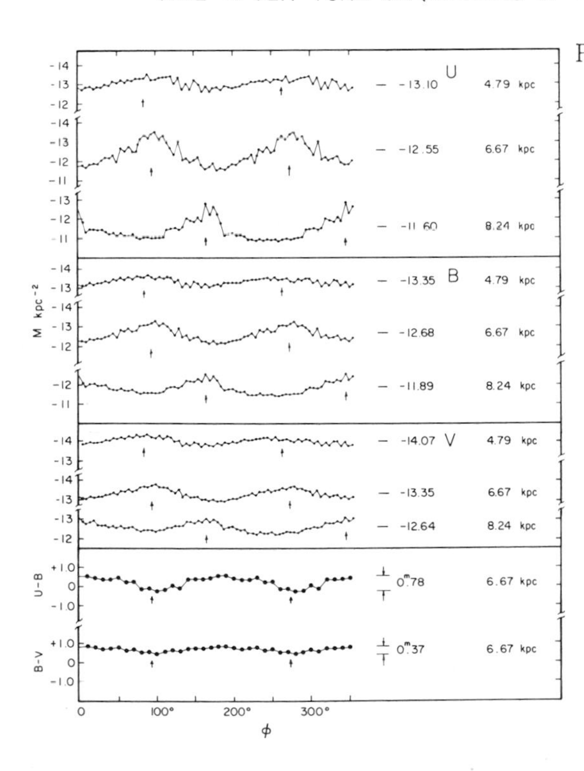

Figure 2. The UBV surface brightness of the model of the Galaxy in annuli of radius 4.79, 6.67 and 8.24 kpc from the center. The galactocentric azimuth, ϕ, increases from the Sun-galactic center line in the direction of galactic rotation. The average surface brightness around each annulus is shown just to the right of each trace and the colors around the R = 6.67 kpc annulus are shown at the bottom. The arrows mark the places where the annuli cross the centers of the density-wave arms.

The older model, reported by Bash (1978), gave similar agreement with Schweizer's (1976) observations. However, the colors of the spiral arms in our older model were bluer by about $0\overset{m}{.}5$ than Schweizer's observed ones. The average colors of the spiral arms in the current model are (B-V) = +0.53 and (U-B) = -0.14. These colors are very similar to values Schweizer has observed for the galaxies in his sample. He finds that the average colors of the spiral arms in M81 are (U-B) = -0.17, (B-V) = +0.74. However, the average (B-V) color of the spiral arms in M99, M51 and M101 is +0.50. This color agreement has been achieved by including giant and supergiant phases in the evolution of the model cluster stars and by assuming that the disk stars exhibit a linear density wave of amplitude 0.2 ≈20% of the disk brightness.

The stellar birthrate averaged over the region where the model applies, 4 kpc < R < 11.4 kpc, required to fit the observations of M81 is 0.08 $M_{\odot}$ yr^{-1}. The birthrate is the product of the mass of each model cluster, the number of model birthsites and the number of clusters born at each per year.

5. SUMMARY

We have attempted to find observational evidence and a model which connects the density wave theory, molecular clouds and star formation. The model, which is constructed for the Galaxy, uses the density wave parameters fit to HI observations and stellar orbits and by assuming that molecular clouds are launched from the TASS wave and emit observable CO spectral lines for 30 million years, it is largely consistent with CO survey results. The model predicts that the CO radiation abruptly cuts off at 30 million years and such a cut-off is seen by observing CO associated with young star clusters. The model predicts the UBV surface brightness of the Galaxy but the stellar birthrate must be adjusted. It is fit to the observed average surface brightness of external spiral galaxies. The model can be adjusted to agree with observations of M81 and other galaxies; however, caution needs to be exercised in applying the stellar birthrate which this fit implies to our Galaxy since, for example, according to Rots (1975), M81 is well fit with a spiral of pitch angle 15° while the density wave fit to the Galaxy has a pitch angle of $6\overset{\circ}{.}65$. However, the model gives spiral arms whose brightness, width and color agree well with values observed for Sb and Sc galaxies.

REFERENCES

Bash, F. N. and Peters, W. L.: 1976, Astrophys. J., 205, pp. 786-797.
Bash, F. N., Green, E., and Peters, W. L.: 1977, Astrophys. J., 217, pp. 464-471.
Bash, F. N.: 1978, Astrophys. J., in press.
Cameron, A. G. W. and Truran, J. W.: 1977, Icarus, 30, pp. 447-461.
Davis, R. J.: 1977, Astrophys. J., 213, pp. 105-110.
Elmegreen, B. G. and Lada, C. J.: 1977, Astrophys. J., 214, pp. 725-741.

Gerola, H. and Seiden, P. E.: 1978, Astrophys. J., in press.
Herbst, W. and Assousa, G. F.: 1977, Astrophys. J., 217, pp. 473-487.
Iben, I.: 1967, Ann. Rev. Astr. Astrophys., 5, pp. 571-626.
Johnson, H. L.: 1966, Ann. Rev. Astron. and Astrophys., 4, pp. 193-206.
Loren, R. B.: 1977, Astrophys. J., 218, pp. 716-735.
Mathewson, D. S., van der Kruit, P. C., and Brouw, W. N.: 1972, Astron. and Astrophys., 17, pp. 468-486.
Morton, D. C. and Adams, T. F.: 1972, Astrophys. J., 151, pp. 611-622.
Ögelman, H. B. and Moran, S. P.: 1976, Astrophys. J., 209, pp. 124-129.
Panagia, N.: 1973, Astron. J., 78, pp. 929-934.
Rots, A. H., and Shane, W. W.: 1975, Astron. and Astrophys., 45, pp. 25-42.
Rots, A. H.: 1975, Astron. and Astrophys., 45, pp. 43-55.
Scalo, J. M.: 1978, to appear in Protostars and Planets, ed. T. Gehrels (Tucson: University of Arizona Press).
Schweizer, F.: 1976, Astrophys. J. Suppl., 31, pp. 313-332.
Shu, F. H., Milione, V., Gebel, W., Yuan, C., Goldsmith, D. W. and Roberts, W. W.: 1972, Astrophys. J., 173, pp. 557-592.
Shu, F. H., Milione, V., and Roberts, W. W.: 1973, Astrophys. J., 183, pp. 819-842.
Stothers, R.: 1963, Astrophys. J., 138, pp. 1074-1084.
__________: 1964, Astrophys. J., 140, pp. 510-523.
__________: 1965, Astrophys. J., 141, pp. 671-687.
__________: 1966a, Astrophys. J., 143, pp. 91-110.
__________: 1966b, Astrophys. J., 144, pp. 959-967.
Stothers, R. and Chin, C.: 1968, Astrophys. J., 152, pp. 225-232.
Taff, L. G.: 1974, Astron. J., 79, pp. 1280-1286.
Wheeler, J. C. and Bash, F. N.: 1977, Nature, 268, pp. 706-706.
Wielen, R.: 1973, Astr. and AStrophys., 25, pp. 285-297.
Woodward, P. R.: 1976, Astrophys. J., 207, pp. 484-501.
Yuan, C.: 1969a, Astrophys. J., 158, pp. 871-888.
______: 1969b, Astrophys. J., 158, pp. 889-898.
Zwicky, F.: 1955, Pub. Astr. Soc. Pacific, 67, pp. 232-236.

DISCUSSION

Solomon: The "CO emitting lifetime" of 3×10^7 years which you give suggests that during most of a galactic rotation of 2×10^8 years the gas is in some form other than molecules. However, the CO surveys of the Galaxy clearly show that 70%-90% of the gas is in molecules, particularly at 4-7 kpc where you are matching the observations. Where, and in what form, is the gas during the other 2×10^8 years in your model? The connection between CO and HII regions in your model is a result of matching the CO ℓ,v diagram. This diagram maps temperature; hot molecular clouds are correlated with the presence of HII regions, as has been known for the past seven years. However, most of the mass in molecular clouds is not in hot regions, and most molecular clouds do not have strong heating sources. Therefore, your result appears to be a confirmation of the correlation of hot molecular clouds with HII regions, not a correlation of all molecular clouds with such regions.

Bash: Our model assumes that CO-emitting molecular clouds are seen for only $3x10^7$ years after they leave the spiral shock wave. They may just mean that such clouds are only observable there (say due to higher CO temperatures) and not that CO clouds only exist there. However, because the inferred presence of a large population of very long-lived CO molecular clouds from your work depends on assumed C/H ratios, the number of such clouds must be uncertain.

Kaufman: First, a question: How do you account for the excitation of hot CO during the $15x10^6$ years before the stars turn on?

Second, a comment: The 30-million-year width of the observed hot CO distribution is comparable to the 1-kpc apparent width of the HII region spiral arms of Georgelin and Georgelin. So it seems that you and the Georgelins may be detecting the same type of galactic features. By restricting your analysis to star formation by spiral shocks, you must work very hard to make your theoretical arms as wide as the observed hot CO and HII spiral arms. This is especially so in view of the fact that Carson's opacities and stellar winds suggest that massive stars evolve towards the red even faster than the conventional O-star lifetimes. However, one can easily account for the width of the HII arms if spiral shocks are not the only star-forming mechanism in our Galaxy. Because there are a number of observations linking star formation to expanding HII regions and expanding SNR's, a model based solely on star formation in galactic shocks is incomplete. The observed spiral features are probably a composite of, first, stars formed by the spiral shock and, then, stars formed by shock waves from high-mass stars. The massive stars formed by the spiral shock act as a trigger for further star formation.

Bash: First, I believe that only some, perhaps a minority, of all stars are directly formed by the action of the spiral-shock wave. However, our model, which assumes (for convenience) that all stars form from the spiral-shock wave is capable of producing spiral arms as wide as the observed ones.

Second, I imagine that the CO is excited during the $15x10^6$ years before the stars turn on basically by the collapse of the cloud.

Wielen: I understand that you favor the post-shock version of initial velocities. This is in contrast to our results that the pre-shock velocities give a better description of the observational results, especially in M51. How did you obtain the initial velocities, e.g., in M81?

Bash: The model discussed here is for our Galaxy and the predicted surface brightness was then compared to Schweizer's observations of M81. Post-shock velocities are required to give a model ℓ,v diagram for CO which agrees with the observed one.

OBSERVATIONS OF CO IN THE PERSEUS ARM

C. Yuan
City College of New York

R. L. Dickman
Aerospace Corporation, Los Angeles

Observations of the 2.6-mm line of $C^{12}O^{16}$ were made in the galactic plane from l=105° to 150° at 1° intervals. The total velocity coverage is 83.2 km s^{-1} centered at -40 km s^{-1} with resolution equal to 0.65 km s^{-1}. The observations were made first in June 1976 and confirmed in October 1977 after some improvement of the observing facilities at the Aerospace Corporation. The integration time in each direction is 15 minutes, giving 30-noise dispersions of 0.9K. Among the 46 directions, there are 18 CO detections. Twelve of them have CO components moving with the velocities typical of those expected for the Perseus Arm. Most of them are situated in or near the O associations or HII regions in the Perseus Arm, while the others lie in the regions rich in young star clusters. All the velocities observed are in excellent agreement with those expected in the post-shock region in the context of the shock formation model of the spiral density-wave theory. These molecular clouds may be viewed as being formed in the compression phase of the cloud medium as the individual clouds enter the shock front of the intercloud medium. Their line-of-sight velocities are in general a few km s^{-1} more negative than those of the HII regions, the young star clusters, and the O associations in the Perseus Arm. This suggests that these young objects may be formed prior to the observed CO clouds. The CO velocities are in agreement with those of the HI ridge. One CO cloud is observed at l=148°, although it is well known that all the tracers of the spiral arm come to an abrupt stop at l=140°.

W. B. Burton (ed.), The Large-Scale Characteristics of the Galaxy, 173.

GASEOUS RESPONSE TO BAR-LIKE DISTORTIONS

William W. Roberts, Jr.
University of Virginia

In this review the large-scale dynamics of the gas in a model disk galaxy which has an oval, bar-like distortion (or bar structure) in the inner parts is reviewed from the standpoint of recent gas dynamical studies, both steady state and time evolutionary.

GALACTIC SHOCKS AND LARGE NONCIRCULAR MOTIONS

In the early 1960s Prendergast already suspected that the narrow dark dust lanes observed along the bar structure in barred spirals, such as NGC 1300 shown in Figure 1, may be related to galactic shocks formed within the gas (Prendergast, 1962). This suspicion together with the early work of Lin and Shu (1964, 1966) on the density wave picture of spiral structure motivated the steady state gas dynamical studies of the late 1960s (Fujimoto, 1968; Roberts, 1969; Roberts and Yuan, 1970) in which the gaseous response to a spiral perturbation gravitational field underlying a spiral density wave pattern was found to be strong and capable of inducing the formation of large-scale shock waves along the spiral arms. These studies identified the dark dust lanes along spiral arms as possible tracers of shocks and the galactic shock itself as a possible triggering mechanism for the formation of young stellar associations and H II regions "strung out like pearls along the arms" (Baade, 1963).

Figure 1. Photograph of the barred spiral galaxy NGC 1300 (The Hale Observatories).

W. B. Burton (ed.), The Large-Scale Characteristics of the Galaxy, 175–186.

The steady state response of the gas to a mild bar-like distortion rotating at a small angular speed Ω_p was considered by Roberts (1971). Figure 2 shows the equipotential contours of the nonaxisymmetric gravitational field. The remarkably strong gaseous response driven by this weakly perturbed field is shown in Figure 3. Galactic shocks form in lanes slightly offset along the major axis of the bar-like distortion. These offset shocks together with the highly oval gas streamlines, large noncircular motions, and strong radial gas streaming along the bar, all driven in the presence of only weak forcing, provided a nice preview as early as 1971 of what was to come in later work.

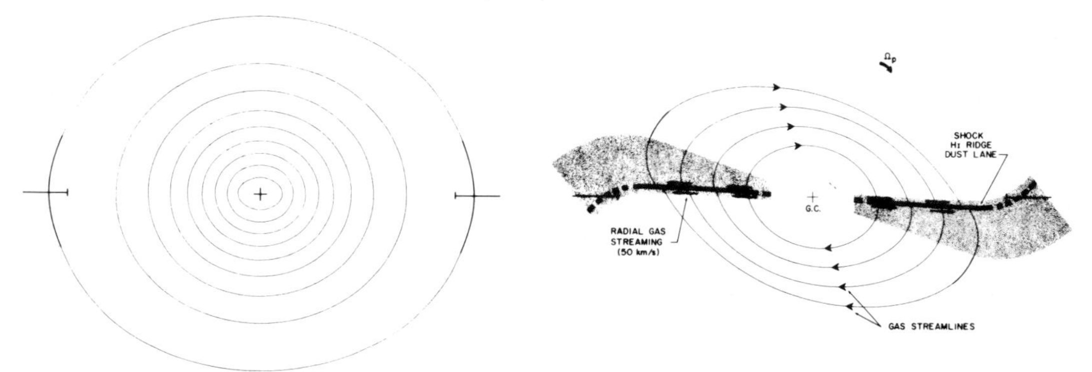

Figure 2. Equipotential contours in a model disk galaxy with a mild (10%) bar-like distortion (Roberts, 1971).

Figure 3. Oval steady state gas circulation driven by the gravitational field of the bar-like distortion in the model disk galaxy (Fig. 2). Galactic shocks and large noncircular motions characterize the flow (Roberts, 1971).

THE "3 KPC ARM" FEATURE

Evidence for the existence of a weak bar structure in the inner parts of our Galaxy has been discussed by de Vaucouleurs (1964, 1970) and by Kerr (1967). Figure 4 shows the observed 21-cm velocity longitude map as derived by Peters (1975) from Kerr's (1969) 21-cm survey, galactic equator scan. The dark curve labelled (1) traces the "3 kpc arm" feature. The first inkling that the "3 kpc arm" feature could result from the gaseous response to a bar-like distortion comes from Peters (Peters and Roberts, 1972) who takes the dynamical model of Roberts (1971, see Figure 3) without any changes and views it at various orientations with respect to the galactocentric line. The 21-cm line profiles which would be observed in our Galaxy from neutral hydrogen having the velocity and density distributions predicted in the model are determined over the longitude range: -20° to 20°. The results give a "3 kpc arm" feature for an orientation of the oval streamlines at about 135° measured counterclockwise from the galactocentric line. What is remarkable: the comparison with the observed "3 kpc arm" feature shows surprisingly good agreement in view of the fact that the orientation is the only parameter adjusted. What is striking: this "3 kpc arm" feature is produced not from gas of high density, as in a real spiral arm, but rather from less dense gas integrated over large distances along the line of sight over the oval annular region along and between streamlines. Figure 5 shows the predicted 21-cm velocity longitude map derived in the subsequent work of Peters (1975) in which he relaxes his dependence on the dynamical model in an attempt to account for additional features, to the extent of kinematically modelling with smoothed ellipses and allowing freedom in the choice of two parameters - the axial ratio and the mean angular momentum/unit mass - as functions of radius. Other attempts to account for the "3 kpc arm" feature are made by Simonson and Mader (1972, 1973) with an elliptical dispersion ring model and by Sanders and Prendergast (1974) with a nuclear explosion model.

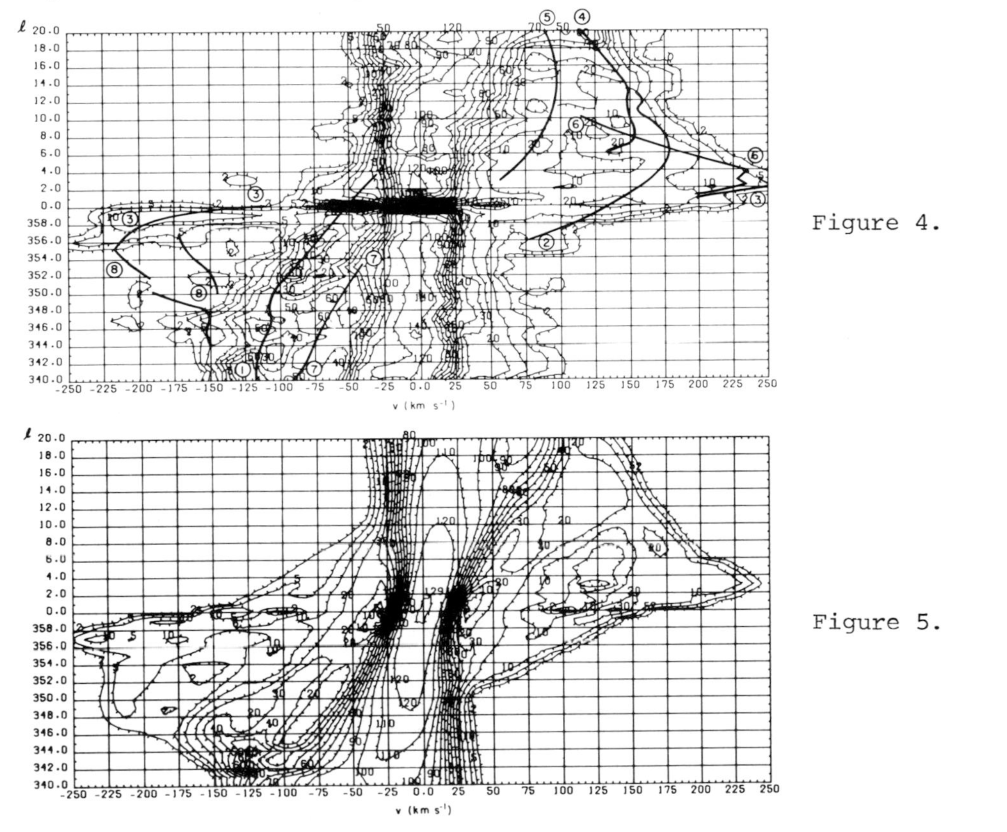

Figure 4.

Figure 5.

Figure 4. Observed 21-cm velocity longitude map derived from Kerr's (1969) 21-cm survey, galactic equator scan. The solid curves, also from Westerhout's (1969) survey. Curve ① - "3 kpc arm" feature (Peters, 1975).
Figure 5. Predicted 21-cm velocity longitude map derived from a kinematic elliptical flow model not greatly different from the dynamical, oval circulation model driven by a bar-like distortion (Figure 3). (Peters, 1975)

STEADY STATE GAS FLOW

Roberts and Huntley (1978; also see Roberts, Huntley, and Lin, 1977) generalize the steady state gas dynamical studies of the late 1960s for tightly-wound normal spirals to include normal spirals with open spiral arms and barred spirals with bar structures in the inner parts. The response of the gas to a perturbing potential that is bar-like in the interior and spiral-like in the exterior is calculated by means of an analysis which enables the two-dimensional flow to be broken up into two physical regimes, illustrated in Figure 6. In regime I near and within the bar (and spiral arms), the flow is determined through an asymptotic approximation that neglects the small variation of the velocity, density, and pressure along a shock with respect to that variation normal to the shock. In regime II where the flow is highly supersonic, the flow is determined through an asymptotic approximation that neglects secondary terms proportional to the square of the dispersion speed, such as the transverse gradient of pressure. The composite picture for the steady state flow of gas is constructed by joining the two regimes of flow in the transition layer between regimes.

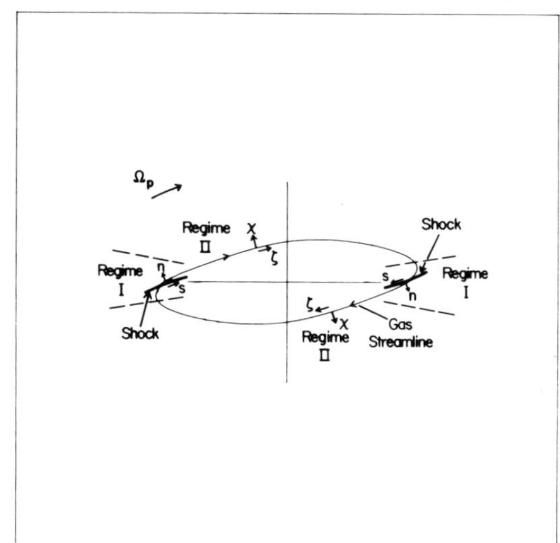

Figure 6. Schematic of the steady state flow along a streamline, and the locations of regimes I and II (Roberts and Huntley, 1978).

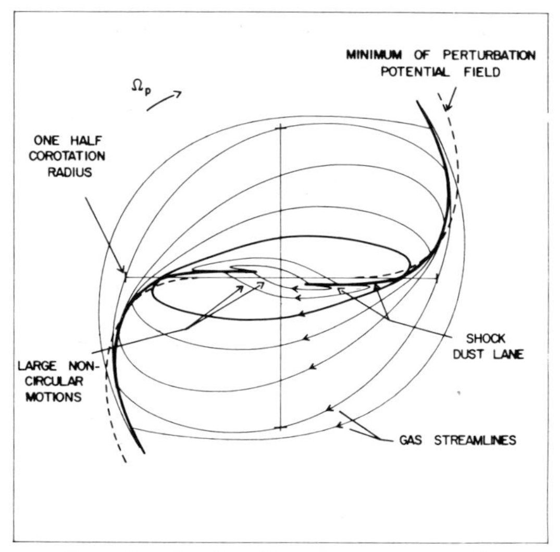

Figure 7. Oval steady state gas circulation in a Toomre model with a spheroidal component, driven by a mild bar-like distortion. In the inner parts the response shows offset shocks, large noncircular motions (50-100 km/s), and radial gas streaming (Roberts and Huntley, 1978).

This composite picture is shown in Figure 7 for the case of a thin disk Toomre (1963) model with a spheroidal component superposed in the inner parts and a perturbing potential representing a 6% perturbation. Galactic shocks are present, and highly oval gas streamlines characterize the circulation in the inner parts where large noncircular motions occur. Strong velocity gradients are present across the major axis of the gas bar near the shock. Because gas, in passage through an oblique shock, must leave the shock at a more oblique angle than that angle at which it entered, postshock gas in the bar region in the inner parts is directed outwards whereas postshock gas in the spiral arm region in the outer parts is directed inwards. This shock-focusing phenomenon, which focuses gas in the inner parts outwards and gas in the outer parts inwards, leads to enhanced concentrations of compressed gas in the region of convergence where the spiral arm bends from the bar and may account in part for the enhanced star formation activity observed at the ends of the bar structure in many barred spirals.

Figure 8 shows why shocks form. The cusp-transition curve, for any one of the four models shown, represents the critical level of forcing, at any radius, for which the gas flow undergoes a transition from flow without shocks to flow with shocks. The transition curve delineated with solid round dots (—●—●—) represents the case of the Toomre model with spheroidal component shown in Figure 7. The strong gaseous response and the formation of shocks in this model result from the moderately-strong (6%) forcing amplitude, here indicated by the horizontal dashed line at A = 1.2, at a level above the transition curve.

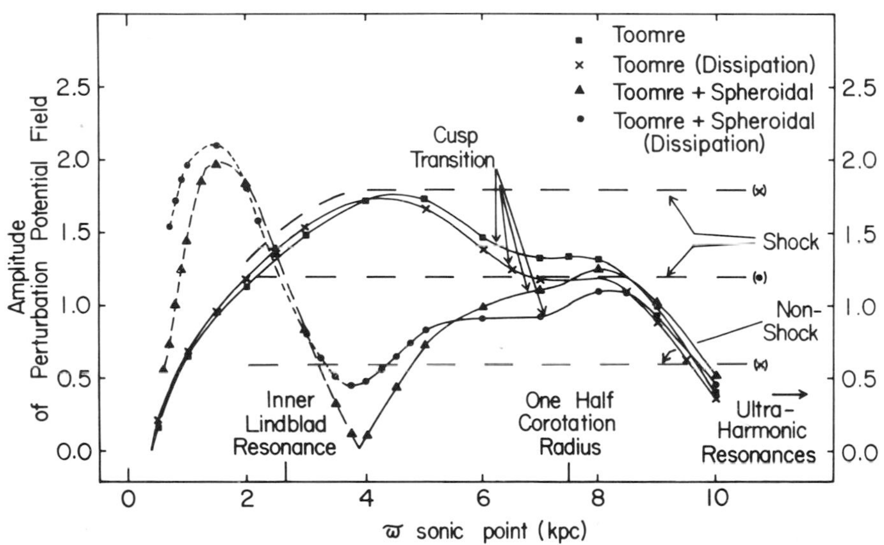

Figure 8. Amplitude of the perturbation potential field of the bar-like distortion which, at each radius, characterizes the transition from steady state flow without shocks to steady state flow with shocks. A = 2.0 represents a 10% perturbation (at $\varpi = 5$ kpc). (Roberts and Huntley, 1978)

TIME EVOLUTIONARY GAS FLOW

Time dependent, two-dimensional, numerical hydrodynamical calculations are carried out by Sanders and Huntley (1976), Sorensen, Matsuda, and Fujimoto (1976), Huntley, Sanders, and Roberts (1978), and Liebovitch (1978) for the time evolutionary response of gas to bar-like perturbations. Sanders and Huntley for a power law disk model with a mild oval distortion and Sorensen, Matsuda, and Fujimoto for an Eggen, Lynden-Bell, and Sandage disk model with a strong prolate spheroidal distortion both found in 1976 the result that the time evolutionary gas response shows open trailing

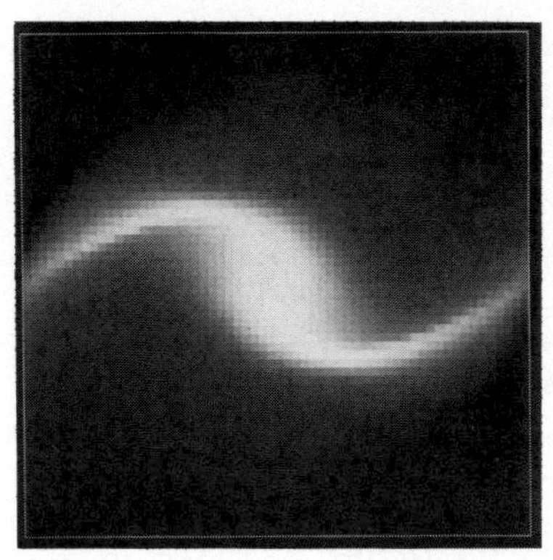

Figure 9. Photographic simulation (a snapshot) of the time evolutionary gas response in a power law disk model to a mild bar-like distortion for two cases: left - inner and outer Lindblad resonance and corotation are contained within the gaseous disk; right - only inner Lindblad resonance is contained within the gaseous disk (Sanders and Huntley, 1976, as re-calculated in the inertial frame by Huntley, 1977).

spiral density waves rotating with the angular speed of the distortion. Figure 9 from Sanders and Huntley shows a photographic simulation of the gas density distribution for the power law disk model as a snapshot in time for two such cases. The shape of the spirals can be interpreted through the coherent shifting in orientation of periodic particle orbits between and across the resonances and corotation.

Figure 10 from Huntley, Sanders, and Roberts for a Toomre model with a mild bar-like distortion shows a gas response in the form of a prominent central gas bar, with open trailing gaseous spiral density waves. Sanders (1977) finds that the response of an ensemble of non-interacting test particles to a bar-like distortion is bar-like, without appreciable spiral arm formation; whereas when he repeats the calculation in a way in which the test particles take on the character of a gas with an artificial viscosity, he finds the response to be bar-like inside inner Lindblad resonance, with two trailing spiral density waves outside. This result indicates that the gaseous spiral density waves form through the dissipation provided by the artificial viscosity in the numerical codes. Viscous effects also induce a strong perennial inward drift of gas toward the center.

Sorensen, Matsuda, and Fujimoto in their model with strong prolate spheroidal distortion find the time evolutionary response to be offset toward the leading edges of the bar. With higher resolution, Huntley, Sanders, and Roberts consider a similar case of a Toomre model with a prolate spheroidal distortion, and a photographic simulation of this is shown in Figure 11. The gas response in the inner parts is quite strong and shows distinctly-offset lanes along the leading edges of the underlying bar-like prolate spheroid. Of course there is still a problem of resolution, directly related to the smoothing effects of artificial viscosity in the numerical code; nevertheless, it is tempting to identify these offset features as tracers of large-scale shocks in the gas.

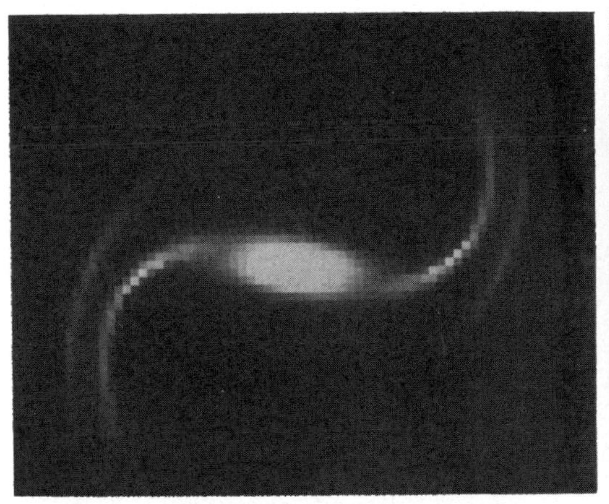

Figure 10. Photographic simulation (a snapshot) of the time evolutionary gas response in a Toomre model to a mild bar-like distortion (Huntley, Sanders, and Roberts, 1978).

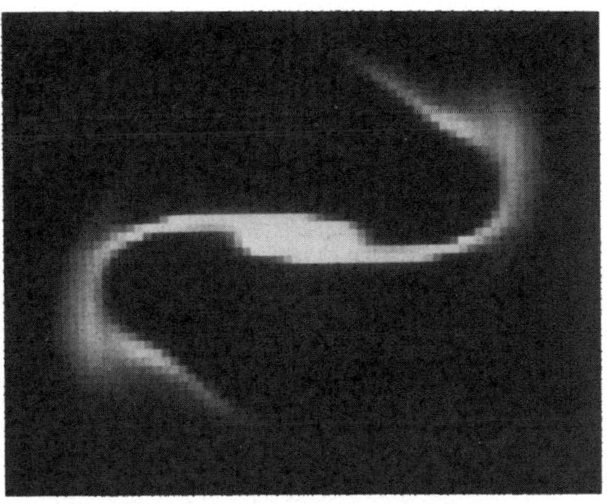

Figure 11. Photographic simulation (a snapshot) of the time evolutionary gas response in a Toomre model to a prolate spheroidal distortion (Huntley, Sanders, and Roberts, 1978).

In order to focus in more detail on these features, we will return to the steady state model (Figure 7) where viscous smoothing does not enter. Figure 12 shows the velocity field for this steady state model in a display which plots velocity vectors at approximately 6400 cell positions over an 80 by 80 grid, 1 cell/300 pcs here, which is the identical grid size used in the time evolutionary calculations. Numerous arrows delineate the strong outward preshock flow approaching the shock and also the strong inward postshock flow at distances downstream from the shock. Only two arrows are apparent which delineate the outward, highly oblique postshock flow just adjacent to the shock. Corresponding velocity field displays for the time evolutionary calculations show a strikingly similar character; except in the narrow postshock region just adjacent to the shock, not even one arrow can be found which delineates outward, highly oblique postshock flow. The artificial viscosity is no doubt playing a role in the time evolutionary calculations to smooth such fine features. On the other hand the resolution of even the best, current observations is coarser; and to that degree of resolution, such differences in the theoretical displays would not even be detectable.

Recent time evolutionary, two dimensional, numerical calculations by Liebovitch (1978) on the gas response to an imposed perturbing potential that is bar-like in the interior and spiral-like in the exterior, on a polar coordinate grid, provide important confirmations of many of the results found in these earlier studies. Liebovitch confirms such effects as the offsetness of the gas features along the bar. Figure 13 shows his determination of the clockwise phase shift toward 90° of the gas bar relative to the (horizontal) driving bar, for the case of a disk model with one inner Lindblad resonance when the amplitude of the driving bar is

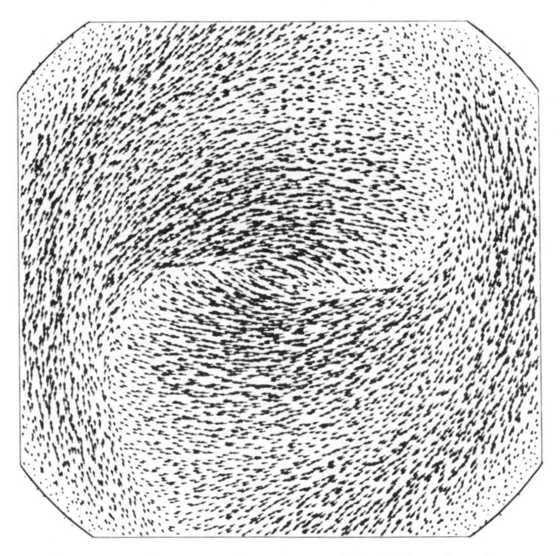

Figure 12. Velocity field of the steady state model in Figure 7, delineated by some 6400 velocity vectors. To this degree of resolution, shocks are detectable (Roberts and Huntley, 1978).

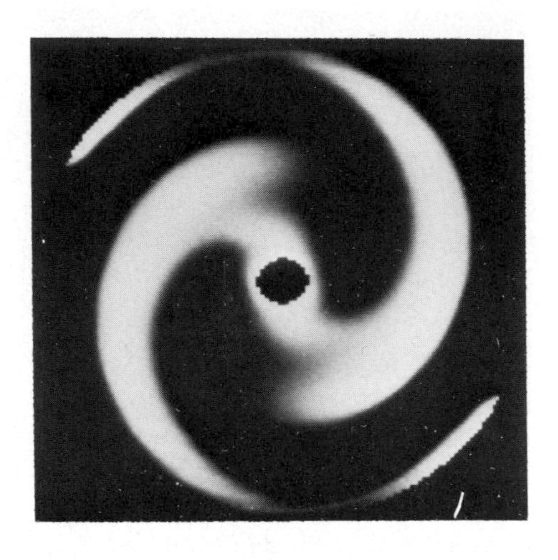

Figure 13. Photographic simulation of the gas density distribution for a case in which the gas bar response shifts to approximately 90° out of phase with the driving bar (Liebovitch, 1978).

reduced toward zero. In addition, he finds broad appendages of high gas density that extend past the bend where the central bar bends into spiral arms (most likely related to the shock-focusing phenomenon discussed earlier); and these he associates with the bright appendages to the spiral arms seen in many barred spirals.

A SAMPLE BARRED SPIRAL

One of the most detailed velocity field maps for barred spirals available to date in the literature is that of NGC 5383 (photograph shown in Figure 14) derived from spectroscopic observations by Peterson,

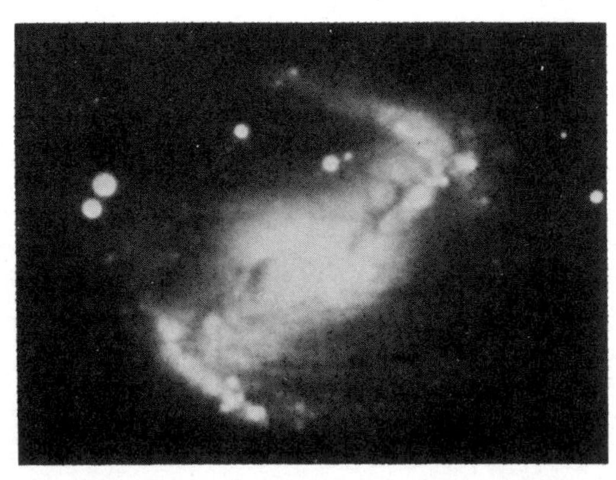

Figure 14. Photograph of the barred spiral galaxy NGC 5383 (The Kitt Peak National Observatory).

Rubin, Ford, and Thonnard (1978). Figure 15 shows this observed velocity field map. Probably what is most striking: the strong velocity gradient across the bar, as evidenced by the crowding of isovelocity contours there. Such a strong velocity gradient may be directly related to the bar-driven flow of gas in NGC 5383.

In order to investigate this possibility we will adopt the steady state model (Figures 7 and 12) and view it in the same orientation as NGC 5383 is viewed in the sky. Figure 16 from Roberts and Huntley (1978) shows the velocity field map (solid lines) predicted from the steady state model; independently Huntley (1978) has followed a similar procedure for a time evolutionary model. As in the observed map, what is most striking in Figure 16 is the strong gradient of velocity across the bar, which in the model is a result of the highly oval gas circulation driven by the bar-like potential field. Overall there seems to be general agreement between these isovelocity contours and those in the observed map. Perhaps of further interest is the trough in the -50 and -100 km/s isovelocity contours in the NW quadrant where these contours cross the shock and spiral arm, and also the corresponding trough in the 50 and 100 km/s isovelocity contours in the SE quadrant. Similar troughs show up in the observed map in the -100 and 100 km/s isovelocity contours in the regions where these contours cross the arms near the ends of the bar. If further observations of this type could be made, particularly near and along the bar structure and spiral arms, corresponding troughs and other detailed variations in more of the observed isovelocity contours might be detectable and help to further reflect the underlying dynamics.

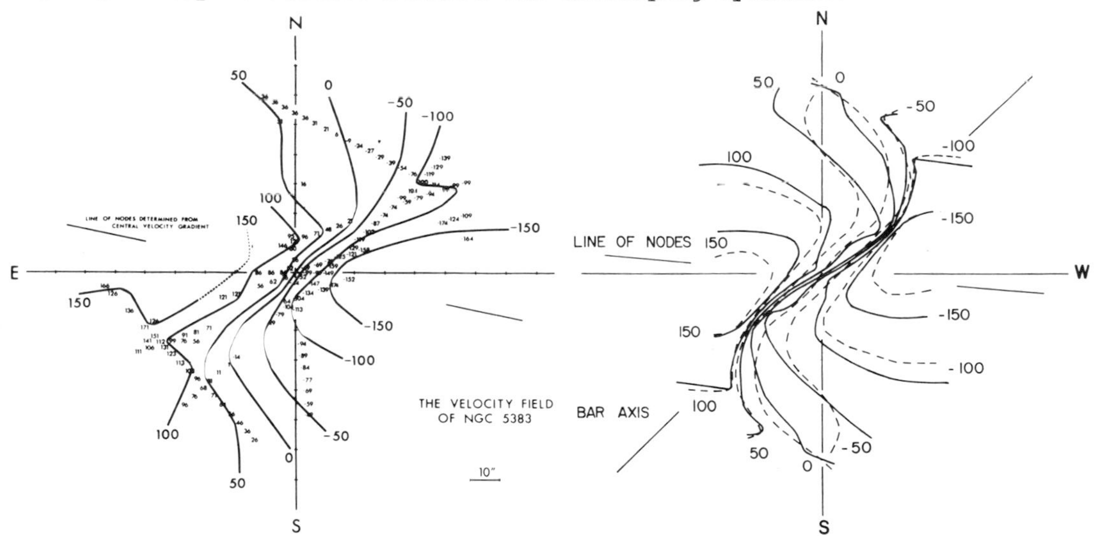

Figure 15. Observed velocity field of the barred spiral galaxy NGC 5383 characterized by a strong gradient in velocity across the bar (Peterson, Rubin, Ford, and Thonnard, 1978).

Figure 16. Predicted velocity field of the steady state model in Figures 7 and 12, with the same orientation as that observationally-derived for NGC 5383 (Roberts and Huntley, 1978).

OUR GALAXY - NOT A BIRD'S EYE VIEW

Does our Galaxy contain a bar-like distortion or bar structure in the inner parts? If so, how prominent is it and what characteristics does it share in common with the bar structures in NGC 1300 or NGC 5383 or other external barred spirals which we see with a bird's eye view? Perhaps the results and implications of the studies reviewed herein will motivate future studies, both observational and theoretical, to help us focus further on this intriguing question, toward a deeper understanding of our Galaxy.

I would like to thank Jim Huntley, Larry Liebovitch, Bill Peters, and Charles Peterson for providing photographs included in this review. Portions of this work were supported in part by the National Science Foundation under grant AST72-05124 A04.

REFERENCES

Baade, W., 1963, "Evolution of Stars and Galaxies," Harvard U., Cambridge.
de Vaucouleurs, G., 1964, in IAU Symp. No. 20, p. 195.
de Vaucouleurs, G., 1970, in IAU Symp. No. 38, p. 18.
Fujimoto, M., 1968, in IAU Symp. No. 29, p. 453.
Huntley, J. M., 1977, Ph. D. Thesis, U. of Va., Charlottesville.
Huntley, J. M., 1978, preprint.
Huntley, J. M., Sanders, R. H., Roberts, W. W., 1978, Ap. J., 221, p. 521.
Kerr, F. J., 1967, in IAU Symp. No. 31, p. 239.
Kerr, F. J., 1969, Australian J. Phys., Ap. Suppl., No. 9.
Liebovitch, L. S., 1978, Ph. D. Thesis, Harvard U., Cambridge.
Lin, C. C., and Shu, F. H., 1964, Ap. J., 140, p. 646.
Lin, C. C., and Shu, F. H., 1966, Proc. Natl. Acad. Sci., 55, p. 229.
Peters, W. L., 1975, Ap. J., 195, p. 617.
Peters, W. L., and Roberts, W. W., 1972, B.A.A.S., 4, p. 265.
Peterson, C. J., Rubin, V. C., Ford, W. K., Thonnard, N., 1978, Ap. J., 219, p. 31.
Prendergast, K. H., 1962, in Interstellar Matter in Galaxies, L. Woltjer (ed.), Benjamin, N. Y.
Roberts, W. W., 1969, Ap. J., 158, p. 123.
Roberts, W. W., 1971, B.A.A.S., 3, p. 369.
Roberts, W. W., and Huntley, J. M., 1978, Ap. J. (submitted for publ.).
Roberts, W. W., Huntley, J. M., Lin, C. C., 1977, in IAU Symp. No. 77.
Roberts, W. W., and Yuan, C., 1970, Ap. J., 161, p. 877.
Sanders, R. H., 1977, preprint.
Sanders, R. H., and Huntley, J. M., 1976, Ap. J., 209, p. 53.
Sanders, R. H., and Prendergast, K. H., 1974, Ap. J., 188, p. 489.
Simonson, S. C., and Mader, G. L., 1972, B.A.A.S., 4, 266.
Simonson, S. C., and Mader, G. L., 1973, Astr. and Ap., 27, p. 337.
Sorensen, S. A., Matsuda, T., and Fujimoto, M., 1976, Ap. Space Sci., 43, p. 491.
Toomre, A., 1963, Ap. J., 138, p. 385.
Westerhout, G., 1969, Maryland-Green Bank Galactic 21-cm Line Survey, 2nd ed.

DISCUSSION

van den Bergh: Radio observations show that there is a small area of very active star formation in the center of our Galaxy. In this respect the Galaxy resembles external galaxies with nuclear hot spots. About two-thirds of the objects with such hot spots are barred spirals. This suggests that there is perhaps a 60% to 70% chance that the Galaxy also contains a nuclear bar.

Contopoulos: I noticed (in one of your slides) that the streamlines inside the bar are very elongated. According to my calculations this should happen mainly near the Inner Lindblad Resonance. Where is the Inner Lindblad Resonance in your model?

Roberts: In the steady-state model shown in Figure 7 of my paper, highly oval gas streamlines occur in the vicinity of the Inner Lindblad Resonance and in fact pass across this resonance region. In the (kpc) units of Figures 7 and 8 the half co-rotation radius occurs at 7.5 kpc (marked, and just outside the bend of the bar into spiral arms); the Inner Lindblad Resonance, at 2.6 kpc; and the maximum and minimum radii of the second innermost streamline, at 4.0 and 1.0 kpc, respectively.

Sanders: It seemed to me that in the steady-state calculations made by you and Huntley the co-rotation radius was near the end of the bar; that is, the bar is a fast bar. I know that with respect to the time dependent calculations, it is certainly the case that co-rotation must lie near the end of the bar in order to get reasonable offset shocks. Now we lie in a region of spiral structure. This would suggest--if our galaxy is a barred spiral like NGC 1300--we lie beyond co-rotation. On the other hand, the fact that active star formation is occurring in the region of the Sun implies that we lie inside co-rotation. This apparent contradiction suggests to me that our Galaxy is not a barred spiral like NGC 1300.

Roberts: I agree that a fast bar would result if co-rotation occurs as far inwards as the radius at the ends of the bar. However, the bar in NGC 1300 may not be so fast a rotator. First, the co-rotation radius in the steady-state model, shown in my Figure 7, actually lies at the outer edge of the disk outside the spirals; the bar is a slowly-rotating bar, and the shocks occur offset with about the same offsetness as the dark dust lanes observed along the bar in NGC 1300. Second, in the observed photograph of NGC 1300 the dust lanes on the inner edges of the spiral arms can be traced to a considerable radius; this implies that co-rotation occurs outside this radius. These considerations do not support your chain of arguments. I believe that the evidence available at this time does not rule out the possibility that our Galaxy may have a bar or bar-like distortion like that of NGC 1300. However, I do agree that such a bar or bar-like distortion in our Galaxy would probably be a slow rotator.

Mark: Dr. Roberts showed that even a weak driving force in the form of a bar plus open spiral would result in strong bar-like gas shocks. A relevant question is whether or not the gas would react with strong shocks to the modes we calculated which sometimes have the appearance of an open spiral in the inner parts of the galaxy model. My guess is that the gaseous response might not look like NGC 1300 because the perturbation densities and potentials do not reach the galactic center in our modes. The situation could be different if the amount of bulge matter is somewhat less than that used in our N-body simulation with Bob Berman. In that case a weak bar or oval distortion forms in the central regions and the gaseous response might indeed behave as in Dr. Roberts' simulation.

Sanders: Let me elaborate on the point I made a minute ago about fast bars. We have recently been investigating the gas response to bars by means of time-dependent hydrodynamical calculations. The model for the gravitational field consists of a homogeneous prolate spheroid embedded in a disk. We have been exploring the parameter space of this model and find that the gas response looks most similar to a barred spiral like NGC 1300 whenever co-rotation lies just beyond the ends of the bar. This is consistent with calculations made several years ago by Prendergast which suggest that the location of the straight dust lanes depend critically upon the presence of Lagrangian points near the ends of the bar.

van Woerden: The pile up of gas just outside a bar may be seen well in the barred Magellanic irregular NGC 4449. A Westerbork map of neutral hydrogen in this galaxy (by Bosma and myself, in preparation) shows a very strong ridge of hydrogen at the NE end of the bar, coinciding with a chain of bright HII regions and associations.

de Vaucouleurs: I should like to call your attention to the on-going program of Fabry-Perot interferometry at McDonald Observatory, including many barred spirals. The first results will appear in the dissertation of W. Pence; this is a study of the velocity field in NGC 253 from about 10,000 individual velocities.

Roberts: I look forward to the publication of this high-resolution study which perhaps can be compared with high-resolution theoretical velocity-field maps, such as the one shown in my Figure 12, to help guide future theoretical studies.

THE FOUR-ARMED RESPONSE NEAR THE LINDBLAD RESONANCES IN GALAXIES

G. Contopoulos
Astronomy Department, University of Athens, Greece
and
Astronomy Program, University of Maryland, USA

ABSTRACT. We show that there is an important four-armed term in the response of a flat galaxy to an imposed two-armed spiral field near the Inner and Outer Lindblad Resonances.

We will show how an important four-armed component arises near the Lindblad Resonances of a two-armed spiral, or a bar. By "important" we mean that it is of the same order as the two-armed component, although it may be numerically smaller.

We will describe the main steps leading to this result, while the detailed calculations are included in a forthcoming paper (Contopoulos 1978).

We assume that the potential of a (flat) galaxy is composed of an axisymmetric background and a two-armed spiral perturbation

$$V = V_o(r) + A(r) \cos [\Phi(r) - 2\theta]. \tag{1}$$

The azimuth θ is calculated in a frame of reference rotating with angular velocity Ω_s. In the case of a bar the phase Φ is a constant.

We know now that the spiral perturbation can be analysed into components of the form (Kalnajs 1971)

$$V_{\ell m}(I_1, I_2) \begin{matrix} \cos \\ \sin \end{matrix} (\ell\theta_1 - m\theta_2), \tag{2}$$

where $(I_1, I_2, \theta_1, \theta_2)$ are action-angle variables and ℓ, m integers. Namely I_1 is the radial action, I_2 the azimuthal action (essentially the angular momentum), θ_1 the epicyclic angle and θ_2 the azimuth of the epicyclic center; θ_2 is close to the azimuth of the star, θ.

The most important terms of the form (2) are (Contopoulos 1973; paper I):

W. B. Burton (ed.), The Large-Scale Characteristics of the Galaxy, 187–190.

$$A_o \cos\ (2\theta_2 + \text{const.}), \qquad (3)$$

$$A_1 \cos\ (\theta_1 - 2\theta_2 + \text{const.}), \qquad (4)$$

and

$$A_2 \cos\ (\theta_1 + 2\theta_2 + \text{const.}), \qquad (5)$$

where A_o, A_1, A_2 are of the same order as the amplitude A.

The orbit of each star in the potential (1) is an oscillation around a periodic orbit. In general the periodic orbits are almost circles, therefore the orbits of stars are slightly perturbed epicycles. However near the main resonances of the galaxy the periodic orbits change considerably. Near the Particle Resonance the most important term in the potential is the term (3), while near the Inner and Outer Lindblad Resonances the most important terms are (4) and (5) respectively.

In the lowest approximation the periodic orbits at a resonance are given by setting the argument of the corresponding term (3), (4), or (5) equal to zero, or to π. E.g. at the Inner Lindblad Resonance the periodic orbits are given approximately by

$$\theta_1 - 2\theta_2 + \text{const.} = 0, \pi. \qquad (6)$$

These two orbits are nearly ellipses, perpendicular to each other. The non-periodic orbits form, in general, rings around one of the periodic orbits (6).

In order to find the density response we must superpose all the orbits oscillating around the periodic orbits (6).

However, the fact that we have two perpendicular populations of orbits is not sufficient to produce a four-armed component in the density distribution. If the density around each periodic orbit has a sinusoidal form, similar to (4), then the maxima of density of one population coincide with the minima of density of the other population, and thus the contributions of the two populations tend to cancel each other. If the amplitudes of the two populations are equal, only the axisymmetric background remains, but if one population is stronger, its amplitude is simply reduced.

On the other hand we may think of a situation where only the immediate neighborhoods of the periodic orbits are populated with stars. In such a case the four-armed component is clearly seen, but the distribution of orbits is not sinusoidal, but indudes several higher harmonics.

One may argue that such a distribution contains higher order terms in A, therefore it is not a first order effect. However we will prove that even if we limit our discussion to first order terms in A we find the same effect. In order to do that we examine the perturbations of the orbits near the Inner Lindblad Resonance due to the term (5). If we introduce into (5) the values (6) (appropriate for the periodic orbits) we find terms of the form

$$A_2 \cos (4\theta_2 + \text{const.}). \qquad (7)$$

Thus the perturbation is close to a four-armed term

$$A_2 \cos (4\theta + \phi), \qquad (8)$$

where A_2 and ϕ are functions of r.

If we follow now the effects of the term (8) we find that the response contains a four-armed component, which is strongest near the Inner Lindblad Resonance (Contopoulos 1978). The 4θ-response is, in fact, of the same order as the 2θ-response, namely of $O(\sqrt{A})$. The amplitude of the four-armed component as a function of r in the models of paper I is given in Figure 1, both for a bar and for a somewhat tight spiral of inclination $\sim 16°$.

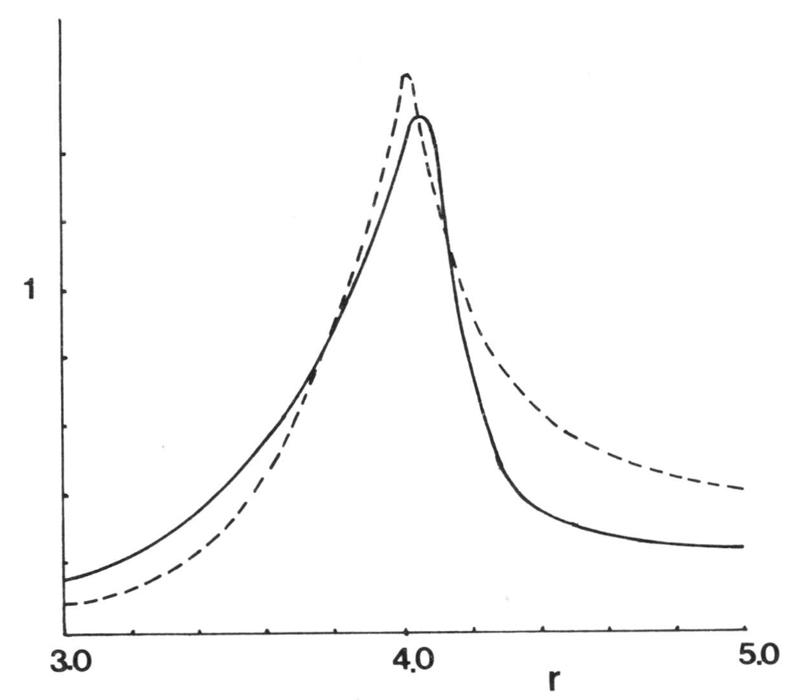

Figure 1. The amplitude of the four-armed term in the models of Paper I. (——) Spiral with k=-10/(r+4); inclination angle $\simeq$ 16°. (---) Bar. Resonance at r=3.73kpc.

This form of the response seems to be quite general. Namely the four-armed term is strongest near the inner Lindblad resonance, but it is small inside and outside it.

This result is consistent with the empirically found 4θ-term in the case of the SB0 galaxy NGC 2950 (Crane 1975). Crane found that the amplitude of the 4θ-term in this galaxy has a localized maximum, as in Figure 1, while the 2θ-component has a broader maximum in the same general region. This can be explained if both maxima are associated with the Inner Lindblad Resonance (Contopoulos and Mertzanides 1977).

Another example of a conspicuous four-armed component in the inner parts of a galaxy was found by P.O. Lindblad (Figure 2).

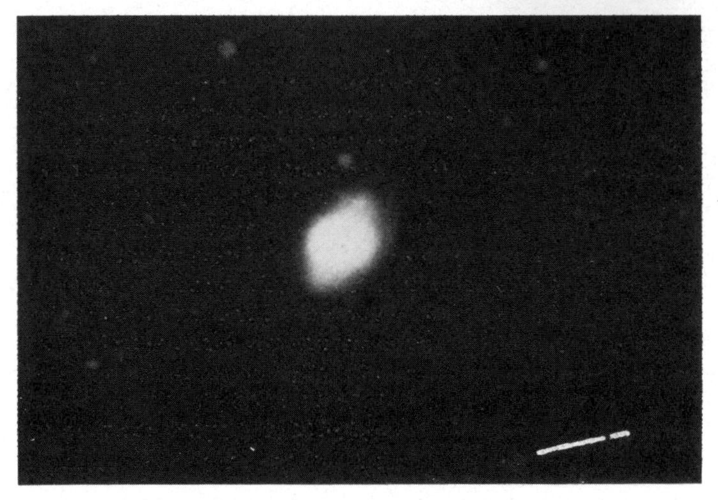

Figure 2. An anonymous southern galaxy that shows one strong and one weak bar (ESO photograph).

A similar four-armed component should be expected near the Outer Lindblad Resonance. It remains to be seen if the various models that introduce a four-armed component in the outer parts of our Galaxy can be explained by such a mechanism, or whether it is necessary to consider a separate four-armed mode, which forms a new grand design of its own.

REFERENCES

Contopoulos, G.: 1975, Astrophys. J. 201, pp. 566-584.
Contopoulos, G.: 1978, Astron. Astrophys. (in press).
Contopoulos, G., Mertzanides, C.: 1977, Astron. Astrophys. 61, pp. 477-485.
Crane, P.: 1975, Astrophys. J. 197, pp. 317-328.
Kalnajs, A.J.: Astrophys. J. 166, pp. 275-293.

ON THE USE OF THE WKBJ DENSITY WAVE THEORY AT THE INNER LINDBLAD RESONANCE REGION OF OUR GALAXY

E. Athanassoula
Observatoire de Besançon

I have tested the reliability of certain approximations involved in the asymptotic WKBJ density wave description of the inner Lindblad resonance (=ILR) of our galaxy.

To calculate the induced density one has to integrate along the orbit of the star. Since this is difficult to do exactly, epicyclic orbits and series expansions are used instead. This necessarily introduces the problem of ordering the small parameters involved and, after such an ordering has been adopted, the neglect of higher order terms. At the ILR such higher order terms include the azimuthal dependance of the potential and the derivatives of the wavenumber. It would be desirable to know the errors thus introduced.

I have done so using the integral equation for the reduced potential, which does not rely on the above mentioned approximations and does not introduce any further approximations not included in the WKBJ treatment. The test is in no way complete since it does not raise either the question of nonlinearity, or of how adequate a description of real orbits epicycles can be, since the integral equation is linear and uses epicycles. Also I have applied it only to the short waves.

For the trailing waves, in all cases tested so far, the error introduced by all other approximations except for the neglect of the azimuthal dependance of the potential, does not exceed 15-20%. If one includes this also, one reaches 30-40%. Unfortunately this does not hold for leading waves, where in all cases tested the error was larger than the solution itself. This result shows an inadequacy of the WKBJ resonant theory, at least in one example, and thus raises a doubt as to its unquestioned applicability.

DISCUSSION

Burton: Would you remind us of the observational data which you have in the back of your mind when you are thinking about the Inner Lindblad Resonance in our Galaxy.

W. B. Burton (ed.), The Large-Scale Characteristics of the Galaxy, 191–192.

Contopoulos: Near the Inner Lindblad Resonance the orbits are rather elongated along a certain direction, and the response density is also elongated. In a barred galaxy, like NGC 2950, the ILR should be near the maximum ellipticity of the isophotes, and this is supported by the fact that the 4θ-term is also maximum there. In the case of our Galaxy we also expect the orbits of stars (and the streamlines of gas) to be most elongated near the ILR. This effect may produce a streaming along the 3-kpc arm. Thus there may also be an appreciable four-armed component in this region.

IV. GALACTIC KINEMATICS AND DISTANCES

THE GALACTIC DISTANCE SCALE

J.A. Graham
Cerro Tololo Inter-American Observatory *

Recent work on the distance scale of the Galaxy has largely been in the direction of refining previously established methods. The RR Lyrae variable stars appear to be better distance indicators than was once thought and they have been used in determining R_o, the distance to the Galactic center. R_o is probably somewhat less than 10 kpc but greater than 7 kpc. Most methods point to a value near 8.5 kpc.

1. INTRODUCTION

In reviewing the present state of the Galactic distance scale, it is encouraging to find that there have been no major revisions over the past decade or so. This is not to imply that the field has become inactive but rather to note that several major investigations have verified with improved precision, results obtained earlier from more fragmentary data. There has been little development of new methods but considerable refinement of those already established.

2. CALIBRATIONS WITHIN THE SOLAR NEIGHBORHOOD FOR YOUNG STELLAR POPULATIONS

The distance to the Hyades cluster, an important base for several determinations of the distance scale, has undergone revision. Van Altena (1974) discussed the several methods employed for this and adopted in summary a distance modulus of $3^m.21$ (± .03). More recently, a new evaluation has been made by Hanson (1978) using a new set of absolute proper motions. The distance modulus of $3^m.30$ (± .06) is independent of previous determinations. This 25% increase over the value in use prior to 1970 does not affect distance scales in general use by a corresponding amount. Van den Bergh (1977) has pointed out

* Cerro Tololo Inter-American Observatory is supported by the National Science Foundation under Contract No. AST 74-04128.

W. B. Burton (ed.), The Large-Scale Characteristics of the Galaxy, 195–200.

that the change should be only about half of this because 1) some calibrations of distance indicators bypass the Hyades completely and 2) the distance scale for the Cepheid variables is sensitive to the assumptions made in the sequence fitting of the calibrating clusters to the Hyades.

Other recent work on the calibrations of young stellar populations in the solar neighborhood is independent of the Hyades modulus. Crawford (1975) (1978) has just completed definitive calibrations of *uvby*, *β* photometry in terms of intrinsic color and absolute magnitude for B and F - type stars. The zero points of the calibrations are based on the parallaxes of nearby F stars. Crawford has also recalibrated the luminosities of some of the MK spectral classes and finds good agreement with the calibration published by Blaauw (1963). For individual O and early B stars, the average error in absolute magnitude derived from the hydrogen line index remains large. In a calibration of Hγ measurements, Balona and Crampton (1974) find an average uncertainty of $0^m.6$ for O, early B and supergiant stars while the expected error for late B stars is much smaller, probably no more than $0^m.3$.

3. CALIBRATIONS OF OLD STELLAR POPULATIONS

Variable stars of the RR Lyrae type are important representatives of the oldest stellar populations because they can be easily identified to faint apparent magnitudes. They appear to be better indicators of distance than was once thought. Observationally, absolute magnitudes for RR Lyrae stars come from 3 sources: 1) stellar systems whose distances are known. 2) statistical parallaxes. 3) application of the Baade-Wesselink method. Some results are summarized in the following Table.

THE ABSOLUTE MAGNITUDE OF RR LYRAE STARS

Method	Object	$\langle M_V \rangle$
Stellar systems	globular clusters: Sandage (1970)	$0^m.6 \pm 0^m.2$
	Magellanic Clouds	
	LMC : Graham (1977)	0.7 ± 0.2
	SMC : Graham (1975)	0.5 ± 0.2
Statistical parallaxes	Heck (1973)	0.5 ± 0.2
	Hemenway (1975)	0.5 ± 0.4
Baade-Wesselink	Metal weak stars: McDonald (1977)	0.55 ± 0.2
	Metal strong stars: McDonald (1977)	0.85 ± 0.2
	McNamara and Feltz (1977)	0.9 ± 0.2

The Magellanic Cloud values depend on assumed distances moduli of $18^m.5$ and $19^m.0$ for the Large and Small Clouds respectively. If, as

Eggen (1977) believes, a modulus of $18^m.25$ is more appropriate to the Large Cloud then the corresponding value above would need to be increased accordingly. While there is some sign of a dependence of absolute magnitude on metal line strength, it may not affect the Galactic calibration. Tamm et al. (1976), Butler et al. (1976), and Rodgers (1977) have shown that the RR Lyrae stars encountered in the Galaxy away from the immediate vicinity of the Sun, generally have weak metal lines even in the crowded region of the Galactic bulge. A time averaged mean visual magnitude $< M_V > = 0^m.6 \pm 0^m.2$ is probably appropriate for the RR Lyrae stars. Oort and Plaut (1975) note that a dispersion greater than $0^m.2$ in the mean magnitude will give unreasonable space densities within the fields that they studied.

Because of their high integrated luminosities, globular clusters are important probes in studies of our own and other galaxies. The most basic method of calibration is the sequence fitting procedure described by Sandage (1970). Harris (1976) has recently written a comprehensive paper on the subject. He discusses various methods of estimating distances for known globular clusters and concludes that the most reliable method of distance determination is based on the assumption of a mean absolute magnitude for RR Lyrae stars in globular clusters. Adopting $< M_V > = 0^m.6$, he derives distances for 111 globular clusters in this way.

4. THE DISTANCE TO THE GALACTIC CENTER

Dense obscuration, much of which is in thick dust clouds within 2 kpc of the Sun, prevents direct estimates of the distance to the Galactic center. Proceeding less directly, methods generally fall into one or other of two classes: 1) observation through relatively clear "windows" a little above or below the plane to determine the distance of maximum density or symmetry for objects known to be concentrated towards the Galactic center. 2) study of the dynamics of stars at the same distance from the center as the Sun.

Studies of the space distribution of RR Lyrae stars are typical of the first category. An important investigation of this type was made by Oort and Plaut (1975). In 5 relatively clear fields, 980 variable stars were detected and measured. Because of the coarse nature of the photographic magnitudes, interstellar absorption could only be taken into account in an average way. Assuming an average photographic magnitude $\overline{Mpg} = \frac{1}{2}\ (M_{max} + M_{min}\ ) = 0^m.7$, a mean distance to the Galactic center, R_o, of 8.7 ($\pm$ .23) kpc was derived.

A similar method, based on the newly determined globular cluster distances has been followed by Harris (1976). As with the RR Lyrae stars, one must look away from the Galactic plane to obtain a sample which is reasonably complete with distance. Harris finds that the most reliable solution for R_o is obtained by restricting consideration to globular clusters more than 2.5 kpc from the plane. This yields 8.5 kpc with an uncertainty of ± 1.6 kpc. Harris notes that his estimated error includes an uncertainty of ± $0^m.3$ in the absolute magnitude of the RR Lyrae stars which corresponds to a distance error of ± 1 kpc. The results in the first Table suggest that this estimate may be excessive.

The distance to the Galactic center can also be found by observing remote OB stars with nearly the same distance from the Galactic center as the Sun. One version of the method consists of determining the galactic longitude at which stars with a known distance have zero radial velocity with respect to the Sun. The method was applied most extensively at Radcliffe Observatory. Balona and Feast (1974) determine $R_o = 9$ kpc with a range of 7.7 - 10.9 kpc. The method is unfortunately very sensitive to streaming motions whose systematic effects are difficult to evaluate. In a rediscussion with improved Northern Hemisphere data, Crampton et al. (1976) find $R_o = 8.4$ (±1) kpc.

An interesting new method has come from Toomre (1972) who combines observations of local mass density with theoretical disk models and investigates the conditions under which the disk can remain stable. He concludes that $R_o = 7$ kpc is too small and $R_o = 10$ kpc is too large to satisfy these constraints. Rybicki et al. (1974) report a development of this method which leads to a distance of 9.0 kpc. Eggen (private communication) also points out that with R_o as low as 7 and plausible values of the Oort constants, there would be an uncomfortably large number of high velocity stars with unbound orbits at present in the Galaxy.

In summary, it can be seen from the following Table that recent work favors a distance to the Galactic center, R_o, somewhat less than 10 kpc but greater than 7 kpc. A straight mean of 8.7 kpc is found. The quoted errors are not independent and reflect uncertain systematic effects. A mean error of 1 kpc therefore remains in this average value.

DISTANCE TO THE GALACTIC CENTER

RR Lyrae Stars:	Oort and Plaut (1975)	8.7 (±0.6) kpc
Globular Clusters:	Harris (1976)	8.5 (±1.6)
Solar distance O and B stars:	Crampton et al. (1976)	8.4 (±1)
Galactic disk:	Toomre (1972) Rybicki et al. (1974)	9.0 (±1)
	Average	8.7 (±1)kpc

REFERENCES

Balona, L.A. and Crampton, D.: 1974, Monthly Notices Roy.Astron.Soc., 166, 203.

Balona, L.A. and Feast, M.W.: 1974, Monthly Notices Roy.Astron.Soc., 167, 621.
Blaauw, A.: 1963, Stars and Stellar Systems, 3, 383.
Butler, D., Carbon, D. and Kraft, R.P.: 1976, Astrophys. J., 210, 120.
Crampton, D., Bernard, D., Harris, A.D. and Thackeray, A.D.: 1976, Monthly Notices Roy.Astron.Soc., 176, 683.
Crawford, D.L.: 1975, Astron. J., 80, 955.
Crawford, D.L.: 1978, Astron. J., 83, 48.
Eggen, O.J.: 1977, Astrophys. J. Suppl., 34, 1.
Graham, J.A.: 1975, Publ.Astron.Soc.Pacific, 87, 641.
Graham, J.A.: 1977, Publ.Astron.Soc.Pacific, 89, 425.
Hanson, R.B.: 1978, Bull.Am.Astron.Soc., 9, 585.
Harris, W.E.: 1976, Astron. J., 81, 1095.
Heck, A.: 1973, Astron. Astrophys., 24, 313.
Hemenway, M.K.: 1975, Astron. J., 80, 199.
McDonald, L.H.: 1977, Ph.D. thesis, University of California, Santa Cruz.
McNamara, D.H. and Feltz, K.A.: 1977, Publ.Astron.Soc.Pacific., 89, 699.
Oort, J.H. and Plaut, L.: 1975, Astron.Astrophys., 41, 71.
Rodgers, A.W.: 1977, Astrophys. J., 212, 117.
Rybicki, G., Lecar, M. and Schaefer, M.: 1974, Bull.Am.Astron.Soc., 6, 453.
Sandage, A.R.: 1970, Astrophys. J., 162, 841.
Tamm, R.E., Kraft, R.P. and Suntzeff, N.: 1976, Astrophys. J., 207, 201.
Toomre, A.: 1972, Quart. J. Roy.Astron.Soc., 13, 241.
van Altena, W.F.: 1974, Publ.Astron.Soc.Pacific, 86, 217.
van den Bergh, S.: 1977 Astrophys. J. Letters, 215, L103.

DISCUSSION

Bok: What role can the Magellanic Clouds play in future calibration problems?

Graham: In the Magellanic Clouds we can now observe stars all at the same distance but with a large variety of ages, to luminosities as faint as that of the Sun. They are ideal places to check the distance indicators of various types. Chemical abundances in the Large Magellanic Cloud appear nearly normal compared to those in the Sun, but the abundances of some elements in the Small Cloud are found to be comparatively low. In the Small Cloud it will be possible to examine the effects of these anomalous abundances on the various calibrating objects and thus enable us to extend the Galactic distance scale with some assurance to more remote stellar systems.

Schmidt-Kaler: Although the most recent stream parallax of the Hyades yields the modulus m_0-M = 3.4, the trigonometric parallaxes (including some new Van Vleck values) yield the old value, 3.0, exactly. This is due to a systematic correction of +0$''$004 of the Jenkins Catalogue relative parallaxes which I find necessary in a recent study.

Muzzio: I would like to make a comment on the use of H_β photometry: although the β-index is independent of the star's color if both the narrow and wide filters have the same mean wavelengths, such is not always the case. In fact, published transmission curves for the two filter sets used to establish the β-system show that one of them should yield β values strongly dependent on the star's color (about $0^m.022$ increase in β with every 1^m increase in color excess). This effect explains also most of the difference between the calibration curves obtained for B and for A-F stars. Because the third set of filters used to establish the standard system also requires the use of those different curves, it is not unlikely that that set is influenced by the same problem too (there are no published transmission curves for it). Present calibrations may thus need some revision and β observations should be corrected by including a color term.

Graham: Observers using interference filters should be careful of effects such as these. Much of Crawford's work was done with a single set of filters, so I doubt that the calibration work will need revision.

de Vaucouleurs: I should like to call your attention to the circular argument involved in "calibrating" the RR Lyrae by way of the Magellanic Clouds using a distance derived from Cepheids calibrated by one method only (and using incorrect absorption corrections).

Graham: The Magellanic Cloud distances are derived from a number of distance calibrators, not just from the Cepheids (Westerlund, B. E.: 1974, in "Galaxies and Relativistic Astrophysics", B. Barbanis and J. D. Hadjidemetrion, eds. (Berlin: Springer-Verlag), p. 39.)

Kraft: I hope colors can be obtained for individual RR Lyrae, so that E(B-V) can be estimated directly. One can also pray that finding charts will be published for the individual variables.

Graham: The Blancos are planning to determine colors for individual RR Lyraes.

RR LYRAE VARIABLES IN BAADE'S WINDOW

B. Blanco and V.M. Blanco
Cerro Tololo Inter-American Observatory*

In his pioneer study of the RR Lyrae variable stars near NGC 6522, in the relatively unobscured window close to the galactic center ($l = 0.9$, $b = -3.9$), Baade (1963) was limited by the high zenith distance of the galactic center as seen from Palomar, and suggested that southern hemisphere observations would be of value. Subsequent studies of the region have been based on the variables found in Baade's original search. Southern hemisphere plates were taken by Hartwick et al. (1972), and their re-analysis of a sample of Baade's variables showed many periods to be in error. They did not, however, search the plates for new variables. Plaut (1973) re-analyzed all the variables but no search for new variables was done.

B. Blanco and V.M. Blanco have obtained a new set of 82 plates with the 1.5m telescope at CTIO on 7 nights, including series on successive ones. Blue and visual plates were alternated to permit a redetermination of absorption in the region as well as an analysis of mean B or V magnitudes. A new search for RR Lyrae variables is now in progress. In order to optimize the selection of plate pairs for blinking, the Monte Carlo method was used. The blinking of the first three plate pairs has confirmed that the method strongly favors the discovery of RR Lyraes.

The uniformity of obscuration is also being examined. The area selected by Baade for his statistical study appears not to be of uniform transparency. However, there does appear to be an extension or wing of relatively uniform and high transparency that extends the window toward the galactic center. This new wing is being searched along with the original area. After blinking only three plate pairs, the rate of discovery is as follows:

	No. vbles.	Area (sq. arc min)
Original window	17	483
Wing	10	176

* Cerro Tololo Inter-American Observatory is supported by the National Science Foundation under Contract No. AST 74-04128.

W. B. Burton (ed.), The Large-Scale Characteristics of the Galaxy, 201–202.

Thus there seem to be at least as many if not more variables per unit area in the wing. Of the ten in the wing, four were not previously known. The photometric sequence in this region is being improved, in view of the existence of appreciable differences between the photoelectric sequencies which have been derived previously by Arp (1965) and by van den Bergh (1971). The discrepancies are significant after mag 16.0 in both B and V, thus affecting the maximum of the period-frequency distribution, found by Baade to be around $m_B = 17.5$.

We are using neither the random sky method used by Arp nor the fixed preselected sky-regions used by van den Bergh. (See also Oort and Plaut 1975). Instead, at least two sky areas near each star are being individually matched as closely as possible to the area around the program star in question, in order to minimize the effects of the inevitable contamination by background stars in this very crowded region. The use of TV acquisition techniques at the 4m telescope is being supplemented by sensitometric analysis of stellar images on direct plates and by 2-dimensional photometry of the field with a vidicon detector. Similar studies in two other galactic nucleus "windows" (Sgr I and II) (Oosterhoff *et al.*, 1967) are also being initiated.

REFERENCES

Arp, H. C.: 1976, *Astrophys. J.* 141, pp. 43-72.
Baade, W.: 1963, Evolution of Stars and Galaxies, Harvard University Press, Chapter 21.
Bergh, S. van den: 1971, *Astron. J.* 76, pp. 1082-1098.
Hartwick, F. D. A., Hesser, J. E., and Hill, G.: 1972, *Astrophys. J.* 174, pp. 573-582.
Oort, J. H. and Plaut, L.: 1975, *Astron. Astrophys.* 41, pp. 71-86.
Oosterhoff, P. Th., Ponsen, J., and Schuurman, M. C.: 1967, *Bull. Astron. Inst. Neth. Suppl.* 1, pp. 397-413.
Plaut, L.: 1973, *Astron. Astrophys.* 26, pp. 317-319.

DISCUSSION

Bok: It is not generally known that on his last big night of observing (at Mount Stromlo Observatory, on the 74-inch telescope) Walter Baade took 12 one-hour plates of his "window". I was his night assistant on that night, along with Gerrit Oom. Walter Baade told me that he planned to give the plates to his collaborator Henrietta H. Swope.

van den Bergh: The "Wing" of Baade's window is closer to the galactic center than is NGC 6522. Its high surface brightness might therefore be due both to high stellar density and to relatively low absorption.

Kerr: Quantitatively, how much of a "window" does this region provide? I once did a sweep across the region in the 21-cm line, and did not find much variation in the HI column density in crossing the "window".

van den Bergh and Feast: Photometry of Mira variables show that $A_V \sim 1.6$ in the Baade window, in good agreement with van den Bergh's value.

QUANTITATIVE CLASSIFICATION OF THE GALAXY FROM NEW DATA ON THE PHOTOMETRIC PROPERTIES OF ITS SPHEROIDAL AND DISK COMPONENTS

G. de Vaucouleurs
The University of Texas at Austin

Abstract. After a brief review of previous attempts at identifying the morphological type of our Galaxy by optical or radio methods, a new approach from surface photometry is described. A two-component model consisting of an $R^{\frac{1}{4}}$ spheroid and an exponential disk is fit to the local galactic disk brightness inferred from star counts and to new observations of the brightness distribution in the bulge along the galactic prime meridian. All parameters are in close quantitative agreement with corresponding quantities for Sbc II galaxies and confirm the AB(rs) morphology first proposed in 1963 (IAU Symposium No. 20). An average of NGC 1073, 4303, 5921 and 6744 described in 1969 (IAU Symposium No. 38) closely approximates the photometric properties of our Galaxy.

1. INTRODUCTION

Identification of the morphological type of our Galaxy in the classification system of galaxies (de Vaucouleurs 1959, Sandage 1961, 1975) requires determination of its class (spiral from all evidence), family (ordinary A, transition AB, or barred B), variety (ringed r, mixed rs, or spiral s), and stage (a to m) along the Hubble sequence. The latter characteristic is the most important because it is most closely related to physical parameters such as bulge to disk ratio, hydrogen ratio, mass-luminosity ratio, etc. while the other two seem to be mainly determined by minor, and perhaps transient, dynamical details. In addition, we need to determine the luminosity class, or better, the absolute magnitudes, color indices, scale lengths, and other photometric parameters of the bulge, the disk and the Galaxy as a whole.

Optical observations are best suited to define the stage and scale parameters while radio observations are essential (but perhaps not sufficient) to analyze the spiral structure of the disk which defines the family and variety characteristics. Optical and infra-red observations of the galactic bulge by comparison with M31 have suggested a galaxy type Sb (Baade 1951, Arp 1965) or more precisely Sb to Sb^+, luminosity

W. B. Burton (ed.), The Large-Scale Characteristics of the Galaxy, 203–209.

class I-II (Schmidt-Kaler and Schlosser 1973, Maihara et al. 1978). The spiral patterns derived from 21 cm observations or from models of the radio continuum distribution are generally consistent with types Sb or Sc and both have been suggested (Mills 1959, Oort, Kerr and Westerhout 1958, Becker 1964, Kerr and Westerhout 1965, Kerr 1969, 1970, Simonson 1976). However, radio models have either an excessive amount of detail due to confusion between distance and velocity differences or too little due to gross oversimplification of the model. They look unrealistic, in particular the spirals in the 2-arms models make too many turns. Optical studies of the multiplicity of the spiral pattern, by analogy with M101, have suggested a type as late as SAB(rs)cd (Courtès 1972). However, multiplicity of the spiral pattern is correlated with family and variety, not with Hubble stage (de Vaucouleurs 1959, 1963; Sandage 1975). A better approach combining radio and optical data on HII regions has been recently developed by the Georgelins (1976 and references therein). It leads to a more plausible 4-arms spiral pattern and suggests a type closer to Sc than Sb.

The radio and optical evidence for an incomplete ring of giant HII regions in the inner regions of the galaxy, the "3-kpc arm" and the radial outflow of gas in the direction of the galactic center have led to the suggestion that a bar and ring structure are present in the central regions of the galaxy (de Vaucouleurs 1964, Kerr 1967, 1969). Computer models of the HI kinematics in the inner regions of the Galaxy tend to support this view (Simonson and Mader 1973, Peters 1975, Simonson 1976). A statistical evaluation of all available criteria (de Vaucouleurs 1970) has lead to the proposal that SAB(rs)bc is the most probable morphological type of our Galaxy. A good example of this type is NGC 4303. Examples of SAB(r), SB(rs) and SB(r) which are other possible types include NGC 6744, 1073 and 5921 (de Vaucouleurs 1964, 1970). The quantitative comparisons presented below strengthen the similarities.

2. QUANTITATIVE CLASSIFICATION OF GALAXIES

Since I last reviewed the evidence on the morphological type of our Galaxy at the Basel symposium in 1969 a great deal of progress has been made toward a quantitative classification of galaxies (de Vaucouleurs 1977a). In particular the concept of bulge to disk ratio -- one of the two fundamental criteria of the Hubble classification system -- can be precisely defined by the decomposition of the luminosity profile $I(r)$ into two major components, (I) a spheroidal component obeying the $r^{\frac{1}{4}}$ law, and (II) a disk component having an exponential distribution (de Vaucouleurs 1959, 1962, 1974; Freeman 1970; Schweizer 1976). Then $k_I = L^I/L_T$, the fraction of the total luminosity L_T contributed by the spheroidal component and the ratio r_e^I/r_e of its effective radius to that of the whole galaxy are quantitative measures of the bulge to disk ratio.

A number of important physical and kinematical parameters are closely related to the Hubble stage index (T = 1 to 9 from Sa to Sm) and to the luminosity index $\Lambda = (T + L)/10$, where L is the luminosity class (L = 1 to 9 from SI to SV) as reported to the Yale Conference last year (de Vaucouleurs 1977a).

3. A TWO-COMPONENT MODEL OF THE GALAXY

A spheroid + disk model is completely determined by the effective parameters r_e, I_e of each component. The effective radius r_e^I of the galactic spheroid can be estimated from the distribution of globular clusters which obey closely the $r^{1/4}$ law (de Vaucouleurs 1977b) in both the Galaxy and M31, and in the latter with precisely the same scale factor as the luminosity distribution in the spheroid (de Vaucouleurs and Buta 1978). Photometric observations of the galactic bulge along the prime meridian provide a test of the validity of the model and give the brightness scale factor I_e^I. Then the total luminosity of the spheroid is $L_I = 7.268\pi I_e^I(r_e^I)^2$. The effective radius r_e^{II} of the exponential disk can be derived from the surface brightness of the disk near the sun $\mu^{II}(r_o) = 24.16$ mag sec^{-2} calculated from star counts and from the assumption that its apparent face on surface brightness at the center is $\mu_B^{II}(0) = 21.65 \pm 0.3$ mag sec^{-2} (Freeman 1970). The validity of this assumption was verified a posteriori by comparison of the derived standard isophotal diameter D_o (at $\mu_B = 25.0$ mag sec^{-2}) with independent estimates. Then $L_{II} = 3.803\pi I_e^{II}(r_e^{II})^2$ and the total luminosity is $L_T = L_I + L_{II}$.

A model based on these principles was recently constructed in collaboration with W. D. Pence. This model assumes a solar galactocentric distance $r_o = 8.0$ kpc; it is consistent with the face on total surface brightness (disk + spheroid) inferred from star counts $\mu_B(r_o) = 23.93$ mag sec^{-2} (for an absorbing layer of constant optical depth $A_B = 0.4$ mag) and gives a good representation of new observations of the luminosity distribution along the galactic prime meridian with the 0.9-m reflector at McDonald Observatory. Two cases were considered: (a) a spherical bulge of effective radius $r_e^I = 2.67$ kpc, suggested by the distribution of globular clusters (de Vaucouleurs and Buta 1978), and (b) an ellipsoidal bulge of effective radius $a_e^I = r_e^I/(0.6)^{1/2} = 3.45$ kpc, if the axis ratio is $c/a = 0.6$ as suggested by infra-red photometry of the central regions (Maihara et al. 1978). The results differ little, except that the fractional luminosity k_I of the spheroidal component is 28% in case (a) and 40% in case (b). Since the true situation is intermediate between these two extreme cases the averages of the two solutions are adopted in what follows.

4. GALACTIC PARAMETERS

For a distant observer the face-on total magnitude of the Galaxy is $M_T^o(B) = -20.08$; the corresponding luminosity is $L_T(B) = 1.6 \cdot 10^{10} \mathcal{L}_\odot$ of which half is emitted within the effective radius $r_e = 5.1$ kpc. The fraction $k_I = L^I/L_T = 0.34 \pm 0.04$ contributed by the spheroidal component is in close agreement with corresponding values for the Sbc (T = 4) galaxies NGC 5194 (k_I = 0.32) and NGC 6744 (k_I = 0.33); it is definitely less than in M31 (Sb, T = 3, k_I = 0.45), but more than in NGC 253 (Sc, T = 5, k_I = 0.15) (de Vaucouleurs 1958, Pence 1978). This is a strong indication that the Hubble stage of our Galaxy is Sbc, or T = 4 ± 0.5.

The calculated face-on isophotal diameter of the Galaxy at the μ_B = 25.0 level is D_o = 23.0 kpc. This is in close agreement with three independent estimates from statistical relations for spiral galaxies: (a) if $V_M = V(R_M)$ is the maximum rotational velocity, a sample of 18 well-observed galaxies shows that $\log V_M + \log(2R_M/D_o) = 2.18 \pm 0.03$ is a constant independent of morphological type and luminosity class (de Vaucouleurs 1977a). In our Galaxy V_M = 255 kms^{-1} and R_M = 7.0 kpc (for r_o = 8.0) imply D_o = 23.4 kpc, (b) if D_1 is the diameter of the largest ring-shaped HII region, a sample of 10 nearby spirals shows that $\log D_o/D_1 = 1.98 - 0.135(M_T^o + 20)$ for all types Sb and later (de Vaucouleurs 1978c). In our Galaxy $D_1 \geq 220$ pc (the Gum nebula) and $M_T^o = -20.1$ imply $D_o \geq 21.5$ kpc, (c) both V_M and D_o are statistically related to the luminosity index Λ by $\log V_M = 2.15 - (\Lambda - 1)$ and $\log D_o = 4.18 - 0.6(\Lambda - 1)$ (de Vaucouleurs 1977a, 1978a). For V_M = 255 kms^{-1} these two relations imply D_o = 21.6 kpc.

The three independent estimates of D_o are in good agreement with each other and with the value derived from the two-component model. This proves that the mean luminosity gradient of the disk between 8 and 11.5 kpc is very nearly the same as between 0 and 8 kpc. Conversely this can be taken as evidence that the "Freeman constant" $\mu^I(0) = 21.65$ applies to our Galaxy.

The luminosity index of the Galaxy derived from the two equations above are $\Lambda(V_M) = 0.75$ and $\Lambda(D_o) = 0.70$; a third estimate from the correlation between Λ and absolute magnitude (de Vaucouleurs 1978d) $M_T^o = -19.15 + 3.0(\Lambda - 1)$ gives $\Lambda(M_T^o) = 0.69$, if $M_T^o = -20.08$. The three determinations are in remarkably close agreement with $\langle\Lambda\rangle = 0.71 \pm 0.03$. Since the most probable combination of morphological type T and luminosity class L is T = L + 1, with a dispersion σ = 1.4 (de Vaucouleurs 1977a, 1978d), the previous conclusion that T = 4 ± 0.5 implies that L = 3 ± 1. In conventional notation this means a classification Sbc II.

The mean intrinsic color of the spheroidal component derived from the new observations of the bulge (corrected for interstellar extinction and scattering) is $\langle B - V\rangle_I = 0.65 \pm 0.05$, in close agreement with previous estimates (Arp 1965); the mean color of the disk stellar population derived from local luminosity functions is $\langle B - V\rangle_{II} = 0.40$ (after Holmberg 1950, transformed to B,V system). In the face-on view a uniform absorbing layer with A_B = 0.4 mag produces a color excess E(B - V) = 0.039 mag and the total color index of the Galaxy for an external observer is $(B - V)_T = 0.53 \pm 0.05$, in fair agreement with the average values 0.57 for Sbc and 0.51 for Sc galaxies (de Vaucouleurs 1977a).

The hydrogen index HI implied by the luminosity index $\Lambda = 0.71$ and the statistical relation <HI> = 1.65 - 1.30(Λ - 1) (de Vaucouleurs 1977a) is HI = 2.03 which by definition of the hydrogen index corresponds to a logarithmic ratio of the neutral hydrogen mass $\mathfrak{M}_H$ to the face on B luminosity (both in solar units) $\log \mathfrak{M}_H/\mathfrak{L}_B = -0.02 - 0.4(HI) = -0.83$. With $M_T^o = -20.08$, this implies a total HI mass $\mathfrak{M}_H = 2.32.10^9 \mathfrak{M}_\odot$.

The HI mass in the disk within $r = 1.5\ r_o$ is $M_H(r < 1.5\ r_o) = 1.66.10^9 \mathfrak{M}_\odot$ or 71% of the total (after Baker and Burton 1975, scaled down from $r_o = 10$ kpc to $r_o = 8$ kpc); this leaves $0.66.10^9 \mathfrak{M}_\odot$ of atomic hydrogen or 29% of the total to be distributed in the warped disk at $r > 1.5\ r_o = 12$ kpc and in the galactic corona, which appears very reasonable.

With the total mass of the Galaxy derived from the HI rotation curve (Schmidt 1965) and from the velocity dispersion of globular clusters (de Vaucouleurs 1977b) $\mathfrak{M}_T \simeq 2.10^{11} \mathfrak{M}_\odot$ (for $r_o = 8.0$ kpc), the mass-luminosity ratio is $\mathfrak{M}_T/\mathfrak{L}_B = 12.5$ (solar units), and the HI mass fraction is $\mathfrak{M}_H/\mathfrak{M}_T = 0.012$. The total hydrogen fraction (including ionized and molecular) may be about 2 to 2.5%.

In conclusion, our Galaxy appears to be a normal giant, multi-arm spiral of type SAB(rs)bc II, total absolute magnitude -20.1(B), -20.6_5(V) and effective diameter 10 kpc, if $r_o = 8.0$ kpc. It is in many respects similar to NGC 1073, 4303, 5921 and 6744 (for illustrations see de Vaucouleurs 1970). The average of these four galaxies is in remarkably close agreement with the photometric properties of the Galaxy. The main results are summarized in Table I; a more detailed report will appear in the Astronomical Journal. This work was supported in part by the National Science Foundation under Grant AST 75-22900.

TABLE I
Photometric Parameters of the Galaxy and of Average of Four Spirals[†]

Parameter		Average	Galaxy
Isophotal diameter, D_o	(kpc)	22.7	23.0
Effective diameter, D_e	"	10.3	10.2
Inner ring diameter, D(r)	"	5.75	6.:
Absolute B magnitude, M_T^o		-20.0	-20.1
Color index, $(B - V)_T^o$		0.51	0.53
Spheroidal fraction, $k_I(B)$		(0.33)[+]	0.34
Luminosity index, Λ		0.67	0.71
Mean effective surface brightness*, m_e'		22.08	22.06

[†]NGC 1073, SB(rs)c II; NGC 4303, SAB(rs)bc I; NGC 5921, SB(rs)bc I-II; NGC 6744, SAB(r)bc II. [+]NGC 6744 only. *B mag sec^{-2}.

REFERENCES

Arp, H. C. 1965, Astrophys. J., 141, 43.
Baade, W. 1951, Publ. Univ. Michigan Obs., X, 16.
Baker, P. L., Burton, W. B. 1975, Astrophys. J., 198, 281.
Becker, W. 1964, Z. Astrophys., 58, 202.
Courtès, G. 1972, in *Vistas in Astronomy*, 14, 140.
de Vaucouleurs, G. 1958, Astrophys. J., 128, 465.
de Vaucouleurs, G. 1959, in *Handbuch der Physik*, Vol. 53, 275.
de Vaucouleurs, G. 1962, in IAU Symp. no. 15, 1.
de Vaucouleurs, G. 1963, Astrophys. J. Suppl., 8, no. 74, 31.
de Vaucouleurs, G. 1964, in IAU Symp. no. 20, 88.
de Vaucouleurs, G. 1970, in IAU Symp. no. 38, 18.
de Vaucouleurs, G. 1974, in IAU Symp. no. 58, 1.
de Vaucouleurs, G. 1977a, in *The Evolution of Galaxies and Stellar Populations*, Yale Univ. Obs., 43.
de Vaucouleurs, G. 1977b, Astron. J., 82, 456.
de Vaucouleurs, G. 1978a, Astrophys. J., 223, in press.
de Vaucouleurs, G. 1978b, Astrophys. J., 224, in press.
de Vaucouleurs, G. 1978c, Astrophys. J., 224, in press.
de Vaucouleurs, G. 1978d, Astrophys. J., submitted.
de Vaucouleurs, G., Buta, R. 1978, Astron. J., submitted.
Freeman, K. C. 1970, Astrophys. J., 160, 811.
Georgelin, Y. M., Georgelin, Y. P. 1976, Astron. Astrophys., 49, 57.
Holmberg, E. 1950, Medd. Lunds Obs., II, no. 128, 44.
Kerr, F. J. 1967, in IAU Symp. no. 31, 239.
Kerr, F. J. 1969, in Ann. Rev. Astron. Astrophys., 7, 39.
Kerr, F. J. 1970, in IAU Symp. no. 38, 95.
Kerr, F. J., Westerhout, G. 1965, in *Stars and Stellar Systems*, V, 167.
Maihara, T., Oda, N., Sugiyama, T., Okuda, H. 1978, Publ. Astron. Soc. Japan, 30, 1.
Mills, B. Y. 1959, in IAU Symp. no. 9, 441.
Oort, J., Kerr, F. J., Westerhout, G. 1958, Monthly Notices Roy. Astron. Soc., 118, 379.
Pence, W. D. 1978, Univ. Texas Ph.D. thesis.
Peters, W. L. 1974, Astrophys. J., 195, 617.
Sandage, A. 1961, *The Hubble Atlas of Galaxies*.
Sandage, A. 1975, in *Stars and Stellar Systems*, IX, 1.
Schmidt, M. 1965, in *Stars and Stellar Systems*, V, 513.
Schmidt-Kaler, T., Schlosser, W. 1973, Astron. Astrophys. 29, 409.
Schweizer, F. 1976, Astrophys. J., 31, 313.
Simonson, S. C. 1976, Astron. Astrophys., 46, 261.
Simonson, S. C., Mader, G. L. 1973, Astron. Astrophys., 27, 337.

DISCUSSION

van den Bergh: Recently Tammann used $L = 4 \times 10^{10}\ L_\odot$ for the Galaxy to obtain a mean interval τ = 20 yr between galactic supernova outbursts. Lowering the galactic luminosity to $L = 1.7 \times 10^{10}\ L_\odot$, as you have just suggested, would increase τ to a (perhaps more comfortable!) 50 years.

Schmidt-Kaler: I completely agree with the type Sbc in your system of classification for our Galaxy. But I wonder about the luminosity class: from the size, we find about $R_H \cong 15$ kpc or a little more (if $R_\odot$ = 8 kpc); from the maximum rotation velocity, $v_m \cong 250$ km/s.

de Vaucouleurs: The Holmberg radius is 17 kpc in our mode; the luminosity class is derived mainly from the total absolute magnitude and from our re-calibration of luminosity classes. With all the uncertainties in photometry, calibrations, etc., a 12% difference is not significant.

Basu: What is the distance of the Inner Lindblad Resonance in your model of our Galaxy.

de Vaucouleurs: I don't know. Can one of the theorists present tell us?

Contopoulos: It is about 3.5 kpc (for R_o = 10 kpc), essentially independent of the model adopted.

Miller: I should like to comment on the interpretation of numbers such as the distance to the Galactic center, and to pose a question for the observers. Is our Galaxy axisymmetric, as we usually assume in interpreting observations, or might it be elongated, possibly as far as a 2:1 axis ratio? The background to this question is as follows. In self-consistent N-body experiments we typically find a prolate bar that rotates end-over-end in space. An elliptical galaxy flatter than E2 is probably prolate. The same situation must occur with S0's and ordinary spirals. With spirals, gas should settle into the equatorial plane and assume an elongated form. Gas streaming and the local standard of rest would move on elliptical tracks about the center of the galaxy. The velocity dispersion in these prolate forms is anisotropic, with the long axis parallel to the long axis of the elongated form. We can tentatively locate the Sun in such a system by noting the vertex deviation: the long axis of the velocity ellipsoid leads the motion of the local standard of rest. This would place the Sun some 10-15° away from the long axis in the direction of streaming. There would be a small K-term, but within observational limits. The streaming takes place within this elongated pattern as the pattern itself rotates. A spiral pattern might exist within this elongated outline just as it could within a circular outline. This possibility complicates Galactic models appreciably, and is unfortunate from that point of view. I would like to ask observers to tell me of observational evidence that forces us to treat the Galaxy as circular, or as nearly circular.

Burton: Deviations from axial symmetry are illustrated by systematic differences in the total extent of HI spectra measured at corresponding longitudes on either side of the Sun-center line. I have given such a plot in Figure 3 of P.A.S.P. 85, 679, and in Figure 4.7 of "Galactic and Extragalactic Radio Astronomy" (Verschuur and Kellermann, eds.). There are trends that could be interpreted in terms of an overall elongated form. In addition, it is interesting in this regard that the $b = 0°$ v,ℓ diagram is pinched closer to $v = 0$ km s^{-1} at $\ell = 186°$ than at $\ell = 180°$.

ROTATION CURVES OF HIGH-LUMINOSITY SPIRAL GALAXIES AND THE ROTATION CURVE OF OUR GALAXY

Vera C. Rubin
Department of Terrestrial Magnetism
Carnegie Institution of Washington

ABSTRACT. Rotation curves of high luminosity spiral galaxies are flat, to distances as great as r=49 kpc. This implies a significant mass at large r. Rotational velocities increase about 20 km/s across a spiral arm, as predicted by the density wave theory. By analogy, it is suggested that our Galaxy has a flat rotation curve out to $r \sim 60$ kpc, with $V \sim$ constant at near the solar rotational velocity, and $\mathcal{M} \sim 7 \times 10^{11} \mathcal{M}_\odot$. Values of A and B imply that the sun is not located in a spiral arm.

Knowledge of the structure and dynamics of our Galaxy has come not only from the study of stellar and gas motions within the Galaxy, but also from a comparison of the properties of our Galaxy with those of neighboring spiral galaxies. Until recently, this comparison was hampered by very incomplete observations of rotation curves of external galaxies. Optical observations generally determined velocities only across the nucleus and inner regions; velocities at large nuclear distances were rarely obtained. Radio 21-cm line observations generally integrated all the neutral hydrogen into a single profile; all spatial information was lost. Recently, both optical and radio instrumentation has developed sufficiently so that detailed rotation curves of high accuracy can be obtained across most of a galactic disk. An outstanding recent study of rotation curves from 21-cm line observations is due to Bosma (1978). I shall present results of our optical studies; briefly discuss their implications for spiral dynamics and kinematics, and from them infer an extended rotation curve for our Galaxy. This curve is necessarily speculative. However, unless our Galaxy is very different from the high-luminosity sample which we have studied, the major characteristics of its rotation curve are reasonably well defined.

ROTATION CURVES FOR HIGH LUMINOSITY SPIRAL GALAXIES

For a sample ($n \sim 15$) of high luminosity spiral galaxies, Hubble types (HT) Sa through Sc, we have obtained accurate rotation curves which extend to about 80% of the deVaucouleurs radii. The galaxies were

W. B. Burton (ed.), The Large-Scale Characteristics of the Galaxy, 211–220.

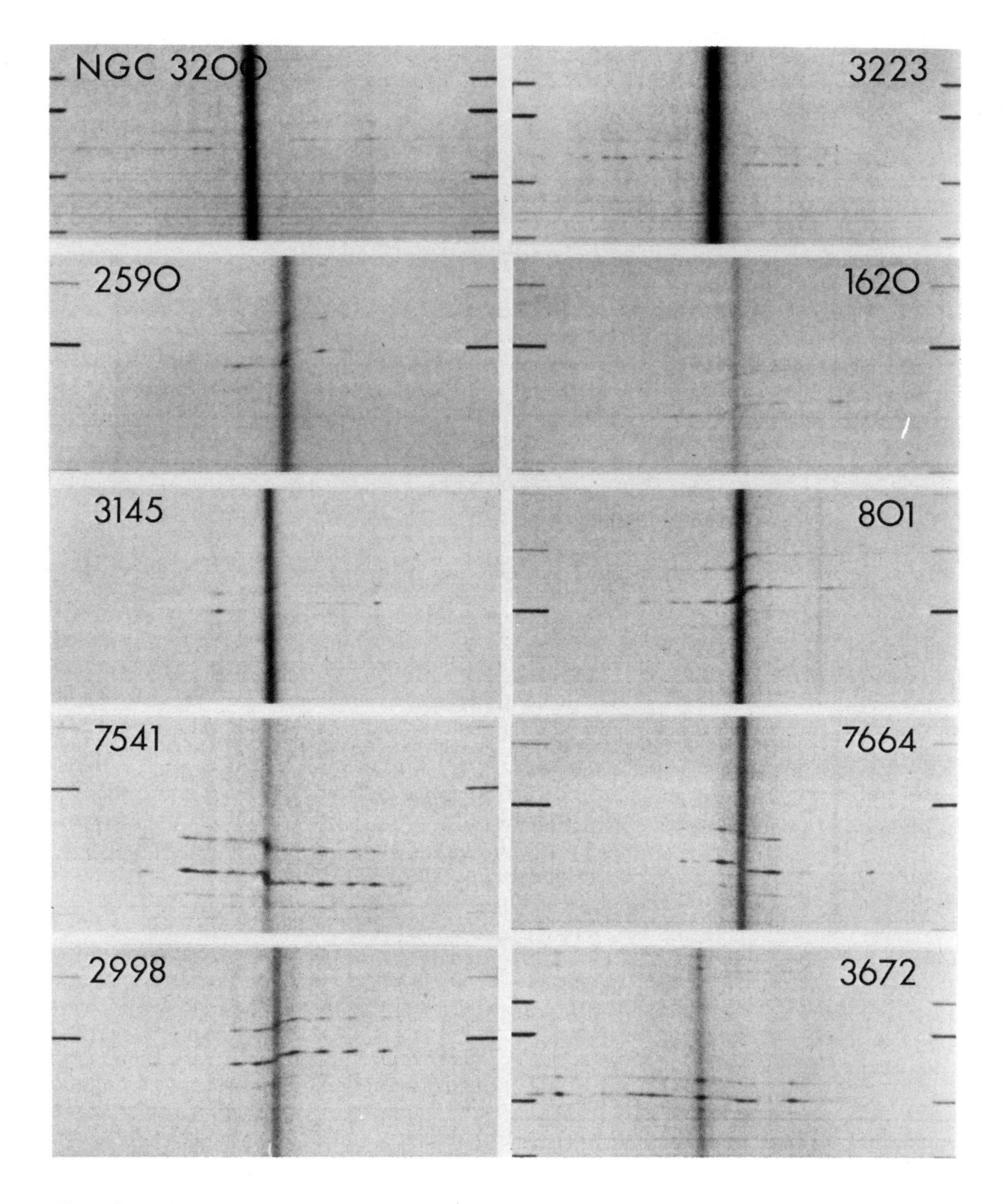

Fig. 1. Hα region of spectra for galaxies of different Hubble types, taken with the KPNO or CTIO 4-m spectrographs plus Carnegie image tube. Exposure times are 90 min. to 200 min., dispersions are 25 or 52 A/mm. Linear extent of emission varies from a radius of 17 kpc (NGC 2590) to 49 kpc (NGC 801). We adopt H = 50 km/s per Mpc.

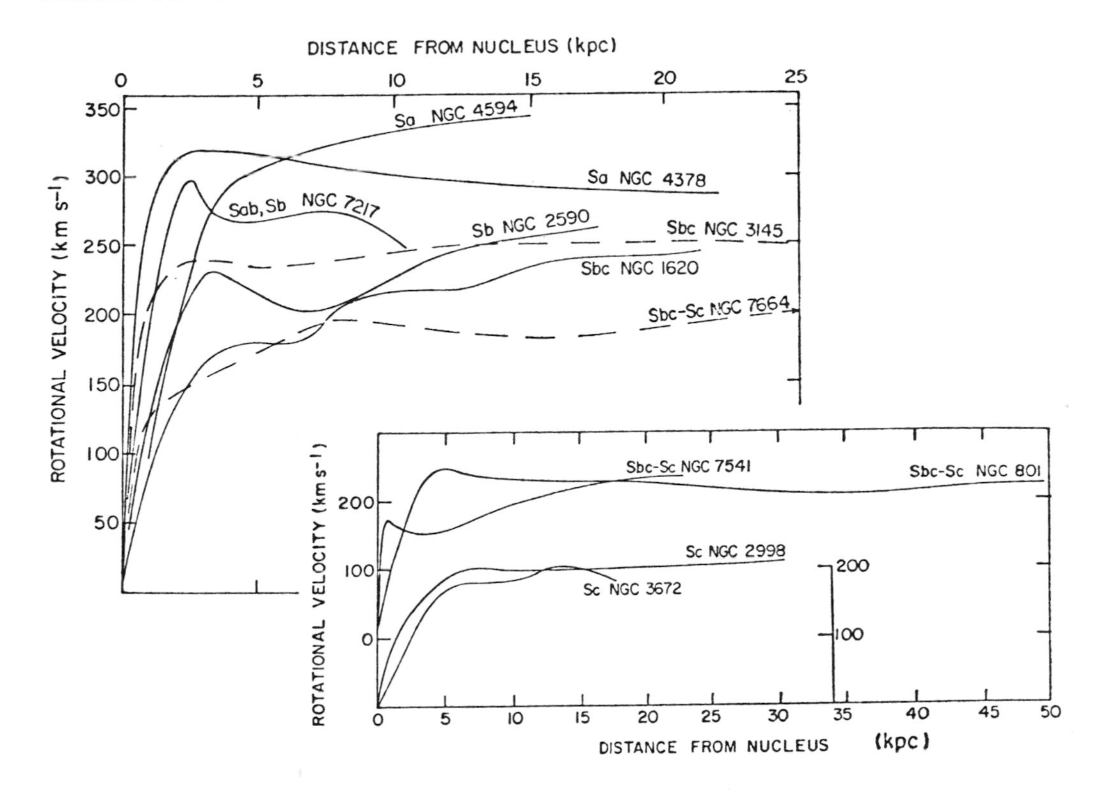

Fig. 2. Rotational velocities for 11 galaxies, as a function of distance from nucleus, Curves have been smoothed to remove velocity undulations across arms and small differences between major axis velocities on each side of nucleus. Early-type galaxies have higher peak velocities than later types.

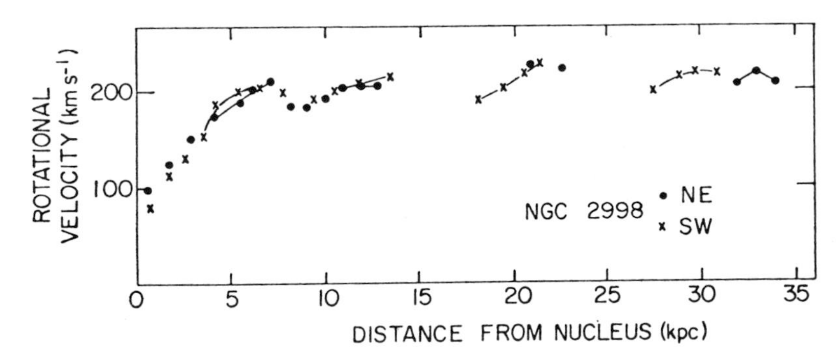

Fig. 3. Rotational velocities in NGC 2998, as a function of distance from nucleus. Velocities for strongest emission are connected with lines. Note fairly good velocity agreement between velocities from NE and SW major axes, and positive velocity gradient across arms.

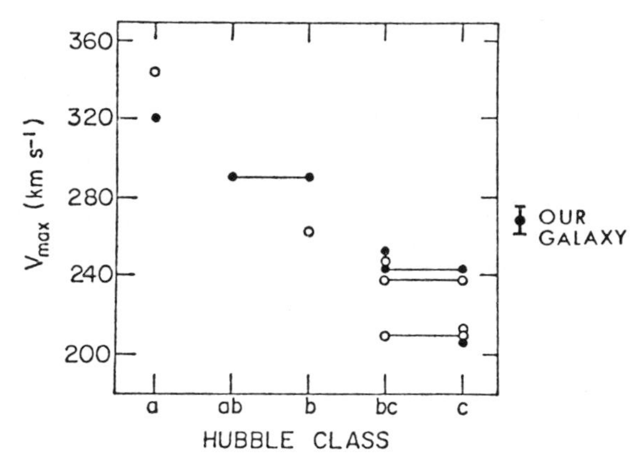

Fig. 4. Maximum rotational velocity, in the plane of the galaxy, as a function of Hubble type. Galaxies with 2 classifications are entered twice with a connecting line. Open circles denote peak velocity at the last measured velocity (i.e., rising rotation curve).

chosen with extreme care: to have angular diameters near 3' or 4' to match the KPNO and CTIO spectrograph slit lengths; to be of high inclinations so that uncertainties in inclinations produce little effect on rotational velocities and hence masses; to be of high luminosity as indicated by the widths of their 21-cm profiles; and to have large linear diameters.

Optical spectra were obtained with the Kitt Peak and Cerro Tololo 4-m spectrographs plus Carnegie image tube, generally at a dispersion of 25A/mm. Errors in the rotational velocities (measuring errors plus projection uncertainties) are generally less than 8 km/s per point. Reproductions of 10 spectra are shown in Fig. 1. Rotation curves are drawn in Fig. 2; data for NGC 4594 come from Schweizer (1978).

The following conclusions come from analysis of these data (Rubin, Ford, and Thonnard 1978): (1) All rotation curves are nearly flat to distances as great as 50 kpc radius. Eight of these 11 galaxies have their maximum velocity at r > 10 kpc. Secondary velocity undulations indicate that rotational velocities are lower by about 20 km/s on the inner edges than on the outer edges of spiral arms. This is especially apparent in NGC 2998, whose velocities are plotted in Fig. 3. This observation confirms a prediction of the density wave theory, and is a major result of our study. While velocity gradients would exist also in any gravitational model with mass concentrations in the arms, the density enhancement would have to be enormous to produce the observed effect; (2) There is a pronounced increase in the maximum rotational velocity, Vmax, with earlier HT; Fig. 4 shows this tight correlation. A correlation between Vmax and HT found earlier by Brosche (1971) lies about 50 km/s below the relation indicated in Fig. 4, and is defined principally by galaxies of types later than Sbc. This suggests that Fig. 4 represents an upper envelope defined by high luminosity galaxies.

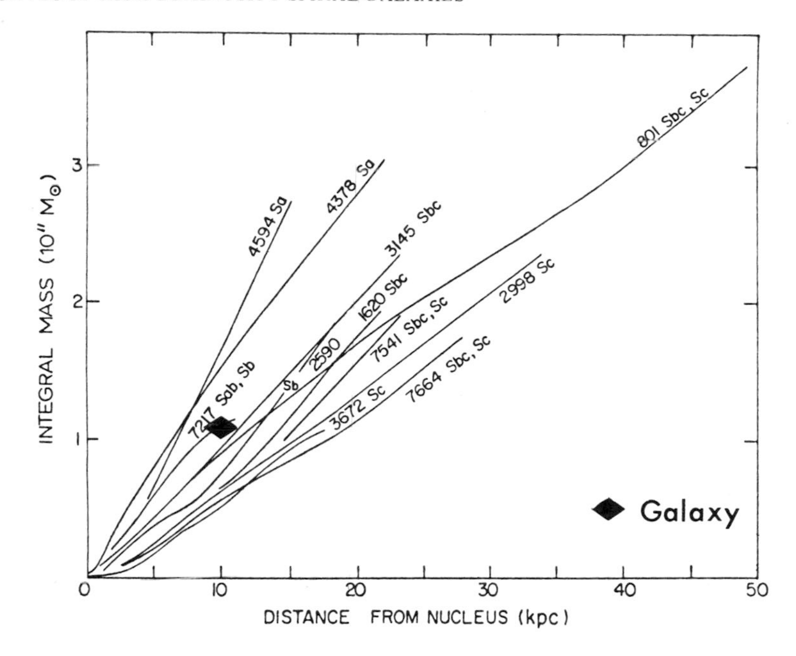

Fig. 5. Integral mass within a disk of radius r, as a function of r, for galaxies types Sa through Sc. Scale gives mass for disk models; masses for spherically modeled galaxies are 1.4 times larger.

A larger sample of rotation curves for galaxies of all luminosity classes must be available before the true scatter of Vmax versus HT and luminosity can be determined; (3) Masses out to the deVaucouleurs radius ($25^m/\square''$) have been calculated, and are generally accurate to about 25%. Masses range from $1<\mathcal{M}<7\times10^{11}\mathcal{M}_\odot$ for spherical modeled galaxies. Thus some spiral galaxies will have masses approaching $10^{12}\mathcal{M}_\odot$ to their Holmberg radii ($26\overset{m}{.}5/\square''$). Integrated mass as a function of radius is plotted in Fig. 5. The linear increase in mass with r is a consequence of the flat rotation curves, for $\mathcal{M}\propto V^2 \cdot r$.

SOME IMPLICATIONS FOR GALACTIC DYNAMICS AND THE DENSITY WAVE THEORY

Spiral galaxies with rotation curves of the forms shown in Figs. 2 and 3 possess interesting dynamical properties. Values of Oort's constants A and B calculated locally for a generally flat but undulating rotation curve (Fig. 3) will oscillate, with A = -B at the local maxima and minima, but with A and $|B|$ slowly decreasing with r. Thus values of A and B determined for stars within any few kiloparsec radius may not indicate the large scale characteristics of the rotation field, but merely the sign of the gradient of the rotation curve locally.

Rotational periods at large r are long, $\sim10^9$ years, and V/r is only a very slowly decreasing function of r. Hence the old worry of winding up of spiral arms will be less relevant. Within the framework of the density wave theory, rotating gas at large r will encounter a two-armed spiral pattern only at widely separated intervals; stars will form and die long before the next passage through the density wave. Lacking alternative mechanisms for massive star production, such galaxies would have two widely spaced arms, of the ∽ type. One alternative mechanism is proposed by Seiden and Gerola (1977), who suggest that self-propagating star formation in a differentially rotating disk can produce persistent large scale spiral features. Their computer pictures indicate features of a feathery nature. Perhaps different mechanisms of star formation produce galaxies of different morphological types; in some galaxies a combination of mechanisms may participate.

Problems with corotation continue to exist. If corotation is placed at the outer HII regions or at the disk limits, then it occurs at r = 49 kpc in NGC 801, near 35 kpc for other program galaxies. No single pattern speed can account for spiral structure over this distance; solutions involving several spiral modes are required. Theoretical progress (Bertin *et al*. 1977) will have to be rapid to keep pace with observational results.

ARE THERE SPIRAL GALAXIES WITH FALLING ROTATION CURVES?

What was the reason for the earlier belief that disk galaxies had Keplerian velocities at moderate nuclear distances? A search of the literature now reveals little to support this belief. Early velocity measures by the Burbidges and colleagues generally show a large scatter near Vmax, followed by sometimes falling, sometimes rising, velocities. But it is almost impossible to identify a galaxy with a falling optical rotation curve. NGC 4321 may be an example (van der Kruit 1973), although measured velocities cover less than 1/2 the optical disk. From Bosma's 21-cm compilation, M81, M51, and M101 appear to have falling rotation curves. For M81, the intergalactic gas enveloping both M81 and M82 confuses the determination of the rotation curve; for M51 the 21-cm velocities are only marginally falling (Shane 1975; r = 12 kpc) and conflict with the optical velocities which are constant (Burbidge and Burbidge 1964; r = 14 kpc). We conclude that nearly constant velocities and significant mass at large r are the rule, at least for high luminosity spiral galaxies. Exceptions have yet to be confirmed.

THE EXTENDED ROTATION CURVE OF OUR GALAXY

Based on observations of high luminosity spiral galaxies, and on the premise that our Galaxy is of luminosity class I or II, we suggest the following rotational properties for our Galaxy. Incomplete justification is given for most of the assumptions; this is due to a lack of space and knowledge. We adopt $R_{\odot}$ =10kpc and $V_{\odot}$=250km/s for the solar neighborhood, although recent work suggests that Ro is less than 10kpc

(Oort and Plaut 1975, Harris 1976) and Vo is greater than 250 (Lynden-Bell and Lin 1977) and less than 250 (Knapp, this Symposium).

Current values of Oort's constants are $A = 15.6 \pm 2.8$, $B = -11.4 \pm 2.8$ km/s per kpc (Fricke and Tsioumis 1975), although O-B2 stars produce values as discrepant as $A = +26$, $B = -37$ (Asteriadis, 1977). At the 1σ level, $A = -B$ is not excluded, i.e., the rotation curve could be flat in the solar vicinity. More likely, there is a negative velocity gradient at the position of the sun. Because spiral arms show positive velocity gradients, the sun is probably not located in a spiral arm. Lin et al. (1977) conclude that the sun is situated between spiral arms, from a comparison of stellar motions with predictions from the density wave theory.

Interior to the solar circle, HI (Sinha 1978) and HI plus CO (Burton and Gordon 1978) velocities at the tangent points are used to derive new rotation curves similar to the Schmidt (1965) rotation curve, $r > 4$kpc. (Fig. 6). Differences between HI velocities in the 1st and 4th quadrants of $\sim$10 km/s are attributed by Sinha to streaming motions. Exterior to the solar circle, HII regions (Georgelin and Georgelin 1976), OB stars (Rubin 1965), OB stars with Walborn distances (unpublished), stellar aggregates associated with Sharpless regions (Jackson, this Symposium) and HII region recombination line velocities (calculated from Silverglate and Terzian 1978) suggest generally flat or rising velocities, $12 < r < 18$ kpc, with perhaps a shallow minimum, $r \sim 11$kpc. (Fig. 7).

How far does the galaxy extend? The surface brightness at the sun compared with appropriate external galaxies, the hydrogen radii of late-type galaxies which are often twice the optical, and the extent of the globular cluster system ($r \sim 40$ kpc, Harris 1976) all suggest that our Galaxy could have a radius $\sim$50 kpc. Faint anticenter blue stars

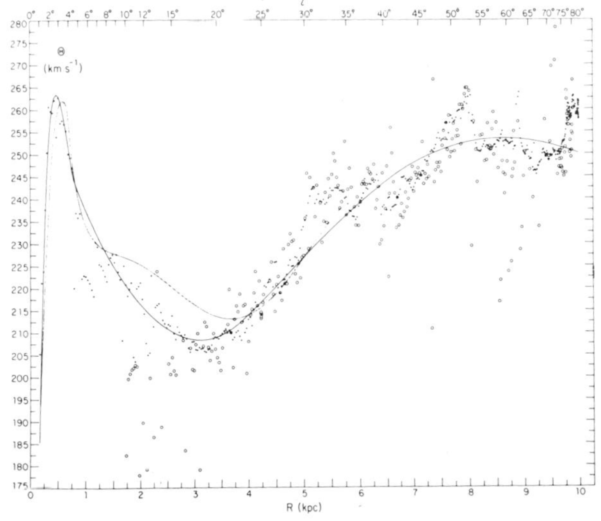

Fig. 6. Linear velocity as a function of distance from the center of our Galaxy, from terminal velocities of CO and HI (Burton and Gordon 1978).

identified by Rubin and colleagues (1971, 1974) show no proper motions (Cudworth 1974, 1975, 1977), and have photometric distances ∿20 - 30 kpc from the center (Chromey 1978). Spectroscopic data confirm these distances; one O5 star (R152, ℓ = 219°, b = +7.6°) is located 48 kpc beyond the sun, 56 kpc from the center, with z = 6 kpc. Such stars may belong to the 50% of all early O stars which are located in regions free of nebulosity (Torres-Peimbert et al. 1974) and may constitute a fraction of the high velocity halo population. If part of a more normal disk population, they will be valuable for determining the rotation curve at large nuclear distances. However, lack of HII regions at large galactocentric distances constitutes a problem for the extended galaxy model.

For a rotation curve, flat at near the solar velocity to r = 60 kpc, then $\mathcal{M} \sim 7 \times 10^{11} \mathcal{M}_\odot$, which is just the Galaxy mass out to r = 60 kpc derived by Hartwick and Sargent (1978) from velocities of intergalactic tramp clusters and satellite galaxies (assuming an isothermal velocity dispersion). If V is constant out to r = 75 kpc, $\mathcal{M} \sim 10^{12} \mathcal{M}_\odot$, close to the mass at r_1 = 75 kpc derived by Einasto et al. (1976). A small Galaxy, $\mathcal{M} \sim 3 \times 10^{11} \mathcal{M}_\odot$ with $\mathcal{M} \propto r$ out to r = 28 kpc and no significant mass beyond, would have a Keplerian velocity decrease to V∿165 km/s at r = 60 kpc, and still be at the lower limit of the Hartwick-Sargent mass determination. But such a galaxy would be atypical. We believe that we live in a high luminosity massive galaxy which extends to r = 60 or 75 kpc. Studies of the extent of star formation and the spiral pattern of the outer regions should occupy observers for many years.

I thank the directors of KPNO and CTIO for making telescope time available, and Drs. W.K. Ford, Jr., N. Thonnard, and numerous colleagues for valuable discussions.

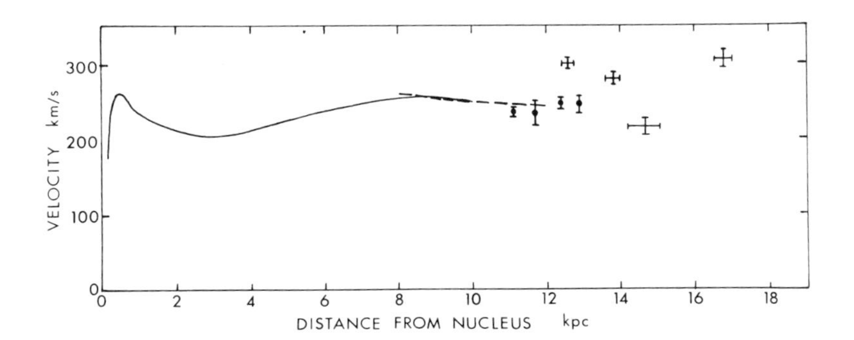

Fig. 7. Schematic rotation curve for the Galaxy, compiled from Burton and Gordon (1978, r < 10 kpc): Georgelins (1976, 8 < r < 12): Rubin (1965, r ∿ 12); Walborn (unpublished, r ∿ 12); Silverglate and Terzian (1978, r ∿ 12); and Jackson (this Symposium, 13 < r < 17kpc).

REFERENCES

Asteriadis, G.: 1977, Astron. Astrophys. 56, pp. 25-38.
Bertin, G. Lau, Y.Y., Lin, C.C., Mark, J. W. -K., and Sugiyama, L.: 1977, Proc. Nat. Ac. Sci. U.S. 74, pp. 4726-4729.
Bosma, A.: 1978, Ph.D. dissertation, Rijksuniversiteit te Groningen.
Brosche, P.: 1971, Astron. Astrophys. 13, pp. 293-297.
Burbidge, E. M. and G. R.: 1964, Astrophys. J. 140, pp. 1445-1461.
Burton, W.B. and Gordon, M.A.: 1978, Astron. Astrophys. 63, pp. 7-27.
Chromey, F.R.: 1978, Astron. J. 83, pp. 162-166.
Cudworth, K.M.: 1974, Astron. J. 79, pp. 979-980; Astron. J. 80, pp. 826-827; Astron. J. 82, pp. 516-517.
Einasto,J., Haud,U., Joeveer,M., & Kaasik,A.: 1976, MNRAS 177,pp.357-375.
Fricke, W. and Tsioumis, A.: 1975, Astron. Astrophys. 42, pp. 449-455.
Georgelin, Y.M. and Y.P.: 1976, Astron. Astrophys. 49, pp. 57-79.
Harris, W.E.: 1976, Astron. J. 81, pp. 1095-1116.
Hartwick, F.D.A., and Sargent, W.L.W.:1978, Astrophys.J. 221, pp. 512-520.
Lin, C.C., Yuan, C., and Roberts, W.W.: 1977, preprint.
Lynden-Bell, D. and Lin, D.N.C.: 1977, MNRAS 181, pp. 37-57.
Oort, J.H. and Plaut, L.: 1975, Astron. Astrophys. 41, pp. 71-86.
Rubin, V.C.: 1965, Astrophys. J. 142, pp. 934-942.
Rubin, V.C., Ford, W.K. Jr., and Thonnard, N.: 1978, in preparation.
Rubin, V.C. and Losee, J.M.: 1971, Astron. J. 76, pp. 1099-1101.
Rubin,V.C., Westphahl,D., & Tuve,M.: 1974, Astron.J. 79, pp. 1406-1409.
Schmidt,M.: 1965, in Galactic Structure (U. of Chi. Press), pp.513-530.
Schweizer, F.: 1978, Astrophys. J. 220, pp. 98-106.
Seiden, P.E. and Gerola, H.: 1977, Bull. Am. Ast. Soc. 9, p. 641.
Shane,W.W.: 1975, in La Dynam. des Gal. Spir., CNRS #241, pp. 217-227.
Silverglate, P.R. and Terzian, Y.: 1978, preprint.
Sinha, R.P.: 1978, to appear Astron. Astrophys.
Torres-Peimbert, S., Lazcano-Aranjo, A., and Peimbert, M.: 1974, Astrophys. J. 191, pp. 401-410.
van der Kruit, P.C.: 1973, Astrophys. J. 186, pp. 807-813.

DISCUSSION

Greyber: Is there evidence to indicate how the nature of the rotation curves changes going towards types with enhanced emission, such as Markarian and Seyfert galaxies and radio galaxies like Centaurus A and Fornax A?

Rubin: Only very limited observations are available for these exotic objects. For a few Seyfert galaxies, rotational properties seem relatively normal. For radio galaxies, velocities are seldom available beyond the nucleus.

Innanen: A comment: The velocity dispersions of the points in the rotation curves appear remarkably small.

Rubin: Except for the nuclear regions, all line widths are probably instrumental, and thus less than 50 km s^{-1}.

van Woerden: Could one of our spectroscopic experts estimate the luminosity of HD 46150, and its uncertainty? I note that, at $\ell \sim 260°$, $b \sim +8°$ as given by Dr. Rubin, HD 46150 would not be in the well-known warp. (However, z(R) in warped disks may be oscillating.)

Rubin: Fred Chromey will discuss his results on "Spectroscopy of Distant Blue Stars near the Galactic Anticenter" at the Madison AAS meeting; details of the work will be available shortly (B.A.A.S., 1978).

Sanders: Several of the rotation curves which you show have two principal peaks. Indeed, the rotation curve of our own Galaxy has a conspicuous inner peak. Is the existence of an inner peak in the galaxies which you observe associated with any other galactic morphological characteristic, such as a conspicuous bulge?

Rubin: Unfortunately, we do not have direct large-scale plates of most of these galaxies. Some of them are not even in the revised de Vaucouleurs Catalogue. The prints from the Palomar Sky Survey are not adequate to answer your question.

van Woerden: Because M51 and M101 are both only little inclined to the sky, the rotation curves derived for both these galaxies are quite sensitive to a possible warp in their disks. Both Sancisi and Bosma have found that major warps occur frequently. Therefore, we should not attach much significance to the finding that the rotation curves of M51 and M101 are falling outwards.

Rubin: You are correct; I meant to suggest that M51, M81, and M101 are all slightly peculiar galaxies.

Pişmiş: I would like to know how you used the rotation curves to obtain masses of those galaxies where the rotation curves show wiggles or "waves". Did you take the upper envelope of the curve, an average curve, or a polynomial representing the details of the curve?

Rubin: The masses shown were derived from the observed velocities, smoothed only to take out local undulations, and from the assumption of an infinitely thin (Kuzmin) disk scaled by the factor 1.1. However, because the rotation curves are so flat, an expression of the form $M = 1.5 \times 10^5 \, V^2 r$ (V in km s^{-1}, r in kpc) reproduces all 11 masses to within 20%, and 7 of them to within 10%. Spherically modeled galaxies have masses well described by $M = 2.1 \times 10^5 \, V^2 r$.

RECENT EVIDENCE ON THE ROTATION CURVE OF OUR GALAXY FOR $R > R_0$

P. D. Jackson
University of Maryland

M. P. FitzGerald
University of Waterloo

A. F. J. Moffat
Université de Montréal

Studies of the rotation curve of our Galaxy at galactocentric radii, R, greater than the solar distance, R_0, from the center require the use of conventional optical techniques since the distances to as well as the radial velocities of Population I objects are needed.

HII regions are the best objects to use for a study of the rotation curve over large distances because (i) they can be detected easily on, say, Palomar red prints, (ii) they have high intrinsic luminosity and (iii) they have low (~ 8 km/sec) dispersion from circular galactic rotation. Y. M. Georgelin's thesis (Georgelin 1975) provides us with H α radial velocities for HII regions which are sufficiently accurate that the measurement error is usually less than the actual velocity dispersion. However, she usually relied on MK spectral classification of presumed exciting stars in order to determine the distance to the regions. This procedure is subject to the relatively large errors in MK luminosity determination for a single star as well as possible misidentification of the exciting star. Nevertheless, Georgelin's work clearly showed a discrepancy between her spectrophotometric distances and kinematic distances based on the Schmidt (1965) mass model in the sense that the former were greater than the latter.

The present authors have undertaken extensive UBV observations (Moffat, FitzGerald, and Jackson 1978 - in preparation) in order to determine more accurate distances by ZAMS fitting to de-reddened color-magnitude diagrams for stars in the neighborhood of distant (mostly Sharpless) HII regions, in the longitude range $\ell = 150°$ to $260°$. Many of the HII regions have such stellar 'aggregates' (clusters or associations) surrounding them. Image tube slit spectrograms for MK classification and (some) for radial velocity determinations were taken for many of the brighter member stars. The distances we determined by ZAMS fitting were not systematically different from spectrophotometric distances we determined for the same aggregates, nor were they systematically different from Georgelin's (1975) distances for stars studied in common. Our

W. B. Burton (ed.), The Large-Scale Characteristics of the Galaxy, 221–224.

stellar radial velocities (Jackson, FitzGerald and Moffat 1978 - in preparation), while showing a larger scatter (σ = 14 km/sec) were not systematically different from Georgelin's H α velocities.

Table 1 gives a summary of our results for regions having both distances (given in kpc with estimated errors) and velocities available. The radial velocity, referred to the LSR, is the mean of stellar and H α velocities if both were available for a given region.

Table 1. Distances and velocities for HII regions

Name	ℓ	b	d		V_r
S206 = NGC 1491	150.°6	-0.°9	3.0 ±	0.7	-22
{S207	151.2	+2.1}	7.6	0.8	-30
S208	151.3	+2.0			
Wat 1	151.4	+1.9	4.4	0.4	-30
S212 = NGC 1624	155.4	+2.5	6.0	0.6	-48
S217	159.2	+3.3	5.2	0.8	-37
S219	159.3	+2.6	4.2	0.6	-19
S224	166.2	+4.4	2.4	0.6	-89:
S225	168.1	+3.1	3.7	0.9	-22
S241	180.9	+4.1	4.7	1.2	-4
S247	188.9	+0.8	3.5	0.9	+6
S253 = Bo 1	192.4	+3.2	4.4	0.4	+6
S254-8 = IC 2162	192.6	-0.1	2.5	0.4	-4
S269	196.4	-1.7	3.8	1.0	+6
{S271	197.8	-2.3}	4.8	0.5	+15
S272	197.8	-2.3			

Name	ℓ	b	d		V_r
S284 = Dol 25	211.°9	-1.°3	5.2 ±	0.8	+27
Bo 2 (cluster)	212.3	-0.4	4.8	0.7	+49
S285	213.9	-0.6	6.9	0.7	+38
S287	218.1	-0.4	3.2	0.8	+21
S289	218.8	-4.6	7.9	0.8	+47
{S299	231.0	+1.5}	4.4	0.6	+37
S300	231.1	+1.5			
S301	231.5	-4.4	5.8	0.9	+50
S305	233.7	-0.3	5.2	1.3	+29
S306 = RCW 10	234.3	-0.4	4.2	0.4	+47
S307 = RCW 12	234.6	+0.7	2.2	0.5	+35
S309 = Bo 6	234.8	-0.2	5.5	0.8	+32
{RCW 19	253.8	-0.3}	3.0	0.7	+30
RCW 20 = NGC2579	254.4	-0.1			
Ru 44	245.7	+0.5	6.6	0.6	+20

Figure 1, from Moffat et al. (1978), shows the positions of HII regions plotted in the galactic plane, as well as the positions for O-B2 galactic clusters (see Vogt and Moffat 1975).

Figure 2 shows the galactic rotation curve in the form where $\omega - \omega_o$ is plotted against $R - R_o$ and where we have assumed circular galactic rotation for which $\omega - \omega_o = V_r / (R_o \sin \ell)$. We have also assumed the IAU values of 10 kpc for R_o and 250 km/sec for Θ_o , the circular velocity at the Sun, whence ω_o = 25 km/sec/kpc. Lines of constant circular velocity, Θ , equal to 250 km/sec (lower line) and 300 km/sec (upper line) are also shown on figure 2. We have used $\omega - \omega_o$ as ordinate in figure 2 instead of $\Theta = R\,\omega = (R/R_o)\,(\Theta_o + V_r / \sin \ell)$ because (i) errors in $\omega - \omega_o$ depend only on errors in V_r and (ii) it is easy to see the effect of assuming different values for Θ_o and R_o.

In figure 2, we see an initial drop in Θ from R = 10 kpc outwards to near R = 11.8 kpc where Θ is about 230 km/sec, but there is then a rise to about Θ = 275 km/sec near R = 14 kpc, followed by, perhaps, a drop back to near 250 km/sec at greater R. Thus, the rotation curve for our Galaxy seems flat overall at large galactocentric distances, but with large-scale regional deviations of about 15 km/sec.

Dr. Rubin, at this Symposium, showed evidence for overall flat rotation curves in the outer regions of most external spiral galaxies, but with clear rises across spiral arms. Note that the rise in Θ near R = 14 kpc shown in figure 2 corresponds to the distance of a general maximum in the HI density near R = 14 kpc as reported at this Symposium by Dr. Henderson.

The non-circular motions closer to the Sun shown in figure 2 have been discussed by Crampton and Georgelin (1975). Note that the value of $d\omega / dR$ in the neighborhood of the Sun is steeper than average and thus values for the rotation constant, A, near 15 km/sec/kpc determined by many workers is not inconsistent with the value for A of 12.5 km/sec/kpc required for a completely flat rotation curve with no local variations.

The effect of different values for R_0 can be easily seen in figure 2, since reasonable changes in R_0 have negligible effects on $R - R_0$. A smaller value for R_0 , say R_0 = 8.5 kpc, causes a steepening in the lines of constant Θ (after rescaling the ordinate to keep the positions of the data points the same) which worsens the fit in the outer parts

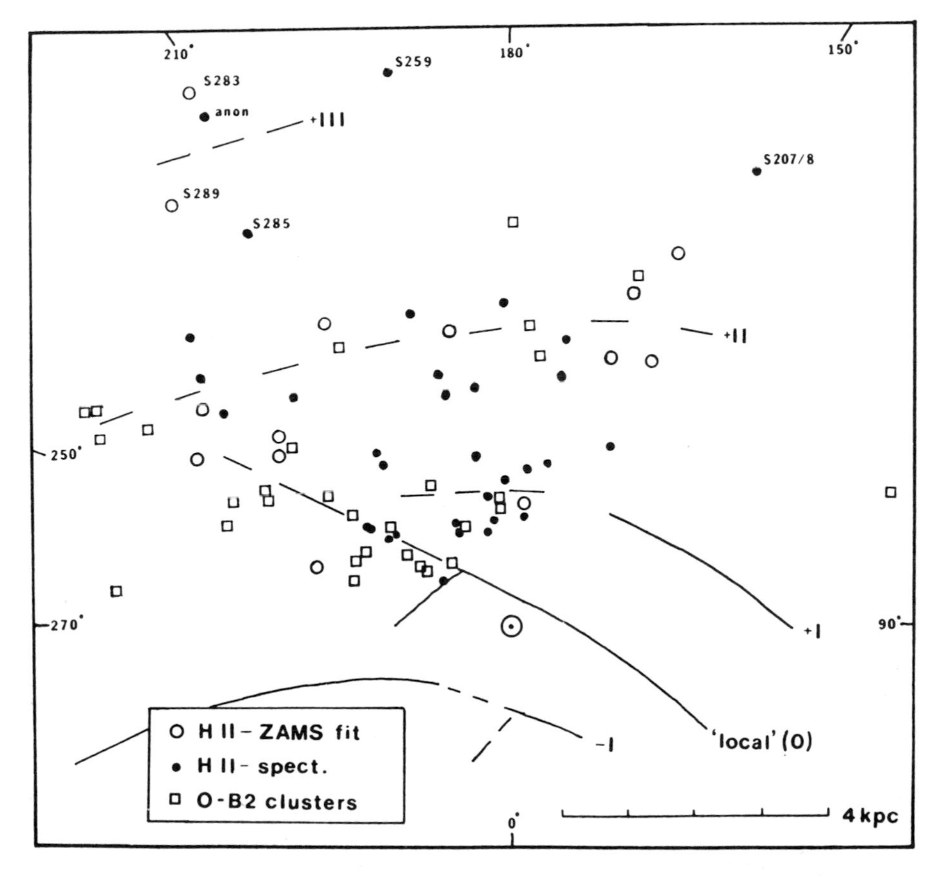

Figure 1. Positions of HII regions and O-B2 clusters plotted on the galactic plane in the longitude range 150° to 260°. Well determined spiral features are shown as solid lines while tentative features are shown as dashed lines.

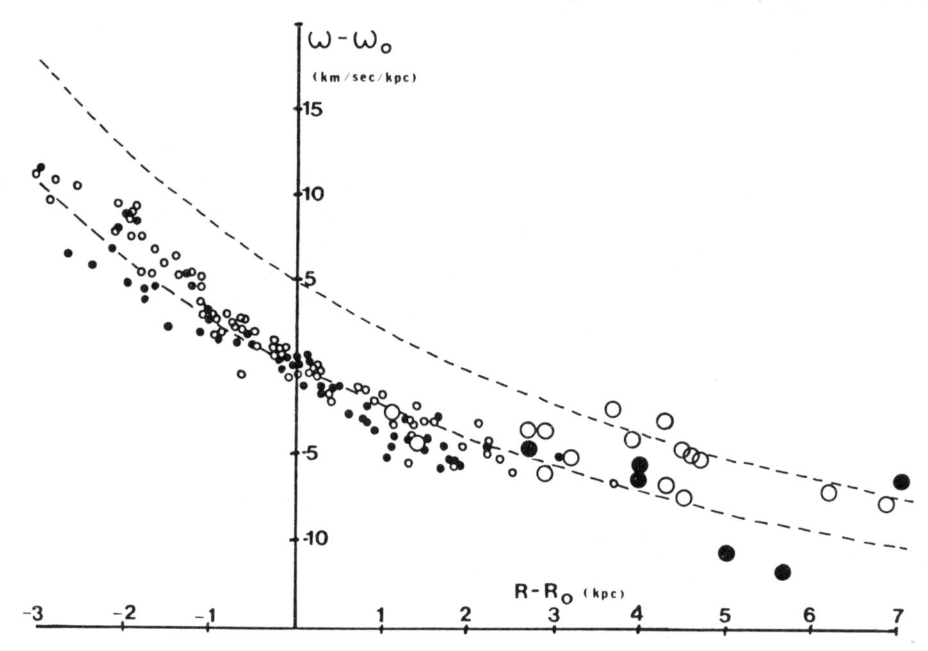

Figure 2. Plot of $\omega - \omega_0$ versus $R - R_0$ for HII regions whose distances have been determined by the present authors (large circles) and for HII regions for which data have been taken from Georgelin (1975; small circles). In both cases, filled circles represent the first and second quadrants of ℓ and open circles represent the third and fourth quadrants of ℓ . The dashed lines represent constant rotation speeds of 250 km/sec (lower line) and 300 km/sec (upper line), assuming R_0 = 10 kpc and ω_0 = 25 km/sec/kpc.

of the Galaxy.A smaller value for Θ_0 (say 220 km/sec, favored by Drs. Knapp and Einasto at this Symposium) flattens the lines of constant in figure 2, improving the fit in the outer parts of the Galaxy. A lower value for Θ_c would also be indicated in view of Dr. Rubin's results that no Sb or later galaxies examined have rotation curves which exceed 250 km/sec.

The authors wish to acknowledge the support of the National Research Council of Canada and PDJ wishes to acknowledge the support of U.S. National Science Foundation grant AST-77-26898, to Prof. F. J. Kerr.

REFERENCES

Crampton, D. and Georgelin, Y. M. 1975, Astron. and Astrophys. 40, 317.
Georgelin, Y. M. 1975, Thèse de Doctorat, Marseille.
Schmidt, M. 1965, in Stars and Stellar Systems, Vol. V- 'Galactic Structure', (Univ. Chicago Press: Chicago), p513.
Vogt, N. and Moffat, A. F. J. 1975, Astron. and Astrophys. 39, 477.

THE CIRCULAR VELOCITY AT THE SUN

G. R. Knapp
Owens Valley Radio Observatory, Caltech

The galactic rotation velocity at the Sun, Θ_0, can be derived several ways, none of them direct and unambiguous - (1) the solar velocity can be found relative to the halo population (the RR Lyrae stars, globular clusters etc.), but may contain an unknown contribution from possible systematic rotation of the halo system (2) the product $R_0\ \omega(R_0) = R_0\ (A-B)$ can be calculated but is uncertain because of large uncertainties in each of these three quantities (3) the motion of the Sun with respect to the center of the Local Group can be found but includes the motion of the galactic center of mass and (4) the velocity-longitude dependence of the outer HI boundary can be examined to deduce the most likely value of Θ_0. The incorporation of new data into analyses using methods (1) and (3) gives essentially the same answers as older studies. Examination of the accumulated current evidence suggests that the best values for the solar rotation velocity Θ_0 and the galactocentric distance R_0 are 220 km s^{-1} and 8.5 kpc respectively.

The value of the velocity of the LSR about the galactic center, Θ_0, is of fundamental importance for any studies of objects at more than a few kiloparsecs from the Sun, both those of galactic structure and of the motions of external galaxies. Unfortunately, there is no simple and direct way to measure this quantity. Its value must therefore be deduced in the context of the other galactic motions and scales, viz. the distance R_0 to the galactic center, the Oort constants A and B which describe the differential galactic rotation in the solar vicinity, the galactic dynamics and rotation curve, the motions of the halo population, and the peculiar motions of our nearest neighbors in extragalactic space, the members of the Local Group. All of the lines of evidence must be examined to find a self-consistent set of values for the galactic parameters, including Θ_0.

There are the following ways, more or less indirect, to estimate the value of Θ_0.
(1) Measure the peculiar motion of the LSR with respect to the halo population, viz. the globular clusters. This method gives $\Theta_0 \sim 200$ km s^{-1}.

W. B. Burton (ed.), The Large-Scale Characteristics of the Galaxy, 225–230.

(2) Measure the peculiar motion of the LSR with respect to the members of the Local Group. This method gives $\Theta_0 \sim 300$ km s^{-1}.
(3) Analyse the motions of the high-velocity stars. There are no stars traveling in the $\ell=90^{\circ}$ direction at velocities greater than ~ 65 km s^{-1} with respect to the LSR, which must measure the local escape velocity. This method suggests $\Theta_0 \sim 300$ km s^{-1}.
(4) Find Θ_0 by measuring both ω_0 (R) and R_0, where ω_0 is the local angular velocity. R_0 is measured by the distance to the center of mass of the globular cluster or RR Lyrae systems, or by heliocentric distances to stars at galactocentric distances R_0. Such stars are characterized by having radial velocities relative to the LSR of 0 km s^{-1}, and a value of R_0 follows from simple geometry. The angular velocity ω is measured via the Oort A and B constants:

$$A = -\frac{1}{2} R_0 \left\{\frac{d\omega}{dR}\right\}_{R_0} \quad \text{km s}^{-1}\ \text{kpc}^{-1} \tag{1}$$

$$B = -\omega_0 - \frac{1}{2} R_0 \left\{\frac{d\omega}{dR}\right\}_{R_0} \quad \text{km s}^{-1}\ \text{kpc}^{-1} \tag{2}$$

so that $\Theta_0 = R_0\,(A-B)$
The constants A and B are measured by observations of stellar radial velocities, distances and proper motions [see the descriptions by Schmidt (1965) and Mihalas (1968)]. The values of these constants must further be in agreement with the value of $2AR_0$ found from 21-cm observations of the HI disk. A careful discussion of all the evidence by Schmidt (1965) led to: $A=15$ km s^{-1} kpc^{-1}; $B=-10$ km s^{-1} kpc^{-1}; $R_0=10$ kpc; $\Theta_0 = 250$ km s^{-1}. The resulting dynamical model for the Galaxy produces a rotation curve which rises approximately linearly from ~ 200 km s^{-1} at $R \sim 1$ kpc to 250 km s^{-1} at 10 kpc then falls slowly ($\sim R^{-0.2}$) beyond R_0. This model for the Galaxy was adopted by the I.A.U. in 1963, and these values used in galactic studies ever since.

There has been a lot of activity of late using each of the above methods; for the first three methods the numerical values found agree well with older values.

(1) <u>The Motions of the Halo Population</u>

The LSR motion may be measured by a 'solar motion solution' relative to the halo population. The disadvantage of this method is that there may be an (unknown) net rotation of the halo objects <u>as a system</u> which adds to the calculated value of Θ_0.

Woltjer (1975) from an analysis of globular cluster motions, finds $\Theta_0 = 200$ to 225 km s^{-1}. Hartwick and Sargent (1978) have measured the velocities of members of the distant halo population (the dwarf spheroidal galaxies and the 'Palomar' interloper globular clusters) and find $\Theta_0 = 220$ km s^{-1}. It has been suggested (Lynden-Bell 1976) that these latter objects do belong to a system, i.e. the Magellanic Stream, but the radial velocities measured by Hartwick and Sargent do not bear this out. Thus, except for the unlikely eventuality that both subsets of the halo objects belong to the same systematically rotating system, this straightforward observation leads to a lower value of Θ_0 ($\sim$215±20 km s^{-1}) than the standard value.

(2) Motion with respect to the Local Group of Galaxies.

This method suffers from several disadvantages, the most serious being that the resulting solar motion is compounded of the galactic rotation and of the peculiar motion of the Galactic Center with respect to the barycenter of the Local Group, and there is no way intrinsic to the method to disentangle the two. Other disadvantages, which are not so intractable since they can be dealt with reasonably by careful analysis, are the non-uniform distribution of local group members about the Galaxy, the concentration of most of the mass in the Galaxy and M31, the fact that several of the members are satellites of one or the other of the big galaxies, and the question of membership. Two such analyses, those of Lynden-Bell and Lin (1977) and Yahil, Tammann and Sandage (1977) confirm the value of the solar motion of $\sim$ 300 km s^{-1} (towards $\sim \ell=105^o$, $b=-8^o$). From an accompanying analysis of the Magellanic Cloud motions (Lin and Lynden-Bell 1977), Lynden-Bell and Lin suggest that essentially all of this motion is galactocentric rotation of the LSR. The value of $\sim$ 300 km s^{-1} is also found by de Vaucouleurs (1972) from observations of nearby galaxies outside the local group, when a distance-dependent K-term (Hubble expansion) is included.

(3) The orbits of the high-velocity stars.

Recent analyses of the motions of the high-velocity stars have been made by Isobe (1974), who finds $\Theta_0 \sim 275$ km s^{-1}, and by Greenstein and Toomre (1978) who suggest $\Theta_0 \sim 300$ km s^{-1}. However, this method contains a hidden assumption; that essentially all of the mass of the Galaxy is inside the solar orbit. This is certainly true of the visible mass, but there is increasing evidence (e.g. Hartwick and Sargent 1978; Jackson 1978 this conference) for a large component of the galactic mass at distances $R > R_0$. The high-velocity stars may thus be affected by a much larger potential than previously assumed.

(4) Value of $\Theta_0 = (A-B)\ R_0$

This determination is probably on the shakiest ground at the moment because of uncertainties in the quantities A,B and R_0. Recent measurements of R_0 all give values less than 10 kpc. R_0 has been measured using globular clusters by Harris (1976), RR Lyrae stars by Oort and Plaut (1975) and using OB stars on the solar circle by Crampton et al. (1976). These measurements give values of R_0 between 8 and 9 kpc.

Many recent measurements of A and B have likewise been made, amongst them those by Fricke and Tsioumis (1974), Dieckvoss (1978), Balona and Feast (1974), and Crampton and Georgelin (1975). Values of A from $\sim$ 11 to 17 km s^{-1} kpc^{-1} and B from -7 to -15 km s^{-1} kpc^{-1} have been found; no attempt will be made in the present paper to suggest "correct" values of A and B from these numbers. Finally, the quantity $2AR_o$ has recently been re-evaluated from 21-cm data by Gunn et al. (1978); we find AR_o = 110±2 km s^{-1}, suggesting that both R_o and A are smaller then their "standard" values. All of this suggests that the value of Θ_0 found from equation (3) is significantly smaller then 250 km s^{-1}.

In the above work, we have also investigated the galactic rotation curve and its interaction with the value of Θ_0. It is already known that the rotation curve flattens outside R_o (Jackson 1978, this symposium), and further, it has been known for a long time that if $\Theta_0 \sim 220$ km s^{-1}, the galactic rotation curve is flat inside R_o also, to $R/R_o \sim 0.5$ (see Kwee, Muller and Westerhout 1954). This rotation curve shape agrees with those determined both optically and by 21 cm observations for almost all other galaxies (Rubin 1978, this symposium). If this is the case, then the rotation curve itself gives the value of Θ_0, since the tangent point velocity is

$$V_M = \Theta_0 (1-\sin\ell) \quad (4)$$

so long as $\Theta_R = \Theta_0$. Thus we find a formal value of Θ_0 = 220±3 km s^{-1}.

Thus I think the best determination of the velocity of the LSR about the galactic center, Θ_0, is that given by the globular motions, i.e. 220 km s^{-1}. This value also provides a good fit to 21 cm observations both at distances $R > R_o$, and at $R < R_o$. With this value, the 21 cm observations lead to a flat rotation curve for the Galaxy, consistent with those of every other large galaxy observed.

The implied flatness of the rotation curve and the large extent of the (invisible) Galaxy requires a high total mass for the Galaxy ($\sim 10^{12} M_\odot$). This value is consistent with the derived using distant halo objects by Hartwick and Sargent (1978), with the large mass implied by the relative approach velocities of the Galaxy and M31 ($\sim$ 100 km s^{-1}), and with the large mass required by the orbits of the high-velocity stars.

REFERENCES

Balona, L.A., and Feast, M.W. 1974, M.N.R.A.S. 167, p 621.
Crampton, D., Bernard, D., Harris, B.L., and Thackeray, A.D. 1976, M.N.R.A.S. 176, p 683.
Crampton, D., and Georgelin, Y.P. 1975, Astron. Astrophys. 40, p 317.
de Vaucouleurs, G. 1972, in I.A.U. Symposium 44, ed. D.J. Evans, D. Reidel Publishing Co., Dordrecht, Holland.
Dieckvoss, W. 1978, Astron. Astrophys 62, p 445.
Fricke, W. and Tsioumis, A. 1975, Astron. Astrophys. 42, p 449.
Greenstein, J.L., and Toomre, A.R. 1978, in preparation.
Gunn, J.E., Knapp, G.R., Tremaine, S.D. 1978, (in preparation)

Harris, W.E. 1976, A.J. 81, p 1095.
Hartwick, F.D.A., and Sargent, W.L.W. 1978, Ap. J. (in press)
Isobe, S. 1974, Astron. Astrophys. 36, p 327.
Jackson, P.D. 1978, preceding paper.
Kwee, K.K., Muller, C.A., and Westerhout, G. 1954, B.A.N. 12, p 211.
Lin, D.N.C., and Lynden-Bell, D. 1977, M.N.R.A.S. 181, p 59.
Lynden-Bell, D. 1976, M.N.R.A.S. 174, p 695.
Lynden-Bell, D., and Lin, D.N.C. 1977, M.N.R.A.S. 181, p 37.
Mihalas, D. 1968, 'Galactic Astronomy' W.H. Freeman Co., San Francisco.
Oort, J.H., and Plaut, L. 1975, Astron. Astrophys. 41, p 71.
Rubin, V.C., 1978, preceding paper.
Schmidt, M., 1965, in 'Galactic Structure', ed. A. Blaauw and M. Schmidt, University of Chicago Press.
Woltjer, L. 1975, Astron. Astrophys. 42, p 109.
Yahil, A., Tammann, G., and Sandage, A.R. 1977, Ap. J. 217, p 903.

DISCUSSION

Oort: I feel some hesitation to accept as low a value as 220 km s^{-1} for Θ_o (as you suggest), because it leads to a disturbingly high relative velocity of the Galaxy and the Andromeda nebula. This velocity is already difficult to account for if Θ_o = 250 km s^{-1}. Commenting also to Dr. Jackson, I want to draw attention to the fact that there are apparently quite high radial motions in the outermost arms of the Galaxy. These were first discovered by Miss Kepner from 21-cm observations at Dwingeloo, and were later confirmed by Verschuur from NRAO observations. In particular there is an arm-like feature extending over some 20° around the longitude of the anticenter, where it has a radial velocity of -100 km s^{-1}. The presence of such large radial motions may seriously affect our determination of the rotation velocity for points in the second and third quadrants of longitude.

Knapp: The terminal-velocity vs longitude curve for the outer HI envelope (defined as the velocity of the 1 K contour in the Weaver-Williams and Kerr-Harten-Ball Surveys) appears to reflect circular motion except in the ℓ = 80°-180° region. The negative-velocity field in the anticenter direction is certainly not in circular motion. In any case, the fact that $\Theta_o = 220 + \beta$ from the rotation curve at $R < R_o$ where β is the gradient of Θ_R in km s^{-1} per 8.5 kpc strongly suggests that $\Theta_o < 250$ km s^{-1}, because the largest value of β observed for an external galaxy (NGC 4596) is 30.

Jackson: In fact we see a difference between the rotation curves for the second and third quadrants which is consistent with a radial outward motion of 5 km s^{-1} for the LSR or an inward motion of 5 km s^{-1} for the outer parts of the Galaxy. Nevertheless, because we included both outer quadrants of the Galaxy, our mean rotation curve results are independent of a uniform radial motion.

Schmidt-Kaler: In a joint work with Dr. Maitzen we obtained RV's of more open clusters. We now have spectra of 90 early-type open clusters; in the classical Johnson and Svolopoulos paper there were only 29. What we see is exactly in line with your results: $\omega(R)$ is steeper than Schmidt's model going inside and flatter going outside.

Jackson: Indeed, we are familiar with your extensive cluster work which essentially extends to $R-R_o \simeq 3$ kpc. However, we are going out to $R-R_o \simeq 7$ kpc, where the flattening, or even the increasing, of the rotation curve becomes more evident.

de Vaucouleurs: One well-known difficulty with Θ_o = 200-220 km s^{-1} is the large residual velocity of the Galaxy relative to the Local Group compared with the velocity dispersion of the other members, which is less than 50 km s^{-1}.

Knapp: With Θ_o = 220 km s^{-1}, the center-of-mass velocity of the Galaxy is 115 km s^{-1} towards ℓ = 145°, b = 20°. The direction is only $\sim$ 20° away from M31. It appears that M31 and the Galaxy are currently approaching each other: the timing argument suggests that the mass of the M31/Galaxy system is $\sim 3 \times 10^{12}$ $M_\odot$. The Local Group is perhaps a little unusual in having much of its mass in these two members.

THE GALACTIC CIRCULAR VELOCITY NEAR THE SUN

J. Einasto, U. Haud, and M. Jôeveer
W. Struve Astrophysical Observatory, Tôravere, Estonia, USSR

The conventional value of the galactic circular velocity near the Sun is V_o = 250 km s^{-1} (Schmidt 1965). Recently both lower (Mathewson) et al., 1974; Einasto et al., 1976) and higher (Lynden-Bell and Lin, 1977) values have been suggested. The following summarizes our dynamical determinations of V_o. Altogether six independent methods have been used.

Recently Illingworth (1977) has measured the rotation velocities of a number of elliptical galaxies. Rotation velocities are considerably smaller than expected. This result is confirmed by the radial velocities of the globular clusters of M31 determined by van den Bergh (1969). Using these data, we adopt the rotation velocity of extreme halo population objects as 40 km s^{-1}. The heliocentric centroid velocity of these objects near the Sun, corrected for the solar motion in terms of the circular velocity, is 172 km s^{-1}. Thus from these data V_o = 212 km s^{-1} has been obtained.

Using the data on centroid velocities and density gradients of disk population objects, one can determine the solar motion which brings centroid velocities and density gradients into mutual agreement. The result is V_o = 230 km s^{-1}.

The solar velocity can be determined by a trial-and-error procedure, demanding that apogalactic distances derived for objects moving with tangential velocity $V_o + \Delta(\Delta \sim 65$ km s^{-1} is the Oort limiting velocity) should coincide with the boundary of the Galaxy. This method gives the value of V_o = 225 km s^{-1} for the circular velocity.

The dynamics of the nearby companions of the Galaxy (galactocentric distances between 30 and 250 kpc (open circles in Fig. 1) yields V_o = 220 km s^{-1}. More distant galaxies (R between 500 and 1500 kpc; dots in Fig. 1) give $V_o + V_{gal}$ = 380 km s^{-1}. Radial velocities have been taken from Yahil, Tammann, and Sandage (1977) and from Hartwick and Sargent (1978).

W. B. Burton (ed.), The Large-Scale Characteristics of the Galaxy, 231–232.

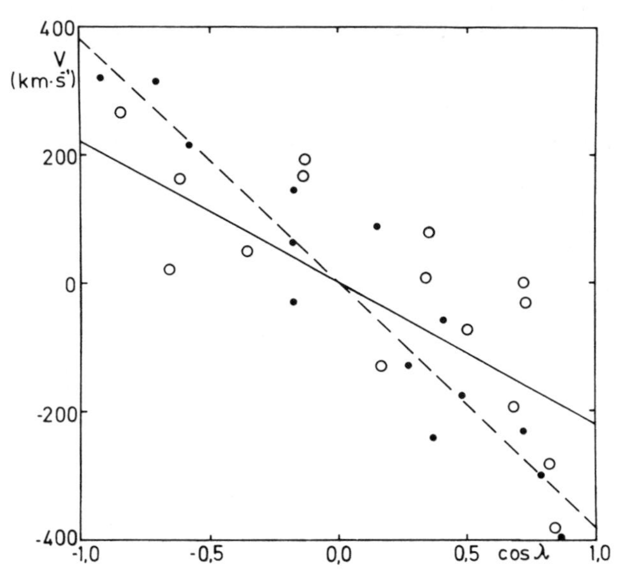

For five galaxies (NGC 224, 1068, 3031, 4594, and 5194) both maximum rotation velocities and velocity dispersions in the center (and in the halo in case of NGC 224) have been measured. The ratio V_{max}/σ depends slightly on the morphological type of the galaxy; for Sb (supposed type of the Galaxy) we adopt $V_{max}/\sigma=2$. The velocity dispersion of galactic halo objects is $\sigma=115$ km s^{-1}, which yields $V_{max} = 230$ and $V_o = 225$ km s^{-1}.

Finally, the circular velocity can be determined from the system of galactic constants. Adopting for the galactic constants and their rms errors the values given by Einasto (1978), one obtains $V_o = 217 \pm 10$ kms.

Adopting for all determinations, excluding the last one, an rms error of 20 km s^{-1}, we found the weighted mean circular velocity to be $V_o = 220 \pm 7$ km s^{-1}, in good agreement with the scatter of individual estimates.

REFERENCES

Einasto, J.: 1978, IAU Symp. 84, invited report.
Einasto, J., Jôeveer, M., and Kaasik, A.: 1976, Tartu Astron. Obs. Teated, No. 54, 3.
Hartwick, F. D. A., and Sargent, W. L. W.: 1978, Astrophys. J. 221, 512.
Illingworth, G.: 1977, Astrophys. J. Lett., 218, L43.
Lynden-Bell, D., and Lin, D. N. C.: 1977, Monthly Notices Roy. Astr. Soc. 181, 37.
Mathewson, D. S., Cleary, M. N., and Murray, J. D.: 1974, Astrophys. J. 190, 291.
Schmidt, M.: 1965, Stars and Stellar Systems 5, Univ. Chicago Press, p. 513.
van den Bergh, S.: 1969, Astrophys. J. Suppl. 19, 145.
Yahil, A., Tammann, G. A., and Sandage, A.: 1977, Astrophys. J. 217, 903.

STELLAR KINEMATICS AND INTERSTELLAR TURBULENCE

Richard B. Larson
Yale University Observatory

1. INTRODUCTION

It is well known that the velocity dispersion of nearby stars increases systematically with age, and this fact is conventionally explained by postulating that stars are randomly accelerated by encounters with massive gas clouds (Spitzer & Schwarzschild 1951, 1953; Wielen 1977). However, a strong constraint on the possible role of random accelerations is provided by the apparent existence of fairly old moving groups of stars (Eggen 1969, Boyle & McClure 1975). An alternative explanation (Tinsley & Larson 1978) is that the age dependence of the velocity dispersion of stars older than 10^9yr is produced by a gradual decay with time of interstellar turbulent motions, as predicted by plausible collapse models. This effect cannot account directly for the variation of velocity dispersion with age observed for stars younger than 10^9yr, but Tinsley & Larson suggested that this could be explained if the velocity dispersion of the youngest stars reflects only the local turbulent motions of the gas, while the velocity dispersion of older stars reflects in addition larger-scale non-circular motions in the galactic gas layer. If the interstellar medium possesses a hierarchy of motions whose velocity dispersion increases with the size of the region considered, older stars which have traveled farther since their formation will sample the gas motions over a larger volume of space, and thus will have a larger velocity dispersion than the younger stars.

Here we summarize some data on the velocity dispersions of interstellar gas and young stars which show that the increase of velocity dispersion with region size is, in fact, sufficient to account for the age dependence of the stellar velocity dispersion for ages up to $\sim 10^9$yr. Further details and references are given by Larson (1978).

W. B. Burton (ed.), The Large-Scale Characteristics of the Galaxy, 233–237.

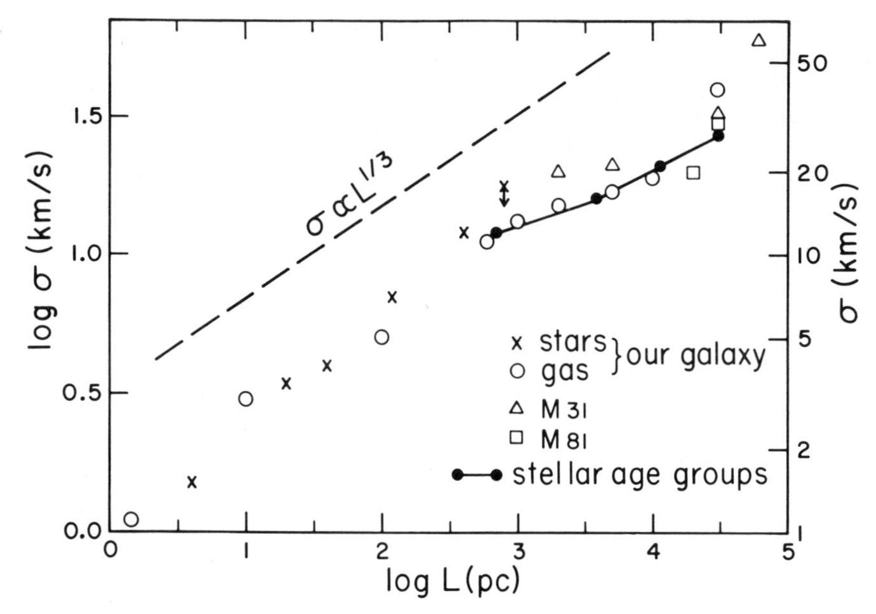

Figure 1. The 3-dimensional velocity dispersion σ versus the diameter of the region in which σ is measured.

2. VELOCITY DISPERSIONS OF STARS AND GAS

Data have been collected from a variety of sources for the velocity dispersions of young stars and gas in our galaxy and in M31 and M81, as measured in different ways and in regions of different size. The observations used are listed below; given in parentheses for each case are the values of (σ, L), where σ is the estimated 3-dimensional RMS random velocity (km/s) and L is the diameter (pc) of the region in which σ is measured. The resulting dependence of σ on L is plotted in Figure 1.

Data on the velocity dispersions of young stars in regions of different size include the following: the velocity dispersion of a typical young open cluster (1.5, 4); the velocity dispersion of a typical T association (3.4, 20); a typical subgroup of an O association (4, 40); a whole O association (7, 120); OB stars within 200 pc of the sun (12, 400); and OB stars with distances between 200 and 400 pc (≲ 18, 800). These data are plotted as crosses in Fig. 1, and a regular increase of σ with L is evident.

Data for cool interstellar gas or gas clouds include: the internal velocity dispersion of a typical dark cloud (1.1, 1.4); a typical HI or CO cloud (3, 10); a typical HI or CO cloud complex (5, 100); and the velocity dispersion of HI clouds within ∿ 300 pc of the sun (11, 600). For larger length scales, velocity dispersions can be estimated from observations of non-circular gas motions, if isotropic motions are assumed and if the velocity dispersions of the smaller-scale motions that are not included are added in. Such observations include: local gas flows associated with the Gould belt (13, 1000); irregularities in the galactic rotation curve (15, 2000); vertical motions associated with warps or corrugations of the galactic gas layer (17, 5000); the large-scale asymmetry of the rotation curve (19, 10000); and non-circular motions and warps in the outermost part of the galaxy (∿40, 30000). These data are plotted as circles in Fig. 1.

Data for non-circular motions in other galaxies include: the small-scale velocity dispersion of HI in M31 (20, 2000); the velocity dispersion of HII regions in M31 (21, 5000); non-circular motions on the minor axis of M31 (32, 30000); motions associated with the outer warp of M31 (∿60, 60000); minor axis motions in M81 (20, 20000); and non-circular motions in the outer part of M81 (∿30, 30000). These data are shown as triangles and squares in Fig. 1. Note that, while M31 seems to show somewhat larger non-circular motions than our galaxy or M81, all three galaxies show non-circular velocities of comparable magnitude which increase in a similar systematic way with increasing length scale. Also, the results for young stars and gas agree well where they overlap, supporting the assumption that the stars form with the same velocity dispersion as the gas.

3. IMPLICATIONS FOR STELLAR AND GAS DYNAMICS

The observed (σ, L) relation for the interstellar motions shown in Fig. 1 can be compared with that required to explain the age dependence of the stellar velocity dispersion by plotting the velocity dispersions of Wielen's (1977) stellar age groups versus the diameter of the region of origin for the stars in each group. Assuming that these stars disperse principally in the azimuthal direction, since their ages are mostly greater than one-quarter of an epicyclic period, the diameter L of the region of origin for stars of age τ and velocity dispersion σ is $L \sim 2\sigma\tau/\sqrt{3}$. The resulting ($\sigma$, L) relation for the groups with ages up to 10^9yr is indicated by the dots joined by solid lines in Fig. 1.

It is seen that the agreement between the velocity dispersions of gas and stars as a function of effective region size is quite close for region sizes up to at least 10 kpc, corresponding to ages up to 5×10^8yr, and remains good for scales up to 20 kpc, or ages of nearly 10^9yr. This agreement means that most of the age dependence of the stellar velocity dispersion is due to the increase of σ with L for interstellar motions, and it sets an upper limit on the possible importance of random accelerations; in terms of Wielen's diffusion coefficient, a value at most 1/3 as large as that derived by Wielen is allowed by these data. This is consistent with the existence of the Hyades moving group, which suggests a diffusion coefficient not more than 1/10 of Wielen's value (Larson 1978).

The (σ, L) relation shown in Fig. 1 also has implications for the origin and dynamics of the observed gas motions. Clearly no single mechanism, e.g. supernova explosions, that produces motions with only a limited range of velocities or length scales can account for the full spectrum of interstellar motions. A broad spectrum of motions could,however, be produced in a turbulent flow in which large-scale motions decay through various instabilities into smaller-scale motions, which thus derive their energy from the larger-scale motions. The dashed line in Fig. 1 shows the slope of the Kolmogoroff spectrum $\sigma \propto L^{1/3}$ for incompressible turbulence, and the observations bear at least a superficial resemblance to this relation. Thus it is at least energetically feasible for the smaller-scale interstellar motions to derive part of their energy from larger-scale motions via a turbulent cascade process.

The short time ($<10^9$yr) required for the decay of non-circular motions implies a continuing energy input for at least the largest-scale motions. The existence of the Gould belt and other vertical displacements of the galactic gas layer suggests that mechanisms internal to the galactic disk are not entirely adequate, and that external perturbations act on the galactic gas layer. An attractive possibility for explaining some of the observed random or non-circular motions of the gas layer is infall of gas from outside the galactic disk (Larson 1972, Saar & Einasto 1977).

4. CONCLUSIONS

The observed non-circular interstellar motions are sufficient to explain the age dependence of the stellar velocity dispersion for ages up to $\sim 10^9$yr; thus it appears possible to understand the kinematics of stars in terms of the initial turbulent motions of the gas, without invoking random accelerations. It is therefore important for both stellar dynamics and interstellar gas dynamics to understand the origin of the observed gas motions on various scales. Interactions with material outside the galactic plane may play a role in producing some of these motions.

REFERENCES

Boyle, R.J., & McClure, R.D., 1975. Publ.Astr.Soc.Pacific,87,p.17.
Eggen, O.J., 1969. Astrophys.J., 155, p.701.
Larson, R.B., 1972. Nature, 236, p.21.
Larson, R.B., 1978. Submitted to Mon. Not. Roy.Astr. Soc.
Saar, E., & Einasto, J., 1977. In "Chemical and Dynamical Evolution of Our Galaxy", ed.E.Basinska-Grzesik&M.Mayor,p. 247. Geneva Observ.
Spitzer, L., & Schwarzschild, M., 1951. Astrophys.J., 114, p.385.
Spitzer, L., & Schwarzschild, M., 1953. Astrophys.J., 118, p. 106.
Tinsley, B.M.,& Larson, R.B., 1978. Astrophys.J., 221, p. 554.
Wielen,R., 1977. Astr. Astrophys., 60, p. 263.

DISCUSSION

de Vaucouleurs: I am curious to find out what would be the slope of your σ_V,L diagram if you reinterpreted it as a density, L diagram using the virial theorem.

Larson: If you adopt the relation $\sigma \propto L^{1/3}$, which is roughly satisfied empirically, and associate a density ρ with each L and σ via the virial theorem, you obtain $\rho \propto L^{4/3}$. This gives the density which regions of different size would have to have in order to be gravitationally bound. Some representative densities predicted this way are $n \sim 10^4$ cm^{-3} for L = 1 pc, $n \sim 10^3$ cm^{-3} for L = 10 pc, $n \sim 2$ cm^{-3} for L = 1 kpc, and $n \sim 0.1$ cm^{-3} for L = 10 kpc. I leave it to the reader to decide whether the rough coincidences between these numbers and the observed densities of interstellar structures or regions of different sizes are meaningful.

Innanen: The existence of a third "quasi" integral of motion demands a strongly anisotropic velocity distribution function. Consequently its operation will assure the persistence of the anisotropy and no "collisional accelerating " mechanism is therefore required.

Larson: Actually I assumed isotropic velocity dispersions throughout, but this makes no significant difference to the conclusions. You are right that an anisotropic velocity distribution would be preserved by the third integral if random accelerations are not important, and therefore the anisotropy might also be a result of initial gas motions. If the gas motions are maintained by a balance between an energy input and turbulent dissipation, the observed anisotropy might reflect the fact that the oscillation period of vertical motions is shorter than the epicyclic period, perhaps resulting in a faster dissipation of vertical gas motions and a smaller velocity dispersion in the vertical direction than parallel to the galactic plane.

Wielen: (1) Typical disk stars have a total velocity dispersion of about 60 km s^{-1}. Do you suggest that this was the typical non-circular velocity of the gas during the past history of the galactic disk? (2) What are the typical time scale (age) and the typical length scale over which a sufficient mixture has occurred among the younger disk stars in order to show the large-scale velocity dispersion according to your theory?

Larson: (1) Yes, I suggest that this was the velocity dispersion of the gas motions at the time when the "typical" disk stars in the solar neighborhood formed, perhaps some 5×10^9 years ago. (2) It takes $\sim 10^9$ years for stars to disperse completely around an annulus of radius 10 kpc, so only stars older than this can be considered to be "completely mixed". If the velocity dispersion of the gas increases monotonically with length scale, it is necessary to integrate the characteristics of younger stars over a complete annulus of radius 10 kpc in order to compare their properties with older stars and study the long-term effects of galactic evolution.

AGE VARIATIONS OF THE MEAN ANGULAR MOMENTUM

P.J.Grosbøl
Copenhagen University Observatory, Denmark

The mean angular momentum of stars in the solar vicinity with respect to the galactic center shows a significant change as a function of age. In figure 1a the variation is given for B stars using the Strömgren index [u-b] as abscissa since it depends nearly linear on the logarithm of the age (i.e. the interval $0^m.6 < [u-b] < 1^m.4$ corresponds approximately to $7.7 < \log(\text{age}) < 8.2$). This can be explained by a change in the birthrate as a function of place, velocity and time and/or an azimuthal variation in the galactic potential. To investigate if the presence of a density wave potential could cause the observed distribution a number of models were made. In these models stars were formed close to the spiral arms according to the density wave theory and traced back to the present epoch. The mean angular momentum of stars being within 300 pc from the sun is shown in figure 1b where the amplitude of the wave was 5 percent, the inclination angle $-7^\circ.2$ and one arm coincided with the Sagittarius arm. For a pattern speed $\Omega_p = 13.5$ km/s/kpc the observed values could be reproduced quite well which indicats the existence of a density wave in our Galaxy.

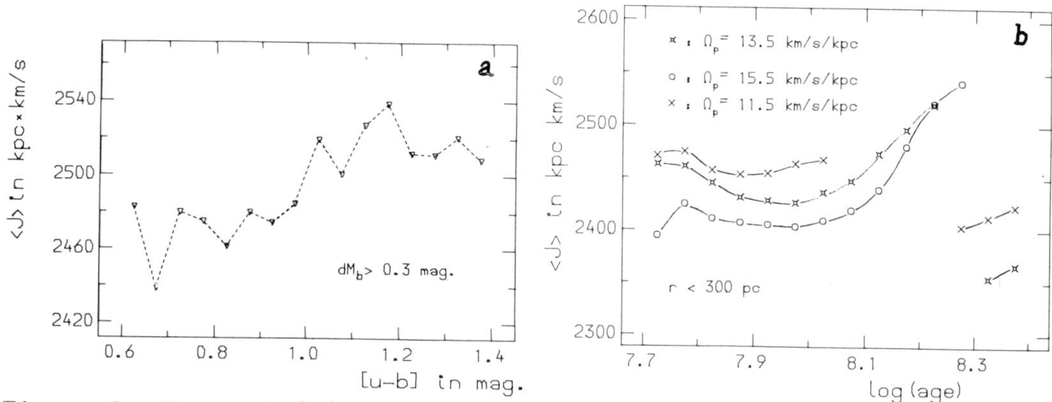

Figure 1. Observed (a) and theoretical (b) mean angular momentum <J>.

W. B. Burton (ed.), The Large-Scale Characteristics of the Galaxy, 238.

A THEORY OF GALACTIC EVOLUTION

S. V. M. Clube
Royal Observatory, Edinburgh, Scotland

..."there exist many galaxies where most of the light originates in an apparently flat rotating disk, and where the random stellar motions appear to be small compared to the systematic circular motion i.e. they are apparently "cold." Our own Galaxy is such a system, and it does not seem to suffer any large scale, large amplitude, short-time-scale instability"... Ostriker and Peebles, Ap. J. 186, 467, 1973.

The generally agreed foundation of galactic theory could not be more clearly expressed. Yet the second sentence is no longer true. That many radio-astronomical kinematic observations are still interpreted in a way that gives substance to such theory directly contradicts the evidence of nearby stars (Clube 1978). The pre-1950 analysis of stellar motions in the solar neighbourhood was seriously distorted by failure to recognise the existence of a young disc population (Pop $I^{I} \lesssim 5.10^{8}$y) and an old disc population (Pop I^{II}) with different kinematic properties. The origin of the error can be traced to an erroneous assumption that Kapteyn's two drifts and Schwarzchild's velocity ellipsoid were completely equivalent. Drift I represents the predominant Pop I^{I} and has an outward motion in the Galaxy relative to Drift II representing Pop I^{II}. This outward motion of 40 kms^{-1} is now reflected in observations of Sgr Aw in the galactic centre and the symmetrically placed inner spirals known as the 3 kpc arm and the +135 kms^{-1} arm. Detailed study of the spiral arms shows that the observed structure is temporary, unstable and probably the result of very energetic processes in the central region of the Galaxy. That spiral arms are such a common feature of galaxies suggests they originate in a recurrent galactic process. Unusual violent behaviour in the centres of other systems suggests galaxies evolve by quasi-periodically producing material that first moves out and then comes in, all the time spreading out under the influence of differential rotation. The clue to the origin of this process may well lie in galactic nuclei themselves (cf. van der Kruit et al. 1972), but there is no proven physical mechanism.

W. B. Burton (ed.), The Large-Scale Characteristics of the Galaxy, 239–241.

REFERENCES

Clube, S. V. M.: 1978, Vistas in Astronomy 22.
van der Kruit, P. C., Oort, J. H., and Mathewson, D. W.: 1972, Astron. Astrophys. 21, 169.

DISCUSSION

Oort: There is one class of objects for which there can hardly be doubt that they form an old, and presumably well-mixed, population, vis. the planetary nebulae near the galactic center. They show a velocity of the LSR close to zero and not +40 km s^{-1} as you suggest. The random motions are high and the mean errors is therefore about 10 km s^{-1}, so that unfortunately there is considerable uncertainty in this result.

Clube: The planetaries are in principle well suited to a determination of the motion of the LSR, but absorption obscures our view of the nuclear bulge at positive b, and furthermore the situation is complicated by the negative motion of objects between us and the galactic center. Allowing for these effects (Clube, 1978), the motion of the center is 74±36 km s^{-1}. As you say, there is considerable uncertainty, but the result is not inconsistent with the expanding model.

Bok: A similar test to the one just suggested by Prof. Oort for the velocity of the LRS relative to an old component could be based on the system of globular clusters. Have you made this test?

Clube: The radial motion of the "nearby" globular clusters appears to be similar to that of disc stars. The few objects near the center have a large velocity dispersion and do not give a definitive result.

Burton: There is a whole collection of observational material, comprised of the spectral line data of HI, H recombination lines, CO, and other molecules, which shows a pronounced tendency to accumulate near zero radial velocity near $\ell = 0°$. This accumulation represents gas lying along a transgalactic path. That it is centered near 0 km s^{-1} rules out, in a way which seems to me very definite, the possibility that the local standard of rest determination could be 30 km s^{-1} in error. Braes (B.A.N. 17, 132) used this sort of argument in his test of Kerr's suggestion of a general galactic expansion.

Clube: Because of limitation on time, I discussed only a small amount of the optical evidence, but obviously no determination of the motion of the LSR can be made without reference to the important data which you mention. Most of the zero velocity emission near $\ell = 0°$ is contributed by material which has a shallow velocity gradient in galactic longitude and is evidently nearby. The remainder cannot be presumed to be uniformly weighted along a transgalactic path unless the optical depth is small. There is good evidence that this is not the case for receding material on the far side of the Galaxy (Clube, 1978). Braes' test is therefore defective. Although my model of diminishing radial motion with galactic radius is similar to Kerr's and originates from much the same observed features, the specific difference in our motions of the LSR can in fact be attributed to entirely different treatments of absorption.

Stark: There is a narrow, very deep absorption feature seen by several observers in the direction of Sgr A in the HCO^+ and HCN $J = 0 \rightarrow 1$ transition, which deviates from a LSR velocity of 0 km s^{-1} by less than 2 km s^{-1}. Thus, there are in fact narrow velocity molecular features at 0 km s^{-1} towards the galactic center.

van Woerden: The very narrow HI absorption component near zero velocity seen in absorption against Sgr A arises in a very nearby cloud, at $\sim$ 100 pc distance, as shown by Riegel and Crutcher (Astr. and Ap. 12, 43) and earlier workers (e.g., Heeschen, 1955). Because of its distance, this absorption component carries no information about possible outward motion of the local standard of rest.

DISCUSSION (after Dr. Grosbøl's talk)

Contopoulos: Have you considered the case of a much larger value of the angular velocity of the spiral pattern, e.g., Ω_s = 32 km s^{-1} kpc^{-1}?

Grosbøl: Not yet, but it will be done shortly.

DISCUSSION (after Dr. Upgren's talk)

Wielen: I am puzzled that you do not find the classical asymmetric drift in the V-component of the solar motion for objects with increasing velocity dispersion.

Upgren: These stars do not include halo members, nor stars with very high velocity. For disk stars only, the drift of the V-component is not noticeable.

AN OUTWARD MOTION OF YOUNG STARS WITH RESPECT TO OLD STARS

Arthur R. Upgren
Van Vleck Observatory

Space motions are calculated for 145 dK2-M2 stars with radial velocities and with parallaxes and proper motions determined and published at the Van Vleck Observatory. The stars are divided into young and old disk components kinematically and also according to the age-sensitive CaII emission intensities. Rigorous solutions for the solar motion and velocity ellipsoid were calculated for each population group using three methods of weighting errors in parallax, proper motion and radial velocity. All methods show a mean motion of the young stars outward away from the galactic center of about ten km/sec when referred to the old stars. Details are presented by Upgren (1978). The conclusion appears to confirm an outward motion suggested by Kerr (1962) from 21-cm observations. It is reasonable to inquire whether a similar outward motion can be seen in the space motions of other stars. Unlike the Van Vleck parallaxes combined with the most rigorous of the weighting methods used here, earlier results may be too insensitive to measure such a motion. Nonetheless a search of some of the existing literature shows that a small outward motion is consistent with sources so far examined. Of the motion solutions listed by Delhaye (1965) only two involve space motions of stars covering most of the main sequence. For stars with velocity dispersions similar to ours, both studies show a small outward motion of the order of five km/sec (for the A dwarfs relative to the generally older F and G dwarfs). The A dwarfs brighter than $5^{m}_{\cdot}5$ (Eggen 1965) have a planar motion distribution very similar to our young dK-M stars and both groups possess a mean motion close to the basic and standard solar motions. These last are mostly based on young stars and the solar motion should be reexamined relative to stars of all ages. It might be worthwhile to redetermine the mean motions of stars for which age-dependent parameters are now available, such as the F-stars with uvby measures or the giants with DDO photometry, since the local standard of rest appears to be a function of stellar age. This study was supported by grant AST77-26554 of the National Science Foundation.

Delhaye, J. 1965, Stars and Stellar Systems 5, pp. 61-84.
Eggen, O. J. 1965, Stars and Stellar Systems 5, pp. 111-129.
Kerr, F. J. 1962, Monthly Notices Roy. Astron. Soc. 123, pp. 327-345.
Upgren, A. R. 1978, Astron. J. 83, pp. 626-635.

W. B. Burton (ed.), The Large-Scale Characteristics of the Galaxy, 242.

V. PHYSICAL PROPERTIES OF THE INTERSTELLAR MEDIUM

INTRODUCTION TO THE SESSION

Carl Heiles, University of California, Berkeley

Five years ago, at the previous IAU Symposium (No. 60) on the galaxy, most of us pictured the interstellar medium (ISM) as a mainly quiescent medium which evolved by orderly processes--the "steady state" ISM. Radhakrishnan reviewed the "intercloud medium" which was warm (some few thousand degrees), rarefied (some few tenths cm^{-3}), uniformly - distributed, partly ionized, and responsible for the smoothly-distributed pulsar dispersion measures and Faraday rotations. Heiles reviewed the "clouds," which had densities of tens cm^{-3}, temperatures of tens K, and nonspherical shapes. These two components were in approximate pressure equilibrium; Grewing, in his review of heating mechanisms, emphasized pervasive, nonvarying processes.

We have since learned that this picture is drastically incomplete. Observationally, the discoveries of interstellar OVI (Jenkins and Meloy, 1974; Jenkins, 1978) and of diffuse X-ray emission from large areas of sky (see the review by Tanaka and Bleeker, 1977) showed that most, or at least much, of the volume relatively close to the sun is filled with hot ($\sim 10^6$ K), highly rarefied (~ 0.003 cm^{-3}) gas. Theoretically, Cox and Smith (1974) showed that type II supernovae generate cavities which expand to radii of order 100 pc and which remain in existence for about 10^7 yr. The total volume occupied by such cavities is a significant fraction of the total interstellar volume in the galaxy. The insides of these cavities contain the hot, rarefied gas of the sort observed. The first steps towards a comprehensive theoretical picture have begun (Smith, 1977; McKee and Ostriker, 1977).

We conclude that much of the volume is occupied by this hot, rarefied gas, and that supernovae have a dominant effect on the ISM.

W. B. Burton (ed.), The Large-Scale Characteristics of the Galaxy, 243–244.

We note that this is true even in the "Great Galactic Ring" where the dense molecular clouds are concentrated; they occupy only about 1% of the volume.

This picture is hardly complete and provides a great many fundamental questions for us. Some of my favorites include the following. Where are the warm HI, the electrons, and the magnetic fields which produce the smooth distributions previously so well accounted for by the intercloud medium? How do spiral density wave shocks fit into this picture? And, of course, the usual question: how do the molecular clouds, which are so dense and massive, form?

REFERENCES

Tanaka, Y., and Bleeker, J. A. M.: 1977, Space Sci. Rev. 20, 815.
Jenkins, E. B.: 1978, Ap. J. 220, 107.
Jenkins, E. B., and Meloy, D. A.: 1974, Ap. J. 193, L21.
Cox, D. P., and Smith, B. W.: 1974, Ap. J. 189, L105.
McKee, C. F., and Ostriker, J. P.: 1977, Ap. J. 218, 148.
Smith, B. W.: 1977, Ap. J. 211, 404.

ENERGY BALANCE IN THE INTERSTELLAR MEDIUM

E. E. Salpeter
Astronomy and Physics Depts., Cornell University

Abstract: A large number (approximately 7) of different components or phases are needed to describe the interstellar medium. The neutral intercloud medium is probably a composite of (a) "lukewarm, substandard" clouds (heated by grain photoeffect and shockwaves), (b) the interfaces between clouds and coronal gas and (c) some "phase 2" gas heated by soft X-rays. Ionizing UV photons are mainly produced by OB-stars and are responsible for most of the average electron density. Bulk kinetic energy for "stirring" the medium and soft X-rays are mainly produced by supernova remnants, less by O-star stellar winds.

1. INTRODUCTION

The energy balance in a molecular cloud is a special case and I will leave this topic for Turner's review. Before discussing energy balance I list (in Sect. 2) modern views on the various components, or "phases", of the interstellar medium. Sect. 3 deals with mass balance, or turnover rates, and Sect. 4 with energy balance itself.

Although the energy input takes on different forms, the primary energy source mainly resides in individual stars. Spiral density waves, powered by rotational and gravitational energy of the galactic disk, are an exception. These density waves--in producing shocks which may initiate star formation (see, in particular, Wielen's and Woodward's contributions)--are triggering mechanisms more than primary energy sources. There is controversy between different rival stellar sources, such as supernova remnant versus O-star stellar wind and nuclei of planetary nebulae versus B-stars. Part of the controversy concerns the precise threshold mass for various processes, so I want to review just what is sensitive to mass and what is not:

For the stellar population I, which applies to the overall Galactic Disk, there is a turnover in the mass-function near 0.2 $M_{\odot}$ and a main sequence break-off near 1 $M_{\odot}$. As regards the present-day luminosity function ψ, massive stars are rare and get rarer rapidly

W. B. Burton (ed.), The Large-Scale Characteristics of the Galaxy, 245–252.

with increasing mass. However, for energy balance considerations it is not ψ itself that matters but mass and energy "fluxing rates", which are related more directly to the birthrate function or "initial mass function" (IMF). A key feature of star formation in the galactic disk is the fact that stellar mass M times (IMF) is a very slowly varying function of M, roughly $\propto M^{-1/3}$ for $M>0.2\ M_\odot$. As a consequence, the fraction of total mass used in star formation which goes into stars of mass $M>1\ M_\odot$ has a total nuclear energy output proportional to M, the fraction of total integrated luminosity $\int L dt$ which comes from stars with mass larger than M is also of order $(M/M_\odot)^{-1/3}$.

Initial main sequence masses larger than $\sim 10\ M_\odot$ are required for (a) the main sequence luminosity to be mainly in the Lyman continuum, (b) the eventual production of a supernova and (c) the generation of a high-velocity stellar wind. In fact, supernova statistics (Maza and v.d. Bergh 1976; Tammann 1977) suggest that the threshold for (b) may be $\sim 5\ M_\odot$ rather than $10\ M_\odot$; Copernicus data (Snow and Morton 1976; Lamers and Morton 1976) suggests that the threshold for (c) may be $\sim 20\ M_\odot$ rather than $10\ M_\odot$. However, because of the weak dependence of $(M/M_\odot)^{-1/3}$ this difference matters little. What matters more is the energy output per star: The main sequence integrated luminosity (msil) is $\sim 10^{-3}\ Mc^2$ and the supernova energy release ~ 0.05 (msil). The gravitational energy content of a main sequence star is only $\sim 10^{-3}$ (msil), O-star stellar winds usually flow at $\lesssim 3$ escape velocity, so that the bulk kinetic energy released in such a stellar wind is $\lesssim 9\times$(gravitational energy) ~ 0.01 (msil). Hence O-star stellar winds are likely to be a smaller primary energy source than supernova remnants, but not by a large factor.

Another comparison concerns central stars of planetary nebulae versus OB-stars: Most stars above $1\ M_\odot$ can produce a planetary nebula (Weidemann 1977), whereas OB-stars are massive, but the important question is what fraction of the (msil) is emitted in the far UV during and after the planetary nebula stage. There was enough theoretical uncertainty (Salpeter 1978) so that fraction might have been appreciable. However, recent UV studies of surface temperatures (Pottasch et al 1978) indicate that this fraction is only ~ 0.01 and planetary nebulae are a minor UV source.

2. THE VARIOUS "PHASES" OF THE INTERSTELLAR MEDIUM

Spitzer (1956) pointed out that the different components of the interstellar medium (ISM) should be in rough pressure equilibrium with each other, much (but not all) of the time. The "two-phase model" (Field et al 1969, Dalgarno and McCray 1972) of the ISM was an elegant application of this principle. However, more recently a larger variety of ISM components (and "transient" material far from pressure equilibrium) have been discovered.

I reiterate first the most obvious components of "phases" of the

ISM in an "average" galactic disk. I assume a radius $\sim$13 kpc, half-height at half-density $\sim$130 pc, a hydrogen mass $M_H \sim 3\times10^9\ M_\odot$ and average pressure (divided by k) of (1500 to 3000) $cm^{-3}\ ^oK$:

Phase 0: "Molecular cloud-OB-star complexes": Typical internal density $n \sim 10^3$ (H) cm^{-3}, temperature $T \sim 10$ K and an overall contribution to mean density of $\bar{n} \sim 0.4\ cm^{-3}$ (volume filling factor $f \sim 4\times10^{-4}$). I will not discuss this phase further.

Phase 1: The "Standard" HI clouds: Internal density $n \sim 40\ cm^{-3}$, $T \sim 70$ K, $\bar{n} \sim 0.2$, $f \sim 0.005$. This component accounts for about half of the neutral atomic hydrogen.

The "Old" Phase 2: A hypothetical, ubiquitous intercloud medium, heated and partially ionized by a hypothetical, ubiquitous flux of X-rays or cosmic rays with ionizing rate per H-atom of $\zeta \sim 10^{-15} s^{-1}$: This, now partially abandoned, component would have had $n \sim \bar{n} \sim 0.2\ cm^{-3}$; $T \sim 7000$ K, $f \sim 1$ and an electron density $n_e \sim \bar{n}_e \sim 0.03\ cm^{-3}$. It would have accounted for the full $\bar{n}_e$ indicated by pulsar dispersion measures (Gómez and Guelin 1974) and the half of the neutral atomic hydrogen which is not strongly absorbing in the 21cm-line. The total flux of soft X-rays is not sufficiently large for a uniform phase 2, but there should be some of it with a slightly smaller internal n and a smaller filling factor (the temperature depends on ζ/n).

Phase 4-: The OVI-containing "coronal gas" (I label components in order of increasing temperature): This component is suggested by satellite observations of absorption by the OVI ion which occurs in gas in a narrow temperature range around 3×10^5 K. An extrapolation from OVI column densities (Jenkins 1978) gives a contribution to the average density $\bar{n}$ from this component of $\sim 3\times10^{-4} cm^{-3}$, but the filling factor is not known.

The evidence for further components is less direct and the following is my personal selection. I start with two components (1+ and 2-) which I feel are needed (together with 2) to account for the neutral atomic hydrogen which is "not strongly absorbing" (with $\bar{n} \sim 0.2\ cm^{-3}$):

Phase 1+: "Lukewarm" and "substandard" clouds: Evidence has been accumulating that the clouds display a wide range of temperatures. The latest absorption-emission survey (Dickey et al 1978a) confirms such "lukewarm" clouds with $T \sim (10^2$ to $10^3)$ K, contributing $\sim$0.1 cm^{-3} to the average density. There is some anticorrelation between the temperature and the 21cm optical depth for these clouds. A measured temperature is only a harmonic mean along the line of sight, not necessarily a single physical temperature, but it is clear (Dickey et al 1978b, Baker 1978b) that mere blending of phase 1 and a uniform phase 2 is not sufficient.

Phase 2-: Interfaces: As mentioned, some phase 2 (neutral, but

$T > 10^3$ K) is produced by soft X-rays, but not sufficient to account for all the "not-strongly-absorbing" material. Some very hot, fully ionized components are discussed below. Theoretical investigations (McKee and Ostriker 1978, Cox 1978) show that the interface between an interstellar cloud and phase (3 or) 4 produces some material similar to phase 2.

Phase 3: Medium-density Strömgren spheres: The phases described above are predominantly neutral and cannot account for the mean electron density of $\bar{n}_e \sim 0.03$ cm^{-3}. Central stars of planetary nebulae and B-stars (Elmegreen 1976) could give some low-level, but widespread ionization. However, as emphasized by Mezger (1978), OB-stars contribute by far the largest amount of ionizing UV to the general ISM (see Sect. 4). With ζ_{UV} the ionizing rate per H-atom, the RMS electron density is given by

$$f^{-1} \sim \langle n_e^2 \rangle/(0.03)^2 \sim \zeta_{UV}/10^{-15}\text{s}^{-1} \qquad (1)$$

The average of 0.03 cm^{-3} can come from Stromgren spheres ($T \sim 10^4$K) with internal densities up to $n \sim 3$ cm^{-3} and filling factors as low as 0.01 (although there is some evidence for larger filling factors, Reynolds 1977).

Phase 4: Coronal gas in pressure equilibrium: Gas containing OVI is easily recongized but there could be more coronal gas at slightly higher temperatures still. It is theoretically likely that coronal gas in pressure equilibrium ($n \sim 3\times10^{-3}$cm^{-3}, $T \sim 10^6$ K) has an appreciable filling factor, say $f \sim 0.2$ to 0.8.

3. MASS TURNOVER RATES

With r the rate in $M_\odot$/year for some process to flux mass through the ISM, the corresponding turnover time is $\sim r^{-1}$ (3×10^9 years). The ejection of planetary nebulae and the formation of white dwarfs are now both estimated (Weidemann 1978) to have $r \sim 1$. Thus, an appreciable fraction of all star deaths proceed via the planetary nebula stage. The corresponding turnover time is only a few times shorter than the present age of our Galaxy, which fits the fact that the ISM is a few powers of 2 less massive than the stars.

This contrasts with some processes which give the appearance of leading to star formation: Galactic spiral shocks set in about every 10^8 years, so that (if all parts of the ISM were affected) $r \sim 30$. Giant molecular clouds present an even bigger puzzle: A large fraction of the total mass of the ISM is in this form (Solomon 1978) and if these clouds were undergoing free gravitational collapse the turnover time would be only $\sim 10^6$ years and r would be enormously large, $r \sim 3000$. Obviously these clouds are not in gravitational collapse, but we don't know if rotation (Field 1978) magnetic pressure (Baker 1978a) or something else is balancing gravitation.

While discussing high-velocity clouds, Oort (1969) pointed out that more intermediate-velocity clouds are approaching the galactic plane than receding from it. More recent 21cm observations at high galactic latitudes have confirmed this trend for clouds and also established it for the "not-strongly-absorbing" neutral hydrogen (Dickey et al 1978b). If interpreted as net infall to the galactic plane (or fluxing through, with the outward flow ionized, the inward neutral), this velocity asymmetry corresponds to $r \sim 3$.

4. ENERGY BALANCE

Regarding the energy input into the ISM we have to distinguish sources for bulk kinetic energy from sources of photons. For photons in turn we have to distinguish between (i) "near UV" photons below the Lyman-edge, which can heat but not ionize the medium, (ii) ionizing UV photons and (iii) penetrating X-rays. I will give photon rates ζ expressed in units of $10^{-15}s^{-1}$ per H-atom. For the non-ionizing UV in HI-regions: Lyman-α is unimportant (Spitzer 1978, Draine and Salpeter 1978) but continuum stellar photons contribute $\zeta \sim 300$.

For the ionizing UV ($h\nu > 13.6$ eV), I have already mentioned "absolutely free" B-stars (Elmegreen 1976) and the lowered estimates (Pottasch et al 1978) for emission from central stars of planetary nebulae. These two sources contribute $\zeta \sim$ (2 to 3) each to the ionizing UV, with the sources rather widely distributed. Supernova remnants contribute comparable a comparable amount of UV. There is still some slight controversy about the exact value of ζ from OB-stars: Older estimates, based on direct counts in the solar neighborhood (Terzian 1974, Torres-Peimbert et al 1974), give $\zeta \sim 25$. The average value for the whole galactic disk (including active spiral arms) should certainly be larger than the local value; working back from the observed diffuse radio-emission (both in free-free continuum and recombination lines) Mezger (1978) estimates $\zeta \sim 80$, even after allowing for the photons which are "wasted" in the dense, immediate vicinity of the star. At any rate, the uncertainty in the OB-star contribution to the general ISM is relatively small and this contribution is greater than that of any other primary source. Most of the average electron density ~ 0.03 cm^{-3} thus comes, not from a uniform phase 2, but from phase 3 with an appreciable "clumping factor" f^{-1} (see eqn. 1).

The flux of soft X-rays (100 eV $< h\nu <$ 300 eV, say) cannot compete in total ionizing rate with that of ionizing UV, but it is more penetrating and can contribute to the coexistence of neutral and ionized hydrogen which is a characteristic of phase 2. Most of the soft X-ray flux probably comes from the coronal gas in phases 4 and 4-; most of that was probably produced by supernova remnants (Jenkins 1978) and somewhat less by "blastwave bubbles" (Weaver et al 1977) from O-star stellar winds. No accurate estimates have been made to date, but $\zeta \sim 0.2$ is probably a reasonable guess for the soft X-rays (there is, in any case, no sharp dividing line between UV and X-rays).

As a primary source of the bulk kinetic energy, required for "stirring" the interstellar clouds, I am particularly fond of blast-waves produced by supernova remnants (Salpeter 1976). The dynamics is complicated because of the inhomogeneity of the ISM (Cox 1978), McKee and Ostriker 1978, Spitzer 1978), but $\sim 10^{-14}$ eV s^{-1} (per H-atom) is a reasonable estimate. Qualitatively, at least, the observational evidence for large-scale effects of supernova remnants is compelling (Weaver 1978). "Blastwave bubbles" (Weaver et al 1977) are also blown by stellar winds emanating from O-stars at speeds exceeding 1000 km s^{-1}. They have qualitatively similar effects to those of supernova remnants, but I estimate their total energy input to be somewhat lower, probably by a factor of about 2 to 10. The stirring rate required to keep up the velocity dispersion of interstellar clouds against "cloud-cloud collisions" is somewhat uncertain because hydromagnetic phenomena and magnetic fields affect the "collision cross section" of a cloud in an unknown way (Spitzer 1978). The estimate for this dissipation rate is $3 \times 10^{-15 \pm 1}$ eV s^{-1}, which comfortably overlaps the (also rather uncertain) above production rate from supernova remnants.

Regarding the temperature balance between heating and cooling, the situation is fairly clear for most of the components: The heating is maintained by the grain photoeffect and by carbon ionization for phase 1 and by hydrogen ionization for phase 3. Phase 4 material was originally heated by shocks and in any case cools rather slowly. Only the "warm, neutral intercloud medium" with $\bar{n} \sim 0.2$ cm^{-3} seems to present a problem in the sense that no single heat source can provide sufficient heating for all this material:

I hope that this puzzle will be solved by the fact that a number of very different heat-sources are at work and that the "intercloud medium" is itself a composite: As discussed in Sect. 2, some part is contributed by "phase 2" where neutral gas in some fraction of the ISM is kept at a few thousand degrees by soft X-rays from some nearby source. Some part is contributed by "phase 2-", neutral material at the edge of a cloud which is heated by thermal electron conduction from phase 3 to 4 (McKee and Ostriker 1978). The exact amount of this contribution is uncertain (partly because plasma oscillations and magnetic fields make the conduction coefficient uncertain), but it certainly helps raise the harmonic mean temperature of a cloud. Finally, a number of heat sources contribute to interstellar clouds, especially to the "substandard clouds" represented by "phase 1+":

Besides the heating produced by the ionization of carbon-atoms, heating by the photoejection of electrons from grains (using the "near UV" stellar emission) is quite important (Watson 1972, Jura 1976, deJong 1977, Draine 1978). Furthermore, the bulk kinetic energy provided by the "stirring rate" I discussed above is transformed into heat via shockwaves or hydromagnetic waves (Cesarsky 1975, Silk 1975) produced by "cloud-cloud collisions". The actual temperature reached is somewhat history-dependent because molecular hydrogen, if present, is an efficient cooling agent.

I finally return to scaling arguments for giant molecular cloud complexes which, as mentioned, cannot normally be under gravitational collapse: This fact is particularly striking in the 5 kpc "molecular ring" where these complexes make up most of the mass of the ISM and densities and luminosities are enhanced. Essentially, one such complex is more like a "mini-galactic-disk" than like a single, unstable cloud. Luminosity L and gas mass M are both high there and controversy exists (Mezger 1978, Puget et al 1978) whether $\zeta \propto L/M$ is the same there as in our "local disk" or slightly larger. However, for most aspects of the heating-cooling balance it is not ζ which counts but ζ/n. Whether ζ is up slightly or not, the density n is certainly up by an enormous factor in one of these "mini-galactic-disks" and cooling must certainly have the upper hand over heating!

This work was supported by the U.S. National Science Foundation under Grant AST 75-21153.

REFERENCES

Baker, P. L.: 1978a, preprint.
Baker, P. L.: 1978b, this volume.
Cesarsky, C. J.: 1975, Proc. 14 Cosmic Ray Conf. 12, pp. 4166.
Cox, D. P.: 1978.
Dalgarno, A., and McCray, R. A.: 1972, Ann. Rev. Astron. Astrophys. 10 pp. 375.
deJong, T.: 1977, Astr. Ap. 55 pp. 137.
Dickey, J. M., Salpeter, E. E., and Terzian, Y: 1978a, Ap. J. Suppl. 36, pp. 77.
Dickey, J. M., Salpeter, E. E., and Terzian, Y.: 1978b, NAIC Report No. 95, Cornell Univ.
Draine, B. T.: 1978, Ap. J. Suppl. 36, pp. 595.
Draine, B. T., and Salpeter, E. E.: 1978, Nature 271, pp. 730.
Elmegreen, B. G.: 1976, Ap. J. 205, pp. 405.
Field, G. B.: 1978, in "Protostars and Planets".
Field, G. B., Goldsmith, D. W., and Habing, J. J.: 1969, Ap. J. (Letters) 155, pp. L149.
Gómez-Gonzalez, J., and Guelin, M.: 1974, Astr. Ap. 32, pp. 441.
Jenkins, E. B., 1978, Ap. J. 220, pp. 107.
Jura, M.: 1976, Ap. J. 204, pp. 12.
Lamers, H. J., and Morton, D. C.: 1976, Ap. J. Suppl. 32, pp. 429.
Maza, J., and v. d. Bergh, S.: 1976, Ap. J. 204, pp. 519.
McKee, C. F. and Ostriker, J. P., 1977, Ap. J. 218, pp. 148.
Mezger, P. G., 1978, Astro. Ap. (in press).
Oort, J. H.: 1969, Nature 224, pp. 1158.
Pottasch, S. R., Wesselius, P. R., Wu, C. C., Fieten, H. and v. Duinen, R. J.: 1978, Astr. Ap. 62, pp. 95.
Puget, J. L., Serra, G., and Ryter, C.: 1978, this volume.
Reynolds, R. J.: 1977, Ap. J. 216, pp. 433.
Salpeter, E. E., 1976, Ap. J. 206, pp. 673.

Salpeter, E. E.: 1978, in Planetary Nebulae (ed. Y. Terzian) D. Reidel, Dordrecht.
Silk, J.: 1975, Ap. J. 198, pp. L80.
Solomon, P.: 1978, this volume.
Snow, T. P. and Morton, D. C.: 1976, Ap. J. Suppl. 32, pp. 429.
Spitzer, L.: 1956, Ap. J. 124, pp. 20.
Spitzer, L.: 1978, Physical Processes in the Interstellar Medium, John Wiley, New York.
Tammann, G. A.: 1977, in Supernovae (ed. D. Schramm), D. Reidel.
Terzian, Y.: 1974, Ap. J. 193, pp. 93.
Torres-Peimbert, S., Lazcano, A., and Peimbert, M.: 1974, Ap. J. 191, pp. 401.
Watson, W. D.: 1972, Ap. J. 176, pp. 103.
Weaver, H. F.: 1978, this volume.
Weaver, R., McCray, R., Castor, J., Shapiro, P., and Moore R.: 1977, Ap. J. 218, pp. 377.
Weidemann, V.: 1977, Astr. Ap. 61, pp. L27.

DISCUSSION

Verschuur: The absence in the symposium of a major paper on magnetic fields is regrettable. This may reflect the complexity of the field data and its interpretation. We should bear in mind that supernovae may act to destroy spiral structures. In some of the photos of other galaxies we have seen so far one often notices complete disruption of an arm somewhere along its length and in some cases arms simply terminate within the Galaxy. Magnetic field data may be very important in helping us understand what forces capable of destroying spiral arms are operating. Perhaps by recognizing the disruptive influences, and the disrupted regions, we might be able to derive a clearer view of the underlying "grand design".

Kerr: The lack of direct discussion on magnetic fields is partly due to the fact that the intended speaker on this subject was unable to come, and partly because many of the things we know about them are not "large-scale".

Baker: I should like to comment on the high rate of fragmentation and collapse in magnetically supported clouds as derived by Langer (preprint) and by Nakano (PASJ 28, 355; 29, 197). Both assume that the coupling of field and gas is due solely to ions. As these are rare, their rates are high. The coupling is actually dominated by dust grains (Baker, A.&A. 50, 327) which are charged by electron collisions (Spitzer, Ap. J. 93, 369). The rate remains significant but must compete with turbulent diffusion which homogenizes field and gas. Thus a quiescent cloud might be unstable on a timescale of 10^7 years, but not clouds of the sort which we actually observe.

A STATISTICAL STUDY OF LOCAL INTERSTELLAR MATTER BASED ON THE NANCAY 21-CM ABSORPTION SURVEY

J. Crovisier and I. Kazès
Département de Radioastronomie, Observatoire de Meudon

The survey of 21-cm galactic absorption towards 819 extragalactic sources, recently carried out with the Nancay radio telescope (Crovisier *et al*. 1978) provides a sample of absorbing neutral hydrogen clouds which is an order of magnitude larger than those resulting from other available surveys (Hughes *et al*. 1971, Radhakrishnan *et al*. 1972, Lazareff 1975, and Dickey *et al*. 1978). A statistical study of this sample offers insight into the properties of local interstellar matter.

We will refer to the results of a kinematical analysis of nearby clouds, and present a preliminary analysis of the internal velocity dispersion of the clouds and of their 21-cm optical depths. In these studies, account must be taken of the observational bias caused by the blending of absorption features coming from clouds at nearby velocities, and by the presence of spurious features resulting from an incomplete elimination of 21-cm HII emission when measuring the absorption profiles. Hydrogen spin temperatures may in principle be derived from comparison of emission and absorption observations at the same direction. However, the small-scale structure of the hydrogen distribution, which is now indicated by several observations (*e.g.* Greisen 1976, Schwarz and Wesselius 1978), precludes accurate determination of emission profiles corresponding to the same clouds as those responsible for absorption along the line of sight. Up to now, the closest approach to real spin temperature is perhaps the one made with the Arecibo antenna by Dickey *et al*. (1978).

The kinematical analysis was made on 299 clouds detected in absorption at $|b| > 10°$(Crovisier, 1978). The main results are: (i) The systematic motion relative to the Sun of cold neutral local hydrogen is the same as that of nearby stars defined by the standard solar motion. (ii) It has been possible to separate statistically the local differential galactic rotation from random motion in the radial velocities of the clouds. This was done by assuming a plane-parallel distribution of local HI clouds, symmetrical with respect to the galactic plane. Although large discrepancies from this model exist (*e.g.* gas associated with Gould's Belt) it was possible to determine the average distance from the galactic plane of HI cold clouds: $<|z|> = 107 \pm 29$ pc. (iii) The

W. B. Burton (ed.), The Large-Scale Characteristics of the Galaxy, 253–256.

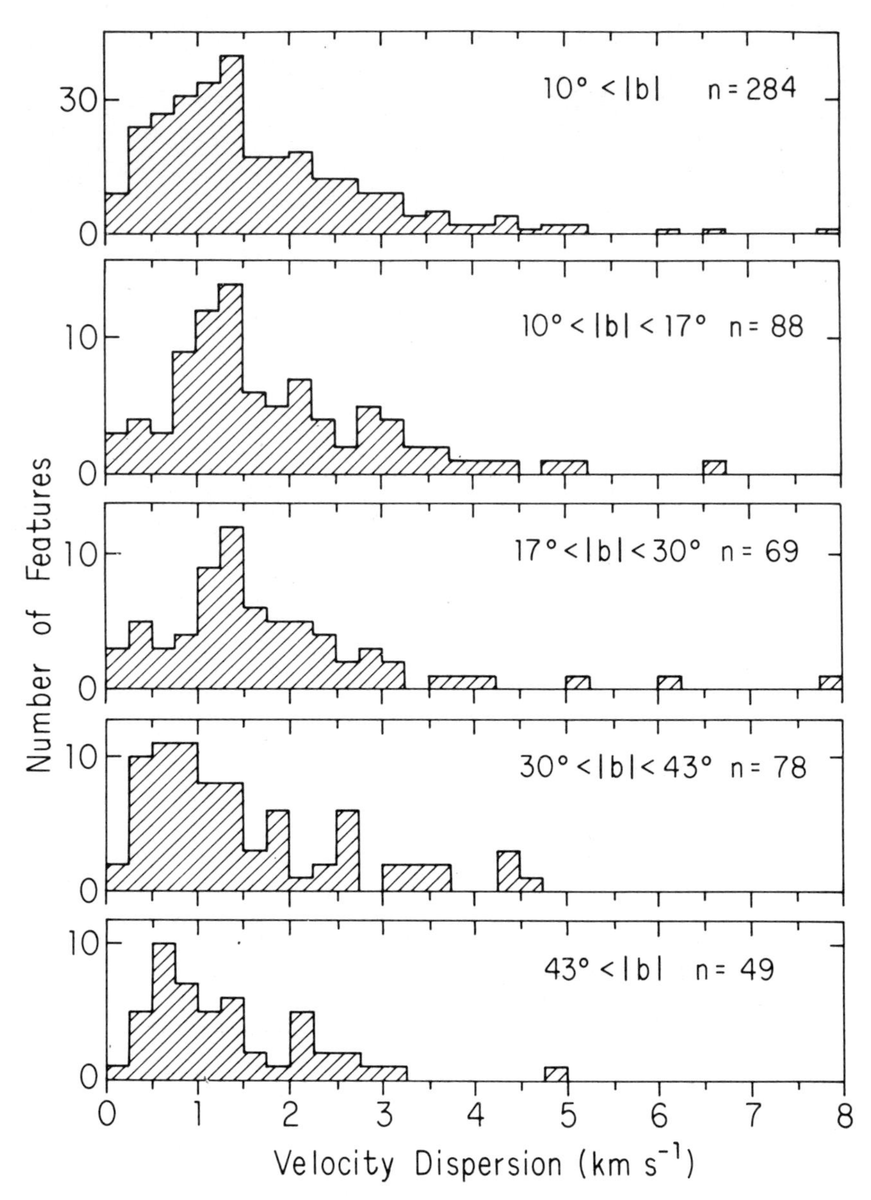

Figure 1. Histograms of the internal velocity dispersion of HI clouds for various latitude ranges. n is the number of clouds in each sample.

radial velocity dispersion, after subtraction of differential galactic rotation, is $<V^2>^{1/2} = 5.7 \pm 0.9$ km s^{-1} and corresponds to a three-dimensional velocity dispersion of about 10 km s^{-1}. No anisotropy greater than 30% was found in the velocity dispersion.

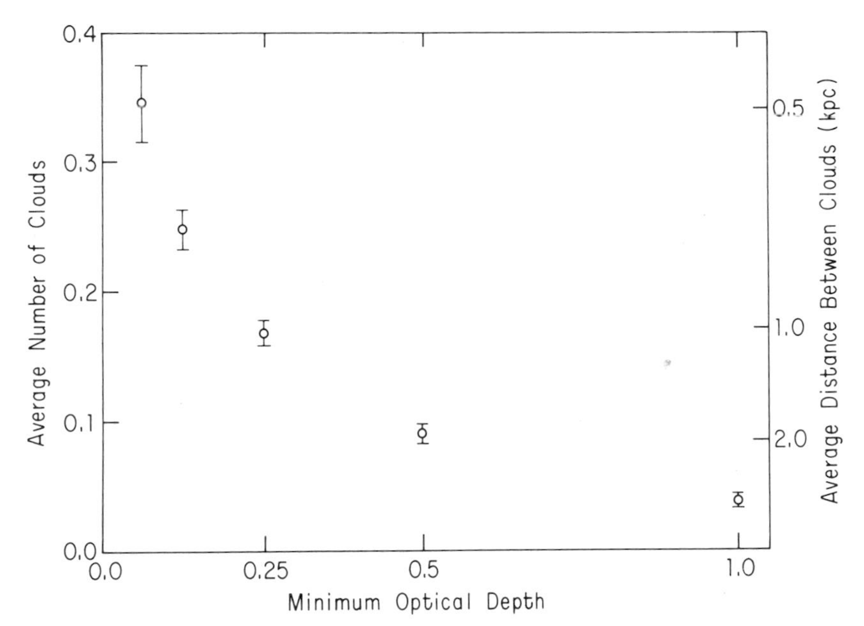

Figure 2. Average number P (left-hand ordinate) of clouds of optical depth at 21 cm greater than τ in a standard line of sight towards $b = 90°$, and average distance L (right-hand ordinate) between these clouds assuming a plane-parallel distribution of equivalent half-thickness equal to 170 pc.

Histograms of internal velocity dispersions, deduced from the observed widths of the absorbing features, are shown in Figure 1 for different latitude ranges. We do not think these histograms are biased against low velocity-dispersions owing to the spectral resolution of the Nancay survey (full half-power width of the channels was 1.3 km s^{-1}); they are, however, affected by blends which cause the feature widths to be overestimated. We believe that the difference between the histograms at $|b| < 30°$ and $|b| > 30°$ is due to the presence of blends, which are more liable to occur at low latitudes, and that the high-latitude histograms are closer to the real distribution of the internal velocity dispersions of the clouds. The distribution peaks around 0.6 km s^{-1}, but the average velocity dispersion is 1.7 km s^{-1}.

In order to derive objectively the distribution of 21-cm optical depths of HI clouds, it is necessary to take into account the sensitivity of each absorption spectrum, which depends on the intensity of the background source. We have calculated the average number per line of sight, $P(\tau)$, of features with optical depths greater than τ, reduced to the standard direction $b = 90°$ using a consecant law (plane-parallel model). A correction was made to account, in a statistical way, for the presence of spurious features. $P(\tau)$ is shown in Figure 2. Note that with increasing τ, $P(\tau)$ decreases more rapidly than might have been expected from the exp ($-\tau$) distribution of optical depths assumed up to now on the basis of the observed histograms of optical depths (Clark 1965, Hughes et al. 1971, and Radhakrishnan and Goss 1972). $P(\tau)$ is related to the average distance L (τ) between clouds of optical depth greater than through $L(\tau) = H/P(\tau)$, where H is the equivalent half-thickness of the cloud distribution in a plane-parallel model. The right-hand ordinate in Figure 2 is scaled in L assuming a Gaussian distribution for clouds along the z-direction, and the average distance from the galactic plane found in our kinematical study. There is for instance one cloud of $\tau > 0.062$ every 500 pc, which may be compared with three clouds per kpc found by Radhakrishnan and Goss (1972)--who did not consider the limit of detection for τ--and one cloud of $\tau > 0.01$ every 80 pc derived recently from a smaller sample by Dickey et al. (1978).

REFERENCES

Clark, B. G.: 1965, Astrophys. J. 142, 1388.
Crovisier, J.: 1978, Astron. Astrophys. 70, 43.
Crovisier, J., Kazès, I., Aubry, D.: 1978, Astron. Astrophys. Suppl.32,20.
Dickey, J. M., Salpeter, E. E., Terzian, Y.:1978, Astrophys. J. Suppl.36,7.
Greisen, E. W.: 1976, Astrophys. J. 203, 371.
Hughes, M. P., Thompson, A. R., Colvin, R. S.: 1971, Astrophys. J. Suppl. 23, 323.
Lazareff, B.: 1975, Astron. Astrophys. 42, 25.
Radharkrishnan, V., Goss, W. M.: 1972, Astrophys. J. Suppl. 24, 161.
Radhakrishnan, V., Murray, J. D., Lockhart, P., Whittle, R. P. J.: 1972, Astrophys. J. Suppl. 24, 15.
Schwarz, U. J., Wesselius, P. R.: 1978, Astron. Astrophys. 64, 97.

DISCUSSION

Rubin: Did you adopt a value for Oort's constant A, or did you estimate a distance for each cloud?

Crovisier: I assumed the standard value A = 15 km s^{-1}.

GENERAL PHYSICAL CHARACTERISTICS OF THE INTERSTELLAR MOLECULAR GAS

B. E. Turner
National Radio Astronomy Observatory*
Green Bank, West Virginia, U.S.A.

The interstellar medium may be characterized by several physically rather distinct regimes: coronal gas, intercloud gas, diffuse clouds, isolated dark clouds and globules (of small to modest mass), more massive molecular clouds containing OB (and later) stars, and giant molecular clouds. Molecules first appear in the denser diffuse clouds, and occur everywhere that $A_V \gtrsim 1^m$. Values of temperature, density, ionization fraction, mass, size, and velocity field are discussed for each regime. Heating and cooling mechanisms are reviewed. Nearly all molecular clouds exceed the Jeans criteria for gravitational instability, yet detailed models reveal no cases where observations can be interpreted unambiguously in terms of rapid collapse. The possibility that clouds are supported by turbulence, rotation, or magnetic fields is discussed, and it is concluded that none of these agencies suffice. Comments are made about fragmentation and star formation in molecular clouds, with possible explanations for why only low mass stars form in low mass clouds, why early-type stars form only in clouds with masses $\gtrsim 10^3\ M_\odot$, and why O-stars seem to form near edges of clouds. Finally, large-scale interactions between molecular clouds and the galactic disk stellar population are discussed.

I. MORPHOLOGY OF THE INTERSTELLAR MEDIUM

Although the range of physical parameters characterizing the interstellar medium may well form a continuum of values, it is convenient (and present observations permit us) to describe the interstellar medium in terms of a few, rather distinct regions.

1) "coronal" gas refers to a hot, diffuse component revealed by absorption lines of atoms in high stages of ionization seen in the far-UV spectra of OB stars; the most conspicuous of these is the O VI ion, whose abundance reaches a peak in the temperature range $5.3 \leq \log T \leq 5.9$ which, along with $-2.3 \leq \log n \leq -1.5$, characterizes these regions (Jenkins, 1978a,b). The filling factor for coronal gas is 0.2 to 0.5 (Myers 1978) and the pressure in these regions is $p/k \lesssim 10^4$ cm^{-3} K.

* Operated by Associated Universities, Inc., under contract with the National Science Foundation.

W. B. Burton (ed.), The Large-Scale Characteristics of the Galaxy, 257–270.

2) intercloud gas is seen in the form of broad, low-intensity emission features in the 21-cm spectra of HI observed toward extragalactic sources at high latitudes (Davies and Cummings 1975; Lazareff 1975). This gas is somewhat cooler ($2.9 \leq \log T \leq 4$) and denser ($-1.0 \leq \log n \leq 0$) than the coronal gas, although the current observations are insensitive to gas hotter than $\sim 10^4$ K. This gas is well described by a pressure $p/k = nT \simeq 10^3$ cm^{-3} K.

3) diffuse interstellar clouds are defined here as those with $2(19) \leq N \leq 2(21)$ cm^{-2}, where $N = N_H + 2N_{H_2}$, the total column density, is inversely correlated with temperature: $\log N = a - b \log T$. The lower limit on N is the currently lowest detectable value (by 21-cm absorption techniques; Davies and Cummings 1975; Lazareff 1975), while the upper limit on N represents the value above which T becomes roughly independent of density (A_V becomes $>1^m$ and the UV field becomes unimportant). At the low-N range, these clouds have no molecular content, while the higher-N clouds have appreciable amounts of H_2, observed optically (Spitzer and Jenkins 1975; Savage et al. 1977), and CO, observed at $\lambda 2.6$ mm (Knapp and Jura 1976). The distribution in sizes, based on statistical analyses of selective extinction, indicates two groups of clouds: large ones with radii ~ 35 pc, and "standard" ones with radii of ~ 5 pc (Spitzer 1978). The latter, ~ 8 times more numerous, are characterized in Table 1. Direct measures of a few cloud sizes have been made by Greisen (1973, 1976). A representative pressure for these clouds is $p/k \simeq 3700$ cm^{-3} K, although some are much lower.

4) "isolated" dark clouds refer to well-defined regions of extinction, $\sim 1 \leq A_V \leq 25^m$ (the upper value representing the limit of star-count techniques) which are not associated with emission or reflection nebulosity (i.e., regions containing early-type stars). These clouds are largely, if not completely, molecular, except for possibly a shell of HI around the larger ones. The temperature is nearly always 10 ± 3 K, and the linewidths are much more constant over these clouds than over other cloud types. While the parameters in Table 1 characterize these clouds as a whole, several of them are now recognized to contain denser cores ($n \sim 10^4$ cm^{-3}) not related to the presence of embedded stars.

5) "large" globules as defined by Bok et al. (1971) and Bok (1977) have recently been found from molecular studies (Dickman 1976; Martin and Barrett 1978) to have densities and temperatures very similar to the denser isolated dark clouds. The mass and size distributions of these globules seem to fit onto the lower end of the corresponding isolated dark cloud distributions. Bok (1977) estimates there are 25,000 large globules in the galaxy.

6) dark clouds associated with AB star formation have long been recognized by the presence of emission or reflection nebulosity associated with the obscured regions. CO maps of these regions have shown that the attendant dark clouds are statistically larger and more massive than the isolated dark clouds, and that, as expected, sizable regions of enhanced temperature and density accompany the locations of embedded stars or shock fronts. As judged from linewidths, these clouds are generally more quiescent than those associated with O star formation.

7) molecular clouds associated with O star formation are discussed here as a separate category because it is often felt that they are larger and more massive than other clouds. While the two or three most massive clouds known do contain O stars, the mass distributions of categories 6) and 7) almost completely overlap; some clouds with O stars are smaller than some clouds with only later-type stars.

8) molecular cores in clouds forming A,B,O stars do not appear to show a wide range of masses or sizes. While parameters describing cores are subject to large uncertainties of selection (e.g., limited spatial resolution) and of definition, it does appear that no great distinction can be made between cores associated with regions of O stars, and later-type stars.

9) giant molecular clouds have recently been mapped in many regions, primarily with the Columbia and Texas telescopes, and include the prominent examples M17 (Elmegreen 1977), Cep OB3 (Sargent 1977), Sgr B2 (Scoville *et al*. 1975), Orion (Kutner *et al*. 1977), W49 (Mufson and Liszt 1977), Ser OB1 (Elmegreen and Lada 1976), Cyg OB1, Cyg OB2, Cyg OB8, Cyg OB9 (Cong 1977), Cas OB6, IC1848/W3 (Lada *et al*. 1978), Per OB2, Gem OB1 (Baran 1978), and Mon OB1, Mon OB2, CMa OB1 (Blitz 1978). A clear-cut definition of a giant cloud is not presently possible, as the edges of many of them are defined by observational sensitivity, and because some giant clouds have recently been found to merge with others (e.g., Orion-Monoceros, Taurus-Perseus clouds). A tentative definition is $L > 30$ pc in one dimension, $M > 5(4)\ M_\odot$. Recent survey work of Scoville and Solomon (this volume) finds the galaxy containing $\sim$4000 such clouds. A rather general cutoff in size and mass seems to be $L \sim 100$ pc, $M \sim 3(5)\ M_\odot$, probably as a result of tidal disruption caused by the differential rotation of the galaxy (Stark and Blitz 1978). (Note, however, that M17 has $L \simeq 170$ pc in one direction.) These clouds are often quite elongated and the majority of them seem to be oriented roughly parallel to the galactic plane. All giant clouds studied so far contain or are associated with regions of early-type star formation (a selection effect: see e.g., Blitz 1978b). As might be expected, the non-core component of giant clouds, comprising virtually all of the volume and mass, consists of gas similar in properties to that of dark clouds, or even diffuse clouds at the outer regions.

These categories of the molecular gas are phenomenological only. The next steps are to quantify the properties of these classes, and then to seek physical explanations.

II. PHYSICAL PROPERTIES OF THE MOLECULAR MORPHOLOGICAL COMPONENTS

The above categories divide rather naturally into the low-pressure (coronal, intercloud, and diffuse cloud) components, and the high-pressure (various molecular) components of the interstellar medium. Methods for determining T and n have been described by Myers (1978).

The low-pressure components are not the subject of this review, and we will only remark on them briefly. They are usually assumed to be in pressure-equilibrium (although there may be significant pressure differences between coronal and intercloud components), and the pressure may be taken as $\sim$3700 cm^{-3} K, as predicted by McKee and Ostriker (1977; "MO") in their recent 4-component model of the interstellar medium. With this assumption, Myers (1978) is able to derive filling factors of $\sim$0.5 and 0.3 for coronal and intercloud components, respectively. The latter may be identified with MO's "warm ionized" plus "warm neutral" media.

Molecules make their appearance in the onset of the high-pressure components which clearly cannot be in pressure equilibrium with the interstellar medium as a whole. All categories of molecular components are gravitationally bound (the question of collapse is dealt with later), and, taken together, they correspond with the "cold neutral medium" of MO, who deduce a filling factor of f $\gtrsim$ 0.02 for it. This factor is probably consistent with observations; Scoville and Solomon (1975) find f $\gtrsim$ 5(-3), a lower limit owing to incomplete sampling. Of course, the MO theory is not concerned with the details of the molecular component.

Table 1. Physical Properties

Type	T(K)		n (10^3 cm^{-3})		Mass ($\odot$)		Size (pc)	
	typical	range	typical	range	typical	range	mean	range
diffuse clouds	80	40-150	0.03	0.02-1.0	4(2)	?	5	$\lesssim$1 - ?
isolated dark clouds	10	6-15	1	0.2 -7.4	2.6(2)	5-1.3(3)	0.9	0.2-2.3
large globules	10	7-14	8	2.5 -14.	20	0.3-70	0.3	0.1-1.1
dark clouds + AB stars	see text		see text		1(4)	2.5(2)-7.6(5)	8	1-60
mol. clouds + O stars	see text		see text		2(4)	1.5(3)-2.5(6)	30	3-170
molecular cores	36	20-150	40	3-1(3)	1(3)	1(2)-5(4)	1	0.2-5.4
giant molecular clouds	see text		see text		$\geq$1(5)	7(4)-2.5(6)	60	30-170

Table 1 gives the physical properties of the molecular components. Several points should be noted.

1) Masses are estimated from observations of ^{13}CO, using the relation $^{13}CO/H_2$ = 2(-6) or CO/H_2 = 8(-5) (Dickman 1976). This relation has been questioned. Leung and Liszt (1976) find $CO/H_2 \gtrsim$ 3(-5) for warm (T $\gtrsim$ 40 K) clouds, and $\sim$1(-5) for cold (T $\gtrsim$ 15 K) clouds. Wootten et al. (1978) find CO/H_2 = 4(-6) in warm clouds, 2(-5) in cold clouds. Guélin et al. (1977) and Turner and Zuckerman (1978) find $CO/H_2 \leq$ 3(-6) in certain cold clouds, but this result is not necessarily general. Thus CO/H_2 should possibly be reduced from Dickman's value by 3 to 20 times in warm clouds, and $\sim$4 times in cold clouds. In addition, LTE analyses, used here, may underestimate or overestimate N(CO) depending on n and T; probably (Leung and Liszt 1976) N(CO) is overestimated by factors of $\sim$1.5 in warm clouds and $\sim$4 in cold clouds. Thus the above masses for the warmer sources may need revision upward by a factor of 2 to 13 times.

2) There is an almost complete overlap in the mass distribution of dark clouds with associated AB stars, and molecular clouds with O stars. On the other hand, the mass distribution of these two categories shows only a small overlap with the mass distributions of the lighter isolated dark clouds and large globules. (These latter two categories have highly overlapped mass distributions also, although the lowest-mass objects tend to be globules). Thus it appears that a threshold mass, $\sim 1000\ M_\odot$, is required before ABO star formation can occur, and that once this mass is attained, O-stars and AB stars may both form, although far fewer such clouds contain O stars.

3) For dark clouds with AB stars, T and n range from ≤ 10 K and $\leq 1(3)\ cm^{-3}$ at the outer extremities to values characterizing the cores, namely, $25 \leq T \leq 50$ K and $5(3) \leq n \leq 5(4)$. Many of these clouds are small enough that the cores noticeably heat the entire cloud, over its presently mapped extent. For molecular clouds with O stars the same remarks apply except that the most intense cores have values of T up to $\gtrsim 100$ K and n up to $1(6)\ cm^{-3}$. These values are defined as averages over a 1' beam and therefore involve selection effects. The presence of masers indicates densities as high as $1(9)\ cm^{-3}$ in some cores.

4) For dark clouds with AB stars, Knapp et al. (1977) find observationally that the presence of $\sim$A0 to B0 stars is a necessary but not sufficient condition for heating of the molecular gas over even a small region ($\gtrsim 1'$) readily detectable with current instrumental resolutions. On the other hand, some clouds are known to be significantly heated over extensive regions by B stars (e.g., ~ 3 pc in S255; Evans et al. 1977). O stars always reveal their presence by heating extensive volumes of molecular clouds, although in cases where the core density is not high, the temperature increase may be less than for several B-star cores.

5) Velocity widths of the molecular gas appear to be indicators of both the presence of stars or protostars, and of their type. For isolated dark clouds, Dickman (1975) finds widths of 2.5 ± 2 km/s. For dark clouds with AB stars, the distribution in widths peaks at 2.5 km/s but has a tail out to values of ~ 10 km/s, arising in the core regions; however the majority of cases show values typical of isolated dark clouds. For molecular clouds with O stars, line profiles are often much wider and of complex shape, variously indicating core rotation (Ori (KL)), multiple core components (W51, W49, Sgr B2?), a high degree of unspecified "turbulent" motion (M17, M8), or mass flow (Ori (KL)). The question whether enhanced velocity widths are also indicative of collapse near molecular cores is controversial (see below). The best case for collapse is probably made for globules (Martin and Barrett 1978), but this does not obviously apply to other types of clouds.

Fractional ionization are: 2(-4) to 1(-3) for diffuse clouds, $\leq 1(-8)$ but $>3(-10)$ for most dark clouds and dense molecular cores, and 3(-10) to 1(-9) for some isolated dark cloud cores with enhanced deuterium fractionation.

III. HEATING AND COOLING OF MOLECULAR CLOUDS

1) Diffuse Clouds (Jura 1978) are cooled by collisional excitation of the fine structure levels of C^+, whose fractional abundance according to Copernicus observations ranges from 4(-5) to 4(-4) depending on the type of determination. If $T > 100$ K, rotational levels of H_2 also become important. Heating by cosmic rays is inadequate, even if the ionization rate ζ_o is as high as $1(-16)\ \mathrm{sec}^{-1}$, which appears unlikely. Starlight seems to be the only viable heat source. Energy released by the photodissociation of H_2 (and its reformation) is probably insufficient, but continuum absorption of starlight by grains, with the attendant photo-ejection of electrons, probably suffices especially if the grains are rather small. A combination of cooling by C^+ and heating by photo effect on grains leads to an expected $T \propto n^{-1}$ dependence, which is consistent with observations and with the hypothesis of constant pressure for all diffuse clouds (i.e., pressure equilibrium). If the grain photo effect indeed dominates heating, then a minimum mass for clouds to become self-gravitating (against thermal pressure) can be estimated as a function of the external intercloud pressure. Jura (1976) finds critical masses of $10^2 - 10^4\ M_\odot$ depending on several uncertainties.

2) Isolated Dark Clouds, along with the outer regions of star-containing clouds, have a well defined temperature, $T \approx 10 \pm 3$ K, apparently independent of density over the range $1(2) \lesssim n \lesssim 3(3)$. Cooling is via collisional excitation of CO rotational transitions (cooling by grains is unimportant at these densities). Heating by starlight is negligible owing to the large attenuation. Heating by cosmic rays, by gravitational collapse, and by ion-slip (ambipolar diffusion) are all possible (Myers 1978) in the formal sense that they fit the observed T *vs* n relationship within rather sizable uncertainties. However gravitational collapse is not favored because there is generally no evidence for it in the observed lineshapes, and because the fraction α of gravitational potential energy loss rate that must go into heating (0.3) is much higher than the expected fraction (0.02) corresponding to uniform compression. Ion-slip heating requires magnetic field strengths of $\gtrsim 100\ \mu$G, probably not consistent with current observed upper limits in two of these clouds (Crutcher *et al*. 1975). The required fields would marginally offer support against gravitational collapse. However, the cosmic-ray heating model probably best fits available data.

The same picture applies to globules. Here, however, gravitational collapse appears inconsistent with observed line profiles according to detailed models of Leung and Liszt (1978).

3) Clouds with stellar-type sources here include clouds associated with A, B, and O stars, HII regions, and protostars. Outside the T = 10 K contours of these clouds, the processes described for isolated dark clouds apply. Inside these contours the same heating processes (except cosmic rays) operate, in addition to heating by the embedded or adjacent stellar/protostellar sources. Within the 10 K contours, the observed narrow range of $T_B(^{12}CO)/T_B(^{13}CO)$ implies $T \propto N(^{13}CO)^p$ with $p \approx 0.2$ to 0.3 (Myers 1978). Also, with more scatter, $T \propto n^p$. In the presence of CO as the dominant coolant, heating solely by gravitational collapse

yields the observed (T,n) relation only for $\alpha \gtrsim 0.3$ and $M \gtrsim 1(5)\ M_\odot$. Again, this process is not to be favored, as α is much higher than the value 0.01 corresponding to uniform compression, and because many warm clouds do not have $1(5) M_\odot$. Ion-slip heating appears satisfactory except for the hottest core regions. (A magnetic field strength of $\sim 270\ \mu G$ is required (Myers 1978), comparable with present relevant observational limits.) In these cores, or anywhere that $n \gtrsim 10^4 - 10^5\ cm^{-3}$, gas collisions with warm grains will dominate the heating. For the specific grain-heating model given by Goldsmith and Langer (1978) the slope of the (T,n) relation is predicted as ~ 0.3 for $n = 10^4 - 10^5\ cm^{-3}$ (in agreement with observations) and ~ 0.5 for $n = 10^3 - 10^4\ cm^{-3}$. A grain temperature $T_{gr} \gtrsim 100$ K is needed for most sources, and ~ 200 K for a few particularly hot cores. $T \gtrsim 0.5\ T_{gr}$ at $n = 10^5\ cm^{-3}$. A rather simpler model by Knapp *et al.* (1977) also provides satisfactory agreement with observations of warm cores in clouds heated by B0 and later stars. The observed fact that A0 stars are necessary but not sufficient to heat cores above 10 K is explained in terms of a threshold density of $\sim 5(3)\ cm^{-3}$. A detailed model of the S255 molecular cloud (Evans *et al.* 1977) shows that three B0 stars at the cloud edge are able to heat all of the 7 x 2 pc region ($M \sim 4(3) M_\odot$) that lies above 10 K. By contrast, Blair *et al.* (1978) find that a corresponding 6 x 3 pc region in the S140 cloud cannot be entirely heated by two nearby B0 stars plus an embedded IR source (latent B0 star); heating mechanisms other than heated grains are required here, although the core region is adequately heated in this way by the IR source. The unusually hot cores and extended warm regions associated with many O-star molecular clouds seem readily explained by the observed high grain temperatures and unusually high gas densities (e.g., Orion: Harvey *et al.* 1974; Werner *et al.* 1974).

In this section we have omitted several other possible heating mechanisms for molecular clouds, including conversion of differential rotation via magnetic fields (Hartquist 1977), chemical heating (Dalgarno and Oppenheimer 1974), and hydromagnetic waves (Arons and Max 1975). All of these appear inadequate.

IV. THE ENERGETICS AND EVOLUTION OF MOLECULAR CLOUDS

Are molecular clouds collapsing, or are they quasi-static? What is the nature of the fragmentation of clouds prior to star formation? What determines whether massive (O) stars form, as distinct from only B-type or later stars? We touch briefly on these presently unresolved questions.

A. Conditions in Collapsing Clouds

Obviously clouds both collapse and fragment since stars exist. Furthermore, the critical Jeans mass for gravitational instability is exceeded by *all* clouds except diffuse ones and a few of the smallest dark clouds and globules. The question is whether *any* clouds contract in a free-fall mode, or only much more slowly as a result of support by turbulence, magnetic fields, or rotation. If all clouds were free-falling, the rate of star formation would exceed the observed rate in the galaxy by a factor of 100 unless the efficiency of forming stars were only 1%; this seems unlikely based on recent studies of fragmentation (Larson 1978) and on observations (Vrba 1977) which indicate that the mass of stars formed is typically 10% of the parent cloud mass. Arguments in favor of free-fall collapse based on observed line profiles have been made for a few reasonably massive clouds (e.g., Loren 1977a,b; Snell and Loren 1977) and for several globules (Martin and Barrett 1978). However more detailed interpretation of these observations (Leung and Brown 1977; Leung and Liszt 1978) show that this conclusion is unwarranted. There appears to be no case at present of uncontroverted free-fall collapse.

According to Goldreich and Kwan (1974) turbulence cannot support clouds against rapid collapse because the required condition that the turbulent stress greatly exceed the thermal pressure (which gives negligible support) means that the turbulence must be supersonic. Then the turbulence will generate shock waves which can be shown to radiate away the shock energy in a time $\lesssim$ free-fall time. Hence turbulence at most increases the cloud lifetime against gravitational collapse by a factor of order unity.

Field (1978) has considered the possibility that rotation stabilizes clouds against collapse. Only a small minority of clouds so far studied via molecular lines can be interpreted as rotating. On the other hand, Hopper and Disney (1974, 1975) examined over 200 fairly compact dark clouds, many of which could be globules, and found that a significant fraction (44%) are elongated roughly parallel (within 30°) with the galactic plane (but not correlated with the magnetic field direction). Hopper and Disney believe that these clouds are probably disks, formed as a result of the gradient in gravitational force perpendicular to the galactic plane. Heiles (1976) suggests instead that these disks are rotating, and that they formed via rapid compression of diffuse clouds. The large-scale galactic magnetic field should keep diffuse clouds in corotation with the galaxy; this initial angular momentum, if conserved during compression to dark clouds, should correspond to a rotation of the latter of $\sim$1 km $s^{-1}pc^{-1}$, roughly what is necessary to rotationally stabilize them. Field points out that the thinner members of these disk-like objects are unstable to formation of bars, and that even bar-like clouds may be gravitationally unstable on a smaller scale. Examples of such clouds have probably been seen (Clark et al. 1977). To stabilize against such effects in most clouds, which do not show them, magnetic fields may be important. While the Hopper-Disney clouds have not been observed in molecular lines to test the rotation hypothesis, Heiles and Katz (1976) fail to observe rotation in several elongated dark clouds and conclude that their elongations arise from other factors. The fact that only three or four clouds of any type show signs of rotating with plane parallel to the galactic plane, as expected for this picture, suggests that rotation in general may not be a very important stabilizing effect.

The role of magnetic fields in molecular clouds is still speculative. Observationally, little is known about such fields in dense clouds. Heiles (1976) cites best values of a few μG and n $\sim$ 1 cm^{-3} for the general intercloud medium. If we scale $|B|$ by $n^{1/2}$as suggested by Mouschovias (1976), values of 10^{-4} to 10^{-3} gauss can be expected in dense clouds. Vrba et al. (1976) have demonstrated the presence of aligning fields in five dark clouds, but their magnitudes cannot be determined. Crutcher et al.'s (1975) upper limits for two dark clouds are 5(-5) gauss. As Field (1978) points out, magnetic fields do influence the distribution of gas on a large scale in the galaxy. But (amplified) magnetic fields may well diffuse out of molecular clouds in time scales short compared to their evolutionary times. This would not occur for diffuse clouds, but may well do so for the more highly condensed cloud types. For typical dark clouds, the diffusion time is only $t_d \approx 6(7)$. $[B/3(-6)\ \mu G]^{-2}$ yr for an ionization fraction of 1(-8) (Guélin et al. 1977). If the field has not diffused out, then radiation of Alfvén waves into the surrounding medium will brake the rotation of a cloud in a time which Field shows is less than the rotation period of the galaxy for diffuse clouds (thus they corotate with the galaxy); however for dark molecular clouds it is not clear whether magnetic diffusion or braking is more rapid. Despite the Hopper-Disney hypothesis, the general lack of observed rotation of expected magnitude might suggest that magnetic braking has been effective for many dark clouds, and therefore that magnetic fields are present which could in principle stabilize against gravitational collapse. In giant molecular clouds, the picture is equally unclear; condensations in M17 are not rotating (Elmegreen et al. 1978) while the Mon R2 cloud appears to rotate at $\sim$2 km s^{-1} pc^{-1} (Kutner and Tucker 1975). Magnetic fields seem possibly to be effective in some, but not all massive clouds.

The influence that an internal magnetic field has on cloud evolution is by no means clear. Mouschovias (1976a,b) finds in a 3-dimensional numerical analysis of low-density (diffuse) clouds that quasi-static solutions can be found, corresponding to slow collapse, and to only modest flattening of the cloud. Dissipation (CO cooling, turbulence) is not included; its effect would be to encourage faster collapse. In an effectively 1-dimensional study of the MHD equations of motion for a gas of ions and neutrals, which does not include dissipation effects, Langer (1978) finds that, for relevant densities and field sizes, only rather small-mass clouds have their collapse times significantly increased over the Jeans ($\sim$ free-fall) time. In Table 2, we compare Jeans time and actual collapse time from Langer's work, for our typical morphological types as listed in Table 1. We conclude that, with the exception of diffuse clouds and probably globules, the expected magnetic field strength doesn't slow the collapse rate significantly over the Jean's time. If dissipative mechanisms or any density gradients are included, the effect of the field would be decreased even further. For the relevant ranges of mass and density, these conclusions do not depend significantly on the magnetic field strength (in the range 1(-4) to 1(-3) gauss). Collapse times do depend strongly on the fractional ionization; unless this is considerably greater than believed, magnetic fields do not seem able to explain why molecular clouds do not collapse at roughly the free-fall rate.

Table 2. Collapse Time in the Presence of Magnetic Fields

Type	Jeans Time (yr)	Collapse Time (yr) B = 1(-4)	Collapse Time (yr) B = 1(-3)	Assumed Ionization Fraction
diffuse clouds*	4(6)	> 1(9)		1(-4)
isolated dark clouds	6(5)	1.5(6) to 2(6)	1.5(6) to 2(6)	1(-8) to 1(-7)
large globules	2(5)	∿4(6)	∿4(6)	1(-7)
dark clouds with AB stars†	6(5)	7(5) to 1(6)	7(5) to 1(6)	1(-8) to 1(-7)
mol. clouds with O stars†	6(5)	7(5) to 1(6)	7(5) to 1(6)	1(-8) to 1(-7)
molecular cores	1(5)	1(5) to 2(5)	1(5) to 3(5)	1(-8) to 1(-7)

* B = 1(-5) gauss

† overall density n = 1(3) cm^{-3}

B. Comments on Fragmentation and Star Formation

The salient observational facts seem to be these:

i) only low mass stars (e.g., T Tauri) have formed in the lower mass isolated dark clouds.

ii) early-type stars (A,B,O) form only in clouds whose masses exceed ∿1(3) $M_\odot$. Such masses are a necessary but not sufficient condition for the presence of O stars. There is no clear distinction in the mass distribution of clouds that form O stars, and those that form only B or later stars.

iii) O stars (and possibly early B stars) appear preferentially near the edges of clouds (cf. e.g., Lada 1978).

Myers (1977) surveyed in the continuum molecular clouds that do not show obvious presence of O stars. Although deeply embedded stars as early as B1 were found, no O stars were detected. This result suggests that O stars appear at cloud edges because they tend to form there. Conversely, lower-mass stars form anywhere in a cloud.

Elmegreen and Lada (1977) and Elmegreen (1977) have explained iii) in terms of "stimulated" collapse--a sudden compression of a cloud's outer layers by a shock front associated with the spiral density wave, a nearby expanding HII region or supernova, or (on occasion) a cloud-cloud collision. These processes do not discriminate against formation of less massive stars, and in fact it is theoretically unclear whether massive stars are even favored. (In particular, Elmegreen (1977) has shown that the maximum fragment mass that will collapse under a shock depends on the 4th power of the time scale of the shock; HII region shocks have time scales ∿10 times those of supernovae). Two interesting questions arise from this picture. 1) Perhaps the massive stars formed earlier near the cloud edge, and the shock merely swept away obscuring material. 2) If shocks induce O-star formation in massive clouds, why not in low-mass clouds? A possible answer to 1) is that OB associations are apparently seen lying along a single line of increasing age away from the cloud. If this is really the case, it does suggest that successive formation of OB clusters was triggered by the previous cluster via HII shocks. A possible answer to 2) lies in a calculation of Jura (1976) which shows that clouds less massive than $\sim 10^3$ $M_\odot$ will not collapse under external pressures $p/k \sim 1500$ cm^{-3} K typical of spiral density waves, while values of $p/k \sim 10^5$ cm^{-3} K near supernovae and HII regions will induce collapse of clouds with $M \gtrsim 10$ $M_\odot$. Thus low-mass, isolated dark clouds do not suffer induced collapse because the spiral density wave is too weak, and because they never encounter HII regions or supernovae. Massive clouds, conversely, are susceptible to stimulated collapse by all of these means. We note that the threshold mass, $\sim 10^3$ $M_\odot$, for induced collapse by the spiral density wave, is observationally about the mass above which OB star formation occurs. This is a possible explanation of observational point ii).

To explain point i) we note that gravitational instability can certainly occur anywhere, anytime, inside a cloud as a result of local loss of internal energy by enhanced molecular cooling, magnetic diffusion, or turbulent viscosity. In the presence of a magnetic field and for sufficiently small ionization fraction ($X_i \lesssim 10^{-7} - 10^{-8}$), Langer (1978) finds that for a given density all fragment sizes have the same (i.e., Jeans) collapse time. In this case, multiple fragmentation appears possible, smaller fragments continuing to separate from larger ones. Suppression of massive stars may result. However, for larger X_i, smaller fragments collapse much more slowly than massive ones. In this case, multiple fragmentation is inhibited, and more massive stars tend to be produced. In isolated dark clouds, $X_i < 10^{-7}$ is indicated by observations, thus explaining lack of massive stars. In more massive clouds, the presence of massive stars (triggered by shocks) heats the surrounding gas, increases the ionization fraction, and therefore possibly inhibits formation of smaller stars.

Returning to point iii), alternative explanations to shock-induced formation of massive stars have been suggested. The critical mass for gravitational instability, with or without a magnetic field, can be written in terms of mass density ρ as $M_{cr} \propto \rho^{-\alpha}$. When coupled with a cloud of non-uniform density given by $\rho \propto r^{-\beta}$, the result is that massive stars tend to form preferentially in the outer regions (for α, $\beta > 0$ as expected in most cases). This suggestion, due to Silk (1978), does not appear to explain the lack of O stars in low-mass clouds.

Another long-standing hypothesis (cf. Mezger 1977) is that low-mass stars form first in a cloud; massive stars later. If so, then the more massive clouds, containing O stars, are older than the low-mass clouds (they could have become more massive by accretion, for example). This idea may be contradicted by recent work by Wootten et al. (1978) who believe that cold, lower-mass clouds have distinctly different chemical composition (in particular, proportionately more CO) than warmer, more massive clouds. Based on our present understanding of cloud chemistry, this result, if confirmed, would imply that the CO is closer to achieving equilibrium abundance in the cold, low mass clouds, which would therefore be the more evolved.

One might argue that in low-mass clouds O stars are simply not expected on the basis of the Salpeter mass function and the relatively few stars of all other types present. This argument is quantitatively untenable. We might also note that Larson's (1978) simulated calculations (no magnetic field) indicate that the largest fragments in a fragmenting cloud should have masses independent of the total mass of the cloud. The number of fragments is roughly the total cloud mass divided by the Jeans mass ($\sim$6 for large globules, $\sim$20 for isolated dark clouds, $\gtrsim$2000 for large molecular clouds).

V. MOLECULAR CLOUDS AND THE LARGE-SCALE CHARACTERISTICS OF THE GALAXY

Based on a galactic plane survey of CO, Gordon and Burton (1976) estimated the total mass of galactic H_2 at 2(9) $M_\odot$, somewhat greater than that found in giant clouds (4000 clouds x 3(5) $M_\odot$ per cloud = 1.2(9) $M_\odot$), but comparable to the total mass in HI (2.3(9) $M_\odot$). While the mass of the molecular component is only $\sim$5% of the mass of the galactic disk, giant molecular clouds are individually the most massive objects in the disk.

Stark and Blitz (1978) have discussed two important interactions that occur between molecular clouds and the galaxy as a whole. The first is the effect on large molecular clouds of tidal forces arising from the differential rotation of the disk. Assuming that the clouds are gravitationally bound, a typical giant cloud of radius 50 pc must have a mass in excess of 2(5) $M_\odot$, when at a typical galactocentric radius R of 6 kpc. The larger value of the tidal acceleration for R < 4 kpc could account for the absence of molecular clouds here (clouds in this region would have to be more massive or more compact than clouds at larger R). Alternatively, if the mass spectrum of clouds is everywhere the same, then masses $\geq$ 3(5) $M_\odot$ are implied. Note that some giant clouds are larger than 50 pc, implying even larger possible masses. (If gravitationally bound, Sgr B2 has a mass of $\gtrsim$5(7) $M_\odot$).

If even a small fraction of the total disk mass is concentrated in objects much more massive than stars, the dynamical relaxation time of the disk stellar system can be significantly reduced. Spitzer and Schwarzschild (1953) suggested that the observed increase in stellar velocity dispersions toward later main-sequence spectral types could be explained by a partial relaxation of the disk population, brought about by large-scale inhomogeneites in the interstellar medium. Stark and Blitz (1978) find that the relaxation time in the presence of the giant molecular clouds is $t_R \gtrsim 7(8)$ yrs. Stars (O,B,A) with main sequence lifetimes $t_\ell << t_R$ do not relax from their initial velocity dispersion of $\sim$10 km s^{-1}, whereas late-type stars with $t_\ell > t_R$ relax toward higher dispersions. The predicted values of velocity dispersion as a function of t_ℓ/t_R are quantitatively in agreement with observations. This implies that there cannot be many giant clouds with M > 3(5) $M_\odot$, or the velocity distribution of disk stars would be more relaxed than is observed.

REFERENCES

Arons, J., and Max, C. E.:1975, Ap. J. (Letters) 196, L77.
Baran, G. P.:1978, Ph.D. dissertation, Columbia U., in preparation.
Blair, G. N., Evans, N. J., vanden Bout, P. A., and Peters, W. L.:1978, Ap. J. 219, 896.
Blitz, L.:1978a, Ph.D. dissertation, Columbia U., in preparation.
Blitz, L.:1978b, Conf. Proceedings on the Massive Molecular Clouds Workshop, Gregynog, Wales.
Bok, B. J.:1977, P.A.S.P. 89, 597.
Bok, B. J., Cordwell, C. S., and Cromwell, R. H.:1971, in Dark Nebulae, Globules, and Protostars, ed. B. T. Lynds, U. of Arizona Press.
Clark, F. O., Giguerre, P. T., and Crutcher, R. M.:1977, Ap. J. 215, 511.
Cong, H. I. 1977, unpublished Ph.D. dissertation, Columbia U.
Crutcher, R. M., Evans, N. J., Troland, T., and Heiles, C.:1975, Ap. J. 198, 91.
Dalgarno, A., and Oppenheimer, M.:1974, Ap. J. 192, 597.
Davies, R. D., and Cummings, E. R.:1975, M.N.R.A.S. 170, 95.
Dickman, R. L.:1975, Ap. J. 202, 50.
Dickman, R. L.:1976, unpublished Ph.D. dissertation, Columbia U.
Disney, M. J., and Hopper, P. B.:1975, M.N.R.A.S. 270, 177.
Elmegreen, B. G.:1977, Conf. Proceedings of the Massive Molecular Clouds Workshop, Gregynog, Wales.
Elmegreen, B. G., and Lada, C. J.:1976, A. J. 81, 1089.
Elmegreen, B. G., and Lada, C. J.:1977, Ap. J. 214, 725.
Elmegreen, B. G., Lada, C. J., and Dickinson, D. F.:1978, in preparation.
Evans, N. J., Blair, G. N., and Beckwith, S.:1977, Ap. J. 217, 448.
Field, G. B.:1978, in Proc. of Conf. on Protostars and Planets, Tucson, ed. T. Gehrels.
Goldreich, P., and Kwan, J.:1974, Ap. J. 189, 441.
Goldsmith, P. F., and Langer, W. D.:1978, Ap. J. in press.
Gordon, M. A., and Burton, W. B.:1976, Ap. J. 208, 364.
Greisen, E. W.:1973, Ap. J. 184, 379.
Greisen, E. W.:1976, Ap. J. 203, 371.
Guélin, M., Langer, W. D., Snell, R. L., and Wootten, H. A.:1977, Ap. J. (Letters) 217, L165.
Hartquist, T. W.:1977, Ap. J. (Letters) 217, L45.
Harvey, P. M., Gatley, I., Werner, M. W., Elias, J. H., Evans, N. J., Zuckerman, B., Morris, G., Sato, T., and Litvak, M. M.:1974, Ap. J. (Letters) 189, L87.
Heiles, C.:1976, Ann. Rev. Astron. and Astrophys. 14, 1.
Heiles, C., and Katz, G.: 1976, A. J. 81, 37.
Hopper, P. B., and Disney, M. J.:1974, M.N.R.A.S. 168, 639.
Jenkins, E. B.:1978a, Ap. J. 219, 845.
Jenkins, E. B.:1978b, Ap. J. 220, 107.
Jura, M.:1976, A. J. 81, 178.
Jura, M.:1978, in Proc. of Conf. on Protostars and Planets, Tucson, ed. T. Behrels.
Knapp, G. R., and Jura, M.:1976, Ap. J. 209, 782.
Knapp, G. R., Kuiper, T.B.H., Knapp, S. L., and Brown, R. L.: 1977, Ap. J. 214, 78.

Kutner, M. L., and Tucker, K. D.:1975, Ap. J. 199, 79.
Kutner, M. L., Tucker, K. D., Chin, G. C., and Thaddeus, P.:1977, Ap. J. 215, 736.
Lada, C. J.:1978, Conf. Proceedings of the Massive Molecular Clouds Workshop, Gregynog, Wales.
Lada, C. J., Elmegreen, B. G., Cong, H. I., and Thaddeus, P.:1978, in preparation.
Langer, W. D.:1978, preprint.
Larson, R. B.:1978, preprint.
Lazareff, B.:1975, Astron. and Astrophys. 42, 25.
Leung, C. M., and Liszt, H. S.:1976, Ap. J. 208, 732.
Leung, C. M., and Brown, R. L.:1977, Ap. J. (Letters) 214, L73.
Leung, C. M., and Liszt, H. S.:1978, in preparation.
Loren, R. B.:1977a, Ap. J. 215, 129.
Loren, R. B.:1977b, Ap. J. 218, 716.
Martin, R. N., and Barrett, A. H.:1978, Ap. J. Suppl. 36, 1.
McKee, C. F., and Ostriker, J. P.:1977, Ap. J. 218, 148.
Mezger, P. G.:1977, in Star Formation, I.A.U. Symp. #75, ed. T. De Jong and A. Maeder (Reidel).
Mouschovias, T. Ch.:1976a, Ap. J. 206, 735.
Mouschovias, T. Ch.:1976b, Ap. J. 207, 141.
Mufson, S. L., and Liszt, H. S.:1977, Ap. J. 212, 664.
Myers, P. C.:1977, Ap. J. 211, 737.
Myers, P. C.:1978, Ap. J., in press.
Sargent, A. I.:1977, Ap. J. 218, 736.
Savage, B. D., Bohlin, R. C., Drake, J. F., and Budich, W.:1977, Ap. J. 216, 291.
Scoville, N. Z., and Solomon, P. M.:1975, Ap. J. (Letters) 199, L105.
Scoville, N. Z., Solomon, P. M., and Penzias, A. A.:1975, Ap. J. 201, 352.
Scoville, N. Z., and Solomon, P. M.:1978, this volume.
Silk, J.:1978, in Proc. of Conf. on Protostars and Planets, Tucson, ed. T. Gehrels.
Snell, R. N., and Loren R. B.:1977, Ap. J. 211, 122.
Spitzer, L.:1978, Physical Processes in the Interstellar Medium, J. Wiley & Sons.
Spitzer, L., and Schwarzschild, M.:1953, Ap. J. 118, 106.
Stark, A. A., and Blitz, L.:1978, preprint.
Turner, B. E., and Zuckerman, B. 1978, Ap. J. (Letters), submitted.
Vrba, F. J., Strom, S. E., and Strom, K. M. 1976, A. J. 81, 958.
Vrba, F. J. 1977, in Star Formation, I.A.U. Symp. #75, ed. T. De Jong and A. Maeder (Reidel), pp. 243-245.
Werner, M. W., Elias, J. H., Gezari, D. Y., and Westbrook, W. E. 1974, Ap. J. (Letters) 192, L31.
Wootten, A., Evans, N. J., Snell, R. N., and vanden Bout, P. 1978, Ap. J. submitted.

STATISTICAL MODELING OF CO EMISSION IN THE GALAXY

M. A. Gordon and W. B. Burton*
National Radio Astronomy Observatory[†], Green Bank

We summarize here some of the distribution properties of CO clouds in our galaxy as determined by observation and by computer simulations of these observations. The simulations adequately account for the number of clouds, their spatial separation, and their relative velocities with respect to one another. The principal shortcoming of the axisymmetric simulations is their failure to describe adequately the clustering of clouds seen in the velocity-longitude domain.

The 2.6-mm emission line of carbon monoxide has been extensively used to explore the optically inaccessible regions of the galactic plane. The first major survey, by Wilson et al. in 1974, dealt primarily with known discrete sources of continuum radio emission. Subsequent surveys, listed as references for Table 1, deal with the general morphology of the cold interstellar gas irrespective of the discrete sources.

Although by themselves the CO spectra give information about the distribution of cold material in the galaxy, the construction of computer simulations give additional insight. The Monte Carlo procedure which we have followed involves construction of a model of the CO distribution within the galactic disk, "observation" of this model by means of a simulated telescope, and subsequent comparison graphically and quantitatively of features found in the observed and modeled data. The initial constraints to the model are those determined unambiguously from the observations. Iterative trial and error guided changes in the simulation.

Specific results on the distribution characteristics of CO clouds

*Present address: Department of Astronomy, University of Minnesota, 116 Church Street S. E., Minneapolis, Minnesota 55455.

[†]Operated by Associated Universities, Inc., under contract with the National Science Foundation.

W. B. Burton (ed.), The Large-Scale Characteristics of the Galaxy, 271–276.

CHARACTERISTICS OF CO IN THE GALAXY

Property	Value	Source	Reference
Kinematics	like HI	Terminal velocities	2,7
Galactic radial extent	$4 < \varpi < 10$ kpc	Longitude extent, kinematic distances	1,2,3,6,7
Abundance maximum	$\varpi = 5.7$ kpc	Longitude extent, kinematic distances	1,2,3,6,7
<z extent>	$h = 50$ pc, $n(z) = \exp(-z^2/2h^2)$	Latitude extent, kinematic distances	4,6
Mean layer	like HI	Latitude extent, kinematic distances	4,6,9
Uniformity of distribution	heavily clumped	Velocity-longitude diagram	1,2,7
<Cloud separation>	1 kpc, increasing away from $\varpi = 5.7$ kpc	Computer simulation	2,7
Cloud - cloud velocity	$\sigma = 4$ km s^{-1}	Computer simulation	2,7
Number of clouds	$< 10^6$	Computer simulation	2,7
Cloud occultation	barely unimportant	Computer simulation	2,7
Cloud density within layer	$< 5 \times 10^{-5}$ pc^{-3}	Computer simulation	7
Fractional volume of layer	$< 0.3\%$	Computer simulation	7
Total mass of CO	$> 10^6$ $M_\odot$	Computer simulation	3,7
Cloud diameters	Most between 5 and 30 pc	Observation	7,8
Cloud masses	3×10^4 to 10^6 $M_\odot$	Virial theorem	1,7,8
Internal turbulence	$\sigma = 2.5$ km s^{-1}	Terminal velocities, isolated clouds	2,7
Cloud lifetimes	$> 2 \times 10^7$ y	Computer simulation	5

[1] Scoville and Solomon 1975
[2] Burton et al. 1975
[3] Gordon and Burton 1976
[4] Burton and Gordon 1976
[5] Bash et al. 1977
[6] Cohen and Thaddeus 1977
[7] Burton and Gordon 1978
[8] Gordon and Burton 1978
[9] Lockman 1977

in the galaxy are summarized in the table. Initial results on the galactic morphology of clouds were based on observations made with an uncooled receiver by Scoville and Solomon (1975) and by Burton, Gordon, Bania, and Lockman (1975). Subsequent papers by various authors generally are based on improved observations made with a cryogenic receiver. In our series of papers, Paper I (ref. 3 in the table) deals with the distribution of cold interstellar gas as a function of galactocentric radius. Paper II (ref. 4) deals with observations and simulation of the latitude extent of the CO layer. Paper III (ref. 7) summarizes results from a rather extensive survey and its computer simulation. Related papers include one on the sizes of molecular clouds (ref. 8) and one on the determination of the pattern speed of the density wave in our galaxy (Gordon, 1978). An attempt has been made by Roberts and Burton (1977) to incorporate the precepts of the density-wave theory in the Monte Carlo simulations. A parallel effort at incorporating these precepts into computer simulation of CO observations has been made by Bash, Green, and Peters (1977). In both attempts, the CO clouds behave as ballistic objects after passing through the galactic density wave. The Bash et al. approach results in a determination of a cloud lifetime.

REFERENCES

Bash, F. N., Green, E., and Peters III, W. L.: 1977, Astrophys. J. 217, pp. 464-472.

Burton, W. B., Gordon, M. A., Bania, T. M., and Lockman, F. J.: 1975, Astrophys. J. 202, pp. 30-49.

Burton, W. B., and Gordon, M. A.: 1976, Astrophys. J. Letters 207, pp. L189-L193.

Burton, W. B., and Gordon, M. A.: 1978, Astr. Astrophys. 63, pp. 7-27.

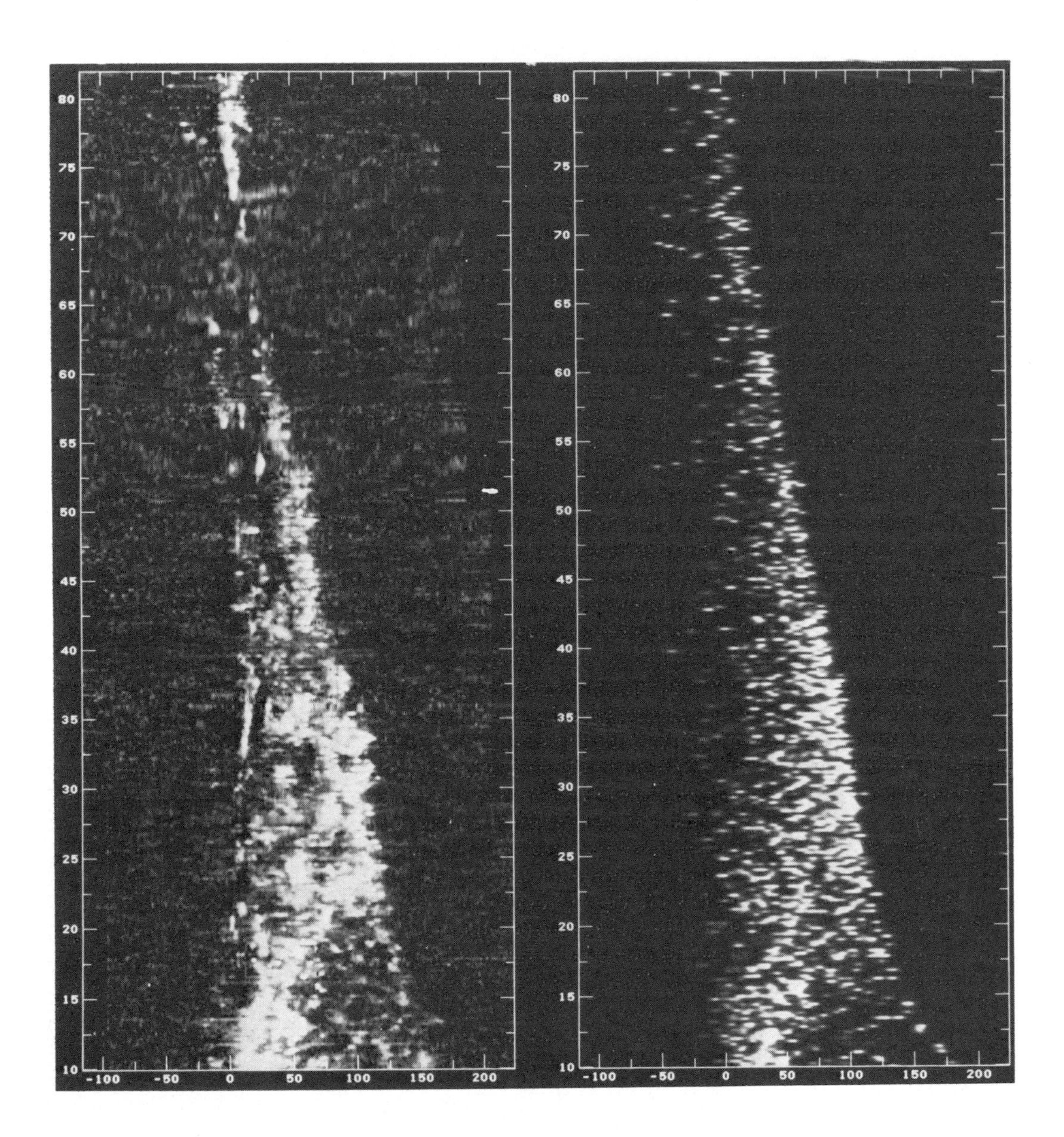

Figure 1. (left) Grey-scale representation of the longitude-velocity arrangement of ^{12}CO emission observed at 0°2 intervals along the galactic equator. (right) Longitude-velocity arrangement of emission inherent in synthetic spectra representing stochastically distributed discrete CO clouds (Burton and Gordon, 1978).

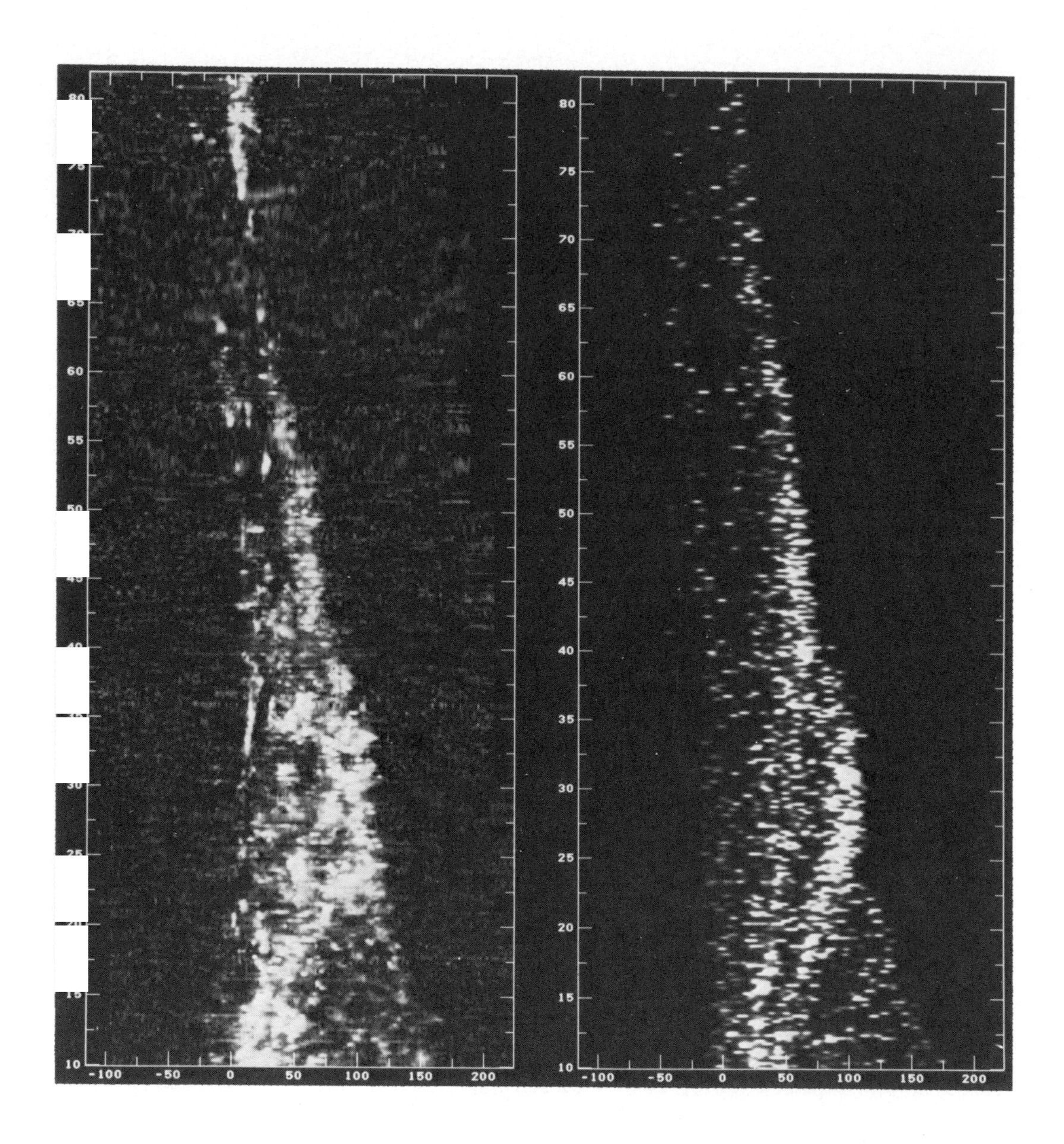

Figure 2. Comparison of the longitude-velocity arrangement of ^{12}CO emission observed at b=0° with the situation representing CO distributed stochastically in discrete clouds in a spiral model. The clouds' kinematics and probability of occurrence follow the predictions of the density-wave theory (Roberts and Burton 1977).

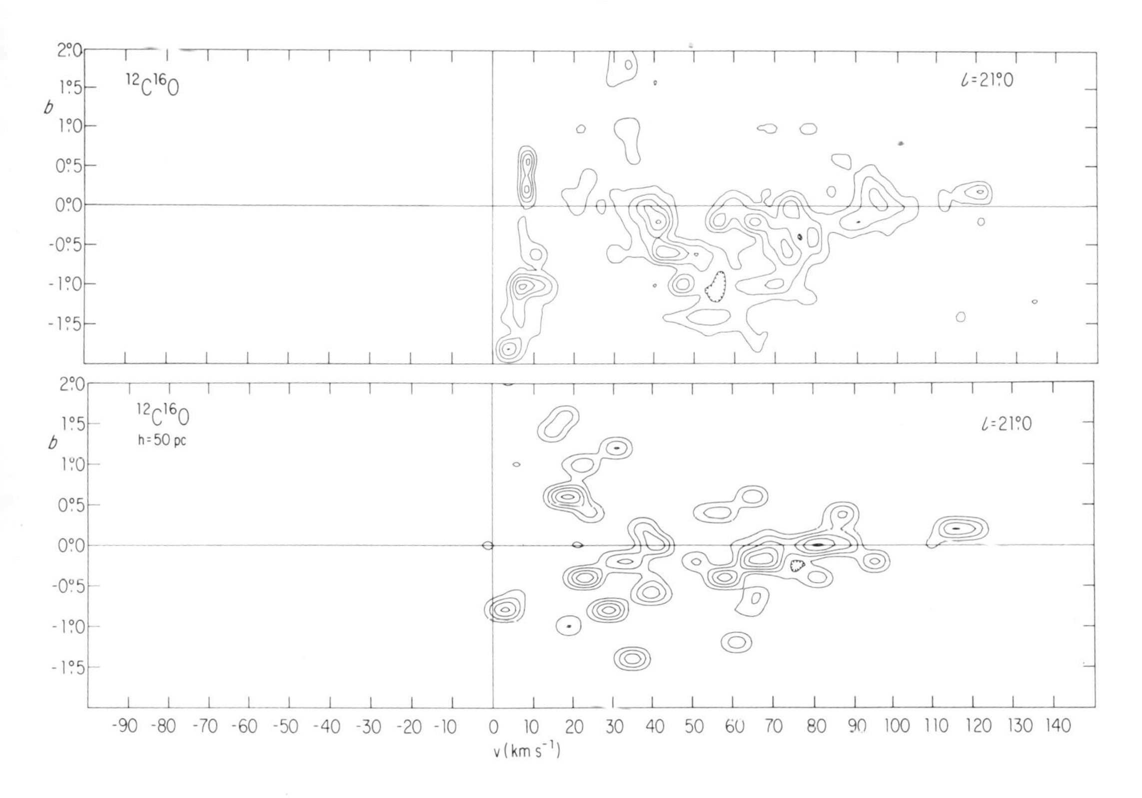

Figure 3. (top) Observed latitude-velocity distribution of CO emission in the direction ℓ=21°. (bottom) Computer simulation of this CO layer. The scale height in the form $n(z)=\exp(-z^2/2h^2)$ is 50 pc in this simulation.

Cohen, R. S., and Thaddeus, P.: 1977, Astrophys. J. Letters 217, pp. L155-L159.

Gordon, M. A.: 1978, Astrophys. J. 222, pp. 100-102.

Gordon, M. A., and Burton, W. B.: 1976, Astrophys. J. 208, pp. 346-353.

Gordon, M. A., and Burton, W. B.: 1978, Proc. Gregynog Workshop, in press.

Roberts, W. W., and Burton, W. B.: 1977, in "Topics in Interstellar Matter," H. van Woerden (ed.), pp. 195-205.

Scoville, N. Z., and Solomon, P. M.: 1975, Astrophys. J. Letters 199, pp. L105-L109.

Wilson, W. J. et al.: 1974, Astrophys. J. 191, pp. 357-374.

DISCUSSION

Sinha: What is your definition of the outer boundary of the CO distribution? Chi Yuan described his detection of a substantial number of CO clouds in Perseus; others have observed CO clouds to farther distances in the anticenter direction.

Burton: The CO abundance distribution continues to fall off gradually at $\mathcal{R} > 8$ or 10 kpc. The accumulated emission which remains is *much* less than at smaller distances. The CO surveys have until recently been principally confined to $b = 0°$. If the CO layer were to be strongly warped, some emission would remain undetected, and the outer-region abundances would need revision. However, there is sufficient data at $b \neq 0°$ to indicate that this effect is not substantial.

Shuter: What value do you use for the random velocities of the clouds? I suggest that you check the consistency of this with the new determinations of scale height because the scale height and random velocities are simply related.

Burton: The one-dimensional dispersion is 4 km s^{-1}, a value derived principally from the scatter in the terminal velocities. We have not made the check you suggest.

van den Bergh: Photographs of external galaxies show that the densest dusty regions have a vein-like structure. Is there any evidence to indicate that galactic CO clouds also have such a structure?

Burton: The sampling interval of the CO data is still too coarse to provide an answer to this.

Bash: Regarding the lifetime of CO-emitting molecular clouds, I would just like to make the point that one can fill the CO v,ℓ diagram, similar to the way it is observed to be filled, by assuming that molecular clouds are launched from the spiral shock wave and live 3×10^7 years.

Burton: The cloud-cloud velocity dispersion, which is measured in one dimension to be 4 km s^{-1}, is not large enough to fill the v,ℓ space if the starting distribution is a simple pattern. (Unfortunately, we do not know much at all about the underlying fundamental pattern.)

CO OBSERVATIONS OF SPIRAL STRUCTURE AND THE LIFETIME OF GIANT MOLECULAR CLOUDS

N. Z. Scoville
Department of Physics and Astronomy
University of Massachusetts, Amherst, Massachusetts

P. M. Solomon and D. B. Sanders
Space Sciences Department
State University of New York, Stony Brook, New York

Observations of CO emission at ℓ=0 to 70°, $|b| \leq 1°$ are analyzed to give a map of the molecular cloud distribution in the galaxy as viewed from the galactic pole. From the fact that this distribution shows no obvious spiral pattern we conclude that the giant molecular clouds sampled in the CO line are situated in both arm and interarm regions and they must last more than 10^8 years. A similar age estimate is deduced from the large mass fraction of H_2 in the interstellar medium in the interior of the galaxy. An implication of this longevity is that the great masses of these clouds may be accumulated through cloud-cloud collisions of originally smaller clouds.

1. INTRODUCTION

One of the great expectations of the 2.6-mm CO line is that its distribution will delineate the spiral structure of the Milky Way with a clarity not obtained in either 21-cm line studies or radio observations of HII regions. The line is highly selective towards the densest interstellar clouds where star formation must occur; it is ubiquitous and easily detected in these regions; and the angular resolution of 1 arcminute which is available with existing telescopes is smaller than typical cloud sizes even for clouds on the far side of the galactic center.

The giant molecular clouds revealed by CO surveys of the galactic plane (Scoville and Solomon 1975; Burton et al. 1975; Burton and Gordon 1978) are a new class of interstellar cloud which constitutes the dominant mass component of the interstellar medium in the interior of the galaxy. Individual masses are typically 10^5 - 10^6 $M_\odot$. Their total number is estimated to be about 5000; most of them are situated in an annulus at 4 to 8 kpc galactic radius (cf. Solomon - this volume).

Our picture for the evolutionary history of the giant molecular clouds and the role played by the spiral shock in precipitating star formation will be strongly influenced by whether the clouds are situated

W. B. Burton (ed.), The Large-Scale Characteristics of the Galaxy, 277–283.

only within spiral arms or whether they are also found just as frequently in interarm areas. The evidence I shall present here indicates that they are found in both regions. A consequence of this is that the clouds must exist as entities for very long times - perhaps 10^9 years rather than 10^7 years as previously assumed, and their large masses can be plausibly accounted for by growth occurring as a result of cloud-cloud collisions. Since the protostellar gas flowing into the spiral arm shock is therefore already composed of high density clouds; the role of the shock in provoking star formation may be substantially less than previously imagined.

2. A "FACE-ON" PICTURE OF THE GALAXY

By virtue of its broad latitude coverage, the survey of Solomon, Scoville, and Sanders (1978) affords an excellent opportunity to trace large scale galactic features such as spiral arms or cloud clusters which do not necessarily maintain a constant latitude over their full longitude extent. Approximately 1100 points in the inner galactic plane at longitudes -5° to 90° were observed with the NRAO 36-foot antenna and the recently completed 45-foot telescope at the University of Massachusetts. Spectra were taken both in the plane and out to $b = \pm 1.5^\circ$ so as to cover at least one scale height of the cloud distribution for any gas at greater than 2 kpc distance.

In transforming the CO data in the ℓ/v plane into galactocentric coordinates of radius and azimuth, there results an ambiguity in the azimuthal angle for clouds interior to the solar circle. In the present survey the extended latitude coverage enables us to resolve the ambiguity for many cloud complexes. Since the scale height deduced for molecular clouds is 70 - 100 pc, most of the features we observe at $|b| > 0^\circ\!.35$ are probably at the near-point azimuth.

The "face-on" distribution of CO emission at high latitudes and at low latitudes are shown separately in Figure 1. In the first figure no attempt was made to resolve the distance ambiguities; each CO feature is shown at both its near and far point locations. The mapping of the observed emission from ℓ/v coordinates to $\tilde{\omega}/\phi$ employed the Schmidt rotation law which has been found to apply within 5 $km \cdot s^{-1}$ to the molecular gas.

In Figure 2 we have attempted to resolve the distance ambiguities by comparison of high and low latitude data in Figure 1 and by identification of particular features with galactic HII regions for which the ambiguity has been resolved from absorption line studies. The ambiguities were resolved for all strong features, but for many of the weak features where the assignment was unclear both positions remain.

There is excellent correlation between the positions of giant HII regions (cf. Georgelin and Georglin 1976) and giant molecular clouds. The correlation is in the sense that nearly all HII regions for which

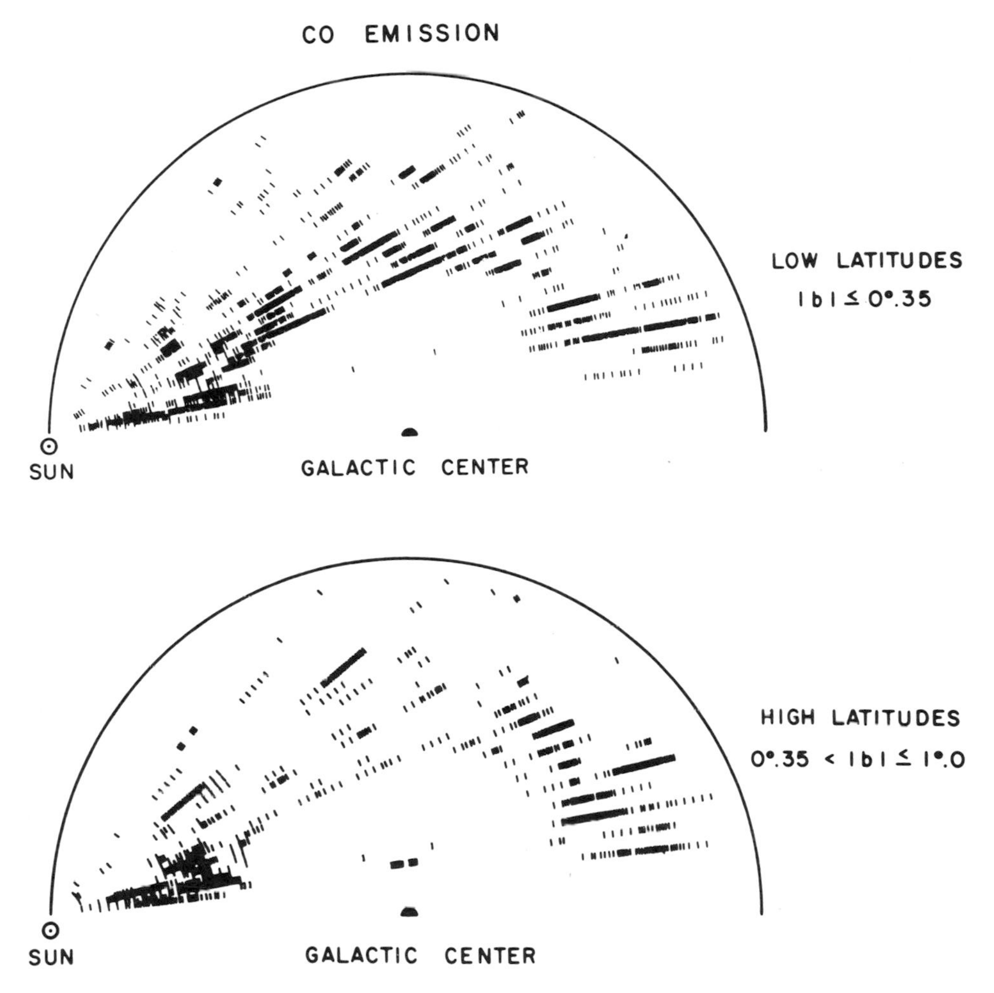

Figure 1: The relative distribution of CO emission in the galaxy at ℓ=0 to 70° as viewed from the North galactic pole. Only strong features with T_A^*/η > 5°K have been plotted. Each region is shown at both near and far points along the line-of-sight calculated from the Schmidt rotation law. High and low latitude data are shown separately to enable distance determinations.

we have CO data within 1/4° have a CO counterpart with similar radial velocity. It is clearly not the case that every giant molecular cloud has an associated HII region since we count in this survey alone many more separate clouds than there are known giant HII regions.

The two major spiral features deduced from 21-cm line observations of the northern hemisphere (ℓ=0 to 90°) are the Scutum and Sagittarius arms which are tangential to the line of sight at ℓ=33° and 50° with radial velocities of 100 and 60 km·s^{-1}. Concentrations of CO emission

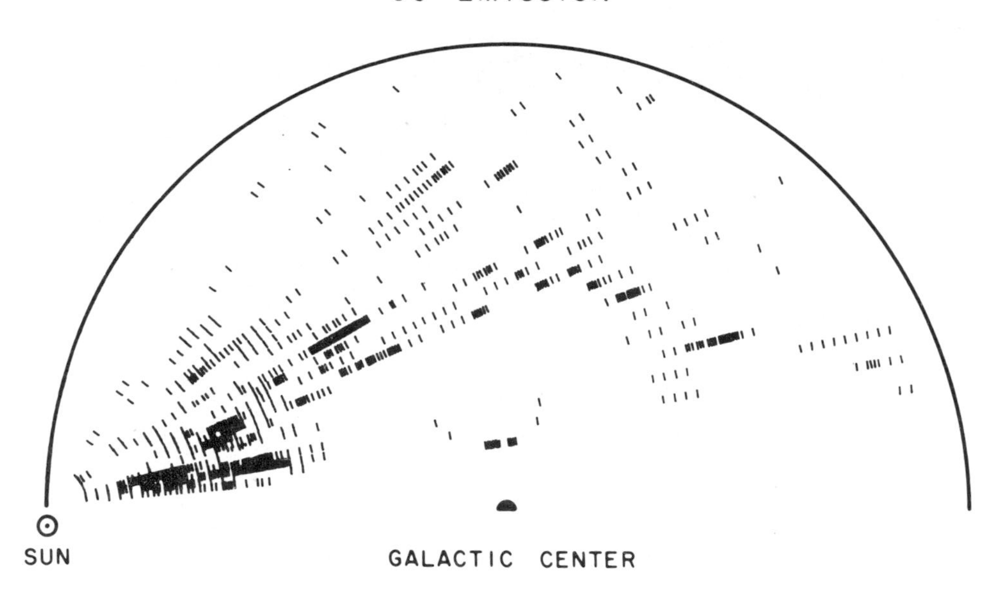

Figure 2: The relative distribution of strong CO features ($T_A^*/\eta > 5°K$) at $|b| \leq 1°$ is shown in a face on view of the galaxy. Distance ambiguities are resolved where possible by comparison of the high and low latitude emission data in Figure 1 and by identification of features with HII regions at known distances.

corresponding to the tangential directions of these two arms are clearly seen in Figure 2 at a distance about 7 kpc from the sun. In the CO data neither of these features may be traced over more than 10° of longitude.

It is clear from the absence of a recognizable spiral form in Figure 2 that most of the clouds cannot be situated within a regular pattern of spiral arms. It is difficult to estimate the percentage of clouds in arms as compared to that between arms for two reasons. First, there is not any strong agreement among previous observers of spiral structure as to precisely how the 21-cm features and HII regions should be linked up into arms. Secondly, with two arms running through the observed ℓ/v space, a good fraction of the CO emission regions should fall at ℓ/v locations close to the presumed spirals even if they were randomly situated in the galactic plane. Examples of features definitely not in the spiral arms are, however, found at $\ell=40°$, $v = 85\ km\cdot s^{-1}$; $\ell=24°$, $v = 105\ km\cdot s^{-1}$; and the low velocity features and $\ell > 20°$.

3. THE LIFETIMES OF MOLECULAR CLOUDS

Based upon the detection of molecular clouds in interarm regions we must abandon the notion that the molecular clouds represent a brief compression phase through which the interstellar medium passes each time a galactic shock passes by. Since the clouds last well into the interarm regions we conclude that their ages are at least 10^8 years. Perhaps even more persuasive is an argument based upon the observational determination that the greatest fraction of the interstellar hydrogen in the interior of the galaxy is molecular rather than atomic or ionic. If the interstellar medium constantly cycles gas through the H_2, HI, and HII phases and the relative abundances are maintained in a steady state, then continuity demands that the mass flow rate out of the H_2 phase (M_{H2}/τ_{H2}) must equal the flow rate out of the HI and HII phases into H_2 ($M_{HI + HII}/\tau_{HI + HII}$). Since M_{H2} is found to be greater than $M_{HI + HII}$, continuity requires $\tau_{H2} > \tau_{HI + HII}$. The timescale for a typical low density atomic or ionized hydrogen region to exist is probably at least 10^7 years and generally approaches 10^8 years in the interarm regions so that τ_{H2} must be at least 10^8 years. It is thus deduced from two independent lines of reasoning that the giant clouds last at least 10^8 years and may well last 10^9 years since it is unclear what process is capable of disrupting them when they are so massive.

4. COLLISIONAL GROWTH OF MASSIVE CLOUDS

Once it is recognized that these clouds are extremely old, it becomes plausible that their very large masses, typically 4×10^5 $M_\odot$ may be built up through the mechanism of cloud-cloud collisions (Scoville and Hersh 1978). The timescale for cloud collisions in the interior of the galaxy ($\tilde{\omega}$ = 4 to 8 kpc) is 3×10^8 years using current estimates for the cloud parameters. A high efficiency for binding of the two colliding clouds is expected isasmuch as their mean velocities are typically comparable with the 10 km·s^{-1} escape velocity from a single cloud. They will still probably coalesce even in cases where the collision velocities are large since the excess kinetic energy can be dissipated rapidly via radiation in the H_2 rotational transitions. In the absence of a mechanism decreasing the mass of H_2 clouds, we therefore expect that they can double their masses within the period of each collision which is presently comparable with <u>lower limits</u> to the cloud lifetime.

It is clear that the mass buildup via accretion cannot go unchecked for more than 10^9 years lest all the clouds become too massive. There must be one or more processes which remove H_2 from the clouds at an average rate equal to that provided by collisions. Though various mechanisms such as supernova explosions may be capable of completely dispersing a cloud of 10^3 - 10^4 $M_\odot$ in a single event this is not possible for the very massive giant molecular clouds. For such a cloud of 4×10^5 $M_\odot$ the escape velocity is about 10 km·s^{-1}; a single event would therefore need to supply a momentum of 4×10^6 $M_\odot$ km·s^{-1} or it must ionize 4×10^5 $M_\odot$ at an average density of 10^3 H nucleii per cc. Since these requirements far exceed the capabilities of supernova or common O-stars, the process of removing H_2 from clouds is more likely in the form of "chipping away" at the cloud rather than catastrophic occurrences.

Formation of massive stars which subsequently ionize an HII region or undergo supernova explosion while in the cloud can lead to a decrease in the mass of H_2 within the giant clouds. The mass-loss rates of these processes must be added to that associated with steady conversion of H_2 gas into stars. Of these processes, the most important are probably the ionization by O-stars and the absorption of H_2 into newly formed stars. The former rate is very difficult to estimate with reliability; the latter is estimated at $\sim 2\ M_{\odot}\ yr^{-1}$ for the entire galaxy (Scoville and Hersh 1978). If this galactic star formation rate is evenly divided amongst the total of about 5000 giant molecular clouds in the galaxy, we find that the rate of collisional growth of molecular clouds is very nearly balanced by the rate of star formation.

5. CONCLUSIONS

The evolutionary model suggested here represents a radical departure from the notion that dense star-forming clouds are the result of compression by a galactic density-wave travelling through a low density medium. Except for the higher rate of star formation which clearly does occur when the clouds pass through a spiral arm, the large scale cloud structure may be effected very little. If the removal of gas from the clouds is piecemeal through star formation, rather than catastrophic, we should then picture a cloud as an entity which survives even longer than 10^9 years though it does not retain a typical gas element longer.

This is contribution number 279 of the Five College Observatories. This research is supported by National Science Foundation grant AST76-24610.

REFERENCES

Burton, W. B., and Gordon, M. A.: 1978, Astron. Astrophys. 63, 7.
Burton, W. B., Gordon, M. A., Bania, T. M., and Lockman, F. J.: 1975, Astrophys. J. 202, 30.
Georgelin, Y. M., and Georgelin, Y. P.: 1976, Astron. Astrophys. 49, 57.
Scoville, N. Z., and Solomon, P. M.: 1975, Astrophys. J. (Letters) 199, L105.
Scoville, N. Z., and Hersh, K.: 1978, submitted to Astrophys. J.
Solomon, P. M., Scoville, N. Z., and Sanders, D. B.: 1978, in preparation.

DISCUSSION

Woodward: Is the observed number of giant molecular clouds in interarm regions consistent with a dynamic model of cloud motions and with the long cloud lifetime you are suggesting? If clouds are long lived, one might expect to see about half of them in interarm regions.

Heiles: From observational data can one appeal to large-scale kinematics (such as rotation) to keep the star formation rate down?

Scoville: We have not yet analyzed the data to determine if many clouds show significant rotational velocity gradients. It will be difficult to discriminate between rotational gradients and the random velocity changes which occur quite generally within these clouds.

Rohlfs: If you argue that the giant molecular clouds are old ($\tau > 10^8$ y), and on the other hand you show that they are gravitationally bound, what prevents these clouds from collapsing?

Scoville: It is clear that there is some mechanism restraining the observed clouds from free-fall collapse. It is also clear from observations of external galaxies that although dust clouds exist between the spiral arms, the spiral arm shock does provoke rapid formation of massive stars. One property we have not been able to measure yet is the magnetic field strength; it is possible that there is a significant magnetic pressure restraining the clouds from collapse. (This appeals to the well-known astronomical maxim that if you can't measure an effect, it could be important.)

Van Woerden: I might remind Dr. Scoville that it takes only 10^7 years, at an expansion velocity of 10 km s^{-1}, for an association to grow from 0 to 200 pc diameter.

Shuter: It seems to me that the best way to treat CO data in an attempt to observe spiral structure is to average profiles that are far enough apart for the individual clouds to be uncorrelated, but near enough that the galactic rotation has not changed significantly.

Cesarsky: Could it be that molecular clouds can exist for some time without making stars, so that the age of molecular clouds is even larger than the age you quoted, based on destruction by an OB association?

Elmegreen: There are no observations for the lifetimes of giant clouds before the onset of massive star formation. One must make a distinction between the lifetime of the molecules themselves, which may be quite long (10^9 years) and the lifetimes of the cloud complexes as physical self-gravitating entities, for which I suggest a maximum of $\sim$ 50 million years. This long time is already 100 times the cloud's dynamical (free-fall) timescale. It is difficult to imagine a delay in star formation for a time much longer than this. Of course, once massive star formation occurs, the cloud destruction is relatively rapid.

IMPLICATIONS FOR STAR FORMATION IN SPIRAL GALAXIES FROM OBSERVATIONS OF NEARBY MOLECULAR CLOUD COMPLEXES

Bruce G. Elmegreen
Harvard-Smithsonian Center for Astrophysics

I want to make three points about star formation in spiral galaxies that follow from consideration of the internal structure of giant molecular cloud complexes (GMCC). The first point comes from pressure considerations. The total pressure inside the star-forming core of a GMCC may be written 10^6k)v/3kms^{-1})^4(17pc/D)^2 for virial theorem line width v and cloud diameter D; the pressure from a spiral density wave shock (SDWS) is 10^5 k(n_s/1cm^{-3})(v_s/20kms^{-1})^2 and the thermal pressure in the cloud is 10^4 k(n/10^3cm^{-3})(T/10K) for Boltzmann constant k. These three pressures differ by factors of 10. An SDWS has too low a pressure to affect a cloud core; the only way an SDWS could influence a GMCC is if it interacted with the low thermal pressure in the cloud, i.e., the SDWS could propagate into a cloud along the direction of a magnetic field which may be the source of large scale pressure in a transverse dimension. The second point is that the density and mass of a GMCC are so large that the cloud will enter an SDWS like a cannon ball and will not be readily deflected. GMCC in other galaxies would then look like spurs on the spiral pattern and not like dust lanes. The alternative to these two points is that an as yet undiscovered (or uncommon) population of low density (100cm^{-3}) clouds exists involving GMCC-type masses, or that smaller clouds coalesce at the SDWS. This implies that the star-forming clouds studied by molecular observers would be post-SDWS and post-gravitational collapse objects. Finally, the maximum age of a GMCC in the solar neighborhood is probably less than 50 million years. Its destruction is a result of pressure forces from the stars which it creates. Destruction in this sense does not necessarily imply that the molecules are converted into atoms -- only that the cloud is pushed around. In the solar neighborhood, some clouds may, in fact, turn into 21-cm features; e.g., an HI half shell with a radius of 100 pc and a visual extinction through the shell of 0.2 mag. contains 3×10^5 $M_\odot$, the mass of a GMCC. However, in the 5-kpc ring of the Galaxy, there is too much H_2 relative to HI to allow any cycling between H_2 and HI that is in phase with an SDWS unless the cloud remains molecular for 80% of the cycle. More likely, the cloud will be "destroyed" before that time. The implication is that cloud destruction at 5 kpc must produce molecular shells in addition to some atomic shells. This could be observed.

W. B. Burton (ed.), The Large-Scale Characteristics of the Galaxy, 284.

THE COLDEST MOLECULAR GAS IN OUR GALAXY

B. Zuckerman
University of Maryland

Galactic CO emission surveys such as those carried out by Gordon and Burton (Ap. J., 208, 346, 1976) and references cited therein are sensitive only to CO with J = 1 → 0 excitation temperature $\gtrsim$ 5 K. "Cold" molecular (H_2) gas with CO excitation temperature $\lesssim$ 5 K would go undetected in such surveys. Low T_{ex} could be due to either low kinetic temperature or low H_2 density. Evidence for the possible existence of substantial amounts of cold gas ($T_{ex} \sim 3$ K) towards the galactic center is apparent in spectra of Liszt *et al*. (Ap. J. 213, 38, 1977 and 198, 537, 1975) in the form of narrow gaps, suggestive of self-absorption, at velocities of 0, -30 and -55 km/s with respect to the LSR.

We have performed two experiments to determine the fraction of the interstellar medium that is too cold to have been accounted for in existing CO surveys. Drs. N. J. Evans, R. H. Rubin, and myself have compared 6-cm H_2CO absorption spectra against distant continuum sources with CO emission spectra in the same directions. The 100-m Bonn and the 5-m Texas antennae were used so that the beamwidths were comparable ($\sim 2\overset{\prime}{.}6$). Cold molecular clouds should show up well in the H_2CO spectra but not in the CO spectra. A preliminary analysis of the data for about 40 clouds detectable in H_2CO does *not* suggest the existence of a substantial population of cold CO clouds.

The other experiment, carried out with the NRAO 36 ft telescope, involved E.N. Rodriguez Kuiper, T.B.H. Kuiper and myself. We carefully checked a number of the Liszt *et al*. positions to better determine τ_{co} and T_{ex} for the CO self-absorption features towards Sgr A and vicinity. By taking great care with the "off" source positions we were able to establish reliable baselines and found $T_{ex} \gtrsim 7$ K for all CO features in all five of the directions we observed.

In summary, there is, at present, no indication of substantial amounts of molecular gas in our galaxy with temperatures colder than the $\sim$ 8 K expected from cosmic ray heating of molecular clouds (Elmegreen B. G. *et al*., Ap. J., 220, 853, 1978).

W. B. Burton (ed.), The Large-Scale Characteristics of the Galaxy, 285–286.

DISCUSSION

Stark: The millimeter-wave radio astronomy group at Bell Laboratories, Crawford Hill, has observed an absorption feature in Sgr A and Sgr B, in a number of molecules. This feature absorbs from several of the strong continuum sources in this region, in a velocity range 0 → -150 km s^{-1}. The feature is optically thick in HCO^+ and HCN, but is optically thin in $H^{13}CO^+$. The excitation temperature of this gas does not differ significantly from 2.8 K in these molecules. Part of this gas is coincident in position and velocity with the "300-pc expanding ring" seen in CO. The existence of this feature in the galactic nuclear region does not necessarily imply the existence of very cold gas in the galactic disk.

Burstein: Heiles and I have reanalyzed the interrelationships among Shane-Wirtanen galaxy counts, HI column densities, and reddenings, and have resolved many of the problems raised by Heiles (Astrophys. J. 204, 379). These problems were caused by two factors: subtle biases in the reddening data and a variable gas-to-dust ratio in the galaxy. Details of this work have appeared in Burstein, D. and Heiles, C.: 1978, Astrophys. J. 225, 40.

Lyngå: If I had had your preprint before my review talk, this could have been updated by your sound interpretation of galaxy counts. However, there is still a disagreement between the values of E_{B-V} at the poles where some determinations give $E_{B-V} = 0\overset{m}{.}05$ and others give considerably lower values. Possibly, this can be due to the cloud structure. In any case we cannot disregard the stellar statistical determination that give high E_{B-V} values.

Felten: What can you say about the ratio of visual extinction to reddening, and its variation over the sky?

Burstein: Our value of $R = A_{pg}/E(B-V)$ derived from log NGAL is of course dependent on the value of $\gamma(= -d(\log NGAL)/d(A_{pg}))$ one assumes or derives. If $\gamma = 1$ (Heiles, Ap. J. 204, 379), $R \approx 3.6$; i.e., approximately the standard value of $R = 4$.

de Vaucouleurs: I agree with you that galaxy clustering and patchy extinction combine to present a clear-cut distinction between a constant log N or a cosec b law in the range $|\mathrm{cosec}\ b| < 1.20$, but the current argument is mainly with the hypothesis of dust-free polar caps extending down to $|b| \simeq 50°-45°$ (cosec $|b| < 1.35-1.4$) where the galaxy counts are adequate to show that the $|\mathrm{cosec}\ b|$ law holds (for counts to different limiting magnitudes).

An unpublished study of colors of galaxies from the second Reference Catalogue shows that the color excess E(U-V) also obeys the cosec b law right up to both galactic poles within the postulated polar windows.

We find no support in the extragalactic data for the hypothesis that $A_B = 0.0$ at $|b| > 50°$ and no evidence against the adopted value $A_B \cong 0.2$ mag. The possible apparent disagreement with the evidence from stellar data suggesting perhaps $A_B \cong 0.05$ to 0.1 mag at the poles indicates the need for further work.

GLOBAL PHYSICAL CHARACTERISTICS OF THE HI GAS.

P.L. Baker
Max-Planck-Institut für Radioastronomie

The interstellar HI is best discussed in temperature categories because the requirement of pressure equilibrium leads to gross differences between hot and cold gas. The temperature of the hot (6000K) neutral gas has been measured by five techniques while the cold (60K) gas is visible mainly in 21cm absorbtion. There is evidence for warm (200K to 1000K) gas. Because of emission line blending, the small scale structural morphology is undetermined; however, there is evidence it is filamentary. With few exceptions, spatial sizes observe a lower limit of about 5 pc. On a larger scale, the gas is organized into sheets which reflect the recent history of cloud formation.

I. THE CLASSIFICATION OF NEUTRAL HYDROGEN.

The study of HI gas is an interesting field of Astronomy because the amount of data which can be obtained is so large that the observations tend to demolish the assumptions used to interpret them. As a result, 21cm studies force us to face the hermeneutical question: where does one begin with the derivation of physical characteristics from the emission of a single spectral line? The first interpretations were based on models of randomly-placed independently-moving spherical clouds of gas. This historical assumption was critically reviewed by Heiles in the previous symposium in this series (Heiles 1974). The situation in brief is that the gas elements do not move independently, nor are they spherical.

To formulate a correct interpretation of the data, it may be fruitful to make an initial generalization from observations. It appears that the gas is approximately in pressure equilibrium, a circumstance which is also theoretically comfortable. If this is the case, the physical characteristic of temperature should assign the gas to remarkably distinct thermal classes or phases. Decreasing temperature will correlate with increasing gas density; so that cold gas, no matter how abundant, can occupy only a tiny fraction of the interstellar volume. Most of the volume must be filled with hot tenuous material. The hot gas is therefore exposed to mechanical disturbances and radiation which main-

W. B. Burton (ed.), The Large-Scale Characteristics of the Galaxy, 287–294.

tain its elevated temperature, while the cold material tends to shield itself and present a low interaction cross section. These differences of density, volume filling factor and exposure mean that the dominent physical processes change with temperature. Moreover, gas elements with different temperatures have certainly experienced separate histories.

It is useful to retain the term cloud in a temperature classification as a name for formations in the cold gas. Such formations are distinct because the small volumes containing cold material are clearly defined by a temperature-density contrast. The objection that observers have made to theoretical cloud models in the past concerns not the existence of cloud material but the inference that the cloud material forms quasipermanent and isolated entities that move randomly. Later, I will try to justify the view that this seemingly pedantic difference in interpretation is connected with a rather gross difference in the physical state of the interstellar gas. First however, let us review some measurements of the gas temperature.

II. NEUTRAL HYDROGEN TEMPERATURES

The coldest HI gas has a temperature of about 10K, but since it occurs in clouds that are primarily molecular (Burton et al. 1978 and references therein), it will not be discussed here. The cold atomic hydrogen clouds have higher temperatures on the order of 60K. One may set upper limits on the temperature of this cloud material from the width of its emission line, but an accurate average temperature can be determined from absorption measurements against background continuum sources (Shuter and Verschuur 1964, Hughes et al. 1971, Radhakrishnan et al. 1972). Only an average temperature is available because a statistical correction must be made for foreground emission from the hotter gas. The most recent high-sensitivity absorbtion experiments confirm the older work for high opacity features but have also detected a type of low-opacity HI which one might call warm gas (Davies and Cummings 1975, Lazareff 1975, Dickey et al. 1977, 1978). The individual temperature measurements for this gas range from a few hundred to a thousand Kelvin, without correction for foreground emission. One would be tempted to ascribe the unusual temperatures simply to the absence of an emission correction were it not for the common occurrence of the same velocity structure in both emission and absorbtion. The similarity of opacity and emission profiles could arise either from gas at the indicated intermediate temperature or from spatial volumes containing a heterogeneous mixture of hot and cold gas. A few instances of warm gas have also been located by independent techniques such as the 590K HI measured by Baker (1973a) and the 570K gas measured in NaI by Hobbs (1976). These detections would argue against the heterogeneous mixture explanation. At the moment, observations are insufficient to quantify the amount of warm gas in the Galaxy, or to prove or exclude a connection with the cold gas.

The highest measured HI temperatures are those associated with the diffuse, neutral, intercloud medium. For this gas, we are fortunate to have five independent temperature sensitive techniques. To indicate the degree of consistency of the results, the measurements are summarized in Figure 1. Hobbs (1976) used NaI optical lines to obtain measurements or lower limits whose average is displayed as the lower end of the NaI range in Figure 1. The upper end is an average of Hobbs's upper limits established by line widths. Figure 1 also displays the HI absorbtion temperatures obtained by Mebold and Hills (1975). In this case, the range shows the spread of temperatures for distinct spectral components. Apparently, the intercloud temperature is not a constant (Davies and Cummings 1975). These published values are systematically high due to stray radiation; however, the open circle in Figure 1 shows the average temperature of the components after correction for this effect (Kalberla 1978). On the whole, the stray radiation effect is not serious in this case. For a few samples of gas, there are temperature measurements from emission data that were analysed by one of two methods. In principle, both methods seek to establish the thermal broadening pertinent to the smallest discernable spatial structure. The presumption is that no significant nonthermal broadening occurs on this small spatial scale because turbulence would dissipate too quickly. The analysis is complicated by the fact that if one has N structures spread over a velocity range V and each emits with a linewidth of v, blending of the spectral lines from these structures becomes serious when N exceeds V/v. For small structures in the intercloud medium with a large thermal broadening v, blending will always occur. Consequently, the analysis must

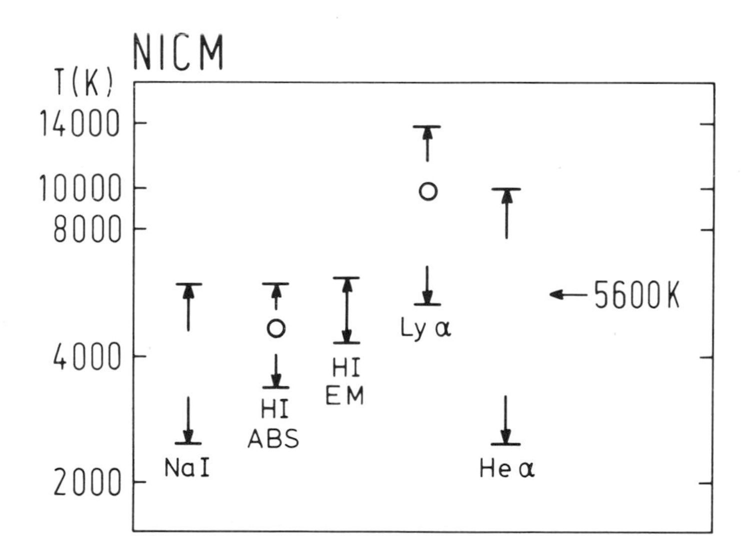

Figure 1. A comparison of temperature measurements of neutral intercloud gas obtained by 5 techniques. A temperature of 5600K is compatible with all of the available data (references are given in the text); however, the range for HI absorption is clearly due to point to point temperature variations.

exploit the spatial fluctuations of blended emission profiles. Baker (1973a) analysed primarily the velocity autocorrelation function of the fluctuations to derive the thermal line width, while Mebold, Hachenberg, and Laury-Micoulaut (1974) relied on a Gaussian model of turbulence that allowed a temperature estimate based on the variance profile. It should be noted that if the assumptions of the 21cm methods fail, the absorbtion measurements of the spin temperature of the hot medium yield estimates of the kinetic temperature that are too low (1) because some of the opacity may be due to cooler material and (2) because the collision rate may be insufficient to thermalize the spin temperature (Field 1958, Davies and Cummings 1975). On the other hand, the emission measurements may be too high if there exists a form of subparsec scale turbulence that mimics thermal broadening. The agreement of the two 21cm techniques is therefore especially satisfying.

While the preceding measurements refer to the gas within a few hundred to several kiloparsecs from the sun, the backscatter of solar Lyman alpha photons yields a measurement of temperature for just those neutral hydrogen atoms entering the solar wind cavity. To indicate the uncertainty in the individual experiments, Figure 1 contrasts the average of the lowest temperatures compatible with the data of three independent observing groups with the average of the highest compatible temperatures (Fahr 1974, Bertaux et al. 1976, Adams and Frisch 1977). The open circle shows the average of all the values. A similar experiment can be performed using the backscatter of 584 Å photons from interstellar helium. The helium result in Figure 1 (Weller and Meier 1974) serves as a check on the backscatter analysis because hydrogen and helium atoms follow different trajectories through the solar cavity.

There is a range of overlap for the temperature determinations which lies at about 5600 K. This result is somewhat lower than theorists had expected for the neutral ICM. In addition, the existence of temperature variation is also not well understood.

III. THE SPATIAL DISTRIBUTION OF HI GAS

Because the volume filling factor of cloud material is small, it was expected that the sun would lie in an intercloud region as proven by the backscatter measurements. However, the fact that the interstellar gas entering the solar system is neutral could not have been predicted so readily. The large exposure of the hot medium hints that a significant fraction of its volume may be ionized by stars (Grewing and Walmsley 1971, Torres-Peimbert 1974). Indeed, large ionized regions must exist near the sun because of the small Lyman alpha opacity in the direction of certain nearby stars (Bohlin 1975) and the detection of stellar EUV radiation (Lampton et al. 1976). The spatial distribution of the remaining neutral intercloud material should be obtainable from 21cm observations. However, reliable information is nearly nonexistent because one needs the brightness temperature to derive the column density. Most observers process 21cm antenna temperatures in a manner recommended by the IAU (van Woerden 1971) in answer to a demand for standardization (van Woerden 1964), and frequently denote the result as brightness temperature. However, the standard procedure is simply a scale change, not a correction for antenna response. The method given by Westerhout et al. (1973) is correct for their special case; but in general, the antenna response far from the main beam must be known and the differential doppler shift of the Earth's motion in these directions must be taken into account. The feasibility of exactly measuring the response of a given antenna by including it in an interferometer was demonstrated by Hartsuijker et al. (1972). Recently, Kalberla derived the response of the Effelsberg telescope without an interferometer and demonstrated a reproducible brightness temperature calibration (Kalberla 1978). The aspect of this problem that is relevant here is that the derivation of the brightness temperature of emission from a low intensity, intercloud field requires the removal of the stray radiation received from other, usually more intense areas on emission. We observe the dim intercloud regions only through a haze of scattered emission and we cannot readily measure the actual column density or its variations. This technical problem also prevents us from defining the boundaries of galactic gas (Baker 1976) and confuses the comparison of 21cm column density against interstellar reddening (Kalberla 1978) and stellar spectra (Giovanelli et al. 1978).

The 21cm lines from the cold and warm gas are bright and relatively narrow in velocity; consequently, they are not seriously confused with contributions from stray radiation or from the neutral ICM. Nevertheless, confusion due to emission line blending complicates the study of small common structures. While it is straight forward to construct a map of blended emission or to assign its fluctuations to putative spatial entities by Gaussian fitting, the result is without significance unless there is a one-to-one and not a one-to-many correspondence between emission structure and spatial structure. One can minimize the blending problem by examining high galactic latitudes where the line of sight leaves the galactic disc after a short distance. Small scale structure seen here is largely filamentary (Heiles and Jenkins 1976). Blending is also avoided if one picks out rare features that are exceptional in velocity or intensity. These structures again appear to be filamentary (Baker 1973b). These results cannot conclusively establish the small scale morphological structure of the gas, but they suggest that filamentary structure is very common (Heiles 1974).

The typical length scale of spatial structure can be measured with some confidence even in the presence of blending because the emission scale size should reflect the underlying spatial scale through the decorrelation length of the chance arrangements in space that - when blended in emission - produce the observed line. Observations relating to the length scale seem to show that structure grows more common with decreasing size until a lower limit is reached. It is rare to find smaller features than this limit. There has been no systematic attempt to determine whether the size cutoff is just what it seems, namely a limit to or turnover in the size spectrum, or whether it results from some interaction between the spectrum of sizes and the sensitivity of the observations. Nevertheless, it is striking that improvements in angular resolution and receiver sensitivity have left the observational situation little changed. The minimum size seen in emission appears to be 5 pc according to a number of surveys tabulated by Verschuur (1974). The main uncertainty seems to be the distance of the observed gas. A similar size is required by emission observations pertaining selectively to the cold cloud material (Baker and Burton 1975); however, absorption measurements of the same material indicate a smaller cutoff at 1 pc (Griesen 1976). These results are not necessarly contradictory because Baker and Burton tested the size of gas patches showing significant opacity at one velocity. Their definition allows smaller size enhancements in opacity at the same velocity; furthermore, it does not exclude a spatial extension of the gas patch at a different velocity.

It has become increasingly common for observers to interpret the intermediate size scale morphology of HI gas - that between 20pc and 100pc - in terms of sheet distributions. This term was invoked by Heiles (1967) to describe the distribution of gas in one large area where the emission showed rifts. Without an unlikely preferential alignment, rifts cannot be seen if the gas is more extended in the line of sight than the width of the rift. The sheet geometry also appears to describe a large area of saturated 21cm emission in the galactic anticenter region (Baker 1974a). For this region, the measured spin temperature implies a number density necessary for pressure equilibrium. This density together with the column density yields a depth of a few parsecs for gas covering tens of parsecs on the sky i.e. the gas occurs in a thin layer. The 21cm opacity of this sheet has also been measured at the location of 5 continuum sources by means of the NRAO 3-element interferometer (Baker 1974b). While the absorption was present at the expected velocity at every point, the opacity varied greatly. This variation confirms the existence of substructure in the sheet.

If one speaks of cloud-like morphologies for HI, one should have in mind the terrestrial cirrocumulus or altocumulus clouds which consist of small filaments or cloudlets organized into a much larger coherent layer. Because both terrestrial and interstellar clouds move, one might think of the sheets as a coherent pattern of motion common to the gas fragments composing the sheet. In this respect, cloud gas imitates star streams - those groups of young stars that move through the solar neighborhood with a coherent velocity determined by their common point of origin. These three velocity groupings - terrestrial clouds, star streams, and HI sheets - must all consist of relatively young objects. For example, if terrestrial clouds were old, the atmosphere would mix them and every day would bring a new arrangement of standard clouds, but a statistically homogeneous one. Because the velocity groups show little mixing, the objects cannot have travelled far. The conclusion must be that HI clouds are continually formed and destroyed and that the temperature-density contrast which distiguishes clouds observationally is transient.

The impermanence of structure in the cloud medium may derive from the weakness of the selfgravitation. Cloud material cannot resist turbulent motion in the interstellar medium and it will be advected into a large volume by turbulent diffusion. The requirement of pressure equilibrium maintains the density contrast but the difference in exposure between hot and cold material is quickly lost. As the cloud material is drawn into ever more extended shapes, its surface to volume ratio increases dramatically and the energy fluxes that maintain the hot medium can then reheat the cloud material.

The cloud hypothesis concerning HI has one correct implication, we are studying a kind of interstellar weather; namely, organized, time-dependent changes of phase and form in the atmosphere of the Galaxy. Most of the results discussed here were won from observations near the sun; and given the nature of the phenomenon, changes in position in the Galaxy should be expected. The global characteristics that can be abstracted from our experience with the local gas are the coexistence of temperature-density phases and the continual reprocessing of material between the phases. In the future, it should be possible to extend measurements to greater distances using high-sensitivity, high-angular resolution instruments. It may then be possible to speak of climatic conditions in the HI and not just of todays weather in the local HI swimming hole.

REFERENCES

Adams, T.F., Frisch, P.C.: 1977, Astrophys. J. 212, pp. 300.
Baker, P.L.: 1973a, Astron. and Astrophys. 23, pp.81.
Baker, P.L.: 1973b, Astron. and Astrophys. 26, pp. 2o3.
Baker, P.L.: 1974a, Astrophys. J. 187, pp. 223.
Baker, P.L.: 1974b, Astrophys. J. 194, pp. L109.
Baker, P.L.: 1976, Astron. and Astrophys. 48, pp. 163.
Baker, P.L., Burton, W.B.: 1975, 198, pp. 281.
Bertaux, J.L., Blamont, J.E., Tabarié, N., Kurt, W.G., Bourgin, M.C., Smirnov, A.S., Dementeva, N.N.: 1976, Astron. and Astrophys. 46, pp. 19.
Bohlin, R.C.: 1975, Astrophys. J. 200, pp. 402.
Burton, W.B., Liszt, H.S., Baker, P.L.: 1978, Astrophys. J. 219, pp. L67.
Davies, R.D., Cummings, E.R.: 1975, Mon. Not. Roy. Astron. Soc. 170, pp. 95.
Dickey, J.M., Salpeter, E.E., Terzian, Y.: 1977, Astrophys. J. 211, pp. L77.
Dickey, J.M., Salpeter, E.E., Terzian, Y.: 1978, Astrophys. J. Suppl. 36, pp. 77.
Fahr, H.J.: 1974, Space Sci. Rev. 15, pp. 483.
Field, G.B.: 1958, Proc. Inst. Radio Eng. 46, pp. 240.
Giovanelli, R., Haynes, M.P., York, D.G., Shull, J.M.: 1978, Astrophys. J. 219, pp. 60.
Grewing, M., Walmsley, M.: 1971, Astron. and Astrophys. 11, pp. 65
Griesen, E.W.: 1976, Astrophys. J. 203, pp. 371.
Hartsuijker, A.P., Baars, J.W.M., Drenth, S., Gelato-Volders, L.: 1972, IEEE Trans. on Antennas and Propagation Ap20, pp. 166.

Heiles, C.: 1967, Astrophys. J. Suppl. 15, pp. 97.
Heiles, C.: 1974, in F.J. Kerr and S.C. Simonson III (eds) "Galactic Radio Astronomy", IAU Symp. 60, pp. 13.
Heiles, C., Jenkins, E.B.: 1976, Astron. and Astrophys. 46, pp. 333.
Hobbs, L.M.: 1976, Astrophys. J. 206, pp. L117.
Hughes, M.P., Thompson, A.R., Colvin, R.S.: 1971, Astrophys. J. Suppl. 23, pp. 323.
Kalberla, P.M.W.: 1978, Ph. d. Dissertation, University of Bonn.
Lampton, M., Margon, B., Paresce, P., Stern, R., Bowyer, S.: 1976, Astrophys. J. 203, pp. L71.
Lazareff, B: 1975, Astron. and Astrophys. 42, pp. 45.
Mebold, U., Hills, D.Ll.: 1975, Astron. and Astrophys. 42, pp. 187.
Mebold, U., Hachenberg, O., Laury-Micoulaut, C.A.: 1974, Astron. and Astrophys. 30, pp. 329.
Radhakrishnan, V., Murray, J.D., Lockhart, P., Whittle, R.P.J.: 1972, Astrophys. J. Suppl. 24, pp. 15.
Shuter, W.L.H., Verschuur, G.L.: 1964, Mon. Not. Roy. Astron. Soc. 127, pp. 387.
Torres-Peimbert, S., Lazcano-Araujo, A., Peimbert, M.: 1974, Astrophys. J. 191, pp. 401.
Van Woerden: 1964, Trans. IAU XIIB, pp. 359.
Van Woerden: 1971, Trans. IAU, XIVB, pp. 217.
Verschuur, G.L.: 1974, Astrophys. J. Suppl. 27, pp. 65.
Weller, C.S., Meier, R.R.: 1974, Astrophys. J. 193, pp. 471.
Westerhout, G., Wendlandt, U., Harten, R.H.: 1973, Astron. J. 78, pp. 569.

LARGE SUPERNOVA REMNANTS AS COMMON FEATURES OF THE DISK

Harold Weaver
Department of Astronomy and
Radio Astronomy Laboratory
University of California, Berkeley

The distribution of the column density of HI over the sky shows strongly organized features of large angular extent. In particular, there is (1) a conspicuous set of arching concentric filaments extending more than 80° on either side of $\ell = 330^\circ$, and (2) a prominent extended region of low column density above $\ell \sim +40^\circ$ and centered on $\ell \sim 130^\circ$. The changes is column density in the b-direction in this region of low N_H is almost step-like at $\ell \sim +40^\circ$.

These distinctly organized structural forms in the two dimensional N_H distribution on the sky imply that there are organized, statistically regular structural forms in the three dimensional space distribution of the gas in the larger neighborhood of the Sun. In this paper models of these statistically regular space structures are described, and the interactions between the ISM and stellar winds, and between the ISM and supernovae that produced the space structures are discussed.

The region of the sky centered at $\ell = 330^\circ$. The arching pattern shown in the HI is also shown by the position angles of optical polarization vectors of nearby stars. In the presentation it was demonstrated that to a very high order of accuracy the HI filaments and the optical polarization vectors are aligned. The magnetic fields that orient the polarization-producing grains lie along or within the elongated space-density structures that we observe as the HI filaments. Position-angle data for the optical polarization vectors serve to determine the center of curvature of the arching structure: $\ell = 331^\circ.3 \pm 1^\circ.3$, $b = +14^\circ.0 \pm 1^\circ.4$. Distances of stars whose light is polarized by the aligned grains in the space structures that we observe as arching filaments indicate that the structure extends over the distance range approximately 20 to 400 pc.

Loop I, the large SN shell strongly visible in synchrotron radiation overlaps the arching HI filaments. In the presentation it was shown that the limb of Loop I is remarkably well represented by the projections of a sphere of radius $55^\circ.5$, centered at $\ell = 336^\circ.0$, $b = +24^\circ.0$. The estimated uncertainty of this location is $< 2^\circ$. The centers of the arching filaments and Loop I differ by more than 6σ; Loop I does not

W. B. Burton (ed.), The Large-Scale Characteristics of the Galaxy, 295–300.

appear to be the source that created the space-density structure we observe as the HI filaments and that is revealed by optical polarization.

The Sco-Cen Association overlaps the HI filaments. The center of the association is at $\ell = 330^\circ$, $b = +15^\circ$, a point 1.2σ from the center of curvature of the filaments. Sco-Cen is 170 pc distant. It contains more than three dozen stars in the mass range 10-20 $M_\odot$. It is $1 - 2 \times 10^7$ years old. At a distance of 170 pc, Sco-Cen is centrally embedded in the space-density structure we observe as the curved HI filaments. Sco-Cen must be strongly involved with that structure.

<u>A model of the $\ell = 330^\circ$ region</u>. All these observations can be drawn together in a simple physical model. Between 1 and 2×10^7 years ago the rich and rather large Sco-Cen group formed. A considerable amount of left-over interstellar material remained after the stars were formed. This unused material, clumpy and striated in form, distributed throughout the general region surrounding the association, and acted upon by differential galactic rotation over a substantial time interval, had been drawn out into line-like formations generally parallel to the galactic plane and along the local spiral arm or spur. A magnetic field, oriented along the arm, was embedded in these striations. The numerous massive stars in the newly formed association produced strong stellar winds. These inflated a bubble of gas and dust concentric with the Sco-Cen Association. As the striations were swept up on the expanding bubble of wind-driven gas, they were compressed in the radial direction, stretched on the surface of the bubble into the filaments we see today. The near side of the bubble is very close, witness the fact that it subtends an angle of $\sim 170^\circ$. Perspective and shape together produce the observed arching form of the filaments, which are in the surface of the expanding bubble. HI observations indicate that the current rate of expansion is ~ 2 km s^{-1}. Preliminary calculations indicate that the moving bubble contains about 10^6 $M_\odot$ of gas, and represents about 10^{50} ergs of energy; it is approximately 300 pc in diameter.

In this bubble model, Loop I is a SN shell produced by the explosion of one of the most massive members of Sco-Cen. The explosion occurred <u>inside</u> the bubble. Since the medium into which the shell expanded was hot, uniform, and of low density (the bubble has been essentially evacuated by the stellar winds that inflated it) the shell would be expected to be large and closely spherical in form as is observed. The SN shell (Loop I) is just beginning to encounter, and hence to interact with, the inside surface of the HI bubble. Where the SN shell interacts with the bubble, higher velocity features (up to 50 km s^{-1}) are observed. An extensive interactive feature of this type has been found in the Southern Hemisphere by M. Cleary.

The bubble model provides many testable predictions. For example: (1) Given the direction of the center of Sco-Cen and the angular diameter of the bubble (170°) as input data, the form of the arching filaments (or, what is the same thing, the pattern of the optical polarization vectors) can be predicted. The calculations reproduce the observations remarkably well. (2) Given the observed expansion velocity of the HI bubble, ~ 2 km s^{-} , one predicts that pictures of the HI gas in the narrow velocity ranges 0 to -5 km s^{-1}and 0 to +5 km s^{-1}should show the front and back sides of the bubble. The two pictures should be similar in character (arching filaments) but different in detail (different filaments) since the structures are far apart in space (~ 300 pc) and are uncorrelated. Pictures of the HI in the narrow velocity ranges 0 to -5 km s^{-1}and 0 to +5 km s^{-1}do, indeed, show precisely the behavior predicted.

Observational features of the ℓ = 130° region. The most outstanding HI feature in the region of low N_H centered at ℓ ~ 130°, is a velocity disturbance which is of large amount in velocity and of great angular extent in the sky. This disturbance, which occurs throughout the longitude range ℓ ~ 60° to ℓ ~ 220°, was discovered by the early group at the Department of Terrestrial Magnetism, by the Dutch observers, and others. It was described by Blaauw as existing in both the N and S galactic hemispheres (more pronounced in the N), extending over a large range of longitudes, and being like an approaching stream of gas centered at ℓ = 115°. The extensive and detailed sky coverage now available in the HI show clearly that this disturbance is not a stream, but is the nature of an extensive shell-like structure having its greatest velocity in the range -50 to -60 km s^{-1}. The shell structure is so pronounced over considerable areas on the North side of the galactic plane that there is little zero-velocity gas.

Loops II and III are conspicuous observational features of the ℓ = 130° region visible in synchrotron radiation; they overlap the HI shell structure just described. It has generally been assumed that these loops had their origins in two supernovae which exploded at moderately large z distances (z = 100 to 200 pc), one N of the plane, the other S, at essentially the same time since the loops are similar in size and character. There are no clusters, associations, bright early stars, or any unusual objects in the vicinity of the centers of Loops II and III individually. The unusual requirement of having two simultaneous supernovae on opposite sides of the plane, together with the lack of interesting objects near the Loop centers have always raised doubts about the source or origin of these particular loops.

An important feature of the ISM directly relevant to an explanation of the ℓ = 130° region is the great irregularity in the space density distribution of the ISM in the larger neighborhood of the Sun. Lyα observations indicate marked fluctuations in HI density over short distances. The Sun is located in a "hole" of generally low gas density as discussed by Fejes and Wesselius, and others. The nature of the hole can be shown most dramatically with the extensive Hat Creek HI data; the region of minimum gas space density lies in the longitude range 130° to 140°. However, at distances greater than 175 to 200 pc in the second quadrant of longitude there is a steep positive density gradient; in a short distance the gas density becomes greater than the value in the general solar neighborhood by an order of magnitude or more. These general features of the local ISM density distribution (here described for the gas) have long been known to the optical observers.

A model of the $\ell = 130^\circ$ region. The distribution of the intermediate negative velocity gas is statistically regular over the ℓ-range of interest, and forms an expanding bilobed shell. The observed velocity distribution cannot be accounted for by expansion from two centers (two SN, Loops II and III); it demands for its explanation expansion from a single center in or very near the galactic plane.

In the model proposed, then, a single SN event occurred in or near the galactic plane at $\ell \sim 130^\circ$. Because of the nature of the density distribution of the ISM (density greatest in the plane, decreasing with $|z|$), a bilobed structure with axis in the z-direction has been produced. Loops II and III together form the bilobed structure as seen in synchrotron radiation. A picture of HI in a restricted velocity range (which enhances the shell structure and makes it more visible) likewise clearly shows the bilobed structure in the HI gas surrounding the synchrotron radiation.

In this simple explosion model the direction assigned places the star that exploded in the α Per Association. The α Per Association, age $\sim 10^7$ years, is ~ 150 pc distant; it contains about a dozen stars more massive than 10 solar masses, the mass of the moving intermediate negative velocity gas is $\sim 10^3$ $M_\odot$; the kinetic energy of the gas is $\sim 10^{50}$ ergs.

The proposed bilobed model is a very rich one; it provides many predictions open to test and, together with the observed density distribution of the ISM, provides explanations of many phenomena observed in this region of the sky as, for example, the expanding "ring" of HI discovered by P. O. Lindblad, and the classic high velocity gas discussed by Oort. Lack of space precludes discussion of these and other topics; they will be considered in detail elsewhere.

DISCUSSION

Oort: If high-velocity clouds are connected with huge expanding shells, the approaching (negative-velocity) parts should be very nearby, at distances not much larger than 100 pc. In the few cases where interstellar absorption lines have been observed, the distances appear to be rather larger. In particular, the so-called feature A is inferred from the lack of interstellar absorption lines at its velocity (of about 200 km s^{-1}) in two stars at estimated distances of 500 and 1500 pc, respectively, to be much too distant to be part of a shell of the type you have discovered. The absorption-line data are unfortunately still extremely meager, and it is quite possible that part of the intermediate velocity gas and the high-velocity gas in the arc-like structure formed by the high-velocity features CI, CII, CIII which you showed should be interpreted in the way you have suggested.

Weaver: The question of distance of the high-velocity clouds (or of features clearly associated with them) remains as a very important observational problem. One of the strengths of the model I have presented is that it suggests new approaches to this problem--approaches that will involve optical polarization and various other observational data. I believe that feature A is one of the objects that can be investigated by new methods.

van den Bergh: You say that these structures have ages of $\sim 10^7$ year. During such a time period one would expect $\sim$ 500 supernovae per kpc^2 to explode. This would seem to make it a bit difficult to associate your arcs with a single supernova outburst.

Weaver: The arcs are not associated with a supernova outburst. They are parts of the bubble of gas blown by the stellar winds originating from the Sco Cen Association. The stars in Sco Cen are now in process of going off as supernovae--at least those of the stars that have sufficient mass to become supernovae--and they very likely will change or destroy the bubble eventually. In my model, Loop I is a shell from a supernova that originated from one of the members of the Sco Cen. It exploded inside the bubble in a very hot, tenuous atmosphere. The SN shell is now starting to interact with the inside of the bubble. It and the other SN shells that will in the future develop from the Sco Cen stars will probably destroy the bubble, but we see the bubble now because of the stage of development of Sco Cen.

It is, I believe, an error to use statistics as implied in this question. In any specific 10^7 years not every kpc^2 of the Galaxy is equally likely to produce SN. We do know where clusters and associations around us are located, and we know which ones are likely to produce SN. As an analogy to these broad statistics just quoted, the average birthrate in the U.S. is one child per 2 square miles per year. But not all areas of 2 square miles--the top of Pike's Peak or the middle of Lake Erie--are not, at least during the present geologic era, likely to produce the average number of births.

Cesarsky: An alternative explanation, existing in the literature, of the observed arches which are aligned to the magnetic field, is that they are the end result of the Parker instability in the galactic disk. As our calculations imply that the end result of the Parker instability is not a collection of stable arches, I am pleased that another, more convincing explanation of these features has been presented.

Weaver: I have also worried about the Parker instability explanation of the arching filaments seen on the sky. I was very pleased to be able to propose an alternative explanation.

Burke: I just checked with Dr. Salpeter, and he agrees that supernova shells do not ordinarily last 10^7 years. A million years is a better number and so multiple SN remnants are not as common as Dr. van den Bergh suggested. By 10^7 years, the shells are well thermalized.

Weaver: The shell around the Sco Cen Association I described was not produced by a supernova; it was produced by stellar winds from the Sco Cen Association. The SN shell within the Sco Cen bubble and the SN shell producing the intermediate velocity and high velocity gas are both much younger than 10^7 years. They are probably not more than 10^5 to 10^6 years old. It is certainly probable that they will become thermalized in 10^7 years.

Felten: You have rather definite kinematical models for the three structures you discussed: Can you say something about the ages of these structures, on the basis of the model kinematics? Is your estimate of 10^7 years for the age of the large structure based upon the kinematics, or simply upon the notion that is started when the SCO-CEN association was formed?

Weaver: It will be possible to make some reasonable age estimates by means of the kinematics of the models. It has not yet been done. The age I quoted for the Sco Cen shell is based only on the age of the association.

Assousa: In primordial times of 1975, in a lecture at Berkeley discussing your early results observationally linking HI shells with known Galactic supernova remnants, Professor Minkowski was skeptical! We have come a long way through our understanding of such structures in HI surveys, typified by the work of Professors Weaver and Heiles. Noting that these shells provide coherent large-scale structures in our Galaxy, having masses $\sim 10^4\ M_\odot$ and lasting $\gtrsim 10^6$ years, I suggest that we begin to think of such objects as important "unit" components--in the same way we treat Giant Molecular Clouds of mass $\sim 10^5\ M_\odot$.

Weaver: I quite agree. I think we must come to realize that giant star clusters and associations can and will be organizers of the kinematics and structure of the surrounding gas.

A NEW CLASS OF EXTRAORDINARY HI SHELL

Carl Heiles
Berkeley Astronomy Department
University of California

I consider an ordinary shell to be one which can be produced by ordinary means such as a supernova. Good examples are the HI shells associated with the prominent radio loops, discussed today by Prof. Weaver, and the large HI shell in Eridanus (Heiles, 1976). Figure 1 shows the shell which is associated with radio Loop I, made from a consolidation of data from Martha Cleary of Australia, Raul Colomb and Wolfgang Poppel of Argentina, and myself. It changes size with velocity in the manner expected for an expanding shell. The distance is only 100 pc or so. Similar shells should be easily resolvable in the galactic plane, even with a modest telescope, because differential galactic rotation eliminates the confusing effects of foreground and background gas.

I used the galactic plane survey of Weaver and Williams (1973) to make a large number of photographs equivalent to those of Figure 1. They reveal a wealth of structure, much of which is filamentary. In fact one has the impression that, if only the angular resolution were somewhat better, nearly all of the structure would be clearly resolved into filaments. Many filaments form circular arcs, some of which change size with velocity. The majority of these shells are of the ordinary class discussed above. But there is a small number, located mainly at large galactic radii, which are extraordinarily large in size and have swept up a huge amount of matter.

It is these shells, "supershells," which compose the new class of shell. An example is shown in Figure 2. This supershell has a galactic radius of 17 kpc, a distance of 13 kpc, a radius R of 800 pc, a kinetic energy of 4×10^{52} erg, and has swept up 2×10^{7} $M_{\odot}$ of gas while expanding through an average density of $n = 0.2$ cm^{-3}. If our supershells were produced by the sudden ejection of energy E_e, as would occur with a supernova, we have (Chevalier, 1974)

$$E_e = 5.3\ 10^{43}\ n_{cm^{-3}}^{1.12}\ R_{pc}^{3.12}\ V_{km/sec}^{1.4}\ \text{erg}$$

W. B. Burton (ed.), The Large-Scale Characteristics of the Galaxy, 301–305.

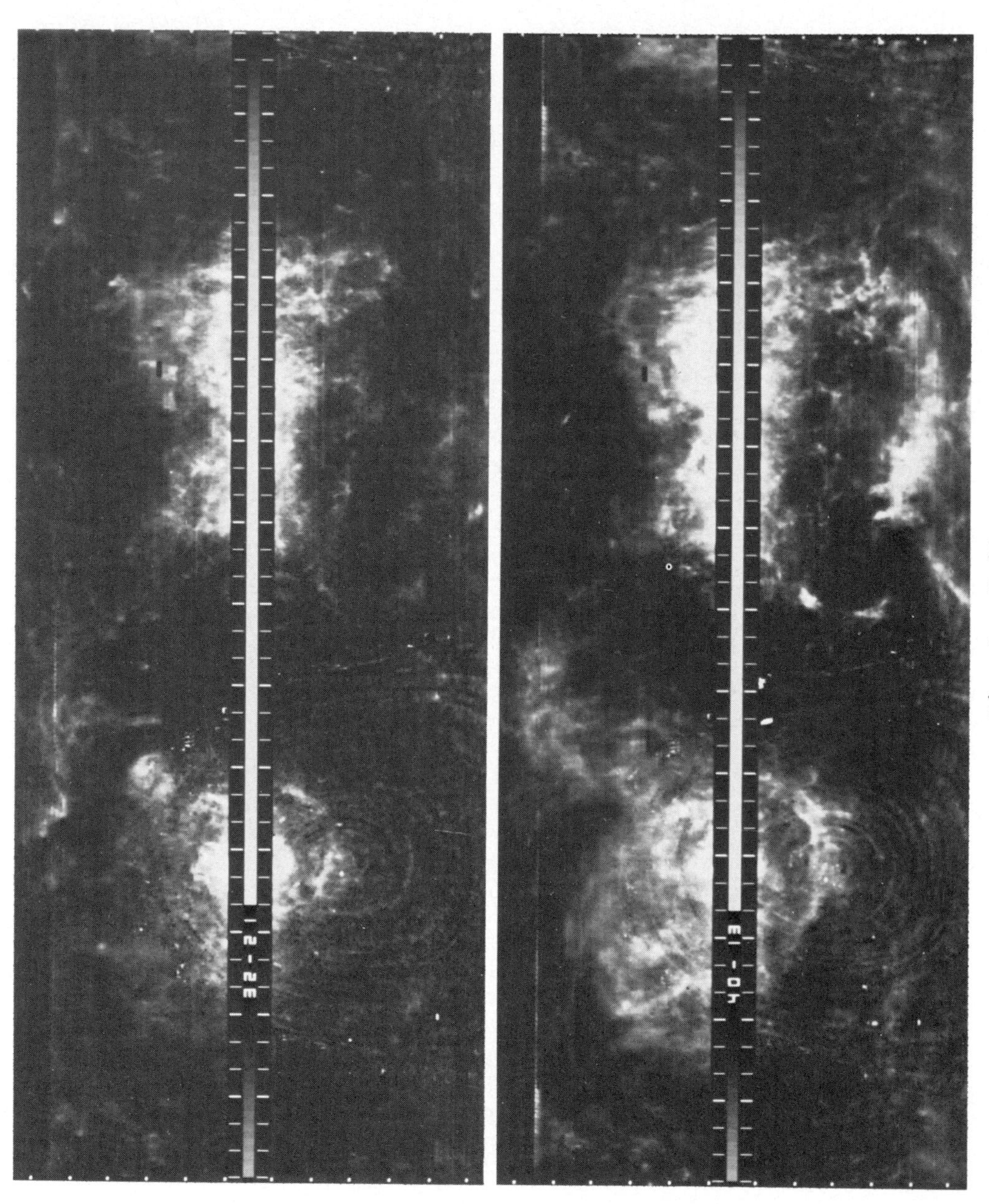

Fig. 1. Heiles

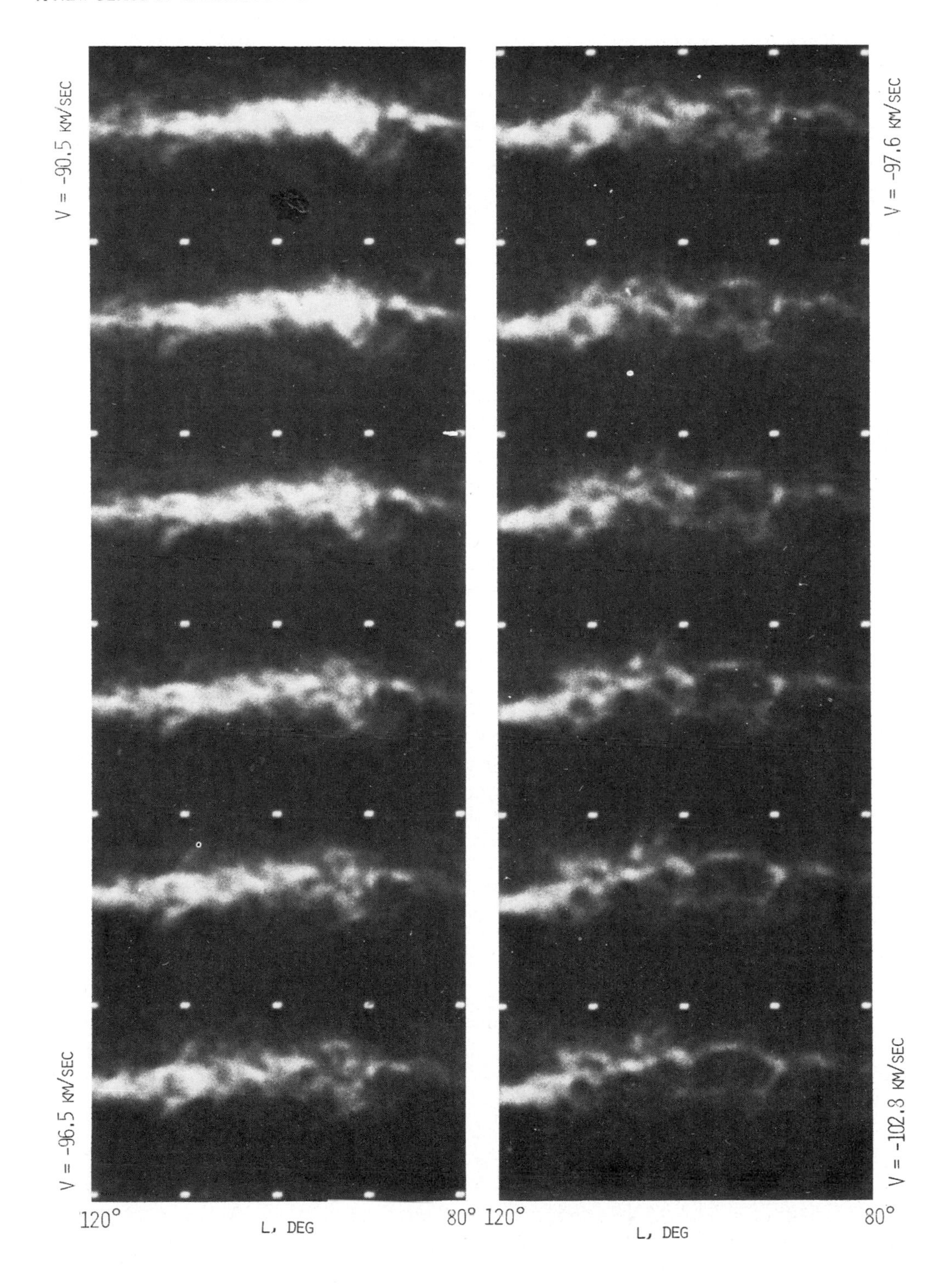
V = -90.5 KM/SEC
V = -97.6 KM/SEC
V = -96.5 KM/SEC
V = -102.8 KM/SEC
120°
L, DEG
80°
120°
L, DEG
80°

Fig. 2. Heiles

where V is the expansion velocity. For the shell in Figure 2, $E_e = 3 \times 10^{53}$ erg. We define supershells as having $E_e > 3 \times 10^{52}$ erg; the largest value for our sample is 6×10^{53} erg.

HI supershells have probably been observed in other galaxies. Hindman (1967) discovered three in the SMC. Several large holes in the HI distribution of M101 appear very prominently in the beautiful photograph of Allen et al (1978). Westerlund and Mathewson (1967) discovered a large ring of HI, bright blue stars, HII regions, and supernovae remnants in the LMC.

A shell will no longer be discernable when interstellar clouds, or other shells, penetrate the shell and fill the volume. For cloud velocities of 10 km/sec the time scale for this process is R/10 million years, about 10^8 years for the supershell in Figure 2. If the galaxy contains 10 supershells, their formation rate is of order 10^{-7} yr^{-1}.

The production agent must release vast quantities of energy into the interstellar medium. Supernova searches of external galaxies should be well-suited for discovering the production agent. They have been in progress for about 40 years; during this time interval about 250,000 galaxies would have had to be searched to accumulate a reasonable probability of seeing such an infrequent event. The Palomar Supernova Search encompasses only 3003 galaxies (Sargent et al, 1974), and searches by other groups increase this number by a modest amount. Thus it is extremely unlikely that the production agent has ever been seen. One possible candidate is type III supernovae, which have expansion velocities of 12,000 km/sec and have very optically thick shells, probably containing hundreds of solar masses (Zwicky, 1964). Unfortunately type III supernovae have not been well-studied, since only two have ever been observed (Sargent et al, 1974).

Figure 1. HI at -21 and -13 km s^{-1} in upper and lower photographs, respectively. For each photograph, ℓ=0° at the righthand edge, increasing through 360° to 60° at the left; latitude increases from -65° at the bottom to +65° at the top, with HI between -10° and +10° omitted; fiducial marks for both coordinates are given every 10°. Note the shell located at $270° < \ell < 350°$, $b < -10°$, and its change of size with velocity.

Figure 2. 12 photographs showing HI at 12 velocities separated by 1.1 km s^{-1}, ranging from -90.5 at the upper left to -102.8 at the bottom right. Each of the 12 pictures covers the range b = -10° to +10°. Note the supershell centered near ℓ=95°, b=+4°, and its change of size with velocity.

REFERENCES

Allen, R.J., van der Hulst, J.M., Goss, W.M., and Huchmeier, W.: 1978, Astron. Astrophys. 64, 359.
Chevalier, R.A.: 1974, Astrophys. J. 188, 501.
Heiles, C.: 1976, Astrophys. J. 208, L137.
Hindman, J.V.: 1967, Aust. J. Phys. 20, 147.
Sargent, W.L.W., Searle, L., and Kowal, C.T.: 1974, in Supernovae and Supernovae Remnants, ed. by C.B. Cosmovici, p. 33.
Weaver, H.F., and Williams, D.R.W.: 1973, Astron. Astrophys. Suppl. 8, 1.
Zwicky, F.: 1964, Ann. d' Astrophys. 27, 300.

DISCUSSION

Rubin: Some years ago at one of these symposia, Paul Wild pointed out large circular rings (∿ few kpc) in galaxies. Can you form "spiral" structure" from these structures?

Heiles: One might think that large explosions would punch holes in the HI distribution, destroying spiral structure. On the other hand, they might enhance the spiral pattern by providing density contrasts whose shapes would be drawn out into a spiral pattern by differential rotation.

CHEMICAL EVOLUTION OF THE GALACTIC INTERSTELLAR MEDIUM: ABUNDANCE GRADIENTS

Manuel Peimbert
Instituto de Astronomía, Universidad Nacional Autónoma de México

Abstract. Recent abundance determinations of galactic H II regions and planetary nebulae are reviewed. The presence of O/H and N/H abundance gradients is well established; there is observational evidence indicating the presence of N/S, He/H and C/H abundance gradients. Some implications of these results are discussed.

1. INTRODUCTION

The presence of abundance gradients of O/H, N/H and N/S in external galaxies is now well established (e.g. Searle 1971, Benvenuti et al. 1973, Shields 1974, Comte 1975, Peimbert 1975, Smith 1975, Sarazin 1976, Jensen et al. 1976, Collin-Souffrin and Joly 1976). Recent observations of the interstellar medium indicate that a similar situation prevails in the Galaxy. It is the purpose of this review to discuss these observations.

2. PLANETARY NEBULAE, PN

A review of PN abundance determinations is presented elsewhere (Peimbert 1978a). Barker (1974, 1978), D'Odorico et al.(1976), Aller (1976), Torres-Peimbert and Peimbert (1977) and Kaler (1978) have studied the presence of abundance gradients in the Galaxy. To study the presence of abundance gradients in the disk of the Galaxy from PN it is necessary to select Type II PN which are of population I and that apparently have not been affected by considerable helium enrichment due to their own stellar evolution (Peimbert 1978a). Based on these considerations the results for PN of Type II by Torres-Peimbert and Peimbert (1977) are presented in Table I and Figure 2.

Kaler (1978) has rediscussed the He/H abundance ratios from a larger sample of PN than those used in previous studies and confirms that there are systematic abundance differences as reported by D'Odorico et al. (1976) and Torres-Peimbert and Peimbert (1977), he finds that the He/H

W. B. Burton (ed.), The Large-Scale Characteristics of the Galaxy, 307–316.

abundance ratios diminish with increasing distance above the galactic plane, with increasing radial velocity and with increasing distance to the galactic center. Kaler attributes the radial gradient as due mostly to an excess of population II PN in the anticenter direction and not to the presence of an interstellar gradient.

In the solar neighborhood the N/O abundance ratio in PN of Type II is larger than in H II regions by about a factor of four while the O/H abundance ratio is similar (Torres-Peimbert and Peimbert 1977), this result agrees with the idea that PN are producing nitrogen without affecting their O/H ratio and consequently that their O/H ratio can be used as a tracer of Interstellar medium abundances at the time the parental star was formed. In Figure 2 we present the N/O versus O/H plot for PN of Type II.

There is some controversy regarding the carbon abundance in PN, the faint permitted lines in the visual indicate carbon overabundances with respect to the solar value (Torres-Peimbert and Peimbert 1977, Aller 1978, Shields 1978), while the ultraviolet C III and C IV lines indicate more normal abundances (Bohlin et al.1978, Pottasch 1978). In gas rich regions where star formation has not been appreciable one would expect the C/H and O/H gradients to be similar, as more gas condenses into stars the C/H gradient should become less steep than the O/H gradient because some C is transformed into N, nevertheless if stars like PN are ejecting carbon rich material produced in them then the C/H gradient could be steeper than the O/H one. Panagia et al. (1977), Tinsley (1978) and Mallik (1978) have studied the relevance of the PN enrichment of the interstellar medium under various assumptions regarding the N and C overabundance as well as the rate of PN formation.

3. H II REGIONS

Recently several investigations devoted to the study of abundance gradients across the disk of the Galaxy have been carried out, the results are presented in Table I. We decided to combine the results of all the observers and redetermine the N/H, O/H and N^+/S^+ abundance gradients considering only those objects in which the electron temperature had been determined observationally, the results are presented in Table I and Figure 1. The reduction procedure, that includes ionization correction equations, spatial temperature fluctuations (t^2= 0.035 was adopted) and atomic data, was the same for all the abundance determinations. The logarithmic abundance gradients from this sample were computed and are presented in Table I, equal weight was given to all the points and in those cases where the same object was observed by different groups we considered the results as independent. To compute the N/H gradient the results for NGC 2359 were not considered because the central star, WR of type N5, is losing mass at a very high rate, and this material is probably nitrogen rich.

TABLE I

Solar Neighborhood Abundance Gradients*

Object	He/H	C/H	O/H	N/H	N^+/S^+	Objects with observed T_e	Reference
PN	-0.02		-0.06	-0.18		15	(1)
H II					-0.04	0	(2)
H II	-0.02		-0.13	-0.23	-0.09	5	(3)
H II	0.00		-0.04	-0.10	-0.05	5	(4)
H II			-0.11	-0.11	-0.06	4	(5)
H II		-0.09	-0.09	-0.12	-0.05	4	(6)
H II			-0.10	-0.14	-0.05	18	(7)

* Given in d log(X/Y)/dR kpc^{-1}.
(1) Torres-Peimbert and Peimbert 1977, (2) Sivan 1976, (3) Peimbert <u>et al</u>.1978a, (4) Hawley 1978, (5) Talent and Dufour 1978, (6) Peimbert and Torres-Peimbert 1978, (7) this paper.

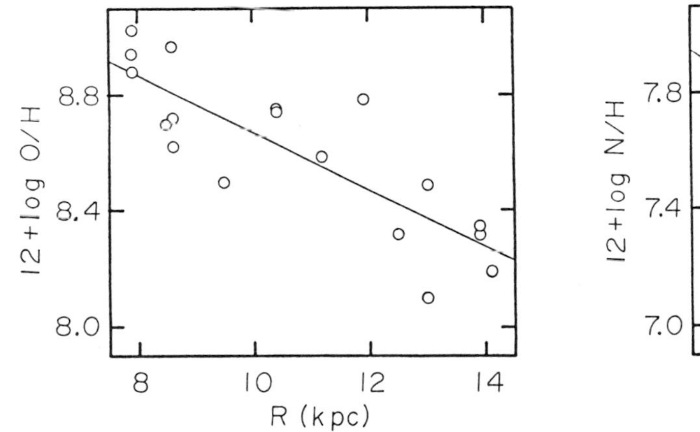

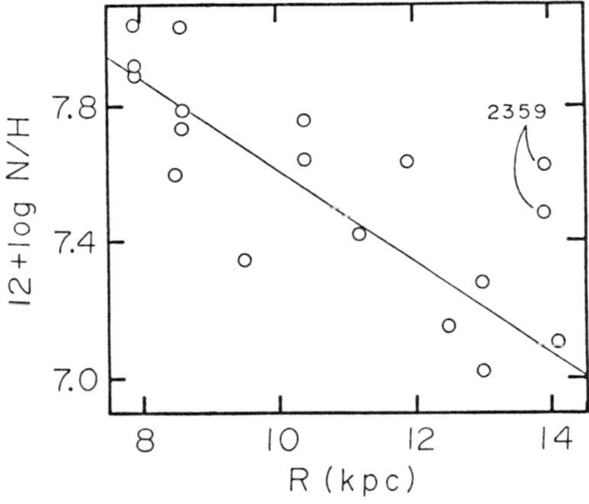

Figure 1. Oxygen and nitrogen abundances versus galactocentric distance for galactic H II regions. The solid lines represent the least-squares solution. For the nitrogen solution NGC 2359 was not included.

There are three possible explanations for the scatter in the abundances at a given R present in Figure 1: a) observational errors due to the faintness of the auroral lines needed to determine the electron temperature, b) inadequate abundance determination procedures, <u>i.e</u>. improper ionization correction factors or temperature distribution scheme, c) real abundance differences. In Figure 1 some of the points represent the same object observed by different groups which implies that the scatter due to observational errors amounts to ~0.2 dex. We cannot rule out the absence of real abundance differences between H II regions at the same galactocentric distance, these differences, if present, cannot amount to more than a factor of two. Chevalier (1978) has studied the problem of element mixing in the interstellar medium and possible causes for abundance differences at a given galactocentric distance.

If instead of fitting an exponential curve to the observations we do a linear fit with the data presented in Figure 1, it is found that O/H= 0 for R= 15.77 kpc and N/H= 0 for R= 14.25 kpc. These fits are reasonably good and imply that very little star formation activity has taken place at galactocentric distances larger than R= 15 kpc. In this review it has been assumed that $R_{\odot}$= 10 kpc.

It is relatively easy to observationally determine N^+/S^+ abundance ratios in H II regions and therefore to derive its gradient in the Galaxy and other galaxies, this ratio is almost independent of electron temperature, reddening, and instrumental correction. Theoretical models (Peimbert et al. 1974, Hawley and Grandi 1977) predict that in those regions where N is once ionized not only S^+ but also S^{++} is present and consequently to derive the N/S gradient from N^+/S^+ observations it has to be shown that N/S is proportional to N^+/S^+. Since the S/O ratio is almost the same in the Orion nebula, the SMC and the LMC H II regions (Pagel 1978) we would expect the O/H and S/H gradients in the Galaxy to be similar. Consequently the validity of the d log (N^+/S^+)/dR= d log(N/S)/dR relation depends on the similarity of the N^+/S^+ and N/O gradients. From Tables I and II it follows that this is indeed the case and that the N^+/S^+ gradient is a good indicator of the N/S gradient.

To obtain the He/H abundance ratio in H II regions it is necessary to estimate the amount of neutral helium present in them. This estimate is based on the ionization degree of other elements that can be observed in several stages of ionization. The best results are obtained from objects of relatively high degree of ionization with a negligible amount of He^0 and from observations of different points with varying degrees of ionization within the same H II region. Peimbert et al. (1978a) have found a log(He/H)/dR= -0.02±0.01 while Hawley (1978) does not find such a gradient. Shields and Searle (1978) from their observations and independently from those of Smith (1975) obtained a similar He/H gradient in M101 to that derived by Peimbert et al. for the Galaxy (see Table II). The helium abundance differences found in PN and H II regions in the Galaxy and other galaxies correspond to $2 \lesssim \Delta Y/\Delta Z \lesssim 3.5$.

Based on the λ4267 line of C II, transition $4f^2F^0$-$3d^2D$, Peimbert and Torres-Peimbert (1978) have obtained the carbon abundance from three H II regions and derive the gradient given in Table I. It is very difficult to obtain the C/H abundance in H II regions from observations in the optical region since λ4267 is typically about 400 times fainter than Hβ in H II regions of the solar neighborhood.

TABLE II

Abundance Gradients in M101*

He/H	O/H	N/H	N^+/S^+	Reference
	-0.06	-0.09	-0.03	Smith (1975)
+0.01	-0.15	-0.14	-0.04	Hawley (1978)
-0.02	-0.08	-0.15	-0.07	Shields and Searle (1978)

* Given in d log(X/Y)/dR kpc^{-1} and evaluated at a distance comparable to that of the solar neighborhood.

The very large C/N ratio in the solar neighborhood (Peimbert and Torres-Peimbert 1977, Lambert 1968) coupled to the similarity between the C/H and O/H abundance gradients implies that: a) only a small amount of carbon has been converted into nitrogen by a secondary mechanism and b) stars like PN, have ejected to the interstellar medium a small amount of carbon, produced in them, as compared to that ejected by massive stars. Furthermore the C/H abundance gradient implies that if the fraction of carbon embedded in CO molecules is the same in all the Galaxy the estimated amount of H_2 molecules for the Galaxy based on the CO distribution should be reduced by a factor of two to three (Gordon and Burton 1976, Peimbert 1978b).

In Table II we present the abundance gradients for M101 they are in good agreement with the galactic ones particularly those derived by Shields and Searle (1978). From an analysis of radio recombination lines Churchwell and Walmsley (1975) and Churchwell et al. (1978) have found a positive electron temperature gradient in the Galaxy extending from 5 to 13 kpc, from their model calculations an increase in Z of about a factor of 2 from R= 13 to R= 5 kpc is required; this variation corresponds to d log(O/H)/dR $\sim$ -0.03 which is in good agreement with the results by Hawley (1978) but is significantly smaller than the other values given in Table I. Considerable work has been done on CNO isotopic variations in the interstellar medium (see Linke 1977 and references therein), significant differences have been found mainly between the solar neighborhood and the galactic center but no clear-cut results on isotopic abundance distributions across the disk of the Galaxy have been found with the exception of a practically constant $^{12}C/^{13}C$ ratio. Audouze et al.(1976) have computed simple models of galactic chemical evolution that are in agreement with the $^{12}C/^{13}C$ observations and that predict a substantial N/O gradient.

4. DISCUSSION

A relationship of the type $[N/O]= \alpha[O/H]$ with α= 1 is predicted by simple models with instant recycling approximation assuming nitrogen to be of secondary origin (Talbot and Arnett 1973). In Figure 2 we show the N/O versus O/H plot for the galactic H II regions, the data show a value of α close to 0.4 in poor agreement with the theoretical prediction. Furthermore a very poor fit is provided by more precise models that drop the instant recycling approximation (Talbot and Arnett 1974) presented in Figure 2.

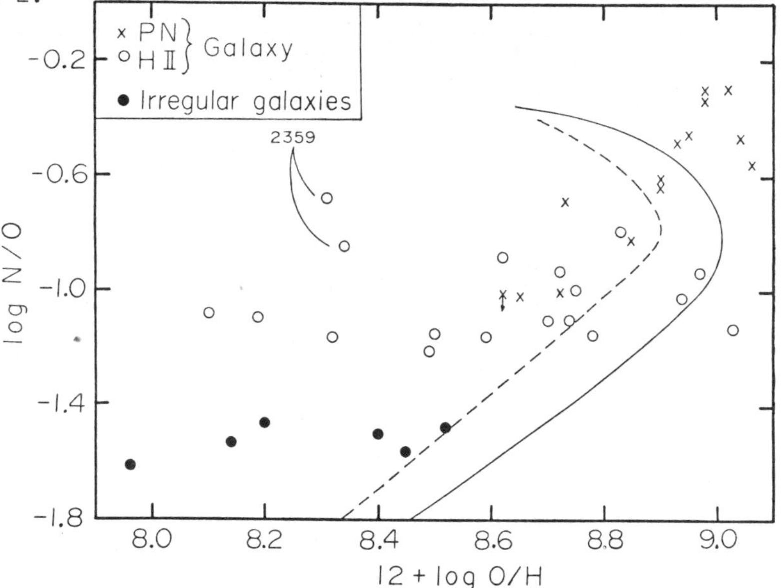

Figure 2. Relative nitrogen abundance N/O versus oxygen abundance. Broken and full curves show theoretical predictions by Talbot and Arnett (1974) for MESF and simple models respectively.

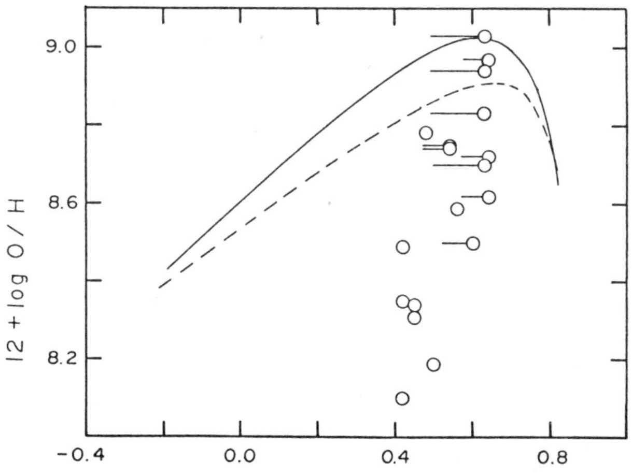

Figure 3. Oxygen abundances plotted against the astration parameter, log(ln $1/\mu$), for H II regions in the Galaxy. Horizontal bars represent the correction for H_2 molecules after Gordon and Burton (1976). Broken and full curves show theoretical predictions by Talbot and Arnett (1974) for MESF and simple models respectively.

Some support for the galactic result is given by Smith (1975) who obtained a value of $\alpha \sim 0.5$ for the H II regions in M33 and M101, however Shields and Searle (1978) find $\alpha = 0.8 \pm 0.3$ for M101. In Figure 2 we have also plotted the H II region values derived for the irregular and dwarf blue galaxies: NGC 6822 V, LMC, SMC, II Zw40, II Zw70, and NGC 4449 (Peimbert and Spinrad 1970, Peimbert and Torres-Peimbert 1974, 1976, Dufour 1975, Dufour and Harlow 1977, Aller *et al*. 1977, Pagel *et al*. 1978, Peimbert *et al*. 1978b). The N/O versus O/H relationship is almost flat with $\alpha \sim 0.2$ again in disagreement with the predictions for simple models. A possible explanation, not the only one, is that in these objects nitrogen is mostly of primary origin with a primary production close to log N/O ~ -1.7. This is a very modest ratio about 1/5 of the Orion nebula value.

Another interesting difference present in Figure 2 is that H II regions in irregular and dwarf blue galaxies have a smaller N/O ratio for a given O/H ratio than galactic H II regions. A possible explanation is that the galactic initial mass function (IMF) is different to those of irregular and dwarf blue galaxies or that infall in the Galaxy of material with pregalactic abundances has reduced the O/H ratio of regions with substantial amounts of nitrogen produced by secondary mechanisms.

An excellent discussion on the implications of abundance gradients for galactic enrichment models has been made by Pagel *et al*. (1978). They have tested simple models of galactic chemical evolution in the instantaneous recycling approximation (Schmidt 1963, Talbot and Arnett 1971, 1973, Searle and Sargent 1972, Pagel and Patchett 1975, Audouze and Tinsley 1976) by means of the equation

$$\log Z = \log p + \log(\ln 1/\mu) \qquad (1)$$

where it has been assumed that gas changes into stars according to a constant IMF in an isolated zone, Z is the mass-fraction of heavy elements in the interstellar gas, p the yield and μ the ratio of the remaining mass of gas to the total mass of the zone. Pagel *et al*. find that equation (1) is not in contradiction with the observations of the Magellanic Clouds, on the other hand for the H II regions in the Galaxy observed by Peimbert *et al*. (1978a) they find that the fit is very poor since for very different O/H ratios μ is almost the same.

In Figure 3 we present the O/H ratio versus log(ln $1/\mu$) for those H II regions in Figure 1 and Table I. The values of μ were obtained from Gordon and Burton (1976) and Innanen (1973), in the galactocentric range of interest the correction due to molecular hydrogen is small. The additional data confirms the result of Pagel *et al*. (1978) that the O/H ratio is independent of μ and that there is no agreement with simple models of galactic chemical evolution.

It is a pleasure to acknowledge several discussions as well as communications prior to publication by R.J. Dufour, S.A. Hawley, J.B. Kaler, B.E.J. Pagel, L. Searle, G.A. Shields, J.P. Sivan, R.J. Talbot, D.L.

Talent, S. Torres-Peimbert and B.L. Webster.

REFERENCES

Aller, L.H.: 1976, Publ. Astron. Soc. Pacific 88, pp. 574-584.
Aller, L.H.: 1978, in Y. Terzian (ed.), "Planetary Nebulae", IAU Symp. 76, Reidel, Dordrecht.
Aller, L.H.,Czyzak,S.J.,and Keyes,C.D.: 1977, Proc. Natl. Acad. Sci. 74, pp. 5203-5206.
Audouze,J.,Lequeux,J.,and Vigroux,L.: 1976, RGO Bull. No. 182, R.J. Dickens and J.E. Perry (eds.), Herstmonceux, pp. 81-86.
Audouze, J. and Tinsley,B.M.: 1976, Ann.Rev.Astron.Astrophys. 14, pp. 43-79.
Barker,T.: 1974, Ph.D. Thesis, Univ. California, Santa Cruz.
Barker,T.: 1978, Astrophys. J. 220, pp. 193-209.
Benvenuti, P.,D'Odorico,S.,and Peimbert, M.: 1973, Astron. Astrophys. 28, pp. 447-455.
Bohlin,R.C.,Harrington,J.P.,and Stecher,T.P.: 1978, Astrophys. J. 219, pp. 575-584.
Chevalier, R.A.: 1978, Mem. Soc. Astron. Italiana, in press.
Churchwell,E.,Smith,L.F.,Mathis,J.,and Mezger,P.G.: 1978, preprint.
Churchwell,E.and Walmsley,C.M.: 1975, Astron. Astrophys. 38, pp. 451-454.
Collin-Souffrin,S.and Joly,M.: 1976, Astron. Astrophys. 53, pp. 213-225.
Comte,G.: 1975, Astron. Astrophys. 39, pp. 197-205.
D'Odorico,S.,Peimbert, M.,and Sabbadin,F.: 1976, Astron. Astrophys. 47, pp. 341-344.
Dufour,R.J.: 1975, Astrophys. J. 195, pp. 315-332.
Dufour,R.J. and Harlow,W.V.: 1977, Astrophys. J. 216, pp. 706-712.
Gordon, M.A. and Burton,W.B.: 1976, Astrophys. J. 208, pp. 346-353.
Hawley, S.A.: 1978, Astrophys. J., in press.
Hawley, S.A. and Grandi,S.A.: 1977, Astrophys. J. 217, pp. 420-424.
Innanen,K.A.: 1973, Astrophys. Space Sci. 22, pp. 393-411.
Jensen,E.B.,Strom,K.M.,and Strom,S.E.: 1976, Astrophys. J. 209, pp. 748-769.
Kaler,J.B.: 1978, preprint.
Lambert, D.L.: 1978, Monthly Notices Roy. Astron. Soc. 138, pp. 143-179.
Linke,R.A.,Goldsmith,P.F.,Wannier,P.G.,Wilson,R.W.,and Penzias,A.A.: 1977, Astrophys. J. 214, pp. 50-59.
Mallik,D.C.V.: 1978, in E. Basinska-Gresik and M. Mayor (eds.) "Chemical and Dynamical Evolution of our Galaxy", IAU Coll. 45, in press.
Pagel,B.E.J.: 1978, Monthly Notices Roy. Astron. Soc. 183, pp. 1p-4p.
Pagel,B.E.J.,Edmunds,M.G.,Fosbury,R.A.E.,and Webster,B.L.: 1978, preprint
Pagel,B.E.J. and Patchett,B.E.: 1975, Monthly Notices Roy. Astron. Soc. 172, pp. 13-40.
Panagia,N.,Bussoletti,E.,and Blanco,A.: 1977, in J. Audouze (ed.), "CNO Isotopes in Astrophysics", Reidel, Dordrecht, p. 45.
Peimbert, M.: 1975, Ann.Rev.Astron.Astrophys. 13, pp. 113-131.
Peimbert, M.: 1978a, in Y. Terzian (ed.), "Planetary Nebulae", IAU Symp. 76, Reidel, Dordrecht, pp. 215-223.
Peimbert, M.: 1978b, in E. Basinska-Gresik and M. Mayor (eds.) "Chemical and Dynamical Evolution of our Galaxy", IAU Coll. 45, in press.
Peimbert,M., Rodríguez,L.F.,and Torres-Peimbert, S.: 1974, Rev. Mexicana

Astron. Astrof. 1, pp. 129-141.
Peimbert, M. and Spinrad, H.: 1970, Astron. Astrophys. 7, pp. 311-317.
Peimbert, M. and Torres-Peimbert, S.: 1974, Astrophys. J. 193, pp. 327-334.
Peimbert, M. and Torres-Peimbert, S.: 1976, Astrophys. J. 203, pp. 581-586.
Peimbert, M. and Torres-Peimbert, S.: 1977, Monthly Notices Roy. Astron. Soc. 179, pp. 217-234.
Peimbert, M. and Torres-Peimbert, S.: 1978,in preparation.
Peimbert, M.,Torres-Peimbert,S.,and Rayo,J.F.: 1978a, Astrophys. J. 220, pp. 516-524.
Peimbert, M.,Torres-Peimbert,S.,and Rayo,J.F.: 1978b, in preparation.
Pottasch, S.R.: 1978, preprint.
Sarazin, C.L.: 1976, Astrophys. J. 208, pp. 323-335.
Schmidt, M.: 1963, Astrophys. J. 137, pp. 758-769.
Searle, L.: 1971, Astrophys. J. 168, pp. 327-341.
Searle, L. and Sargent, W.L.W.: 1972, Astrophys. J. 173, pp. 25-33.
Shields, G.A.: 1974, Astrophys. J. 193, pp. 335-342.
Shields, G.A.: 1978, Astrophys. J. 219, pp. 559-564.
Shields, G.A. and Searle,L.: 1978, Astrophys. J., in press.
Sivan, J.P.: 1976, Astron. Astrophys. 49, pp. 173-177.
Smith, H.E.: 1975, Astrophys. J. 199, pp. 591-610.
Talbot,R.J. and Arnett,W.D.: 1971, Astrophys. J. 170, pp. 409-422.
Talbot,R.J. and Arnett,W.D.: 1973, Astrophys. J. 186, pp. 51-67.
Talbot,R.J. and Arnett,W.D.: 1974, Astrophys. J. 190, pp. 605-608.
Talent, D.L. and Dufour,R.J.: 1978, private communication.
Tinsley, B.M.: 1978, in Y. Terzian (ed.) "Planetary Nebulae", IAU Symp. 76, Reidel, Dordrecht.
Torres-Peimbert, S. and Peimbert, M.: 1977, Rev. Mexicana Astron. Astrof. 2, pp. 181-207.

DISCUSSION

Tinsley: You showed a plot in which O/H does not vary with $\ell n(1/\mu)$ in the way simple models predict, and you mentioned that infall could solve the problem. It is true that with infall there need be no correlation between abundances and gas fraction. However, in order to get an abundance spread, some parameter must vary from region to region. One possibility is that the IMF varies, and another possibility is that the ratio of star formation rate to infall rate varies. In particular, infall leads to a limiting abundance of a given element which is the product of its yield (dependent on the IMF) and the ratio (star formation rate)/(infall rate).

Wollman: As one moves toward the center, is there a preferential settling of heavier material into the plane? Might this account for an abundance gradient? The abundance of neon within the central parsec of the Galaxy is apparently not more than about three times the adopted solar abundance.

Peimbert: Since the observations come from HII regions that are located very close to the galactic plane, $<|z|> \sim$ pc, the effect that you mention will produce a negative gradient. I think that theoretical quantitative predictions should be carried out considering at least the following elements: H, He, C, N, O, Ne and S. The optical results are not in contradiction with the results for the galactic center. There are theoretical models of the chemical evolution of the Galaxy that predict a maximum in the O/H ratio in the 2 to 6 kpc range. More observations in the infrared are needed to obtain the galactocentric distribution of chemical abundances.

A NEW LOOK AT THE GALACTIC MAGNETIC FIELD

P.P. Kronberg and M. Simard-Normandin
University of Toronto; David Dunlap Observatory and
Scarborough College

We have measured the linear polarization of a new large sample of extragalactic radio sources, and by combining these with polarization values already in the literature, we have been able to compute a large number of rotation measures, with improved quality. We have also investigated the depolarization properties of these sources and as a result have been able to identify most sources with a large internally generated Faraday rotation. Figure 1 shows the rotation measures of 475 extragalactic radio sources on an equal-area projection, after "cleaning out" the extragalactic effects to first order.

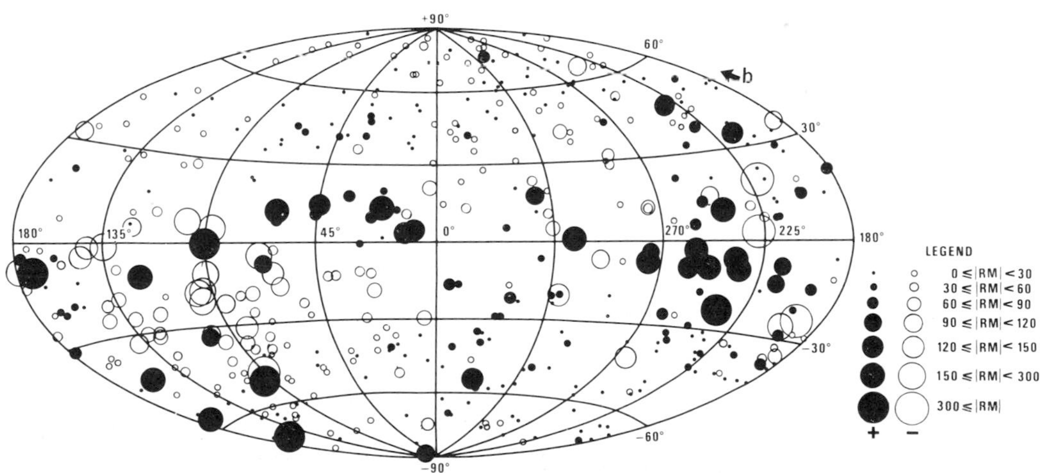

Figure 1 Rotation measures of a subset of 475 extragalactic radio sources that are statistically unlikely to have large internally generated RM's.

W. B. Burton (ed.), The Large-Scale Characteristics of the Galaxy, 317–319.

In the southern galactic hemisphere the RM's suggest a prevailing magnetic field pointing towards $\ell \sim 90^\circ$ and coming from $\ell \sim 255^\circ$. This systematic pattern persists all the way to the south galactic pole, and also agrees with the prevailing trend in the second and third quadrants ($90^\circ \lesssim \ell \lesssim 270^\circ$) in the northern hemisphere. Only above the plane and towards the galactic centre ($270 \lesssim \ell \lesssim 90$) is the general trend of RM's reversed - this is the region occupied by loop I. This "reversal" has previously been observed by Vallée and Kronberg (1975) and Gardner, Morris and Whiteoak (1969).

The most prominent feature in the RM sky is located at $45^\circ \lesssim \ell \lesssim 160^\circ$ and $-40^\circ \lesssim b \lesssim +10^\circ$ and contains surprisingly high negative RM's $-60 \gtrsim RM \gtrsim -200$ rad m^{-2} to quite large negative latitudes (Fig. 1). The predominance of this large feature has not previously been recognized in earlier RM maps with fewer sources.

The vast majority of sources at $|b| > 30^\circ$ have small rotation measures ($|RM| < 30$ rad.m^{-2}), and the large-RM zones are virtually free of small-RM sources. At $\ell \sim 90^\circ$, the RM's are much smaller at $b > 10^\circ$ than $b < -10^\circ$. The influence of loop I on the RM map is much smaller than the large RM features near $\ell \sim 255^\circ$ and $\ell \sim 90^\circ$, and loop III has no corresponding large scale RM feature. Loop II encircles the large feature below the plane, although the physical significance of this positional coincidence is not yet clear.

Another strong feature of positive RM is centered at $\ell \approx 40^\circ$, $b \approx +5^\circ$. Between this feature and the strongly negative zone near $\ell \sim 90^\circ$ below the plane the RM's change abruptly from large positive to large negative values. These features are consistent with a large scale magnetic field directed in the sense of galactic rotation between the Sagittarius and Perseus arms, and in the opposite direction between the Norma-Scutum and Sagittarius arms, in the first two quadrants of galactic longitude.

REFERENCES

Gardner, F.F., Morris, D. and Whiteoak, J.B.: 1969, Aust. J. Phys., 22, 813.

Vallée, J.P. and Kronberg, P.P.: 1975, Astron. and Astrophys., 43, 233.

DISCUSSION

Cesarsky: Previous maps (e.g., Wright) showed much more disorder in the magnetic field directions, and were interpreted as meaning that the galactic magnetic field consists of an ordered and a disordered component, of similar strengths. Measurements of the magnetic field that average over a distance large compared to the scale of the irregularities can only give a measure of the mean, or the possibly ordered component, of the magnetic field. Could, in fact, the disordered component dominate?

Kronberg: The rotation-measure technique is of course more sensitive to ordered fields. It would be premature to give a quantitative value to the ratio B ordered/B disordered. Nevertheless, one can deduce that for the large-scale regions that we see, the absence of small rotation measures is an indication that this ratio must be at least, and probably greater than, one.

Felten: It is worthwhile to emphasize the point that Dr. Cesarsky raised: There may be a sizable random component to the field as well as a uniform component. People should not assume from these studies that the field is smooth, uniform, and directed along spiral arms. If you took a cloud containing a perfectly tangled field, and stretched it along one dimension, you would get a field which would still be tangled. It would have a random component, but it would still give high radio polarization. This could be a realistic model for an interstellar cloud. Until Dr. Kronberg can give us a ratio of the random component to the uniform component, we cannot say how tangled the field is.

Berkhuijsen: At a wavelength of 11 cm, using the Bonn telescope, we have detected polarized emission from the southern part of the bright ring in M31. With a beamwidth of 4!5 ($\simeq$ 1x4 kpc in the plane of M31), we find percentages of polarization of 10 to 20%, with some peaks of more than 30%. A magnetic field along the spiral arm is consistent with out data, but observations at other wavelengths are needed to confirm this. A letter has been submitted to A.&A. about this work by Beck, Berkhuijsen, and Wielebinski.

STABILITY OF AN INTERSTELLAR MEDIUM WITH CURVED MAGNETIC FIELD LINES

Catherine J. Cesarsky
Centre d'Etudes Nucléaires de Saclay

Parker (1966) has considered a simple equilibrium state of the interstellar gas and fields system, such that the magnetic field lines are parallel to the galactic plane; he has shown that this state is subject to the Rayleigh-Taylor instability. The gravitational pull of the stars on the gas and the buoyancy of the magnetic field tend to bend the field lines, allowing the gas to slide down towards the plane. The stability of equilibria with curved magnetic field lines remained to be considered.

In collaboration with E. Asséo, M. Lachièze-Rey and R. Pellat, I have applied the extended energy principle of Bernstein et al. (1958) to two-dimensional, periodic, curved equilibrium configurations of the interstellar medium. We find that, in addition to the Rayleigh-Taylor effects brought about by gravity, a second cause of instability appears: the curvature of the field lines. Sufficient instability criteria have been obtained, assuming that the adiabatic index of the gas is $\gamma=1$, for two types of perturbations: (i) "flute" perturbations, where the shape of the lines of force is conserved; (ii) anti-symmetric perturbations, where the volume of each tube of flux is conserved, and the perturbation extends over several cycles of the equilibrium configuration. We can show that the type of curved equilibria constructed by Mouschovias (1974),which are related through flux-freezing to an initially unstable equilibrium of the Parker type, are stable to "flute" perturbations, but unstable according to the second criterion. The unstable modes have short wavelengths in the direction perpendicular to the planes containing the magnetic field lines. Thus, that type of equilibrium cannot represent the present state of the interstellar medium.

The interaction of cosmic rays with curved magnetic field lines is another source of instability, and probably the dominant one, especially at some height above theplane. The time of growth of this type of instability can be much shorter than the Rayleigh-Taylor time, which is $\sim 10^7$ years.

W. B. Burton (ed.), The Large-Scale Characteristics of the Galaxy, 321–322.

REFERENCES

Bernstein, I. B., Frieman, E. A., Kruskal, M. D., and Kulsrud, R. M.: 1958, Proc. Roy. Soc. (London), A244, 17.
Mouschovias, T. Ch.: 1974, Astrophys. J. 192, 37.
Parker, E. N.: 1966, Astrophys. J. 145, 811.

DISCUSSION

Lockman: Would your conclusions be changed if the interstellar medium were filled almost entirely with coronal gas?

Cesarsky: These studies, following Parker, consider the interstellar medium as a "gas of clouds"; the "sound velocity" which I quoted would then be approximately the dispersion velocity of clouds, say 7 km s^{-1}. The same analysis could be applied to an interstellar medium of hot gas; but then the sound velocity would be higher by a factor of order 10.

Kronberg: There seem to be systematic shearing motions on the insides of spiral arms. This must surely have implications for the large scale magnetic field structure in the galactic disk?

Cesarsky: Probably. But this is very far from the problem I just discussed.

Greyber: An old observation by Morris and Berge using Faraday rotation of distant radio sources shows that the magnetic field tends to be along the spiral arm but in one direction above the galactic plane and in the opposite direction below. This magnetic field configuration in spiral arms, predicted on theoretical grounds (Greyber, Liège Symposium, 1966) is the same as observed in the earth's magnetotail.

Verschuur: When we discuss cooling of clouds we should bear in mind that there are neutral hydrogen clouds, as cold as 10 K, that appear devoid of dust and molecules such as OH and CO. How do clouds get that cold without dust or molecules to help cool them? Until we can explain that, we should remain cautious about cooling mechanisms for interstellar clouds.

Cesarsky: Two comments: (1) Cloud models where clouds are supported by magnetic fields always exhibit the "wrong" curvature, and thus could easily be unstable because of the kind of processes I just discussed in connection with the galactic disk. (2) The rate of angular momentum you quoted is derived for the case of a uniform field; possibly distortions of the magnetic field, brought about by the contraction associated with cloud formation, could alter this rate. Thus one has to be careful when applying the Eberts et al. formula to a contracting, or already contracted, cloud.

VI. THE GALACTIC NUCLEUS

INTRODUCTION TO THE SESSION

J. H. Oort, Sterrewacht, Leiden

Observational data:

There has been a great development in the past few years of observational data of various kinds:

(a) HI observations of high resolution and sensitivity (Cohen and Davies, Burton *et al.*, Kerr *et al.*)
(b) Observations of molecules, in particular high-resolution observations of CO (Burton, Gordon, Bania, Liszt, Solomon *et al.*)
(c) Infrared and far-infrared
(d) Ionized gas: hydrogen recombination lines, and in particular NeII (with 4" beam) (Mezger *et al.*, Townes, Wollman *et al.*)

Discoveries:

(a) and (b) Many new features having large radial motions in addition to rotation. These features lie in a plane tilted ∿22° relative to the galactic plane. Probably most, if not all, of the gas within R ∿ 1.3 kpc lies in a tilted disk (Shane, van der Kruit, Cohen and Davies, but in particular Burton and Liszt). Very high gas density within R ∿ 200 pc (Bania).
(c) Distribution of old population; gravitational field. Infrared nuclear disk of radius ∿10" with discrete "10-μ sources."
(d) High random velocities of 10-μ sources, which are probably compact HII regions.
(e) Nucleus of about 0.″001 (20 A.U.) diameter.

Problems of Interpretation:

Are the radial motions due to expulsion or can they be explained by gasflow along a small inclined bar? Can the tilted disk be maintained?

W. B. Burton (ed.), The Large-Scale Characteristics of the Galaxy, 323.

HI IN THE INNER FEW KILOPARSECS OF THE GALAXY

W. B. Burton* and H. S. Liszt
National Radio Astronomy Observatory†, Green Bank, W.Va. U.S.A.

After reviewing the available observational material, we describe here a simple model of the distribution and kinematics of HI gas within 1.5 kpc of the galactic center. According to this model, most of the inner-Galaxy gas is smoothly distributed in a tilted disk, within which the perceived kinematics are consistent with axisymmetric rotation and expansion of approximately equal magnitude. The model subsumes in a coherent way many observed spectral features which were previously studied separately, without requiring important density enhancements or anisotropic ejection from the nucleus.

PERSPECTIVE AND EARLIER WORK

The evidence for peculiar kinematics in the inner regions of our Galaxy and of other nearby spiral galaxies is well established. The disturbances extend to z-distances of at least several hundred parsecs, to radii of several kiloparsecs, and involve non-rotational velocity components of substantially more than 100 km s^{-1}. Most earlier papers interpreting the wide variety of anomalous-velocity, non-planar phenomena observed toward the central region of the Galaxy have considered them as ejecta produced by violent activity in the nucleus (see Oort's 1977 review). Here we suggest an alternative interpretation of these features which acknowledges little evidence of anisotropic ejection from the nucleus.

For the case of our own Galaxy, the most definite and extensive observational data come from radio observations at λ21 cm of atomic hydrogen. The Galaxy is to a large extent transparent to this radiation, allowing access to the nuclear region, and the emission is intense, allowing more extensive sampling than in the much weaker molecular lines.

* Present address: Department of Astronomy, University of Minnesota, 116 Church Street S.E., Minneapolis, Minnesota 55455.

† Operated by Associated Universities, Inc., under contract with the National Science Foundation.

W. B. Burton (ed.), The Large-Scale Characteristics of the Galaxy, 325–336.

In Table 1 we have compiled the observational parameters of the most recent 21-cm surveys of the general region of the galactic center. Each body of data represented in the table is unique in terms of at least one of its observational parameters and satisfies at least in part the requirements for a general investigation of the nuclear phenomena: extensive angular and velocity coverage and high sensitivity. Additional material, suitable for special-purpose investigations, is tabulated by Simonson (1974) and by Heiles and Wrixon (1976).

Table 1. Still-current general HI line surveys of the galactic center

Reference	Telescope	ℓ-coverage (degrees)	b-coverage (degrees)	v-coverage (km s^{-1})	Sensitivity (K)	Form of Display
Burton *et al*. 1977	43 m	349 to 12, $\Delta\ell = 1$	-10 to 10, $\Delta b = 1$	-500 to 500, $\Delta v = 5.5$	0.2	$(b,v)\|_\ell$ maps
Burton and Liszt 1978	43 m	349 to 13, $\Delta\ell = 1$	-10 to 10, $\Delta b = 0.5$	-320 to 320, $\Delta v = 5.5$	0.1	$(\ell,v)\|_b$ & $(b,v)\|_\ell$
Cohen 1975	30 m	355 to 10, $\Delta\ell = 1$	5 to 5, $\Delta b = 0.25$	-300 to 300, $\Delta v = 7.3$	0.3	$(b,v)\|_\ell$ maps
Kerr 1969	64 m	300 to 60, $\Delta\ell = 1$	-2 to 2, $\Delta b = 0.2$	-250 to 120, $\Delta v = 7$	2.0	$(b,v)\|_\ell$ maps
Lindblad 1974	43 m	339 to 12, $\Delta\ell = 3$	-10 to 15, $\Delta b = 0.3$	-120 to 120, $\Delta v = 1$	(2.0)	$(b,v)\|_\ell$ maps
Mirabel 1976	30 m	355 to 5, $\Delta\ell = 1$	-5 to 5, $\Delta b = 1$	-1000 to -300, $\Delta v = 25$	0.3	none
Sanders and Wrixon 1972a and Sanders *et al*. 1972	6 m	350 to 12, $\Delta\ell = 2$	-10 to 0, $\Delta b = 2$	-340 to -40, $\Delta v = 16$	0.1	special purpose
Sanders and Wrixon 1972b	6 m	355 to 5, $\Delta\ell = 2$	-5 to 5, $\Delta b = 2$	-300 to 300, $\Delta v = 16$	0.1	special purpose
Sanders *et al*. 1977	100 m	357 to 3, $\Delta\ell = 0.1$	0	-300 to 300, $\Delta v = 3$	0.5	$(\ell,v)_b$ maps
Simonson and Sancisi 1973	25 m	356 to 2, $\Delta\ell = 0.5$	0 to 5, $\Delta b = 0.5$	-120 to 120, $\Delta v = 3.4$	2.0	$(\ell,v)\|_b$ & $(b,v)\|_b$
Sinha 1978	43 m	339 to 11, $\Delta\ell = 0.5$	-2 to 2, $\Delta b = 0.25$	-260 to 300, $\Delta v = 4$	0.1	$(\ell,v)\|_b$ maps
Wrixon and Sanders 1973	43 m	357 to 3, $\Delta\ell = 0.3$	-3 to 3, $\Delta b = 1$	-300 to 300, $\Delta v = 5.5$	0.2	$(\ell,v)\|_b$ maps

Note: The sensitivity refers to the lowest-level contour plotted or to the quoted 3σ value, in the temperature units of the survey rounded to one decimal. The velocity coverage refers to the published data, if any.

The observations show many apparently isolated HI features with anomalous velocities. To account for these features, separate, highly directional nuclear events have been invoked, with the epoch and imparted initial velocity of each event being tailored in accordance with the individual feature's observed properties. It has been recognized for some time that the anomalous material occurs mainly in the two opposed quadrants $\ell > 0°$, $b < 0°$ and $\ell < 0°$, $b > 0$. This has been shown by Kerr (1967), van der Kruit (1970), Cohen (1975), and others using maps of the HI integrated-intensity first moment. Kerr and Sinclair (1966) have shown that perturbations in the distribution of the non-thermal continuum radiation at λ20 cm also occur in these opposed quadrants. Figure 1 shows two examples of HI moment maps; by suitable choice of the range of integration, the inner-Galaxy material may be separated to a large degree from other line-of-sight material. The obvious preference for two quadrants has been taken as evidence of a favored collimation axis for the violent events. Such explanations are unsatisfying because of the number of unrelated events which they require and because they address directly neither the nature of the ejection mechanism nor the focussing. Also puzzling are the general lack of disruption of the nuclear-region gas, which remains predominantly neutral and conducive to molecule formation, and the confinement of the anomalous-velocity features within a well-defined velocity envelope.

We suggest here (see also Burton and Liszt, 1978, and Liszt and

Burton, 1978) an alternative model of the distribution and kinematics of the gas within 1.5 kpc of the galactic center which accounts in a simple way for many of the phenomena observed. Lacking a dynamical foundation, the principal use of the model is the constraint of other interpretations of the inner-Galaxy gas.

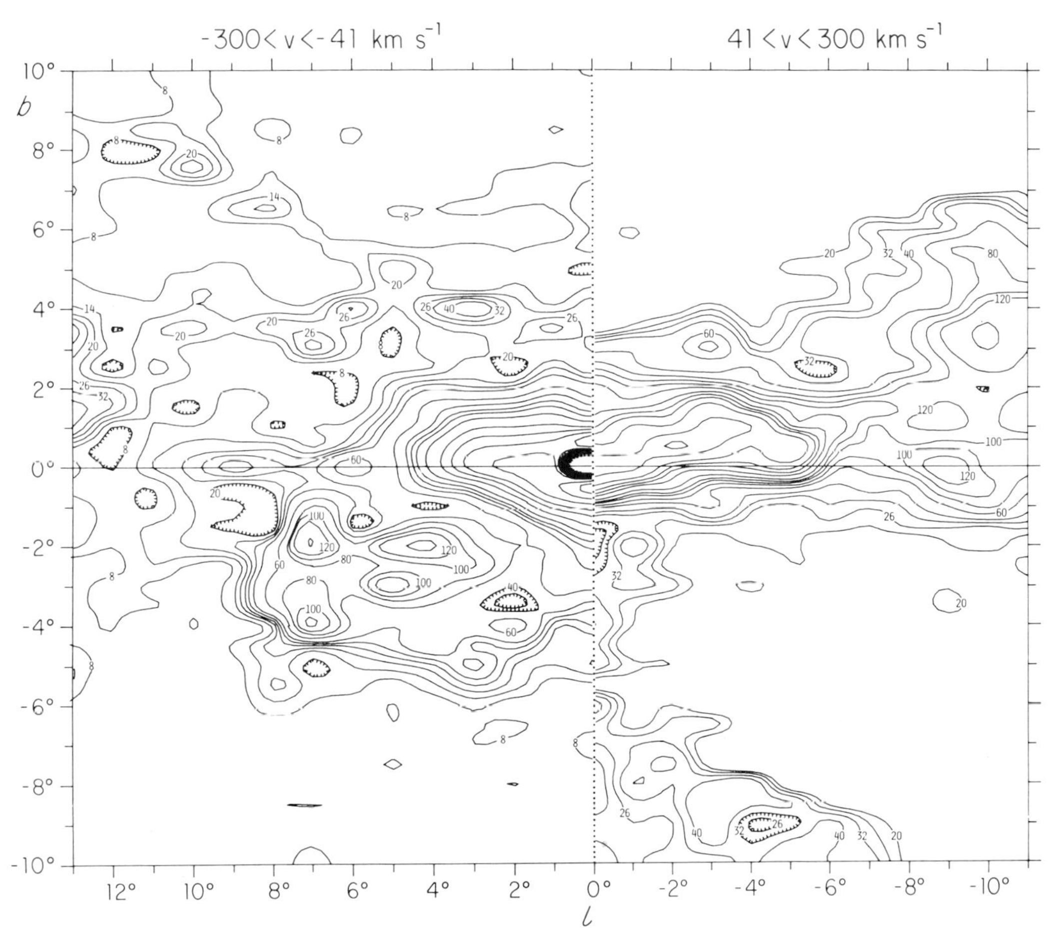

Figure 1. Contours of observed T_A integrated over the velocity ranges $-300 < v < -41$ km s^{-1} (left) and $+41 < v < +300$ km s^{-1} (right). The choice of the velocity intervals excludes most of the emission from the Galaxy at large.

TILTED-DISK MODEL OF THE INNER-GALAXY GAS DISTRIBUTION

The model which we suggest (see Figure 2) confines the gas in a layer of 0.1 kpc scale height to a disk of 3 kpc diameter. This disk is tilted $\alpha = 22°$ with respect to the plane $b = 0°$ and $i = 78°$ with respect to the plane of the sky; it is the plane of this disk, not the plane $b = 0°$, which is fundamental to the gas distribution in the inner Galaxy. Within this disk the kinematics are axisymmetric and density varies only with distance from the equatorial plane.

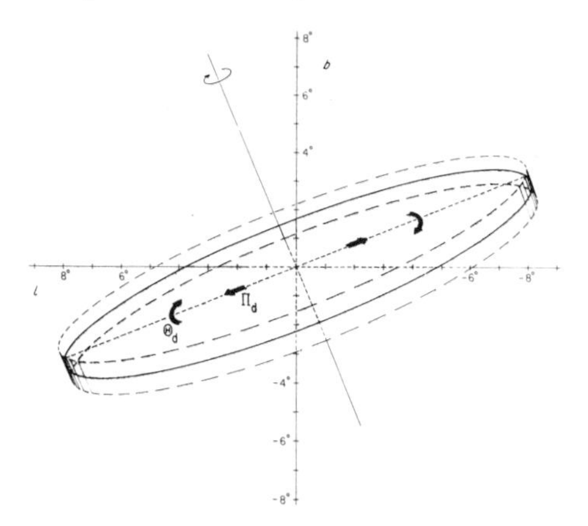

Figure 2. Appearance of the model tilted gas distribution as projected onto the plane of the sky. The solid-line approximate ellipse represents the equatorial plane of a disk of radius 1.5 kpc tilted through the angles $\alpha = 22°$, $i = 78°$. The vectors indicate schematically the model rotation and expansion functions.

The free parameters of the model are specified--and the success of the model judged--by requiring agreement of synthetic with observed spectra. The synthetic spectra are generated in a way which includes the consequences of the radiative transfer inherent in the model velocity field and density distribution. In practice, the choice of parameters defining the model required iterative comparison between observed and simulated profiles.

We specify perpendicular distance from the disk axis by ϖ_d, velocities in the ϖ_d-direction by Π_d, perpendicular distance from the central plane of the disk by z_d, and rotation velocities orthogonal to both the ϖ_d- and z_d-directions by Θ_d. The disk has azimuthal symmetry, so that Π_d and Θ_d depend only on the radius ϖ_d. Necessary for the generation of synthetic data is the velocity measured with respect to the local standard of rest at a point at distance r from the Sun on a line of sight in the direction (ℓ, b):

$$v_d = \Pi_d(\varpi_d) \cdot \{\varpi_d^2 - \varpi_o \cdot (\varpi_o - r \cdot \cos b \cdot \cos \ell - z_d \cdot \cos i)\} \cdot (\varpi_d \cdot r)^{-1} - \Theta_d(\varpi_d) \cdot \varpi_o \cdot \sin i \cdot (\sin b \cdot \sin \alpha - \cos b \cdot \sin \ell \cdot \cos \alpha)/\varpi_d - \Theta_o \cdot \sin \ell \cdot \cos b. \qquad (1)$$

Kinematics, rather than the gas density or temperature distributions, dominate the appearance of long-line-of-sight spectra of ubiquitous galactic tracers like HI or CO. The kinematic functions which follow from the model/observation comparison, utilizing in addition to the HI the CO data described by Liszt and Burton (1978), are

$$\Pi_d(\varpi_d) = 170\ (1-\exp\ (-\varpi_d/0.07))\ \text{km s}^{-1}, \qquad (2)$$

and

$$\Theta_d(\varpi_d) = 180\ (1-\exp\ (-\varpi_d/0.20))\ \text{km s}^{-1} \quad \text{if } \varpi_d \leq 0.85 \text{ kpc}$$

$$= 180\ (1-\exp\ (-(1.7 - \varpi_d)/0.20\ \text{km s}^{-1} \quad \text{if } \varpi_d > 0.85 \text{ kpc} \qquad (3)$$

COMPARISONS OF OBSERVED AND MODELLED SPECTRA

Integrated-intensity moment maps are convenient at the beginning of the modelling process because they reflect directly the size and orientation of the disk. Figure 3 shows the arrangement on the plane of the sky of intensities integrated in synthetic spectra over the indicated velocity ranges. The model moment maps show the same general trends as those observed and plotted in Figure 1, including the zero-longitude crossing at $b \neq 0°$ as well as the tilted nature and extent of the moment distributions.

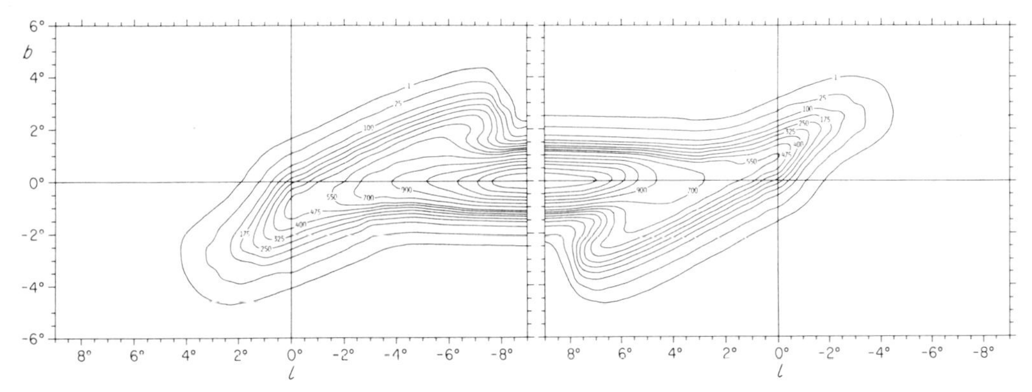

Figure 3. Arrangement on the plane of the sky of intensities integrated over the indicated velocity ranges in synthetic spectra representing HI in the modelled inner-Galaxy distribution. Some emission from the Galaxy at large also enters.

The principal disadvantage of the moment maps is their insensitivity to the available kinematic information which position, velocity maps do not suppress. Spectral data from the central region of our Galaxy are usually displayed in ℓ,v maps taken parallel to the plane $b = 0°$ or in b,v maps taken parallel to $\ell = 0°$. For gas distributed as in Figure 2, such cuts are not the most revealing. The tilted disk projects approximately to an ellipse whose major axis lies approximately on the line $b = -\ell \tan \alpha$. A position, velocity map constructed along this line should, according to the model, show a high degree of kinematic symmetry (because the projections of the functions Π_d and Θ_d onto the line of sight are very nearly antisymmetric about $\ell = b = 0°$, $v = 0$ km s^{-1}) and so should yield straightforward information about these functions in the inner Galaxy. Such a map should also reveal the maximum extent of the tilted distribution.

Figure 4 shows an observational ℓ,v map on the line $b = -\ell \tan 22°$. To avoid distortion of the contours caused by absorption at the position of the galactic center, we have replaced the spectrum at that point by the average of those at $\ell,b = +0.°0, +0.°0$, and $0.°0, -0.°5$. As predicted, and in great contrast to the ℓ,v arrangement taken at $b = 0°$, the observations show a high degree of kinematic symmetry about the map origin. That the envelope of the emission is not pinched toward $v = 0$ km s^{-1} at $\ell = 0°$ indicates radial motion; that the pattern is skew indicates

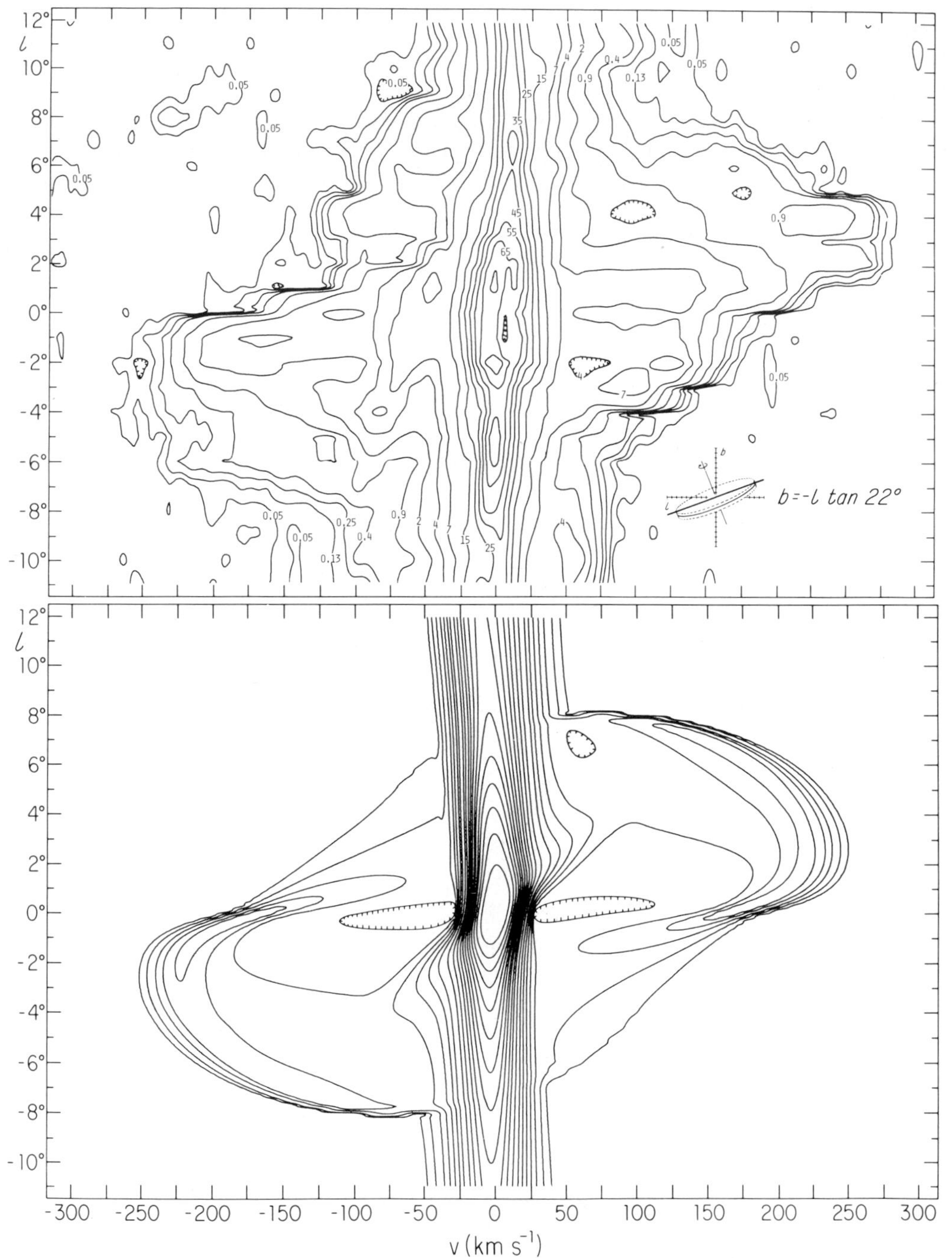

Figure 4. Longitude, velocity arrangement of emission from directions with b = − ℓ tan 22°. This cut through the inner-Galaxy reveals the approximate extent of the tilted fundamental distribution of gas, and shows its kinematic symmetry. (top) Observed profiles. (bottom) Synthetic profiles.

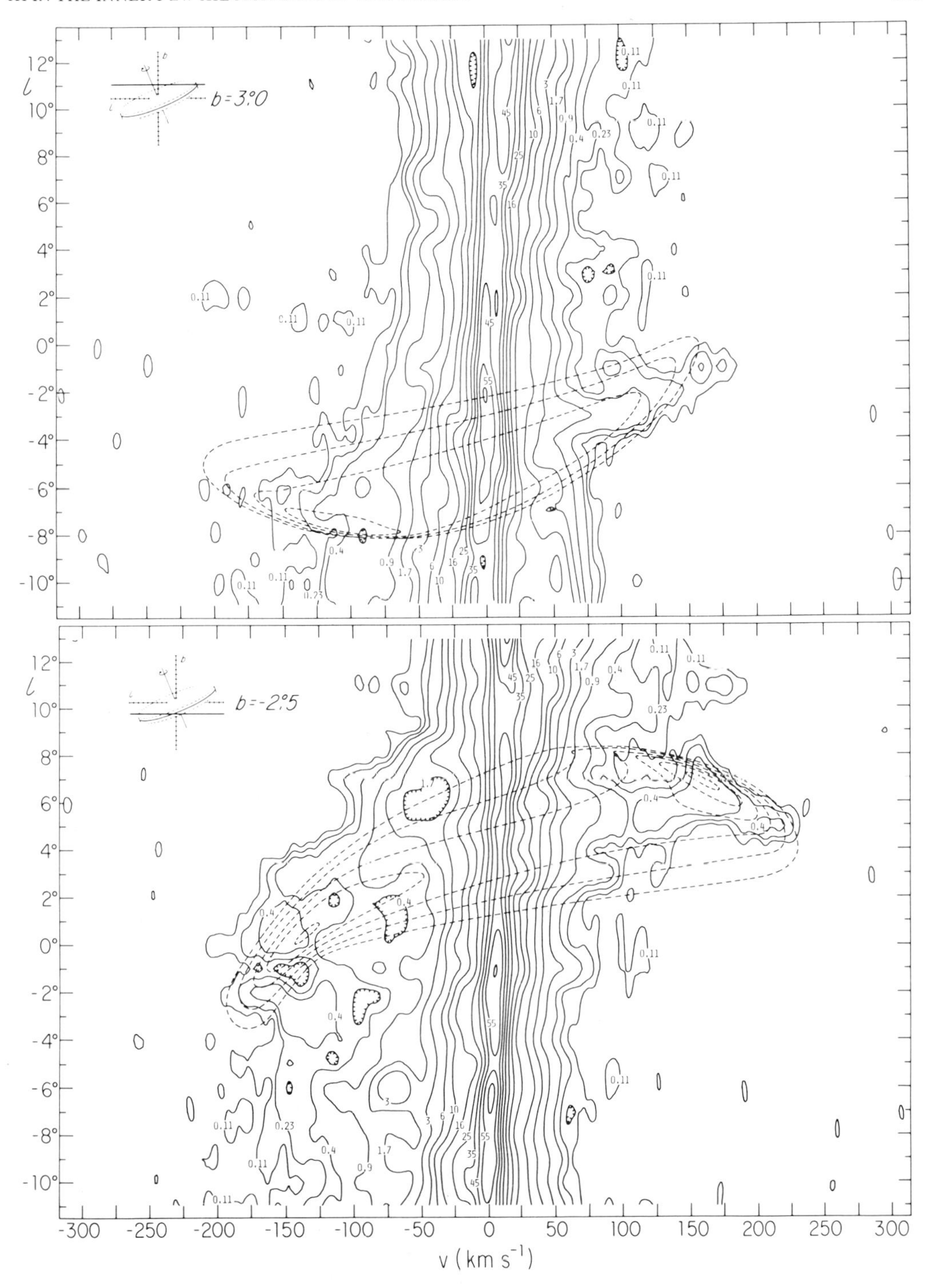

Figure 5. Emission in the ℓ,v planes at the indicated latitudes. The dashed contours represent synthetic emission from the modelled disk;

the Galaxy at large does not enter these spectra. As these sample comparisons show, the model accounts for a number of apparently isolated features.

rotation. The negative-velocity HI absorption (see Figure 1) against the Sagittarius continuum sources shows directly that the radial motion is expansion. Detailed specification of the Π_d- and Θ_d-functions in accordance with the observational material is described by Burton and Liszt (1978).

A synthetic map generated along the line $b = -\ell \tan 22°$ is also shown in Figure 4. The basic agreement of the outer envelopes of the observed and modelled distributions indicates that the kinematics are successfully modelled. Obviously the presence of large general expansion velocities forces our rotation function to be much smaller in magnitude than those derived on the basis of pure rotation.

The data displays which most usefully show the previously studied and apparently isolated anomalous features are ℓ,v intensity - contour maps. Figure 5 shows two examples of these, drawn at $b = 3\overset{\circ}{.}0$ and at $b = -2\overset{\circ}{.}5$. Superimposed on the observed material are dashed contours representing synthetic spectra. The model profiles exclude the Galaxy at large, which, in the observations, contributes the broad band of emission at $|v| \lesssim 50$ km s^{-1}. In both ℓ,v maps, the characteristic signature of the disk encompasses the principal perturbations observed. Thus, at $b = 3\overset{\circ}{.}0$, the model subsumes Cohen's (1975) feature J2 at high positive velocities near $\ell = -1°$ as well as the feature near $\ell = -3°$, $v = 100$ km s^{-1}, discussed by Sanders and Wrixon (1972b). At $b = -2\overset{\circ}{.}5$, the positive-velocity pattern near $\ell = 6°$ is Cohen's feature J4. The negative-velocity portion of the observed ℓ,v plane shows the combined emission from features X and XII of van der Kruit (1970) and from feature E of Sanders _et al_. (1972).

Other model/observation comparisons show that many of the apparently isolated observed spectral features occur along the position, velocity loci predicted by the model. Because the density of gas in the model varies smoothly, and because the velocity fields are axially symmetric and simply described, we see no compelling evidence for important density enhancements or kinematic perturbations associated with particular observational features. In this respect the model avoids some of the most troubling aspects of earlier interpretations of the emission from the inner Galaxy. Like them, however, it lacks a dynamical foundation.

REFERENCES

Burton, W. B., Gallagher, J. S., and McGrath, M. A.: 1977, Astr. and Astrophys. Suppl. 29, pp. 123-138.
Burton, W. B., and Liszt, H. S.: 1978, Astrophys. J., in press.
Cohen, R. J.: 1975, M.N.R.A.S. 171, pp. 659-696.
Heiles, C., and Wrixon, G. T.: 1976, in "Methods of Experimental Physics", ed. M. L. Meeks, 12C, pp. 58-77.

Kerr, F. J.: 1967, in "Radio Astronomy and the Galactic System", ed. H. van Woerden (London: Academic Press), pp. 239-251.
Kerr, F. J.: 1969, Australian J. Phys. Astrophys. Suppl. 9, pp. 1-147.
Kerr, F. J., and Sinclair, M. W.: 1966, Nature 212, p. 166-167.
Lindblad, P. O.: 1974, Astr. and Astrophys. Suppl. 16, pp. 207-236.
Liszt, H. S., and Burton, W. B.: 1978, Astrophys. J., in press.
Mirabel, I. F.: 1976, Astrophys. Space Sci. 39, pp. 415-417.
Oort, J. H.: 1977, Ann. Rev. Astr. Astrophys. 15, pp. 295-362.
Sanders, R. H., and Wrixon, G. T.: 1972a, Astr. and Astrophys. 18, pp. 92-96.
Sanders, R. H., and Wrixon, G. T.: 1972b, Astr. and Astrophys. 18, pp. 467-470.
Sanders, R. H., Wrixon, G. T., and Penzias, A. A.: 1972, Astr. and Astrophys. 16, pp. 322-326.
Sanders, R. H., Wrixon, G. T., and Mebold, U.: 1977, Astr. and Astrophys. 61, pp. 329-337.
Simonson, S. C.: 1974, in "Galactic Radio Astronomy", ed. F. J. Kerr and S. C. Simonson, pp. 511-519.
Simonson, S. C., and Sancisi, R.: 1973, Astr. and Astrophys. Suppl. 10, pp. 283-364.
Sinha, R. P.: 1978, in preparation.
van der Kruit, P. C.: 1970, Astr. and Astrophys. 4, pp. 462-481.
Wrixon, G. T., and Sanders, R. H.: 1973, Astr. and Astrophys. Suppl. 11, pp. 339-345.

DISCUSSION

<u>Contopoulos</u>: About two years ago Mr. Sinha (University of Maryland) sent me a letter, suggesting a tilted disk near the center of our Galaxy, and asked me if I could provide a dynamical explanation. Well, I have no ready-made dynamical explanation, thus I would like first to have a feeling how certain is the kinematical model provided. May I ask, therefore, whether you exclude a model without expansion and for what reasons?

<u>Burton</u>: Although our kinematic model cannot be defended dynamically, we do believe that any dynamically consistent model must provide line-of-sight motions which are not very different from those given by the combined effects of our Θ_d and Π_d functions. No pure-rotation situation would do that. However, motions in elliptical streamlines might suffice. Liszt and I are testing this now. If such a solution could be found, it would avoid the problem of net outward mass flux implied by our Π_d function.

<u>Sanders</u>: Have you produced moment maps of the neutral hydrogen surface density distribution at very high velocities--say $|v| > 200$ km s^{-1}, where the contribution from transgalactic hydrogen is certainly excluded? Do you see the tilt in such moment maps?

It seems to me that the tilt does not show up in other conspicuous tracers of the gas density distribution, such as the extended non-thermal continuum source and the extended far-infrared emission. Concerning the far-infrared, this is almost certainly thermal radiation of star light by dust in the inner 100-200 pc and thus should be a tracer of the gas density in the inner region. Can you comment on this?

Burton: We have produced moment maps over a large number of velocity ranges, and find in all cases the tilted distribution in all the inner-galaxy material, whether at permitted or forbidden velocity. Regarding the highest velocities, it is an important fact that the gas is confined within very definite kinematic boundaries. Very high velocities, which might be expected (Oort 1977) for ejection from the nucleus, are not found (Mirabel 1976, Burton et al. 1977).

In two respects the non-thermal continuum and infrared data are not well suited for a search for the tilted distribution: the observed angular extent (especially in b) is small, and the measurement technique is a differential one, making the results rather insensitive to a weak, extended background. In addition, because kinematic isolation of the inner-galaxy gas is not possible, one must somehow separate the contribution from a ∿500 pc intersection of the tilted disk from the ∿30 kpc intersection of the general galactic layer.

Davies: In contrast to Dr. Sanders, R. J. Cohen and I found in our 1976 paper that the locus of the moments of the HI distribution in fixed velocity ranges made an inclined line to the plane; the distribution at each velocity was also inclined. The distributed ionized hydrogen over a scale of 10° shows no evidence for the inclined disk. This is not in contradiction to the HI result because this ionized gas, as measured by the 166α recombination line, is near zero velocity and is evidently foreground material.

I have a comment about the distribution within the inclined disk. R. J. Cohen and I found a number of features within the central region--too many to be explained by velocity crowding in a uniform disk.

A study by K. Grape of the observed terminal-velocity hydrogen has shown that the inner region of the Galaxy has a distributed HI component with a density of about 0.1 atoms cm^{-3}.

Sinha: The area under the profile integrated up to 100 km s^{-1} from the permitted-velocity edge shows the inclined feature very clearly.

Ostriker: Have you made any progress in assessing the suggestion that there is neither explosion or even expansion but that, rather, the apparent expansion is caused by motions along elliptical streamlines. Even if the gravitational field were approximately axisymmetric so that $j = rv_{\perp}$ were constant along a streamline, it would appear from a superficial examination of your results that a good fit to observations would be possible and the large rates of mass and energy outflow (implied by your present model) could be avoided.

Burton: Liszt and I are pursuing this, motivated by the desire to find a dynamically plausible model. Our kinematic model shows the sort of restraints which an elliptical-streamline model will have to satisfy.

Oort: It is hard to imagine how a smooth expansion and rotation could exist simultaneously throughout your tilted disk. In a physically possible model one should either have distinct expanding features, or a bar-like structure strongly differing from azimuthal symmetry. However, I understand and appreciate that your model was based on the wish to have as simple a model as possible.

Menon: Your model implies that the disc is transient. What is the time scale for its appearance and disappearance?

Burton: Such a time scale is implied by the parameters of the kinematic model, but probably does not merit much discussion until a dynamically satisfying model can be found.

Tinsley: What mass outflow rate would be predicted by your expansion model?

Burton: The expansion flux across the outer boundary of the tilted disk is 4 $M_{\odot}$ per year. Because this refers only to HI, the total flux would be much greater. This is of course uncomfortably large, and provides one of our motivations for searching for a closed-streamline elliptical model which can still satisfy the restraints indicated by our kinematic model.

de Vaucouleurs: Your schematic map of the nuclear region agrees well with what should be expected in a barred spiral having its bar in the position angle suggested by the major axis of the inner ring of the Simonson map and the Georgelins' map of the spiral pattern. Models with pure circular symmetry are not likely to lead to realistic pictures of the gas distribution.

Burton: I am bothered by the insistence that the evidence indicates important high-density features in the central region. Our model shows that a smooth, axisymmetric distribution of density and velocity results nevertheless in intensity concentrations in position, velocity maps. These concentrations occur at the locations of features E, J2, J4, J5, VII, X, and XII, as well as at other locations. Among these other features which we believe adequately accounted for in these terms is the "connecting arm" feature of Rougoor, also identified by van der Kruit (feature III) and by Cohen (feature IIIa). It plays an important role in the interpretations of Rougoor, Kerr, and Cohen and Davies, where it is identified as a steeply inclined arm, or bar-like feature. Would you care to comment on this identification in view of our opinion that it is adequately accounted for by the vagaries of radiation transport through a smooth density in a rotating disk?

Cohen: The velocity-longitude diagram, Figure 1 of my paper, indicates some large-scale symmetry in the velocity field of gas in the central region, since the outermost velocities at which emission is detected are symmetrical through $\ell = 0°$, $v = 0$ km s^{-1}. However the gas density distribution must be rather irregular. For example, the "connecting arm" you mentioned is a major feature in the map at positive longitudes, but it has no symmetric counterpart at negative longitudes. In general it is very hard to find symmetry in the ridge lines of emission, although the overall extent of the emission is symmetric. This is very difficult to understand in terms of your axisymmetric model. Another point is that for a given line of sight such kinematic models give rise to only a single emission peak as a rule, at the terminal velocity. The observations however show wavy peaks at lower velocities. Because of the concentration of molecules to these HI peaks we believe they represent real density enhancements.

Burton: Kinematic models can give rise to the multiple peaks; in addition, the subcentral point (terminal velocity) region is not necessarily favored in models which deviate from pure circular rotation. Furthermore, molecules--being kinematic tracers--respond to the velocity field in the same way HI does.

THE HI STRUCTURE OF THE NUCLEAR DISK

R. J. Cohen
University of Manchester, Nuffield Radio Astronomy Laboratories, Jodrell Bank, Macclesfield, Cheshire, SK11 9DL England

It is nearly twenty years since Rougoor & Oort (1960) reported the discovery of neutral hydrogen (HI) emission from the galactic nucleus, and I would like to summarize some of the progress that has been made since then in observing and interpreting this emission. One of the most surprising properties to be found is the tilt of the HI distribution relative to the galactic equator. This was noted originally by Kerr (1967), and recent measurements indicate that the tilt is present from within 200 pc of the centre out to a distance of some 2 kpc. (Cohen & Davies 1976, 1978). Because of the tilted distribution the properties of the nuclear emission cannot always be deduced from observations at zero galactic latitude. For example, Rougoor & Oort (1960) deduced a sharp cut-off in the negative-velocity "nuclear disk" emission at $\ell \simeq -4^{o}$, from their measurements near zero latitude, whereas the more extensive observations by Kerr (1967) show that the cut-off is only an apparent effect. In reality the emission continues to more negative longitudes, but at higher galactic latitude because of the tilt.

The nuclear emission at the highest "permitted" velocities has very broad line profiles over the range of longitude where the tilt is seen, increasing in velocity dispersion from 8 km s^{-1} outside the tilted region to over 20 km s^{-1}. Such broad HI profiles occur only in the "nuclear disk" emission feature discovered by Rougoor & Oort (1960) and in a number of weaker HI features seen within a few degrees of the galactic centre (Cohen 1975, Cohen & Davies 1978).

An overall view of the emission can be taken from Fig. 1, which is a velocity-longitude map of the 21 cm emission averaged in galactic latitude between latitudes $\pm$ 3°. By averaging in this way we properly represent the emission which lies away from zero latitude because of the tilt. Within 5° longitude of the centre 34% of the high-velocity emission ($|V|$ > 50 km s^{-1}) shown in Fig. 1 has velocities which are "forbidden" in terms of normal galactic rotation. At zero longitude there is detectable HI emission at velocities up to $\pm$ 200 km s^{-1}, indicating that the non circular motions are as large as the rotational

W. B. Burton (ed.), The Large-Scale Characteristics of the Galaxy, 337–340.

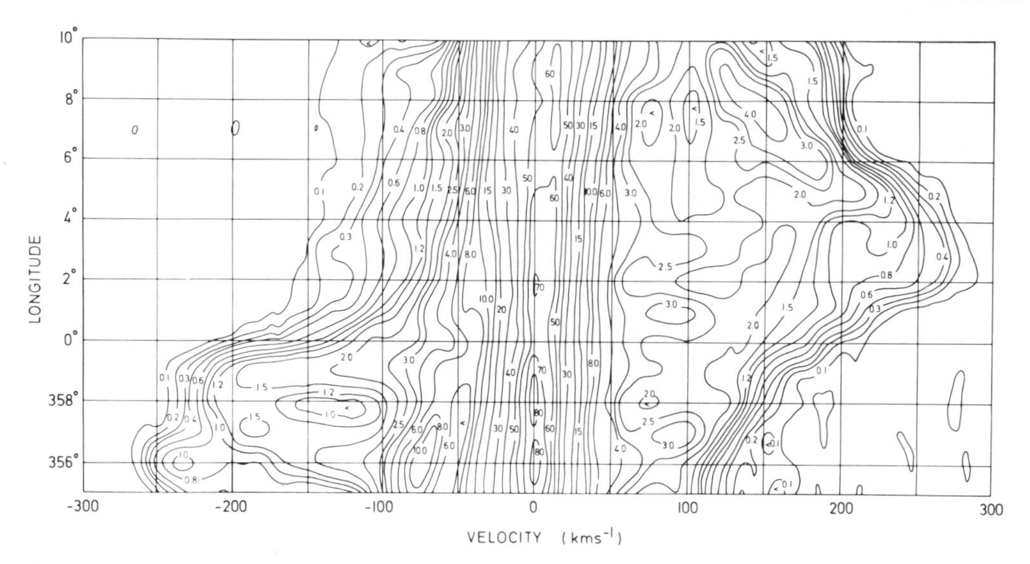

Fig. 1. Velocity-longitude diagram showing the mean HI emission from the nuclear region of the Galaxy, averaged between galactic latitudes $\pm 3^{\circ}$. The observations were taken at Jodrell Bank with the Mark II radio telescope (beamwidth 31 x 35 arcmin) and have a velocity resolution of 7.3 km s^{-1} (Cohen 1975). The region was fully sampled in latitude but undersampled in longitude ($\Delta \ell = 1^{\circ}$). Contours give line temperature in degrees K.

components of motion. The noncircular motions increase on the whole towards the centre, since the emission features with the largest "forbidden" velocities are seen over the smallest range in longitude. We know from absorption measurements of the strong radio continuum, source Sgr A near the galactic centre that in this direction most of the HI with "forbidden" velocities is moving away from the centre. However, the transverse velocities are not directly measurable, so we can derive the velocity field and distribution of the gas only by making further assumptions.

Three simple kinematical models are often used to describe the large scale motion of the gas: pure rotation, rotation plus expansion, and motion about a central bar. Rougoor & Oort provisionally assumed that the motion within 800 pc of the centre was purely rotational, and that all the noncircular motions lay outside this region and were directed away from the centre. Using these assumptions Rougoor (1964) was able to construct a number of axially symmetric velocity models which gave the various "expanding arms" the appearance of spiral arms when they were located according to their measured radial velocities. Grape (1978) has extended this analysis, allowing the possibility of expansion nearer the centre, and has shown that the extreme "permitted" and "forbidden" velocities observed strongly constrain such axially symmetric models. A slightly different approach has been taken by

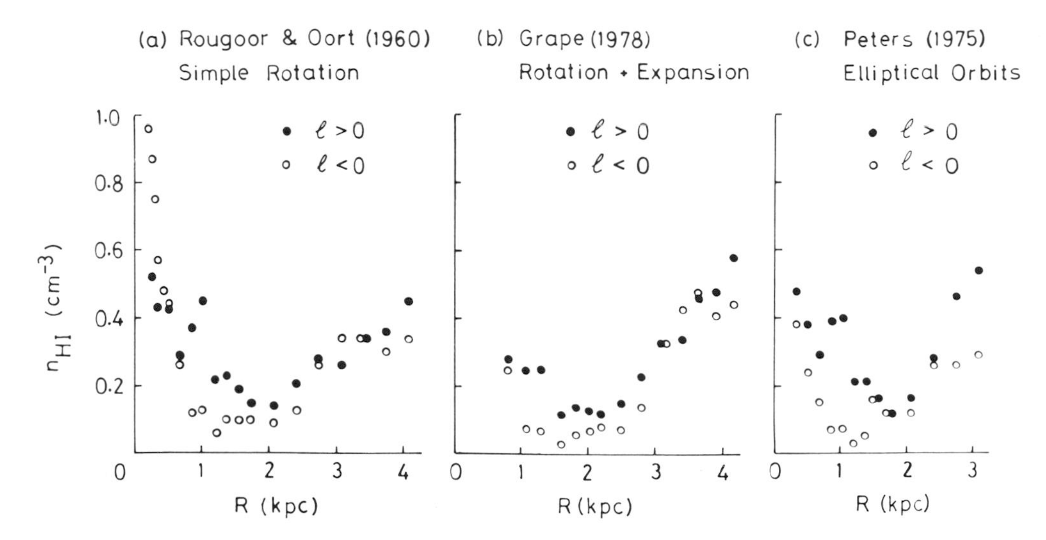

Fig. 2 Radial distribution of HI density in the nuclear disk, estimated from the terminal velocity emission using three different kinematical models: (a) pure rotation throughout the disk according to the Rougoor & Oort (1960) rotation curve, (b) axially symmetric rotation plus expansion (Grape 1978), and (c) motion in elliptical orbits about a central bar (Peters 1975, his case $\theta = 150^{\circ}$). Observational data were taken from Weaver & Williams (1973), Cohen (1975) and Kerr & Sinha (private communication).

van der Kruit (1971) and Sanders & Wrixon (1973), who treat the "expanding arms" as singular regions in an otherwise normally rotating galactic disk. Other workers have noted that it is possible for non-circular motions about a central bar to reproduce the measured radial velocities for these arms without there being any net outflow of gas from the galactic nucleus. Shane (1972), Simonson & Mader (1973), and Peters (1975) have constructed kinematical models of this kind in which gas moves in elliptical orbits which are closed when viewed from a frame co-rotating with the bar.

The three kinematical models can be used to estimate the radial distribution of HI in the nuclear disk. To avoid the problem of distance ambiguities we consider only the HI emission peak at the highest "permitted" velocity, for which there is no distance ambiguity in any particular model. In general this extreme velocity peak occurs at a non-zero galactic latitude. The path lengths contributing to this emission peak at a given longitude occur at different places along the line-of-sight in the three models and are different in length. Nevertheless the general character of the derived HI distribution is similar in all three cases, as shown in Fig. 2, the mean HI density having a minimum at a distance of 1.5 to 2 kpc from the centre and rising smoothly towards the centre. No evidence is found for the ring of enhanced density at 700 pc from the centre suggested by Rougoor & Oort (1960). Fig. 2 also illustrates the considerable differences between the emission seen on opposite sides of the galactic centre. These differences are as great as the detailed differences between the three estimates of HI density, and make it impossible to distinguish between the kinematical models on this basis alone.

New insight into the structure of the nuclear disk has come from the observations of molecular lines, which sample the regions of higher gas density. The structures which appear most prominent in molecular observations are often inconspicuous in the HI data (and vice versa) because of the great increase in molecular density in the nucleus. When this is allowed for then generally good agreement is found between the HI and molecular data (cf. Cohen & Few 1976, Bania 1977). The molecular data emphasize that most of the neutral gas in the galactic nucleus has strong noncircular motions. Thus it is becoming very difficult to support the Rougoor & Oort model of a nuclear disk in simple rotation: the real situation appears to be much more complex.

REFERENCES

Bania, T.M., 1977. Astrophys.J., 216, pp.381-403.
Cohen, R.J., 1975. Mon.Not.R.astr.Soc., 171, pp.659-696.
Cohen, R.J. & Davies, R.D., 1976. Mon.Not.R.astr.Soc., 175, pp.1-24.
Cohen, R.J. & Few, R.W., 1976. Mon.Not.R.astr.Soc., 176, pp.495-523.
Cohen, R.J. & Davies, R.D., 1978. Mon.Not.R.astr.Soc., in press.
Grape, K. O., 1978. Mon.Not.R.astr.Soc., in press.
Kerr, F.J., 1967. IAU Symp. No. 31, pp.239-251.
Kruit, P.C., van der, 1971. Astr.Astrophys., 13, pp.405-425.
Peters, W.L., 1975. Astr.J., 195, pp.617-629.
Rougoor, G.W. & Oort, J.H., 1960. Proc.Nat.Acad.Sci., 46, pp.1-13.
Rougoor, G.W., 1964. Bull.astr.Inst.Nethl., 17, pp.381-441.
Sanders, R.H. & Wrixon, G.T., 1973. Astr.Astrophys., 26, pp.365-377.
Shane,W.W., 1972. Astr.Astrophys. 16, pp.118-148.
Simonson, S.C. & Mader, G.L., 1973. Astr.Astrophys., 27, pp.337-350.
Weaver, H. & Williams, D.R.W., 1973. Astr.Astrophys.Suppl., 8, pp.1-503.

SOME RESULTS FROM AN HI SURVEY OF THE CENTRAL REGION OF THE GALAXY

R. P. Sinha
Astronomy Program
University of Maryland, College Park, Md 20742, USA

A complete, high-sensitivity survey of HI (Sinha 1978, in preparation) has been examined for model-independent morphological properties of the distribution of HI in the central region of the Galaxy. The dominant symmetry of the kinematics of HI is evident in Figure 1, where equivelocity contours of the permitted velocity edge (corresponding to 1 K antenna temperature) have been plotted on the longitude-latitude plane. The axis of symmetry is coincident with the Kerr-Sinclair ridge of 20-cm continuum radiation from this region. The positions of the maxima of the absolute velocity in the first and the fourth quadrants are aligned along a line inclined to the line of symmetry indicated in Figure 1. The HI gas, therefore, must have noncircular motion. A similar but not identical symmetry is evident in the distribution of the forbidden velocity edge of the profiles. The absolute magnitude of the velocity peaks and the area under the profiles differ in the two quadrants. The nuclear disk appears to be symmetrical in an angle-velocity diagram in which the angle is measured along the line of symmetry in Figure 1.

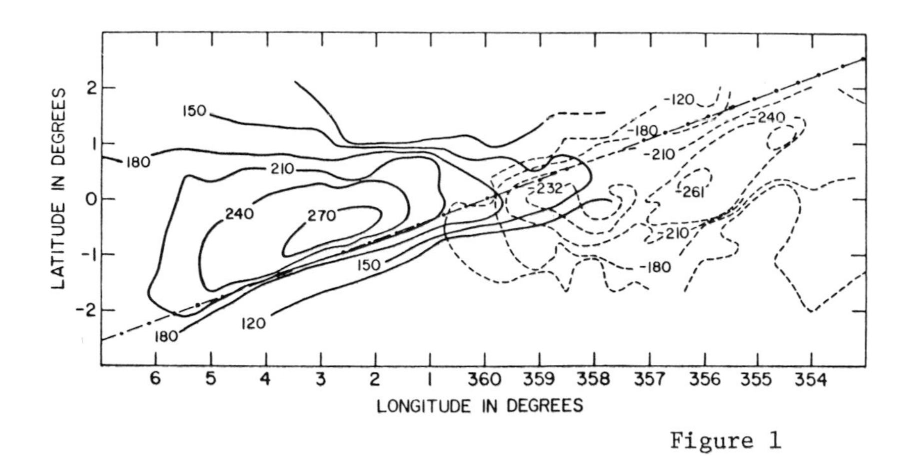

Figure 1

W. B. Burton (ed.), The Large-Scale Characteristics of the Galaxy, 341–342.

DISCUSSION

Baker: You showed a tilted line that reproduced well a tilt in the HI, but you said the tilt was derived from continuum measurements. Which measurements were you using?

Sinha: The tilted structure is visible in the 20-cm continuum map of Kerr and Sinclair (1966) and in the 327 MHz lunar occultation maps of Gopal-Krishna *et al*. (1974).

MOLECULES IN THE INNER FEW KPC OF THE GALAXY

H. S. Liszt and W. B. Burton*
National Radio Astronomy Observatory†, Green Bank, W.Va.,U.S.A.

INTRODUCTION

Emission and absorption spectra of several molecules have been used to trace the kinematic patterns of molecular material in the inner Galaxy. For a variety of reasons, the great majority of effort in this area has focussed on the region $357 \lesssim \ell \lesssim 3°$ very near the plane $b = 0°$. Extending the observational coverage of the molecular distribution away from this plane and over a wider longitude range is a tedious process only now beginning and severely hampered by relatively small beamwidths and long integration times. Our knowledge of the arrangement of molecular material in the inner few kpc of the Galaxy is primitive and quite incomplete compared to that of the atomic gas sampled at λ21 cm.

Because of its ubiquity and because it appears in emission in a suitable atmospheric window, carbon monoxide (CO) is the most readily accessible tracer of molecular material throughout our Galaxy. Comparison of inner-Galaxy spectra shows clearly that other molecular species exhibit only a limited subset of the behaviour apparent in carbon monoxide and, for this reason, our discussion is heavily oriented toward CO observations. Our discussion concentrates on the behaviour of material within about 2 kpc of the galactic nucleus but does not deal with the complicated behaviour of material within the Sagittarius source complex, which is essentially a local phenomenon.

THE MOLECULAR RING

The earliest molecular maps of the inner Galaxy were of the absorption spectra of H_2CO and OH. Instead of revealing a clear signature ascribable to the "rotating nuclear disk" of Rougoor and Oort (1960), they presented an entirely new picture of cold, dense inner Galaxy material

*Present address: Department of Astronomy, University of Minnesota, 116 Church Street S.E., Minneapolis, Minnesota 55455.

†Operated by Associated Universities, Inc., under contract with the National Science Foundation.

W. B. Burton (ed.), The Large-Scale Characteristics of the Galaxy, 343–350.

moving with large outward radial velocities with respect to the galactic center. These absorption measurements culminated in the interpretation of Scoville (1972) and Kaifu, Kato, and Iguchi (1972) which postulated the existence of an expanding ring of molecular material encircling the center of the Galaxy at a distance of ∿250 pc, presumably formed of matter ejected from it. At $\ell = 0°$, $b = 0°$, the ring feature has a velocity (LSR) of -135 km s^{-1}.

Because these absorption measurements necessarily sampled only gas in front of the Sagittarius source complex, the absence of a positive velocity branch of the ring did not itself indicate a lack of circular symmetry. However, the rear portion of the ring was also not present in CO emission spectra (Solomon et al. 1972) and it was not until later that a high - positive-velocity "expanding" molecular feature was detected crossing the plane $\ell = 0°$ at +165 km s^{-1} by Sanders and Wrixon (1974) and by Scoville, Solomon, and Jefferts (1974); these latter authors detected for the first time molecular emission at velocities $\pm$ 200 km s^{-1}, perhaps corresponding to the nuclear disk, and presented a two-arm spiral model to account for their observations.

Both anomalous-velocity molecular features are shown in Figure 1, a longitude-velocity diagram of CO emission in the plane $b = -3'$ extending over the region $358° \leq \ell \leq 2.5°$ with a 2' sampling interval (similar but less extensive or less densely sampled data from this region have been presented by Bania 1977, by Liszt et al. 1977, and by Scoville et al. 1974). Although the similar perceived velocities of the two expanding molecular features are certainly indicative of a fair degree of front-back symmetry in the molecular distribution, the larger-scale kinematic pattern cannot easily be reconciled with the existence of a single ring. The high-positive-velocity feature extends from $\ell \sim -50'$ to $\ell > 145'$ and is not even approximately symmetrically positioned about $\ell = 0°$. The high-negative-velocity gas is more difficult to isolate but probably crosses 0 km s^{-1} at $\ell \lesssim 100'$ which, in the oval locus of an expanding, rotating ring would necessarily be the maximum positive longitude excursion of any emission arising from it (cf. Scoville 1972).

As an alternative explanation of the expanding molecular features, we show in Figure 2 the kinematic arrangement of molecular emission in the plane $b = -3'$ which is generated synthetically by the tilted disk model of the inner Galaxy HI distribution described by Burton and Liszt (1978) and discussed by them in a separate paper in this volume. This model, schematically represented in Figure 3, is characterized by a density variation depending only on distance from its midplane of the disk and by kinematic rotation and expansion motions depending only on perpendicular distance from the disk axis. A more complete discussion of the applicability of this model to the large-scale CO distribution is given by Liszt and Burton (1978). It is clear that the synthetic longitude-velocity arrangement of Figure 2 reproduces most of the observed "ring" characteristics: These include the dissimilar longitude placements of the positive and negative velocity emission branches and their quite different slopes $dv/d\ell$ seen very prominently at positive longitudes. We believe that the presence of two discrete expanding features does not

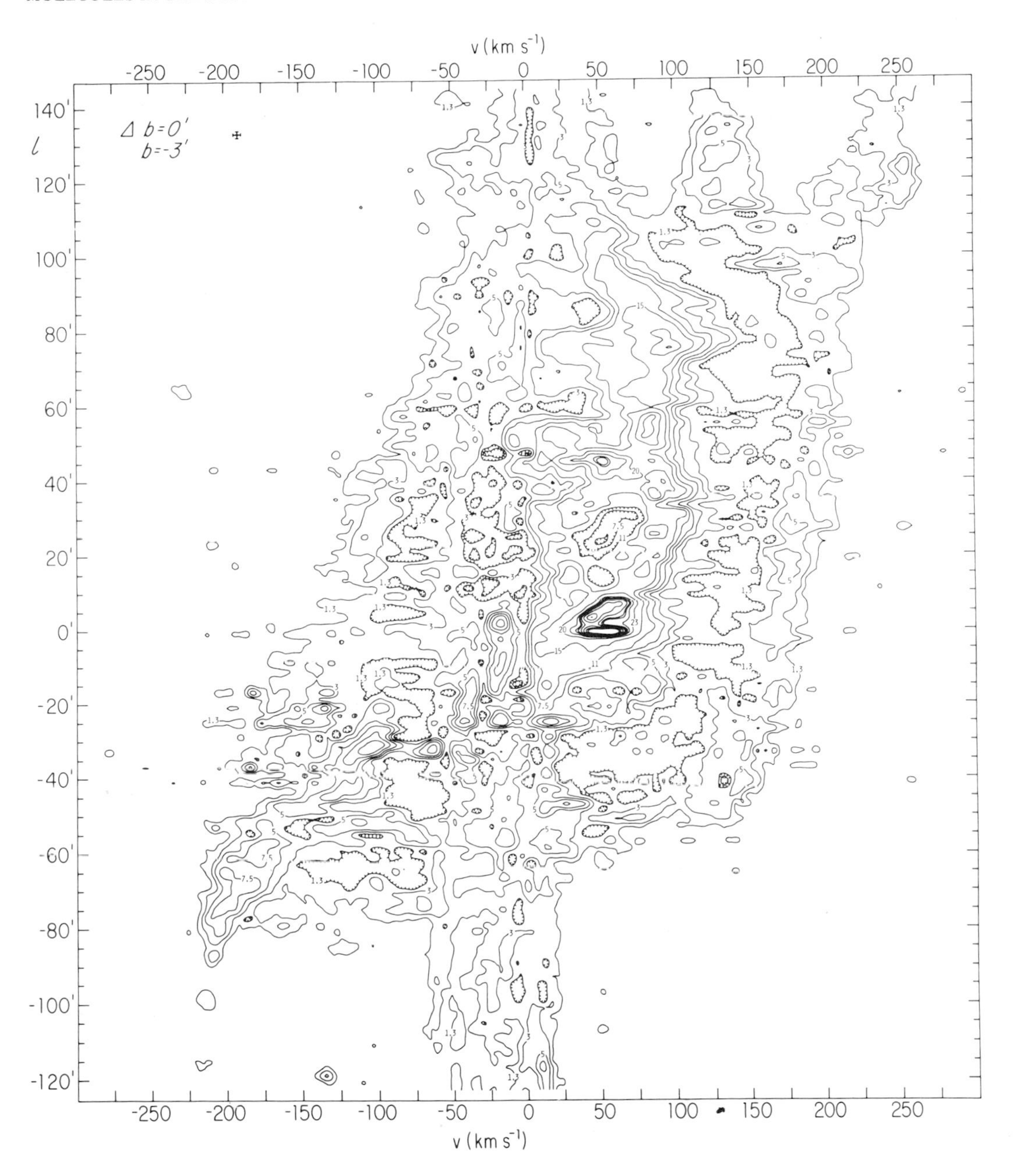

Fig. 1. The longitude-velocity arrangement of λ2.6 mm CO emission at b = -3' taken with a 1' beam at 2' intervals. Intensities are expressed in terms of antenna temperature corrected for atmospheric attenuation, telescope losses and a nominal beam efficiency 0.65 (as also in Figs. 3-5).

necessitate the existence of locatable material bodies, nor do they imply that molecular kinematics are anomalous compared to those of the HI. Rather, projection effects in space and velocity are sufficient to explain their presence in terms of a model which reproduces and lends coherence to a large body of HI observations. One characteristic of the observations which is not accounted for in the synthetic diagram is the non-zero average velocity of the two molecular features near $\ell = 0°$, about +15 km s^{-1}. In terms of the disk model, such a result is obtained only if the kinematic center of the disk is displaced slightly further (10'-15') into the fourth longitude quadrant than is Sgr A.

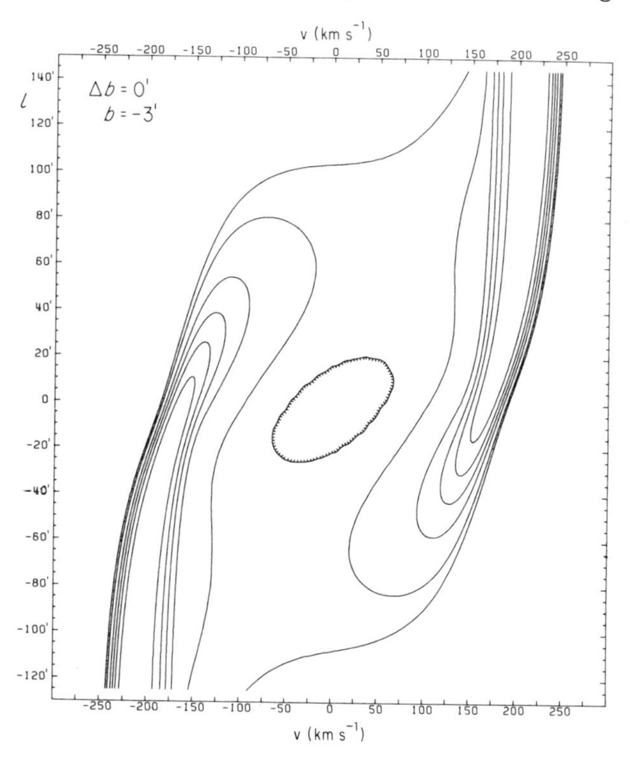

Fig. 2. Synthetic longitude-velocity arrangement of λ2.6 mm CO emission at b = -3' generated by the tilted disk model. Intensity contours are drawn at levels 0.6 K, 1 K, 2 K, 3 K, 4 K, 4.5 K.

THE TILTED NATURE OF THE MOLECULAR DISTRIBUTION IN THE INNER GALAXY

In Figure 4 we show the latitude-velocity arrangement of molecular emission at the longitude of Sgr A (West), ℓ = -3'. It can be seen there that the high positive and negative velocity emission features are symmetrically placed about b = -3' but are not contiguously distributed. Such behaviour is predicted to occur in the context of a flattened general gas distribution (containing expansion motions) whose symmetry axis is tilted out of the plane of the sky in the sense indicated in Figure 3. Below the apparent major axis of such a distribution outwardly directed motions will be perceived as negative and above it as positive. The latitude extent of the molecular emission in Figure 3 results from this tilt, not from the scale height of the molecular distribution. This latter property is only manifested in the small overlap in latitude of the positive and negative velocity features. The "inverted integral sign" signature of Figure 4 gives a clear indication that the molecular distribution is tilted as predicted by the model. If this aspect of the geometry were ignored, inplausibly large shearing effects would be required to reproduce such a pattern. The presence of the predicted model signature in latitude-velocity maps constructed over a range of galactic longitude supports the conclusion that the tilted distribution is filled rather than annular.

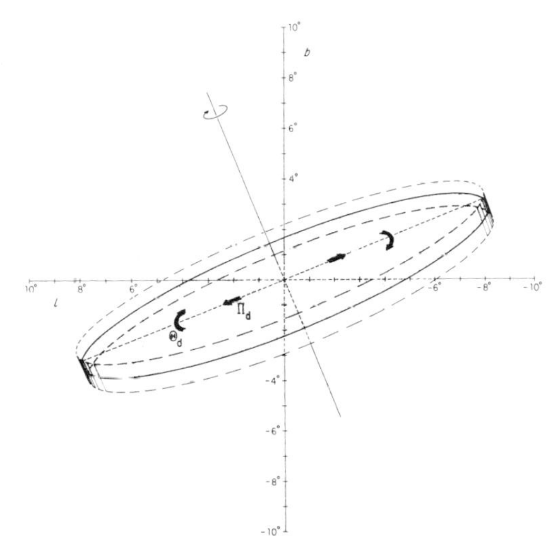

Fig. 3. Appearance of the model tilted gas distribution described by Burton and Liszt (1978) as projected onto the plane of the sky. The solid-line approximate ellipse represents the equatorial plane of a disk of radius 1.5 kpc, whose axis has been tilted through the angles 22° and 12°, respectively, in and out of the plane of the sky. The vectors indicate schematically the model expansion and rotation functions.

Although molecular emission from the inner Galaxy is largely confined at b = 0° and to the region $357 \lesssim \ell \lesssim 3°$, our model predicts that other molecular features should be present at $\ell > 3°$ or $\ell < 357°$ when $b < 0°$ and $b > 0°$, respectively. Figure 5 shows the longitude-velocity arrangement of CO emission at b = -1° made with 20' sampling intervals, along with a similar HI map made with a coarser spacing. The molecular feature observed at high velocities with a large negative slope $dv/d\ell$ is the familiar HI "connecting arm" of Rougoor (1964). As discussed by Burton and Liszt (1978), this arm is well accounted for by our tilted disk model, again without requiring the existence of a separately locatable material body.

Quite generally, CO emission follows the ridge lines present in HI observations and is readily detectable (antenna temperatures above one

Kelvin) whenever the antenna temperature at λ21 cm exceeds ∿5 K. These ridges are formed by projection effects and often represent but a small portion of the total distance over which a given line of sight remains in the disk. Alternatively, in other directions, such as ℓ = 3° in Figure 5, the perceived velocity gradient is large over the entire line of sight and only a very weak and broad (200 km s^{-1}) HI feature is observed. Corresponding emission is not apparent in most of our molecular observations only because the signal-to-noise ratio required to detect it cannot be obtained except after several hours' integration. Nonetheless, it is present in sufficiently long integrations and argues strongly for a molecular distribution which fills fairly uniformly the entire disk volume.

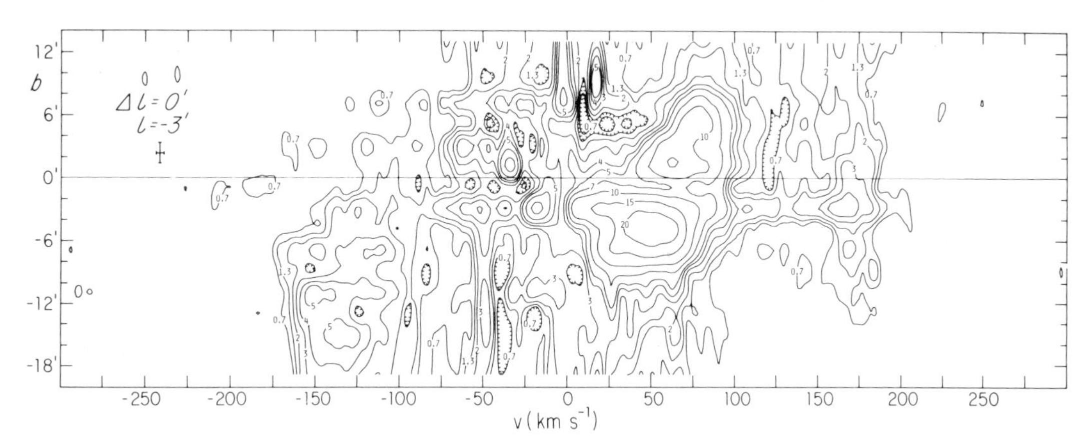

Fig. 4. The latitude-velocity arrangement of λ2.6 mm CO emission at ℓ = -3' taken with a 1' beam at 2' intervals.

PHYSICAL CONDITIONS IN THE INNER GALAXY MOLECULAR GAS

Because the more intense CO emission is optically thick and the relative CO abundance ill-determined, the mass of the molecular constituent of the inner Galaxy gas distribution is necessarily very uncertain. With a radius 1.5 kpc and scale-height 0.1 kpc as for the HI, the molecular mass is related to density in the midplane of the model by $M \sim 7x10^7 \, n_{H_2} \, M_\odot$. Thus even for relatively low molecular densities ∿100 cm^{-3}, the mass of the inner Galaxy material will be large. The crudest estimates of the required densities arise from the condition that collisional excitation of the CO rotation ladder alone be sufficient to produce excitation temperatures as large (5-7 K) as those needed to produce emission lines at levels 2-4 K. At a kinetic temperature of 100 K, this density is of order 100 cm^{-3}. Photon trapping could lower the required density by perhaps as much as a factor 3-4 at this temperature, but no more if the relative CO abundance is limited to $[CO]/[H_2] \lesssim 5x10^{-4}$. Alternatively, one could uniformly clump the molecular gas, but these clumps would be rather different from ordinary galactic molecular clouds which at their most copious have an intercloud spacing of order 1 kpc (Burton and Gordon 1978). Unless the clump separation were much less than even 200 pc, velocity projection effects would not be able to concentrate molecular emission in the ridges that are observed.

As discussed by Liszt and Burton (1978), a consistent clumped model could perhaps lower the mass of gas in the disk to $\sim 10^9$ $M_\odot$. Within 1 kpc of the disk axis, such a model would contain $\sim$4% of the total (stellar) mass derived by Oort (1977) from analysis of the nuclear disk in terms of pure rotation. The inner Galaxy gas is, however, unique in the extent to which the molecular component dominates the atomic gas. Because the HI profiles are well modelled by an optically thin gas (Burton and Liszt 1978), the mass in atomic hydrogen probably does not exceed $\sim 10^7$ $M_\odot$, or 1% of the total gaseous mass.

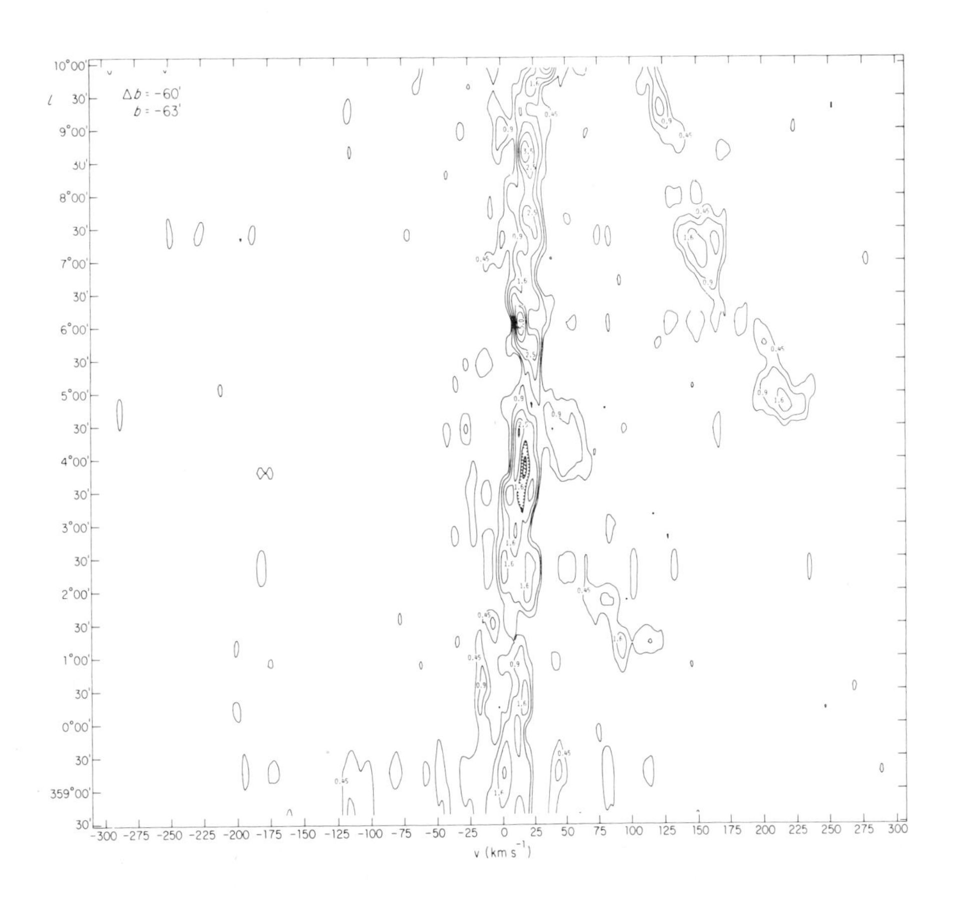

Fig. 5a. The longitude-velocity arrangement of λ2.6 mm CO emission at ℓ = -63' taken with a 1' beam at 20' intervals.

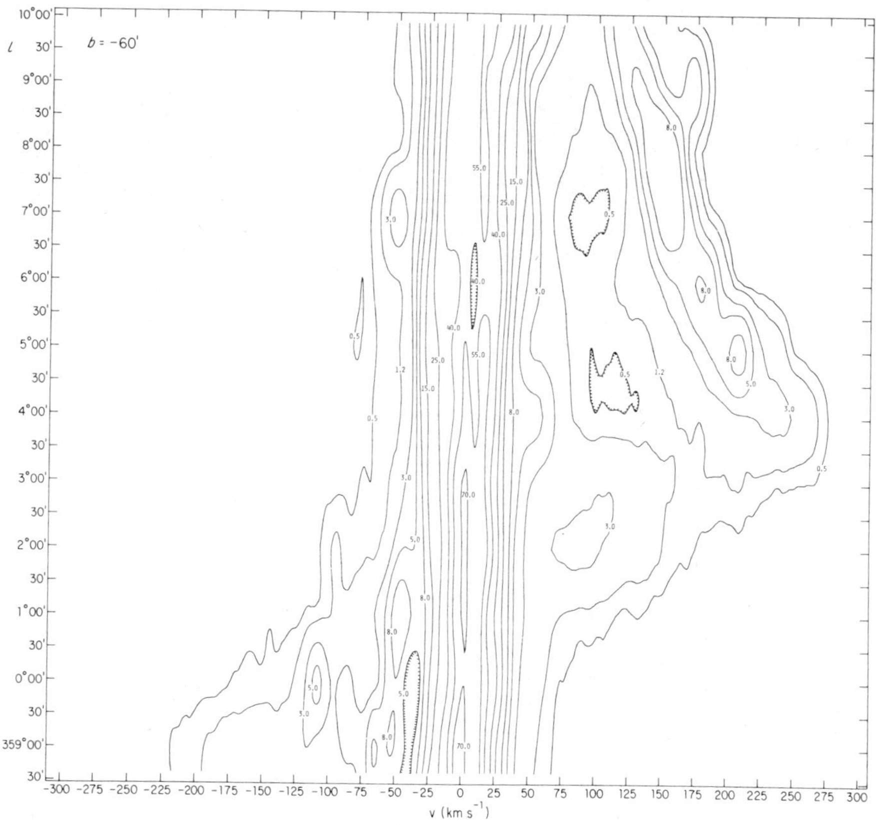

Fig. 5b. The longitude-velocity arrangement of λ21 cm HI emission taken with a 21' beam at 1° intervals. Intensities are expressed in terms of antenna temperature corrected only for losses.

REFERENCES

Bania, T. M.: 1977, Astrophys. J. 216, pp. 381-403.
Burton, W. B., and Gordon, M. A.: 1978, Astron. Astrophys. 63, pp. 7-27.
Burton, W. B., and Liszt, H. S.: 1978, Astrophys. J., in press.
Kaifu, N., Kato, T., and Iguchi, T.: 1972, Nature Phys. Sci. 238, pp. 105-107.
Liszt, H. S., Burton, W. B., Sanders, R. H., and Scoville, N. Z.: 1977, Astrophys. J. 213, pp. 38-42.
Liszt, H. S., and Burton, W. B.: 1978, Astrophys. J., in press.
Oort, J. H.: 1977, Ann. Rev. Astron. and Astrophys. 15, pp. 295-362.
Rougoor, G. W.: 1964, Bull. Astr. Inst. Neth. 17, pp. 381-441.
Rougoor, G. W., and Oort, J. H.: 1960, Proc. Nat. Acad. Sci. 46, pp. 1-13.
Sanders, R. H., and Wrixon, G. T.: 1974, Astron. Astrophys. 33, pp. 9-14.
Scoville, N. Z.: 1972, Astrophys. J. Lett. 175, pp. L127-L132.
Scoville, N. Z.: Solomon, P. M., and Jefferts, K. B.: 1974, Astrophys. J. Lett. 187, pp. L63-66.
Solomon, P. M., Scoville, N. Z., Jefferts, K. B., Penzias, A. A., and Wilson, R. W.: 1972, Astrophys. J. 178, pp. 125-130.

CARBON MONOXIDE AND THE 3-KPC ARM FEATURE

T. M.Bania
Arecibo Observatory, Box 995, Arecibo, Puerto Rico 00612

First isolated using the 21-cm HI spectral line, the 3-kpc arm exhibits the largest non-circular motion of any large-scale gaseous structure in the Galaxy (see e.g. Rougoor, 1964). Carbon monoxide observations best delineate the 3-kpc arm because the CO features are both narrower in velocity and also less confused with emission from more local gas than are HI observations. Because CO is a tracer of molecular hydrogen, the HI and CO data can in principle provide a good estimate of the total gaseous mass of the 3-kpc arm. A previous CO study of the inner Galaxy (Bania, 1977) suggested that, unless it is significantly tilted with respect to b=0°, the 3-kpc arm cannot be a continuous ring structure because emission at extreme positive velocities which should be produced by ring segments lying farther than 10 kpc from the Sun is absent. Consequently, the latitude distribution of CO in the 3-kpc arm has been studied by surveying ^{12}CO emission over the region $350° \leq \ell \leq 25°$ at b=0° and $\pm 0.°33$. The survey positions are separated by $\Delta\ell=1°$. This coarse angular sampling resolution is deemed sufficient because the typical 3-kpc arm CO cloud has a linear dimension of 100 pc (Bania, 1977). Such clouds will subtend ~0.°4 at the maximum line-of-sight distances expected for the extreme positive velocity 3-kpc arm emission (~14 kpc).

CO emission from the 3-kpc arm feature is clearly seen in Figure 1 starting at $(\ell, v)=(-10°, -100\ \text{km s}^{-1})$ and continuing until (~14°, ~20 km s^{-1}). In HI the feature is lost due to confusion with low velocity emission at ℓ~6°. The Figure 1 data are projected onto a common plane by sandwiching the b=0° line profile for each longitude with the $b=\pm 0.°33$ profiles at that longitude. The resulting map will best show any continuity of the 3-kpc feature in (ℓ, v) space but unfortunately will mask any latitude structure. Examination of the individual latitude maps reveals that the 3-kpc arm CO emission is more evident, i.e. both more intense and more continuous, for $\ell<0°$ at b=0.°33; for $\ell>0°$ it is more evident at b=-0.°33. Although suggestive of a tilt, these CO data are too scanty to make any definitive statement.

The average emission characteristics of a 3-kpc arm CO feature imply

W. B. Burton (ed.), The Large-Scale Characteristics of the Galaxy, 351–356.

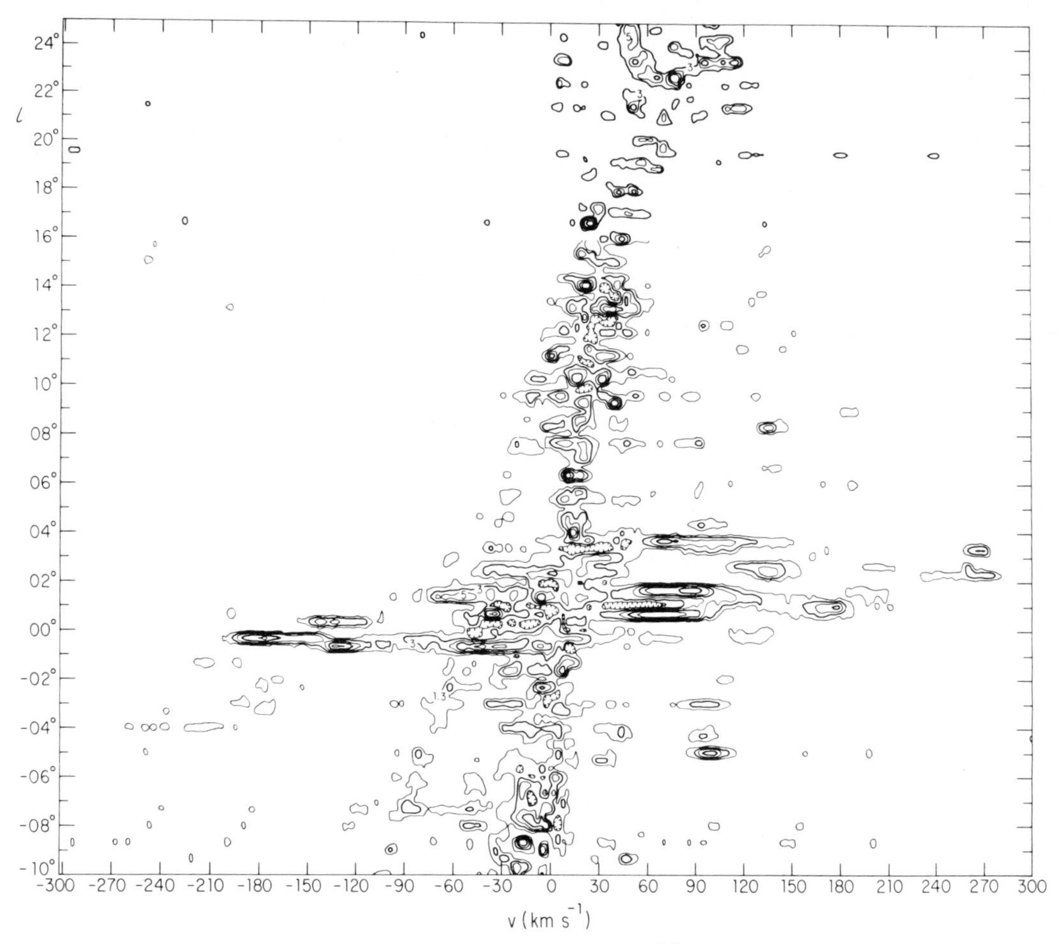

Figure 1: Longitude-velocity contour map of ^{12}CO emission for the region 350°≤ℓ≤25° and for latitudes b=0° and ±0°.33 (see text). The angular sampling resolution of the survey is Δℓ=1° and the LSR velocity resolution is Δv=2.6 km s^{-1}. Contours are drawn at T_A*= 1.3, 3, 5, 7.5, 10, 15, ... K. The lowest contour drawn is 3 times the mean rms noise level. The data quality for ℓ>16° was degraded by weather so for clarity the 1.3 K contour level has been suppressed in this region. The survey data were taken with the N.R.A.O. 11-m telescope on Kitt Peak. The N. R. A. O. is operated by Associated Universities, Inc., under contract with the National Science Foundation.

an H_2 column density of 2.4 x 10^{21} cm^{-2} using the methods given in Bania (1977). Assuming constant density spherical clouds and a 4 kpc radius for the arm, the observed angular size of the CO features, ~0°.8 implies an H_2 cloud mass of 4.7 x 10^5 $M_\odot$ and an H_2 mass of 2.4 x 10^7 $M_\odot$ for a 90° galactocentric arm segment. This is an upper limit because it assumes that the arm is comprised of an unbroken string of clouds. Even

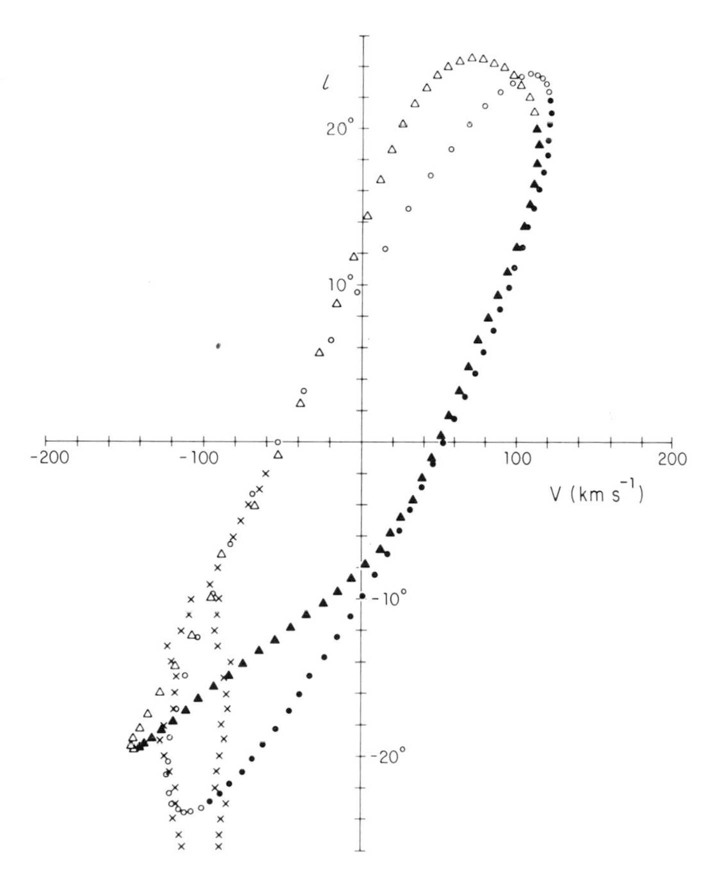

Figure 2: Longitude-velocity loci predicted by two kinematic models: circles show the simple kinematic ring of Cohen and Davies (1976) and triangles display the Simonson and Mader (1973) dispersion ring. Filled symbols represent positions lying farther than 10 kpc from the Sun. Crosses trace the HI emission maxima attributed to the 3-kpc arm. The HI data are from the DTM survey of Burke and Tuve (1963). Note that the observations do not show an obvious tangent point near $\ell \sim 334° - 340°$.

so, the total HI mass, 3.6×10^7 $M_\odot$ (Cohen and Davies, 1976), still exceeds the H_2 mass. This arm segment has a total neutral hydrogen mass, $M(HI) + M(H_2)$, of $\sim 6 \times 10^7$ $M_\odot$ and an expansion energy, $\frac{1}{2}Mv^2$, of $\sim 1.7 \times 10^{54}$ ergs. If the HI and H_2 are uniformly mixed in clouds, the implies cloud mass is 10^6 $M_\odot$. Yet the total mass of the 3-kpc arm still appears to be too small to constrain the explosion models that have been invoked to produce the observed kinematics (van der Kruit, 1971, and Sanders and Prendergast, 1974).

The (ℓ, v) loci predicted by two simple kinematic models for the 3-kpc arm are shown in Figure 3. In each case, the plotted points were calculated for constant intervals in galactocentric azimuth. Thus any crowding in (ℓ,v) space is due solely to the coordinate transformation involved, and, for complete structures of constant density, enhanced emission would be expected in these regions. It is evident from a comparison of Figures 1 and 2 that while the models fit the data well for the region $(\ell,v)<(0,0)$, neither succeeds very well in predicting the CO emission locus anywhere else. In particular, the enhanced emission expected at extreme positive velocities for $\ell>0°$ is still not observed even though the $b=\pm0.^\circ33$ data sample a $|z|$ of ~80 pc at a distance of 14 kpc from the Sun. The HI 3-kpc arm has an observed z extent to half intensity of 90 pc (Cohen and Davies, 1976).

The simple kinematic ring model becomes a tangent to the line-of-sight at $\ell\sim23°$ where there in fact is an intense feature in the CO data. However, the counterpart southern tangent point is not present at $\ell\sim-23°$ as can be seen by the crosses in Figure 2 which trace the HI maxima in the survey of Burke and Tuve (1963). These authors point out that one cannot fit both the southern HI tangent point and the negative velocity (ℓ,v) slope of the 3-kpc arm feature simultaneously.

The CO and HI data suggest, therefore, that the 3-kpc arm is not a uniform circular feature, and in fact may not be at 3 kpc at all. It should be emphasized that we really only know that the 3-kpc arm lies somewhere between the Sun and Sgr A. Its "tangent points" do not occur at symmetric longitudes about $\ell=0°$ and in fact they lie dangerously close to regions in the (ℓ,v) diagram where pseudo-features are naturally produced by the observing geometry. Even a constant density, differentially rotating disk of optically thin gas could produce arcs of enhanced emission in these regions!

REFERENCES

Bania, T. M.: 1977, Astrophys. J. 216, pp. 381-403.
Burke, B. F. and Tuve, M. A.: 1963, I.A.U.Symps. No. 20, pp. 183-186.
Cohen, R. J. and Davies, R. D.: 1976, M.N.R.A.S. 175, pp. 1-24.
Kruit, P. C. van der: 1971, Astron. Astrophys. 13, pp. 405-425.
Rougoor, G. W.: 1977, Bull. Astron. Instit. Neth. 17, pp. 381-441.
Sanders, R. H. and Prendergast, K. H.: 1974, Astrophys. J. 188, pp. 489-500.
Simonson, S. C., III and Mader, G. L.: 1973, Astron. Astrophys. 27, pp. 337-350.

DISCUSSION

de Vaucouleurs: From CO observations only, could you precisely define the longitude of the tangential point (or points) of the "Rougoor (3-kpc) arm"? Does it coincide with the HI tangential point?

Lockman: Can you be certain that the expanding arm is actually seen tangent to the line of sight at any longitude?

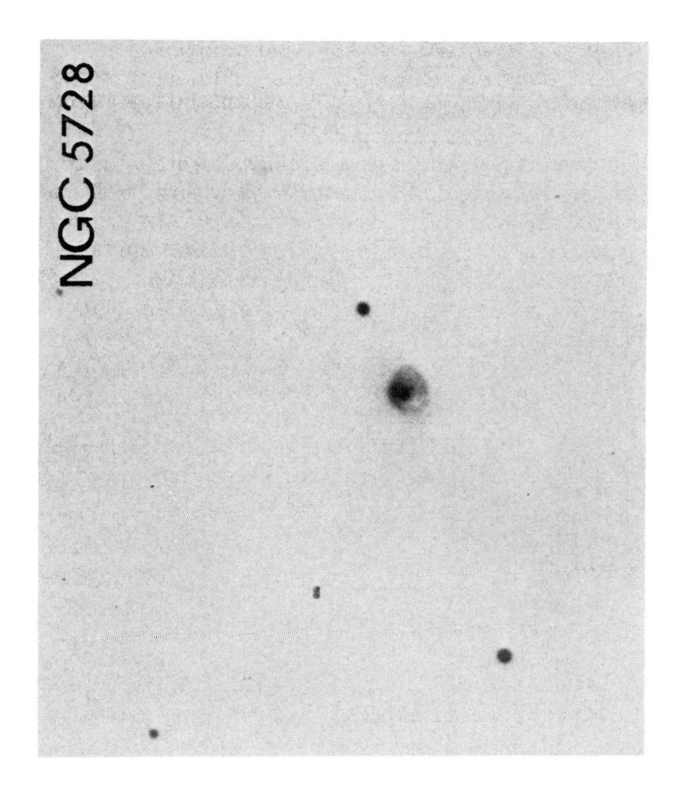

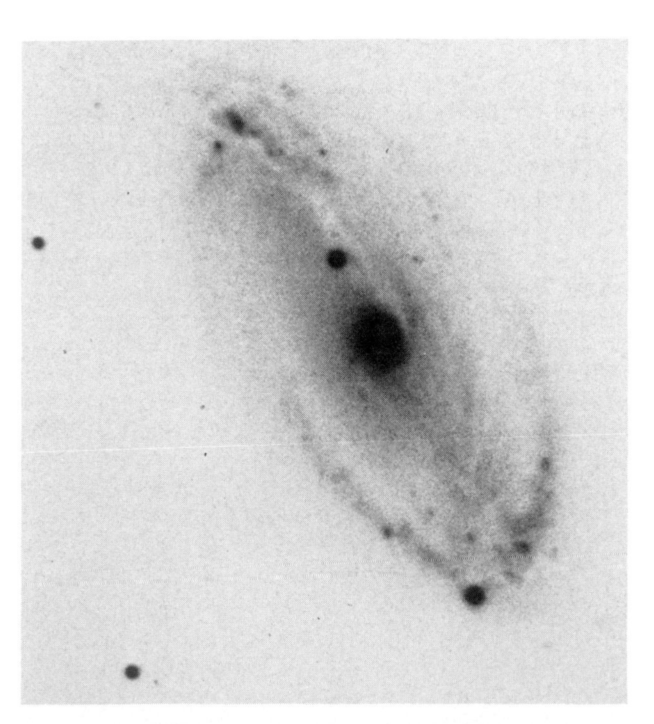

Figure caption: Reproductions of Las Campanas 2.5-m plates of NGC 5728, kindly made available by Sandage. Original scale 10".8/mm; 103a-0 emulsion plus GG 385 filter. (left) exposure 45 min.; (right) exposure 5 min.

Bania: No. Pseudo features in the ℓ,v diagram produced by, for example, a constant density, differentially rotating disk of optically thin gas naturally form arcs in the range $\ell = 15°-25°$ which could be mistaken as tangent points. The proposed tangent points for the 3-kpc arm are not convincing both for this reason and also because these points do not occur at longitudes symmetric about $\ell = 0°$.

Maihara: Do you have any idea about the actual location of the so-called 3-kpc arm?

Bania: It certainly lies somewhere between the Sun and Sgr A. It cannot be much closer than R = 1-2 kpc, though, because the feature subtends too large an angle. One might apply Occam's Razor and suggest that it is associated with the molecular annulus at R = 5.5 kpc. In short, the distance to the 3-kpc arm feature is unknown.

van Woerden: The dispute about the name of the "3- or 4-kpc expanding arm" could be resolved by calling it "Rougoor's arm", after our deceased young colleague who made the pioneering study of it.

Ostriker: The "dispersion" of stellar velocities in the interior part of the Galaxy may be more apparent than real. If our Galaxy has, as I consider likely, a bar of several kpc extent, then streaming motions along the bar (or caused by its rotating non-axisymmetric potential) projected onto the line of sight, might be interpreted as a dispersion. In fact, the young stars and gas might move with the local centroid in these regions as they do near the sun.

Oort: This appears quite possible. It is unfortunate that no proper model based on this idea has been worked out.

Sanders: The "molecular ring"--200 pc from the galactic center--has an oscillation period of about 5×10^6 years (or would have if its motion were just that of particles in a gravitational field). So to keep the molecular ring oscillating, we might require a burst of 10^{56} ergs every 5×10^6 years. Now, if all of this energy came out in a short time, say 10^5 years, and some fraction were radiation, then our Galaxy might be a Seyfert ($L \sim 10^{44}$ ergs s^{-1}) for 2% of the time. All of this, of course, presumes an explosive origin for the non-circular motion of the inner-Galaxy molecular clouds.

Rubin: NGC 5728 is a southern spiral galaxy with a very complex optical nuclear spectrum. Direct blue plates show a nuclear ring, radius $\sim$7", with the nucleus apparently tangent to the ring. Adopting H = 50 km s^{-1} Mpc^{-1}, the radius of the ring is about 2 kpc. I thought these prints would be of interest, because of the likelihood that our Galaxy has a complex spatial structure on this scale.

HII REGIONS AND STAR FORMATION IN THE GALACTIC CENTER

P. G. Mezger and T. Pauls
Max-Planck-Institut für Radioastronomie, Bonn,
Federal Republic of Germany

The centimeter wavelength continuum radiation seen toward the Galactic center (Figure 1) is a mixture of thermal (free-free) and nonthermal (synchrotron) radiation which originates in the nucleus and along the line-of-sight. In this review we discuss only the thermal emission (also see Mezger 1974 and Oort 1977). High-frequency radio continuum and recombination line observations show that the thermal radiation comes from extend, low-density (ELD) HII, and a number of giant "radio HII regions" (see Mezger 1978 for definitions). The approximate half-power contour of the ELD HII (labelled EI in Fig. 1), probably represents a superposition

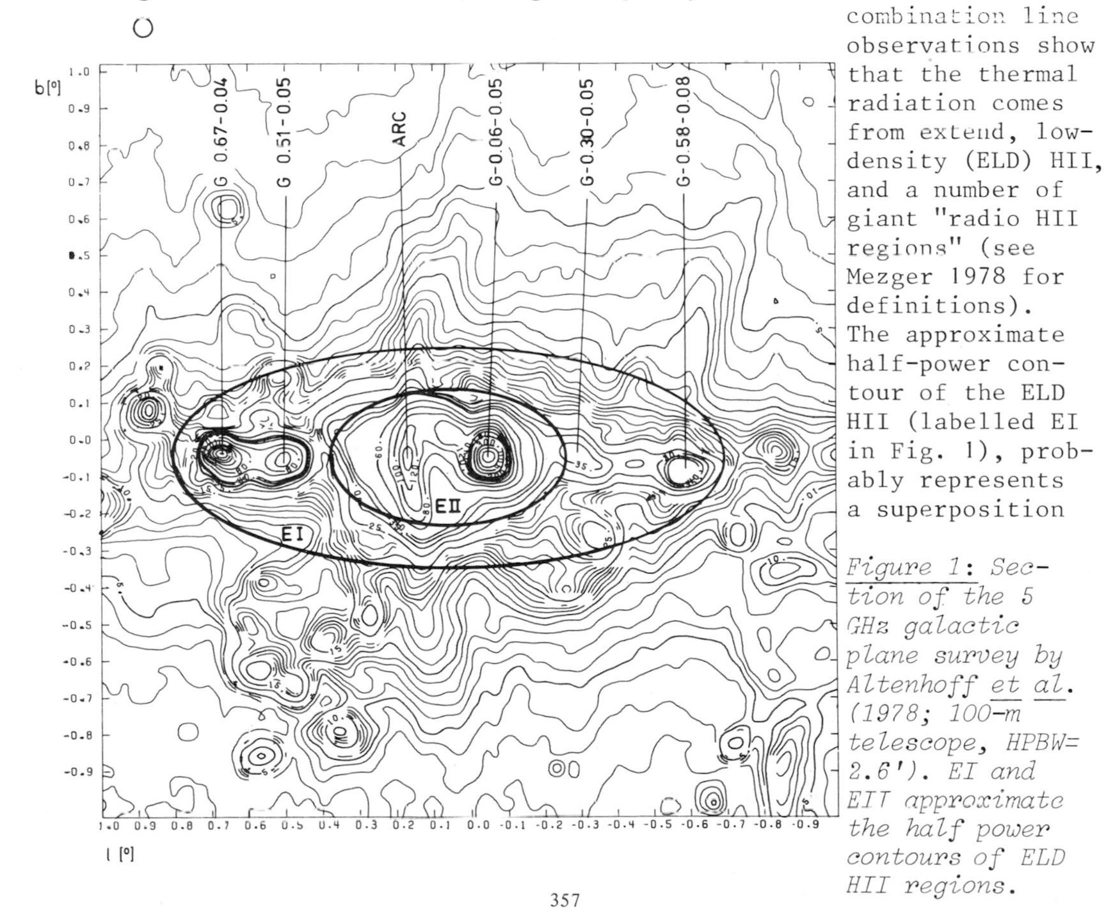

Figure 1: Section of the 5 GHz galactic plane survey by Altenhoff et al. (1978; 100-m telescope, HPBW= 2.6'). EI and EII approximate the half power contours of ELD HII regions.

W. B. Burton (ed.), The Large-Scale Characteristics of the Galaxy, 357–366.

of evolved and expanded HII regions. Thermal radiation outside EI comes predominantly from along the line-of-sight (see Pauls and Mezger 1975).

In the Galactic center, as in the spiral arms, HII regions, IR emission from dust, and molecular clouds are closely associated. These three quantities are plotted in Figure 2 for b = 0° and $|\ell|<5^o$. We see that the free-free and far infrared (FIR) emission correlate well, but the CO is stronger for $\ell>0^o$ and is larger in extent. The latitude distribution is reversed, with the neutral gas forming a thin layer (HPW ∿40 pc (CO), ∿80 pc (HI)) and the FIR and free-free emission more extended (HPW ∿120 pc). This suggests that the ELD HII may be ionization bounded along the galactic plane, but density bounded perpendicular to it.

The Extended, Low Density Ionized Gas. Schmidt (1978a, 1978b) has used continuum observations with the 100-m telescope at 1.7 GHz (HPBW = 7.6'), 2.7 GHz (4.4') and 4.9 GHz (2.6') to separate the thermal and nonthermal radiation in the Galactic center. Assuming S_ν (thermal + nonthermal) $\propto \nu^{-0.1+\alpha}$, Schmidt finds, at 5 GHz, S_5 (thermal) $\sim S_5$ (nonthermal) ∿950 Jy, and α(nonthermal) = -0.9 ± 0.15. Then, from a map of just thermal radiation, Schmidt represents the ELD HII by two spheroidal, constant density components (shown in Fig. 1), whose properties are listed in the Table 1.

The electron temperatures in Table 1 were obtained by comparing (Schmidt 1978b) the thermal continuum with the H166α observations of Kesteven and Pedlar (1977). For ($|b|>0^o$, $\ell = 0^o$) and (b = 0°, $\ell<0^o$) Schmidt (1978b) finds that T_e = 5000 K gives a satisfactory fit to both line and continuum data; but for $\ell>0^o$ the data require T_e = 7500 K. However, since all giant HII regions are at positive longitudes and have T_e ∿6000 - 9000 K, we expect the beam-smoothed electron temperature for $\ell>0^o$ to be higher.

Radio recombination line emission attributed to the ELD ionized gas has been reported by Mezger et al. (1974; 5 GHz), Pauls and Mezger (1975; 5 GHz) and Kesteven and Pedlar (1977; 1.4 GHz). 5 GHz observations at positions away from

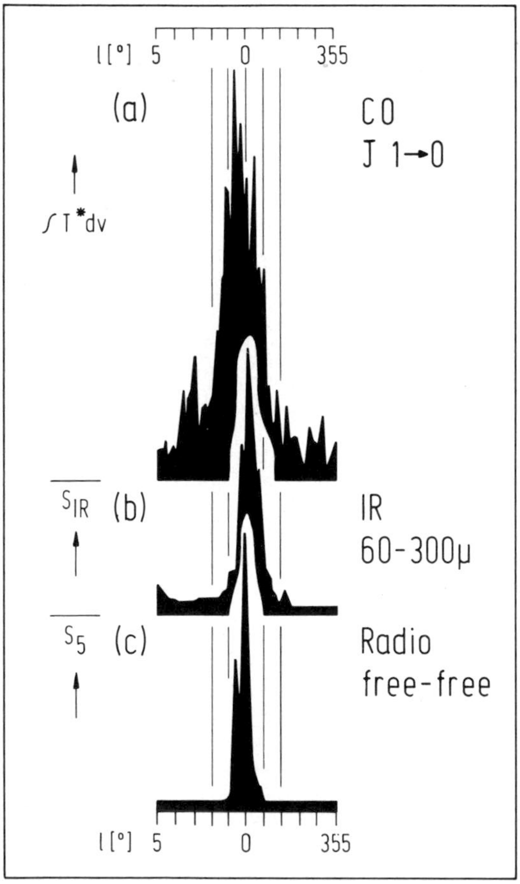

Figure 2: Distribution in the galactic plane of a) integrated CO J=1-0 line profiles (Bania, 1977; HPBW = 1'); b) FIR flux density observed with a 15' beam by Low et al. (1977); c) thermal radio flux density, smoothed to a HPBW of 12' (Schmidt 1978b).

Table 1

	HPW	Diameter of equivalent spheroid	S_5 Jy	T_e K	N'_c s^{-1}	EM cm^{-6} pc	n_e cm^{-3}	M_{HII} $M_\odot$
EI	90'x36'	300 pc x 120 pc	300*	5000	3.6×10^{51}	1.4×10^{4}	7	1.4×10^{6}
EII	38'x22'	130 pc x 80 pc	180	5000	2.2×10^{51}	3.1×10^{4}	16	3.7×10^{5}

* corrected for line-of-sight contribution = ∿150 Jy.

the individual sources yield line widths (FWHM) Δv = 50→100 km s^{-1}, and the systematic longitude variation shown in Figure 3. Also shown in Fig. 3 is the observed rotation curve of the HI nuclear disk (extrapolated to positive longitude). We see that the majority of the ionized gas rotates at a lower velocity than that required for dynamical equilibrium with the gravitational field of the stars near the nucleus. However, the ionized gas at ∿-135 km s^{-1} may be associated with the nuclear disk (see also, Kesteven and Pedlar 1977). For $\ell > 0^o$ the ionized gas lies in the same region of the ℓ-v diagram as the molecular ridge running between Sgr A and Sgr B2 (see Bania 1977, Scoville 1972). This suggests that the extended HII may be simply a collection of evolved HII regions with associated molecular clouds. The large line-widths of the recombination lines, also present in CO and H_2CO, may be a result of rotation and the large path-length over which the emission is observed. Finally, we note that the diffuse ionized gas appears to lie completely inside the "molecular ring" seen in OH, H_2CO and CO (Bania 1977).

The Giant HII Regions shown in Fig. 1 have a total flux density of ∿320 Jy. All compact HII regions are located at positive longitudes and, with the exception of the Arc feature, are similar to giant HII regions observed in spiral arms. The abundance of ionized He in the Galactic center HII regions is very low, typically 1-2% integrated over the source, implying selective absorption of Lyc-photons by dust. This is consistent with an increase of metal abundance and dust-to-gas ratio towards the center of the Galaxy (Churchwell et al. 1978), but the high LTE electron temperatures and nominal infrared excesses (Gatley et al. 1978) do not fit this picture. Apart from the Arc feature, radial velocities of the giant and ELD HII regions correlate well, suggesting a generic relationship as found in the spiral arm region.

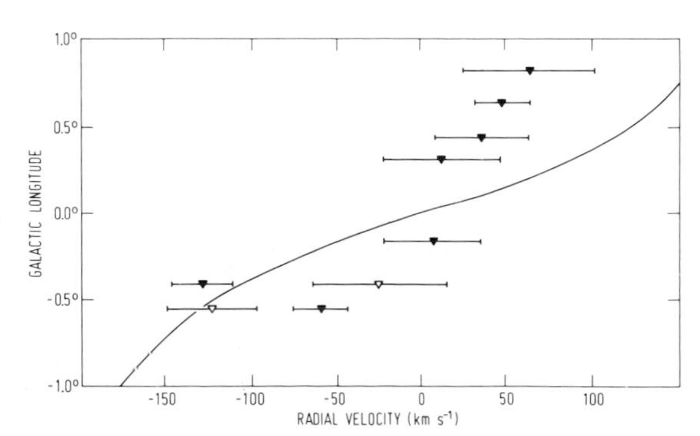

Figure 3: Longitude-velocity diagram of recombination line observations of the ELD HII (filled (dominant component) and open (weaker component) triangles; Mezger et al. (1974, Pauls and Mezger 1975)) and the rotation curve of the HI nuclear disk (extrapolated to positive longitudes; Sanders et al. 1977).

G0.67-0.04 (Sgr B2), G0.51-0.05 and G-0.58-0.08 (Sgr C). These sources are symmetrically located about Sgr A at a projected distance of ∿120 pc. Sgr B2 and G0.51-0.05 are giant HII regions with LTE electron temperatures of 8000 K and 6500 K, respectively. The measured He^+/H^+ ratio in Sgr B2 increases with decreasing beam size (cf. Thum et al. 1978), suggesting different sizes for He^+ and H^+ Strömgren spheres. Sgr C may be a supernova remnant (Downes 1974) since it has a nonthermal spectrum and shows no recombination line emission intrinsic to the source, only lines from the extended HII (Pauls and Mezger 1975). However, Sgr C is a strong far infrared source (Jenning 1975, Low et al. 1977).

The Arc. For $0^o \leq \ell \leq 10'$, there is a complex of radio sources which form an arc-like structure near $b = 0^o$. Radio continuum and recombination line emission from this region have been studied by Pauls et al. (1976), Gardner and Whiteoak (1977) and Pauls and Mezger (1978). The continuum observations of Schmidt (1978a) show an extended region of emission not centered on Sgr A (EII in Fig. 1), while at high frequency (ν>10 GHz) and high resolution (HPBW ∿1') the entire region break ups into discrete sources. One of these, G0.07+0.04, is a strong FIR source (second only in intensity to Sgr A at 69μm) and Gatley et al. (1978) argue that it may be a site of current star formation. Using the 100-m telescope at 5 GHz (HPBW = 2.6'), we have unsuccessfully searched for He^+ in G0.07+0.04 (He^+/H^+ <0.01). This result is consistent with the large infrared excess of the source. At 5 GHz, T_e (LTE) = 8000 K.

The ionized gas emits primarily at $V_{LSR} = -40 \pm 20$ km s^{-1} and most of this emission arises north of $b = 0^o$. Observations of HCN indicate that the molecule clouds are anti-correlated in velocity and position with the ionized gas. The maximum HCN emission is in the range $+15 \leq V_{LSR} \leq +80$ km s^{-1} and lies south of $b = 0^o$ (Fukui et al. 1978).

Sgr A. The radio source Sgr A, associated with the Galactic nucleus, consists of a thermal component (Sgr A West), a nonthermal component (Sgr A East), a halo (diameter ∿6') of thermal + nonthermal emission (see Pauls et al. 1976), and a point-like source (cf. Ekers et al. 1975). Some of the halo emission is thermal because: (1) radio recombination lines are seen within a radius of ∿3' of Sgr A; (2) the low-frequency turnover in the continuum spectrum of Sgr A requires ionized gas with an emission measure of $\sim 4 \times 10^4$ cm^{-6} pc. The sources EI and EII can only provide about 50-60% of this emission measure, suggesting that the remainder comes from HII near Sgr A.

The recombination line emission from Sgr A has recently been summarized by Pedlar et al. (1978). These authors show that there

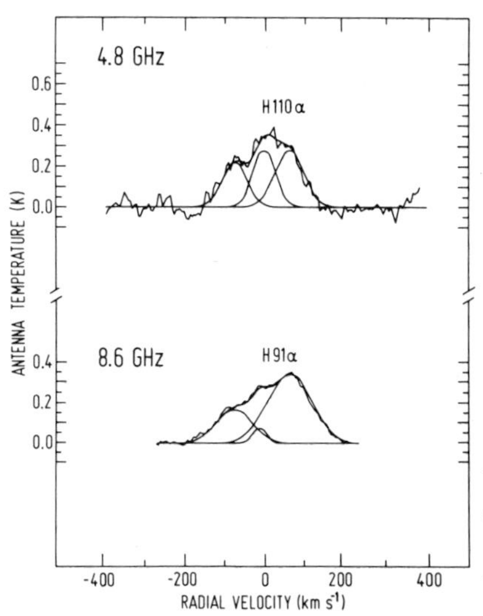

Figure 4: Radio recombination line spectra toward Sgr A West. HPBW (H110α) = 2.6'; HPBW (H91α) = 1.5'.

is weak masering of the line emission which sets in at $\nu \sim 1.4$ GHz and increases as ν decreases (see also Casse and Shaver 1977). Above 5 GHz, Sgr A West emits very broad recombination lines (Pauls et al. 1974). H110α and H91α spectra (Figure 4) show three components with LSR velocities near 0 km s^{-1} and $\sim\pm 75$ km s^{-1}. The line near 0 km s^{-1} probably arises in HII in the halo around Sgr A West and Sgr A East; then, the increased intensity at H110α may be explained by stimulated emission from Sgr A West. The observed line-widths are larger than those from any other HII region in the Galaxy. Similar velocities and velocity dispersions are found from observations of NeII (Wollman et al. 1976, 1977).

Infrared observations show that the inner $\sim$30" of Sgr A West contains a number of compact sources plus more extended emission (Becklin et al. 1978a, b; Willner 1978; Neugebauer et al. 1978; Rieke 1978). Some of the compact sources appear to be O stars or star clusters. The extended emission seen in the middle to FIR seems likely to be dust heated by hot stars. These facts suggest that star formation is currently taking place within the inner parsec of the Galaxy.

The Origin of the FIR Radiation from the nuclear region is generally accepted to be due to dust grains heated by stellar radiation. However, it is still not clear if these dust grains are mainly located in dense molecular clouds or in ionized gas.

Figure 5 shows on overlay of the main contours of a FIR map (40-350μm) by Alvarez et al. (1974) on the 5 GHz map of the thermal radiation constructed by Schmidt (1978a). This map confirms the close correlation already seen in Fig. 2. We have integrated radio and IR maps in circles of radius R centered on Sgr A West; for $\lesssim 8'$ we used the high resolution observations by Ekers et al. (1975), Pauls et al. (1976) and Harvey et al. (1976).

In Fig. 6 we plot, as function of radius R, (1) the IR luminosity L_{IR}; (2) the total luminosity of PopII stars from eq. (5) of Sanders and Lowinger (1972); (3) the total luminosity of PopI stars, estimated by converting S_5 into N_c', the number of Lyc-photons absorbed by gas, and subsequently into $N_c = \{f_{net}(1-f)\}^{-1} N_c'$, the number of Lyc-photons emitted by all O stars contained in a cylinder of radius R. We used $f_{net}(1-f) = 0.33$ (see following section). Then $5.75\ N_c h\nu_\alpha$ is the total luminosity of stars of all masses which have been

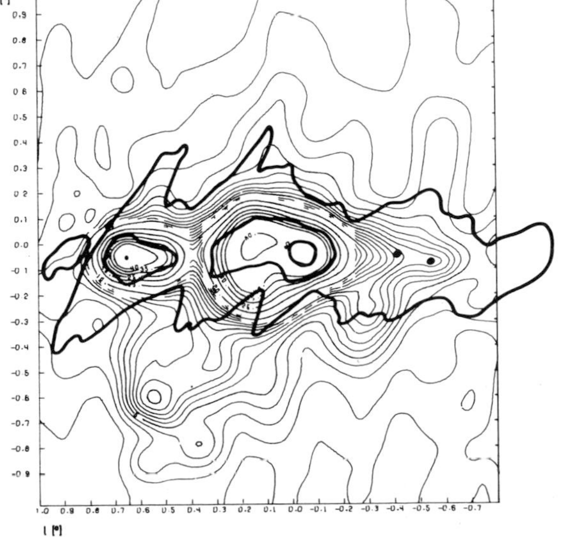

Figure 5: Overlay of the IR map of the galactic center region obtained by Alvarez et al. (1974; 40-350μm; HPBW = 5.6') on the map of thermal radio emission constructed by Schmidt (1978a; HPBW = 7.6').

formed together with the O stars, *if* Salpeter's original luminosity function holds.

Figure 6a shows that both PopI and PopII stars provide enough energy to account for the observed FIR radiation. However, (1) the correlation between thermal radio and IR emission (Figs. 2 and 5); (2) the fact that for well mixed PopI and II stars the effective (Planck mean) absorption optical depths for PopI radiation is larger; and (3) the fact that 3-color photometry by Gatley et al. (1977) shows molecular clouds in the nucleus as "cold" and ionized gas as "hot" regions, indicates that the contribution of PopI stars to the heating of dust is stronger than in the spiral arms. In the spiral arms Mezger (1978) has estimated that old and young stars contribute equally to the heating. Within $R \lesssim 150$ pc we estimate that PopI stars contribute 2/3 of the energy which heats the dust.

The infrared excess, (IRE) = $L_{IR}/N_c' h\nu_\alpha$, is related to the fraction of Lyc-photons absorbed directly by dust. This quantity is shown in Fig. 6b. The low values for $R \lesssim 5'$ show that the ionized gas may be dust depleted beyond Sgr A West (Mezger 1974; Gatley et al. 1977).

The Star Formation Rate. The number of Lyc-photons absorbed per sec by gas, N_c' is related to the Lyc-photon production rate N_c of the ionizing O stars by $N_c' = f_{net} N_c$ for ionization bounded HII regions, and by $N_c' = f_{net}(1-f) N_c$ for density-bounded HII regions. Here $(1-f_{net})$ is the fraction of Lyc-photons absorbed by dust and f is the fraction of Lyc-photons which escape the density bounded HII region.

HII Regions	ΣS_5/Jy	T_e/K	$\Sigma N_c'/s^{-1}$	f_{net}	(1-f)	$\Sigma N_c/s^{-1}$
giant	320	8000	3.1×10^{51}	0.3	1.0	1.0×10^{52}
ELD	480	5000	5.8×10^{51}	0.61	0.57	1.7×10^{52}

The table above gives N_c' for compact and ELD HII regions based on the values S_5 and T_e listed. For the compact HII regions we estimate $f_{net} = 0.3$; an average value derived from IR luminosity, He^+-abundance and the assumptions used by Smith et al. (1978). For the ELD HII region we estimated the values f_{net} and (1-f) given in the table above from an analysis similar to that used by Mezger (1978) for the spiral arm ELD HII region.

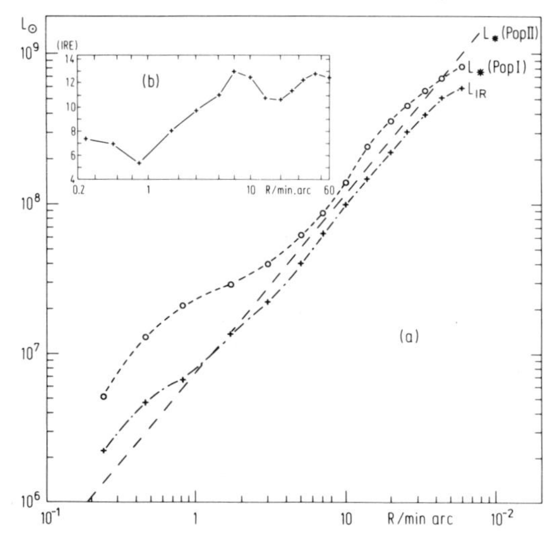

Figure 6: 6a: Luminosities, integrated within cylinders of radius R, of i) the IR emission L_{IR}; ii) the radiation of old (PopII) stars L_* (PopII); iii) the radiation from O star clusters L_* (PopI). 6b: The IR excess, (IRE) = $L_{IR}/N_c' h\nu_\alpha$.

The total Lyc-photon production rate in the nuclear region ΣN_c (giant + ELD HII) = 2.7×10^{52} s^{-1}, or $\sim$8% of the total Lyc-photon production rate in the Galaxy, $(30+2.7)10^{52}$ s^{-1} (Mezger 1978). If the stellar birth rate function is constant within the Galaxy, a similar ratio must hold for the star formation rates. The Arc feature may not be ionized by O stars; in this case the nuclear star formation rate would be $\sim$7%.

Mezger (1974) has shown that the Lyc-photon production rate by nuclei of planetary nebulae is negligible.

Finally, although the estimates of L_{IR} and N_c are not quite independent, it is of interest to compare the integrated luminosity from the map by Alvarez et al., 6×10^8 $L_\odot$, with the IR luminosity of the spiral arm region estimated by Mezger (1978) on the basis of observations by Low et al. (1977), $\sim 6 \times 10^9$ $L_\odot$. The corresponding ratio, $\sim$9%, is also a roug estimate of the star formation rate in the nuclear and spiral arm regions

Summary. Ionized gas extends from the center of the Galaxy out to a radius R $\sim$150 pc, while the molecular clouds extend to R $\lesssim$300 pc. Within this radius, the total stellar mass is $\sim 1.5 \times 10^9$ $M_\odot$, the total gas mass is $\sim 1.2 \times 10^7$ $M_\odot$ of which $\sim 2 \times 10^6$ $M_\odot$ is ionized. Most of the neutral gas is in molecular clouds which probably occupy less than 5% of the interstellar space.

With the possible exception of the Arc feature, the giant HII regions indicate recent, large-scale star formation. The extended, low-density HII appears to be evolved HII regions whose Strömgren spheres have merged. Between 25% (Arc feature not included) and 37% (Arc feature included) of the O stars are contained in compact HII regions; in spiral arms the corresponding ratio is 18%. This may indicate that the present star formation rate is somewhat higher than the star formation rate averaged over the past 5×10^6 yr. O stars appear to contribute most to the heating of dust grains which emit in the FIR.

The nucleus of the Galaxy consists of the compact HII region Sgr A West, which contains a cluster of O stars, surrounded by more extended HII of lower density (halo). Within a radius of $\sim$10 pc the gas appears to be depleted of dust. Sgr A West as well as the thermal halo emit very broad recombination lines; the mechanism for the broadening of the lines is not understood.

The He^+-abundance in nuclear giant HII regions is extremely low. This could be due to an increased metal abundance, which would effect the Lyc-photon radiation field through 1) increased stellar opacities, 2) and/or increased dust-to-gas ratio, and/or 3) selective absorption by dust grains. However, LTE electron temperatures of nuclear giant HII regions are higher than those in spiral arm HII regions; this argues against an increased metal abundance in the center of the Galaxy.

Acknowledgements. We thank N. Panagia for very helpful discussions on the interpretation of the infrared data, and J. Schmidt for making his results available prior to publication.

REFERENCES

Altenhoff, W. J., Downes, D., Pauls, T., and Schraml, J.: 1978, Astron. Astrophys. Suppl., in press.

Alvarez, J. A., Furniss, I., Jennings, R. E., and King, K. J.: 1974, in HII Regions and the Galactic Centre, Proc. Eighth ESLAB Symp., ed. A.F.M. Moorwood (Neuilley-sur-Seine: ESRO), p. 69.

Bania, T. M.: 1977, Astrophys. J. 216, 381.

Becklin, E. E., Matthews, K., Neugebauer, G., and Willner, S. P.: 1978a, ibid 219, 121.

________: 1978b, ibid 220, 831

Casse, J. L., and Shaver, P. A.: 1977, Astron. Astrophys. 61, 805.

Churchwell, E., Smith, L. F., Mathis, J., and Mezger, P. G.: 1978, ibid in press.

Downes, D.: 1974, in HII Regions and the Galactic Centre, Proc. Eighth ESLAB Symp., ed. A.F.M. Moorwood (Neuilley-sur-Seine: ESRO), p. 247.

Ekers, R. D., Goss, W. M., Schwarz, U. J., Downes, D., and Rogstad, D. H.: 1975, Astron. Astrophys. 43, 159.

Fukui, Y., Iguchi, T., Kaifu, N., Chikada, Y., Morimoto, M., Nagane, K., Miyazawa, K., and Miyaji, T.: 1977, Pub. Ast. Soc. Japan 29, 643.

Gardner, F. F., and Whiteoak, J. B.: 1977, Proc. Ast. Soc. Australia 3, 150.

Gatley, I., Becklin, E. E., Werner, M. W., and Wynn-Williams, C. G.: 1977, Astrophys. J. 216, 277.

Gatley, I., Becklin, E. E., Werner, M. W., and Harper, D. A.: 1978, ibid 220, 822.

Harvey, P. M., Campbell, M. F., and Hoffmann, W. F.: 1976, ibid 205, L69.

Jennings, R. E.: 1975, in Proceedings of the Symposium on HII Regions and Related Topics, eds. T. L. Wilson and D. Downes (Springer: Berlin), p. 137.

Kesteven, M. J., and Pedlar, A.: 1977, Mon. Not. Roy. Ast. Soc. 180, 731.

Low, F. J., Kurtz, R. F., Poteet, W. M., and Nishimura, T.: 1977, Astrophys. J. 214, L115.

Mezger, P. G.: 1974, in Proceedings of ESO/SRC/CERN Conference on Research Programs for the Large Space Telescopes, ed. A. Reig (Geneva: ESO), p. 79.

Mezger, P. G.: 1978, Astron. Astrophys., in press.

Mezger, P. G., Churchwell, E. B., and Pauls, T. A.: 1974, in Stars and the Milky Way System, ed. L. N. Marvridis (Berlin - Heidelberg - New York: Springer), p. 140.

Neugebauer, G., Becklin, E. E., Matthews, K., and Wynn-Williams, C. G.: 1978, Astrophys. J. 220, 149.

Oort, J. H.: 1977, Ann. Rev. Astr. Ap. 15, 295.

Pauls, T., Mezger, P. G., and Churchwell, E.: 1974, Astron. Astrophys. 34, 327.

Pauls, T., and Mezger, P. G.: 1975, ibid 44, 259.

Pauls, T., Mezger, P. G.: 1978, in preparation.

Pauls, T., Downes, D., Mezger, P. G., and Churchwell, E.: 1976, Astron. Astrophys. 46, 407.

Pedlar, A., Davies, R. D., Hart, L., and Sh-ver, P. A.: 1978, Mon. Not. Roy. Ast. Soc. 182, 473.

Rieke, G. H., Telesco, C. M., and Harper, D. A.: 1978, Astrophys. J. 220, 556.
Sanders, R. H., Wrixon, G. T., and Mebold, U.: 1977, Astron. Astrophys. 61, 329.
Schmidt, J.: 1978a, dissertation, University of Bonn.
Schmidt, J.: 1978b, in preparation.
Scoville, N. Z.: 1972, Astrophys. J. 175, L127.
Smith, L. F., Biermann, P., and Mezger, P. G.: 1978, Astron. Astrophys. in press.
Thum, C., Mezger, P. G., Pankonin, V., and Schraml, J.: 1978, ibid 64, L17.
Willner, S. P.: 1978, Astrophys. J. 219, 870.
________: 1977, ibid 218, L103.

DISCUSSION

Terzian: Although O-type stars may be present in the galactic nucleus which may contribute a large fraction of the ionizing radiation, we should also include the important contribution of ionizing radiation from the central stars of planetary nebulae, because the number density of planetary nebulae in the galactic central region is very high.

Mezger: We considered this using your own estimates of the characteristics of PN: these contribute less than 10% of the Lyc-photon production rate in the nuclear region.

Sanders: From your recombination line observations, you can isolate, kinematically, those components of the thermal continuum emission which arise in the inner part of the Galaxy. If you do this, can you say whether or not the HII seems to be distributed in a tilted disk?

Mezger: The distribution of the ELD ionized gas shows in the inner part a remarkable spherical symmetry, whose center, however, does not coincide with Sgr A West but is somewhat shifted towards $\ell > 0°$. Further out, the ELD ionized gas is elongated parallel to the galactic plane but its plane of symmetry lies at $b < 0°$.

Puget: In the far-infrared data which you mentioned, there is evidence for variation in the dust temperature which might be relevant to the question of the region of origin of this radiation: in HII regions or in molecular clouds. Except for the central peak the evidence suggests that most of the far-infrared emitted within 300 pc of the galactic center comes from molecular clouds.

Mezger: I do not agree. All the observational evidence which I have presented indicates that the main contribution to the FIR emission at wavelengths $\lambda < 300\mu$ comes from dust particles embedded in ionized gas which are heated by radiation from O stars.

Schmidt-Kaler: In the last passages of your talk you compared the nuclear region with the whole Galaxy. I have three questions: (1) What were the initial mass functions you used? (2) What is the gas turn-over rate ($M_{\odot}$/year) as a consequence of the stellar evolution (and formation) processes which you described? (3) What is the ensuing chemical enrichment in the nuclear region compared to the whole Galaxy?

Mezger: (1) I derived Lyc-photon production rates. If these are to be converted into star formation rates one needs an IMF. (2) If I assume that the IMF is constant throughout the Galaxy and similar to Salpeter's IMF the star formation rate would be about 0.3 $M_{\odot}$/year in the nuclear region and the instantaneous return rate would be about 0.08 $M_{\odot}$/year. (3) At the moment I can only make the qualitative statement that Z(nuclear region) $> Z_{\odot}$, but I cannot say it is three or six times larger.

Sinha: Dr. F. J. Kerr, Dr. R. W. Hobbs, and I have mapped a 5'x5' area around Sgr A at 90 GHz. In the map shown in the adjacent figure, the Sgr A East component can be separated from the Sgr A West source. The ratio of the flux of the two components is equal to the ratio expected from the 6-GHz map of Ekers et al. (A.&A. 43, 159) after proper corrections for the different beam sizes are made. The Sgr A East component does not appear to have a non-thermal spectrum between 6 and 90 GHz.

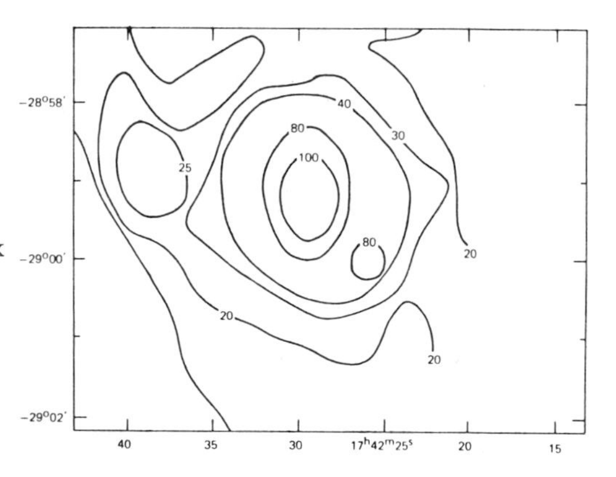

INFRARED OBSERVATIONS OF THE NUCLEAR REGION

E. R. Wollman
Kitt Peak National Observatory[1)]

ABSTRACT

Infrared observations have provided considerable information about the structure and energetics of the galactic center. The stellar bulge dominates the mass and luminosity of the nuclear region. The luminosity of the bulge is strongly peaked toward the center, even within the central parsec. Dust in the nuclear disk absorbs the power output of the central portion of the bulge and reemits it in the far infrared. Near the center, molecular clouds move in a plane apparently tilted toward the Sun. Within the central few parsecs, or core, the inclination may be as large as 45°. The total power output of the core is about twice that of the bulge population alone. The source of excess luminosity is uncertain, but evidence points to ongoing star formation associated with the Sgr A molecular complex.

1. INTRODUCTION

Becklin and Neugebauer (1968) published the first infrared studies of the center of our Galaxy. Since then, virtually every new development in infrared astronomical instrumentation has been applied in the search for hints about galactic nuclear structure and energetics. The results teach us a great deal about both the galactic center and the capabilities of infrared astronomy.

In describing the galactic center, the nomenclature of Oort (1977) is adopted. The large-scale structures of the nuclear region are the stellar bulge and the gaseous nuclear disk. The disk is about 1500 pc in diameter. Strong infrared emission from dust in the nuclear disk is limited to the region inside $r \simeq 200$ pc. Within this radius, the gas is primarily either molecular or ionized. Molecular clouds lie close to the galactic plane ($\Delta z \simeq 30$ pc), whereas the layer of ionized gas is

1) Operated by the Association of Universities for Research in Astronomy, Inc., under contract with the National Science Foundation.

W. B. Burton (ed.), The Large-Scale Characteristics of the Galaxy, 367–375.

much thicker ($\Delta z \simeq 100$ pc at the center). Since molecular clouds occupy a small fraction of the volume of the region, this inner ionized/molecular portion of the nuclear disk will be referred to as the H II disk.

The central few parsecs is called the core. Its most distinguishing infrared feature is 10μ continuum emission from warm dust associated with the dense ionized gas of the Sgr A West thermal H II region. In any volume about the center significantly larger than the core, the stellar bulge accounts for nearly all the luminosity. Within the core, the total power output is about twice that of the bulge population alone. The source of excess luminosity is uncertain, but various evidence points to star formation.

Almost all the infrared radiation from the nuclear region is optically thin thermal emission from dust. This dust enshrouds the central portion of the bulge and soaks up all the energy radiated at wavelengths shorter than about 1μ. This energy is reradiated in the mid- and far-infrared, a wavelength region in which the properties of the emitting dust are not well known. The small amount of undegraded near-infrared radiation that does penetrate the nuclear dust encounters a comparable or larger quantity of dust in the outer disk. By the time it reaches us, a near-infrared spectrum, which might inform us about some hot source at the center, is so heavily reddened that its shape reflects the shape of the extinction more than that of the emission. Fortunately, improvements in sensitivity and resolution during the past five years have allowed the identification of some luminosity sources on the basis of unambiguous spectral features. Examples are CO-band absorption (→ red giant) and ion fine-structure line emission (→ H II region). These identifications in turn provide checks on the assumptions about extinction. The result is a picture with enough internal consistency to be moderately convincing. Some fundamentals and many details are missing, but the past decade of investigations has provided a solid framework within which the next generation of questions can be formulated.

It is impossible to acknowledge here all the contributions that have been made to the present understanding of the infrared emission from the galactic center. However, several recent papers of broad scope deserve special mention. These can direct the reader to earlier literature. A series by Becklin, Neugebauer, and collaborators (Becklin et al. 1978a,b; Willner 1978; Neugebauer et al. 1978) discusses the core in detail. Rieke et al. (1978) review both the near- and mid-infrared structure of the core and the far-infrared emission from the immediate surroundings. Their paper also provides a rather complete guide to the literature. Gatley et al. (1977) investigate the far-infrared emission from the central few tens of parsecs with attention to determining the far-infrared properties of dust in the region. Krügel and Tutukov (1978) have constructed a model for infrared radiation from the galactic center which incorporates the known essential features and serves as a guide to the general distribution of sources and luminosity throughout the region. Finally, a complete review of both radio and infrared observations of the galactic center is given by Oort (1977).

2. THE BULGE

The bulge accounts for nearly all the mass and luminosity in the nuclear region. Because of the contribution of red giants, the bulge is bright at 2μ, where the intervening extinction is only a few magnitudes. The 2μ brightness distribution has been mapped extensively (Becklin and Neugebauer 1968, 1975, 1978; Ito et al. 1977; Maihara et al. 1978; Oda et al. 1978). It is elongated approximately along the plane and is strongly peaked toward the center. Even within the central parsec, the bulge appears to be an identifiable, centrally condensed structure.

From the distribution of 2μ surface brightness, one might hope to extract the mass distribution. However, the brightness distribution is modified to an uncertain extent by dust in the nuclear region. The 2μ optical depth from the center to the edge of the nuclear disk is probably in the range $0.3 \lesssim \tau_{2.2\mu} \lesssim 1.5$. There is also some uncertainty in the absolute and relative calibration of various 2μ brightness measurements. The radial dependence of the true luminosity density in the central kiloparsec is probably between $r^{-1.5}$ and $r^{-2.0}$, assuming the stellar population is independent of radius.

3. THE H II DISK

The central region of the bulge is bathed in the diffuse thermal plasma of the H II disk (Pauls et al. 1976). Within the central few tens of parsecs, the average electron density is $\sim 100 \text{ cm}^{-3}$. Young, hot stars which have recently formed in the central region may account for ionization of the H II disk. The evidence for star formation is mostly indirect, however, and ionization may have a different cause. The matter of star formation will be discussed further in connection with the core.

Dust in the H II disk absorbs the power output of the central portion of the bulge and reemits it in the far infrared. Since reradiation of bulge luminosity accounts for nearly all the far-infrared emission, one can conclude that there is no unidentified, heavily obscured, and dominantly powerful source of energy associated with the core.

Molecular clouds near the center lie close to the galactic plane. They occupy a relatively small fraction of the volume of the region, and probably play a minor role in the thermalization and transport of energy from the bulge. The Sgr A complex is of particular interest, since it is close to the center. Column density estimates and far-infrared observations indicate that this cloud should be opaque at 2μ. If the entire cloud were in front of the center, the 2μ continuum from the bulge would be weak in the direction of any part of the cloud. Since this effect is not seen, most of the Sgr A complex must lie behind the center, as indicated by radio observations (Oort 1977).

4. THE CORE

The central few parsecs, or core, is the site of a variety of phenomena. Background emission from what is apparently the bulge component dominates the 2μ continuum. The compact 2μ source IRS 16 (Becklin and Neugebauer 1975) may be the high-density central condensation of the bulge population. Thus the stellar bulge may be an identifiable structure on a scale as small as 0.1 pc, and IRS 16 may be the dynamical center of the Galaxy. Within the uncertainty of 2", IRS 16 coincides with the radio point source discovered by Balick and Brown (1974).

In addition to the bulge emission there are various compact or point sources bright at 2μ. The detection of CO absorption in several of these identifies them as red giants. One has the luminosity of a supergiant and is probably much younger than the bulge population. Stars have apparently formed recently.

A possible site of star formation is the dense, ionized ridge of the Sgr A West thermal H II region. The southern declination prevents complete high-resolution radio mapping of this cloud. However, warm dust associated with the ionized gas makes the ridge bright at 10μ, and infrared line emission mapping confirms that the ridge is the region of highest emission measure. Hence, the 10μ continuum map (Rieke et al. 1978) is probably a reasonable guide to the distribution of ionized gas in the core.

The Ne II 12.8μ line has been used to map the velocity structure of gas in and near the ridge (Wollman et al. 1977). As in molecular clouds near the center, the velocity dispersion is large and radial motion is indicated. There is also the appearance of rotation with a velocity of about 200 km s^{-1}. The location of the axis of rotation is uncertain by several arcsec but may pass through IRS 16. This is the most direct evidence that IRS 16 is (or is very near) the dynamical center. From these motions, the total mass in the central parsec is estimated to be several times 10^6 $M_\odot$. This roughly agrees with the mass estimated on the basis of the luminosity (Becklin and Neugebauer 1968; Sanders and Lowinger 1972) and is very small compared with the mass of gas which seems to flow in and out of the central region during the lifetime of the Galaxy.

Several bright, compact 10μ sources lie along the H II ridge. They may be the locations of newly formed O stars responsible for local excess heating of the dust. The luminosities of the sources are appropriate for O stars and can account for the ionization of the ridge and immediate surroundings. The total mid-infrared luminosity of the ridge, including the compact sources, is comparable to that of the bulge population within the core.

Since the infrared emission from the compact sources differs in several aspects from that of typical compact H II regions, the identification with O stars is uncertain. The ridge sources are unusually bright relative to the emission measure of the associated gas, and their spectra are abnormally hot. The spectrum of the brightest 10μ source may indicate two components at different temperatures. The ionization excitation of the ridge is rather low considering the luminosities of the stars apparently responsible. Furthermore, ionization could be due to some process unique to the core, eliminating the need for O stars.

On the other hand, many of the observed peculiarities may result from the unique location of the cloud. Because of the large mass of the core, gas dynamics will differ from that normally associated with star formation. Also, the high luminosity density in the core likely results in an unusually high dust temperature. Aitken et al. (1976) have suggested that the low ionization excitation is the consequence of somewhat high metal abundances. Finally, whereas the mid-infrared spectra of compact H II regions are normally affected by dust associated with cool circumnebular gas, the line of sight to the ridge sources seems to avoid most of the associated molecular cloud. Thus, the interpretation of the compact sources as the locations of O stars, while not firm, seems reasonable.

All the 2μ sources identified as cool giants, including the background contribution of the bulge, are reddened by about 30 visual magnitudes (Becklin et al. 1978b). This apparently represents a rather uniform minimum extinction and must be due to dust well outside the core. Patchy additional extinction is seen to correlate in position with the distribution of formaldehyde near the core (Rieke et al. 1978). This finding is significant in light of the conclusion mentioned earlier that most of the Sgr A complex is behind the center.

These observations and the velocity structure of the H II ridge suggest that part of the Sgr A cloud has spiralled around the core and lies in front of and below the center (Fig. 1). Star formation and subsequent ionization is occurring on the inner edge of this toroidal segment. In order to account for the appearance of the ridge, the location of the foreground molecular cloud, and the relatively low and uniform extinction to the 2μ sources, the plane of motion of the cloud near the center must be tilted toward the Sun. The shape of the H II ridge indicates a tilt perhaps as large as 45° within the central few parsecs. According to the observed distribution of formaldehyde, the molecular portion of the cloud depicted in Figure 1 must extend downward at least far enough to cover the radio continuum source Sgr A East ($\sim$6 pc).

5. CONCLUSIONS

The essential conclusions to be drawn from the infrared studies of the galactic center can be summarized as follows: The stellar bulge accounts for virtually all the mass and luminosity of the nuclear region. The luminosity density of the bulge is strongly peaked toward the center. Even within the central parsec, the bulge appears to be a centrally condensed feature. Dust in the nuclear disk absorbs the power output

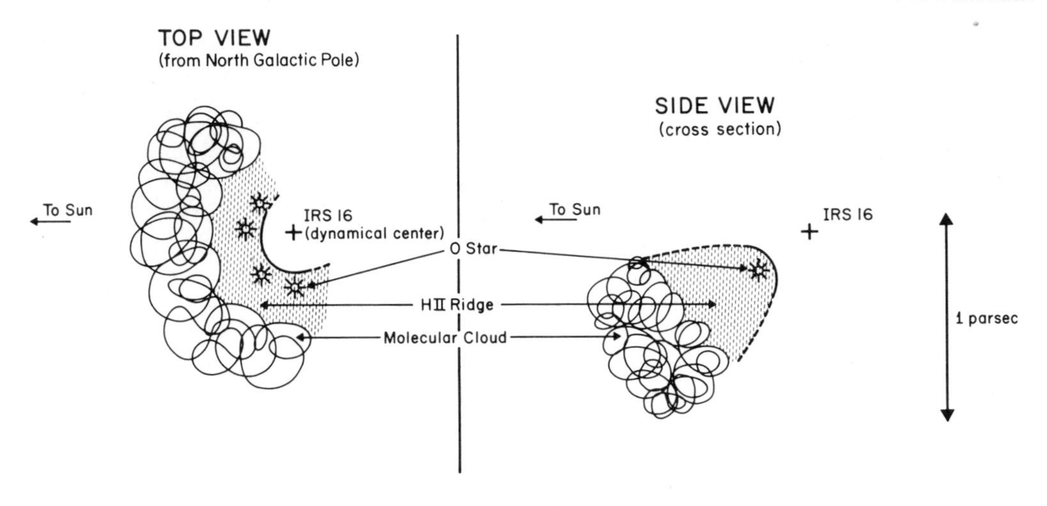

FIG. 1. Suggested location of dense gas in the core.

of the central portion of the bulge and reemits it in the far infrared. The central few parsecs, or core, is not currently a site of violent activity. Nor is it entirely quiescent. The power output of the core is about twice that of the bulge population alone. This excess luminosity may be due to recent or ongoing star formation. Based on the motion of ionized gas in the core, the total mass in the central parsec is estimated to be a few times 10^6 $M_\odot$, in agreement with the mass estimated on the basis of the luminosity. This is much smaller than the total mass of gas that seems to flow in and out of the center during the lifetime of the Galaxy. Various observations suggest that part of the Sgr A molecular cloud is curled around the core and moves in a plane inclined toward the Sun. Within the central parsec, the inclination may be as large as 45°. Star formation on the inner boundary of this cloud can account for the mid-infrared structure of the core.

Among the unanswered questions, that of the mass distribution of the central portion of the bulge may be the most significant. As mentioned earlier, the motion of ionized gas in the core is characterized in part by rotation with a velocity of about 200 km s^{-1}. Approximately the same velocity characterizes rotation throughout the disk, indicating that the radial dependence of the mass density is approximately $\rho(r) \propto r^{-2}$. On the other hand the core may contain unidentified excess mass, and the large-scale density distribution in the nuclear region may be flatter than r^{-2}. Multicolor, near-infrared mapping of the central region of the bulge will help answer this question.

The phenomena that distinguish the core may result from interactions between the disk and the bulge which are not yet understood. More must be learned about the details of star formation and of the compact sources in the core. Because of its unique status as the apparent dynamical center of the Galaxy, IRS 16 warrants special attention. Radio observations can provide a broader multidimentional picture of the nuclear disk. The dynamics of the H II disk is largely undetermined. The motion of molecular clouds in the nuclear region is not yet understood. Both infrared and radio observations indicate that the axis of rotation of gas in the disk may be a function of radius.

Considering the certainty of radial gas flow in the nuclear region, the relatively small mass currently found in the core indicates a roughly balanced circulation of gas in and out of the central region. The ring-like structures observed at radio wavelengths are most likely expanding. The infrared observations suggest that amorphous clouds such as Sgr A and Sgr B are making their way into the center. This situation may be a phase in some as yet unidentified cycle of galactic nuclear activity. It is also interesting to consider the possibility that the center of the Galaxy is in an approximately steady state.

I am grateful to E. E. Becklin, G. R. Gisler, J. H. Lacy, G. H. Rieke, H. A. Smith, and C. H. Townes for helpful discussions.

REFERENCES

Aitkin, D. K., Griffiths, J., and Jones, B.: 1976, *Monthly Notices Roy. Astron. Soc.* 176, pp. 73p-77p.
Balick, B. and Brown, R. L.: 1974, *Astrophys. J.* 194, pp. 265-270.
Becklin, E. E., Matthews, G., Neugebauer, G., and Willner, S. P.: 1978a, *Astrophys. J.* 219, pp. 121-128.
Becklin, E. E., Matthews, G., Neugebauer, G., and Willner, S. P.: 1978b, *Astrophys. J.* 220, pp. 831-835.
Becklin, E. E. and Neugebauer, G.: 1968, *Astrophys. J.* 151, pp. 145-161.
Becklin, E. E. and Neugebauer, G.: 1975, *Astrophys. J. Letters* 200, pp. 71-74.
Becklin, E. E. and Neugebauer, G.: 1978, *Astrophys. J.* (in press).
Gatley, I., Becklin, E. E., Werner, M. W., and Wynn-Williams, C. G.: 1977, *Astrophys. J.* 216, pp. 277-290.
Ito, K., Matsumoto, T., and Uyama, K.: 1977, *Nature* 265, pp. 517-518.
Krügel, E. and Tutukov, A. V.: 1978, *Astron. Astrophys.* 63, pp. 375-382.
Maihara, T., Oda, N., Sugiyama, T., and Okuda, H.: 1978, *Publ. Astron. Soc. Japan* 30, pp. 1-19.
Neugebauer, G., Becklin, E. E., Matthews, K., and Wynn-Williams, C. G.: 1978, *Astrophys. J.* 220, pp. 149-155.
Oda, N., Maihara, T., Sugiyama, T., and Okuda, H.: 1978 (preprint).
Oort, J. H.: 1977, *Ann. Rev. Astron. Astrophys.* 15, pp. 295-362.
Pauls, T., Downes, D., Mezger, P. G., and Churchwell, E.: 1976, *Astron. Astrophys.* 46, pp. 407-412.
Rieke, G. H., Telesco, C. M., and Harper, D. A.: 1978, *Astrophys. J.* 220, pp. 556-567.
Sanders, R. H. and Lowinger, T.: 1972, *Astron. J.* 77, pp. 292-297.
Willner, S. P.: 1978, *Astrophys. J.* 219, pp. 870-872.
Wollman, E. R., Geballe, T. R., Lacy, J. H., Townes, C. H., and Rank, D. M.: 1977, *Astrophys. J. Letters* 218, pp. 103-107.

DISCUSSION

Peimbert: Are there emission line observations in the infrared, in addition to those of the neon 12.8μ line, which can be used to determine the degree of ionization?

Wollman: Yes. The S IV 10.5μ line is not seen to a limit that allows only a small fraction of the gas to be highly ionized. The same conclusion follows from observations of the O III 88μ line, which is quite weak. Comparison between the Ne II 12.8μ line strength and the intensities of hydrogen IR recombination lines also indicates that low ionization states account for most of the ionized gas in the core.

Verschuur: The 2.4μ map of Okuda et al. seems to indicate a tilt at the center in a direction opposite to the tilt which Burton and Liszt talked about.

Wollman: That is a possibility. It is also possible that a band of extinction tilted in the same general direction as the CO could produce the observed effect.

Greyber: Would you explain the nature of the intense 2μ source close to IRS 16?

Wollman: Based on its luminosity and atmospheric CO band absorption, this source (IRS.7) is judged to be an M supergiant.

Bok: We have heard from several sides that there may well be a black hole at the center. There has been no mention this morning of this possibility. If there were a black hole, would one then not expect some spectacular effects of this black hole upon the radio and infrared distribution very near the center?

Wollman: The total mass within the central parsec is approximately 5×10^6 $M_\odot$. The observed radial flow of gas in the nuclear region is quite large--perhaps 1 $M_\odot$ y^{-1}. It seems that under these circumstances, a black hole at the center would grow to be much larger than the observed mass in the core. I consider this evidence against the existence of a black hole at the center.

Oort: It may be confusing to talk about a "dynamical" center of the Galaxy which is quite different from the position of Sagittarius A West. I believe that there is direct evidence for the existence of a center of mass at the position of the ultra-compact radio core. This evidence comes from the motions of the Ne II condensations that are observed in the region of $\sim$ 1 pc radius around the compact core. These motions indicate that there is a mass of about 5×10^6 $M_\odot$ within 0.5 pc.

Burton: Although it seems difficult to argue against the identification of the infrared centroid with the center of the galactic mass distribution, it is puzzling that the central longitude of the large-scale CO patterns is some 10' distant. The plane b = 0° and the direction l = b =10° have earlier been derived from the location of the HI density and velocity centroids. The λ21 cm line provides insufficient resolution to provide the detailed location of the center, but the 1'-resolution CO data are useful for this purpose. Because the CO gas motions are presumably ordered by gravitational forces, and because they show such a clear, symmetric kinematic pattern, it does not seem unreasonable to think that the center of this pattern would coincide with the center of the inner-Galaxy mass distribution.

Wollman: Perhaps the observations are telling us that the distribution of mass does not have a simple azimuthal symmetry. In any event, it is clear that the dynamics of the disk is not yet understood.

Burton: Regarding the location of the dynamical center near Sgr A West: Are you worried that the molecular complexes near there have a large net positive velocity. The molecular complexes are parts of consistent patterns which extend over at least 2°, consequently they cannot be highly localized.

Oort: Couldn't the cloud you refer to have been expelled from the center, and only a small part of it be seen in absorption against Sagittarius A East?

Kerr: I think we should not apply the phrase "dynamical center of the Galaxy" to a particular point until we understand more about the physics of the region.

A STUDY OF THE GALACTIC CENTRE REGION USING MIRA VARIABLES

M.W. Feast
South African Astronomical Observatory, Cape, S.A.

Mira variables are attractive for galactic centre studies because (1) they are numerous there; (2) they have high luminosities ($M_K \sim -7.5$) in the infrared where they can be seen through heavy absorption; (3) good radial velocities can be obtained from emission lines even for very faint objects; (4) at least in the solar neighbourhood and in globular clusters the period seems a good indicator of age and/or chemical composition.

Recent SAAO work covers four inter-related fields.

1. 2 200 sets of JHKL (1.2 - 3.5μ) measures of 220 nearby Miras have been made to establish their basic infrared properties (Feast, Catchpole, Robertson, Carter, Lloyd Evans). Me Miras occupy a limited region in the J-H/H-K diagram allowing reddenings to be determined for those in the galactic centre.

2. Photographic work in V and I has been done (Lloyd Evans) in the galactic centre windows NGC 6522/Sgr I/Sgr II. A wide range of periods is present. Lack of long periods in previous work was due to a selection effect.

3. JHKL photometry of Miras in the windows (Glass) has begun. Scattered observations of 40 stars shows (a) $A_K = 0.14 \pm .02$ ($A_V = 1.62 \pm .17$) for the NGC 6522 field and $A_K = 0.11 \pm .01$ ($A_V = 1.25 \pm .15$) for the Sgr I field; (b) a preliminary absolute magnitude calibration gives a mean distance modulus of 14.54 (8.1kpc). The s.e. of one observation ($\sim 0^m.5$) should be much reduced when full light curves are available.

4. Radial velocities of Miras in the windows have been obtained (Feast). From 23 stars ($\bar{P}$ = 223 days) the dispersion is 113 km/sec. This is the same as that of OH/IR sources in the centre (Baud). These are usually considered as long period (young) Miras. Either (1) OH/IR Miras in the centre have shorter periods than elsewhere or (2) the kinematics of objects in the centre is independent of age or (3) the period-age relation in the solar neighbourhood does not apply in the central region.

The work is extended to include searches for heavily reddened Miras near the centre from which estimates can be made of the density distribution.

W. B. Burton (ed.), The Large-Scale Characteristics of the Galaxy, 376.

OBSERVATIONS OF THE NEAR INFRARED SURVACE BRIGHTNESS DISTRIBUTION OF THE GALAXY

H. Okuda, T. Maihara, N. Oda, and T. Sugiyama
Department of Physics, Kyoto University, Kyoto, JAPAN

Studies of the stellar distribution in the inner region of our Galaxy have been seriously hampered at optical wavelengths by strong interstellar extinction. The extinction decreases considerably at infrared wavelengths, allowing us to look deep into the Galaxy. Motivated by this, we have tried to observe the near infrared brightness distribution of the central region of the Galaxy (Okuda et al.,1977, Maihara et al.,1978, Oda et al.,1978). Similar observations have been carried out by Hayakawa et al.,(1976), Ito et al.,(1977), and Hofmann et al.,(1977). These observations have provided valuable information on the distributions of stars and dust in the inner Galaxy (Hayakawa et al., 1977, Maihara et al., 1978, Oda et al., 1978).

OBSERVATIONS AND RESULTS

The observations were made four times since 1971, using balloon borne telescopes at altitudes near 25 km. The observed wavelength range was limited to a narrow gap in the OH airglow centered at 2.4μm. The field of view adopted in the flights in 1975 and 1976 was 1°x1°; this was improved to 0°.6x0°.6 in the flight of 1977 in order to resolve the fine structure near the galactic center.

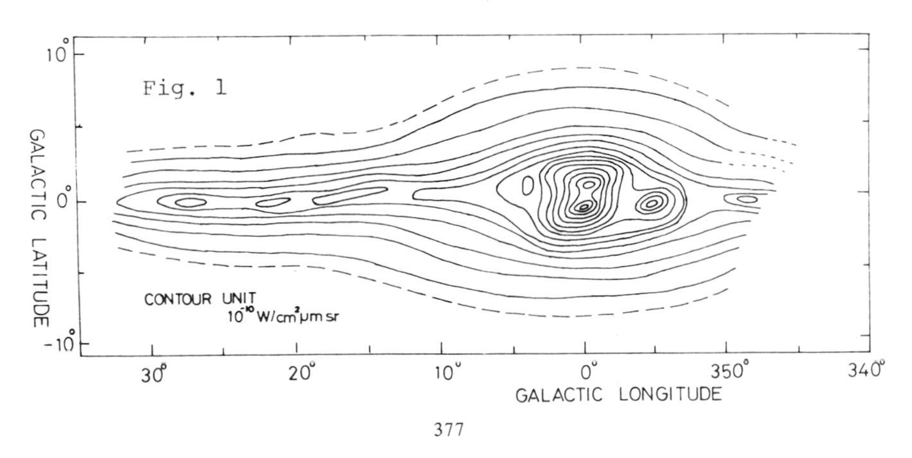

W. B. Burton (ed.), The Large-Scale Characteristics of the Galaxy, 377–380.

The brightness distribution derived from the 1977 flight and partly supplemented by the 1976 data is drawn in Figure 1. The longitudinal dependence of the ridge intensity and the cross-sectional distribution at l=25° are displayed in Figures 2 and 3.

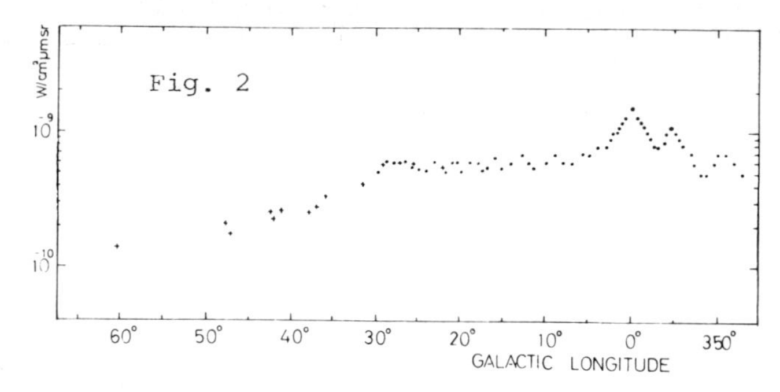

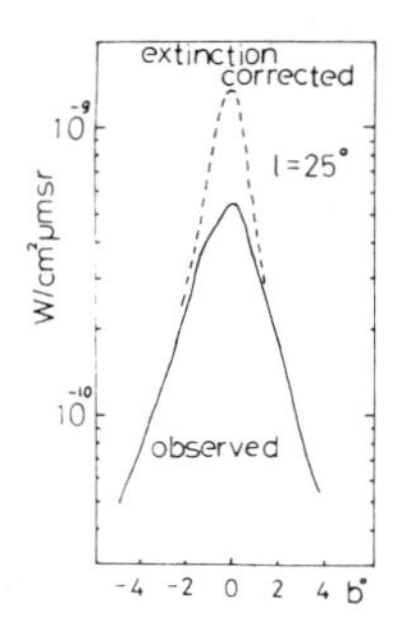

STARS AND DUST IN THE INNER GALAXY

a) Distribution of Dust

The split of the brightness contours close to the galactic center evidently indicates the presence of extremely large interstellar extinction along the galactic plane. Although it is not straightforward to estimate the effect of the extinction from the single-band observations, we have tried to decompose the magnitude of the extinction by assuming that the intrinsic brightness distribution of the bulge is similar to that of M31. The validity of this assumption will be discussed elsewhere.

Magnitudes of the extinction thus derived are shown in Figure 4. Most of the extinction ($A_{2.4}$=1.7 mag, corresponding to 21 mag of visual extinction) is distributed independently of galactic longitude. It may be associated predominantly with the 5-kpc ring of the molecular clouds which has been delineated from CO emission measurements (Scoville and Solomon 1975, Gordon and Burton 1976). In fact, the column density of total hydrogen gas (atomic and molecular), $3x10^{22}/cm^2$ (Burton 1976) is consistent with the extinction, if the relations of $N(H)/E_{B-V}=5x10^{21}$ atoms cm^{-2} mag^{-1} (Savage and Jenkins 1972) and $Av/E_{B-V}=3$ are adopted. The width of the extinction is about 2° in FWHM, or 200pc if the extinction originates in the 5 kpc ring. This is almost comparable to the situation in the solar neighborhood.

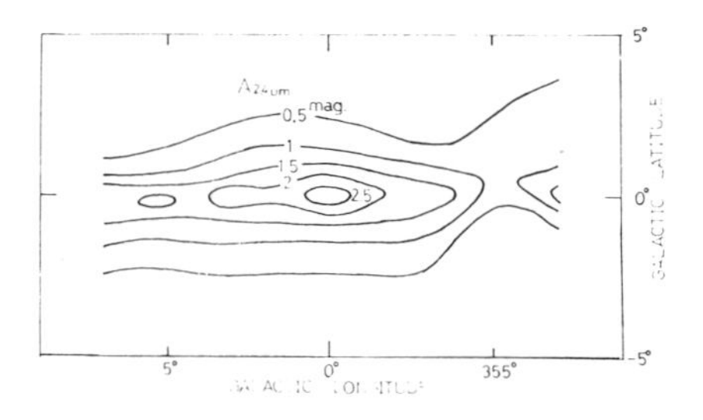

A slight concentration of the extinction ($A_{2.4}$=0.8 mag) toward the galactic center suggests existence of a dust layer in the innermost region of the Galaxy. The distribution corresponds with that of the far infrared emission detected by Hoffmann and Frederick (1969) and Soifer and Houck (1973). The total amount of dust required for the extinction is about 10^5 $M_\odot$, in good agreement with that estimated from the far infrared observations. The dust layer may have something to do with the large CO-clouds in the galactic center (Bania, 1977).

b) Distribution of Stars

The central bulge extends in a spheroid of ±15 in longitude by ±7°5 in latitude, or ±2.5 kpc by 1.3 kpc in linear scale. The total luminosity corrected for extinction amounts to 2×10^{10} $L_\odot$, if 4000 K is assumed for the effective temperature of the constituent stars. This is comparable to the situation in M31.

The most conspicuous feature of the ridge component is its flatness between l=10°and 30°, which suggests an annular distribution of the emitter as proposed for the CO clouds distribution (Burton, 1977). It is also remarkable that the ridge is extremely narrow; its FWHM is about 3°5. This would become much narrower, at most 2° as shown in Figure 3, if we correct the interstellar extinction derived above. The corresponding linear thickness is 300 pc in FWHM, if the annulus is located at a distance of 5 kpc from the galactic center. This would mean that the emitters are extreme Pop. I type objects such as protostars, O or B stars, or late supergiants.

Taking the interstellar extinction into account, a model distribution of volume emissivity of the 2.4 μm radiation is calculated so as to reproduce the observed brightness distribution. That is shown in Figure 5. The radial dependence of the volume emissivity is shown in Figure 6, together with that of the nucleus derived by Becklin and

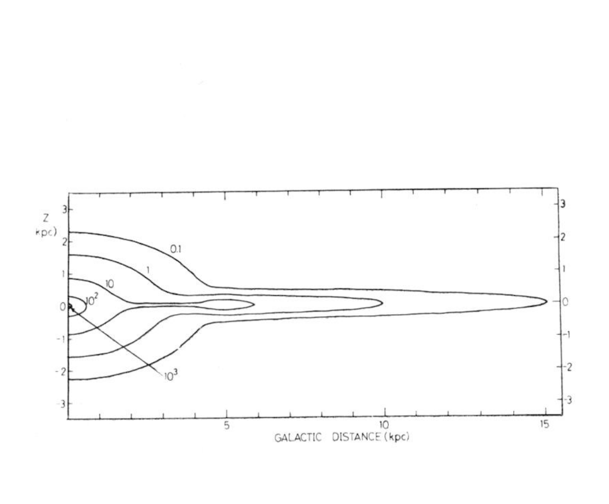

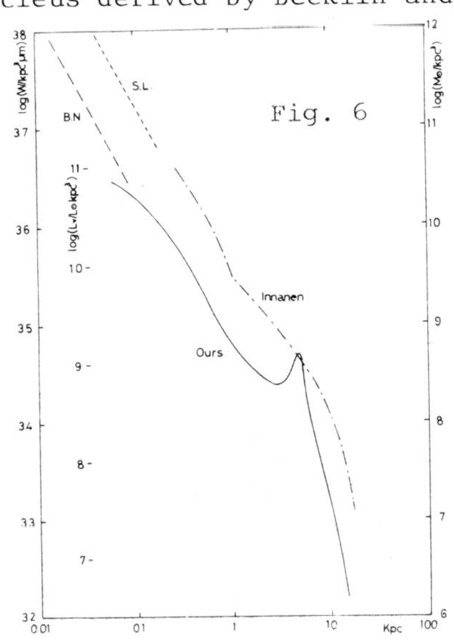

Neugebauer (1968). They are compared with the mass density distribution estimated from the rotation curve of the Galaxy by Sanders and Lowinger (1972) and by Innanen (1973) in the same figure. They are almost parallel to the infrared emissivity distribution, except for an extraordinary enhancement of the latter around 5 kpc. If we assume 4000 K for an effective temperature of the constituent stars, the ratio M/L_{bol} becomes ~2 in the inner Galaxy, while 0.4 in the 5 kpc ring. The relatively small value of the first ratio suggests that giant type stars contribute to the luminosity of the bulge, while much more luminous objects should supply the luminosity in the 5 kpc ring. In this regard, it is worth remarking that CO clouds, thermal radio sources, HII regions, and OH/IR sources cluster in the same region (e.g. Burton 1976). They all indicate that the region is very active in star formation and rich in young generation objects.

Finally, a few words should be added about the anomalous enhancement at $l=355°$, $b=-0°.7$, which has no known identification with any optical or radio sources. The total flux of the excess intensity amounts to 3.5×10^{-10} W/cm^2 μm, or K=-1.5 mag. From the presently available data, it cannot be concluded whether the anomaly is due to a local decrement in the interstellar extinction or to some unknown source hidden from optical detection by the strong interstellar extinction. It is interesting to note, however, that the flux and the size are compatible to those of M32 (Penston 1973), if M32 were put at a distance of 40 kpc.

REFERENCES

Bania, T.M.: 1977 Astrophys. J., 216, pp 381-403.

Becklin E.E., and Neugebauer, G.: 1968, Astrophys. J. 151, pp 145-161.

Burton, W.B.: 1976, Ann. Rev. Astron. Astrophys., 14, pp 275-308.

Gordon, M.A., and Burton, W.B.: 1976, Astrophys. J., 208, pp 346-353.

Hayakawa, S. Ito, K. Matsumoto, T. Ono, T., and Uyama, K.: 1976, Nature, 261, pp 29-30.

Hayakawa, S. Ito, K. Matsumoto, T., and Uyama, K.: 1977, Astron. Astrophys., 58, pp 325-330.

Hoffmann, W.F., and Frederick, C.L.: 1969, Astrophys. J. Letters, 155, pp L9-13.

Innanen, K.A.: 1973, Astrophys. Space Sci., 22, pp 393-411.

Ito, K., Matsumoto, T., and Uyama, K.: 1977, Nature, 265, pp 517-518.

Maihara, T., Oda, N., Sugiyama, T., and Okuda, H.: 1978, Publ. Astron. Soc. Japan, 30, pp 1-19.

Oda, N., Maihara, T., Sugiyama, T., and Okuda, H.: 1978, Astron. and Astrophys., submitted.

Okuda, H., Maihara, T., Oda, N., and Sugiyama, T.: 1977, Nature 265, pp 515-516.

Penston, M.V.: 1973, Monthly Not. Roy. Astron. Soc. 162, pp 359-366.

Sanders, R.H., and Lowinger, T.: 1972, Astron. J., 77, pp 292-297.

Savage, B.D., and Jenkins, E.B.: 1972, Astrophys. J., 172, pp 491-522.

Scoville, N.Z., and Solomon, P.M.: 1975, Astrophys. J. Letters, 199, pp L105-109.

Soifer, B.T., and Houck, J.R.: 1973, Astrophys. J. 186, pp 169-176.

ON THE MASS DENSITY DISTRIBUTION OF THE INNER GALAXY

T. Maihara
Department of Physics, Kyoto University, Kyoto

Based on current 2.4-micron observations of the Galaxy (see Okuda et al. in this Symposium), we have proposed a specific model for the bulge component. This model is a concentric spheroid with an axial ratio of ~0.5; $\rho(a)\propto(a^2+a_c^2)^{-1}\exp(-(a/a_o)^2)$, where a_o = 2.5 kpc and a_c = 0.14 kpc respectively. A constant mass-to-luminosity ratio $M/L_V \simeq 7.6$ is assumed, which yields the relevant rotational velocity in the inner region (Figure 1); the absolute velocity is normalized to 250 km s^{-1}. This value of M/L_V is likely to meet with giant-rich synthetic models for the nuclear bulge of M31.

Other important information from the 2.4-micron observations concerns the existence of a particular inner disk component suggested in the longitudinal brightness distribution (Figure 2). The near-infrared intensity, which should represent the stellar distribution in the Galaxy, shows a prominent excess of radiation at R≃3~5.5 kpc over that of the

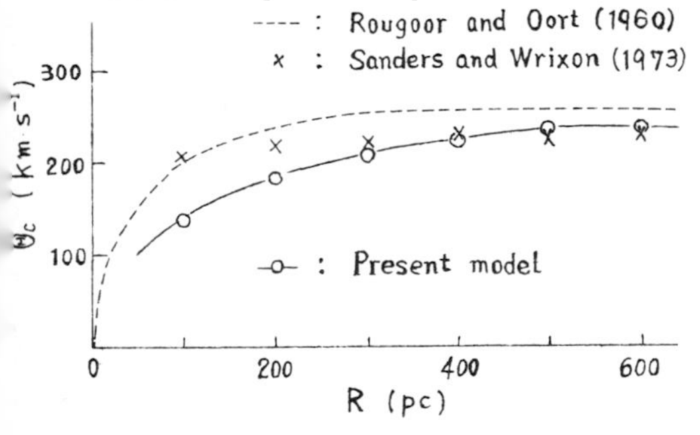

Figure 1. Rotation curves in the central region (left).

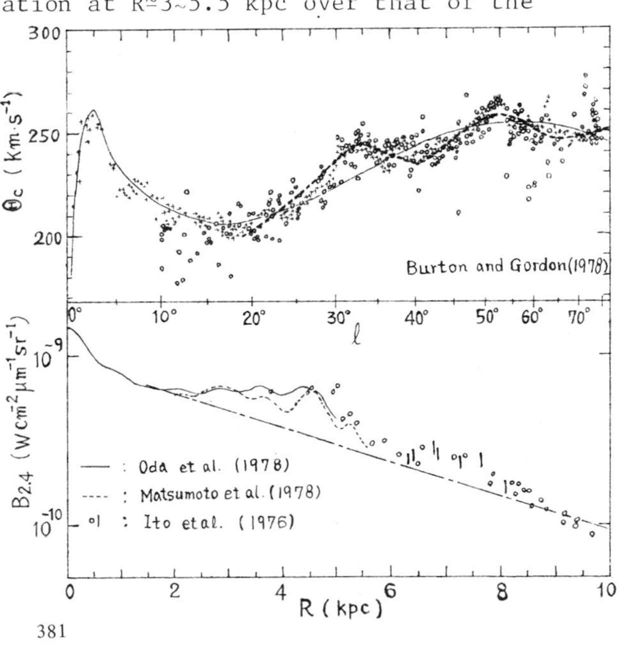

Figure 2. The upper: summarized rotational velocities in the Galaxy (Burton and Gordon 1978), the lower: 2.4-micron surface brightness along the galactic plane (right).

W. B. Burton (ed.), The Large-Scale Characteristics of the Galaxy, 381–382.

so-called exponential disk. One can also notice a small excess at R≃6.5∼8 kpc. Although these features do not directly represent the actual radial distribution of stars, it is interesting to note that the excesses correspond exactly not only to the global behavior of the rotation curve provided by HI and CO observations but also to the wavy flucutations in it, as demonstrated in the figure.

DISCUSSION

Lockman: I notice that your infrared emission is not always symmetric about b = 0°, especially near ℓ = 20°. Do you think that this is a feature of the stellar distribution or of the extinction?

Okuda: Modulation by the interstellar extinction is the most probable explanation.

Puget: What is the extent of the region over which you integrate to get the mass of dust which you give (10^5 $M_\odot$)?

Okuda: It is the mass in a cylinder with radius 300 pc and thickness 150 pc.

Stecker: Would you care to comment about the peak at ℓ ≃ 347°, b ≃ 0° in your observations?

Okuda: The peak may be due to some irregularities in the interstellar extinction or it may reflect a possible arm structure in the inner Galaxy.

Puget: You mentioned a 10μ luminosity for our Galaxy. So far observations have only been done with a beam-switching technique. The extended flux in the far infrared has been underestimated in several cases by about 1 order of magnitude. This could be the same for the 10μ flux.

Viallefond: The spiral M83 has been observed in the far infrared. Using a beam-switching technique, we observe a gradient of the infrared emission in the band 70-95μ along a scan almost parallel to the E-W direction. The signal-to-noise ratio is between 5 and 8. After integration of this gradient profile, we get the distribution of the far infrared emission along the scan. Two maxima are observed, at locations where the scan crosses the arms, which are very rich in HII regions. Because we do not detect a strong signal in the band 115-190μ, we presume that the bulk of the emission comes from HII regions rather than from molecular clouds although a large amount of molecular gas has been observed.

DYNAMICAL INTERPRETATIONS OF THE GALACTIC CENTER REGION

Robert H. Sanders
Kapteyn Astronomical Institute, Groningen

I will define the central region of the Galaxy as being the inner four kiloparsecs. The distinguishing characteristics of this region are:
1) The dominance of a central spheroidal component in the mass distribution -- a bulge.
2) An apparent deficiency of gas, at least between radii of 500 pc and 4000 pc.
3) High non-circular gas velocities.

Now let us consider these characteristics in some detail.

1. THE BULGE

The rotation curve, as observed in the 21-cm line of neutral hydrogen, gives some indication of the form of the gravitational field and, hence, the mass distribution in this region. Figure 1 is the rotation curve in the inner few kiloparsecs (solid line). Inside 1 kpc the curve is essentially that of Rougoor and Oort (1960); and beyond 1 kpc the curve is a fit to the data presented by Simonson and Mader (1973). The existence of an inner centrally condensed component in the mass distribution - a component which is distinct from the more extended disk - is implied by the presence of the inner peak in the rotation curve; specifically, the rise to 260 km s^{-1} at 800 pc.

Independent confirmation of the centrally condensed component came in 1968 with the near infrared observations of Becklin and Neugebauer. They discovered a source of extended emission at 2.2μ which they interpreted as being starlight and, therefore, an indication of the density distribution of stars within the galactic nuclear region. Making a few reasonable assumptions one can easily convert this 2.2μ intensity distribution into a stellar density distribution. This has been done by Becklin and Neugebauer (1968), Oort (1971), and Sanders and Lowinger (1972). In particular, Sanders and Lowinger derive a stellar density distribution in the inner 50 pc described by the formula

$$\rho = \frac{7.6 \times 10^5}{r^{1.8}} \ M_\odot \ pc^{-3} \qquad (1)$$

W. B. Burton (ed.), The Large-Scale Characteristics of the Galaxy, 383–392.

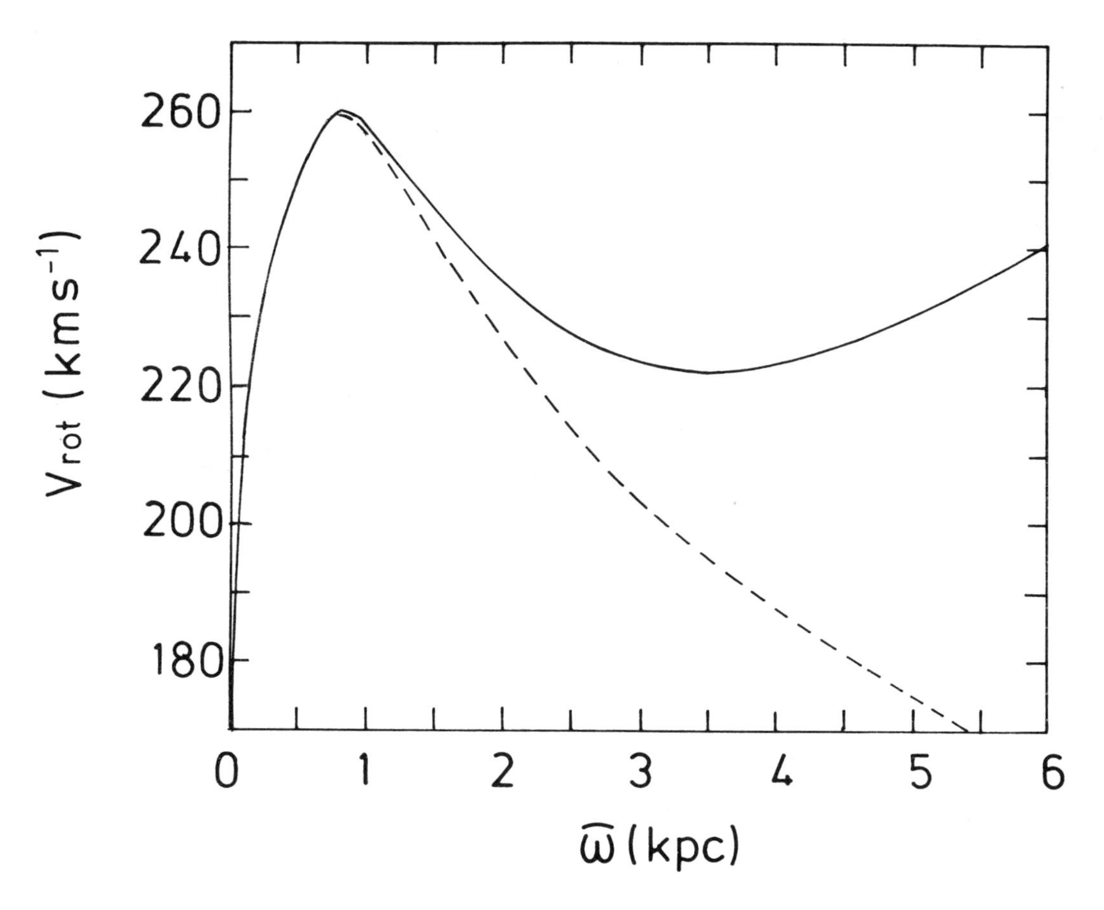

Figure 1. Rotation curve (solid line) in the inner region of the Galaxy and rotation curve (dashed line) of bulge-halo component.

If we extrapolate this density law out to 800 pc (an extrapolation which seems justified by comparison with M31), we may derive a rotation curve which is practically indentical to the observed 21-cm curve at radii less than 800 pc. Thus, there appears to be agreement between two independent observational determinations of the mass distribution in the bulge. This gives us some confidence in both techniques: that the near infrared really does reveal the distribution of old stars near the center and that the neutral hydrogen velocity field inside 800 pc is dominated by circular motion and truly reflects the gravitational field; in other words, that the HI motions are not drastically affected by some presumed general expansion motion inside 800 pc.

Now I would like to emphasize two important aspects of this observed central spheroid:

a) This spheroid, or bulge, could very likely be the central core of an extended spheroidal galactic halo; a halo with a mass sufficient to stabilize the disk. The density law derived from the Becklin-Neugebauer observations (Equation 1) is basically that of an isothermal sphere. But if the bulge is to continue beyond 1000 pc, it must have a steeper density law in order to be consistent with the observed rotation curve; specifically with the dip in the rotation curve between one and six kpc (Fig. 1). The likely tracers of the halo, the low-metal RR Lyrae stars and the globular clusters, would suggest an r^{-3} density law beyond one kpc (Oort 1965, Oort and Plaut 1975). Fitting an r^{-3} law smoothly to the bulge density at 800 pc (that is, the density derived from extrapolation of the near infrared observations), we find

$$\rho = \frac{10^9}{r^3} M_\odot \, pc^{-3} \qquad (2)$$

for r > 800 pc.

The dashed line in Figure 1 is the rotation curve for this bulge-halo component. The difference between the bulge-halo rotation curve and the observed rotation curve is presumably due to an extended disk component, but the significant point is that we can smoothly tie an extended halo onto the bulge without violating the observed rotation law. Moreover, in the context of this model, the mass of the bulge-halo inside 10 kpc is comparable to the mass of the disk, both being about $5 \times 10^{10} M_\odot$. Therefore, the bulge-halo has a mass which is sufficient to stabilize the disk against the violent bar-forming modes discussed by Ostriker and Peebles (1973).

b) The bulge-halo is probably axisymmetric and possibly nearly spherically symmetric. The halo tracers, specifically the low-metal RR Lyrae stars from 1 kpc to 5 kpc, seem to have a spherically symmetric distribution (Oort and Plaut 1975). In M31 the visual isophotes become almost circular within a few hundred parsecs of the center where the bulge dominates the light distribution (Light, Danielson, and Schwarzschild 1975). This supports the hypothesis that the bulges of spiral galaxies are, in general, hot axially symmetric systems.

The probable axial symmetry of the bulge in our Galaxy has implications regarding the interpretation of non-circular gas velocities in the central region. It has been suggested that such high non-circular velocities may be due to the action of a rotating oval or bar-like distortion. If so, then from arguments given above, it is probably the disk component and not the bulge-halo which is ovally distorted. But the disk makes a significant contribution to the gravitational field only at radii greater than about 3 kpc. Therefore, it is possible that an oval distortion of the disk could drive non-circular motions associated with the 3 kpc arm, but an oval distortion is not likely to be responsible for the high non-circular velocities observed within 200 pc of the center.

2. THE OVERALL GAS DISTRIBUTION

The second general characteristic of the galactic center region is the deficiency of gas inside four kiloparsecs. Figure 2 shows the overall radial distribution of gas in the Galaxy (Gordon and Burton 1976). The

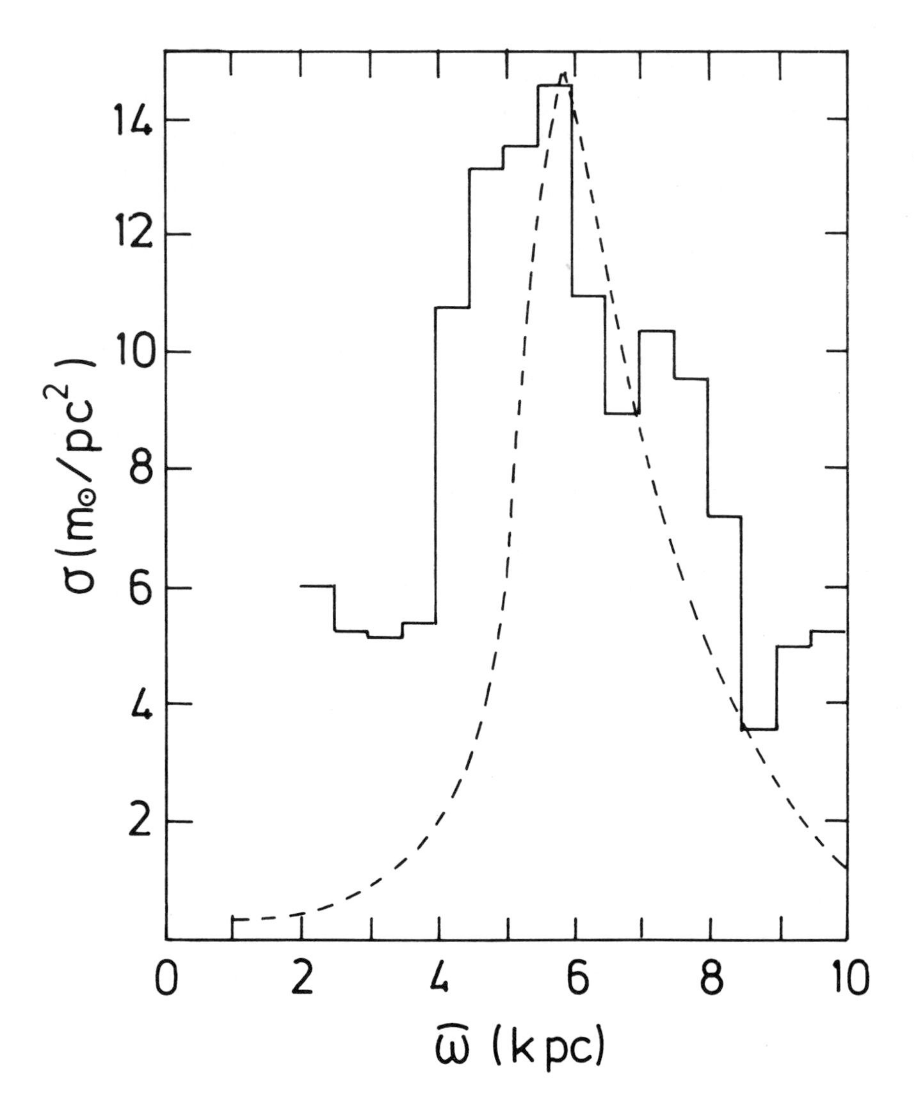

Figure 2. Radial distribution of gas surface density (Gordon and Burton 1976). Dashed line is distribution resulting from turbulent transfer of angular momentum.

gas surface density increases inward to about 5 kpc and then decreases rather abruptly inside 4 kpc; i.e., there is a hole in the total gas density distribution. In fact, there is quite a bit of gas in the central regions but it is all in the form of massive molecular clouds within a few hundred parsecs of the center. Bania (1977) estimates that there may be 7×10^8 $M_\odot$ of molecular gas within 500 pc of the center. If this gas were distributed evenly throughout the inner 4 kpc, it would in some sense fill up the hole. This suggests that inside 4 kpc, the gas has experienced some efficient loss or transfer of angular momentum.

One possible mechanism is an effective breaking of the gas due to interaction with a bar (Matsuda and Nelson 1977). This suggestion is based upon numerical calculations which contain a large artificial viscosity; therefore, it is very likely that this effect has been overestimated.

A second possibility is the outward transfer of angular momentum by a turbulent shear viscosity which is due to large-scale cloud-cloud interactions (Lynden-Bell 1969, Icke 1978). Given such a viscosity, there are two equations for the time development of the gas surface density, the equations of motion and continuity. A numerical solution of these two equations, appropriate for the Galactic rotation curve (or shear), is shown by the dashed line in Figure 2. The initial gas surface density distribution is assumed to be constant, and the distribution shown here develops over a period of 10^{10} years (assuming a constant cloud mean-free-path of 200 pc and a random velocity of 10 km s^{-1}). The general trends in the radial distribution of gas are accounted for in such a picture. In particular there is an accumulation of gas near 5 kpc and severe deficiency in the inner regions. The gas which was originally present inside 4 kpc has been transferred inward, through the inner boundary in this calculation (at $r = 1$ kpc).

It should be emphasized that for a simple single-peaked rotation curve, the shear, and, as a consequence, the rate of transfer of the angular momentum approaches zero. Therefore, if the galactic rotation curve had only one outer peak, the inner regions would not be depleted of gas. In order to deplete the inner 4 kpc by inflow, the galaxy must have a central peak in the rotation curve or, in other words, a massive central spheroid or bulge. It is the effect of the bulge on the rotation curve that creates strong shear in the inner regions. This would seem to be consistent with the observations of Bosma (1978) who has pointed out that central holes in the neutral hydrogen distribution tend to be present in galaxies with conspicuous bulges.

3. NON-CIRCULAR GAS VELOCITIES

Perhaps the most interesting characteristic of the galactic center region is the presence of high peculiar or non-circular gas velocities. (The relevant observations have been extensively reviewed by Oort, 1977.) For simplicity, I would like to divide the galactic center region into two sub-regions. Such a division is necessary because I will propose that the likely mechanism for excitation of high non-circular velocities is different in these two regions. Region I is between radii of 3 kpc and 4 kpc and region II is within 200 pc of the center.

a) Region I: This is the region of the 3-kpc arm, a feature seen in neutral hydrogen and molecular line observations over at least 20° of galactic longitude. At zero longitude the arm has a velocity of 53 km s^{-1} directed radially outward from the center. Now let us consider two suggested mechanisms for the origin of the large peculiar velocity of this feature: explosions and bars.

It is not likely that the 3-kpc arm is caused by expulsions or super explosions at the galactic center as suggested in the work of Van der Kruit

(1971) or Sanders and Prendergast (1974). The energies required to excite radial gas motions so far from the center are enormous -- in excess of 10^{58} ergs for a symmetrical event. More significantly, vast amounts of gas must be ejected, -- 10^8 $M_\odot$ per event. Given 100 such events over the lifetime of the galaxy (such a frequency would be necessary to maintain the observed non-circular motion for a reasonable fraction of the time), the total mass ejected from the central region would be on the order of 10^{10} $M_\odot$. Viscous transfer of angular momentum, the process described above, would fail by a factor of 100 to supply this amount of gas. Apart from these considerations, there is no independent evidence that events of this magnitude occur in the central regions of our own Galaxy.

The most likely explanation of the 3-kpc arm phenomenon is gas flow on elliptical stream lines maintained by a bar or oval distortion of the disk component of the galaxy. This was a suggestion first made by Kerr (1968) and later modelled kinematically by Peters (1975) on the basis of a flow pattern suggested by Roberts (1971). I have recently done time-dependent gas dynamical calculations of flow in the central region of a galactic gravitational field which is ovally distorted. The form of the axisymmetric potential was taken to mock up the central region of the galaxy; i.e., there are two components, bulge-halo and disk, with a rotation curve identical to that shown in Figure 1. The disk component was given a slight oval distortion by adding cos 2Θ variation to the potential. Only the disk was distorted, -- not the bulge. The angular velocity of the distortion is 15 km s^{-1} kpc^{-1}; hence, an inner resonance occurs between 3 kpc and 4 kpc. The strength of the perturbation was chosen such that the Θ force at 4 kpc is about 10% of the mean axisymmetric force at that radius. Within 2 kpc of the center the Θ force is less than 2% of the axisymmetric force due to the dominance of the axisymmetric bulge in this region. Gas motions are numerically followed until a quasi-steady state is reached in the rotating frame of the distortion. The steady state density distribution is that of an open trailing spiral. The steady state gas flow is illustrated in Figure 3. This is a contour map of the radial (or non-circular) component of the gas velocity. Solid lines are outflow, dashed lines are inflow. We see the "quadrupolar" pattern of non-circular velocities which is characteristic of flow on highly elliptical streamlines. In the inner 1000 pc (where the bulge dominates) the motion is predominately circular. The significant point is that a rather weak oval distortion can produce non-circular velocities of the observed magnitude. Moreover, if we view this flow pattern along the indicated line-of-sight (dotted line), we reproduce 21-cm profiles of the 3-kpc arm component along its observed range of longitude. One problem with this model, of course, is that we would also predict an equal positive velocity counterpart to the 3-kpc arm. This is a difficulty of any model with 180° symmetry and may imply the existence of large-scale systematic asymmetries in the gravitational field near the center.

b) Region II: This is the region of the massive molecular clouds within 200 kpc of the center. Here we find the great Sgr A and Sgr B2 complexes as well as the "expanding ring" first noticed by Scoville (1972) and Kaifu et al. (1972). This latter feature has been fit by a kinematic model consisting of an expanding, rotating ring. The ring has a radius of

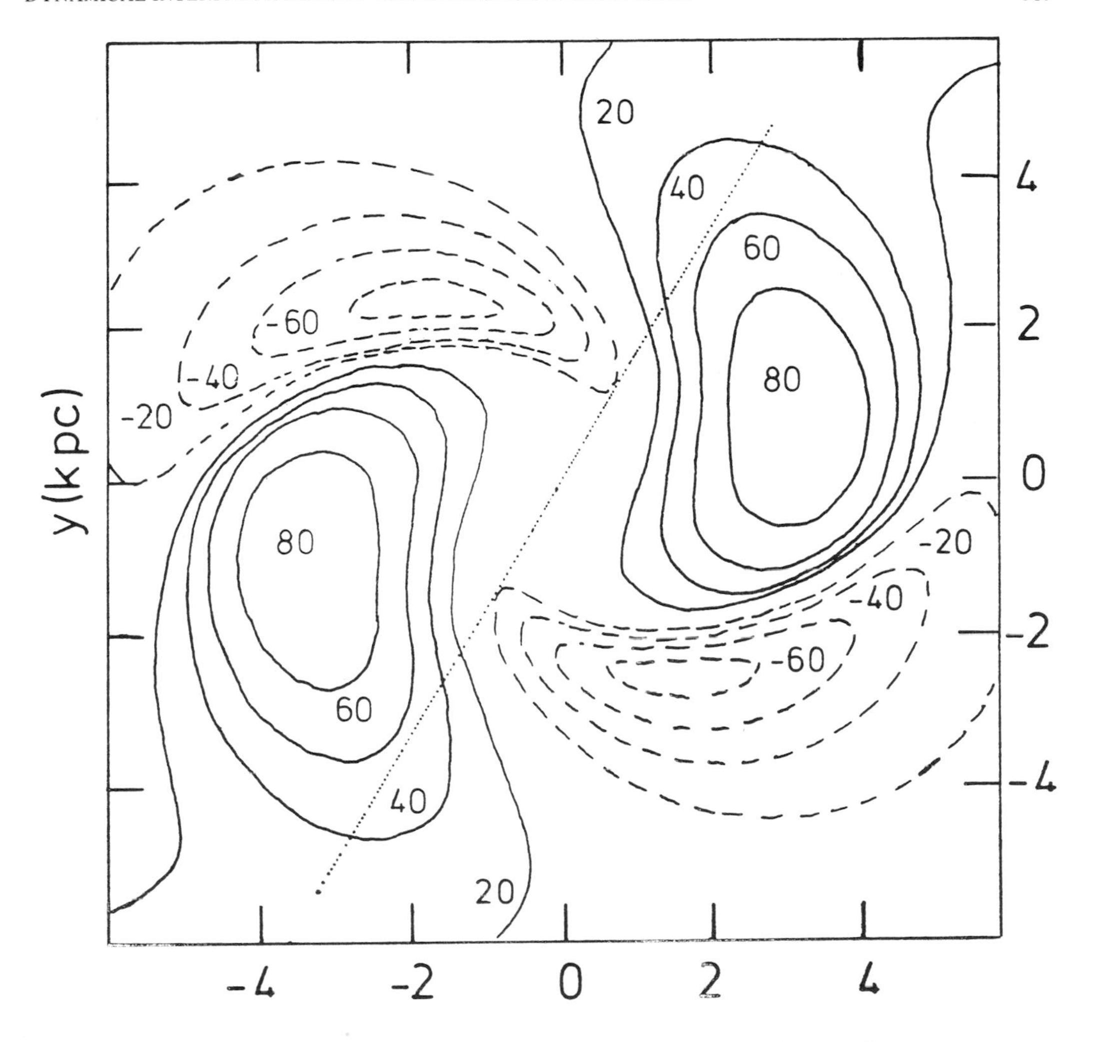

Figure 3. Contour map of non-circular velocity resulting from a rotating oval distortion in the inner 6 kpc of the Galaxy. Velocities are in km s^{-1}.

190 pc, an expansion velocity of 150 km s^{-1}, and a rotational velocity of 65 km s^{-1}. Its mass is on the order of 10^7 $M_\odot$ (Bania 1977) implying a kinetic energy in expansion motion of 10^{54} ergs.

These extremely high systematic non-circular velocities could result either from the effects of a bar or from an explosive process. As discussed above, the overwhelming dominance of the axisymmetric bulge component in this region would seem to rule out a direct gravitational mechanism, such as a rotating bar distortion. Moreover, for a bar to have a strong resonance effect in this region, it would have to be rotating very rapidly, with an angular velocity well in excess of 100 km s^{-1}. Therefore, such an inner bar would be dynamically distinct either from the bar presumably responsible for the 3 kpc arm or from the outer spiral structure. While a little, fast bar in the inner 200 pc cannot be ruled out, the dominance of the hot axisymmetric bulge favors, in my opinion, an explosion hypothesis for the non-circular gas motions in this region.

Above, we have discussed a mechanism by which more than 10^8 $M_\odot$ could accumulate near the center over a galactic time scale. Such an accumulation could provide fuel for a variety of explosive mechanisms. To maintain the observed non-circular motions inside 200 pc, the explosive "events" would have to recur on time scales of 10^7 years and provide at least 10^{55} ergs. Since our galactic nucleus is not spectacularly luminous at the present time ($< 10^{40}$ ergs s^{-1}), the radiation associated with such an event should be quite short lived ($\lesssim 10^5$ years), and what we presently observe is the hydrodynamic relic or fossil of the most recent period of activity.

REFERENCES

Bania, T.M.: 1977, Astrophys.J. 216, 381.
Becklin, E.E., Neugebauer, G.: 1968, Astrophys.J. 151, 145.
Bosma, A.: 1978, Dissertation, University of Groningen.
Gordon, M.A., Burton, W.B.: 1976, Astrophys.J. 208, 346.
Icke, V.: 1978, preprint.
Kaifu, J., Kato, T., Iguchi, T.: 1972, Nature Phys. Sc. 238, 105.
Kerr, F.J.: 1968, in Radio Astronomy and the Galactic System, IAU Symp. 31, ed. H. van Woerden.
Kruit, P.C. van der: 1971, Astron. Astrophys. 13, 405.
Light, E.S., Danielson, R.E., Schwarzschild, M.: 1974, Astrophys.J. 194, 257.
Lynden-Bell, D.: 1969, Nature, 223, 690.
Matsuda, T., Nelson, A.H.: 1977, Nature 266, 607.
Oort, J.H.: 1975, in Galactic Structure, Vol. V of Stars and Stellar Systems, eds. Blaauw and Schmidt, University of Chicago Press, p.455.
Oort, J.H.: 1971, Nuclei of Galaxies, ed. D.J.K. O'Connell, North Holland.
Oort, J.H.: 1977, Ann. Rev. Astron. Astrophys. 15, 295.
Oort, J.H., Plaut, L.: 1975, Astron. Astrophys. 41, 71.
Ostriker, J.P., Peebles, P.J.E.: 1973, Astrophys.J. 186, 467.
Peters, W.L.: 1975, Astrophys.J. 195, 617.
Roberts, W.W.: 1971, Bull. A.A.S. 3, 369.
Rougoor, G.W., Oort, J.H.: 1960, Proc. Natl. Acad. Sci. U.S.A. 46, 1.
Sanders, R.H., Lowinger, T.: 1972, Astron.J. 77, 292.
Sanders, R.H., Prendergast, K.H.: 1974, Astrophys.J. 188, 439.
Scoville, N.Z.: 1972, Astrophys.J. 175, L127.
Simonson, S.C., Mader, G.L.: 1973, Astron. Astrophys. 27, 337.

DISCUSSION

Schmidt-Kaler: (1) The agreement between the density model from the infrared observations and the Sanders-Lowinger rotation curve (if these really agree) may be misleading. The M/L rates need not be constant, because star formation is going on. (2) The rotation curve of the Galaxy inside 4 kpc is certainly far from solid body. To explain just one feature, for just one time period, and from one observer's aspect is the wrong philosophy. Further, you consider viscous shear with slowly ingoing motions. I do not at all object to the existence of that shear but I believe a theory should first explain the whole situation of many features flowing out with high velocities. The philosophy of near-circular orbits for the gas and pure gravitation is probably just wrong in the innermost parts of the galaxy. (3) On the same grounds you obtain the great mass outflow rate, because you assume circular symmetry. Starting from a Riemann instability you find the outflow only at two diametrically opposite points or strips in the plane, and these points represent the onset of the spiral arms. A rate of 0.25 $M_\odot$/year is sufficient to maintain energy and momentum of the density wave in the main body of the Galaxy. That adds up to $\sim 10^9$ $M_\odot$ in those 5×10^9 years, corresponding to the age of the disk stars according to Demarque and McClure.

Sanders: (1) The Sanders-Lowinger rotation curve is the curve for a mass model derived from the near infrared observations. So, of course, they "agree". The statement was that this rotation curve is entirely consistent with the 21-cm observations in the inner few degrees--and that is just a statement of fact. By playing with the assumptions that go into determination of the mass model from the near infrared observations, such as the mass-to-light ratio, one might be able to change the predicted rotation curve by 50 km s^{-1} or so, but not by much more. (2) Your second point escapes me. I did not suggest that the rotation curve inside 4 kpc is solid body; it is far from it. And it is possible to have apparent expansion due to flow on highly elliptical streamlines even in the presence of a low-velocity systematic inflow. (3) In your third point, I presume you are objecting to my arguments against an explosion model for the 3-kpc arm. It is true that the mass outflow rate can be lowered by proposing carefully aimed and highly directional ejection. But, I must admit, I do not understand your model.

van Woerden: There can be no objection to a model requiring two bars. Several external galaxies have two bars; e.g., NGC 1291 has a nuclear bar inside, and not aligned with, the major bar (see de Vaucouleurs 1975, Ap. J. Supp. 29, 193).

Sanders: That is a good point. In the case of our Galaxy, however, both bars would have to lie oriented in such a way that we would observe high-velocity outflow.

de Vaucouleurs: Instead of different ill-defined power laws for the space density distributions in the "nucleus", "bulge" and "halo", it is advisable to recognize that these are parts of a unique spheroidal component pervading the whole Galaxy from the innermost IR nucleus to the outermost globular clusters and obeying throughout the $R^{1/4}$ law in projection. P. Young has given (A.J. 1976) a convenient asymptotic expression for the space density distribution in a spheroid obeying the $R^{1/4}$ law:

$$\rho(s) \simeq \left(\frac{\pi}{8bj}\right)^{1/2} e^{-bj}/2j^3 ,$$

where $s = r/r_e = j^4$ and $b = 7.669$. This is a very good approximation at all $s > 0.2$ ($\sim$ 1 kpc in our Galaxy). Monnet and Simien (A.&A. 1977) have shown how a spheroid and exponential disk combination give good representations of the rotation curves of M31, M81.

Pişmiş: In the dynamical determination of mass one should be aware that the rotational velocity may deviate significantly from circular velocity. The gravitational force, due to the mass interior to the point, is not necessarily balanced by the centrifugal acceleration at the point. The additional force is provided by the dispersion of the velocities. If this dispersion is not negligible, then the masses determined will be underestimated. These statements are based on the hydrodynamical equations of stellar synthesis.

Sanders: The 21-cm profiles in the inner few degrees are not compatible with a velocity dispersion in excess of 20 km s^{-1} or so. The pressure-gradient force is not significant.

Yuan: From the observations of van der Kruit and Davies and Cohen, there are many expanding features in the central region. This seems to contradict the flows moving towards the nucleus, due to the loss of angular momentum, that you suggest in your theory. Could you comment on this?

Sanders: The inward flow of gas due to viscuous transfer of angular momentum is a very low-velocity flow (2-3 km s^{-1} at most), but it is systematic. The high expansion-velocity components which you refer to may not represent a systematic outflow. For example, it has been suggested that we might be observing an apparent expansion due to gas flow on highly elliptical streamlines in the field of a bar. In this case, it would be possible to observe an apparent expansion in the presence of a general low-velocity inflow. A gas "streamline" would look like this, in such a case:

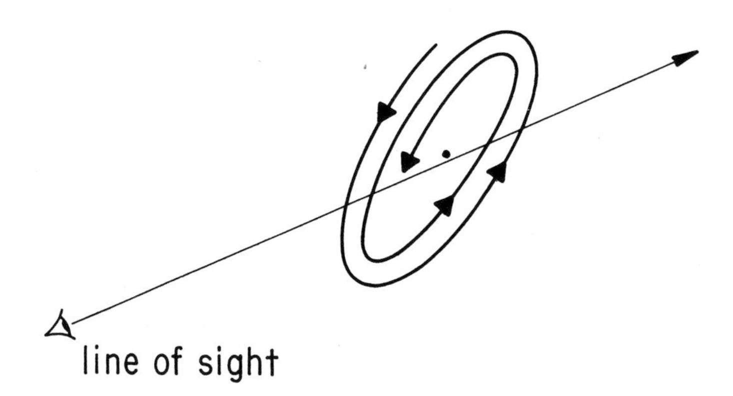

THE BASIC STATE OF THE VELOCITY FIELD IN THE GAS CLOSE TO THE GALACTIC CENTRE

K. Rohlfs and Th. Schmidt-Kaler
Astronomisches Institut, Ruhr-Universität, Bochum, FRG

To explain the features in the range $340° \leq l \leq 22°$ which show expansion velocities, elliptical stream-lines have been proposed that happen to expose expansion components towards the sun. Three general observations, however, make this interpretation doubtful:

i) No case of a contraction velocity with respect to the galactic centre is found (the +40 km s^{-1} feature in the 21-cm absorption data in the direction of the galactic centre is probably closely connected to it (*cf*. Schwarz *et al*. Astron. and Astrophys. 54, 863).

ii) The average radial density distribution of all major observed constituents of the interstellar gas shows a large deficit in the distance range $300 < R < 3500$ pc, while the distribution for $R > 6$ kpc resembles an exponential disk.

iii) Each observed expansion feature requires another special set of parameters for the ellipticity and orientation of the stream-lines.
We therefore propose an axially symmetric expansion velocity field in the gas, which can be approximated by

$$\Pi(R) = 484\ (R/\mathrm{kpc})\ \exp\{-0.93\ (R/\mathrm{kpc})\}\ \mathrm{km\ s^{-1}}$$

This expansion field may be considered as a description of a galactic wind. If it remains time-independent, then this velocity field depletes the gas within 3.5 kpc in about 5×10^7 years.

The observed major features of the gas are compared in a l-V-map (for b=0°) with predicted lines of constant distance from the centre and of constant galactocentric azimuth, for a velocity field with both rotation and expansion (Figure 1). No feature is found at forbidden velocities.

W. B. Burton (ed.), The Large-Scale Characteristics of the Galaxy, 393–394.

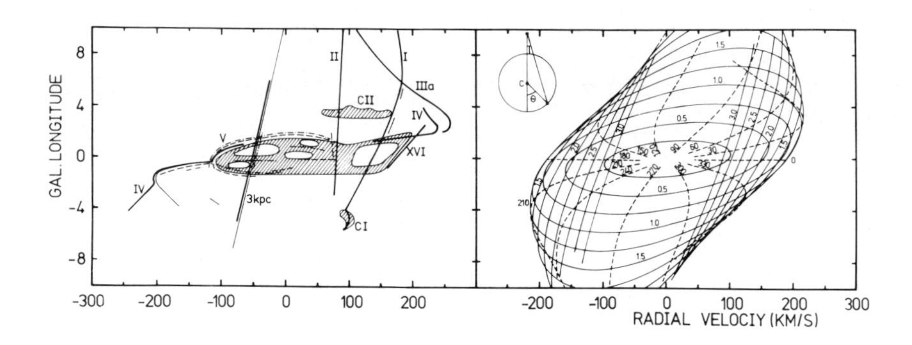

Figure 1: Comparison of the kinematics of observed features close to the galactic centre with a kinematical model containing both rotation and expansion. The observed features are

———	HI data after	Cohen and Davies 1976, MN 175, 1.
------		Cohen and Few 1976, MN 176, 495.
	CO	Bania 1977, Ap. J 216, 381.
-·-·-·	H_2CO	Scoville *et al.* 1974, Ap. J 187, L63

IS THERE A MASSIVE BLACK HOLE AT THE GALACTIC CENTER?

L. M. Ozernoy
Lebedev Physical Institute, Moscow, USSR

1. INTRODUCTION

During the past 10 years an hypothesis about the presence of a massive black hole at the center of our Galaxy (Lynden-Bell, 1969) has been an object of many exciting speculations. This hypothesis is based, firstly, on attempts to explain the nature of the "point radio source" at the galactic center (as well as a presumed much more powerful activity of the galactic nucleus in the remote past), and, secondly, on the opinion that the conditions in the course of dynamical evolution of galactic nuclei are favorable for the formation of massive black holes. However, both these approaches did not succeed in predicting with any confidence the black hole mass at the center of the Galaxy. The estimates available are based on indirect arguments and range from 10^7-10^{11} $M_\odot$ (Novikov and Thorne, 1973) to 10^4 $M_\odot$ (Shklovskii, 1976). A recent dynamical approach using NeII infrared observations of the galactic center (Wollman et al., 1977) has indicated that the black hole mass does not exceed 5×10^6 $M_\odot$ (Oort, 1977), although this value may well be due to a very dense star cluster whose brightest members only are seen in the infrared.

Black hole models are usually based on at least two arbitrary parameters: the black hole mass M_h and the accretion rate $\dot{M}$. As for the galactic center, the situation is fortunately much more definite. Taking into account such an inevitable process as disruption of stars in the vicinity of the black hole by its tidal forces, it is possible to obtain a lower limit on $\dot{M}$ and then (invoking available observational constraints on the luminosity or mass of the point source) an upper limit to the black hole mass M_h.

2. CONSTRAINTS ON M_h FROM LUMINOSITY DATA

The rate of tidal disruption of stars surrounding a black hole at the center of a compact star system has been calculated recently by a number of authors (Hills, 1975; Ozernoy, 1976; Bahcall and Wolf, 1976; Frank and Rees, 1976; Lightman and Shapiro, 1977; Dokuchaev and Ozernoy, 1977a) and may be considered as rather well established. Recently the

W. B. Burton (ed.), The Large-Scale Characteristics of the Galaxy, 395–400.

present author has investigated the character of accretion of gas released from the disrupted stars. Applied to a presumed black hole surrounded by the conditions pertaining in the nucleus of the Galaxy (defined by a core radius $R_c \approx 1$ pc, a stellar concentration inside the core of density $n_c \approx 10^7$ pc^{-3}, and a velocity dispersion v = 200 km s^{-1}) the overall picture is as follows.

As long as the mass of a black hole is comparatively small (less than, say, $3 \times 10^7\ M_\odot$), then the feeding of the hole can be provided by stars from unbound orbits which are disrupted by the tidal forces of the hole with a rate $N \approx 10^{-2}\ (M_h/10^6\ M_\odot)^{4/3}$ yr^{-1}. After 10-100 stars are disrupted, their remnants, forming gaseous disks inclined to each other under different angles, will form as a result of their collisions a more or less spherical cloud which will provide an effective accretion of the gas onto the hole. Afterwards accretion will produce, in addition to a flare (intermittent) component, a steady component of the luminosity. The value of the latter is determined by the rate of star disruption and is equal to

$$L = \varepsilon \dot{M} c^2 \approx 10^{42}\ \varepsilon_{0.1}\ M_6^{4/3}\ n_7 V_{200}^{-1}\ \text{ergs s}^{-1}. \qquad (1)$$

Here ε is the efficiency of mass-to-energy production during accretion and is hardly much smaller than 0.1; the quantities are normalized to 0.1, $10^6\ M_\odot$, 10^7 pc^{-3}, and 200 km/s, respectively.

Although an appreciable part of the luminosity of a massive black hole will be in the optical, UV, and very soft x-ray bands, the spectrum will differ significantly from that in the standard disk accretion model. First of all, dust does re-radiate optical and UV emission into the infrared band. By comparing the recent upper limit to the infrared radiation of the point source, $L \lesssim 10^7\ L_\odot$ (Gatley _et al_., 1977), with eq. (1) it is easily seen that the black hole mass is constrained by the value $M_h < 3 \times 10^4\ M_\odot$ (_cf_. Ozernoy, 1976).

This upper limit may be lowered farther to a much smaller value if one takes into account that an appreciable part of the emission of a massive black hole is in the energy range 1 keV $\lesssim E \lesssim$ 100 keV, according to an "optically thin model" by Payne and Eardley (1977) which is appropriate for the case of interest. Meanwhile the recent high resolution observations of x-ray sources at the galactic center by Cruddace _et al_. (1977) reveal no x-ray emission from the point source in Sgr A West, which yields an upper limit of 1.5×10^{36} ergs/s (2-10 keV) on its x-ray luminosity. Comparing this limit with the x-ray emission of the optically thin model (Eardley _et al_., 1978), one obtains $L/L_{Edd} < 10^{-4}$, where L_{Edd} is the Eddington luminosity of a black hole. The inequality obtained, together with eq. (1), give an extremely low upper limit to the black hole mass

$$M_h < 1\ \varepsilon_{0.1}^{-3}\ n_7^{-3}\ V_{200}^{-3}\ M_\odot. \qquad (2)$$

Evidently, the numerical coefficient is arguable, but the qualitative result seems to be rather significant. True, a confrontation with observational data needs time-dependent models of accretion which should be elaborated to determine details of gas flow near a hole. Because this has not been done, we present in the next Section another method to obtain an upper limit to the black hole mass, without invoking the luminosity arguments.

3. CONSTRAINTS ON M_h FROM SECULAR GROWTH OF A BLACK HOLE

Let us consider the inevitable growth of the black hole mass in the course of tidal disruption of stars surrounding the assumed hole at the galactic center. The calculations of secular growth of the black hole formed presumably 10^{10} years ago in the galactic nucleus were made by Dokuchaev and Ozernoy (1977b) under the following assumptions: (i) stars in the nucleus have a Maxwellian velocity distribution, and the rotation of the nucleus is negligibly small; (ii) most of the gas from disrupted stars is accreted eventually onto the hole; (iii) the main parameters of the galactic core (its radius $R_c \approx 1$ pc and the star density $n_c \approx 10^7$ pc^{-3}) do not change appreciably during secular evolution of the black hole.

The character of the growth of the black hole mass is shown in Figure 1. As long as the black mass is comparatively small, its tidal forces disrupt neighboring stars with a rate $N \approx 6 \times 10^{-7} (M_h/10^3 M_\odot)^{4/3}$ yr^{-1} (Hills, 1975; Dokuchaev and Ozernoy, 1977a). When the mass $M_h \sim 3 \times 10^7 M_\odot$ is reached, disruption of stars due to collisions begins to prevail over tidal disruption. Finally, when the black hole mass becomes comparable with the mass of the core (i.e., when the concentration of stars diminishes noticeably), tidal disruption will again become dominant over star collisions, but will proceed now at the lower rate $N \approx 4 \times 10^{-3}$ yr^{-1} (Dokuchaev and Ozernoy, 1977c).

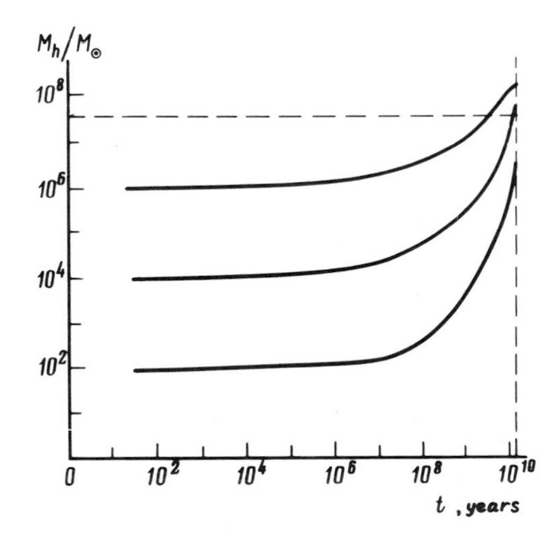

Figure 1. Secular growth, a black hole mass at the Galactic nucleus.

As seen in Figure 1, an appreciable growth of the black hole mass during 10^{10} yr is possible only if the initial mass was greater than $\sim 10^2$ $M_\odot$. An important additional feature is shown in Figure 2, which gives the relation between the initial black hole mass $M_h(0)$ and its expected present value $M_h(10^{10}$ yr) caused by accretion. As one can see, the mass increases during 10^{10} yr to a value $M_h \sim (4\times10^6-10^8)M_\odot$ which depends weakly on the initial mass provided that the latter exceeds 10^2 $M_\odot$ only slightly.

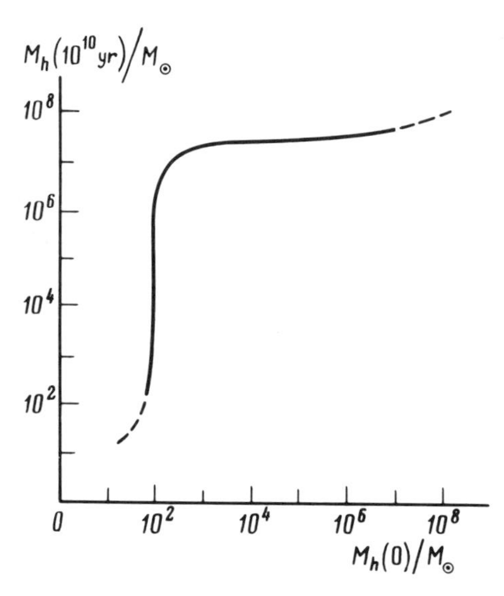

Figure 2. Final mass of a black hole vs its initial mass.

Even the minimal value of the final black hole mass is only marginally consistent with the above-mentioned observational upper limit to the black hole mass 5×10^6 $M_\odot$ (Oort, 1977), if $M_h(0) \gtrsim 10^2$ $M_\odot$. Of course, there is a possibility that a black hole in the galactic center was formed so recently that its present mass lies within the interval $10^2-4\times10^6$ $M_\odot$. However this possibility seems to be rather artificial.

Let us discuss briefly the other assumptions and simplifications listed above which could change the upper limit obtained.

(i) Repopulation of the "loss-cone" by stars due to diffusion of their orbits could depend on the rotation of the core only at a large anisotropy of star velocities. As for the situation in the nucleus of M31 (whose dynamical parameters are very similar to our own), its rotational velocity is much smaller than the velocities of chaotic motions of stars (e.g., Ruiz, 1976). This makes it quite reasonable to neglect rotation, in a first approximation, in estimates of the tidal disruption rate.

(ii) The possibility of partial ejection of gas from disrupted stars out of the sphere of its spreading, as a result, e.g., of thermal flares induced by the tidal forces (Lidskii and Ozernoy, 1978) was not taken into account. On the other hand, the interaction of stars with elongated gas clouds (remnants of disrupted stars) was neglected also. Their collisions lead to a decrease of the orbital angular momentum of the stars and, consequently, to a more rapid filling of the loss-cone. These processes work in opposite directions and compensate each other partially.

(iii) Although the assumption that the main dynamical parameters of the nuclear core are constant during its life is an oversimplification, a detailed analysis indicates (Dokuchaev and Ozernoy, 1977c) that it is not too bad.

4. SOME INFERENCES

A stringent upper limit $M_h \lesssim 10^2\ M_\odot$ to the black hole mass in the galactic center raises an interesting question: Why did the dynamical evolution of the nucleus lead to a small, if any, mass of the black hole? A possible answer, according to results of Dokuchaev and Ozernoy (1977d) may lie in the formation of a large number of close binary systems which may be the main factors preventing both collapse of the core and the formation of a massive black hole there.

Regardless of the eventual explanation of its low value, the upper limit to the black hole mass in the nucleus of the Galaxy appears to be in contradiction

(1) with the hypothesis that nuclei of normal galaxies are dead quasars (Lynden-Bell, 1969);

(2) with the hypothesis that massive black holes serve as sources of activity for Seyfert galaxies unless the Seyfert nuclei belong to some peculiar galaxies rather than to normal giant spirals;

(3) with the hypothesis that relic black holes may be the main factor of galaxy formation, because the mass of a relic black hole must have grown to $\sim 10^7\ M_\odot$ by the present time if it served as a center for the formation of the Galaxy (Ryan, 1972).

These negative results are, in fact, rather positive in the sense that they impose very informative constraints on the dynamical history of the galactic nucleus and, possibly, on the nature of an "engine" in the nuclei of Seyfert galaxies.

5. SUMMARY

During recent years many authors have become so convinced that black holes do exist that they adhere to the statement "Black holes are everywhere until their presence is disproved". Clearly, this claim violates the principle of "presumption of failure to prove". Nevertheless the galactic center appears to be the place where the existence of a massive black hole seems to be inconsistent with observational data coupled with theory, which together impose rather severe constraints on its mass.

REFERENCES

Bahcall, J. N., and Wolf, R. A.: 1976, Astrophys. J. 209, 214.
Cruddace, R. G., Fritz, G., Shulman, S., Friedman, H., McKee, J., and Johnson, M.: 1977, preprint E. O. Hulburt Center for Space Research.
Dokuchaev, V. I., and Ozernoy, L. M.: 1977a, Zh. Exp. Theor. Fiz. 73, 1537.
Dokuchaev, V. I., and Ozernoy, L. M.: 1977b, preprint Lebedev Physical Institute No. 137; Pis'ma Astr. Zh. 3, 391.
Dokuchaev, V. I., and Ozernoy, L. M.: 1977c, Pis'ma Astr. Zh. 3, 391.
Dokuchaev, V. I., and Ozernoy, L. M.: 1977d, preprint Lebedev Physical Institute No. 134, Astr. Zh. 55, 27 (1978).
Eardley, D. M., Lightman, A. P., Payne, D. G., and Shapiro, S. L.: 1978, preprint.
Gatley, I., Becklin, E. E., Werner, M. W., and Wynn-Williams, C. G.: 1977, Astrophys. J. 216, 277.
Hills, J. G.: 1975, Nature 254, 295.
Lidskii, V. V., and Ozernoy, L. M.: 1978, Pis'ma Astr. Zh. (in press).
Lightman, A. P., and Shapiro, S. L.: 1977, Astrophys. J. 211, 244.
Lynden-Bell, D. 1969, Nature 223, 690.
Novikov, I. D., and Thorne, K. S.: 1973, In "Black Holes", eds. C. de Witt and B. de Witt (Gordon & Breach, N.Y.), p. 343.
Oort, J. H.: 1977, Ann. Rev. Astr. Astrophys. 15, 295.
Ozernoy, L. M.:1976, Observatory 96, 67.
Payne, D. G., and Eardley, D. M.:1977, Astrophys. Lett. 19, 39.
Ruiz, M. T.: 1976, Astrophys. J. 207, 382.
Ryan, M. P. 1972, Astrophys. J. 177, L79.
Shklovskii, I. S.: 1976, Pis'ma Astr. Zh. 2, No. 7, p. 3.
Wollman, E. R., Geballe, T. R., Lacy, J. H., and Townes, C. H.: 1977, Astrophys. J. 218, L103.

DISCUSSION

Ostriker: Quite without regard to the ability of the rotating black hole to find and eat stars, the limit on the x-ray luminosity observed sets a rather low limit on the local gas density. Even a 10^6 $M_\odot$ black hole emitting with an efficiency of 10^{-4} would produce too much luminosity if the surrounding gas density were as much as the solar neighborhood value of 1 particle per cm^2.

Trimble: Ozernoy would undoubtedly agree with you, as, of course, do I, provided that the x-rays are emitted isotropically, and not perpendicular to a thick disk which we see edge-on. A creative imagination could probably come up with one or more plausible processes to clear gas out of the immediate black hole environment.

GALACTIC CENTER PULSAR AS A TEST OF BLACK HOLE EXISTENCE AND PROPERTIES

B. Paczyński
Copernicus Astronomical Center, 00-478 Warsaw, Poland

V. Trimble
Astronomy Program, U. Maryland, College Park MD 20742, USA

ABSTRACT: There is a reasonable chance of finding a (probably X-ray) pulsar in a short-period orbit around the galactic center. Such a pulsar can provide a test distinguishing a central black hole from a supermassive object or spinar. It also makes available a good clock in a region of space in which GM/Rc^2 is much larger than solar system values, thus allowing strong-field tests of general relativity.

The existence of expulsive phenomena, short-lived turbulence, IR source 16, excess mass and a compact radio source at or in connection with the galactic center requires a powerful, massive energy source there (Oort, 1977). The source might be a group of small objects (stars colliding in a relativistic cluster, many SNe and pulsars, etc.) or a single large one (supermassive star, spinar, black hole). Observational tests that might distinguish these two classes are discussed by Trimble (1971) and many others. Tests to tell a black hole from an extended object are harder to come by. A pulsar in a relatively short period orbit around the galactic center can provide such a test. It also makes available a good clock in a region of space in which GM/Rc^2 is much larger than solar system values, thus allowing strong-field tests of general relativity.

We need first to assess the probability of finding a usable pulsar. The 5×10^6 $M_\odot$ associated with the galactic center (Wollman et al. 1976) is 2.5×10^{-5} of the total galactic mass. To see a radio pulsar, we need $L_r \sim 10^{32-33}$ erg s^{-1}, implying $P \sim 10$-20 ms (and search frequency $\gtrsim 10$ GHz Davies et al. 1976) and age ~ 100 yr. An X-ray pulsar must have $L_x \sim 10^{34-36}$ erg s^{-1} to be discovered by the HEAO-B imaging camera. This requires a period ~ 0.1 s and age $\sim 10^{3-4}$ yr for a Crab-like pulsar, or an accretion rate $\gtrsim 10^{14}$ g s^{-1} on a 1 $M_\odot$ neutron star. The accretion can be from a binary companion or from high-density regions of the local ISM. If our galaxy has a supernova rate of 0.1 yr^{-1}, 300 weak X-ray binaries, and 10^8 old neutron stars, and the region that scatters radio

W. B. Burton (ed.), The Large-Scale Characteristics of the Galaxy, 401–404.

waves from the central compact source has 10^{-3} of its volume with 10^3 times its average density of 10^3 cm^{-3} (Backer, 1978), then the probabilities of finding one useful pulsar are about 0.0002 for a young radio object, 0.002 for a young X-ray one, 0.02 for an X-ray binary, and 0.2 to 1.0 for an accreting old neutron star. Davies et al. (1976) suggest that the compact radio source is a young pulsar, in which case the problem is solved.

The value of such a pulsar lies in its clock-like nature, allowing us to determine its orbit and probe the nature of space-time in its vicinity. The potential due to a flattened spinar disc at small r is quite different from that of a black hole, so the shape of the pulsar's orbit can distinguish the two cases. In addition, the parameter GM/Rc^2 to which all classic relativity tests are proportional (see, eg., Ohanian, 1976 for the proportionality constants) can be much larger than its maximum solar system and binary pulsar value of 1-2 X 10^{-6}. Table I summarizes the situation for pulsars 10^4 AU (size of the IR source)

Relativistic Parameters for a 1 $M_\odot$ Pulsar a Distance R from a 5 X 10^6 $M_\odot$ Black Hole

R (AU)	$\frac{2GM}{Rc^2}$	V/c (1)	P (2)	Tau (3)	ΔE at 6.6 keV	$\dot{\theta}$ (4)	$\Delta\theta$ (5)	Δt (6)
10^4	10^{-5}	0.003	450 yr	long	-----	0".04 yr^{-1}	4"	700 s
10^3	10^{-4}	0.01	14 yr	long	-----	14" yr^{-1}	40"	800 s
10^2	10^{-3}	0.03	163 d	$6X10^{11}$ yr	-----	1°.2 yr^{-1}	7'	900 s
10	10^{-2}	0.1	5.3 d	$6X10^7$ yr	0.07 keV	1°.1 d^{-1}	1°.1	1100 s
1	10^{-1}	0.3	3.7 h	6000 yr	0.7 keV	14° h^{-1}	11°	1150 s
0.1	1	1.0	-----	----	$\gtrsim$7 keV	-----	2π	1245 s

(1) Orbit Velocity
(2) Orbit Period
(3) Lifetime against gravitational radiation
(4) "perhelion" advance
(5) light deflection for impact parameter R
(6) excess time delay for impact parameter R

to 0.1 AU (Schwarzschild radius for 5 X 10^6 $M_\odot$) from the galactic center. V and P are orbit velocity and period; Tau is the lifetime of the orbit against gravitational radiation; ΔE is the gravitational redshift; $\dot{\theta}$ the "perihelion" advance; and $\Delta\theta$ and Δt the light deflection and excess time delay for em radiation with impact parameter R. Blanks in the table indicate unobservably small values or quantities undefined because there are no stable orbits. Gravitational lens effects will give brightness fluctuations also of order GM/Rc^2, but the details are very sensitive to the angle at which we see the system (Cunningham and Bardeen, 1973).

Requiring observations of two or more of these quantities to give consistent results constitutes perhaps the only strong-field test of general relativity that can be carried out in our lifetimes.

REFERENCES

Backer, D.L. 1968. Astrophys. J. 222, p.L9.
Cunningham, C. and Bardeen, J. 1973. Astrophys. J. 183, p.237.
Davies, R.D., Walsh, D., and Booth, R. 1976. Monthly Notices Roy. Astron. Soc. 177, p.319.
Ohanian, H.C. 1976. "Gravitation and Spacetime" (New York: Norton & Co.)
Oort, J.H. 1977. Ann. Rev. Astron. Astrophys. 15, p.295.
Trimble, V. 1971. Nature 232, p.607.
Wollman, E.R., Geballe, T., Lacy, J., Townes, C., and Rank, C. 1976. Astrophys. J. 205, p.L5.

DISCUSSION

Burke: The existence of apparently separate expanding features in the central region implies recurring activity. How could a black hole produce recurring activity?

Sanders: A black hole might produce recurring activity in the following way: In the central 200 parsecs of the Galaxy the distribution of interstellar material is obviously quite clumpy. If we suppose that within 100 to 200 pc of the center there are 20 to 30 massive molecular clouds and that the velocity distribution of the clouds is completely isotropic with a dispersion of 100-200 km s^{-1}, then a molecular cloud would actually pass through the center every 10^6 to 10^7 years. An encounter between a molecular cloud and a 10^7 $M_\odot$ black hole could produce quite spectacular results even if a very small fraction of the cloud (10^2-10^3 $M_\odot$) is captured by the hole. In this picture accretion is extremely non-steady-state.

Greyber: (1) K. Lo, M. Cohen, et al. point out the possibility of time variations in the flux from the compact radio source coincident with the peak of Sgr A West. (2) The logical possibility of part of the gravitational energy from collapse of the pre-galaxy cloud being stored in coherent relativistic electrons makes possible models other than "spinars" or massive black holes for such radio sources.

Trimble: (1) This means simply that they did not see the 0".001 component, which could have a variety of expanations. (2) You may be right.

Kaufman: The 10% value for the efficiency factor was a value that Fowler pulled out of the air as the best compromise between 1% and 100%. Is there any better justification now for assuming a 10% efficiency?

Trimble: 0.1 mc^2 is approximately the binding energy for a mass m in that last stable circular orbit around a Kerr black hole with the value of a/M (angular momentum per unit mass) that is thought to result from steady-state accretion.

Burke: To the extent that one accepts the conventional interpretation of the 3-kpc expanding arm and to the extent that one accepts Burton and Liszt's model as a temporary phenomenon, there are separated outbursts of large energy from the galactic center, lasting a relatively brief time. This does not seem to be a natural consequence of the kind of black hole models you have discussed here.

VII. COMPARISONS OF OUR GALAXY WITH OTHER GALAXIES

THE GALACTIC NUCLEUS COMPARED TO THOSE OF OTHER GALAXIES

Daniel W. Weedman

Dyer Observatory, Vanderbilt University, Nashville, Tn

ABSTRACT: Observations at various wavelengths are considered for extragalactic nuclei and are compared to how our galaxy would appear at comparable distances. The starlight from our nucleus is similar to that from the spirals in the Virgo Cluster. Our nucleus would show no sign of activity to a distant observer, neither unusual color, nor emission lines, nor excess infrared radiation. For example, the luminosity in Hβ emission is about 10^{38} ergs s^{-1}, which is 100 times fainter than that in the faintest Seyfert galaxy or emission line galaxy. It is also emphasized that there is no evidence from X-ray data for a massive, condensed object in the Galactic nucleus.

I. INTRODUCTION

This review is painful for an observer, because the observations of the nucleus of our own galaxy are so good that most of them have to be thrown out. In order to measure our own nucleus with absolute resolution comparable to that obtainable even for M31, we would need, for example, 21 cm observations with a 5 foot telescope. For optical observations, we have only a few results for the nuclei of other galaxies with 1" resolution; this corresponds to 100 pc at 20 Mpc, the nominal distance of the Virgo Cluster. 100 pc in our own nucleus would subtend 34'. There are excellent observations at many wavelengths that map our nucleus on a much finer scale than this, as reviewed comprehensively by Oort (1977). Because of the vastly different scales observed, it is rarely clear just what is meant by a "galactic nucleus". It won't be clear in this paper, either. Table 1 illustrates the different scales implied by references to various sorts of "galactic nuclei". For conciseness, the nucleus of our Galaxy is referred to as the GN.

II. STARS

One of the more reasonable comparisons that can be made is between the starlight from the GN and that from comparable regions in other galaxies. A substantial amount of small aperture photometry

W. B. Burton (ed.), The Large-Scale Characteristics of the Galaxy, 407–412.

TABLE I

Scale Sizes of 'Galactic Nuclei'

Source	Absolute size
GN(infrared)	1 pc
GN(Sag. A)	10 pc
GN(thermal radio)	100 pc
Seyfert Nucleus	0.1 pc - 1000 pc
M31 Nucleus	10 pc
Nuclear magnitude (Virgo)	1000 pc

exists for galaxies giving what are termed "nuclear magnitudes". From the distribution of 2.2μ light, Oort (1977) deduced a table of the stellar mass contained within various distances from the GN, assuming $M/L_V = 15$. This tabulation can be transformed to give the absolute magnitude of the GN within a given diameter. For comparison, the most complete data set for nearby galaxies is Tifft's (1969) photometry of Virgo Cluster galaxies, done with apertures as small as 9".7. Comparing the GN to those in Virgo depends on the adopted Virgo Cluster distance modulus. Results are given in table 2 for two alternatives. One adopts the Sandage-Tammann modulus of 31.5 (corresponding to $H_o = 55$ km s^{-1} Mpc^{-1}); the other uses the modulus 30.5 ($H_o = 85$ km s^{-1} Mpc^{-1}) preferred by several workers and most recently defended by Hanes (1977). (We have ignored small Galactic absorption corrections). For each modulus, table 2 shows the m_V that would be observed for the GN at that distance using an aperture of 1 kpc projected diameter, taking the mass within a 500 pc radius from Oort's table. The alternative aperture sizes, for the different moduli, are 16" and 10". Fortunately, these correspond to apertures used by Tifft (1969) to observe Virgo galaxies. His results are given in table 2 for all spirals which he observed that are considered by de Vaucouleurs and de Vaucouleurs (1973) to be Virgo Cluster members.

These results show that the nuclear magnitude of our Galaxy is certainly comparable to Virgo spirals, and the values are closer for the lower distance modulus. For now, this is not meant as a meaningful statement on the value of H_o, but it might someday be an interesting approach. In table 2, the GN most closely resembles NGC 4192 (SABab), 4421 (SBa) and 4651 (SAc).

III. GAS AND DUST

We know from 21 cm, high level recombination line, molecular and continuum radio emission that there is a substantial amount of gas in the GN. While there are many interesting details of this radio emission, the radio power of the GN is weaker than that detected from the nucleus of any other galaxy except M31 and M101 (Ekers 1974). The

Table II

Magnitudes for Central kpc of GN and Virgo Spirals

	m - M = 30.5 m_V(16")	m - M = 31.5 m_V(10")
NGC 4192	13.2	13.7
4216	12.1	12.7
4254	13.1	13.9
4321	12.7	13.5
4421	13.5	14.0
4450	12.7	13.2
4501	12.6	13.2
4535	13.9	14.3
4548	13.0	13.6
4569	12.2	12.6
4651	13.2	13.8
4654	14.0	14.7
Virgo mean	13.0±0.6	13.6±0.6
GN	13.4	14.4

most effective technique for optical detection of gas in galactic nuclei is looking for emission lines. In this respect, how would our nucleus appear? To decide, it is necessary to know the amount of ionized gas in the GN. From observations of H109α, Pauls, Mezger and Churchwell (1974) deduced that the highest density ionized gas is in Sgr A West, which they conclude has $M(HII) \simeq 600\ M_\odot$ and $N_e \simeq 1.4 \times 10^3 cm^{-3}$. They also decided that if the extended H 109α emission comes from H II, there is $\sim 10^4 M_\odot$ with $N_e \sim 200$. (Wollman et al. 1977 concluded that $\sim 10^2 M_\odot$ of H II was needed to explain the [Ne II] 12.8μ emission line.) This entire H II complex has an extent of about 4' so it would be unresolved in an extragalactic system. Knowing the volume, density and temperature of an H II region, the total Hβ emission can be calculated (Osterbrock 1974). For a T_e of 6000°K deduced by Pauls et al., Osterbrock's table for case B interpolates to an Hβ emissivity of $1.6 \times 10^{-26} N_e^2$ ergs cm^{-3} s^{-1}. Therefore, the complex of gas responsible for the H 109α emission would produce an Hβ luminosity, L(Hβ), of 5×10^{37} ergs s^{-1}. There could also be Hβ emission from a much lower density but more extended ionized gas producing the thermal radio continuum. This gas fills a volume of 260 x 90 pc with a mean $N_e = 16$, if the 3.75 cm continuum is all thermal in origin (Oort 1977, table 4). Were such gas of comparable temperature to that producing H 109α, it would have $L(H\beta) = 6 \times 10^{38}$ ergs s^{-1}. Therefore, on a scale size of 100 pc - like that observed for the nuclei of other galaxies - the GN would have 5×10^{37} ergs $s^{-1} < L(H\beta) < 5 \times 10^{38}$ ergs s^{-1}.

Compared to many other galaxies, this is very weak emission. The faintest Seyfert nuclei or emission line galaxies from the Markarian and Tololo lists have L(Hβ) ≃ 10^{40} ergs s^{-1} (Weedman 1977); the brightest Seyferts have L(Hβ) ≃ 10^{44} ergs s^{-1}. The faintest emission line measured in the nucleus of another spiral galaxy would correspond to L(Hβ) = 2 x 10^{38} ergs s^{-1}, for M81 (Peimbert 1968). Even here, the Hβ is not visible - only deduced from the Hα strength. Given that there seems to be plenty of H I (about $10^7 M_\odot$) in the GN, the lack of ionized gas has to be caused by a deficiency of ionizing photons. The L(Hβ) limits calculated above could be accounted for by 17 to 170 main sequence O7 stars. (This comes from the relation $N_* = 3.4 \times 10^{-37} L(H\beta)$ in Osmer, Smith and Weedman 1974.) Krugel and Tutukov (1978) recently carried through a detailed synthesis of the GN in order to reproduce the observed infrared data. The infrared radiation is attributed to a combination of that from 4000°K giant stars and that from dust heated by the ultraviolet radiation from O6 V stars; the relative contributions depend on the wavelength observed. Their model requires about 100 such O stars within 50 pc radius from the GN, comparable to that deduced from our L(Hβ). In a review of their observations, Mezger, Churchwell and Pauls (1974) decide that the total thermal radio emission from a 300 pc by 150 pc volume in the GN, including giant H II regions embedded in a lower density extended H II region, could be accounted for by the ionization from 77 O6 stars. All these estimates indicate that at the distances of galaxies beyond the local group, the GN would not show emission lines or an unusually blue continuum in the visible spectrum. The rate of star formation in the GN is orders of magnitude lower than that in objects called "active galaxies".

In the early days of infrared astronomy, radiation at far infrared wavelengths (λ≥10μ) was sometimes attributed to exotic non-thermal mechanisms, but it is now considered that in most cases this radiation is from heated dust. This heating arises primarily from absorption of ultraviolet photons. A small core about 1 pc (20") in diameter stands out at 10μ in the GN (Becklin and Neugebauer 1969). This is superposed on a more extended source whose extent is not well determined because of the limited chopper throw of infrared telescopes. Consequently, it is difficult to compare the 10μ luminosity of the GN with that observed for other galaxies. Aumann and Low (1970) feel that the far infrared luminosity of the GN arises within a diameter of less than 3' and assign the GN a 10μ flux of 10^3 Janskys. The most extensive data for other galaxies is in Rieke and Lebofsky (1978) who observed with a 5".7 beam. An important result is that M31 has an absolute 10μ flux that is a factor of ten fainter than the GN; in fact, all of the 10μ radiation from the nucleus of M31 can be explained as starlight. Because the absolute diameter observed at 10μ for M31 would correspond to 6'.6 at the GN, this result for M31 indicates a real and important difference between the GN and that of M31. However, the difference can be accounted for by the O stars in the GN which are needed anyway to explain the presence of ionized gas. The only galaxy detected with an absolute flux at 10μ comparable to that of the GN is NGC 3031. All of the other normal spiral galaxies detected at 10μ

Table III
Some Normal Spiral Galaxies Detected at 10μ

Galaxy (NGC)	3627	4258	4303	4536	4569	4736	4826	5055
Type	Sb	Sbc	Sbc	Sbc	Sab	Sab	Sab	Sbc
Flux Ratio*	58	40	330	840	400	50	34	64

*This gives the ratio of the absolute 10μ flux of the galaxy listed to that of the GN, using the distance and detection in Rieke and Lebofsky (1978).

have much brighter absolute fluxes, as shown in Table 3. Therefore, while the GN is sometimes cited as part of an "infrared galaxy phenomenon", the activity in the GN is very mild compared to some other spiral galaxies.

IV. BLACK HOLES?

Existing X-ray observations are especially important because they show that there is not an X-ray source in the GN. Though there is an X-ray source centered about 20' from the GN (Giaconni et al. 1974), and known as GCX, this is an extended source (Kellogg et al. 1971). From the Uhuru catalog (Giaconni et at. 1974), it is possible to estimate the limit on luminosity for any discrete X-ray source in the GN. The catalog limit corresponds to about 2×10^{-11} ergs $cm^{-2}s^{-1}$ in the 2-6 kev band for representative spectra. Therefore, any source at the GN has an X-ray luminosity less than 2×10^{35} ergs s^{-1}. This is a factor of 10 to 100 fainter than the X-ray sources identified in some globular clusters (e.g. Cominsky et al. 1977, Clark, Markert and Li (1975). The X-ray sources associated with the nuclei of Seyfert galaxies approach 10^{45} ergs s^{-1} (Tananbaum et al. 1978). Lightman, Giaconni and Tananbaum (1978) analyze the properties needed for a hypothetical black hole to explain the X-ray emission from the Seyfert NGC 4151. They conclude that $0.07L_{44} < M_7$ where L_{44} is the X-ray luminosity in units of 10^{44} ergs s^{-1} and M_7 is the black hole mass in units of $10^7 M_\odot$. For analogous circumstances in the GN, the luminosity limit would require a black hole mass less than $1.4 \times 10^{-3} M_\odot$. A rule of thumb for energy generation by black hole accretion, derived from the Eddington limit whereby the energy released acts against further accretion, is 10^{38} ergs s^{-1} per accreting solar mass. This gives a limit for a GN black hole of $2 \times 10^{-3} M_\odot$. These limits, while very uncertain, do indicate that there is no evidence from X-ray astronomy of a significant accreting object in the GN. In this case, the absence of evidence is meaningful evidence of absence, because there seems to be sufficient interstellar matter in the nucleus to power accretion. Knowing confidently that our galactic nucleus does not contain a dormant black hole would rule out suggestions that all galaxies once went through a quasar or Seyfert stage, powered by a massive black hole.

REFERENCES

Augmann, H. H., and Low, F. J.: 1970, Astrophys. J.(Letters),159, L159.
Becklin, E. E.,and Neugebauer, G.: 1968, Astrophys. J.,151, 145.
Becklin, E. E., and Neugebauer, G.: 1969, Astrophys. J. (Letters),157,L31.
Clark, G. W., Markert, T. H., and Li, F. K.: 1975, Astrophys. J. (Letters) 199, L93.
Cominsky, L., Forman, W., Jones, C., and Tananbaum, H.: 1977, Astrophys. J. (Letters), 211, L9.
de Vaucouleurs, G., and de Vaucouleurs, A.: 1973, Astron. and Astrophys., 28, 109.
Ekers, R. D.: 1974, in IAU Symp. 58 "Formation and Dynamics of Galaxies," ed. J. R. Shakeshaft (Dordrecht: Reidel), p. 257.
Giaconni, R., Murray, S., Gursky, H., Kellogg, E., Schreier, E., Matilsky, T., Koch, D., and Tananbaum, H.: 1974, Astrophys. J. Suppl. 27, 37.
Hanes, D. A.: 1977, M.N.R.A.S., 180, 309.
Kellogg, E., Gursky, H., Murray S., Tananbaum, H., and Giaconni, R.: 1971, Astrophys. J. (Letters), 169, L99.
Krugel, E., and Tutukov, A. V.: 1978, Astron. and Astrophys., 63, 375.
Lightman, A. P., Giaconni, R., and Tananbaum, H.: 1978, Astrophys. J. (in press).
Mezger, P. G., Churchwell, E. B., and Pauls, T. A.: 1974, Proc. 1st European Astron. Meeting, Vol. 2 (Berlin: Springer-Verlag),p. 140.
Osmer, P. S., Smith, M. G., and Weedman, D. W.: 1974, Astrophys. J. 192, 279.
Osterbrock, D. E.: 1974,"Astrophysics of Gaseous Nebulae"(San Francisco: Freeman), p. 66.
Pauls, T., Mezger, P. G., and Churchwell, E.: 1974, Astron. and Astrophys., 34, 327.
Peimberg, M.: 1968, Astrophys. J., 154, 33.
Rieke, G. H., and Lebofsky, M. J.: 1978, Astrophys. J. (Letters), 220, L37.
Tananbaum, H., Peters, G., Forman, W., Giaconni, R., Jones, C., and Avni, Y.: 1978, Astrophys. J. (in press).
Tifft, W. G.: 1969, Astron. J., 74, 354.
Wollman, E. R., Geballe, T. R., Lacy, J. H., Townes, C. H., and Rank, D. M.: 1977, Astrophs. J. (Letters), 218, L103.
Weedman, D. W.: 1977, Vistas in Astron., 21, 55.

DISCUSSION

Sanders: Can you exclude the possibility that our Galaxy is a Seyfert for about 2% of the time?

van den Bergh: Dr. Sanders' suggestion that galaxies might become Seyferts for a short time every $\sim 5 \times 10^6$ years can be excluded because about half of all Seyferts have a peculiar "washed out" spiral structure. This suggests that typical Seyferts have been in their "on" stage for a period comparable to, or longer than, their rotation period ($\sim 10^8$ yr).

MOLECULAR STRUCTURES OF OTHER GALAXIES COMPARED TO THAT OF THE GALAXY

L. J Rickard
National Radio Astronomy Observatory*, Green Bank, W.Va.,U.S.A.

NUCLEAR SOURCES

Of the eleven galaxies with detected CO emission, eight have bright nuclear CO sources: M82, NGC 253, M51, NGC 5236, NGC 1068, Maffei 2 (Rickard et al. 1977a,b), NGC 6946, and IC 342 (Morris and Lo 1978). Two have disk-population CO sources and no detectable nuclear source (M31 and M81, Combes et al. 1977a), and one has no obvious nucleus (LMC, Huggins, et al. 1975). Nuclear maxima thus appear to be the rule for galaxies with extensive molecular components, and such a peak is also seen in our Galaxy (e.g., Bania 1977). In Figure 1, I compare the CO data for the nuclei of M82 and M31 with spectra of the Galactic nucleus as it would be seen at their respective distances. (The Galactic spectra were synthesized from the date of Bania [1977], and assume a uniform z-distribution of 30 pc width.) The Galaxy is roughly intermediate, being about one-fifth the intensity of M82 and more than six times the intensity of M31.

After reasonable assumptions, one infers molecular hydrogen surface densities (corrected to face-on values) for the nuclear sources in the range 40-90 $M_\odot/pc^2$ and total H_2 masses $\sim 10^9$ $M_\odot$ (Rickard et al. 1977a, Morris and Lo 1978). These mass estimates are within a factor of two of those obtained from 390μ, 540μ, and 1 mm observations (Hildebrand et al. 1977, Elias et al. 1978). The ^{12}CO optical depths must thus be $\lesssim 5$, consistent with present upper limits for ^{13}CO emission. The velocity structures of several sources suggest strong non-circular motions, and Rickard et al. (1977a) make a direct comparison of models for M82 and the Galactic center. The inferred kinetic energies of expansion are quite similar, although other gauges of nuclear activity (e.g., infrared luminosity) differ by orders of magnitude.

OH is seen in absorption against the strong nuclear continuum sources in M82, NGC 253, NGC 4945, and NGC 5128 (Weliachew 1971, Whiteoak and Gardner 1973, Gardner and Whiteoak 1975, 1976, Nguyen-Q-Rieu et al. 1976). The total velocity widths of the absorption features in

* Operated by Associated Universities, Inc., under contract with the National Science Foundation.

W. B. Burton (ed.), The Large-Scale Characteristics of the Galaxy, 413–416.

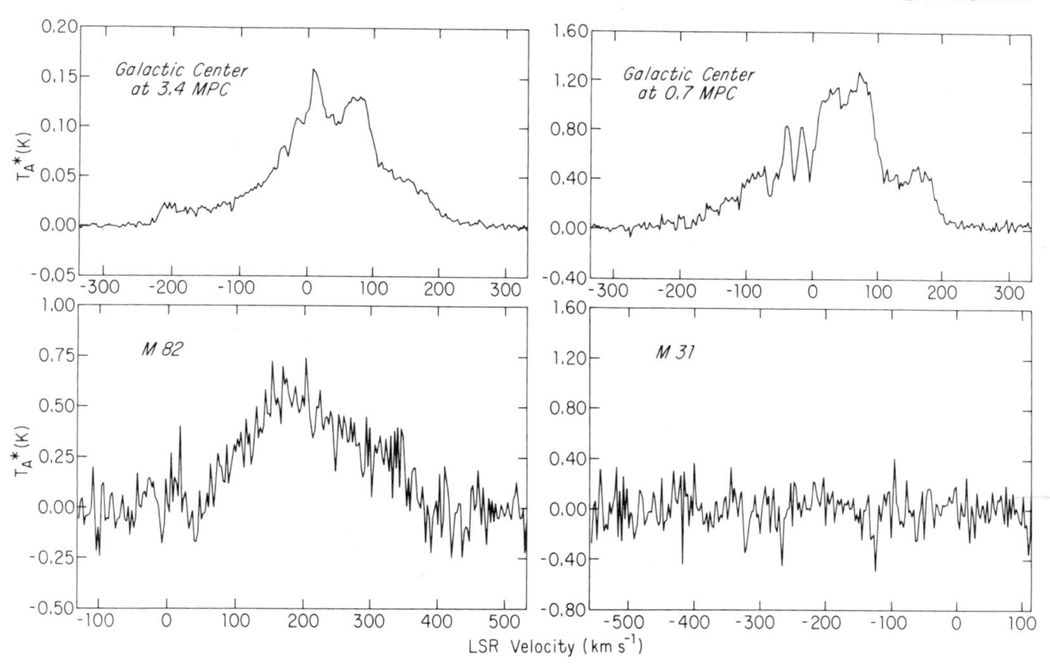

Figure 1. Comparison of CO data for M82 and M31 with synthetic spectra of the Galactic nucleus placed at their respective distances.

M82 and NGC 253 are comparable to the widths of the CO emission features, although the absorption arises from a more narrowly defined region of the galaxy. This also supports the inference of non-circular motions in these sources. Observations of the satellite transitions in NGC 253 and NGC 4945 (Gardner and Whiteoak 1975, Whiteoak and Gardner 1975) show strong hyperfine anomalies, which can be explained by the transport of intense infrared radiation from the nucleus through the OH. The same effect is seen in the Galactic center (Whiteoak and Gardner 1976a).

Also seen in the central molecular sources are HCN emission (Rickard et al. 1977c), H_2CO absorption (Gardner and Whiteoak 1974), and maser emission from H_2O (Lépine and Dos Santos 1977) and OH (Gardner and Whiteoak 1975, Nguyen-Q-Rieu et al. 1976). The OH masers are particularly remarkable, being 10 to 100 times as strong as Galactic masers, and strongest at 1667 MHz, rather than 1665 MHz. They may be amplifying the strong nuclear continuum sources.

DISK SOURCES

Some (but apparently not all) galaxies with bright nuclear CO sources also have detectable CO emission from extended disk components. The radii of the observable disks range from 6 to 10 kpc (Rickard et al. 1977b, Morris and Lo 1978, Rickard et al. 1978). When the data are averaged over azimuth, the resulting radial variation is generally a fairly smooth decline with the suggestion of a plateau for radii > 6 kpc, of intensity $\sim$ 25% of the peak intensity. The implied H_2 surface densities, 10 to 20 $M_\odot/pc^2$, are about twice the mean for the Galactic molecular annulus (Gordon and Burton 1976). There is no evidence for a trough or discontinuity in the radial distribution akin to that at 3-4 kpc in our Galaxy, linked by Gordon (1978) to the inner Lindblad resonance. However, the effect of the coarse angular resolution may be significant here.

Morris and Lo (1978) note that when the inferred H_2 distributions in NGC 6946 and IC 342 are combined with 21-cm data, the total gas density is roughly constant over the inner regions. They then argue that the 21-cm minima in the centers of many spiral galaxies arise from the conversion of atomic to molecular hydrogen. However, there are several cases of galaxies with nuclear HI minima and no compensating molecular sources (e.g., M31). Also there are cases of galaxies with similar H_2 distributions but quite different 21-cm distributions (e.g., NGC 6946 and M51).

H_2O and H_2CO have also been detected specifically towards disk-population sources. Churchwell et al. (1977) and Huchtmeier et al. (1978) have detected H_2O masers associated with several HII regions in M33 and IC 342. The source luminosities are in the range of the stronger Galactic H_2O masers (0.01 $L_\odot$ to $\sim$ 1 $L_\odot$), but a meaningful comparison with the Galactic luminosity distribution cannot be made yet. Whiteoak and Gardner (1976b) reported H_2CO absorption towards N159, in the LMC.

SPIRAL STRUCTURE

Do the molecular components of spiral galaxies themselves show any spiral structure? At present, this question reduces to asking whether there is any evidence for a preferential association of CO emission with spiral structure. One would expect an affirmative answer because of the association of bright Galactic CO sources with HII complexes, which are the sharp delineators of spiral patterns in other galaxies, and because of indirect evidence for such structure in our Galaxy (Roberts and Burton 1977). Unfortunately, present evidence is not compelling.

Combes et al. (1977b) searched for CO emission along the major axis and southern spiral arms of M31. From their detection statistics, they inferred that CO emission was present only on the inner sides of the 21-cm spiral arms within the mean corotation radius--as expected in density-wave models. However, Emerson (1978) has disagreed with this conclusion, noting that the HI distribution on the SW side is too irregular to allow a simple comparison with the under-sampled CO data, and that the CO data do not cover the outer edges of the smoother NE HI arms.

Some of the structure in extended CO disks of galaxies like NGC 6946 and M51 can be associated with particular features of the optical spiral patterns. But global spiral structure is not yet evident in the CO data.

REFERENCES

Bania, T. M.: 1977, Astrophys. J. 216, pp. 381-404.
Churchwell, E., Witzel, A., Huchtmeier, W., Pauliny-Toth, I., Roland, J. and Sieber, W.: 1977, Astron. Astrophys. 54, pp. 969-971.

Combes, F., Encrenaz, P. J., Lucas, R., and Weliachew, L.: 1977a, Astron. Astrophys. 55 pp. 311-314.
Combes, F., Encrenaz, P. J., Lucas, R., and Weliachew, L.: 1978, Astron. Astrophys. 61, pp. L7-L9.
Elias, J. H., Ennis, D. J., Gezari, D. Y., Hauser, M. G., Houck, J. R., Lo, K. Y., Matthews, K., Nadeau, D., Neugebauer, G., Werner, M. W., and Westbrook, W. E.: 1978, Astrophys. J. 220, pp. 25-41.
Emerson, D. T.: 1978, Astron. Astrophys. 63, pp. L29-L30.
Gardner, F. F., and Whiteoak, J. B.: 1974, Nature 247, pp. 526-527.
Gardner, F. F., and Whiteoak, J. B.:1975, M.N.R.A.S., 173, pp. 77P-81P.
Gardner, F. F., and Whiteoak, J. B.: 1976, Proc. Astron. Soc. Aust. 3, pp. 63-65.
Gordon, M. A.: 1978, Astrophys. J. 222, pp. 100-102.
Gordon, M. A., and Burton, W. B.: 1976, Astrophys. J. 208, pp. 346-353.
Hildebrand, R. H., Whitcomb, S. E., Winston, R., Steining, R. F., Harper, D. A., and Moseley, S. H.:1977, Astrophys. J. 216, pp. 698-705.
Huchtmeier, W., Witzel, A., Kuhr, H., Pauliny-Toth, I., and Roland, J.: 1978, Astron. Astrophys. 64, pp. L21-L24.
Huggins, P. J., Gillespie, A. R., Phillips, T. G., Gardner, F. F., and Knowles, S.: 1975, M.N.R.A.S. 173, pp. 69P-71P.
Lépine, J.R.D., and Dos Santos, P. M.: 1977, Nature 270, p. 501.
Morris, M., and Lo, K. Y.: 1978, Astrophys. J., in press.
Nguyen-Q-Rieu, Mebold, U., Winnberg, A., Guibert, J., and Booth, R.: 1976, Astron. Astrophys. 52, pp. 467-469.
Rickard, L. J, Palmer, P., and Turner, B. E.: 1978, in preparation.
Rickard, L. J, Palmer, P., Morris, M., Turner, B. E., and Zuckerman, B.: 1978a, Astrophys. J. 213, pp. 673-695.
Rickard, L. J, Palmer, P., Turner, B. E., Morris, M., and Zuckerman, B.: 1977c, Astrophys. J. 214, pp. 390-393.
Rickard, L. J, Turner, B. E., and Palmer, P.: 1977b, Astrophys. J. (Letters) 218, pp. L51-L55.
Roberts, W. W., and Burton, W. B.: 1977, in "Topics in Interstellar Matter", H. van, Woerden (ed.), Reidel, Dordrecht, pp. 195-205.
Weliachew, L.: 1971, Astrophys. J. (Letters) 167, pp. L47-L52.
Whiteoak, J. B., and Gardner, F. F.: 1973, Astrophys. Lett. 15, pp. 211-215.
Whiteoak, J. B., and Gardner, F. F.: 1975, Astrophys. J. (Letters) 195, pp. L81-L84.
Whiteoak, J. B., and Gardner, F. F.: 1976a, M.N.R.A.S. 174, pp. 627-636.
Whiteoak, J. B., and Gardner, F. F.: 1976b, M.N.R.A.S. 174, pp. 51P-52P.

DISCUSSION

van der Hulst: I have the impression that CO has been detected in primarily late type galaxies. If so, is this a real effect or is this due to observational selection?

Rickard: It is true that the galaxies with bright nuclear CO sources are mainly of Hubble type Sc. However, NGC 1068 is Sb; and M31 and M81, for which only disk-population sources are reported, are Sb also. I think that the present differences in the detection statistics for different Hubble types are not significant.

HI AND CONTINUUM STRUCTURE OF EXTERNAL GALAXIES COMPARED TO THE SITUATION IN THE GALAXY

Arnold H. Rots
Netherlands Foundation for Radio Astronomy
Dwingeloo, The Netherlands

Continuum observations of NGC 6946 strongly suggest the presence of an exponential non-thermal disk component. When applied to galactic 408 MHz observations, the scale length is found to be between 5 and 6 kpc. Continuing the parallel with NGC 6946, a total luminosity of $2 \times 10^{10} L_{\odot}$ is derived. Observations of the edge-on galaxies NGC 891 and 4631 show the existence of flattened halos with a steep spectrum around these galaxies. This is consistent with the constraints that are known for the galactic radio halo. On the basis of the radial distribution of HI in the Galaxy and in M81 three regimes can be defined: a central one which is hyper-deficient in hydrogen; a middle one which is deficient in HI, but not necessarily in total hydrogen; and an outer one which is neither deficient in HI, nor in H. Density wave theory, combined with gas flow dynamics appear a powerful tool in interpreting the kinematical data. No rotation curve is yet known to become Keplerian in the outer parts. Many external galaxies are warped, like our own; one of them, NGC 5907, does not have a visible companion.

1. INTRODUCTION

Since the previous IAU Symposium on the Galaxy (Galactic Radio Astronomy, 1973, Maroochydore) many new data have become available in the fields of galactic and extra-galactic radio astronomy. It is far from true, however, that a unifying picture has emerged, or that it could be put together in a paper like this one. I would just like to bring together some extragalactic examples and various aspects of galactic research.

For more detailed information and references to the existing literature the reader may be referred to the following reviews. For galactic continuum radiation to the ones by Price (1974b) and Baldwin (1977), for "HI and other tracers" in the Galaxy to those by Burton (1976,1977), and to the reviews by Van der Kruit and Allen (1976, 1978) on external galaxies.

W. B. Burton (ed.), The Large-Scale Characteristics of the Galaxy, 417–426.

2. DISTRIBUTION OF CONTINUUM RADIATION

2.1. Radial distribution

Price (1974a,b) showed that the non-thermal base disk component of the galactic radiation at low frequencies can be very well fitted by an exponential disk. I.e., a disk model in which the volume emissivity is proportional to exp (-r/a), where r is the galactocentric radius and a the scale length; he favoured a scale length of 6 kpc. Baldwin (1977), on the other hand, argued that a good fit can be obtained with a fairly uniform disk that rapidly falls off just inside the solar circle.

A suitable external galaxy for studying its non-thermal disk is the spiral system NGC 6946. Van der Kruit et al. (1977) obtained continuum distributions for this galaxy at 0.6, 1.4, and 5 GHz with the Westerbork Synthesis Radio Telescope (WSRT). They conclude that the base disk component shows a nearly exponential behaviour with radius.

This is one more reason to look at the exponential models again. Fig. 1 shows the continuum radiation vs. galactic longitude profile derived by Green.(1974) from her survey of the galactic plane at 408 MHz, done with the Molonglo Cross antenna. The heavy curve shows the profile that would be expected from an exponential disk with a scale length of 5 kpc. If one takes the liberty of having some doubts about the accuracy with which one can determine the zero level of these low frequency surveys, it is possible to get a very good fit for an exponential model with a scale length of 6 kpc as well. The broken curve represents such a model, assuming that the true zero level is 25 K below Green's.

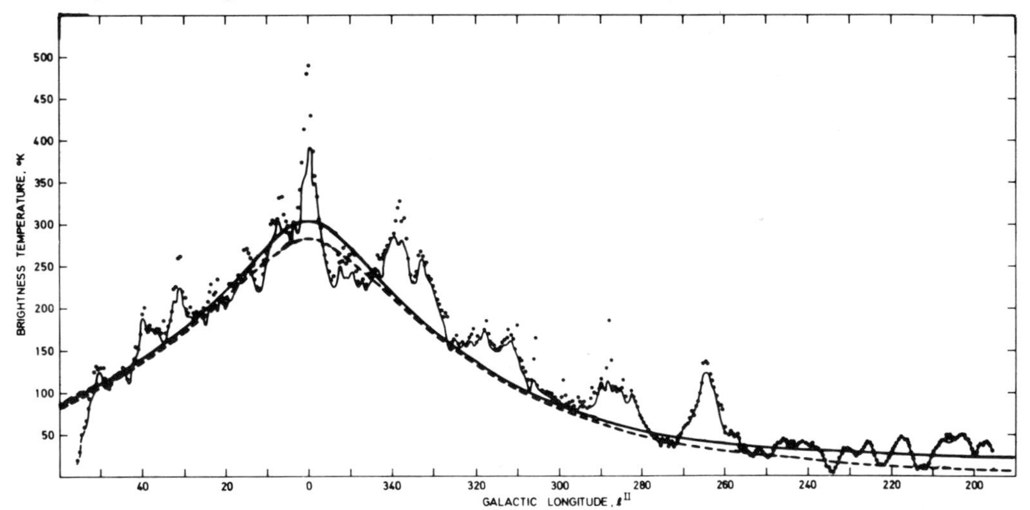

Fig. 1. Brightness temperature at 408 MHz along the galactic plane according to Green (1974). The smooth and broken line are exponential models for the disk with scale lengths of, resp., 5 and 6 kpc (see text).

Both fits (and especially the latter) are in my opinion quite satisfactory and certainly no worse than Baldwin's model. For a determination of the exact scale length one need to separate the disk radiation from the non-disk components in the profile. This is especially difficult in the center and anti-center directions. We can conclude that if the non-thermal base disk has an exponential brightness distribution, the scale length at 408 MHz must be between 5 and 6 kpc (assuming the distance to the galactic center to be 10 kpc).

Comparison with the light distribution proves to be very interesting. Van der Kruit et al. (1977) found that the scale length of the non-thermal disk in NGC 6946 is very close to the scale length of the exponential disk representing the distribution of visible light as derived by Ables (1971). A similar correspondence for M51 had been pointed out by Allen (1975). If this close correlation between the two scale lengths could be well established it would provide an independent method for estimating the total luminosity of the Galaxy (L_{pg}) from:

i. The scale length of the non-thermal disk (a).
ii. The distance of the galactic center (R_o).
iii. The local optical surface brightness in the solar neighbourhood ($\sigma_\odot$ ($L_\odot pc^{-2}$)).

L_{pg} can then be expressed as: $L_{pg} = b \times \sigma_\odot (L_\odot)$. Integrating out to a radius of 30 kpc one obtains $b = 1.15 \times 10^9$ for a = 5.5 kpc and R_o= 10kpc. The uncertainty in a is not important since it results in less than one percent variation in L_{pg}. For R_o = 8.5 kpc one obtains $b = 0.83 \times 10^9$. Hence we shall adopt $b = 10^9$ ($\pm$ 15% due to the uncertainty in R_o). Additional errors arise from:

i. The uncertainty in $\sigma_\odot$.
ii. Neglecting the light contribution from the spiral arms.
iii. Neglecting the light contribution from the nuclear bulge.

The total uncertainty may well be a factor 2. The result and two relevant ratios are shown in the table. The value for $\sigma_\odot$ (21 $L_\odot pc^{-2}$) has been taken from Oort (1965); the total mass from Innanen (1973); and the HI mass from Kerr and Westerhout (1965) and Burton (1976). The M/L averages for the two morphological types Sb and Sbc are from Roberts (1975). One should be warned that the higher weight that is usually attached to M_{HI}/L (since it is a distance independent parameter for external galaxies) is not justified in this particular case. The HI mass of the Galaxy is uncertain because of uncertainties in precisely that parameter: the distance. It seems fair to conclude that the present determination is consistent with the notion of our Galaxy being of morphological type Sbc or slightly earlier.

		Galaxy	<Sb>	<Sbc>
M_{pg}	(mag)	-20.4		
L_{pg}	($10^9 L_\odot$)	20		
M_T	($10^9 M_\odot$)	110		
M_{HI}	($10^9 M_\odot$)	3		
M_T/L	($M_\odot/L_\odot$)	5.5	6.4 ($\pm$2.1)	6.0 ($\pm$1.2)
M_{HI}/L	($M_\odot/L_\odot$)	0.15	0.17($\pm$0.03)	0.22($\pm$0.03)

2.2. z-distribution

I want to emphasize that the use of the word "halo" does not necessarily imply that the feature which it refers to has a spheroidal shape.

In recent years some new data have become available on halos of external galaxies. Allen et al. (1978) show the existence of a flattened halo around the edge-on galaxy NGC 891. At 1.4 GHz the continuum radiation extends at least to a height of 6 kpc above the plane of the galaxy, assuming a distance of 14 Mpc. Observations of NGC 4631 at the same frequency by Ekers and Sancisi (1977) show a vertical extent of at least 7 kpc, at an assumed distance of 5.2 Mpc. The axial ratio of the halo component is larger for NGC 4631 than for NGC 891. In both cases there are strong indications that the spectrum steepens considerably with increasing height above the plane: in the case of NGC 891 from 0.6 at heights up to $z = 2.5$ kpc to 1.0 or 1.5 at $z = 4$ kpc.

These data are consistent with statements by Price (1974b) and Baldwin (1977) that the existence of a radio halo with a steep spectrum around our Galaxy is not precluded by low-frequency observations, and with the steep spectrum halo component derived by Webster (1975). One should note, however, that such a halo may very likely not be spheroidal. A thorough investigation of the change of spectral index with z would be very interesting, but requires higher sensitivity and/or lower observing frequencies then the above mentioned WSRT observations offer.

2.3. Origin of the relativistic electrons.

The question of the origin of the relativistic electrons responsible for the large scale non-thermal radiation of galaxies is actually beyond the scope of this paper. Two short remarks can be made, however.

It is important to note that, apparently, the non-thermal disk (producing the bulk of the non-thermal radiation) mimics the Population II distribution. Whether the supernovae are distributed in the same manner is not entirely clear. The reader be referred to the short discussion on this subject by Van der Kruit et al. (1977).

In trying to understand the mechanisms that are at work, various other tracers may hold important clues. Galactic research has an important advantage here over extra-galactic research. For reasons of sensitivity and resolution several constituents (e.g. γ-radiation, molecular radiation) cannot be observed with enough detail in external galaxies.

3. THE DISTRIBUTION AND KINEMATICS OF NEUTRAL ATOMIC HYDROGEN

3.1. Radial distribution

The two most striking features in the radial distributions of HI in galaxies which have been known for many years, are the deficiency of HI in the central regions of most galaxies, and the fact that the HI seems to extend to much larger radii than all other tracers.

The radial HI distributions for the Galaxy (see Burton, 1976) and M81 (see Fig. 2 which is taken from Rots, 1975) both show a central

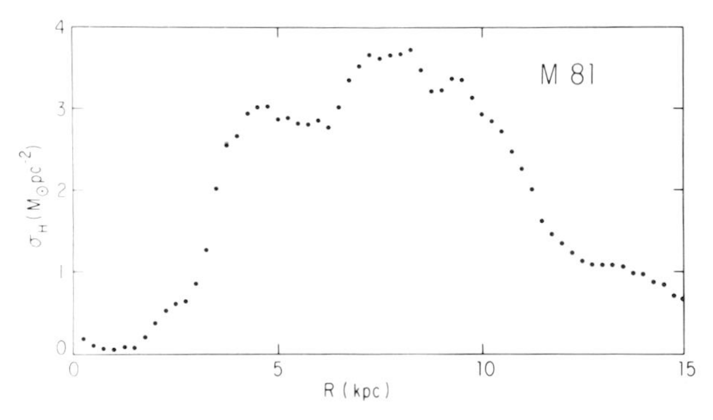

Fig. 2. HI surface density as a function of radius in M81 (from Rots,1975).

deficiency, a steep increase, a plateau of rather constant surface density, and a sudden drop followed by a gradual decline as radius increases. It is remarkable that in both cases the two discontinuities at either side of the plateau region seem to occur at about the same radii as, resp., the inner Lindblad resonance and corotation. However unfortunately located our vantage point in the Galaxy may be, it is here that we have the best observational conditions (in terms of sensitivity and resolution) for studying these resonance regions in detail.

As for the alleged large extent of the HI: Burton (1977, see Fig.3) shows that beyond 6 kpc the total hydrogen surface density (including molecular hydrogen) rather closely follows the total mass distribution as determined by Innanen (1973). The determination of the H_2 surface densities is not very certain, but for the Galaxy as well as for M81 the assertion seems valid that the relative HI surface density -after a gradual increase- remains more or less constant beyond the corotation region. One should bear in mind, however, that there still is considerable uncertainty in the mass models and in the galactic radial distribution of HI. On the other hand, Bosma (1978) finds the same effect in a

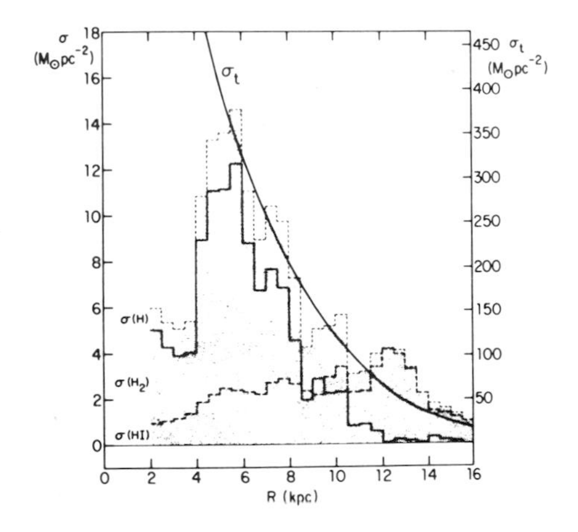

Fig. 3. Surface densities of HI, H_2, HI+H_2, and total mass (Innanen,1973) in the Galaxy, from Burton (1977).

sample of 14 galaxies.

Thus one may tentatively reformulate the "striking features" mentioned at the beginning of this section in terms of the relative HI surface density (μ_{HI}) and discern three areas: (in order of decreasing radius)

i. In the outer regime μ_{HI} is more or less constant. Most of the hydrogen is in the form of HI. There is no deficiency of either H or HI.

ii. In the middle regime there is an increasing deficiency of HI with decreasing radius (i.e.: μ_{HI} decreases). Since the fraction of H_2 increases rapidly in this region there may not be a deficiency of H (i.e.: μ_H may still remain constant)

iii. In the central regime there is a (hyper-)deficiency of both HI and H. μ_{HI} and μ_H quickly drop to (near) zero.

The central regime is obviously dominated by the nuclear bulge; studies of elliptical galaxies may help to understand this region. It is interesting to speculate whether or not the rough coincidence of the regime boundaries and the density-wave-theory resonances is significant.

3.2. Kinematics

Various types of kinematical data for many galaxies are available. We may refer to Van der Kruit and Allen (1978) for a review of these, and to Bosma (1978) for additional data. As far as the dependence on morphological type is concerned, the available data seem to support the notion (Roberts and Rots, 1973) that earlier type galaxies do show a higher degree of mass concentration toward the center.

Intimately related to the problem of determining rotation curves is the matter of spiral structure, if only because the kinematical effects of the latter confuse determination of the former. This has been worked out rather convincingly by Visser (1978) for the case of M81 (see Fig. 4). He corrected the velocity field for streaming motions and removed in that manner the steep decline around 8 kpc that caused Rots (1975) difficulties in fitting a mass model. It should be remarked, however, that in a case like this one, where a reasonably satisfactory rotation curve and mass model have been derived, yet another problem appears: deviations from the crucial assumption of axial symmetry (for M81 a north-south asymmetry in the velocity field; c.f. Rots, 1974, 1975).

This paper will not deal with spiral structure specifically, except to note that models based on density wave theory and gas flow dynamics (including shocks) seem quite capable of reproducing the observed kinematics satisfactorily. Such detailed models have been worked out for the Galaxy by Simonson (1976) and for M81 by Visser (1978). A basic difference in the approaches of these two authors is that Visser corrected the rotational data iteratively for gas flows linked to the spiral structure, while Simonson kept his rotation curve (Schmidt, 1965) fixed.

As far as the rotational data from HI observations in the outer parts of galaxies is concerned, it is fair to say that no rotation curve has yet been derived which turns Keplerian before radio telescopes run out of signal.

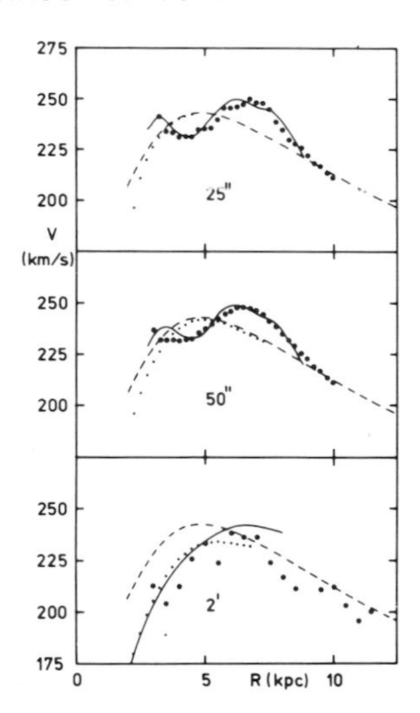

Fig. 4. Rotational data for M81 at different resolutions, from Visser (1978). Filled circles: observed velocities; broken curve: axisymmetric model; full curve: model plus streaming motions; dotted curve: axisymmetric model distorted by the beam.

3.3. HI in the outer parts of galaxies

It is well known that although the HI is generally concentrated in well defined flat disks, the outer parts of many galaxies greatly deviate from the planes of such disks. The Galaxy's HI disk is known to be warped in the outer regions, and several people have tried to explain this phenomenon by gravitational interaction with the Magellanic Clouds (see e.g. Hunter and Toomre, 1969). M31 is now known to be warped (Roberts and Whitehurst, 1975); M33 has a very strong warp (Rogstad et al., 1976); M83 too (Rogstad et al., 1974); IC 342 is conceivably warped (Rogstad et al., 1973); M81 shows HI almost everywhere and it seems unlikely that it would be confined to one plane; and there are many other examples. Quite interesting is the study of Sancisi (1976) who found warps in four out of five edge-on systems. Interesting because of the statistics, but also because one of the warped systems, NGC 5907, does not have a visible companion. This case should renew the interest in alternative warping mechanisms, as has been pointed out by Burton (1976).

The case of M81 that was just mentioned, does not have a proper, clean warp, but actually belongs in the realm of bridges, tails, and Magellanic Streams. Several years ago Toomre and Toomre (1972) produced the first successful gravitational interaction models. Some models have recently proved to be consistent with HI observations (with their added kinematical information): NGC 4038/9 by Van der Hulst (1977), NGC 3628/7 by Rots (1978). In our own Galaxy, however, we are faced with a warp, a Magellanic Stream, and the fact that we are not "outside observers", which seriously restricts our knowledge of the geometry and kinematics. And this yields another parallel with M81: both systems are extremely complicated as far as fitting gravitational interaction models is concerned.

REFERENCES

Ables, H.D.: 1971, Publ. U.S. Naval Obs., Ser. II, Vol. XX, Part IV.
Allen, R.J.: 1975, in L. Weliachew (ed.), "La Dynamique des Galaxies Spirales" (CNRS), p. 157.
Allen, R.J., Baldwin, J.E., Sancisi, R.: 1978, Astron. Astrophys. 62, p. 397.
Baldwin, J.E.: 1977, in C.E. Fichtel and F.W. Stecker (eds.), "Structure and Content of the Galaxy and Galactic Gamma Rays" (NASA), p. 189.
Bosma, A.: 1978, Dissertation, University of Groningen.
Burton, W.B.: 1976, Ann. Rev. Astron. Astrophys. 14, p. 275.
Burton, W.B.: 1977, in C.E. Fichtel and F.W. Stecker (eds.), "The Structure and Content of the Galaxy and Galactic Gamma Rays" (NASA), p.163.
Ekers, R.D., Sancisi, R.: 1977, Astron. Astrophys. 54, p. 973.
Green, A.J.: 1974, Astron. Astrophys. Suppl. Ser. 18, p. 267.
Hulst, J.M. van der: 1977, Dissertation, University of Groningen.
Hunter, C., Toomre, A.: 1969, Astrophys. J. 155, p. 747.
Innanen, K.A.: 1973, Astrophys. Space Sci. 22, p. 393.
Kerr, F.J., Westerhout, G.: 1965, in A. Blaauw and M. Schmidt (eds.), "Galactic Structure", Stars and Stellar Systems V, p. 167.
Kruit, P.C. van der, Allen, R.J.: 1976, Ann. Rev. Astron. Astrophys. 14, p. 417.
Kruit, P.C. van der, Allen, R.J.: 1978, Ann. Rev. Astron. Astrophys. 16, in press.
Kruit, P.C. van der, Allen, R.J., Rots, A.H.: 1977, Astron. Astrophys. 55, p. 421.
Oort, J.H.: 1965, in A. Blaauw and M. Schmidt (eds.), "Galactic Structure", Stars and Stellar Systems V, p. 455.
Price, R.M.: 1974a, Astron. Astrophys. 33, p. 33.
Price, R.M.: 1974b, in F.J. Kerr and S.C. Simonson (eds.), "Galactic Radio Astronomy", I.A.U. Symp. No. 60, p. 637.
Roberts, M.S.: 1975, in A. Sandage, M. Sandage and J. Kristian (eds.), "Galaxies and the Universe", Stars and Stellar Systems IX, p. 309.
Roberts, M.S., Rots, A.H.: 1973, Astron. Astrophys. 26, p. 483.
Roberts, M.S., Whitehurst, R.N.: 1975, Astrophys. J. 201, p. 327.
Rogstad, D.H., Lockhart, I.A., Wright, M.C.H.: 1974, Astrophys. J. 195, p. 309.
Rogstad, D.H., Shostak, G.S., Rots, A.H.: 1973, Astron. Astrophys. 22, p. 111.
Rogstad, D.H., Wright, M.C.H., Lockhart, I.A.: 1976, Astrophys. J. 204, p. 703.
Rots, A.H.: 1974, Dissertation, University of Groningen.
Rots, A.H.: 1975, Astron. Astrophys. 45, p. 43.
Rots, A.H.: 1978, Astron. J. 83, p. 219.
Sancisi, R.: 1976, Astron. Astrophys. 53, p. 159.
Schmidt, M.: 1965, in A. Blaauw and M. Schmidt (eds.), "Galactic Structure", Stars and Stellar Systems V, p. 513.
Simonson, S.C.: 1976, Astron. Astrophys. 46, p. 261.
Toomre, A., Toomre, J.: 1972, Astrophys. J. 178, p. 623.
Visser, H.C.D.: 1978, Dissertation, University of Groningen.
Webster, A.: 1975, Monthly Notices Roy. Astron. Soc. 171, p. 243.

DISCUSSION

Stecker: You mention the interesting possibility that the non-thermal emission in the Galaxy mimics the Population II distribution. If we assume that this reflects the distribution of sources of cosmic-ray electrons and of cosmic-ray nuclei as well, we would expect that the galactic nucleus would be much brighter in γ-rays than is observed. This may mean either that the sources of cosmic-ray electrons in our Galaxy have a different distribution than the sources of cosmic-ray nucleons (which seem to reflect the Population I distribution) or else the sources of cosmic-ray nucleons and electrons in the disk are the same (supernovae or pulsars) and there is an additional source (or sources) of cosmic-rays in the galactic nucleus. This later possibility seems more attractive to me.

Wielebinski: The spectral-index increase of nonthermal emission away from the planes of NGC 881 and NGC 4631 is certainly real. This increase gives us the lifetimes of relativistic electrons. We should go back and try to determine the exact spectrum of the nonthermal emission in the Galaxy, to put limits on lifetimes of our cosmic rays. The measurements will not be easy, but should be considered a challenge.

de Vaucouleurs: Your observation that the mean radial distribution of the radio continuum emission in the disks of spirals is exponential with essentially the same scale length as the optical emission is most significant. However, the radio spectral index varies with radius and so does the optical color index. So which radio wavelength should be compared with which optical wavelength?

Rots: In NGC 6946 there is a variation of spectral index with radius in the sense that the spectrum steepens with increasing distance from the center. However, this can be entirely accounted for by the distributed thermal radiation in the inner parts of the disk. Thus the answer is that one should use a low frequency; 408 MHz seems to be adequate.

Heiles: It seems to me that the facts that (1) the distributions of nonthermal and thermal (HII region) radio continuum have different shapes in any one galaxy, (2) that the relation between these shapes is different for M33, and (3) that HII regions are heated by hot, massive stars which become type II supernovae, show that at least some relativistic electrons are produced by processes other than type II supernovae. But, since type I supernovae also produce relativistic electrons, we cannot then conclude that processes other than supernovae are necessary.

Lockman: I do not agree that your conclusion necessarily follows from the facts you stated.

Heiles: I do not agree that my conclusion does not necessarily follow from the facts I stated.

Oort: With regard to the possibility of producing relativistic electrons in other ways than by supernovae, it should be pointed out that there is now plenty of evidence in radio galaxies that high-energy particles can be produced in shock regions not connected with supernovae.

Bok: So far in this Symposium we have ignored the Magellanic Clouds. For example, we have not mentioned yet that different processes of star formation are at work in our Galaxy and in the Magellanic Clouds. In our Galaxy, star formation is apparently associated with low temperature carbon monoxide and molecular hydrogen. However, at a distance of 60,000 parsec from our Sun there are the two Magellanic Clouds in which star formation is at work on a very big scale, apparently mostly without benefit of large amounts of cold cosmic dust. Evidently excessive amounts of neutral atomic hydrogen are present in the Magellanic Clouds at many positions where star formation is actively in progress. Let us not ignore the Magellanic Clouds whenever they offer to tell us something of interest for consideration in studies of our own Galaxy. May I suggest that everyone who works on the large scale structure of our Galaxy place a picture of the Large or Small Magellanic Cloud on the walls of her or his office. Please bear in mind that the Clouds are at distances one-tenth of the distance to Messier 31!

RADIAL DISTRIBUTIONS OF SOME CONSTITUENTS IN M31, THE GALAXY AND M33

Elly M. Berkhuijsen
Max-Planck-Institut für Radioastronomie, Bonn, F.R.G.

The Figure shows the distributions perpendicular to the planes of the galaxies of: (a) the surface density of HI, $\sigma_{HI}(R)$, (b) the surface density of HII regions, $\sigma_{HII}(R)$, and (c) the radio brightness temperature, $\sigma_{TB}(R)$. Details may be found in Berkhuijsen (A.A. 57, 9, 1977; Proc. IAU Symp. 77, 1978); here $\sigma_{HI}(R)$ of the Galaxy was taken from Gordon and Burton (Ap.J. 208, 346, 1976) and $\sigma_{TB}(R)$ of M31 has been revised. Distances to the centre are scaled with the Holmberg radius R_H (21, 15 and 7.8 kpc for M31, Galaxy, M33).

M31 is of Hubble type Sb, M33 of type Scd and the Galaxy of type Sbc (de Vaucouleurs, this Volume). $\sigma_{HI}(R)$ of the Galaxy is intermediate between that of M31 and that of M33. $\sigma_{HII}(R)$ is typical for the disk component which decreases roughly exponentially from its maximum outwards. Note that only in M31 $\sigma_{HII}(R)$ and $\sigma_{HI}(R)$ peak at the same distance from the centre. It will be intriguing to see whether the difference between $\sigma_{HII}(R)$ and $\sigma_{HI}(R)$ in these galaxies is related to the distribution of CO.

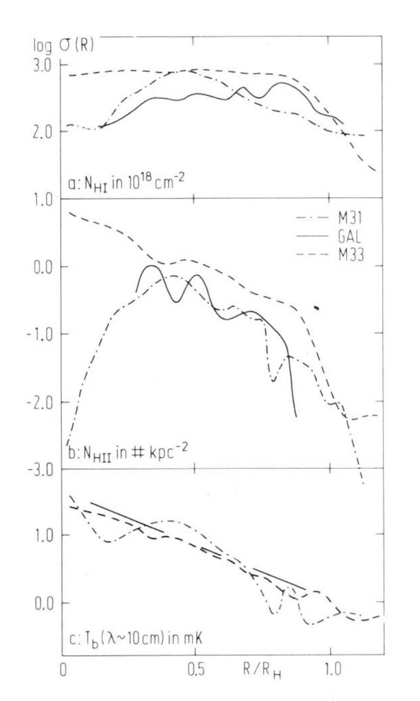

In both M31 and M33 $\sigma_{TB}(R)$ of the disk component is similar in shape to $\sigma_{HII}(R)$ but slightly flatter. Assuming that this also holds for our Galaxy a normalised scale length $L/R_H \simeq 0.3$ may be expected for σ_{TB} ($4.5 < R < 10$ kpc). Subtraction of the thermal contribution to $\sigma_{TB}(R)$ in M33 indicates that the distribution of the nonthermal emission is much flatter than $\sigma_{HII}(R)$ (Berkhuijsen, Proc. IAU Symp. 77, 1978). Therefore we may speculate that also in our Galaxy the nonthermal emission has a flatter distribution than $\sigma_{HII}(R)$ and may have $L/R_H \simeq 0.5$. Supernova remnants may then not be the only sources of relativistic electrons in the Galaxy (van der Kruit, Proc. IAU Symp. 77, 1978).

W. B. Burton (ed.), The Large-Scale Characteristics of the Galaxy, 427.

VIII. THE SPHEROIDAL COMPONENT

THEORETICAL OVERVIEW - INTERACTIONS AMONG THE GALAXY'S COMPONENTS

Beatrice M. Tinsley
Yale University Observatory

ABSTRACT. The structure and evolution of the Galaxy are reviewed in terms of interactions among various components, including its immediate surroundings. Emphasis is given to the large-scale processes responsible for star formation and chemical evolution, which are seen to be controlled by interactions between the ISM and other components and by gas flows. A model for formation of the Galaxy is outlined, in which the spheroidal component results from mergers of former small galaxies, and the disk has been accreted from outlying diffuse matter.

1. INTRODUCTION

Many advances in understanding the Galaxy have been due to a recognition of separate components, either conceptually, as in Baade's division of stars into two populations, or factually, as in the discovery of the interstellar medium (ISM). In trying to see the whole Galaxy in theoretical perspective, we find that much of its structure and evolution can be described in terms of interactions among such components. Figure 1 shows a subdivision of the Galaxy that is convenient in this context.

Disk	galàctic center		old disk stars	young stars, spiral arms	ISM
Spheroidal component		nuclear bulge	stellar halo		
Surroundings			satellite galaxies	invisible corona	IGM

Figure 1. A conceptual division of the Galaxy and its immediate surroundings into components.

W. B. Burton (ed.), The Large-Scale Characteristics of the Galaxy, 431–440.

Most of the borderlines between components in Figure 1 are intended to be only vaguely defined, because physical continuity makes any such divisions rather arbitrary. For example, stars of the nuclear bulge might be described as the innermost halo population, and there is a continuous distribution of ages between the oldest and the youngest disk stars. The intergalactic medium and satellite galaxies are included because they cannot be ignored in a large-scale picture of the Galaxy's structure and evolution; as Saar and Einasto (1977) convincingly argue, the whole hypergalaxy forms an interacting system. The term "halo" refers throughout this paper to ordinary halo stars in the classical sense - a spheroidal subsystem traced by the globular clusters; outlying gas is denoted "intergalactic medium" (IGM); and any extended hidden mass consisting of neither gas nor visible stars is referred to as the "corona".

Many of the processes that are familiar in studies of galactic structure and evolution can be seen as interactions between the components given in Figure 1. For the large-scale picture of interest here, two broad classes of interactions are especially relevant:

i) Exchange of matter between components. Many changes in the structure, stellar content, and chemical composition of the Galaxy are due to exchanges of matter between components, either by local physical processes such as stellar aging, or by bulk motions such as infall. Examples to be discussed below are the formation of stars from the ISM, ejection of matter (after nuclear processing) from stars back to the ISM, formation of the Galaxy itself from IGM and even from smaller satellite galaxies, and continued accretion of IGM onto the disk.

ii) Gravitational interactions. Several current problems in galactic dynamics involve questions of how the mass distribution in one component of the Galaxy affects the form and kinematics of others. Examples are whether density waves in the old disk underly spiral structure (Roberts, 1977), whether an inner halo stabilizes the disk (Ostriker and Peebles, 1973), and whether motions of satellites indicate a hidden corona (Hartwick and Sargent, 1977). This paper will emphasize processes that affect the stellar content and chemical composition of the Galaxy, rather than dynamical problems as such, which are reviewed elsewhere in this Symposium and in IAU Colloquium No. 45 (Basinska-Grzesik and Mayor, 1977). Although the corona may contain most of the mass of the Galaxy, it will be ignored here since the processes of interest are not necessarily affected by the depth of the potential well in which the hypergalaxy lies. Gravity is of course the force behind the large-scale motions that have shaped the Galaxy itself, which could in turn be important in star formation, according to some of the ideas discussed below.

This picture of the Galaxy as an interacting system will be kept in mind in discussing recent ideas on its current evolution (§ 2) and formation (§ 3), as a fairly typical spiral galaxy.

2. SOME EFFECTS OF PRESENT INTERACTIONS

Viewed on large scales, the evolution of the Galaxy now and during most of its past history is controlled by processes involving interactions among its constituents and immediate environment, some of which are reviewed here. The past several billion years are thought to have witnessed only mild changes as a result of such processes: the conversion of some ISM into stars, growth of the disk by accretion, a small fractional increase in the mean metallicities of stars and gas (especially in the outer disk where the time scales for change are longer), and evolution toward redder integrated colors as the ratio of old to new stars has increased (Audouze and Tinsley, 1976; Saar and Einasto, 1977; Tinsley and Larson, 1978a). Spiralling-in of satellites may also be adding sporadically to the Galaxy's mass (Tremaine, 1976). At the galactic center, more dramatic changes may have occurred.

This Section focusses on two related aspects of long-term evolution of the Galaxy: star formation and chemical evolution.

2.1. Causes of Star Formation

Still taking an overview that ignores small-scale physical processes, we can consider star formation to be caused primarily by dynamical compression of the ISM (Larson, 1977). Several different compression mechanisms have been suggested as contributing to star formation in the disk, each of which can be seen as an interaction between the ISM and another component of the hypergalaxy. One set of possibilities involves young stars: dense interstellar clouds can be compressed by blast waves from supernovae (Herbst and Assousa, 1977) or by ionization-shock fronts from hot stars (Elmegreen and Lada, 1977). Another idea is that star formation is due to infall of extragalactic gas (Larson, 1972), in particular to infalling clouds that collide with diffuse interstellar gas (Saar and Einasto, 1977). A third possibility is that interstellar clouds are compressed by shocks due to a galactic density wave (Roberts, 1969). There is at least one major site of star formation in the Galaxy, its center, where a density wave cannot be held responsible. The strongly non-circular or non-coplanar motions observed there suggest, however, that star formation is associated with some kind of dynamical disturbance (see reviews by other authors at this Symposium); further evidence that violent disturbances lead to rapid bursts of star formation is obtained from the colors of interacting galaxies (Larson and Tinsley, 1978; see below).

An important question is whether processes inside the Galaxy itself can sustain star formation in the disk, or whether a continual external impetus is required. If the main mechanism involves short-lived stars, such as supernovae or OB stars, then star formation may spread epidemically; and if old stars form a stable density wave that initiates star formation, the process could again be sustained indefinitely over much of the disk. But if infall is a major source of the necessary interstellar compression, then star formation is strongly linked to the Galaxy's environment.

Of the compression mechanisms mentioned above, only standard density wave theory implies direct involvement of old-disk stars in spiral structure; if instead star formation is due primarily to shock fronts from supernovae or OB stars, or to infall, spiral arms can be a dynamically superficial result of differential rotation (Larson, 1972; Mueller and Arnett, 1976; Saar and Einasto, 1977; Gerola and Seiden, 1978). Of course the differential rotation is itself due to the mass distribution of spheroidal and old-disk stars, so these components still play an indirect role in spiral structure. Moreover, Gerola and Seiden (1978) and Saar and Einasto (1977) have shown how the tightness of spiral winding then naturally increases with the bulge/disk ratio, as observed along the Hubble sequence.

It has often been hoped that the star formation rate could be regarded simply as an increasing function of the mean gas density in a region (Schmidt, 1959, 1963; de Jong and Maeder, 1977, pp. 169-172), but both theory and observation tell against any unique dependence of this form (Larson, 1977). A recent study shows, moreover, that Local Group galaxies with the most efficient present star formation, measured as the rate per unit mass of HI, are those with the *smallest* ratios of interstellar to total mass (Lequeux, 1978). It thus appears that the quantity of gas is determined by the efficiency of star formation, not vice versa; other factors, such as compression and cooling of the gas, must therefore be the chief elements in star formation.

An interesting case arises if infall of extragalactic gas gives the main *dynamical* impetus to star formation, since the infall rate in the Galaxy is thought to be comparable to the star formation rate. In this situation, the consumption and supply of gas in the Galaxy are closely related, and one can understand both how the ISM is not entirely consumed into stars in a few billion years and why chemical abundances change only slowly with time (Larson, 1972; Tinsley, 1977, and references therein).

2.2. Chemical Evolution in the Disk

The histories of chemical abundances in different regions of the Galaxy are thought to depend strongly on both star formation and gas flows, as reviewed by Audouze and Tinsley (1976). If gas flows are negligible, abundance levels depend logarithmically on the relative masses of ISM and stars, and the time scale for enrichment is given by the ratio of interstellar mass to the star formation rate. But if inflow and star formation occur at comparable rates, abundances approach equilibrium values on a time scale given by the ratio of interstellar mass to inflow rate.

For long-term evolution of the disk, the main factors are star formation, gas return from stars of both the disk and spheroidal components, accretion of IGM, radial inflow, and the chemical composition of gas involved in these processes. Many schematic models have illustrated how these factors can help explain salient chemical properties of the Galaxy, and relevant data and models were recently reviewed by many authors at IAU Colloquium No. 45 (Basinska-Grzesik and Mayor, 1977). It has become clear that a realistic picture of evolution in the disk should consider gas inflows from both the spheroidal component and the IGM. When these are included in calculations of chemical evolution, many formerly *ad hoc* model parameters - such as "prompt initial enrichment" and "infall" - appear as natural consequences of dynamical processes. (Of course, small-scale factors, such as nucleosynthesis and mixing of stellar ejecta into the ISM, also play key roles in chemical evolution. They are discussed in the reviews cited above.)

Three examples of possible large-scale dynamical effects on chemical evolution are the following: (1) Supernovae in the halo at its formation time could have produced a rain of metals sufficient to give the disk a substantial fraction of its present metallicity before many disk stars formed; thus there need be no problem in understanding the paucity of metal-poor dwarfs in the solar neighborhood (Ostriker and Thuan, 1975). (2) Further increase of the disk's metallicity would be suppressed by continuing infall of IGM, at a rate comparable to the star formation rate, as is predicted in dynamical models where the disk forms by accretion on a time scale of billions of years; this effect can explain the weakness of the age-dependence of metallicities of nearby disk stars (Tinsley and Larson, 1978a), and some other interesting details such as the near constancy of the interstellar beryllium abundance during the past 5 billion years (Reeves and Meyer, 1978). (3) The time scales for infall and consumption of gas by star formation increase radially outward in the slow collapse models just mentioned, so that the metallicities of stars and gas are predicted to decrease outward in the disk (Tinsley and Larson, 1978a), as observed.

Although these effects have led to plausible solutions to some puzzles concerning chemical evolution of the galactic disk, they show how complicated is the task of understanding other galaxies or the anomalous center of our own. Models that treat galaxies as closed systems, converting a given mass from gas into stars, are probably as misleading for other regions as for the solar neighborhood. In general, no part of any galaxy is likely to be isolated from its surroundings, and gas flows may play important roles both in chemical evolution and in star formation. The colors of a region are, in turn, determined mainly by the ratio of its present star formation rate to the integrated past rate (Larson and Tinsley, 1978), so even the photometric properties of galaxies depend in part on large-scale interactive processes. Finally, the present shape of the galactic disk is another property that could be strongly influenced by the spheroidal component and the IGM (Ostriker, 1977; Saar and Einasto, 1977).

3. FORMATION OF THE GALAXY FROM SUBSYSTEMS

The picture of disk formation reviewed above is one of accretion of IGM (and possibly gas-rich companion galaxies) around the nuclear bulge of the Galaxy, during many billions of years. As a final example of how interactions among various subsystems can be important in galactic evolution, the formation of the spheroidal component will also be considered as due to the coalescence of formerly separate components.

One piece of evidence favoring this view is that a gaseous protogalaxy of a few times 10^{11} $M_{\odot}$ would probably have a Jeans mass of only $\sim 10^{7} - 10^{9}$ $M_{\odot}$, and so consist initially of many lumps (Larson, 1969). Another suggestive argument is that the present frequency of collisions between galaxies is such that most or all of them could have experienced mergers with others during their lifetimes (Toomre, 1977; Vorontsov-Velyaminov, 1977). In particular, Toomre (1977, p. 420) has commented that present-day collisions could be simply the "dregs" of a process that was very common in the past.

It is therefore interesting to replace the conventional view of a rather smoothly collapsing protogalactic cloud with one of an initial cluster of small gas clouds that collide and merge among themselves. Tinsley and Larson (1978b) suggest that spheroidal systems form by a hierarchical sequence of such mergers; the mergers are assumed to induce bursts of star formation, as suggested by an earlier result that many strongly interacting galaxies have colors consistent with recent bursts (Larson and Tinsley, 1978). Each time a given galaxy grows by merging with another of comparable size, some of the residual gas is turned into stars, so that the metallicities of both gas and stars increase with the total mass of the system. Star formation in mergers is cut off before the gas is completely used up, because the gas eventually becomes too hot to cool and form stars between collisions. However, in most cases the violent growth process stops before this stage is reached, because there are no more neighbors to merge with. If the residual gas is swept away by an ambient IGM, an elliptical galaxy remains, but if diffuse outlying gas stays bound to the system, it may be accreted gradually in a disk.

This picture accounts qualitatively for the relatively low frequency of disk galaxies in dense clusters, it explains the observed iron content of intracluster gas as due to residual enriched gas lost by ellipticals, and it predicts that the spheroidal components of spiral galaxies should be identical to elliptical galaxies. An increase of metallicity with mass for ellipticals is a direct consequence of the proposed star formation in mergers; Tinsley and Larson (1978b) find that this relation has approximately the form suggested by the empirical color-magnitude relation of ellipticals if the efficiency of star formation is allowed to increase with successive mergers, i.e. to increase with the mass of the merged system. This requirement is itself found to be plausible in terms of the amount of ISM that would be compressed enough to form stars in a given collision.

The nuclear bulge and halo of the Galaxy are, in this view, simply an elliptical galaxy. Direct evidence for its formation by mergers of subsystems would have been almost obliterated, since tidal disruption and evaporation would have destroyed clumping on scales larger and smaller, respectively, than globular clusters (Rees, 1977). Nevertheless, indirect support for the model is given by Searle's (1977) result that the metallicity distribution of globular clusters can be explained if halo stars formed in isolated subsystems that later merged. A metallicity gradient in the nuclear bulge and inner halo is also consistent with star formation in mergers, since at each stage the gas would cool and condense toward the center before forming new stars, which would always be more metal-rich than the stars formed earlier. Some of the gas that later made the disk would have been enriched during the formation of the spheroidal component, so the initial disk gas could have a significant metallicity, as in the earlier models discussed in § 2.

In many respects, the merger picture is just a close-up view of previous models of galaxy formation, in which spheroidal systems result from rapid star formation during the initial collapse and disks form from dissipative gas with less efficient star formation (Larson, 1976; Gott, 1977). The main innovation is to provide a specific mechanism - collisions between galaxies - for the early efficient star formation needed to produce an elliptical system. Moreover, the main processes envisioned during galaxy formation can be studied directly by interpreting observations of present-day interacting systems: occasional collisions between gas-rich galaxies appear to result in mergers that form elliptical systems and involve bursts of star formation (Toomre, 1977; Larson and Tinsley, 1978); and cD galaxies appear to be growing continually by accretion of other ellipticals at the centers of dense clusters (Ostriker and Hausman, 1977).

As for our own Galaxy, infall and the spiralling-in of satellites represent continuations of the processes of dissipation and dynamical friction that have shaped the system from its start as a swarm of small clouds. It is not difficult to imagine that the bulge and halo are essentially an elliptical galaxy, whose signs of former substructure are the globular clusters and outlying dwarf spheroidal companions, while the disk is still the site of conversion of intergalactic into interstellar matter and ultimately into stars.

ACKNOWLEDGMENTS

It is a pleasure to thank R. B. Larson for many valuable discussions. This work was supported in part by the National Science Foundation (Grant AST77-23566) and the Alfred P. Sloan Foundation.

REFERENCES

Audouze, J., and Tinsley, B.M.: 1976, Ann. Rev. Astron. Astrophys. 14, p. 43.
Basinska-Grzesik, E., and Mayor, M.: 1977, eds., *Chemical and Dynamical Evolution of Our Galaxy, IAU Colloquium No. 45,* Geneva Obs., Geneva.
de Jong, T., and Maeder, A.: 1977, eds., *Star Formation, IAU Symposium No. 75,* D. Reidel Publ. Co., Dordrecht.
Elmegreen, B.G., and Lada, C.J.: 1977, Astrophys. J. 214, p. 725.
Gerola, H., and Seiden, P.E.: 1978, Astrophys. J., in press.
Gott, J.R.: 1977, Ann. Rev. Astron. Astrophys. 15, p. 235.
Hartwick, F.D.A., and Sargent, W.L.W.: 1978, Astrophys. J. 221, p. 512.
Herbst, W., and Assousa, G.E.: 1977, Astrophys. J. 217, p. 473.
Larson, R.B.: 1969, Monthly Notices Roy. Astron. Soc. 145, p. 405.
Larson, R.B.: 1972, Nature 236, p. 21.
Larson, R.B.: 1976, Monthly Notices Roy. Astron. Soc. 176, p. 31.
Larson, R.B.: 1977, in *The Evolution of Galaxies and Stellar Populations,* ed. B.M. Tinsley and R.B. Larson, Yale Univ. Obs., New Haven, p. 97.
Larson, R.B., and Tinsley, B.M.: 1978, Astrophys. J. 219, p. 46.
Lequeux, J.: 1978, Astron. Astrophys., in press.
Mueller, M.W., and Arnett, W.D.: 1976, Astrophys. J. 210, p. 670.
Ostriker, J.P.: 1977, in *Chemical and Dynamical Evolution of Our Galaxy, IAU Colloquium No. 45,* ed. E. Basinska-Grzesik and M. Mayor, Geneva Obs., Geneva, p. 241.
Ostriker, J.P., and Hausman, M.A.: 1977, Astrophys. J. (Letters) 217, p. L125.
Ostriker, J.P., and Peebles, P.J.E.: 1973, Astrophys. J. 186, p. 467.
Ostriker, J.P., and Thuan, T.X.: 1975, Astrophys. J. 202, p. 353.
Rees, M.J.: 1977, in *The Evolution of Galaxies and Stellar Populations,* ed. B.M. Tinsley and R.B. Larson, Yale Univ. Obs., New Haven, p. 339.
Reeves, H., and Meyer, J.P.: 1978, preprint.
Roberts, W.W.: 1969, Astrophys. J. 158, p. 123.
Roberts, W.W.: 1977, in *Chemical and Dynamical Evolution of Our Galaxy, IAU Colloquium No. 45,* ed. E. Basinska-Grzesik and M. Mayor, Geneva Obs., Geneva, p. 11.
Saar, E., and Einasto, J.: 1977, in *Chemical and Dynamical Evolution of Our Galaxy, IAU Colloquium No. 45,* ed. E. Basinska-Grzesik and M. Mayor, Geneva Obs., Geneva, p. 247.
Schmidt, M.: 1959, Astrophys. J. 129, p. 243.
Schmidt, M.: 1963, Astrophys. J. 137, p. 758.
Searle, L.: 1977, in *The Evolution of Galaxies and Stellar Populations,* ed. B.M. Tinsley and R.B. Larson, Yale Univ. Obs., New Haven, p. 219.
Tinsley, B.M.: 1977, Astrophys. J. 216, p. 548.
Tinsley, B.M., and Larson, R.B.: 1978a, Astrophys. J. 221, p. 554.
Tinsley, B.M., and Larson, R.B.: 1978b, preprint.
Toomre, A.: 1977, in *The Evolution of Galaxies and Stellar Populations,* ed. B.M. Tinsley and R.B. Larson, Yale Univ. Obs., New Haven, p. 401.
Tremaine, S.D.: 1976, Astrophys. J. 203, p. 72.
Vorontsov-Velyaminov, B.A.: 1977, Observatory 97, p. 204.

DISCUSSION

Yahil: It is worthwhile pointing out that the buildup of large galaxies from smaller subunits, by means of dynamical friction and subsequent merger, can proceed at a reasonable rate only if the size of the subunit is comparable to their mean separation at the time of merger. Once the mean separation becomes too big, the process will quickly stop. The spectrum of initial perturbations which leads to galaxy formation must therefore have a hierarchial structure which will enable the continued growth of galaxies from small subunits to present-day galaxies. The continued growth of galaxies today in turn depends on the extent of their massive halos (for which there is no secure upper limit) in relation to their mean separation.

Tinsley: The work by Simon White and his collaborators on this point is very relevant. White and Sharp have noted that halos cannot be too extensive, or else a large number of binary galaxies would be within an orbit of merger! The present luminosity function and clustering of galaxies are of course constraints on the initial perturbation spectrum in the merger model, as they are in any view of galaxy formation.

Verschuur: If galaxies form by accretion of smaller units would not the size correlate with age in some way?

Tinsley: This is true, but there would not necessarily be any observable consequences. For one thing, mergers and star formation could go to completion in a few dynamical times of the resulting present-day elliptical galaxy, so even the final star-forming stages of large galaxies could have taken place within 1 or 2 billion years of the start of the process. It would be impossible to detect such age differences among elliptical galaxies, but there does seem to be an age spread among halo stars in the Galaxy suggesting that the very metal-poor globular clusters (e.g., M92) formed before 47 Tuc (Demarque and McClure 1977, at the Yale conference). Another point is that the very biggest galaxies, which might be noticeably young according to their long collapse times, are here seen as forming simply by mergers of stellar systems; there is no further star formation because any gas is too hot. This is just Ostriker's view of the formation of cD galaxies. They were made recently, but of old stars.

Rubin: I would like to call attention to an observation which may be relevant to your model of a spiral as an elliptical plus disk. Some spirals (i.e., M31) have nuclear velocity dispersions of a few hundred km s^{-1}, and in spectra centered in the red, the Na D doublet (separation $\sim$6A) is completely blended. Other spiral galaxies show incredibly narrow, well-resolved nuclear Na D lines, which reproduce the velocity gradient shown by the excited gas. For these galaxies, the disk must extend within a few pc of the center. Why a larger velocity dispersion is not apparent, arising from the bulge stars, is a puzzle. For your model would you expect the disk to extend into the nucleus?

Tinsley: The merger model need not differ from any other in the inward extent of the disk, which we have not tried to predict. In general, the velocity dispersion of the disk need not correspond to that of a spheroidal system in which it is embedded.

Sullivan: How do S0's and other gas-deficient disk galaxies fit into your scheme? What is the relative time scale of disk formation and various possible stripping mechanisms which may occur in the cluster environment?

Tinsley: The origin of gas-deficient galaxies need not be any different in this model than in alternative models of galaxy formation. Perhaps some constraints on the time scale for gas loss or stripping of disks can be obtained from the recent studies by Butcher and Oemler of clusters of galaxies at redshifts ~ 0.4 that are morphologically as regular as Coma: although Coma's galaxies are nearly all red S0's and ellipticals, these clusters are dominated by systems as blue as active spirals. The implication is that many disks were still actively forming stars a few billion years ago, even in cluster environments most favorable to stripping. Butcher and Oemler's results indicate that disk formation can occur on a very long time scale. They are consistent with the apparent youth (mostly $< 10^9$ yr) of disk stars near the Sun, according to Demarque and McClure.

Graham: Dr. Tinsley's talk reminded me of some observations that I have made recently of the radio galaxy NGC 5128. Here, in the middle of a giant, non-rotating elliptical galaxy we see a rapidly rotating but apparently stable gaseous disk in which the stellar density is sufficiently low to enable strong interstellar sodium D-lines to be observed in absorption against the background elliptical stellar component. The rotation of the disk shows some irregularities which suggest that its formation may have occurred relatively recently, less than 10^9 years ago.

Tinsley: Thank you for mentioning this galaxy and your exciting new results. We had also thought of NGC 5128 as a possible example of a disk in formation around an elliptical galaxy, since Larson (1972) suggested that the material could have been recently accreted.

THE MASS AND LIGHT DISTRIBUTION OF THE GALAXY: A THREE-COMPONENT MODEL

J. P. Ostriker and J. A. R. Caldwell
Princeton University Observatory

ABSTRACT

The galaxy is represented schematically by a three-component model: a disc having the form of a modified exponential distribution, a spheroidal (bulge + nucleus) component and a dark halo component which, following the nomenclature of Einasto, we call the corona. The shapes of these components, chosen on the basis of observations of other galaxies, are consistent with imperfect knowledge of the Galaxy; values of the adjustable parameters are chosen by a least square minimization technique to best fit the most accurate kinematical and dynamical galactic observations. The local radius, circular velocity and escape velocity are found to be $(R_0, V_0, V_{esc}) = (9.05 \pm 0.33$ kpc, 247 ± 13 km/s, $550 \pm 24)$ quite close to the values determined from observations directly. The masses in the three components are $(M_D, M_{Sp}, M_C) = (0.78 \pm 0.13, 0.81 \pm 0.09, 20.3) \times 10^{11} M_0$ for a model with coronal radius of 335 kpc. If the quite uncertain coronal radius is reduced to 100 kpc the model is essentially unchanged except that then $M_C = 6.65 \times 10^{11} M_0$. The disc and spheroidal components have in either case luminosities (in the visual band of $(L_D, L_{Sp}) = (2.0 \quad ,0.2 \quad) \times 10^{10} L_0$. The galaxy is a normal giant spiral of type Sb-Sc similar to NGC 4565.

1. MODEL COMPONENTS

1.1 Background

Many investigators have constructed models of the galaxy on the basis of observations made necessarily from the somewhat unfortunate vantage of the sun, and over the years both the modeling techniques and the accuracy of the input data have been steadily refined. But several aspects of the galaxy such as the ratio of the disc to spheroidal components in the inner parts or the mass distribution in the outer parts are essentially unobservable with present techniques. Recently, however, observations of other galaxies have improved so that we now know more about some of the dynamical and kinematic properties of M31 or

W. B. Burton (ed.), The Large-Scale Characteristics of the Galaxy, 441–450.

NGC 4565 than of our own Galaxy. Thus, in this work we have chosen to assume that our galaxy is a normal giant spiral with two mass components following the light distribution observed in two-component fits to the surface brightness of other galaxies (e.g., de Vaucouleurs 1959) and to fix the adjustable parameters by fitting the best determined kinematical and dynamical properties of the Galaxy. While a number of components with varying degrees of flattening (as in the Schmidt, 1965, model) could have been used and may in fact be required, observational constraints on the inner parts of the galaxy do not allow one to subdivide the model into more than a flat and an approximately spherical part (the necessary flattening of the latter due to the gravitational field of the former being simply ignored). In addition, there is dynamical evidence (Ostriker et al. 1974; Turner 1976; Salpeter 1977) for an extended component of dark matter (cf. Spinrad et al. 1978) having a mass exceeding that in the inner parts of the galaxy and a rate of density decline of $\rho \propto r^{-2}$ approximately. Note that the modeling is based only on observations and not on any theoretical preconceptions, although there are reasons (Ostriker and Thuan 1975; Gunn 1977; Fall and Rees 1978) for believing that galaxy formation might occur in distinct stages corresponding to coronal, spheroidal and disc components. Let us now discuss in detail the parameterization of the three components.

1.2 The Disc

Freeman (1970) following de Vaucouleurs (1959) showed that the discs of flattened galaxies obey, in their outer parts, a simple exponential law with the surface brightness (in magnitudes/sec^2) falling linearly with projected radius. But in both its molecular and atomic components the gas content of our galaxy declines within 5 kpc of the center (cf. Scoville and Solomon 1975) after increasing in a more or less exponential fashion inwards towards this radius. Kormendy (1977a) finds that stellar discs also show a substantial decline in their inner parts (< 5 kpc) from that expected on the basis of the exponential law. We thus fit the discs with a surface mass distribution specified by three parameters,

$$\Sigma(\varpi) = \Sigma_D[\exp(-\varpi/\varpi_D) - \exp(-\varpi/\varpi_G)] , \tag{1}$$

where ϖ is the radial coordinate in the disc and (ϖ_D, ϖ_G) are radii characterizing the disc scale length and that of the central gap. A similar form has been used by Einasto (1970).

1.3 Spheroid

The light profile in elliptical galaxies is fairly well fit by a Hubble law (cf. Oemler 1976; Kormendy 1977b) in which the surface brightness falls as r^{-2} and the volume emissivity as r^{-3}. Since in the solar vicinity the population II tracers follow a similar distribution (Oort 1965), we thus adopt the Hubble law for the galactic surface mass density $\Sigma(\varpi) = \Sigma_{Sp}(1 + \varpi/\varpi_{Sp})^{-2}$. There is evidence (Kormendy (1977b) and

Spinrad et al. (1978) that at large radii the surface brightness of other galaxies declines somewhat faster than given by the Hubble law, but a cutoff or correction to the Hubble law is unnecessary since at those radii the mass density in the computed models is determined by the corona, not the spheroidal component and thus the dynamical fitting procedure would be quite insensitive to any outer cutoff of the spheroid component. The three-dimensional ($r = |\mathbf{r}|$) density distribution ρ, determined by inverting Abell's equation is

$$\rho_{Hub}(r) = 3.75\ \rho_{Hub} \times \begin{cases} \left[\dfrac{3-z}{\sqrt{z}}\ \ln\left(\dfrac{1+\sqrt{z}}{\sqrt{1-z}}\right) - 3\right]/z^2, & z < 1 \\ 1/3.75\ , & z = 1 \\ \left[\dfrac{3+z}{\sqrt{z}}\left\{\tan^{-1}\left(\dfrac{1}{\sqrt{z}}\right)+\dfrac{\pi}{2}\right\}-3\right]/z^2, & z > 1 \end{cases} \qquad (2a)$$

where

$$z \equiv |(r/r_{Sp})^2 - 1|\ . \qquad (2b)$$

To this we add a small ($\sim 10^8\ M_{\odot}$) nuclear mass component M_N which is seen in the infrared, and detected dynamically in M31 where a similar infrared nuclear profile is observed. Thus three parameters determine the spheroidal mass distribution:

$$\rho_{Sp}(r) = M_N\ \frac{\delta(r)}{4\pi r^2} + \rho_{Hub}(r)\ . \qquad (2c)$$

1.4 The Dark Corona

We choose the analytically simple two-parameter form

$$\rho_C(r) = \frac{\rho_C}{1 + (r/r_C)^2}\ , \qquad (3)$$

which at large radii $r >> r_C$ approaches the density distribution in an isothermal sphere.

A cutoff radius R_0, applied to all components ($\rho = 0$ for $r > R_0$), is fixed in advance for each model rather than treated as an adjustable parameter to be determined by observations. In all we have nine parameters required to specify a given model.

2. OBSERVATIONAL CONSTRAINTS

Given values of the model parameters, we can calculate the density, gravitational potential and any stellar orbit. Thus the galactic rotation curve, Oort constants, etc. that would be observed from any point in the Galaxy can be computed and compared with observations made locally. On minimizing the difference between observed and computed quantities we determine best values for the model parameters and for the point of observation.

The observational inputs used to constrain the model may conveniently be divided into three groups of four.

2.1 Local Constraints

The most important observed quantity for all the parameters is fortunately fairly well known. From a search of the literature we have arrived at the value of the sun's position $R_\odot = 8.9 \pm 0.6$ kpc. Details of the determination will be given elsewhere but we have essentially combined 18 independent determinations (using R R Lyrae stars, globular clusters, etc.) weighting them in the final result inversely as the square of the individual fractional error.

For Oort's constant A an analysis of seven methods and 45 sources gave 15.2 ± 0.4 km/s/kpc.

As is well known, Oort's constant B is quite poorly known. We examined 18 determinations based either directly on proper motions or indirectly upon the ratio B/A obtained from the velocity ellipsoid axis ratios. Weighting together all determinations from the first method gave -10.4 ± 10 km/s/kpc and from the second gave -9.7 ± 2 km/s/kpc, but we noticed that the more recent studies gave systematically larger (in absolute value) estimates of B as well as larger estimates of the error, although the data had presumably improved! This makes a simple compounding of the weighted results nonsense and we decided to limit consideration to the most recent sources (Fatchikhin 1970; Vasilevskis and Klemola 1971; Fricke and Tsioumis 1975 for proper motions; Erikson (1975) for the velocity ellipsoid). These yielded $B = -11.6 \pm 2.6$ and $B = -11 \pm 2$ respectively. By averaging the results from both methods then assigning an estimated uncertainty of 20% to the mean we obtain our adopted value of $B = -11.3 \pm 2.3$ km/s/kpc.

The adopted values in standard units of ($R_\odot$, A, B) are (8.9 ± 0.6, 15.2 ± 0.4, -11.3 ± 2.3) which are consistent with the IAU (1964) system (10 ± 1, 15 ± 1.5, -10 ± 2) and Oort's suggested revision (8.7 ± 0.6, 16.9 ± 0.9, -9.0 ± 1.5) in Plaut and Oort (1975).

The remaining local constraint is $\mu_\odot$ the local mass per unit area of the galactic plane in the flattened component. On the basis of VanderVoort's (1970) and Toomre's (1972) discussion of Oort's (1960) K_z study we take $\mu_\odot = 90 \pm 9\ M_\odot/pc^2$.

2.2 Rotation Curve Constraints

Following the method of Toomre (1972) we judged that the large scale galactic structure information inherent in the interior region could satisfactorily be incorporated into the model by having it fit the observed maximum recession velocities, Δv, at four longitudes corresponding to $\sin^{-1} \ell$ = (0.0671, 0.3448, 0.5747 and 0.8046) corresponding to R = .58, 3, 5, and 7 kpc, for R_0 scaled to 8.7. For the first of these we take $\Delta v = 250.5 \pm 8.0$ km/s from Rougoor and Oort (1960). For the remaining three velocities, which characterize the main hump of the rotation curve, we take $\Delta v = (131.2 \pm 6.3, 98.1 \pm 6.9, 54.8 \pm 7.1)$ km/s from Tuve and Lundsager's (1973) smoothed rotation curve B_0. At the adopted separation in longitude, the points may be considered to be statistically independent; the uncertainties were set at 3% of the typical corresponding circular velocity following Burton's (1971) suggestion that streaming motions associated with the spiral arms of that order mask the underlying smooth curve.

2.3 Supplemental Constraints

These are additional observations which are necessary to define the model but which do not affect significantly the rotation curve in the vicinity of the sun.

The nuclear mass M_N was taken to be $(1.6 \pm 0.4) \times 10^8$ $M_\odot$ from dynamical studies of M31. Oort (1977) and Tremaine (1976) obtain estimates of $(0.3, 0.6) \times 10^8$ respectively from analysis of the likely number of globular clusters sinking to the center due to dynamical friction so it is possible that our adopted value is somewhat too large. The halo core was fixed at $r_{sp} = 0.11 \pm 0.02$ kpc from comparison of infrared photometry of M31 and the Galaxy's nuclear regions.

The remaining two constraints are designed to determine the characteristics of the coronal component. From an analysis of Eggen's (1964) catalog of high velocity stars, the details of which are to be given elsewhere, we find $V_{esc} = 558 \pm 78$ km/s. This is to be compared with Schmidt's (1965) value of 380 km/s and Innanen's (1973) value of 374. Parenthetically we note that Hesser and Hartwick's (1976) observation of a globular cluster with radial velocity 273 km/s at a distance of R = 26 ± 5 kpc implies a minimum escape velocity of 430 km/s in the framework of this model. Finally, since galaxy rotation curves are typically flat or decreasing (but see the apparent counterexamples NGC 2590 and NGC 1620 found by Rubin et al. 1978) limits on the coronal radius r_c can be set which, with the optimization scheme to be described shortly, determine that parameter.

3. RESULTING MASS MODEL

Given a set of model parameters (including R_0) we calculate the potential by standard analytical and numerical methods and compare with the observational constraints, computing thereby

$$\chi^2 \equiv \sum_{i=1}^{12} (Q_{i,obs} - Q_{i,calc})^2/\sigma^2_{i,obs}$$

TABLE I. SOLUTION FOR MODEL PARAMETERS

Observer's Position	$R_\Theta = 9.05 \pm 0.33$ kpc	Spheroidal Component	$\rho_{Sp} = 145 \pm 63\ M_\Theta/pc^3$
			$r_{Sp} = 0.104 \pm 0.012$ kpc
Disc Component	$\Sigma_D = 7.23 \times 10^4\ M_\Theta/pc^2$		$M_N = 1.6 \pm 0.4 \times 10^8\ M_\Theta$
	$\varpi_D = 2.31711$ kpc		
	$\varpi_G = 2.27969$ kpc	Coronal Component	$\rho_C = 2.19^{+1.90}_{-1.69} \times 10^{-3}\ M_\Theta/pc^3$
	$M_D = 7.81 \pm 1.31 \times 10^{10}\ M_\Theta$		$r_C = 15.4 \pm 5.3$ kpc
	$\langle r^2 \rangle^{1/2} = 7.96 \pm 0.41$ kpc		

TABLE II. THREE-COMPONENT GALACTIC MASS MODEL

R kpc	V_{cir} km/s	V_{esc}	$-\Phi_D$ $10^3(km/s)^2/kpc$	$-\Phi_{Sp}$	$-\Phi_C$	Φ'_D $10^2(km/s/kpc)^2$	Φ'_{Sp}	Φ'_C	Σ_D M_Θ/pc^2	Σ_{Sp}	Σ_C	M_D $10^9 M_\Theta$	M_{Sp}	M_C	ρ_{tid} M_Θ/pc^3
.125	251	913	73	257	86	-19	5070	0.0	61	36300	106	0.0	1.8	0.0	238
.25	274	858	74	208	86	-27	3030	0.1	115	15200	106	0.0	4.4	0.0	83
.5	270	795	74	155	86	-31	1490	0.2	206	5210	106	0.1	8.7	0.0	23
1	245	734	76	107	86	-23	623	0.4	331	1560	105	0.8	14	0.0	5.2
2	219	682	76	70	86	9.0	230	0.8	429	429	105	4.5	21	0.1	.97
3	219	653	74	53	86	35	123	1.2	416	197	104	11	26	0.2	.36
4	228	631	70	43	86	51	78	1.5	359	113	102	20	29	0.6	.19
5	238	612	64	37	86	57	54	1.9	291	73	100	29	31	1.1	.14
6	245	594	58	32	86	58	40	2.2	226	51	98	38	33	1.8	.11
7	249	578	53	29	85	55	31	2.5	170	38	96	46	35	2.8	88 D-3
8	249	563	47	26	85	50	25	2.7	126	29	94	53	37	4.1	74 D-3
9	*247*	*550*	*43*	*24*	*85*	*45*	*20*	*3.0*	*92*	*23*	**91**	*59*	*38*	*5.6*	*62 D-3*
10	243	539	39	22	85	39	17	3.2	66	19	89	63	39	7.4	52 D-3
15	216	497	25	16	83	19	8.4	3.9	11	8.3	76	75	44	20	21 D-3
20	196	472	18	13	81	9.9	5.1	4.1	1.7	4.7	64	77	48	39	9.3 D-3
30	178	441	11	9.1	77	4.0	2.5	4.1	0.0	2.1	48	78	52	85	3.0 D-3
50	170	405	6.8	5.9	69	1.4	1.0	3.4	0.0	0.7	31	78	58	198	0.9 D-3
100	168	356	3.4	3.2	57	0.3	0.3	2.2	0.0	0.2	16	78	67	509	0.2 D-3

We then adjust the values of the model parameters in an attempt to minimize χ^2 using a variation of the standard Levenberg-Marquardt algorithm for nonlinear least square minimization. The model parameters determined by this method are given in Table I. The model itself is presented in Table II; column 1 is the distance from the center (in the galactic plane), 2 and 3 give the local circular and escape velocities, 3-5 the gravitational potential due to the three components, 6-8 the forces, 9-11 the projected mass densities, 12-14 the interior mass in each component and the last column gives ρ_{tid}. The row corresponding most closely to the solar position is shown in italics.

The model has a value of χ^2 equal to 0.90 which for three degrees of freedom indicates a respectable probability of 82% that the model agrees with the observations. The χ^2 test indicates a slightly worse result if the nuclear component is omitted and a much worse fit if the Plaut and Oort local constants are used (due to the considerably smaller value of B).

4. LIGHT DISTRIBUTION

Consistent with the approach we have taken up to this point we need only determine the local light per unit area in the disc and spheroidal components, compare with the model to find the local mass-to-light ratio, and applying that universally in the galaxy determine the light distribution. From Weistrop's (1972) counts of high latitude blue stars we find a local spheroidal visual luminosity of 0.5 $L_\odot/pc^2$ for a local mass-to-light ratio of approximately 40. The disc mass-to-light ratio is approximately 4.0 from Oort (1965) giving total visual luminosity in the two components of 2.0×10^9 and 2.0×10^{10} $L_\odot$, with a total magnitude of -21.0 approximately the same as for the similar edge-on giant spiral NGC 4565.

5. CONCLUSION

We have constructed a mass and light distribution model which, while not greatly different in its details from existing models like those of Schmidt (1966) or Innanen (1973), has the virtues that a) it treats all observations on a comparable footing, b) it is based on our knowledge of external galaxies and c) it accommodates naturally the flat rotation curves found recently by Roberts (1974), Krumm and Salpeter (1977) and Rubin et al. (1978).

ACKNOWLEDGEMENTS

We wish to thank J.H. Oort for helpful discussions and S. P. Bhavsar and R. L. Pariseau for their investigations of the properties of disc/halo galaxy models during an initial phase of this work. Support from National Science Foundation grant AST76-20255 is acknowledged.

REFERENCES

Burton, W.:1971, Astron. Astrophys. 10, 76.
deVaucouleurs, G.:1959, Handbuch der Physik, 53 (Springer-Verlag:Berlin).
Eggen, O. J.:1964, Royal Obs. Bulletin No. 84.
Erikson, R.:1975, Astrophys. J. 195, 343.

Einasto, J.:1970, Teated Tartu Obs.26,1.
Fall, M. and Rees, M.:1978, preprint.
Fatchikhin, N.:1970, Soviet Astron. J.,14,495.
Freeman, K.:1970, Astrophys. J. 160,811.
Fricke, W. and Tsioumis, A.:1975, Astron. Astrophys.,42,449.
Gunn, J. E.:1977, Astrophys. J.,218, 592.
Harris, W. and Hesser, J.:1976, Publ.Astron.Soc.Pacific,88,377.
Innanen, K.:1973, Astrophys.Space Sci.,22,393.
Kormendy, J.:1977a, Astrophys.J.,217,406.
Kormendy, J.:1977b, Astrophys. J.,218,333.
Krumm, N. and Salpeter, E. E.:1977,Astron. Astrophys.,56,465.
Oemler, G.:1976, Astrophys.J.,209,693.
Oort, J.:1960,
Oort, J.:1965,Stars & Stellar Systems V, p.483 (U of Chicago Press: Chicago).
Oort, J.:1977, Astrophys. J.,218,L97.
Ostriker, J.P., Peebles, P.E.J. and Yahil, A.:1974,Astrophys.J.Letters, 193,L1.
Ostriker, J.P. and Thuan, T.X.:1975, Astrophys. J.,202,353.
Plaut, L. and Oort, J.:1975, Astron. Astrophys., 41, 71.
Roberts, M.S.:1974,Rotation Curve of Galaxies, in Dynamics of Stellar Systems, ed. A. Hayli (Reidel Pub.: Dordrecht), 331
Rougoor, G. and Oort, J.:1960, Proc.Nat.Acad.Sci.,46,1.
Rubin, V.C., Ford,W.K. and Thonnard, N.:1978, preprint.
Salpeter, E.E.:1977, IAU Symp. No. 77.
Schmidt, J.:1965,Stars & Stellar Systems, V,513 (U of Chicago Press: Chicago).
Scoville, N.Z. and Solomon, P.M.:1975,Astrophys. J., 199,L105.
Spinrad,H., Ostriker, J.P., Stone, R.P.S., Chiu, G.L-T. and Bruzual, A.: 1978,Astrophys. J., in press.
Toomre, A.:1972, Quart.J.Roy.Astron.Soc.,13, 241.
Tremaine, S.D.:1976,Astrophys. J.,203,345.
Turner, E.L.:1976,Astrophys. J.,208,304.
Tuve, M. and Lundsager, S.:1973, Velocity Structures in Hydrogen Profiles, Carnegie Inst. of Washington Publ. 630.
Vandervoort, P.:1970,Astrophys. J.,162,453.
Vasilevskis,S. and Klemola,A.:1971,Astron.J.,76,508.
Weistrop,D.:1972,Astron.J.,77,366.

DISCUSSION

de Vaucouleurs: Is your escape velocity, which is estimated from rotation curves of other galaxies, sensitive to the assumed distance scale?

Sinha: How sensitively does the determination of the dark halo depend upon observations interior to the Sun? Based on the tangential point velocities of HI, I have derived a rotation curve (Astron. & Astrophys. in press) with three components: a nuclear disk (similar to one of Oort or Sanders and Lowinger) to match IR isophotes, a spherical $1/r^3$ halo to explain excess rotational velocities between 2 and 4 kpc, and a Toomre (n=5) disk. In this model I can explain the steep gradient at

1 kpc without invoking a disk with a hole, and I get a rotation curve, less steep than the Schmidt curve outside the solar circle, which is very similar to your model (for a bulge mass 1/4 x disk mass). The flatness of the rotation curve in our Galaxy found by Moffat and his coworkers and reported by Dr. Jackson, can be fit by introducing an extra halo-like component.

Bok: I hope that Dr. Ostriker will go as far as he can in making specific suggestions for work by observers. Modern techniques make it by now a relatively simple matter to study density and velocity distributions for objects like F stars. Radial velocities can now be measured to 18th magnitude--good spectra for classification purposes to the same limits--and photometry in established color systems (including the near infrared!) can be carried out to 21st magnitude (and fainter if need be!).

As part of theoretical model calculations, people like Dr. Ostriker should provide observers the force law perpendicular to the galactic plane at the Sun, which can then be checked by combined analyses of radial velocity and density distributions perpendicular to the galactic plane at the sun.

There are probably too few globular clusters to permit extensive use of them for studies of the dynamical properties of the halo. However, the search for and study of RR Lyrae variables in high galactic latitudes holds great promise. I hope that theorists in the future will not hesitate to make specific recommendations for observational tests and that in all of their reports on new models they will attempt to give us specific information on the field of force perpendicular to the galactic plane at the Sun.

Ostriker: Here are some recommendations. For the ratio of disc to spheroidal components, better determination of the local force law is critical.

For the mass distribution exterior to the Sun these items come to mind:

a. More observations of velocities of halo objects far from the Sun (like Hartwick and Sargent's work on globular clusters).

b. More and better high-velocity-star searches are important for determination of the local escape velocity.

c. Better measurements of the velocities and masses of the Galaxy's satellites.

Oort: Bok suggested that a strong attempt be made to get information about the structure of the very large halo or "corona" by surveys of distant stars or clusters. A thorough search for distant globular clusters might be feasible; for RR Lyrae variables it might be too time consuming, because they are so rare at the distance to be considered.

Ostriker's mass model seems a very acceptable one. One feature about which I feel some doubt, however, is the gap he assumed in the central part of the disk. It is evident that in many spirals there is a central hole in the gas distribution. The existence of such holes can be a natural consequence of the process of star formation, but I find it difficult to understand that there would be similar holes in the mass distribution. With regard to this point, Ostriker answered that it is difficult to represent the dip in the rotation curve between $R \sim 0.5$ and ~ 3 kpc without a hole in the disk component. Burton drew attention to the fact that this dip may not be real. The gas motions in this region are highly noncircular, and may well be considerably lower than the circular velocities.

Ostriker remarked that our Galaxy would be exceptional in that it has no black hole in its center. However, the NeII radial velocities in the 1-pc infrared nuclear disk indicate the presence of a mass of about 5×10^6 solar masses within $R \sim 1/2$ pc. It seems doubtful whether this can consist of stars.

Yahil: First a comment to Dr. de Vaucouleurs: Because Dr. Ostriker calculates an escape velocity squared from external galaxies, and not a mass, this value is not inversely proportional to the Hubble constant as you suggest.

Then a question to Dr. Ostriker: I am concerned that your rotation curve falls faster at large galactocentric radii than is observed in other galaxies. Which of your data points would have to be different if the rotation curve were indeed flatter?

Ostriker: The calculated escape velocity is only very weakly dependent on the assumed Hubble constant, as you state.

The outer part of the rotation curve is largely dependent on the poorly-known local escape velocity. We chose 558 ± 78 km s^{-1}. A larger value of the escape velocity would be required to have a flatter rotation curve.

Berman: It seems that the inner dip in your final rotation curve is due more to your choice of a model with a cut out disk and a spheroidal component rather than on your use of many measurements of the rotation curve in your least-squares fitting scheme: Is this right? Secondly, even though the final rotation curve falls very rapidly with radius, it seems to satisfy the law that the enclosed mass $M(R) \propto R$, obtained for massive extended galaxies and rotation curves that are constant by Dr. Rubin. Is this law $M(R) \propto R$ more common for flattened galaxies? Or is it due to your model choice of a halo component?

Ostriker: Both Rubin's galaxies and our model's have $M \propto R$ in the outer parts. The reason why our v_{cir} declines slightly (not rapidly) is due to the chosen value of $v_{esc}(R_o)$. Had we chosen a larger value we would have obtained a flatter curve. With respect to your first point, I should note that we also made models (which we shall publish separately in a more complete discussion) without the central hole in the disc. These do not fit the apparent dip in the rotation curve as well as the model we present here. The innermost point made by us (neglecting the "nuclear" point) is at 3 kpc and may be influenced by expansion or circulations.

GALACTIC MASS MODELING

J. Einasto
Tartu Astrophysical Observatory

1. INTRODUCTION

Galactic mass modeling has a long history. The first mass models were designed to represent the galactic attraction force in the radial direction. Considerable progress in galactic mass modeling was made during the fifties when Kuzmin (1952) introduced nonhomogeneous ellipsoids and Schmidt (1956) used a number of ellipsoids to represent various galactic populations. Further progress in galactic mass modeling has followed with the improvement of the system of the galactic constants and with the improvement of our knowledge of the structure of the galactic populations, in particular with the discovery of a massive corona around the Galaxy. In this report we present a new mass model of the Galaxy. It has been constructed using the most recent data available. A preliminary version of this model has been discussed earlier by Einasto, Jôeveer and Kaasik (1976).

2. THE METHOD OF MASS MODELING

By a model of the Galaxy we mean a set of functions and parameters which quantitatively describe the principal properties of the Galaxy and its populations. The main functions needed to describe the Galaxy are the gravitational potential, Φ, and its radial and vertical derivatives K_R, K_z. The structure of the galactic populations is given by the spatial density, ρ_i, the projected density, P_i, velocity dispersions in the cylindrical coordinates, σ_R, σ_θ, σ_z, and the centroid velocity V_i.

The galactic descriptive functions are interrelated by a set of formulae, the form of which depends on the principal properties of the Galaxy. On the basis of the existing data we may assume that the Galaxy is well relaxed, that its populations are physically homogeneous, and that equidensity contours of the galactic populations are similar concentric ellipsoids or can be represented in the form of sums of such ellipsoids. Under these assumptions simple relations hold between all the descriptive functions (Schmidt 1956, Einasto 1974).

W. B. Burton (ed.), The Large-Scale Characteristics of the Galaxy, 451–460.

The most convenient way of determining a model is to use a certain analytic expression for the density of the galactic populations. Schmidt (1956) and Innanen (1966, 1973) have used a polynomial expression for the density. Our expression has shown that a better representation can be obtained by the use of a modified exponential function (Einasto 1974)

$$\rho(a) = \rho(0)\ \exp\{x - [x^{2N} + a^2(ka_o)^{-2}]^{1/2N}\}, \tag{1}$$

where $a = R^2 + z^2/\varepsilon^2)^{\frac{1}{2}}$ is the major semiaxis of the equidensity ellipsoid, ε is the axial ratio of the ellipsoid, $\rho(0) = hM(4\pi\varepsilon\ a_o{}^3)^{-1}$ is the central density, M is the mass of the population, a_o is the harmonic mean radius of the population, x and N are structural parameters of the model, and h and k are dimensionless normalizing constants. The density distribution in the massive corona can be represented by a modified isothermal model (Einasto, Jôeveer and Kaasik 1976). The practical procedure of modeling consists of three steps: determination of the system of the galactic constants; determination of the parameters of the galactic populations; calculation of descriptive functions.

3. THE SYSTEM OF GALACTIC CONSTANTS

In the second Schmidt (1965) model the following principal galactic constants were adopted: $R_o = 10$ kpc, $V_o = 250$ km s^{-1}, A = 15 km s^{-1} kpc^{-1}, $\rho_o = 0.15\ M_\odot$ pc^{-3}. These values have been recommended by the IAU for general use. Recent observational data favor values of R_o, V_o, and ρ_o smaller than the conventional values. Table 1 summarizes the mean values of recent independent determinations of the galactic constants and their estimated rms errors. Here we use the designation

$$W = \frac{1}{2}\frac{dU}{dx} = AR_o, \tag{2}$$

where U is maximum relative radial velocity of rotation in the inner parts of the Galaxy (Fig. 1) and $x = R/R_o$;

$$C^2 = -(\partial K_z/\partial z)_{z=0} \tag{3}$$

and

$$k_z = \sigma_z^2/\sigma_R^2. \tag{4}$$

Some comments on the mean mass density, ρ_o, and the circular velocity, V_o. The galactic mass density adopted in this paper is close to Oort's (1932) first determination. The conventional IAU value is based on Hill's (1960) determination. As demonstrated by Eelsalu (1961), the method used by Hill is not very sensitive and thus his value is of little weight.

Recently Lynden-Bell and Lin (1977) have suggested a very high value, $V_o = 294$ km s^{-1}, for the circular velocity. This value is based on the mean velocity of the Sun with respect to the members of the Local Group. When discussing the dynamics of the Local Group, Lynden-Bell and Lin have treated together both the companions of the Galaxy and the distant members of the Group. To determine the solar velocity with

Table 1. Galactic Constants

Constant	Unit	Observed Value	Smoothed Value	Adopted Value	References
R_o	kpc	8.8 ± 0.7	8.5 ±0.3	8.5	1,2
V	km s^{-1}	220 ±10	221 ±5	225	3
W	km s^{-1}	120 ±15	133 ±4	131.8	4
A	km s^{-1} kpc^{-1}	16 ± 1	15.7 ±0.4	15.5	5-7
Ω	km s^{-1} kpc^{-1}	26. ± 2	26.0 ±0.7	26.5	8-10
k_z		0.282± 0.020	0.285±0.008	0.293	11
ρ	$M_\odot$ pc^{-3}	0.1 ± 0.02		0.097	12,13
C	km s^{-1} kpc^{-1}	70 ± 5		74	14

1. Oort, Plaut: A&A 41, 71 (1975)
2. Harris: A.J. 81, 1095 (1976)
3. Einasto et al. (in this volume)
4. Haud (in press)
5. Crampton, Fernie: A.J. 74, 53 (1969)
6. Balona, Feast: M.N. 167, 621 (1974)
7. Crampton, Georgelin: A&A 40, 317 (1975)
8. Asteriadis: A&A 56, 25 (1977)
9. Fricke: Heidelberg Veröff. 28, 1 (1977)
10. Dieckvoss: A&A 62, 445 (1978)
11. Einasto: Thesis, Tartu (1972)
12. Jôeveer: Tartu Teated 46, 35 (1974)
13. Woolley, Steward: M.N. 136, 329 (1967)
14. Jôeveer, Einasto: Tartu Teated 54, 77 (1976)

respect to the galactic center, one should use only objects moving in space together with the Galaxy. These objects form the galactic subgroup, our Hypergalaxy (Einasto 1977) in the Local Group. All members of the Hypergalaxy are located in a sphere with radius of 250 kpc around the galactic center. Using only these close companions of the Galaxy, we obtain V_o = 220 km s^{-1}. If we omit close companions and consider only distant members of the Local Group, we obtain the sum of the solar and the galactic motion, $V_o + V_{Gal}$ = 380 km s^{-1}. We note that other methods, based on the inner dynamics of the Galaxy, yield also V_o = 215 - 225 km s^{-1}.

The observed values of the galactic constants are subject to random and undetected systematic errors. For this reason they do not exactly satisfy equations connecting individual galactic constants with each other. To remove the role of these errors to some extent, we have found by the method of least squares a smoothed and mutually concordant system of galactic constants where all equations are exactly fulfilled (for details see Einasto and Kutuzov 1964). This system of galactic constants, as well as the adopted system, is presented in Table 1. The last system differs from the previous one by the use of rounded values for principal constants.

4. GALACTIC POPULATIONS

The present model incorporates the following galactic populations: the nucleus, the bulge, the halo, the disk, the flat populations, and the massive corona. Parameters of galactic populations are given in Table 2. The structural parameters N and x of the galactic populations

Table 2. Parameters of galactic populations

Population	ε	a_o (kpc)	M ($10^{10}M_\odot$)	N	x	h	k
Nucleus	0.6	0.005	0.009	1	0.5	2.871	0.48188
Bulge	0.6	0.21	0.442	1	0.5	2.871	0.48188
Halo	0.3	1.9	1.2	4	3.5	101.66	1.2304×10^{-4}
Disk	0.1	4.62	7.68	1	0.5	2.871	0.48188
	0.45	1.026	-0.379	1	0.5	2.871	0.48188
Flat	0.02	6.4	1.00	0.5	0	1.5708	1.1284
	0.025	5.12	-0.64	0.5	0	1.5708	1.1284
Corona*	1	75	110	0.5	25.12	8.3806	0.2575

* For the corona an isothermal model has been used. For the meaning of parameters N and x, see Einasto, Jôeveer and Kaasik (1976).

have been adopted mainly on the basis of the study of external galaxies. Other parameters and properties of populations will be discussed in the following sections.

(a) The nucleus. Its parameters have been adopted by analogy with the Andromeda galaxy.

(b) The bulge. According to Arp (1965), the radius of the bulge of the Galaxy is about half of the radius of the bulge of M31. The adopted mass and radius were found by the method of least squares from the first maximum of the rotation velocity curve (Fig. 1).

(c) The halo. The radius of the halo was determined from the spatial distribution of globular clusters (Kukarkin 1974, Woltjer 1975). The mass of the halo was estimated on the basis of the density determination of the halo population objects in the solar vicinity (Oort 1958, Schmidt 1975). Our halo represents all metal-deficient populations, both the extreme halo and the intermediate population II objects. For this reason a moderate value, $\varepsilon = 0.3$, has been adopted for the axial ratio of the halo.

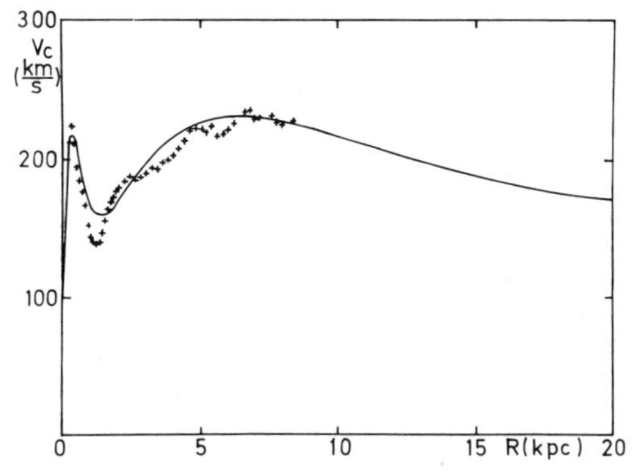

Fig. 1. Adopted curve of the circular velocity and of the observed rotation velocity (crossed).

Table 3. Galactic functions

R (kpc)	V (km s^{-1})	V_{esc}	A (km s^{-1} kpc^{-1})	-B	C	k_z	log ρ ($M_\odot$ pc^{-3})	log P ($M_\odot$ pc^{-2})
0.0	0	874	30168	32166	120740	0.500	5.438	5.476
0.1	150	756	247	1257	1776	0.455	2.059	4.355
0.2	202	727	335	657	1343	0.401	1.663	4.063
0.3	218	702	326	402	961	0.356	1.282	3.760
0.4	217	682	297	246	724	0.312	0.938	3.486
0.5	208	666	257	160	573	0.277	0.660	3.276
1.0	167	628	105	62	321	0.270	0.213	3.007
1.5	160	611	49	58	266	0.351	0.128	2.976
2.0	170	598	30	55	237	0.394	0.047	2.930
2.5	184	587	23	50	213	0.406	-0.038	2.870
3	196.3	575	20.9	44.5	193	0.405	-0.126	2.802
4	215.2	554	19.7	34.1	160	0.388	-0.301	2.656
5	225.8	534	19.1	26.1	133	0.366	-0.470	2.503
6	230.0	516	18.2	20.1	112	0.344	-0.631	2.347
7	229.9	500	17.2	15.6	95	0.322	-0.785	2.190
8.5	225.0	479	15.5	11.0	74	0.293	-1.011	1.951
10	217.1	462	13.8	7.9	57.8	0.268	-1.243	1.707
12	205.0	444	11.5	5.5	40.5	0.245	-1.576	1.372
14	193.6	430	9.6	4.3	27.7	0.236	-1.939	1.029
16	184.0	419	7.9	3.6	19.0	0.240	-2.310	0.686
18	176.6	410	6.5	3.3	13.6	0.252	-2.643	0.360
20	171.1	402	5.44	3.11	10.43	0.267	-2.894	0.063
30	159.4	373	2.91	2.40	5.46	0.311	-3.346	-0.913
50	155.6	336	1.59	1.53	3.17	0.329	-3.746	
75	154.4	304	1.06	1.00	2.09	0.328	-4.117	
100	152.6	283	0.81	0.72	1.54	0.321	-4.409	

(d) The disk. The mass and the radius of the disk were determined by the method of least squares from the rotation velocity curve. The disk represents galactic populations over a wide range of axial ratios between the flat and the intermediate population objects. Its axial ratio, $\varepsilon = 0.1$, is a compromise. There exists evidence indicating that the disk has a ring-like structure. Absence of interstellar hydrogen and young stars near the center of Sb galaxies is a well-known observational fact (Baade and Arp 1964, Roberts 1966, Burton et al. 1975). Stars are formed from the interstellar gas. Thus, if the interstellar gas had been absent in the central region also in the earlier period of Galaxy history, the whole disk should have a minimum in the density distribution near the center. The mass distribution of such a ring-like population can be represented as a sum of two components with identical structural parameters, but with different M, a_o, and ε. If the density of the disk at the center were zero, the parameters of both components should be chosen as follows: $M_2 = -\kappa^2 M_1$, $a_{o_2} = \kappa a_{o_1}$ $\varepsilon_2 = \kappa^{-1}\varepsilon_1$, where $\kappa < 1$ is a parameter which determines the extent of the "hole" in the center of the disk. By introducing the hole into the disk it is possible to represent the deep minimum in the rotation curve of the Galaxy at R = 1.5 kpc. The value of the parameter κ has been derived by trial-and-error to achieve the best possible representation of the rotation curve between the two maxima.

(e) The flat population. This population represents the interstellar gas and young stars. It has a ring-like structure (Fig. 2). The radius a_o, the axial ratio, and the parameter κ were derived from the data available on the distribution of gas and young stars. The mass of this population was determined on the basis of density estimates of gas and young population stars.

(f) The corona. The velocity dispersion of the companions of our Galaxy is only slightly smaller than the velocity dispersion of galactic globular clusters, thus the velocity dispersion remains practically constant over a wide range of distances. This suggests that our Galaxy is surrounded by a massive corona. The harmonic mean radius of the system of galactic companions is a_o = 75 kpc; this value has been adopted for the corona. The mass of the corona has been calculated from the virial theorem, adopting σ_R = 85 km s^{-1} for the velocity dispersion of the coronal objects. Visible elements of the corona (galactic companions) form a flat disk. The form of the invisible corona is at present unknown; in the model we adopt a spherical corona.

5. DISCUSSION

Table 3 contains some descriptive functions. A number of the calculated descriptive functions are shown in Figures 1-6.

The present model differs from the Schmidt and Innanen models in two principal aspects: a new system of galactic constants has been adopted and a massive corona has been added. These changes have been made necessary by the body of available observational data. One consequence of the presence of a massive corona is a very high escape velocity which does not imply a high circular velocity (Table 3).

In our model the velocity dispersion of objects of a homogeneous population slightly decreases with increasing distance from the galactic plane (Figs. 3 and 4). This is a direct consequence of the use of the exponential density law. Schmidt and Innanen have used polynomial laws

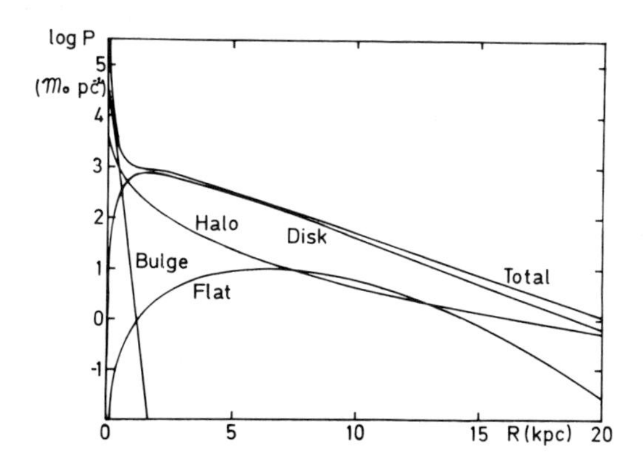

Fig. 2. The surface density of the Galaxy and of its components.

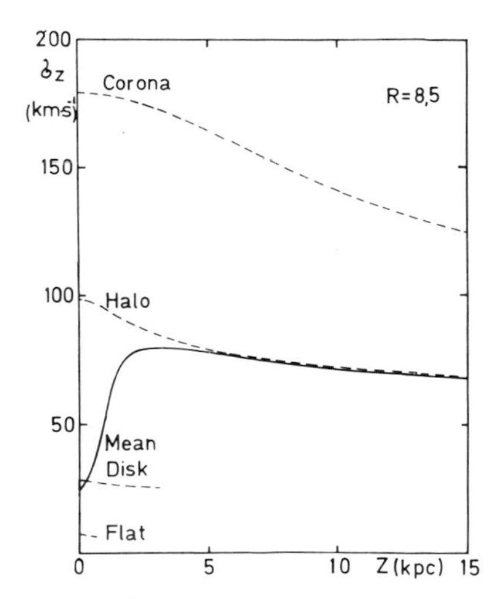

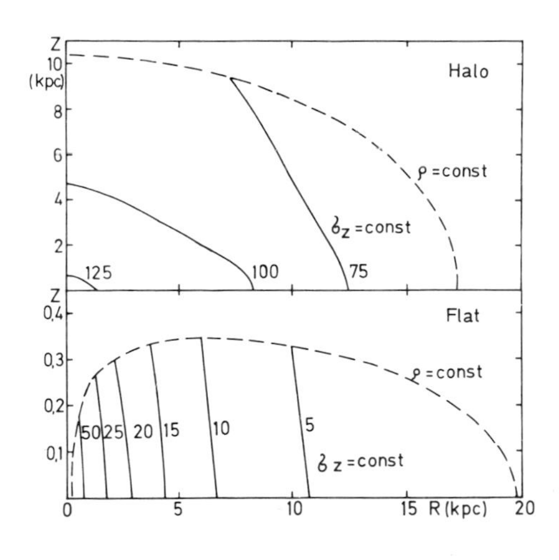

Fig. 3 (left). Velocity dispersion, σ_z, of galactic populations and the mean velocity dispersion versus distance from the galactic plane.

Fig. 4 (right). Meridional sections of the extreme halo (ε = 0.6) and flat (ε = 0.02) populations. Dashed line gives equidensity contours, corresponding to a density of 10^{-4} $M_\odot$ pc^{-3}, solid lines denote lines of constant velocity dispersion.

with a fixed boundary of the populations. In such a case the velocity dispersion would be zero at the boundary. Zero dispersion has not been observed at any distance from the galactic plane. On the contrary, real samples of stars are mixtures of stars of different populations; thus the mean dispersion may even increase with increasing z, as is indeed the case with the overall mean dispersion (Fig. 3).

The model has been checked for stability. As seen from Fig. 5, in general the mean velocity dispersion is larger than the critical Toomre (1964) dispersion, and thus the model is stable against small radial perturbations. However, between 2 and 8 kpc the Toomre dispersion is slightly higher than the calculated mean σ_R. Over a wide region the inner mass of the disk considerably exceeds the inner mass of the spheroidal component (Fig. 6). A similar picture has been found to exist also in other Sb galaxies.

ACKNOWLEDGMENTS

In this report results have been discussed which have been obtained by a team of Tartu astronomers. The most important contributions have been made by Mr. U. Haud, M. Jõeveer, A. Kaasik, P. Tenjes and P. Traat. Mr. L. Kivimägi, Mrs. T. Saar and Miss T. Johanson have assisted in preparing the manuscript. Our sincere thanks are due to all of them.

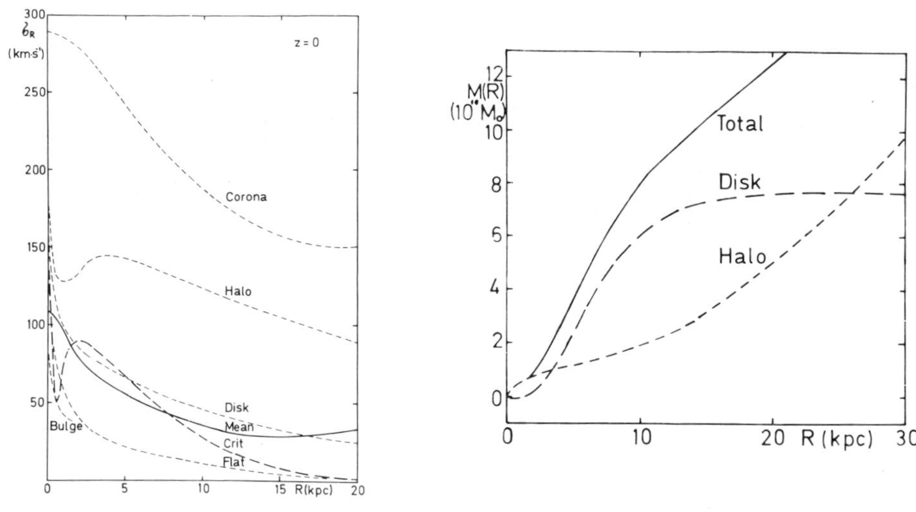

Fig. 5 (left). Velocity dispersions, σ_R, of galactic populations versus distance from the galactic center. The mean velocity dispersion as well as the critical Toomre (1964) dispersion have also been plotted.

Fig. 6 (right). Inner mass of the Galaxy, M(R), of the galactic spheroidal components (including the massive corona) and of disk components. The mass M(R) is defined as the mass of a spherical body which exercises the same attraction force as the model galaxy. Negative M(R) values indicate that the attraction force is directed away from the galactic center.

REFERENCES

Arp, H.: 1965, Astrophys. J. 141, 43.

Baade, W. and Arp. H.: 1964, Astrophys. J. 139, 1027.

Burton, W. B., Gordon, M. A., Bania, T. M. and Lockman, F. J.: 1975, Astrophys. J. 202, 30.

Eelsalu, H.: 1961, Tartu Astron. Obs. Publ. 33, 416.

Einasto, J.: 1974, Proc. First European Astr. Meeting 2, 291.

Einasto, J.: 1977, Tartu Preprint A-3, 3.

Einasto, J., Jôeveer, M. and Kaasik, A.: 1976, Tartu Astron. Obs. Teated 54, 3.

Einasto, J. and Kutzov, S. A.: 1964, Tartu Astron. Obs. Teated 10, 1.

Hill, E. R.: 1960, Bull. Astron. Inst. Neth. 15, 1.

Innanen, K. A.: 1966, Astrophys. J. 143, 150.

Innanen, K. A.: 1973, Astrophys. Space Sci. 22, 393.

Kukarkin, B. V.: 1974, The General Catalogue of Globular Clusters of our Galaxy, Moscow, Nauka.

Kuzmin, G. G.: 1952, Tartu Astron. Obs. Publ. 32, 211.

Lynden-Bell, D. and Lin, D.N.C.: 1977, Mon. Not. Roy. Astr. Soc. 181, 37.

Oort, J. H.: 1932, Bull. Astron. Inst. Neth. 6, 249.
Oort, J. H.: 1958, Ric. Astr. Specola Vaticana 5, 415.
Roberts, M. S.: 1966, Astrophys. J. 144, 639.
Schmidt, M.: 1956, Bull. Astron. Inst. Neth. 13, 15.
Schmidt, M.: 1965, Stars and Stellar Systems, Univ. Chicago Press 5, p. 513.
Schmidt, M.: 1975, Astrophys. J. 202, 22.
Toomre, A.: 1964, Astrophys. J. 139, 1271.
Woltjer, L.: 1975, Astron. Astrophys. 42, 109.

DISCUSSION

Lynden-Bell: There are problems in the determinations of (B-A)/B from local stars. Not only is the vertex deviated but the Hyades group dominates one end--this may well disturb the ratio. Woolley et al. (MNRAS 179, 81) measured radial velocities of K stars at around 600 pc in the direction of galactic rotation and another group in the direction of the anticenter. From about 500 stars they found a ratio of velocity dispersion of $\sqrt{2}$, corresponding to B = -A.

Basu: Could you please comment on the uncertainties that might be introduced into your computed values by using the local values of Oort constants?

Ostriker: The overall uncertainties of the various model parameters are given in the paper. Because quantities enter into the determination with weight inversely proportional to the square of the uncertainty attributed to it, Oort's constant B is relatively unimportant and A is relatively important in contribution to the final result.

Rubin: A comment which is also a question. If the values of Oort's constants A and B are the local values, then the value of $\frac{dV}{dR}$ which they define may not be the correct value for your model. In particular, if the extended rotation curve is flat, then maybe A = -B would be a more appropriate choice. Is this correct? For the 11 rotation curves which we showed earlier, eight of them have their maximum velocity at R > 8 kpc; seven of them have their maximum beyond 15 kpc. In response to Dr. de Vaucouleurs, note that for $H \sim 100$ km s^{-1} Mpc^{-1}, this would be equivalent to rising curves at $R \sim 8$ kpc.

Einasto: The presently adopted Oort constants may indeed reflect only the local behavior of the rotation curve. In order to investigate the consequences of the totally flat rotation curve of the Galaxy at R > 5 kpc we will compute the appropriate version of the model.

Upgren: In response to Bok's comment a few minutes ago and in regard to the last two papers, I want to mention some pertinent observations which are being made. Jurgen Stock and I and others are using the new 1.5-meter Schmidt in Venezuela to get objective-prism radial velocities and spectral types of F stars to 2 kpc or farther. At latitude 9° N, we can reach the entire plane and both poles. We want to determine the force law in z from the F stars which would then not be dependent on the giants because they have been shown by Sturch and Helfer and others to have luminosity calibration problems. I also want to mention that Stock already has objective-prism radial velocities of about 6000 stars in a direction not far from that opposite the direction of galactic rotation. The percentage of high-velocity stars is very small, although a few have velocities of 400 km s^{-1} or more.

Trimble: Concerning your plot of velocity dispersion in the radial direction vs distance from the galactic center; the average stellar velocity distribution is less than Toomre's initial value in the region $\sim$ 4-7 kpc, which is just where we see all sorts of excitement in the form of lumpy molecular clouds, excess star formation, pulsars, and so forth; perhaps there really is an instability?

Einasto: This may be the case. Our attempt to avoid the instability by increasing the mass of the stellar halo and bulge failed because the masses which are needed are too large to be compatible with currently available data.

Burton: In view of the importance of the hole in the disk component at small galactic radii, and its dependence on the dip in the galactic rotation curve, it is worth emphasizing the uncertainty in this curve at R $\lesssim$ 3 kpc. Although the [NeII] data probably give a good indication of the rotation in the inner few pc, at larger distances the pervasive expansion will have caused rotation curves derived on pure-rotation assumptions to be in error. The sense of this influence will exaggerate the appearance of the inner-Galaxy dip.

Einasto: Our adopted rotation curve is corrected for the effect of expansion. Corrected velocities (lower curve) are smaller than the observed ones; in particular the inner dip is considerably lower.

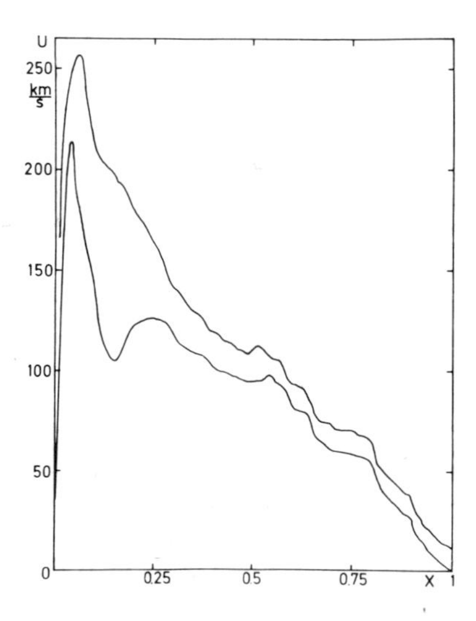

"THREE-INTEGRAL", COLLISION-FREE STATISTICAL MECHANICS AND STELLAR SYSTEMS

K. A. Innanen
Physics Department, York University, Toronto, Canada

In a recent paper, Innanen and Papp (1977) have introduced and discussed the behaviour of a function they called the "super angular momentum," h_t^2. This quantity, although not formally an integral of motion in the classical definition, is nevertheless a straightforward generalization of the classical total angular momentum. The super angular momentum, together with the other two classical invariants, the total energy E, and the Z component of the angular momentum h_Z lead to the introduction for each homogeneous stellar system of a new collision-free distribution function

$$f = f_o \exp(-Q)$$

$$\text{where } Q = 2E - 2\beta h_Z + \gamma h_t^2$$

and f_o, β and γ are constants. The function f satisfies the fundamental equation of stellar dynamics and shows that the velocity dispersions of stellar systems cannot be isotropic, but rather that

(a) the velocity dispersion in the radial direction is constant (i. e. "isothermal") and
(b) the velocity dispersions in the tangential directions follow the law

$$\sigma^2_\theta = \sigma^2_Z \, \alpha(1 + \gamma r^2)^{-1}$$

That is, the tangential components are "isothermal" at the centre of the system, and decay according to the above equation so as to leave a purely radial distribution in the outer parts.

This research, which will be published in detail elsewhere, has been supported by the National Research Council of Canada.

REFERENCES

Innanen, K. A., and Papp, K. A.: 1977, Astron. J. 82, 322.

W. B. Burton (ed.), The Large-Scale Characteristics of the Galaxy, 461–462.

DISCUSSION

Contopoulos: I agree that a third integral of motion is very useful in galactic dynamics, e.g., in explaining the three-axial form of the velocity ellipsoid; this has been done by Barbanis, about 15 years ago. Many other applications have been made by several people since that time. In the case of almost spherical systems a third integral has been calculated by Saaf (a student of Vandervoort), which is a generalization of the total angular momentum.

However your "third integral" includes a function $\psi(t)$, which is an integral in time, calculated along the orbit. Therefore in order to evaluate it you have to calculate the whole orbit of a star. But, according to the standard definition, an integral of motion is a function of local quantities, at any point of an orbit, and remains constant along this orbit. That is, the energy of a star in a time-independent potential is a function of the local position and velocity of the star; we do not have to calculate the whole orbit in order to find it. Therefore what you give as a generalization of the square of the total angular momentum should not be called an integral of motion.

Innanen: The difficulty appears to arise over the semantics of the word "integral". The function we have introduced is not a conservative integral in the "standard" definition. The function does, however, reduce to the square of the total angular momentum integral for spherical systems, and is constant to within 2 or 3% during long numerical experiments. Consequently we have called it the "super-angular momentum". The method makes clear predictions about the variation of velocity dispersions within stellar systems and so is subject to direct observational tests.

Lynden-Bell: For the potential $\psi = \psi_1(r) + \psi_2(\theta)/r^2$ the r and θ motions of one star decouple exactly and the third integral is the square of the angular momentum minus $2\psi_2(\theta)$. Woolley has analyzed the local stars using this integral and has calculated $\psi_1(r)$ and $\psi_2(\theta)$ to fit the Galaxy's potential within some 4 kpc from the Sun (Royal Observatory Annals No. 5, 1971). Dr. Innanen's expression will reduce to this integral because this is a good approximation to the galactic potential. Dr. Innanen's work shows that for a different approximation to the galactic potential an integral of this form still works very well.

Innanen: I am aware of the work mentioned by Dr. Lynden-Bell and agree with his comments.

IS THERE A COMPOSITION GRADIENT IN THE HALO?

Robert P. Kraft, Charles F. Trefzger, and N. Suntzeff
Lick Observatory, Board of Studies in Astronomy and
Astrophysics, University of California, Santa Cruz

We define "metals" as consisting of the elements at and near the Fe-peak, and review several methods by which the metal-abundance gradient of stars in the halo can be obtained. In the inner halo (galactocentric distance R $\lesssim$ 8 kpc), the Basel RGU photometry should allow the derivation of the shapes and dimensions of the iso-abundance contours. For the outer halo to R $\sim$ 30 kpc, we review techniques based on Δs-measurements of RR Lyraes (Lick) and intermediate band-pass photometry of globular-cluster giants (Searle and Zinn, Palomar). Both methods suggest little change in mean [Fe/H] between 10 and 30 kpc; however, both may be biased against the discovery of very metal-poor objects. The conclusion that the outer halo has no abundance gradient may be somewhat premature.

Recent abundance analyses of the classical halo giant HD 122563 (Lambert, *et al*. 1974) and of giants in certain globular clusters (Cohen 1978; Carbon, *et al*. 1978) suggest that in old, metal-poor stars the relative abundances of the α-process and Fe-peak elements, and even of primary elements *inter-alia* are not the same as in the sun. Thus before we can speak of a "metal-abundance gradient in the halo", we need to decide which metals we mean. As a purely practical matter, optical observers require a metal-abundance "indicator" that can be measured with reasonable accuracy in faint cool stars of old stellar populations. Most current photometric techniques depend on the blocking of UV-flux produced by elements in and near the Fe-peak; thus in what follows, "metals" $\equiv$ Fe-peak.

Nevertheless, a moderate decoupling of the C, O group from the Fe-peak is an interesting possibility. In both HD 122563 and the M15 planetary nebula (Hawley and Miller 1978) $[\frac{O}{H}] > [\frac{Fe}{H}]$ (recall the definition $[\frac{A}{H}] \equiv \log (\frac{A}{H})_* - \log (\frac{A}{H})_\odot$). In low mass stars, it is not easy to see how O can be produced by an evolutionary process, and a primordial overabundance of O may be required. Moreover, the strength of the CO-bands in M3 giants is greater than that of M13 giants, even though $[Fe/H]_{M3} < [Fe/H]_{M13}$ (Cohen, *et al*. 1978; Pilachowski 1978).

W. B. Burton (ed.), The Large-Scale Characteristics of the Galaxy, 463–474.

In metal-poor clusters horizontal branch morphology critically depends on the C, O group whereas giant branch morphology depends largely on the Fe-peak (cf. Faulkner 1966, Renzini 1977); these results therefore may bear on the classical problem of the anomalously blue HB of M13 *vis-a-vis* M3.

Returning to the domain of Fe-peak metals, we now review several methods by which the halo gradient problem can be attacked. Space does not permit an exhaustive survey of techniques: what we describe is, however, representative.

A. *Halo Field Dwarfs: Basel Three-color RGU System*. The system (Becker 1972) discriminates metal-abundance in late-type (>F0) stars by means of the ultraviolet excess $\delta(U-G)$ at a given $(G-R)$ in a plot of $U-G$ *vs* $G-R$; $\delta(U-G)$ is similar to the more familiar $\delta(U-B)$ of the *UBV* system. When applied to dwarfs, $\delta(U-G)$ is defined to be 0.0 for metal-rich stars and reaches a value ~0.5 mag in stars with extreme metal deficiency. Compared with *UBV* the system has the advantage that $\frac{\delta(U-G)}{\delta[Fe/H]}$ is almost twice $\frac{\delta(U-B)}{\delta[Fe/H]}$, largely because *B* is affected more by blanketing than is *G*. This large UV-excess sensitivity is particularly favorable when one tries to detect metal-poor dwarfs in photographic Schmidt surveys (Steinlin 1973). The Basel investigators have studied $\delta(U-G)$ for stars as faint as V = 19 in six selected areas lying roughly in a plane containing the sun, the galactic nucleus, and the north galactic pole. The distribution of metal-abundance among dwarfs is thus studied to a distance some 8 kpc from the sun, and the *shapes* of the iso-abundance contours can be derived in principle for comparison with those predicted from galactic collapse models (cf., eg., Larson 1975, 1976). Preliminary results indicate that the "isochromes" of $\langle\delta(U-G)\rangle$ are flattened, corresponding fairly closely to the isodensity contours (Becker 1972) derived from the same surveys. The calibration of $\delta(U-G)$ in terms of [Fe/H] has not yet been carried out, so the precise conversion of "isochrome" to iso-abundance contour is not yet possible.

Since the mean $\delta(U-G)$-values are based on some 200 dwarfs in each field, the *RGU* system carries an enormous statistical weight when compared with other methods (described later), but also has some disadvantages. First, since it is applied to dwarfs, it does not penetrate into the outer halo. Second, since late-type giants exhibit a well-known gravity-induced UV-deficiency relative to dwarfs, one can confuse a metal-deficient giant of the distant halo with an ordinary nearby metal-rich dwarf. The Basel investigators statistically reduce the effect of this contamination by confining their attention to stars with $(G-R)$-colors corresponding to spectral types ~G5 and earlier. By thus limiting the sample to stars blueward of the Hayashi line for giants, they find only about one red horizontal branch "giant" per 100 main sequence stars, per unit volume of space. If the stellar density falloff goes as $\rho \sim R^{-3.5}$ (Kinman, *et al*. 1966; Harris 1976), giant star contamination should be quite negligible. We turn next to methods of deriving metal abundances more applicable to the outer halo (R> 8 kpc).

B. RR Lyrae Stars

Recent work (McDonald 1976; Butler, et al. 1978) bolsters the long-held view that, except for the most metal-rich objects ($\Delta s \leq 2$), RR Lyraes are standard candles with $\langle M_V \rangle = +0.6$. Since they are also F-type stars, objects in which the sources of opacity are well understood, their metal abundances can be reliably estimated. Butler's (1975a) calibration of Preston's (1959) Δs-index, based on the Ca II (K) and hydrogen lines, leads to [Fe/H] = $-0.16\ \Delta s\ -0.23$. Thus measurements of Δs in halo RR Lyraes provides an ideal means of mapping the halo abundance gradient. Fortunately, extensive surveys made with the Lick astrograph have discovered RR Lyraes near the north galactic pole (NGP) and in several other galactic star fields (Kinman, et al. 1965, 1966) to a magnitude limit near m_{pg} = 18.5 (about 30 kpc). Since 1975, measurements of Δs for stars in three of these fields, including the NGP, have been made with the Wampler-Robinson (1972) scanner (IDS) operated at the cassegrain focus of the Lick 3-m (Shane) reflector. Although Δs measurements, generally numbering at least two per star, continue, a preliminary idea of the results can now be stated. As one proceeds above the plane to a distance of about 10 kpc, [Fe/H] on the average declines to a value somewhat smaller than that for RR Lyrae itself; beyond 10 kpc there is little evidence for a gradient, although there is large scatter. Certainly, stars with moderately high metallicity ([Fe/H] $\sim$ -1) exist even at a distance of 20 kpc above the plane.

However, before we can naîvely embrace the evident conclusion that no abundance gradient exists in the halo between 10 and 30 kpc, we must examine two selection effects that discriminate against the discovery of very metal-poor RR Lyraes (By "very metal-poor", we mean [Fe/H] $<$ [Fe/H]$_{M92}$ = -2.2 ± 0.2. M92 has the lowest well-known metal abundance among globular clusters). First, as is well-known (cf. Faulkner 1966, Iben 1974), a decline in metals leads to a bluer and bluer "mapping" of the horizontal branch; thus stellar populations with very low metals may skip the region of RR Lyrae production altogether. If we examine the distribution with [Fe/H] of RR Lyraes on the solar vicinity (Butler 1975), we find no stars with [Fe/H] significantly lower than M92. On the other hand, the giant HD 122563 has [Fe/H]= -2.7, a value reliably established (Wallerstein, et al. 1963), to be 0.5 dex lower than M92, and the existence of two giants with [Fe/H] $\lesssim$ -3.0 has been reported (Bessell 1977). (Lick ITS observations easily show the weakening of metal features in HD 122563 compared with giants of the same T_{eff} in M92.) The number of stars involved is too small to permit definitive conclusions, but the material suggests that RR Lyraes may not in fact be generated in populations with very low metals.

A second selection effect is purely observational. Preston's (1959) work established that Δs is positively correlated with period amongst Bailey a's, but the amplitude of the light curve is in turn negatively correlated with period. This means, on the average, that the stars with lowest metal abundances tend statistically to have the largest periods and smallest light amplitudes. Kinman et al. (1966b) estimated that in the NGP, the sample of Bailey a's with $\Delta m \geq 0.75$ mag is complete to $m_{pg} = 17.0$, corresponding to a distance of 17 kpc. Since this Δm corresponds to [Fe/H] ~ -1.8, it seems likely that beyond 17 kpc, a significant fraction of metal-poor stars may be lost by the survey.

C. Giant Stars in Globular Clusters.

The fact that metal-rich globular clusters such as 47 Tuc and NGC 6171 ($[\frac{Fe}{H}] > -1.0$) are confined to a region in and near the galactic nuclear bulge, whereas metal-poor clusters are found in all parts of the galaxy has been known for some time (e.g., Morgan 1956, Kinman 1959, Arp 1965, Harris 1976, Fig. 5 and 6). What is not so clear is whether in the outer halo beyond the domain of metal-rich clusters $R \gtrsim 8$ kpc), an abundance gradient exists amongst the metal-poor clusters themselves. Several studies (Bell 1976, Canterna and Schommer 1978, Cowley et al. 1978) indicate surprisingly high metallicities ($>[Fe/H]_{M92}$) for a number of remote halo clusters and satellite subsystems (e.g. Draco) of the Galaxy. In a recent comprehensive paper, Searle and Zinn (1978) hereafter "SZ") addressed the halo abundance gradient question, and drew "the tentative conclusion that for $R > 8$ kpc, the distribution over abundance of halo globular clusters is independent of galactocentric distance". Although we suspect that this conclusion is probably valid, we believe the authors are also correct in emphasizing the word "tentative". The main problem is that the method employed by Searle and Zinn is not able to discriminate very well, within the quoted errors, between metal-poor clusters having, for example, [Fe/H] = -2.0 and [Fe/H] = -2.5. The question we raise, therefore, is whether the apparent cut-off in cluster metal deficiency near [Fe/H] ~ -2.0 simply reflects an artifact of the method of analysis; whether a few clusters, for example NGC 5053 and NGC 2419 have been assigned metal abundances that are too large; and whether the apparent absence of a gradient might result from a combination of this lack of discrimination and the fact that the galactic halo contains very few clusters with galactocentric distances $R > 20$ kpc. (We do not quarrel with the conclusion that several remote clusters, e.g., NGC 7006, in fact have [Fe/H] considerably larger than -2.0)

Our point becomes clearer when we examine the SZ method in more detail. Space does not permit an extensive review, but the essential points can be recovered from a study of Fig. 1, in which we plot the energy distributions F_ν expressed in magnitudes, averaged over 160 A intervals, for various metal poor stars, following exactly the treatment by SZ. The energy distributions are plotted against $\psi(\lambda)$, where $\psi = 1.30\ \lambda^{-1} - 0.60$ or $0.75\ \lambda^{-1} + 0.65$, according as $\lambda^{-1} \leq 2.29$ or > 2.29. Note that, as a function of $\psi(\lambda)$ (or $1/\lambda$) the energy distributions are virtually straight lines between λ8000 and 5000 A. The slope of this straight line, i.e., the increase in magnitude between

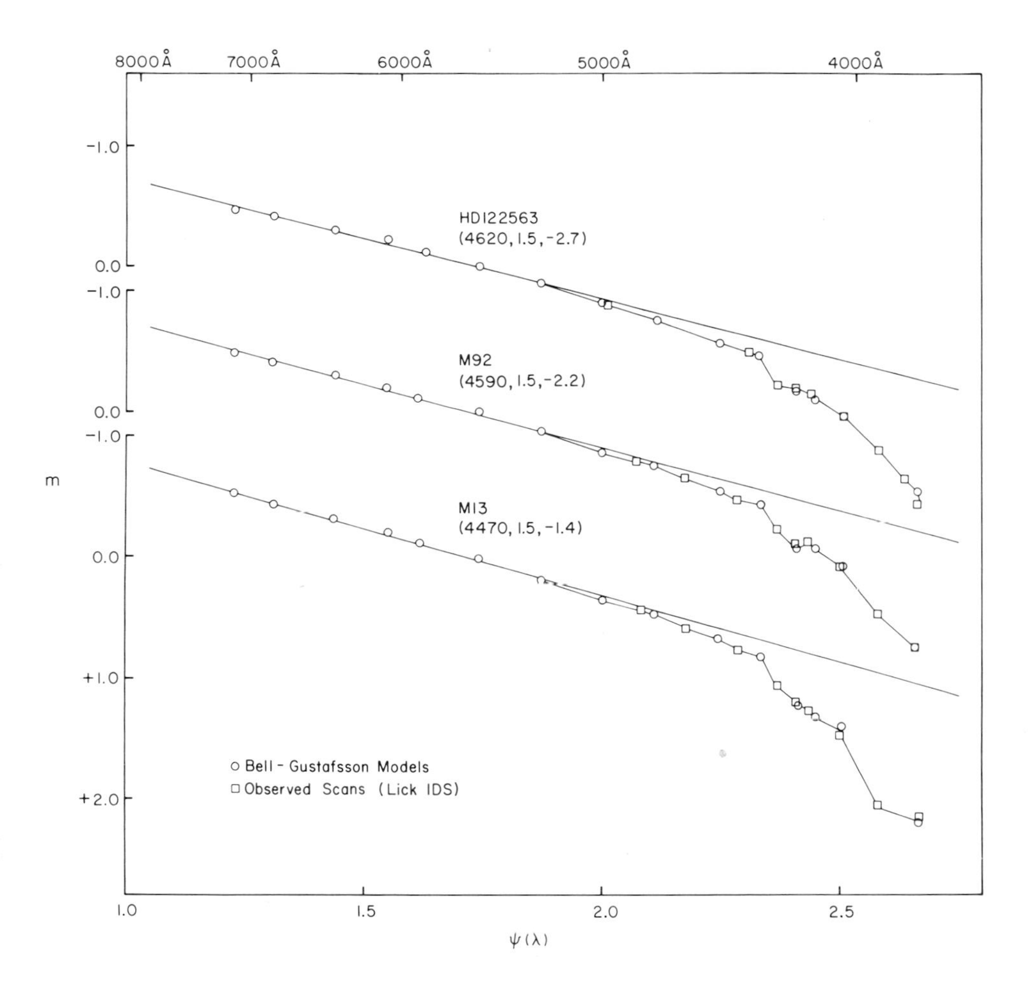

Figure 1. Theoretical and observed scanner fluxes F_ν for metal-poor stars. The latter, averaged over 160 A intervals, are normalized to the former, averaged over 50 A intervals, at λ4850 A; the numbers in parentheses refer to models with parameters (T_{eff}, log g, [Fe/H]).

λ8000 and 5000 A, denoted by ψ_o, depends on the intrinsic energy distribution of the star and on interstellar reddening; however, since the transformation given above is based on the Whitford reddening law (Miller and Mathews 1972) areas between the straight line, extrapolated shortward of $\lambda^{-1} = 2.29$, and the observed stellar flux m_λ, are reddening independent. Such areas, integrated over 160 A intervals and denoted by Q(λ), thus measure the intrinsic spectrum of the star.

In the SZ-method, some dozen red giants typically are scanned in each cluster, the Q(λ) are measured for each star at several wavelengths

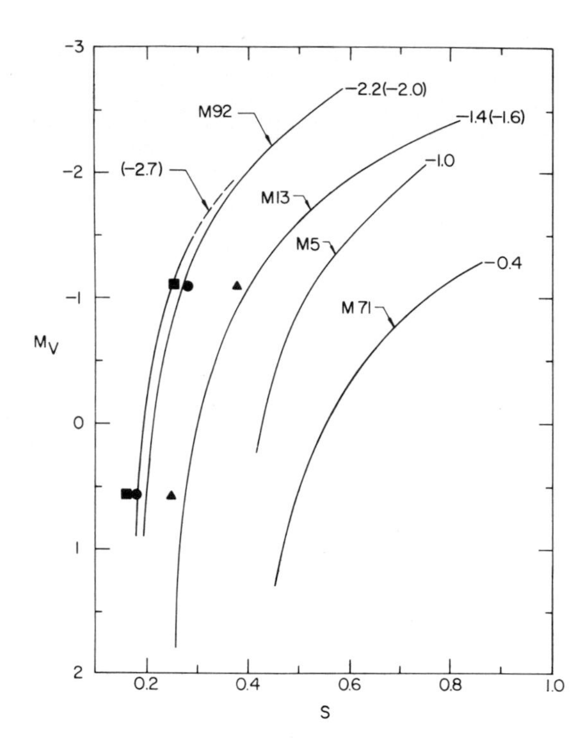

Figure 2. The variation of S with M_V in several globular clusters, after Searle and Zinn (1978). The location of a fictitious cluster with [Fe/H] = -2.7 is shown; filled symbols, reading from left to right, refer to models with [Fe/H] = -2.7, -2.2, and -1.4.

between λ5160 and λ3880, and a weighted mean of these Q's, denoted by S, is formed (the reader is referred to SZ for details). For each cluster star M_V is plotted as a function of S; this is illustrated in Fig. 2, which is adapted from the SZ paper (we do not show the individual observations). A particular S-value, denoted by $\langle S \rangle$, is read out at the convenient luminosity $M_V = -1$ for all clusters; typically the s.d. in $\langle S \rangle$ for a well-observed cluster is ± 0.02. The tight correlation of $\langle S \rangle$ with [Fe/H] determined from RR Lyraes (Butler 1975) provides the basic calibration of cluster abundances.

Consider now a fictitious cluster having [Fe/H] 0.5 dex smaller than that of M92 (for which we adopt [Fe/H] = -2.2). Where would a cluster lie in Fig. 2 if its [Fe/H] were -2.7? We can calculate this with the help of the theoretical scanner fluxes recently computed by Bell and Gustafsson (1978, hereafter "BG") for giants having a range of values of T_{eff}, log g, and [Fe/H]. The idea is that we start with the theoretical energy distribution corresponding to some M92 giant with M_V near -1, calculate S from this distribution, and then ask how S is changed if [Fe/H] is reduced by 0.5 dex and M_V is held constant.

In order to make the calculation, we must do two things. First, we must compare the theoretical with observed energy distributions, since the Bell and Gustafsson theoretical scanner fluxes do not include some wave-length intervals needed to calculate the Q(λ)'s (and therefore S). And second, we must find the change in location of the Hayashi line when [Fe/H] is changed from -2.2 to -2.7. We adopt for this purpose the interior model calculations of Sweigart and Gross (1978, hereafter "SG"). We begin with a real star, HD 122563, for which $(B-V)^\circ = 0.90$ and [Fe/H] = -2.7, and ask where it would lie in the HR diagram of a cluster having [Fe/H] = -2.7 if the turn-off mass (therefore age) and helium abundance were the same as M92. From the BG models, we find that the star has T_{eff} near 4600°K, and for a change in metals of 0.5 dex, it will be hotter than its M92 counterpart by 30°K (cf. SG), a change corresponding to 0.015 in $(B-V)^\circ$. Thus the M92 counterpart has $(B-V)^\circ = 0.915$ and T_{eff} = 4590°K (BG); whence T_{eff} (HD 122563) = 4620°K. For an M92 modulus of 14.5 (Sandage 1970) this puts HD 122563 at $M_V = -1.15$ (log g = 1.5); the "real" stars of interest in M92 are therefore objects such as IV-10 and IV-79.

In Figure 1, we show theoretical scanner fluxes (BG) averaged over 50A intervals, for a star having (T_{eff}, log g, [Fe/H]) = 4620°K, +1.5, -2.7), compared with Lick IDS Scanner fluxes for HD 122563, with Δλ = 160 A (corresponding to the SZ resolution). The agreement between theory and observation is remarkably good (The Lick IDS data also agree very well with those of Christensen (1978) for the same star.). One is therefore encouraged to believe that the Q(λ) can be derived for the "theoretical" star having parameters (4620°K, +1.5, -2.7) using the real energy distribution of HD 122563 interpolated at the appropriate wavelengths. Similarly, in the second panel of Figure 1, we show a comparison between the theoretical model with parameters (4590°K, +1.5, -2.2) and the mean energy distribution of M92 IV-10 and IV-79 also derived from Lick IDS scans. A summary of values of ψ_o and Q(λ) are given in Table 1. The main point is this: when we calculate S for these two energy distributions, we find S(-2.7) = 0.267 (for HD 122563) and S(-2.2) = 0.280 (for the mean of IV-10 and IV-79 in M92); that is, within the error (±0.02) quoted by SZ, the values of S are indistinguishable. Thus we conclude that, with the SZ method, one cannot distinguish a cluster with metal abundance [Fe/H] = -2.2 from one with [Fe/H] 0.5 dex smaller.

The reason for this lack of sensitivity is not hard to find. As one removes metals from a solar abundance giant, the spectrum changes rapidly until [Fe/H] reaches a value near -2, at which point the metal lines are mostly very weak; after that, the spectrum changes slowly and response is confined largely to the region of H and K and the Fe features immediately shortward. Indeed of the Q's, only Q(3880) shows much sensitivity when [Fe/H] < -2.0. But Q(3880) is the least "error-free" of the Q's, and besides, its effect tends to be lost when it is averaged

Table 1

"Star" (ref.)	(B-V)°	ψ_o	Q (4520)	Q (4360)	Q (4040)	Q (3880)	S
(4620,-1.5,-2.7) (present)	0.90	0.98	0.12	0.30	0.43	0.66	0.267
(4590,-1.5,-2.2) (present)	0.915	1.02	0.10	0.33	0.45	0.76	0.280
M92, IV-10 (SZ)	0.935	1.02	0.14	0.23	0.46	0.92	0.28
(4470,-1.5,-1.4) (present)	1.01	1.07	0.14	0.34	0.59	1.10	0.38
M13, A1 (SZ)	1.02	1.07	0.16	0.37	0.61	1.31	0.39

with the other Q's to find S. The Lick scanner observations with resolution 12 Å, easily reveal the differences between HD 122563 and M92, IV-10 or IV-79 at and near H and K; these differences are suppressed at the 160 A resolution employed by SZ.

One may object in our treatment that the absolute magnitude of HD 122563 is unknown: it is true that we know directly only its abscissa in the HR diagram. Nevertheless, we are concerned with its spectrum only as an interpolation device to obtain the Q(λ)'s within a family of theoretical scanner fluxes; this interpolation will be valid if T_{eff} and log g for HD 122563 are close to those of the M92 stars (viz. 4590°, +1.5) used in the comparison. Now the BG-models show that (U-B) is quite sensitive to log g when $T_{eff} \sim 4500$°K mostly because of the change in relative influence of Rayleigh scattering on the opacity. For M92 IV-79, (U-B)° = 0.40 (Cathey 1974) whereas for HD 122563, (U-B)° = 0.38; this difference corresponds to a change of less than 0.1 dex in log g, which is completely negligible.

As a further check on our conclusions, we calculate S in the same way for a star at M_V =-1.15 in a theoretical cluster having M13 abundances. We are concerned only with the difference Δ[Fe/H] between M92 and M13, although the [Fe/H]-value for M13 is somewhat controversial. We summarize in Table 2 the values of Δ[Fe/H] (sense M13 minus M92) given in the literature (Cohen, et al. found [Fe/H] = -2.4 for M92). We adopt Δ[Fe/H] = +0.8, although our conclusion would be little changed for a value as small as +0.6. From SG, we find that the Hayashi line moves to T_{eff} = 4470°K; the corresponding BG-model, with Lick scanner flux points, is shown in the bottom panel of Fig. 1; observational details are given in Table 1. The calculated value of S (M_V = -1.15) is 0.38. The S-values at M_V = -1.15 for the three abundances

Table 2

Method	Ref.	Δ[Fe/H]
Δs (RR Lyraes)	Butler (1975)	+1.1
Echelle, high res. spectro.	Cohen, et al. (1978) Cohen (1978)	+0.8
High res. spectro.	Wallerstein & Helfer (1966)	+0.7
DDO photometry	Hartwick, *et al*. (1977) Bell (1976)	+0.6
	Adopted	+0.8

(Fe/H) = -2.7, -2.2, -1.4) are shown in Fig. 2; except for a slight systematic displacement to the left, the points match the observed clusters well. We have carried out a similar set of calculations at M_V = +0.6. The fit to the "observed" curves is again good; the curve labelled (-2.7) is drawn between the two computed points. The extrapolation to higher luminosities is uncertain, since BG do not compute scanner fluxes for the combination (4000°K, 0.75, -3.0). Examination of their table of theoretical colors suggests, however, that the dashed curve might actually cross over the curve for [Fe/H] = -2.2 as M_V decreases, owing to the influence of Rayleigh scattering.

We conclude by commenting on the use, as an abundance parameter, of $(B-V)_{o,g}$, the unreddened color of the giant branch at its junction with the horizontal branch; this color is achieved near M_V = +0.6. It is well-known that $(B-V)_{o,g}$ is correlated with [Fe/H] in the range -2 < [Fe/H] < 0 (cf. Butler 1975b). But how sensitive is it to abundance changes near [Fe/H] = -2 and lower? $(B-V)_{o,g}$ for metal-poor clusters (Sandage, *et al*. (1977) ranges from 0.67 for NGC 5053 to 0.72 for M68 and M53 and equals 0.69 ± 0.02 for M92; the errors are estimated to range from ± 0.02 to ± 0.05 in the best determined cases. But as we have seen above, if we change our BG model from (5000, 2.25, -2.2) (for M92) to (5030, 2.25, -2.7) (for a fictitious cluster with metals reduced by 0.5 dex), $(B-V)_{o,g}$ changes from 0.71 to 0.685, i.e., $\Delta(B-V)_{o,g}$ = 0.025. Since $\Delta(B-V)_{o,g}$ is of the same order as the typical errors in $(B-V)_{o,g}$, it follows that the isolation of very metal-poor clusters by this technique is also somewhat problematical.

We are indebted to Dr. T. D. Kinman for valuable comments.

This research was supported by NSF contract AST 74-13717. One of us (Ch. F. T.) wishes to acknowledge the support of a grant from the "Schweizerische Naturforschende Gesellschaft".

REFERENCES

Arp, H. 1965, in "Galactic Structure", ed. A.Blaauw and M.Schmidt pp.401-434.
Becker, W. 1972, Quart. Journ. R.A.S., 13, pp. 226-240.
Bell, R. 1976, in I.A.U. Symposium No. 72, ed. B.Hauck and P. Keenan, pp. 49-62.
Bell, R. and Gustafsson, B. 1978, Astron. and Astrophys., (in press).
Bessell, M. 1977, Proc. A.S.A., 3, pp. 144-145.
Butler, D. 1975a, Astrophys. J., 200, pp. 68-81.
Butler, D. 1975b, Pub. Astron. Soc. Pacific, 87, pp. 559-560.
Butler, D., Epps, L., Dickens, R. and Bell, R. 1978, in I.A.U. Symposium No. 80 (in press).
Canterna, R. and Schommer, R. 1978, Astrop. J. Letters, 219, L119-L122.
Carbon, D., Langer, G., Butler, D., Kraft, R., Trefzger, Ch., Kemper, E., Nocar, J. 1978, Astron. J. (in preparation).
Cathey, L. 1974, Astron. J., 79, 1370-1377.
Christensen, C. 1978, Astron. J., 83, pp. 244-265.
Cohen, J. 1978, Astrophys. J. (in press).
Cohen, J., Frogel, J. and Persson, S. 1978, Astrophys. J. (in press).
Cowley, A., Hartwick, F. and Sargent, W. 1978, Astrophys. J., 221. pp. 512-516.
Faulkner, J. 1966, Astrophys. J., 144, pp. 978-994.
Harris, W. 1976, Astron. J., 81, pp. 1095-1116.
Hartwick, F., Hesser, J. and McClure, R., Astrophys. J. Supplements, 33, 471-491.
Hawley, S. and Miller, J. 1978, Astrophys. J., 220, pp. 609-613.
Iben, I. 1974, in Annual Rev. Astron. and Astrophys., 12, pp. 215-256.
Kinman, T. 1959, Monthly Notices Royal Astron. Soc., 119, pp. 559-578.
Kinman, T., Wirtanen, C. and Janes, K. 1965, Astrophys. J. Supplements, 11, pp. 223-276.
Kinman, T., Wirtanen, C. and Janes, K. 1966, Astrophys. J. Supplements, 13, 379-412.
Lambert, D., Sneden, C. and Ries, L. 1974, Astrophys. J., 188, pp. 97-103.
Larson, R. 1975, Monthly Notices Roy. Astron. Soc., 173, pp. 671-699.
Larson, R. 1976, Monthly Notices Roy. Astron. Soc., 176, pp. 31-52.
McDonald, L. 1976, Ph.D. Thesis, UCSC (unpublished).
Miller, J. and Mathews, W. 1976, Astrophys. J., 172, pp. 593-604.
Morgan, W. 1956, Pub. Astron. Soc. Pacific, 68, pp. 509-516.
Pilachowski, C. 1978, Astrophys. J. (in press).
Preston, G. 1959, Astrophys. J., 130, pp. 507-538.
Renzini, A. 1977, Lectures given at the 7th Advanced Course Saas-Fee (Switzerland) March 28-April 2 on "Advanced Stages in Stellar Evolution".
Robinson, L. and Wampler, E. 1972, Pub. Astron. Soc. Pacific, 84, pp. 161-167.
Sandage, A. 1970, Astrophys. J., 162, pp. 841-870.
Sandage, A., Katem, B. and Johnson, H. 1977, Astron. J., 82, pp. 389-394.
Searle, L. and Zinn, R. 1978, Astrophys. J. (in press).
Steinlin, U. 1973, in "Spectral Classification and Multicolour Photometry" ed. Ch. Fehrenback and B. Westerlund, pp. 226-229.

Sweigart, A. and Gross, P. 1978, Astrophys. J. Supplements, 36, pp.405-438.
Wallerstein, G. and Helfer, H. 1966, Astron. J., 71, pp. 350-354.
Wallerstein, G., Greenstein, J., Parker, R., Helfer, H. and Aller, L. 1963, Astrophys. J., 137, pp. 280-300.

DISCUSSION

Peimbert: 1. I agree that from your data there seems to be no abundance gradient in the halo perpendicular to the plane in the solar vicinity. What can be said about regions closer to the galactic center?

2. Can you comment on the state of CNO abundance determinations?

Kraft: 1. Rodgers found evidence for an abundance gradient in a small sample of RR Lyraes nearer the galactic nucleus. I do not remember now the numerical values, but they would need changing to judge by recent work of Bell and Manduca.

2. Bell, Deming, Laird, and myself have obtained high-dispersion (KPNO 4-M echelle) spectra of many field RR Lyrae stars near maximum light--a phase at which the temperature is high enough to permit analysis of atomic lines of the CNO group. The RR Lyraes studied cover the entire range $0 > [Fe/H] > -2$. We have found that moderately substantial carbon over-deficiencies are the rules, and not the exception. Our oxygen analysis involves computation of non-LTE level populations for a model oxygen atom, and is not finished. Preliminary computations indicate slight oxygen under-deficiencies in the most iron-poor stars; results for the iron-rich stars should be available soon. Nitrogen abundances can be determined only for the most iron-rich RR Lyraes, and for those $[N/H] \sim 0$.

Lesh: Do you find a difference in metal abundance between the RR Lyrae stars in globular clusters and those in the disk population, corresponding to the difference in helium abundance between these two groups that you have postulated on other grounds? Would you expect the difference in helium abundance to entail a difference in metal abundance?

Kraft: Yes, in the sense that halo RR Lyraes have [Fe/H] in the range -1 to -2 or so, whereas old disk RR Lyraes have $[Fe/H] \gtrsim -0.7$. But this is opposite to the sense of my proposal that He in the disk RR Lyraes is actually lower than in the RR Lyraes of halo globular clusters. This is, of course, opposite to the generally accepted theoretical picture of chemical evolution in the Galaxy. But direct observational determinations of the He abundance are, of course, lacking.

Bok: How far out in z do you go with your RR Lyrae variables? If the RR Lyraes you have are well mixed perpendicular to the galactic plane, then one would not expect differences in metal abundances for the group.

Kraft: The most distant stars are 25 or 30 kpc above the plane. The question is whether over such a large distance one expects complete mixing in [Fe/H]. Galactic collapse models suggest that a gradient should exist; the present work provides a direct observational test.

Butler: Regarding the question, "What do we do now?", I would like to return to Manduca and Bell, who have recently computed synthetic spectra for RR Lyrae stars, and have shown that synthetic ΔS values are in agreement with observed values for RR Lyrae stars of known calcium abundance. Because it is clear that we are in need of a metal abundance parameter which can be measured efficiently, and one which remains sensitive to abundance in extremely metal-poor G and K-giants, it would certainly be useful to experiment (computationally) with new ΔS-like parameters in which IR colors replace hydrogen lines in removing the temperature dependence of K-line strength. It is my guess that the K-line would remain measurable in a middle G-giant as metal-poor as [Ca/H] = -4.

GAMMA-RAY EVIDENCE FOR A GALACTIC HALO

F. W. Stecker
Laboratory for High Energy Astrophysics
NASA Goddard Space Flight Center
Greenblet, Maryland U.S.A.

Abstract: γ-ray data favor a cosmic-ray propagation halo comparable with the thickness of the primary radio emission disk of our Galaxy and the halo seen around NGC891.

1. DEFINITIONS OF GALACTIC HALOS

We define a galactic halo to be a region containing relativistic particles in significant enough numbers to have observable effects. Such a region must also extend above and below the galactic plane for a distance larger than twice the scale height of the source region. Using these definitions, we distinguish between a cosmic-ray electron halo and a cosmic-ray nucleon halo, the former occupying a volume less than or equal to that occupied by the latter because electrons can suffer significant energy loss by synchrotron emission and Compton interactions during their lifetime whereas nuclei do not. The electron halo is manifested through observations of radio synchrotron emission and x-ray and γ-ray production from Compton interactions. The nucleon halo, which plays a significant role in the dynamics of the interstellar medium and in determining the propagation characteristics of cosmic rays, is not directly observable by γ-ray astronomy because of the extremely tenuous nature of the gas far from the galactic plane. (Cosmic ray nucleons must interact with gas in order to produce observable γ-rays). However, indirect arguments using data on the distribution of γ-radiation in the plane can be used to place limits on the dimensions of the nucleon halo (Stecker and Jones 1977). γ-ray observations provide the only means for studying the nucleon halo.

Both electron and nucleon halos can be separated into two regions. There is (1) a region where particle propagation is dominated by diffusion and there is a significant probability for particles in the region to return to the plane. Since this region will generally have the form of a thick disk, I will call it the diffusion disk. (2) I also define a region where convection or free escape dominates over diffusion. Particles in this region will not return to the propagation disk (Jones 1978). This region, which I will call the exodisk, (Stecker 1977) will generally have a lower particle density but may still have observable radio emission (Webster 1975). We will consider only dynamical (Ipavich 1975, Jokipii 1976,

W. B. Burton (ed.), The Large-Scale Characteristics of the Galaxy, 475–482.

Owens and Jokipii 1977) and diffusion halos to be physically plausible; trapping, closed or "leaky box" models have no physical basis and also contradict evidence of large scale gradients in the galactic cosmic ray distribution (Stecker 1977, Stecker and Jones 1977).

2. EXISTENCE OF A PROPAGATION DISK

The γ-ray data from both SAS-2 (Fichtel et al. 1975) and COS-B (Bennett et al. 1977) require that cosmic-rays not be strictly confined to spiral arms; they must diffuse into a larger propagation region. If cosmic rays were strictly confined to well defined spiral arms with a large arm interarm gas ratio, as in the gas (Roberts 1977), the result would be spiral arm peaks in the γ-ray longitude distribution which would be too pronounced and too intense in comparison to the data (Stecker 1977, Stecker and Jones 1977). As a specific example, we may note the lack of a Sagittarius arm feature at $\ell = 50°$ in the γ-ray data.

Further evidence for a propagation disk comes from analysis of the nonthermal radio continuum data which shows that confinement of cosmic rays to spiral arms is untenable (French and Osborne 1976) and that there is a strong disk component of nonthermal emission (Price 1974). Most measurements of ^{10}Be in the cosmic-rays (e.g. Garcia-Munoz et al. 1977) indicate that cosmic rays have a mean lifetime in the range $(1-2)\times10^7$ yr and have traversed a gas of mean density during that time of 0.15-0.3 cm^{-3}. If this is indeed the case, the cosmic rays within ~ 1 kpc of us must have spent most of their time in regions of quite low density since the mean density in the galactic disk in the solar vicinity is ~ 1 cm^{-3} (Gordon and Burton 1976, Jenkins 1977). The ^{10}Be situation is still not completely settled (see summaries of the data given by Stecker and Jones 1977 and Ormes and Freier 1978). Ormes and Freier (1978) have shown that one can build a model of cosmic-ray propagation which is consistent with the data on cosmic-ray composition as well as the radio and γ-ray data assuming a mean cosmic ray age of $(1-2)\times10^7$ yr. Dynamical considerations (Badhwar and Stephens 1975, Parker 1977) also favor a thick propagation disk model.

3. LATITUDE DISTRIBUTION OF γ-RAYS AND THE ELECTRON HALO

Fichtel et al. (1978) have considered the latitude distribution of γ-rays observed by SAS-2 to be made up of two components of the form $A+BN_{HI}$. Component A has a very steep energy spectrum. It is most likely of cosmological origin (Stecker 1977, 1978a) and is so isotropic as to exclude large quasispherical halos with radii less than 45 kpc (Fichtel et al. 1978). The isotropy and energy spectrum observations rule out the large γ-ray halo models discussed by Worrall and Strong (1977) (although thin halo models considered by Worrall (1977) are consistent with the conclusions of this section). We will thus consider that any extended γ-ray halo ewith dimensions $\sim$ 10-20 kpc be either non-existent or too weak to be observable at present, a conclusion which has interesting implications for high energy cosmology (Stecker 1978b.) We do, however, consider the possibility that the component designated by Fichtel et al. (1978) as BN_{HI}, which they have shown to have roughly linear dependence on N_{HI} for $|b|>10°$, can also represent the sum of a number of disk components which scale roughly as csc|b|. Such a rough dependence has previously been shown by Fichtel et al. (1977). This component can also include a contribution from Compton interactions of cosmic ray electrons in the radio disk (Stecker 1977) as well as bremsstrahlung and π°-decay γ-rays arising from cosmic ray interactions with both HI and H_2 in the gas disk. The electron halo will then be identified with a thick disk-shaped propagation region which extends somewhat further than the radio disk if the magnetic field falls off with distance from the plane. γ-rays from this "electron halo" then arise from Compton interactions between electrons and the various photon fields (starlight, far infrared, universal microwave) in the galactic neighborhood. In support of this hypothesis we note that Fichtel et al. (1978) have also found a good correlation between the γ-ray flux and 150 MHz radio flux for $|b|>10°$. Working from this type of model Schlickeiser and Thielheim (1977) have obtained an effective half-thickness of the electron halo $h \sim 3$ kpc. Stecker and Jones, taking various uncertainties into account have obtained the results $h = 2\pm2$ kpc.

To reexamine this problem here, we first consider the γ-rays arising in the matter disk from π°-decay and bremsstrahlung (π+B). Using the reddening data of Heiles (1976) and the relation between reddening and total hydrogen column density N_{HI+H_2} given by Jenkins (1977), the π°-decay production rate of Stecker (1970), and the bremsstrahlung production rate calculated using the low energy electron spectrum derived by Goldstein et al. (1970), we estimate the integral γ-ray flux above energy E_γ, $J(>E_\gamma)$, to lie within the region in Figure 1, bounded by upper and lower limits

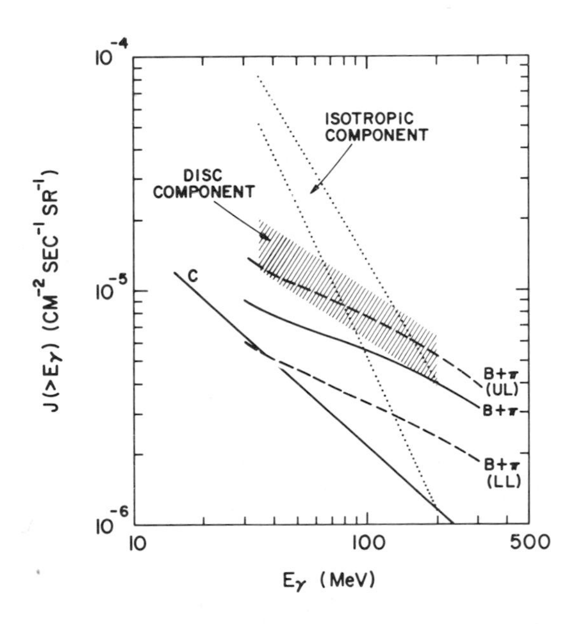

Figure 1. Integral fluxes in the directions of the galactic poles (see text).

designated by (UL) and (LL). Figure 1 also shows the data on the isotropic and disk component (shown shaded) given by Fichtel et al. (1978). All of these results are shown for b = 90°. The remaining curve, marked C, shows the estimated flux of Compton γ-rays from an electron halo of half-thickness h = 1.5 kpc. Such a component when added to the π+B component will provide a better fit to the data, supporting the hypothesis of a thin disk shaped electron halo. This result agrees with analysis of the radio data by Illovaisky and Lequeux (1972) and Baldwin (1977). However, the large uncertainties in both the γ-ray data and the theoretical calculations must be emphasized.

4. LONGITUDE DISTRIBUTION OF γ-RAYS AND THE NUCLEON HALO

Since cosmic ray nucleons do not produce measurable numbers of γ-rays out of the gas disk of the galaxy, we must resort to indirect determination of their propagation based on their deduced radial distribution in the galactic plane. This was the basic idea proposed in the analysis of Stecker and Jones (1977). It is based on using a simple isotropic diffusion model for cosmic ray propagation which seems reasonable if the propagation region does not extend too far from the plane. For non-isotropic diffusion with diffusion coefficients perpendicular and parallel to the plane $D_\perp$ and $D_\parallel$, the derived width of the diffusion disk (halo) need only be scaled by the factor $(D_\perp/D_\parallel)^{\frac{1}{2}}$. Jones (1978) has considered dynamical halos and finds that models with outflow may be replaced by purely diffusive models with smaller effective halo thickness in analyzing the γ-ray results. This brings us back to the picture of a diffusive propagation disk surrounded by an exodisk where outflow dominates over diffusion.

It has been shown that if one analyzes the longitude distribution of γ-ray emission in the plane, taking account of the presence of large amounts of H_2 in the inner galaxy, the implied cosmic ray radial distribution closely resembles the distribution of supernovae (SN) in the Galaxy, implying a galactic origin of cosmic rays (Stecker 1975). The cosmic ray distribution is therefore source dominated on a scale of a few kiloparsecs and we therefore expect the diffusion halo to be at most a few kiloparsecs thick.

Stecker and Jones (1977) have made the conclusion more quantitative by considering various models having boundaries such that at a distance $L \simeq 2h$ from the plane the cosmic ray density drops to 0. The planes defined by $|z| = L$ roughly correspond to the boundary between the diffusion disk and the exodisk. Using SN and pulsar source distributions, γ-ray emissivity distributions were computed for various values of L and compared for probability of fit with the SAS-2 data. The results varied from a $\sim$ 30 to 40 percent probability of fit for L=2h=1 kpc to a $\sim$ 5 to 20 percent probability of fit for L=2h=3 kpc to a probability fit of the order of a few percent for h = 2.5 kpc. Thus, within the framework of the analysis it appears that the cosmic ray nucleon "halo" has an effective half width $h < 3$ kpc. Measurements of the energy spectrum of galactic γ-rays (Bennett et al. 1977, Kniffen et al. 1978) indicate that perhaps up to $\sim$ 50% of the radiation above 100 MeV may be from electron bremsstrahlung. However it is still reasonable to assume that the electrons and nucleons have the same source distribution. Using the results of the previous section which indicate a comparable width for the electron diffusion disk, it follows that the result for the nucleon diffusion disk h < 3kpc, should still be valid.

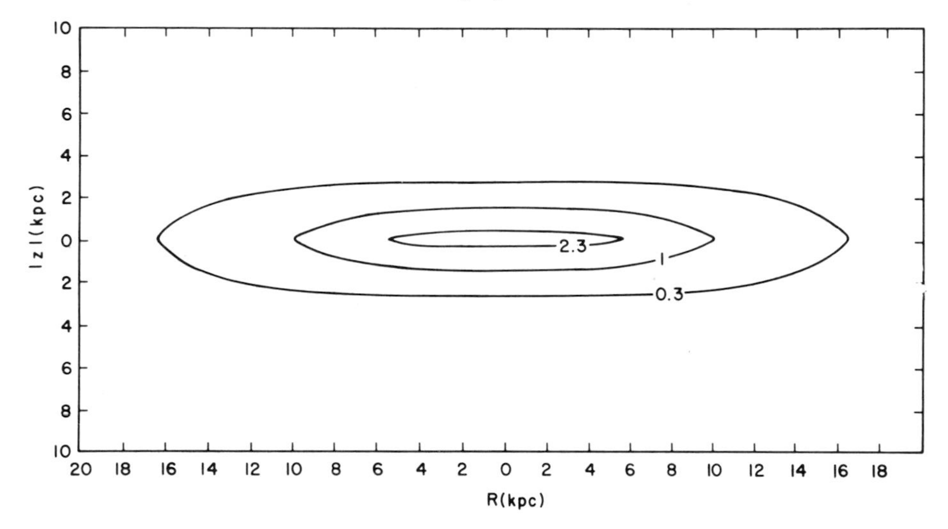

Figure 2. Relative column density of cosmic ray nuclei as seen from outside the Galaxy for the diffusion model of Stecker and Jones with L = 3 kpc.

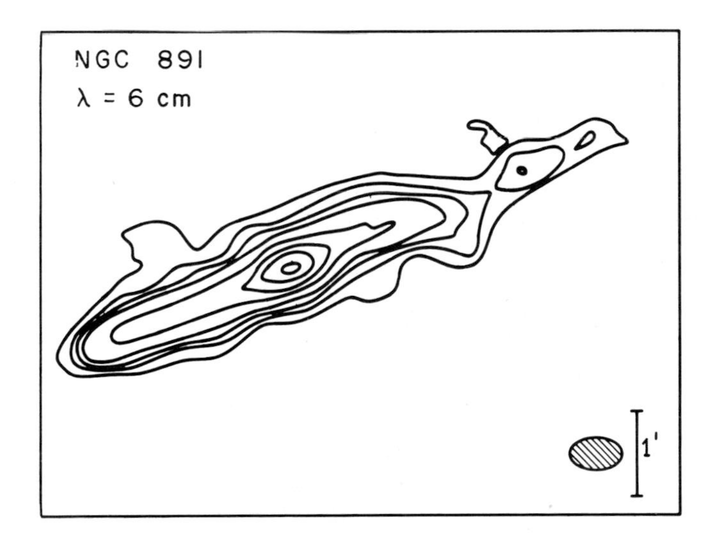

Figure 3. Contour plot of 6cm emission from NGC891, a galaxy similar to ours but with stronger radio emission and a strong nuclear emission component (Allen et al. 1978). Note the similarity to the halo model shown in Figure 2.

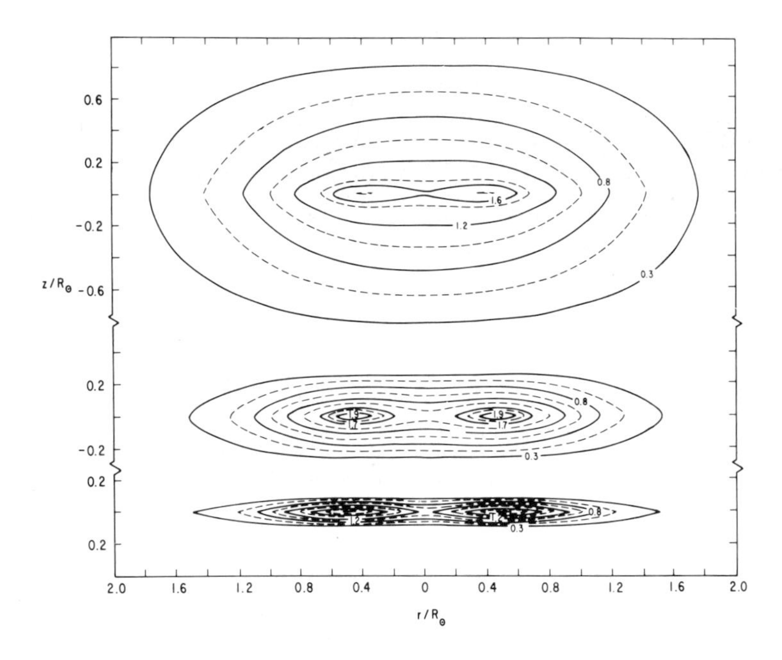

Figure 4. Cross sectional contours of constant cosmic ray intensity in the r-Z plane for diffusion halos with L = 1,3 and 10 kpc. (See Stecker and Jones 1977) The effect of the increased supernova density in the "Great Galactic Ring" at 5-6 kpc is seen in the location of the peaks. Scales are in units of 10 kpc.

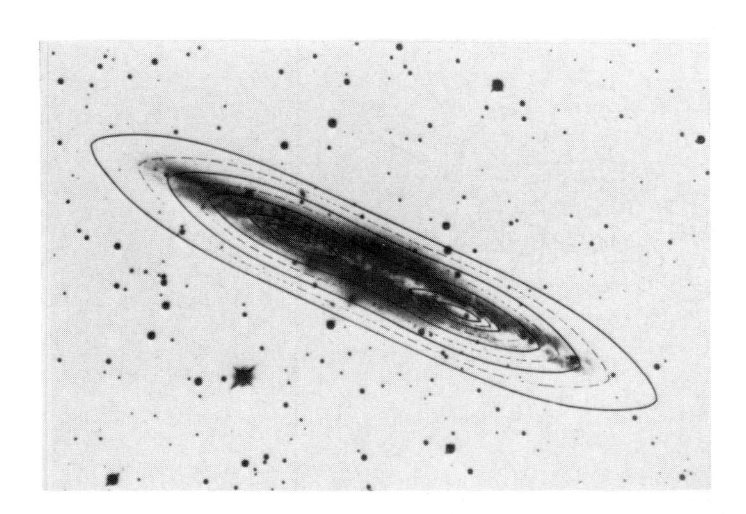

Figure 5. L = 3 kpc contours from Figure 4 superimposed on a photograph of NGC891 (courtesy Hale Observatories) to illustrate the scale of the halo models discussed in the text.

REFERENCES

Allen, R. J., Baldwin, J. E. and Sancisi, R.: 1978, *Astron. and Astrophys. 62*, 397.
Badhwar, G. D. and Stephens, S. A.: 1977, *Ap. J. 212*, 494.
Baldwin, J.: 1977 in *The Structure and Content of the Galaxy and Galactic γ-Rays* (Proc. Greenbelt, Int'l. Symp. NASA CP-002, U. S. Govt. Printing Office, Washington) p. 189.
Bennett, K. et al. (CARAVANE Collaboration) in *Recent Advances in γ-Ray Astronomy* (Proc. Frascati Int'l. Symp. ESA-SP-124; ESA Publications Branch, Noordwijk) p. 83.
Fichtel, C. E., Hartman, R. C., Kniffen, D. A., Thompson, D. J., Bignami, G. F., Ögelman, H., Özel, M. F. and Tümer, T.: 1975, *Ap. J. 198*, 163.
Fichtel, C. E., Hartman, R. C., Kniffen, D. A., Thompson, D. J., Ögelman, H. B., Özel, M. E. and Tümer, T.: 1977, *Ap. J. 217*, L9.
Fichtel, C. E., Simpson, G. A. and Thompson, D. J.: 1978, *Ap. J. 222*, in press.
French, D. K. and Osborne, J. L.: 1976, *Mon. Not. R. Astr. Soc. 177*, 569.
Garcia-Munoz, M., Mason, G. M. and Simpson, J. A.: 1977, *Ap. J. 217*, 859.
Goldstein, M. L., Ramaty, R. and Fisk, L. A.: 1970, *Phys. Rev. Lett. 24*, 1193.
Gordon, M. A. and Burton, W. B.: 1976, *Ap. J. 208*, 346.
Heiles, C.: 1976, *Ap. J. 204*, 379.
Ilovaisky, S. A. and Lequeux, J.: 1972, *Astr. and Ap. 20*, 347.
Ipavich, F. M.: 1975, *Ap. J. 196*, 107.
Jenkins, E. B.: 1977, in *The Structure and Content of the Galaxy and Galactic γ-Rays* (Proc. Greenbelt, Int'l. Symp. NASA CP-002; U. S. Govt. Printing Office, Washington) p. 215.
Jokipii, J. R.: 1976, *Ap. J. 208*, 900.
Jones, F. C.: 1978, *Ap. J. 222*, in press.
Kniffen, D. A., Bertsch, D. L., Morris, D. J., Palmeira, R. A. R., and Ras, K. R.: 1978, *Ap. J.* in press.
Ormes, J. and Freier, P.: 1978, *Ap. J.*, in press.
Owens, A. J. and Jokipii, J. R.: 1977, *Ap. J. 215*, 677.
Parker, E. N.: 1977, in *The Structure and Content of the Galaxy and and Galactic γ-Rays* (Proc. Greenbelt, Int'l. Symp. NASA CP-002; U. S. Govt. Printing Office, Washington) p. 783.
Roberts, W. W., Jr.: 1977, in *The Structure and Content of the Galaxy and Galactic γ-Rays* (Proc. Greenbelt Int'l. Symp. NASA CP-002; U. S. Govt. Printing Office, Washington) p. 119.
Schlickeiser, R. and Thielheim, K. O.: 1977, *Ap. and Space Sci. 47*, 415.
Stecker, F. W.: 1970, *Ap. and Space Sci. 6*, 377.
Stecker, F. W.: 1975, *Phys. Rev. Lett. 35*, 188.
Stecker, F. W.: 1977, *Ap. J. 212*, 60.
Stecker, F. W.: 1978a, *Nature*, in press. (NASA TM 78114)
Stecker, F. W.: 1978b, *Ap. J. 223*, in press. (NASA TM 78045)
Stecker, F. W. and Jones, F. C.: 1977, *Ap. J. 217*, 843.
Webster, A.: 1975, *Mon. Not. R. Astr. Soc. 171*, 243.
Worrall, D. M.: 1977, Ph.D. Thesis, University of Durham.
Worrall, D. M. and Strong, A. W.: 1977, *Astron. and Ap. 57*, 229.

DISCUSSION

Paul: The recent COS-B results show that many gamma-ray sources exist in the Galaxy. Therefore it is quite premature to derive astrophysical conclusions on the basis of pure diffuse-emission models, at least regarding large-scale characteristics of the Galaxy.

Stecker: Fichtel *et al.* have recently shown, in their analysis of the moderate-latitude γ-radiation, that it is not made up of point sources but is diffuse in nature. I believe that you also have recently shown that there is an association of local γ-ray emission with molecular clouds at moderate latitudes. The constancy of the γ-ray spectrum in all parts of the Galaxy also argues for its primarily diffuse origin. I therefore believe that there is strong support for the production of most γ-rays in the inner Galaxy in molecular clouds as I suggested in 1969, and for the diffuse origin hypothesis.

SAS-2 GAMMA RAY OBSERVATIONS RELATED TO A GALACTIC HALO

C. E. Fichtel, G. A. Simpson, and D. J. Thompson
NASA/Goddard Space Flight Center, Greenbelt, Maryland 20771

An examination of the intensity, energy spectrum, and spatial distribution of the diffuse γ radiation observed by SAS 2 away from the galactic plane in the energy range above 35 MeV has revealed no evidence supporting a cosmic ray halo surrounding the galaxy in the general shape of a sphere. The diffuse γ radiation does consist of two components. One component is related to the galactic disk on the basis of its correlation with the 21-cm measurements, the continuum radio emission, and galactic coordinates. Further its energy spectrum is similar to that in the plane, and its intensity distribution joins smoothly to the intense radiation from the plane. The other component appears isotropic, at least on a coarse scale, and has a steep energy spectrum. The degree of isotropy which has been established for the "isotropic" radiation and the steep energy spectrum, which distinguishes it from the galactic disk radiation, place strong constraints on galactic halo models for the origin of this component. Theoretical models involving a galactic halo have generally postulated a halo with dimensions of the order of the Galaxy and hence a radius, at least in the plane, of about 15 kpcs. Since the Sun is about 10 kpc from the galactic center, if such a halo exists and is responsible for the γ rays (through, for example, black body Compton radiation), a very marked anisotropy would be seen, with the γ ray intensity from the general direction of the galactic center being much larger than that from the same latitudes in the anticenter direction. In fact, no such anisotropy is seen; specifically the ratio of the average intensity in the ($300° < \ell < 60°$, $20° < |b| < 40°$) region to that in the ($100° < \ell < 250°$, $20° < |b| < 40°$) region was found to be 1.10 ± 0.19 compared to a calculated value for a model with a uniform cosmic ray sphere with a 15 kpc radius of 2.85. The ratio between the average γ-ray intensity from regions with $|b| < 60°$ to that from $20° < |b| < 40°$ is found to be 0.87 ± 0.09. If the region is assumed to be spherical, but with a larger radius and a uniform cosmic ray density, the upper limit (2σ) set for the anisotropy demands that the radius be at least 45 kpc. An extragalactic origin for the isotropic component currently appears to be a more plausible explanation.

W. B. Burton (ed.), The Large-Scale Characteristics of the Galaxy, 483–484.

DISCUSSION

Verschuur: Conditions in interstellar space at 5 kpc are clearly very different from those in the solar neighborhood. We can therefore question whether the emergence of intelligent life as we know it is even possible over much of the inner parts of the Galaxy. Much work is now being done on the influence of cosmic rays and interstellar gas and dust on planetary atmospheres. In order to make meaningful comments about the possibility of life in the inner parts of the Galaxy we need to have better data on cosmic ray and other high energy radiations as a function of R. Is such data likely to become available? What can we learn from the fact that the γ-ray emissivity is about 10 times greater at 5 kpc than at 10 kpc. Knowing better what the environmental conditions are like as a function of R might show that there is a limited volume of galactic space in which planets could exist with Earth-like conditions.

Stecker: The cosmic γ-ray fluxes even at 5 kpc would be quite insignificant in this context. The cosmic-ray intensity at 5 kpc would be about 2 to 3 times higher than in our vicinity and should not present any real problem to life as we know it on a planet similar to ours, although the genetic mutation rate and consequent evolution rate might be proportionately higher.

Lequeux: A nearly final reduction of the COS-B data for the galactic center shows no excess at $\ell \sim 0°$ in the longitude profile, contrary to what SAS-2 found. The discrepancy might be due to the presence of a transient γ-ray source at the time of the SAS-2 observation. From the COS-B data, taking a mass of interstellar matter of 7×10^7 $M_\odot$ within $R < 300$ pc (rather a minimum), we find that the cosmic ray flux at the galactic center is not more than two times the flux close to the Sun (3σ limit).

Stecker: The result which you give is at variance with the result of the SAS-2 satellite experiment and with the results previously reported by the COS-B collaboration at the γ-ray symposium in Frascati in 1977. Both previous results show a peak in the γ-ray distribution at the galactic center.

Basu: Could your analyses of the γ-ray distribution in the halo be used to make an estimate of the mass of the halo?

Stecker: The cosmic-ray "halo" in the Galaxy, as I have tried to show, is basically a Population I-related phenomenon associated with recent star formation in the galactic disk. The stellar halo, on the other hand, is a Population II phenomenon. Thus, the answer to your question is no. Perhaps the term "halo" as applied to cosmic-rays is confusing.

COSMIC-RAY EVIDENCE FOR A HALO[1]

V.L. Ginzburg
P.N. Lebedev Physical Institute, Acad. Sci. USSR, Moscow, USSR

1. INTRODUCTION

Cosmic rays were discovered in 1912, but it was only about forty years later that they were found to play an important role in astronomy. Firstly, cosmic rays (including the electron component) are an important source of astronomical information, namely the cosmic synchrotron radiation. Secondly, cosmic rays are essential as energetic and dynamical factors in the galaxy and also as a source of heating and transformation of the interstellar gas composition. Suffice it to remember, for example, that near the solar system the cosmic ray energy density is about the same as the thermal energy of the interstellar gas, and the cosmic ray pressure is likewise about the same as the interstellar gas pressure. Thus, there is every reason to believe that galaxies do not consist of stars and gas only, but of cosmic rays as well.

This conclusion is, of course, well known at present but it is emphasized here because the role of cosmic ray astrophysics in galactic astronomy is still rather small except for the case of the synchrotron radiation theory. It seems to me that to a considerable extent this is explained by the difficulties faced by cosmic ray studies and as a consequence by a comparatively slow progress in this field. As a result, a number of basic questions remained vague for a long time. Seeing that there are disputes in the literature even of a galactic vs. metagalactic origin of cosmic rays and whether galaxies have a radio or a cosmic ray halo, an astronomer is naturally apt to be particularly careful with the cosmic ray data.

Meanwhile, the picture has been significantly clarified (at least in my opinion) concerning the two above mentioned questions. These two problems, especially the halo problem, will be discussed here briefly.

2. COSMIC RAY ORIGIN MODELS

The situation with the cosmic ray origin problem as a whole is presented in Refs. 1-3. In the metagalactic models (e.g. Ref. 4) the cosmic

W. B. Burton (ed.), The Large-Scale Characteristics of the Galaxy, 485–490.

rays get into the Galaxy from outside, while in the galactic models the cosmic rays are generated within the Galaxy. The galactic and metagalactic models are so different that without choosing one it is impossible to establish the cosmic ray behavior in the Galaxy. In the case of the cosmic ray electron component, a galactic origin may be considered proved, since Compton and synchrotron losses on the 2.7 K blackbody radiation do not allow electrons of energy $\gtrsim 10^{10}$ - 10^{11} eV formed in other galaxies to reach the Earth or even the Galaxy. However, as far as the proton-nucleon component is concerned, the arguments in the past were for the most part indirect, involving energy considerations, analysis of the charged particle motion, etc. But now with the present data on the intensity of gamma-rays of energies >50-100 MeV in the Galaxy anticenter direction, one can put forward quite direct objections to metagalactic models[1,3,5-8]. (Except perhaps for superhigh energy cosmic rays with energies $\gtrsim 10^{17}$ eV). Meanwhile, no evidence has appeared in favor of these metagalactic models, and so now practically everybody has evidently rejected them and so there is no need to discuss this question further.

3. COSMIC RAY DATA AND THE HALO

In galactic models it is supernovae (including pulsars) that are likely to serve as cosmic ray sources. Even if other active stars or if possible explosions of the galactic nucleus play some role, the sources are in all cases concentrated near the galactic plane, say, within the gas disk with a half-thickness ∿100-150 pc. Now, cosmic rays are confined only by the magnetic field frozen in the interstellar gas. The gas is concentrated near the galactic plane due to gravity. However, investigations of the controlled thermonuclear synthesis problem show how difficult it is to keep charged particles even in special laboratory magnetic traps. In cosmic conditions, and in weak fields it is all the more difficult. Thus, the data on gas clouds far from the galactic plane (at $z \gtrsim 1$ Kpc) and radioastronomical observations also leave no doubt that cosmic rays in the Galaxy do not remain in the gas disk region but occupy some region with a characteristic halfthickness >>100-150 pc. This is the region to be called a "cosmic ray halo".

What conclusions concerning the cosmic ray halo can be made on the basis of the data on cosmic rays near the Earth? The essence of the matter is such that for its analysis one should use various data (often quite indirect) which only in total makes it possible to arrive at more or less definite conclusions. Since we cannot here go into details (see Refs. 1, 2, 3, 5-7) we shall first of all formulate the results. Firstly, there are no indications *against* the assumption that the Galaxy has a large (quasi-spherical) cosmic ray halo with a characteristic scale height ∿10 Kpc, and with cosmic rays at an energy density near that at Earth. Secondly, even beside the radio data there exists some information and arguments in favor of the model with a large halo, although it cannot be considered proved.

Not to touch upon radiodata and the already mentioned cosmic-ray confinement arguments and the presence of gas clouds at large z, one may involve the results of the investigations of cosmic ray anisotropy and elemental and isotopic composition. Cosmic ray isotropy is so high that their anisotropy has even not yet been reliably established. At energies below 10^{12} eV the anisotropy coefficient is $\lesssim 10^{-3} - 10^{-4}$. High isotropy is quite natural in a model with a large halo where the cosmic ray concentration gradients are small. It is obvious, however, that such an argument taken separately is not weighty enough.

The data on the elemental composition of stable nuclei in cosmic rays leads to the conclusion that they pass through a thickness $x \sim 5$ g cm^{-2} in the interstellar medium (evaluated for pure hydrogen). Now, if the cosmic ray "trapping" region is a gas disk, then the density $n \sim 1$ atom cm^{-3}, and the time required to traverse the 5 g cm^{-2} is about 3×10^6 years. If the particles are trapped in a large halo but their passing through the disk is taken into account, then, approximately $n \sim 10^{-2}$ and the lifetime is 3×10^8 years. So, knowing the thickness x, we cannot yet find the cosmic ray lifetime, and so the halo dimension remains unknown. However, the situation is different when one considers secondary radioactive nuclei (e.g., ^{10}Be which decays into ^{10}B + e^- with a mean lifetime 2.2×10^6 years). Knowing x for stable nuclei, and the relative number of ^{10}Be in the cosmic rays, one can already find the cosmic ray lifetime. According to the data available[9] for ^{10}Be the lifetime is 1.7×10^7 years, whence $n = 0.2$ atoms cm^{-3} for the cosmic ray trapping region. This is already proof (though not rigorous) of the fact that cosmic rays leave the gas disk.

At the same time these data by no means contradict the model with a large halo. This is because if the radioactive nuclei lifetime is not large enough, these nuclei have no time to fill up the halo. In other words, for a large halo the radioactive nuclei, the same as relativistic electrons, fill only part of the halo; they pass only to the distance z corresponding to their lifetime. Summarizing, it may be said that at present the direct data on cosmic rays near the Earth only do not contradict the model of a large halo, while they do show that cosmic rays go rather far beyond the gas disk.

4. RADIOASTRONOMICAL EVIDENCE FOR THE HALO

The most reliable of all now available methods of halo study is a radioastronomical one, although it enables one to judge only the halo for the cosmic ray electron component or, as it is often referred to, a radiohalo. The radiohalo is due to synchrotron and Compton losses of the electrons in the magnetic field of the halo region. Unfortunately, the question of a radiohalo of the Galaxy has appeared to be not only difficult to answer, but is has also aroused objections and unpleasant arguments. To my mind several reasons may exist for this situation.

Firstly, being inside a radiating system it is difficult to establish its dimensions and other parameters. Indeed, solutions of the integral equations yielding the radioemission intensity are known to be rather complicated and unstable. And, the presence of discrete sources and various background inhomogeneities complicate the whole picture. Thus, a problem which is simple at first sight is in fact rather complicated, which has aroused errors and misunderstandings and as a result irritation.

Secondly, the radiohalo is often understood as only a spherical or in any case a quasispherical system, and so radiohalo is opposed to radiodisk. However, the difference between halos with scale heights of 1 Kpc vs. 10 Kpc depends to a great extent on the specification of the meaning of the scale height parameter, and so this whole question is of secondary importance. Even so, radiohalo and radiodisk remain opposed in the literature.

A third reason may lie with the desire to solve the radiohalo problem using the minimum of model and theoretical considerations. Such an approach is often, but not always, justified. One cannot make great progress in many radio-astronomical problems when disregarding the synchrotron theory of cosmic radio-emission. The radiohalo problem is not an exception.

I should like to emphasize that I have never (after the work by Pikelner[10]) doubted the existence of a cosmic ray and radio halo, and I believe in it all the more now. The preceding review of reasons for doubting the radiohalo involve problems I consider to be hypothetical. However, I wanted to present them here because I cannot attend the Symposium and discuss this question with colleagues. At the same time I would like to know their opinion.

5. HALOS IN OTHER GALAXIES

Now, the above mentioned difficulties in the study of the galactic radiohalo must to some extent be absent in the observations of other normal edge-on galaxies. What is the observational situation on this question? In NGC4631 such a halo does exist and it is rather bright even at very short wavelengths[11,12]. A radiohalo has also been discovered[13] for edge-on galaxy NGC 891. In fact, I do not know a single case, when a normal spiral edge-on galaxy with a rather high radio-emissivity in the galactic plane had no radiohalo of the type discovered for the above mentioned galaxies.

Summarizing we may state that our Galaxy and similiar ones have a radiohalo, but perhaps this halo is somewhat flattened and less powerful than it was sometimes supposed before. The present report may appear to achieve its goal if it will stop useless arguments concerning the very existence of a radiohalo.

Future work concerning the halo in our own Galaxy should use a broad observational approach. Namely, one should use not only the radio data but also the data on cosmic rays near Earth (elemental and isotopic composition, spectra of protons, nuclei, antiprotons, electrons and positrons, anisotropy) as well as gamma-astronomical information. I am sure that in doing this we should deduce galactic models with a large, or, in any case, a considerable cosmic ray halo. Comparison of all the data will make it possible to specify these models and select the best one.

NOTES

1. This version of V.L. Ginzburg's paper was condensed by G.M. Mason, Dept. of Physics, University of Maryland, College Park, MD USA. The full text has been submitted to Astrophysics and Space Science.

REFERENCES

1. 15th Int. Cosmic Ray Conf., Conf. Papers, Plovdiv, Bulgaria, 1977.
2. Ginzburg, V.L. and Ptuskin, V.S.: 1976, Rev. Mod. Phys. 48, p. 161; 1976, Sov. Phys. Uspekhi 18, p. 931.
3. Ginzburg, V.L.: 1978, Uspekhi Fiz. Nauk 124, p. 307; Engl. transl. Sov. Phys. Uspekhi (1978).
4. Burbidge, G.R.: 1975, Phil. Trans. Roy. Soc. A277, p. 481.
5. Ginzburg, V.L. and Syrovatskii, S.I.: 1964, *Origin of Cosmic Rays*, Pergamon Press, Oxford.
6. Ginzburg, V.L. and Syrovatskii, S.I.: 1967, IAU Symp. No. 31, (ed. H. vanWoerden), Academic Press, London, p. 411.
7. Ginzburg, V.L.: 1975, Phil. Trans. Roy. Soc. A277, p. 463.
8. Strong, A.W., Wolfendale, A.W., Bennett, K. and Wills, R.D.: 1977, Proc. 12th ESLAB Symp. on Astron., Frascati, Italy, p. 167.
9. Garcia-Munoz, M., Mason, G.M. and Simpson, J.A.: 1977, Ap.J. 217, p. 859.
10. Pikelner, S.B.: 1953, Doklady (C.R.) Akad. Nauk USSR 88, p. 229.
11. Ekers, R.D. and Sancisi, R.: 1977, Astron. and Ap. 54, p. 973.
12. Wielebinski, R. and von Kapherr, A.: 1977, Astron. and Ap. 59, L17.
13. Allen, R.J., Baldwin, J.E. and Sancisi, R., 1978: Astron. and Ap. 62, p. 397.

DISCUSSION

Felten: It appears that Prof. Ginzburg still feels that a cosmic-ray halo with a scale height of 10 kpc is compatible with the data, whereas Stecker says that it must be more like 3 kpc. Can a halo as thick as 10 kpc preserve the correlation between cosmic-ray source positions and cosmic-ray densities which Stecker claims is present in the gamma-ray data? Can a halo as thin as 3 kpc account for the high isotropy of the cosmic rays?

Stecker: Prof. Ginzburg's arguments concerning the cosmic-ray evidence for a halo arc, I believe, basically correct. I do, however, believe that the γ-ray evidence is not as ambiguous concerning the size of the halo as the cosmic-ray evidence. The γ-ray evidence favors a flattened halo. It appears to me from the tone of Prof. Ginzburg's remarks that he would not strongly oppose this new result. He is arguing more for the existence of a halo on the basis of the cosmic-ray data. I would have liked to hear his response to your question. Because the gyroradius of cosmic rays in the galactic magnetic field is much less than 1 kpc at 1-10 GeV energies, either halo type would be compatible with the isotropy data.

Cesarsky: It is true that the galactic gamma-ray observations discussed during this Symposium rule out the possibility that cosmic rays in the energy range $\sim$ 1 GeV--i.e., the bulk of the observed cosmic rays--can be extragalactic. But the problem is still alive for the higher energy cosmic rays, especially for $E \gtrsim 10^{17}$ eV. The low value of the cosmic-ray anistropy ($\sim 10^{-4}$) mentioned in the paper was measured at energies of a few hundred GeV; it is believed that, at such energies, cosmic ray trajectories suffer considerable deflections while transiting in the solar cavity. Thus, such measurements are, at best, very difficult to interpret, and, at worst, not relevant to the question of cosmic ray isotropy in the galactic disk.

I want to remark that the discussion of a halo from the point of view of γ-ray data presented by Dr. Stecker, as well as that made on the basis of the observed composition of cosmic rays as elluded to by Ginzburg, only refer to a diffusive halo. This type of argument cannot exclude the presence of a halo made up of particles that are leaving the Galaxy, which would still emit radio-synchrotron radiation. The argument presented by Dr. Fichtel excludes the presence of a strong spherical halo of 100 MeV gamma rays. Such γ-rays had been predicted as arising from inverse Compton interaction of cosmic ray electrons and the 3° black body radiation. But we note that such electrons must have an energy $\gtrsim$ 50 GeV; most cosmic ray observers agree now that the electron spectrum at the Sun is very steep beyond such energies. The steepness is attributed to energy losses of the electrons, and so I suspect that any moderately diffusive model would predict stronger lines, and thus even lower fluxes of high energy electrons in the halo--if there is a halo.

Wielebinski: The halo which has aroused so much theoretical discussion is an intense spheroidal object. When a well-calibrated all-sky survey is taken, the "halo" is what is left over after all the other components have been subtracted. There are, of course, several arbitrary assumptions involved in establishing what is local and what is large-scale. Southern radio continuum surveys are particularly vulnerable because there are relatively few foreground features present. The 408-MHz survey of Haslam should allow a good determination of the halo component. If one accepts the results from edge-on galaxies which indicate a weak ellipsoidal halo with increasing spectral index away from the plane, then a halo of this type should be found around the Milky Way.

IX. THE GALACTIC WARP

OBSERVATIONAL DESCRIPTION OF THE WARP IN OUR GALAXY

A. P. Henderson
University of Maryland, College Park, MD.
and Manhattan College, New York City

In 1975 F. J. Kerr and P. F. Bowers of the University of Maryland made a full coverage survey of the neutral hydrogen in the southern hemisphere between $\pm 10°$ latitude from 240° to 350° in longitude. This survey taken with the CSIRO 18 meter telescope is still in the reduction stage but when completed it will provide an ideal complement to the full $\pm 10°$ coverage of the northern sky by H. Weaver and D. R. W. Williams (1973). The possibility of a unified analysis of the outer region of the Galaxy has inspired this present study. The plan here is to use the $\pm 10°$ Weaver-Williams survey in the region $10° < \ell < 130°$; the $\pm 30°$ extension to this survey (H. Weaver and D. R. W. Williams, 1974) in the region $115° < \ell < 245°$ and finally the Kerr-Bowers survey, $240° < \ell < 350°$. Since the latter survey is incompletely reduced at this time we have only used profiles at 5° intervals in longitude and thereby have produced a preliminary determination of the plane in the southern hemisphere. The northern hemisphere determination is complete.

REDUCTION PROCEDURE

Since this paper is only concerned with the large scale aspects of the outer part of the Galaxy, we assume overall circular symmetry considering "streaming" and "explosive" events as perturbations on this symmetry. We start with an assumed rotation curve and with this, transform the intensity contours at constant longitude which initially measures brightness temperature as a function of velocity and latitude into contours of density as a function of galactocentric radius and z position. By integrating in z, four quantities can be determined: a) the surface brightness σ, b) the half-thickness Δz which is measured as the distance from the $\sigma/4$ to the $3\sigma/4$ position, c) the median position of the plane, $\bar{z}$, which is the $\sigma/2$ position, and finally d) the position of the hydrogen layer as indicated by the z position where the density is maximum. In each case the latter three quantities are only carried out to the radius where $\sigma = 1.25 M_{\odot}/pc^2$.

W. B. Burton (ed.), The Large-Scale Characteristics of the Galaxy, 493–499.

Two rotation curves were assumed in this study: a) the Schmidt curve and b) a flat, θ = 250 km sec^{-1}, curve. The results are similar since the angular velocity, Ω, is monotonic in each curve. The only difference is in the resultant size of the Galaxy such that the $\sigma = 1.25 M_{o}/pc^2$ contour is 30% larger using the flat curve. Using the Schmidt curve we could compare the position of the median z as found by the above reduction procedure with that determined by taking 5° intervals in longitude, integrating at constant velocity (Henderson, 1967). The results are quite compatible, justifying the use of this simple procedure for the preliminary analysis of the southern hemisphere data included in this paper. Optical thinness has been assumed in this outer part of the Galaxy. It will be shown later that this also seems justified.

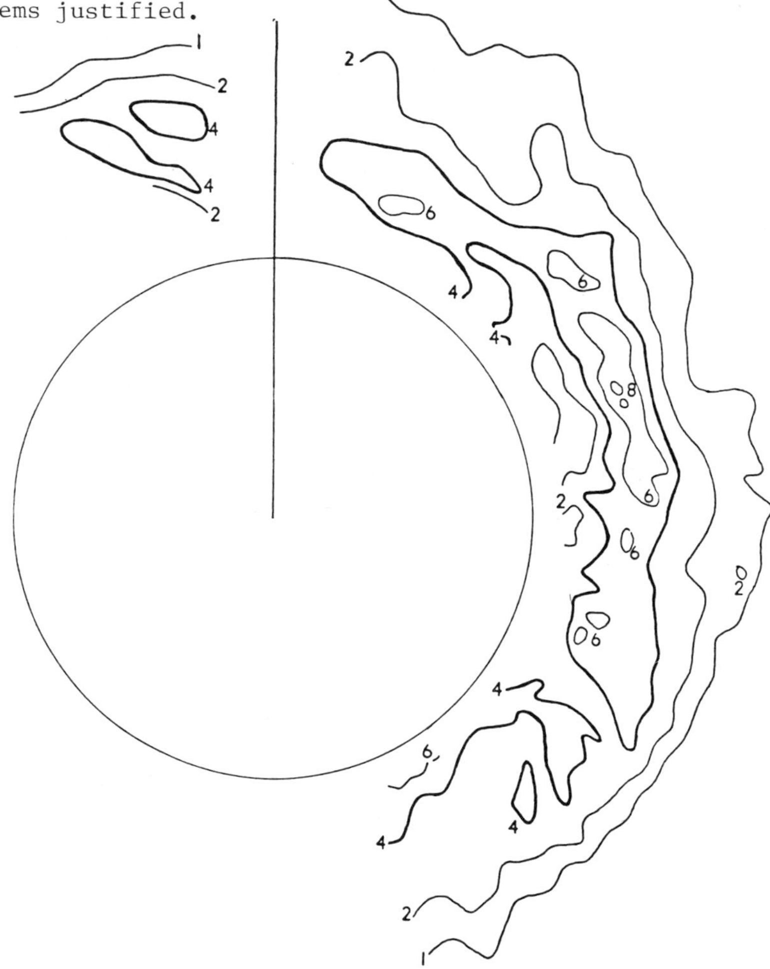

Figure 1. Surface brightness, σ, of the neutral hydrogen determined from the Weaver-Williams surveys using a flat rotation curve. Contours are in units of $1.25 M_{o}/pc^2$, the solar radius (10 kpc) is indicated by the circle and the line from the galactic center through the solar region is also indicated.

RESULTS

As can be seen in figure 1, the increased spatial and velocity resolution of the Weaver-Williams surveys have changed little the over-all structure as represented by the original work of van de Hulst, Muller and Oort (1954). There is a hint of a spiral appearance but it is much weaker than the ring-like structure at 14 kpc. This appearance is not altered when optical depth effects are taken into account. Some of the strong surface brightness features are enhanced but the overall picture remains unchanged. Figure 2 shows the average surface brightness at constant galactic radii averaged between 20° and 160° galactic azimuth, the value calculated on the assumption of a 125° spin temperature is represented by the solid line and the value calculated on the assumption of optical thinness is represented by the dashed line.

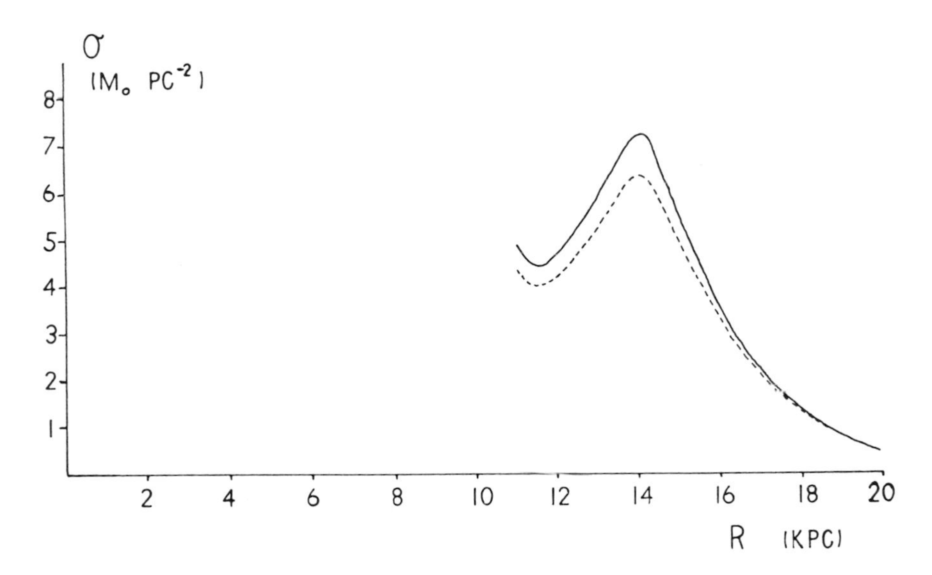

Figure 2. The average surface brightness in the northern hemisphere (dashed line). Optical depth effects were included in producing the solid line.

The half-thickness Δz is shown in figure 3. It does not uniformly increase with radius; in some regions it increases while it decreases in others. One must be careful because there seems to be a geocentric effect present as seen by the fact that Δz is smallest near the solar position. Nevertheless it seems that Δz has its greatest extent just on the outer edge of the regions where σ is maximum, but not coincident with the maximum. This result differs from the findings of Jackson and Kellman (1974) who report that in the region R = 4-10 kpc enhanced values of half-thickness are associated with regions occupied by major spiral features.

Figure 3. The half-thickness Δz of the galactic HI layer. Units represented are kiloparsecs.

The next property of the outer region of the Galaxy to be represented is the median z position of the HI. Here, the preliminary data from the Kerr-Bowers survey is added to the values of the median z determined from the Weaver-Williams surveys and shown in figure 4. It can be seen that the position of maximum warp is at 80° galactocentric azimuth in the north and approximately 260° in the south. However the extent of the warp is much greater in the north. Inside 14 kpc the deviations from the plane are roughly equal; beyond 14 kpc the distortion increases somewhat uniformly in the north making an angle of 18° but in the south the distortion rapidly levels off. In this way the north is apparently much more warped than the south.

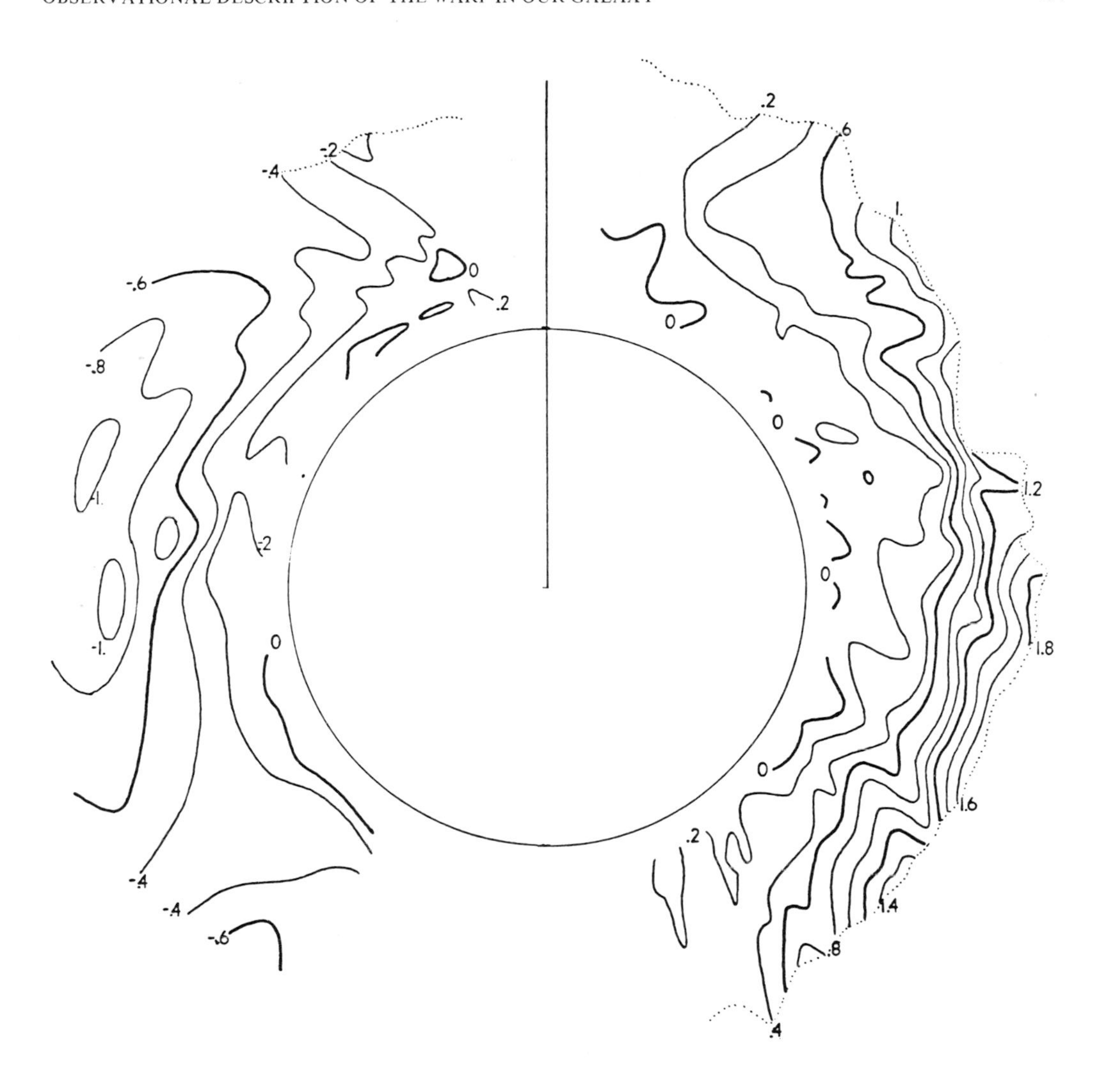

Figure 4. The median position of the HI layer. The results from the southern hemisphere are incompletely sampled.

When the position of the plane in the north is calculated by using the position of the density maximum rather than the median, there is little measurable difference in the contours. Peak density positions show more random fluctuations but the average is quite similar. This is consoling because both methods have been used in the past and this indicates that the results should be consistent. Probably the $\pm 10°$ cut-off in latitude as well as the fact that measurements were only carried out to the $\sigma=1$ contour level have eliminated any of the great latitude asymmetry in the contour diagrams found by Burton and Verschuur (1973).

It is possible to determine a mean plane for the northern hemisphere data by assuming the Galaxy to be in the form of concentric annuli and fitting this model to the data by the method of least squares, calculating the azimuth and extent of the distortion. These are listed in the table. Note that the slope of

Radius	11	12	13	14	15	16	17	18
Max z(kpc)	.04	.18	.28	.40	.58	.87	1.22	1.54
Position(deg)	51	62	64	73	80	80	79	79

the increase of z with radius is a rather constant 18° from R=18 kpc and the position is constant at 80° galactic azimuth. This mean value as determined from the set of rings can now be subtracted from the median z position and in this way large scale, consistent deviations from the plane can be found. At R=14 kpc these regions vary as much as 200 pc above or below the mean whereas in the inner part of the Galaxy (Henderson, 1967) similar regions of large scale deviation are found but usually no more than 50 pc from the plane. Nevertheless, regions which are above the plane in the outer part correlate approximately with regions above the plane in the inner part and vice-versa.

Full analysis awaits completion of the southern hemisphere data. At that time we can make more accurate comparison between the neutral hydrogen on the north and that on the south. As indicated we plan to calculate the surface brightness, the half-thickness, the position of the plane as well as possible large scale deviations from the mean position.

This work has been funded in part by grant AST-77-26898 from the National Science Foundation. The author acknowledges the help of Dr. P. D. Jackson in analyzing the southern hemisphere survey and in general discussions.

REFERENCES

Burton, W.B. and Verschuur, G.L., 1973, Astron. Astrophys. Suppl., 12, 145.
Henderson, A.P., 1967, Ph.D. thesis, Univ. of Maryland, College Park.
Jackson, P.D. and Kellman, S.A., 1974, Astrophys. J., 190, 53.
van de Hulst, H.C., Muller, C.A. and Oort, J.H., Bull. Astron. Inst. Neth., 12, 117.
Weaver, H. and Williams, D.R.W., 1973, Astron. Astrophys. Suppl., 8, 1.
Weaver, H. and Williams, D.R.W., 1974, Astron. Astrophys. Suppl., 17, 251.

DISCUSSION

Verschuur: Having done some work on the warping of the plane, I am keenly aware that the presence of cloudiness in the outer parts of the Galaxy severely confuses estimates of both the z thickness and the deviation from the plane. If you remove the clouds, you get a different picture of the warp and thickness. I believe that it is essential to consider the presence of cloudiness as a biasing factor. Are you taking this into account?

Henderson: There was no attempt to put any model into the reduction in addition to "circular symmetry". Many of the clouds seen by Burton and Verschuur are biased out of the reduction by the $\pm$ 10° cut-off that exists between ℓ = 10° and ℓ = 115° because they were at higher latitudes. Clouds at more extended radii are biased out by the fact that z and $\Delta\bar{z}$ are only carried out to the radius where $\sigma = 1.25\ M_\odot/pc^2$.

van Woerden: Could you give a number for the (maximum) inclination in the warped region?

Henderson: Using the model whereby rings at constant R were fit to the data by the method of least squares, the slope in the outer region from R = 16 to R = 18 is rather constant at 0.32 kpc per kpc.

Pişmiş: In your earlier work you had found indications that adjacent spiral arms were alternately up and down with respect to the average galactic plane. Do your recent results support such an arrangement of the spiral forms?

Henderson: The data show that there are consistent regions below the plane and also consistent regions above the plane. These regions follow an approximate spiral pattern. It is not clear that the center of these regions corresponds to the central region of the spiral arm.

Toomre: How far out do you detect gas unambiguously?

Henderson: Using a flat rotation curve, and to the sensitivity of the Weaver-Williams survey, the extent of surface brightness is approximately 22 kpc.

WARPING AND THICKNESS OF GALACTIC GAS LAYERS

Hugo van Woerden
Kapteyn Astronomical Institute, University of Groningen,
Groningen, The Netherlands

Summary The evidence for warps in the gas layers of galaxies is reviewed. Both the 21-cm line intensity distribution on the sky in edge-on systems and the hydrogen velocity fields in other systems indicate that warped gas layers are common, more common than close companions. Hence, warps probably persist for several times 10^9 years.

The thickness of the gas layer in NGC 891 appears to increase outward, similarly to that in our Galaxy.

INTRODUCTION

The standard model for spiral and lenticular galaxies consists of a spheroidal bulge and a thin, flat disk. The model can, of course, best be tested in galaxies where the disk is seen "edge-on". Among 15-20 such galaxies in the Hubble Atlas, most have an undisturbed, flat disk; NGC 5866 has a slightly (2°) tilted inner dust lane; and only in one, the lenticular galaxy NGC 4762 (Hubble Atlas, page 8), do deep exposures show the outermost parts of its wafer-thin disk to be warped, and probably even corrugated. Arp's Atlas of Peculiar Galaxies illustrates more deviations from flatness, and Arp (1964) noted a 5° bending in the plane of the Andromeda Nebula; but to the optical observer warped disks remain an exception.

THE WARP IN OUR GALAXY

The early surveys of neutral hydrogen in our Galaxy (Muller and Westerhout, 1957; Westerhout, 1957; Schmidt, 1957; Kerr, Hindman and Gum, 1959; Oort, Kerr and Westerhout, 1958) demonstrated that the gas layer is very flat indeed in the inner parts of the system: inside the solar circle, deviations of the midplane from a flat "principal plane" did not exceed 100 pc. At greater distances from the centre, however, the gas layer proved to curl away from the principal plane (Burke 1957, Kerr 1957) up to heights $|z| \sim 1000$ pc (Oort et al., 1958) at $R \sim 16$ kpc*, in a symmetric, integral-sign (or "hat-brim") pattern. Much greater deviations, of several

* All heights and distances in this section have been recalibrated to a sun-centre distance, $R_o = 10$ kpc.

W. B. Burton (ed.), The Large-Scale Characteristics of the Galaxy, 501–510.

kpc, were later reported at greater R by Kepner (1970), Davies (1972) and Verschuur (1973), cf. also Habing (1966).

Several attempts were made to explain this warp. Burke (1957), Kerr (1957) and many others estimated the tidal effects exerted by the Magellanic Clouds, but only Hunter and Toomre (1969) succeeded in showing that these galactic satellites might indeed have caused the warp in the gas layer on a recent, close passage. The intergalactic-wind model by Kahn and Woltjer (1959) and the free-precession model by Lynden-Bell (1965) did not require the influence of such companions. Whether or not the galactic warp was peculiar, remained an Open question.

OTHER GALAXIES

Some fifteen years later, Gordon (1971) and Wright, Warner and Baldwin (1972) observed anomalous features in the distribution and velocity field of HI in Messier 33. They noted that these features could be explained by a warped gas layer; however, both the required amplitude (∿6 kpc) of the warp and the great distance (∿150 kpc) of the nearest neighbour, Messier 31, caused these authors to wrap this suggestion in considerable hesitation.

Several years passed before Rogstad, Lockhart and Wright (1974) showed convincingly that the pronounced asymmetries in the gas distribution and velocity field of Messier 83 could be well represented by a set of concentric rings, progressively inclined with respect to the main plane of the galaxy (Figure 1). This suggested strongly that the gas layer in this galaxy was severely warped (by about 30°). Subsequently, the same authors (Rogstad, Wright and Lockhart, 1976) represented Messier 33 by a similar model. In either case, Rogstad et al. agreed that neither a tidal nor a primordial origin of the warp seemed likely, and they wondered whether recent infall of gas might be responsible.

EDGE-ON GALAXIES

The nearest edge-on systems are an obvious place to look for phenomena similar to those in our Galaxy. As soon as Westerbork had a line receiver

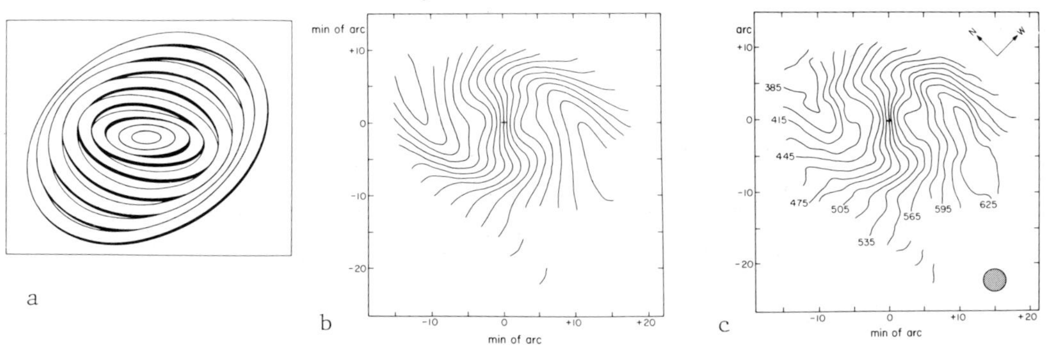

Figure 1. The warp in M83 (Rogstad, Lockhart and Wright, 1974). (a) Tilted-ring model. (b) Model velocity field. (c) Observed velocity field.

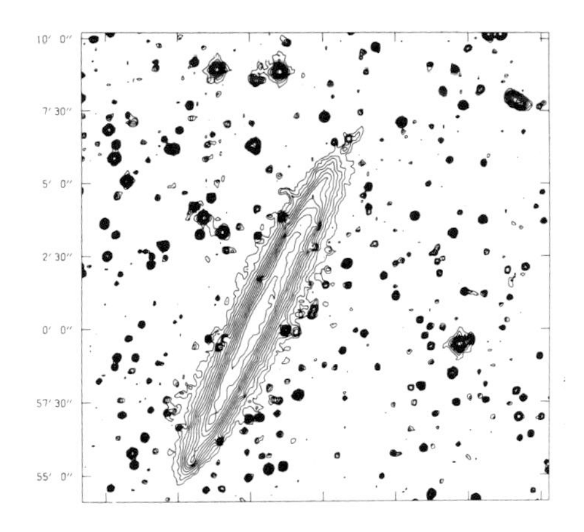

Figure 2. Isophote map made from two deep IIIa-J plates of NGC 5907 taken by Van der Kruit and Bosma with the 48-inch Palomar-Schmidt telescope (Van der Kruit, 1978). Contour values range down from 27.5 mag arcsec^{-2} in steps of 0.5 and are accurate to $\pm$0.2. Note flatness of disk out to faintest contour.

(1971), Sancisi and Allen observed NGC 891. The gas layer in this system turned out to be perfectly flat (Sancisi, Allen and Van Albada, 1975), in keeping with the lack of a companion.

The next object observed, NGC 5907, however, turned out to have a very strong warp (Guélin et al., 1975; Sancisi, 1976). By now, of the five edge-on galaxies studied (Table 1), four are warped. One of these, NGC 4631, is surrounded by a complex distribution of neutral hydrogen (Weliachew et al., 1978); this galaxy has two near companions, and indeed Combes (1978) has shown that the distribution and kinematics of the gas around NGC 4631 can be fairly well represented by a tidal-interaction model. For the other three warped galaxies, the nearest major neighbour is at a projected distance of six or more De Vaucouleurs diameters; this implies that *either* 1) the relative velocity is strongly hyperbolic, hence any interaction would have been brief and inefficient; *or* 2) any close encounter would have occurred several times 10^9 years ago - too long (Hunter and Toomre, 1969) for a warp to maintain its organized shape. The statistics suggest that warps are not induced by major, nearby companions but have some other cause, and/or that they must be able to survive much longer than we think. Tubbs and Sanders (1978) now appear to have found a mechanism that allows warps to persist for long.

Published photographs suggest that the warps in the neutral hydrogen mainly occur in regions outside the optical disk. Isophotometry on deep IIIa-J plates of three edge-ons by Van der Kruit (1978) confirms this, and further shows (cf. Figure 2): a) the optical disks remain flat out to an (edge-on!) brightness level as low as J = 27 mag arcsec^{-2}; b) the intensity gradient over the last three magnitudes is very steep. This indicates that the warps probably occur in regions of quite low mass density, in agreement with the suggestions by Tubbs and Sanders (1978).

"KINEMATIC WARPS"

In his recent thesis, Bosma (1978) has compiled detailed HI velocity fields (Figure 3) for about 20 galaxies, from measurements made at Westerbork and (by others) at several other observatories. In most of these galaxies, Bosma finds evidence for large-scale deviations from axial symmetry in the distribution and motions of the gas. In a number of galaxies (notably: M31, M33, M83, and NGC 300, 2805, 2841, 5033, 5055 and 7331) these deviations are well represented by a tilted-ring model similar to that (cf. Figure 1) first proposed by Rogstad et al. (1974) for M83; the orientation of the rings varies progressively outward, but all rings are presumed to be in circular motion about the centre. Bosma has also developed tilted-ring models for the warped edge-on galaxies discussed in the previous section.

The success of these tilted-ring models suggests that all of the galaxies mentioned in the preceding paragraph have warped gas disks. Bosma calls these "kinematic warps". The high frequency of kinematic warps

TABLE I

Warps in edge-on galaxies

Name	Relative resolution: θ/D_o (a)	Warp?	z/R (max.) (b)	Nearest major companion: distance/D_o (c)	Notes and references
Milky Way	<0.02-0.002	yes	∿0.05	∿2	
NGC 891	0.045	no	<0.03	>20	(e)
NGC 4244	0.10	yes	0.05	8.5	(f)
NGC 4565	0.11	yes	∿0.12	6.1	(f)
NGC 4631	0.14	yes	0.10	3.0	(d) (g)
NGC 5907	0.11	yes	0.2	8.0	(f) (h)

Notes

(a) θ = beamwidth perpendicular to galactic plane;
D_o = De Vaucouleurs diameter, from Second Reference Catalogue.
(b) Highest value of z, height of warp above principal plane, relative to distance R from galaxy centre.
(c) "Major companion" ≡ magnitude difference <2. Velocities (Second Reference Catalogue) taken into consideration.
(d) Elliptical companion, $\Delta m \sim 4$, at 0.2 D_o. Tidal interaction with both companions, see Combes (1978).

References for warp data: (e) Sancisi and Allen (1978), (f) Sancisi (1976), (g) Weliachew, Sancisi and Guélin (1978), (h) Sancisi (1977).

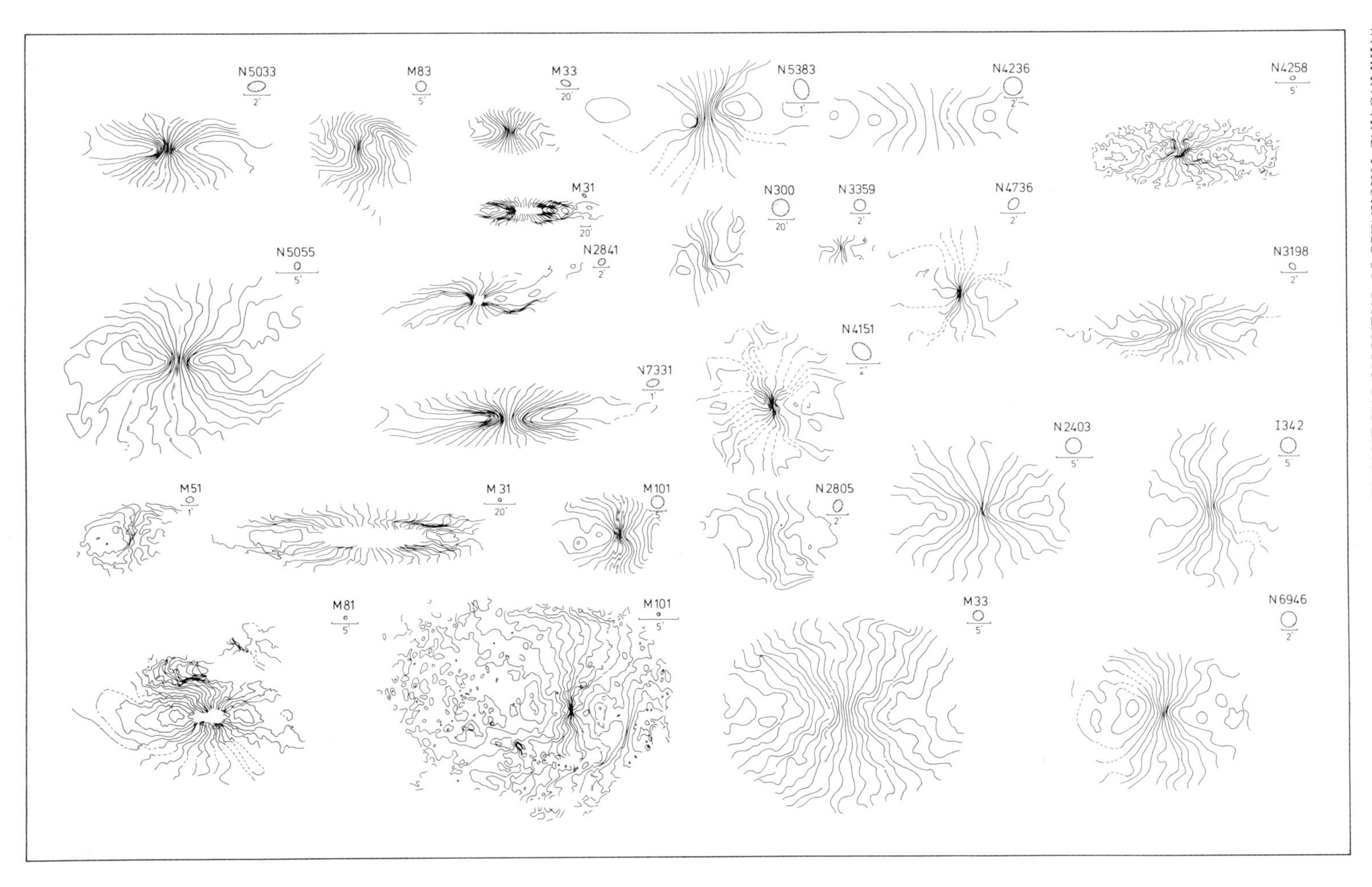

Figure 3. Velocity fields of 22 spiral galaxies, from high-resolution studies by various observers (Bosma 1978). Note the frequent, strong deviations from axial symmetry.

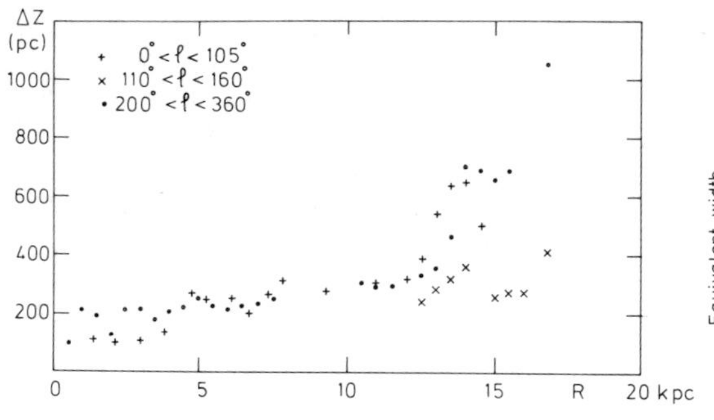

Figure 4. Layer thickness Δz (between half-maximum-density points) of neutral hydrogen in our Galaxy, plotted as function of distance R from the centre. Data taken from Jackson and Kellmann (1974, Tables 1-5): plus signs, regions with $0^o < \ell < 180^o$; circles, regions with $180^o < \ell < 360^o$.

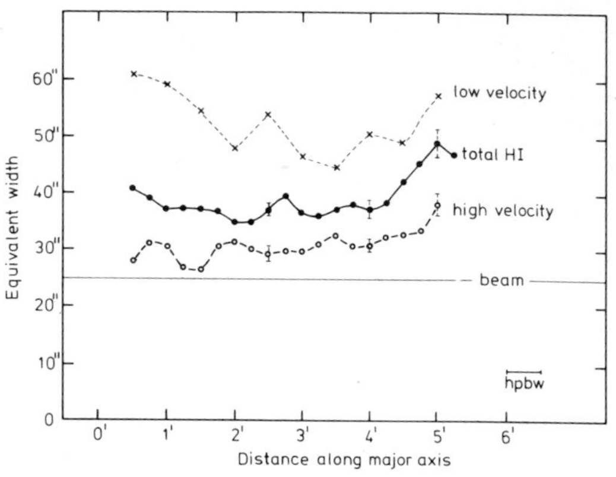

Figure 5. Equivalent width of intensity distribution observed across major axis for neutral hydrogen in NGC 891 (Sancisi and Allen 1978). Separate curves plotted for "high" and for "low" velocities (see text).

supports and amplifies the findings of the previous section. The general conclusion appears justified that warped gas disks are common among galaxies.

THICKNESS OF GAS DISKS

For the thickness of the gas layer in our Galaxy, Schmidt (1957) found Δz ∿ 270 pc between points of half-maximum density at 4 < R < 8 kpc. (Again, we recalibrate thicknesses and distances to a sun-centre distance R_o = 10 kpc.) The thickness varied hardly with R, although the force K_z perpendicular to the plane was estimated to vary by an order of magnitude over this range. The near-constancy of Δz in the annulus 4 to 8 kpc temporarily led to the belief that the thickness was similar throughout the Galaxy (Oort, 1959) except in the central region where it decreases to about 100 pc (Rougoor and Oort, 1960). In the outer parts, Lozinskaya and Kardashev (1963) found the layer thickness to increase strongly, to Δz ∿ 1000 - 2000 pc at R ∿ 15 - 20 kpc. Values Δz ∿ 700 pc at R ∿ 15 kpc were reported by Van Woerden (1967) and by Kerr (1969). A new study by Jackson and Kellmann (1974), based on high-resolution surveys, confirmed the earlier findings: Δz ≃ 250 pc, and almost constant, at 4.5 < R < 10 kpc; Δz ∿ 100 - 200 pc at R < 4 kpc, and ∿600 pc at R < 14 kpc. Figure 4 is a plot of the layer thicknesses tabulated by Jackson and Kellmann. On the whole, Δz appears to rise gradually with R, with a steep increase beyond R = 12 kpc.

Little is known for other galaxies. The 1-arcmin resolution used for most objects listed in Table 1 corresponds to several kpc in most cases! The only edge-on galaxy sofar observed with 0!5 resolution is NGC 891;

for a Hubble constant of 100 km s^{-1} Mpc^{-1}, this corresponds to 1 kpc. The analysis of the observations (Sancisi and Allen, 1978) is difficult: the observed intensity distribution across the major axis is influenced by 1) the z-distribution (and layer thickness) of the gas, 2) the inclination (and possible warping) of the gas layer to the line of sight, 3) optical-depth effects, 4) beam broadening. Only the last effect is accurately known. Figure 5 shows the equivalent width W(y) of the intensity distribution across the major axis, as a function of position x along the major axis. Separate curves are given for gas with "high" and with "low" velocity. Gas located close to the line of nodes (of galactic plane and sky plane) will be observed with its full rotational velocity; gas far from the line of nodes has a much smaller rotation component in the line of sight. The equivalent width W(y) of the "high-velocity" gas is little affected by the inclination; hence the run of W(y) with x suggests that the layer thickness does increase with radius R. This conclusion is strengthened by the fact that optical depth will raise W(y) for small x.

Sancisi and Allen place a lower limit $i > 87^\circ\!.5$ on the inclination, from a comparison of the intensity distributions I(y) and I(x) along minor and major axes. From an analysis of I(y) they then conclude that the layer thickness is unresolved, hence $\zeta < 15'' \sim 0.5$ kpc at radii $R < 2' = 4$ kpc; for R between 5' and 6' (10 and 12 kpc), $\zeta \sim 0.5 - 1$ kpc. These results, though necessarily rough, appear in good agreement with those for our own Galaxy; Sancisi and Allen show that also the hydrogen size and distribution, hydrogen mass, total mass and rotation velocity are all quite similar in both galaxies.

WHAT NEXT?

Further work on disk thicknesses will clearly require observations at $0'\!.5$ resolution (or better), and improved sensitivity. For a better understanding of the dynamics, origin and evolution of warps, highly sensitive observations at high resolution will have to provide extended rotation curves and detailed gas distributions in the far outer parts of edge-on galaxies.

ACKNOWLEDGEMENTS

I thank A. Bosma and P.C. van der Kruit for the use of unpublished information. In particular, I am grateful to Renzo Sancisi for several discussions. The Westerbork Radio Observatory is operated by the Netherlands Foundation for Radio Astronomy, with financial support from ZWO.

REFERENCES

Arp, H.C.: 1964, Astrophys. J. 62, 1045.
Bosma, A.: 1978, "The distribution and kinematics of neutral hydrogen in spiral galaxies of various morphological types", thesis, Groningen.
Burke, B.F.: 1957, Astron. J. 62, 90.
Combes, F.: 1978, Astron. Astrophys. 65, 47.
Davies, R.D.: 1972, Mon. Not. Roy. Astr. Soc. 160, 381.

Gordon, K.J.: 1971, Astrophys. J. 169, 235.
Guélin, M., Sancisi, R., Weliachew, L., Van Woerden, H.: 1975, in :La Dynamique des Galaxies Spirales" (ed. L. Weliachew), CNRS Colloq. no. 241, p. 291.
Habing, H.J.:1966, Bull. Astr. Inst. Netherlands 18, 323.
Hunter, C.S., Toomre, A.: 1969, Astrophys. J. 155, 747.
Jackson, P.D., Kellmann, S.A.: 1974, Astrophys. J. 190, 53.
Kahn, F.D., Woltjer, L.: 1959, Astrophys. J. 130, 705.
Kepner, M.: 1970, Astron. Astrophys. 5, 444.
Kerr, F.J.: 1957, Astron. J. 62, 90.
Kerr, F.J.: 1969, Ann. Rev. Astron. Astrophys. 7, 39.
Kerr, F.J., Hindman, J.V., Gum, C.S.: 1959, Austral. J. Phys. 12, 270.
Van der Kruit, P.C.: 1978, Astron. Astrophys. Suppl., in preparation.
Lozinskaya, T.A., Kardashev, N.S.: 1963, Astron. Zh. 40, 209 (Soviet Astr. - AJ 7, 161).
Lynden-Bell, D.: 1965, Mon. Not. Roy. Astr. Soc. 129, 299.
Muller, C.A., Westerhout, G.: 1957, Bull. Astr. Inst. Netherl. 13, 151.
Oort, J.H., Kerr, F.J., Westerhout, G.: 1958, Mon. Not. Roy. Astr. Soc. 118, 379.
Oort, J.H.: 1959, I.A.U. Symp. 9, 416.
Rogstad, D.H., Lockhart, I.A., Wright, M.C.H.: 1974, Astrophys. J. 194, 309.
Rogstad, D.H., Wright, M.C.H., Lockhart, I.A.: 1976, Astrophys. J. 204, 703.
Rougoor, G.W., Oort, J.H.: 1960, Proc. Nat. Acad. Sci. 46, 1.
Sancisi, R.: 1976, Astron. Astrophys. 53, 159.
Sancisi, R.: 1977, in "Topics in Interstellar Matter" (ed. H. van Woerden), Reidel, Dordrecht, p. 255.
Sancisi, R., Allen, R.J., Van Albada, T.S.: 1975, in "La Dynamique des Galaxies Spirales" (ed. L. Weliachew), CNRS Colloq. no. 241, p. 295.
Sancisi, R., Allen, R.J.: 1978, Astron. Astrophys., in press.
Schmidt, M.: 1957, Bull. Astr. Inst. Netherl. 13, 247.
Tubbs, A.D., Sanders, R.H.: 1978, Astrophys. J., in press, and paper at this Symposium.
Verschuur, G.L.: 1973, Astron. Astrophys. 22, 139 and 27, 407.
Weliachew, L., Sancisi, R., Guélin, M.: 1978, Astron. Astrophys. 65, 37.
Westerhout, G.: 1957, Bull. Astr. Inst. Netherl. 13, 247.
Van Woerden, H.: 1967, I.A.U. Symp. 31, 138.
Wright, M.C.H., Warner, P.J., Baldwin, J.E.: 1972, Mon. Not. Roy. Astr. Soc. 155, 337.

DISCUSSION

Verschuur: Based on my measurements of the warp in our Galaxy, I produced a model cross-section for the Galaxy which when overlaid on the HI map for NGC 5907 gives an excellent fit. I reassert that the evidence for a warp of the order of 5 kpc in the Galaxy is very strong, especially in the second quadrant of galactic longitude.

van Woerden: I agree that our Galaxy may be warped up to 5 kpc. Habing (1966, BAN 18) already showed that the outer arms have asymmetric z-distributions reaching up to 3-5 kpc. However, it is not clear that the column densities of hydrogen in these features are comparable to those found by Sancisi in NGC 5907. Our opinions may diverge in the interpretation of the "high-velocity clouds".

Sanders: Concerning the kinematic warps, such as exemplified by M83 or the various cases cited by Bosma: Is it the case that these warps lie outside of the optical disk?

van Woerden: Bosma's tilted rings generally start outside the Holmberg radius.

de Vaucouleurs: NGC 5907 probably provides a better comparison to our Galaxy than NGC 891 which is type Sb (larger bulge/disk ratio). We have recently completed photometry of NGC 4631. It has an extensive thick disk with an exponential z-distribution on a scale roughly similar to the so-called radio "halo".

Pişmiş: Only in the case of an edge-on Galaxy can one be sure of the existence or not of a warp. I showed by geometrical considerations in 1966 that if spiral arms are not logarithmic spirals the projected major axis will show a warp.

van Woerden: It is clear that the case for warped disks can only be unambiguously proven in edge-on galaxies. In inclined galaxies, any distorted observed velocity field may be represented by a planar distribution of gas together with an appropriate arrangement of non-circular motions. However, in his thesis Bosma shows convincingly that a tilted-ring model à la Rogstad gives a good fit to many galaxies with distorted outer velocity fields, as well as to the warped edge-on galaxies. This suggests that warps are indeed a frequent, if not general, phenomenon.

Burstein: In Bosma's figure depicting the HI rotation curves of spirals, it appeared that the "kinematical warp" in M31 extends to the center of the galaxy, and is continuous throughout the disk of M31. Does this imply a difference in the kinematics of the gas and the stars in M31?

van Woerden: I am not sure that the kinematical warp continues so far in. However, note that a similar phenomenon ("twist") in the velocity field may be caused by an elongated inner structure (bar or oval). Bosma finds evidence for such oval distinctions in many galaxies.

Burstein: In my recent photometric study of SO galaxies, I discovered that the brightness distribution in edge-on SO's requires the presence of a third, separate luminosity component, termed a "thick disk", in addition to a thin disk and a spheroid. I searched for the presence of a "thick disk" in spirals, and did not find any evidence of thick disks (as found in SO's) in six spirals, including M31. One can take the major axis profile of M31, assume the disk is thin (a/b $\gtrsim$ 40:1) and axisymmetric, and fit the light distribution (from de Vaucouleurs) perpendicular to the major axis. The thin disk approximation fits the observed light distribution (termed a perpendicular profile) to 27 mag arcsec^{-2} B. Thus, the stellar distribution is not warped to this surface brightness.

Toomre: Could you please summarize again just how far we have gotten with attempts to observe optical counterparts to the HI warps seen in edge-on galaxies like NGC 5907 and 4565?

van Woerden: Van der Kruit has taken deep IIIa-J plates of NGC 5907, 4565, and 4244. Isophotometry of these plates down to 27.5 mag arcsec^{-2} shows no warps within the Holmberg radii (26.5 mag arcsec^{-2}), and only a faint hint of a warp in the outermost contours of one or two of these galaxies.

Burton: In our own Galaxy, doesn't the warp occur outside of the "optical disk"?

van Woerden: Not quite. Our warp starts just outside the solar circle. The Perseus arm is at $z \approx + 0.2$ kpc, and is rich in associations (though HII regions are locally lacking). The Holmberg radius was estimated at 17 kpc earlier in this Symposium. Remember, however, that a warp of 0.2 kpc is very small compared to those measured at Westerbork in other galaxies.

Lequeux: Guibert, Viallefond and I have just made the sort of search for a warp in the young population in the Galaxy that Dr. Burton was mentioning. The results are inconclusive (Astron. & Astrophys., in press).

Kerr: There is some evidence that stars are to be found in the warp of our Galaxy. Graham was one of the first to show this. A search for such objects is an important problem for optical observers.

van Woerden: I could certainly not claim that warped gas cannot form stars.

Jackson: In our work on distant HII regions, we find that the most distant ones, 7 to 8 kpc from us and in the directions near $\ell = 150°$ and 240° show a departure from the plane which is in the same direction but less than the maximum departure in the 21-cm data (i.e., above the plane near $\ell = 150°$ and below it near $\ell = 240°$).

van Woerden: The presence of such stars, of course, implies that star formation occurs in the warped disk. It does not imply that our Galaxy has a warped optical disk brighter than 27 mag arcsec^{-2}, say.

THE PERSISTENCE OF WARPS

Allan D. Tubbs and Robert H. Sanders
Kapteyn Astronomical Institute, Groningen

Hunter and Toomre (1969) have demonstrated that a simple warp of a stellar disk will damp out within one or two galactic rotation periods due to rapid differential recession of stellar orbits. If, however, the galactic gravitational field is more nearly spherically symmetric, the rate of differential recession is decreased and the warp may persist for a longer time. We therefore propose that the observed gaseous warps exist in regions outside of the massive stellar disk; that is, in regions where the gravitational field is essentially spherically symmetric. A necessary condition of this hypothesis is that long-lived warps are present only in a low mass, low random velocity component of the galaxy -- presumably the gas.

We have investigated this hypothesis by numerically integrating the motion of test particles in a model galactic gravitational field (Tubbs and Sanders 1978). The model galaxy consists of two components: a highly flattened disk and a spherical halo. The disk is truncated at some radius, R_d, and the halo extends beyond the disk. The form of the density distribution is chosen to give a flat rotation curve within the disk. A critical parameter in this model is μ, the halo-to-disk mass ratio within R_d. Such a model is entirely consistent with Sancisi's observations of the rotation curve in NGC 5907 (1978) and Van der Kruit's sky-limited photographs of this galaxy (1978).

We examined two sets of initial conditions for our test particles: a primordial warp in which we simply tilt an outer annulus of particles, and a tidal warp, in which the particles in an outer annulus of particles receives a vertical impulse. We allow the particles to move in the given gravitational field, and, after some time, view them edge-on from various azimuthal angles. Below we see the state of a primordial warp after 10^{10} years. This is the distribution of test particles presented in the form of a high-velocity "channel" map (i.e. the column density in a 15 km s^{-1} velocity range around 230 km s^{-1}) smoothed to the resolution of the Westerbork beam. For the warp to persist for 10^{10} years we require that the stellar disk be truncated and that μ = 3. "Primordial" or tidally caused warps beyond a truncated massive stellar disk may persist for 5×10^9 years if $\mu \geq 1.5$.

W. B. Burton (ed.), The Large-Scale Characteristics of the Galaxy, 511–512.

REFERENCES

Hunter, C. and Toomre A.: 1969, Astrophys.J. 155, 747.
Kruit, P.C. van der: 1978, in preparation.
Sancisi, R.: 1978, in preparation.
Tubbs, A.D. and Sanders, R.H.: 1978, Astrophys.J. (submitted).

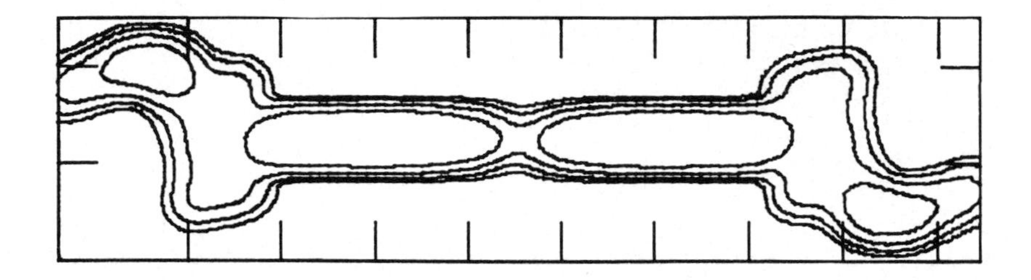

THEORETICAL CONSIDERATIONS OF THE WARP AND THICKENING

E. Saar
Tartu Astrophysical Observatory

1. INTRODUCTION

As a theoretical subject, the study of the bending of the gaseous planes of galaxies has not been extremely popular. It seems that this is due to an excellent and thorough paper on the subject, published by Hunter and Toomre in 1969. At that time there existed three (or four) rival theoretical mechanisms:

(1) vertical oscillations of the galactic disk, considered first by Lynden-Bell (1965);
(2) asymmetrical pressure on the disk due to intergalactic wind (Kahn and Woltjer 1959);
(3) tidal influence of the Magellanic Clouds, from
  (a) the case of Stationary Clouds, and
  (b) the case of Clouds orbiting around the Galaxy (Elwert and Hablick 1965, Avner and King 1967).

The case 3a was, of course, of no interest really, because the Clouds have to orbit somehow. Moreover, it was shown by Burke (1957) and by Kerr (1957) that the amplitude of the effect is inadequate.

Hunter and Toomre eliminated also the first two possibilities and concluded that the observed bending can be explained only by a comparatively recent and close passage of the Magellanic Clouds. This subject could not, in general, be handled analytically, but it seemed that the number of degrees of freedom inherent in the tidal picture could provide an explanation for any observed picture. So the widespread interest in the problem quickly vanished. By the way, the first more or less complete tidal models of the bending appeared only in 1976 (Fujimoto and Sofue).

What has changed now, compared to the year 1969, to justify reinvestigation of the problem? The obvious answer is the discovery of (or rather a sound belief in) the massive outer halos (or coronas) of galaxies. As far as we know the coronas become dynamically dominant just in the outer parts of galaxies, where bending occurs. The introduction of coronas has suggested a few new possibilities. These mechanisms and the re-evaluation of the old ones will be the subject of the next sections.

W. B. Burton (ed.), The Large-Scale Characteristics of the Galaxy, 513–522.

2. SPHERICAL CORONA

The simplest distribution of matter in a corona is a spherical. It reflects, of course, only the fact that we do not have a sufficient number of luminous test particles to determine the detailed density distributions. So, let us assume a spherical corona for this section. The problem is: how do the three mechanisms listed above function if a galaxy is immersed in a corona?

2.1. Vertical Oscillations of the Disk

As the corona becomes gravitationally dominant near the edge of the disk (the phenomenon of constant rotational velocities), the basic approximation is that of test particles in an external gravitational field, moving in ring orbits (cold disk). Rink orbits are stable in a spherical field (e.g. Polyachenko and Fridman 1976), and so the vertical oscillations of our disk are stable too. To obtain a distorted disk we can apply, following Hunter and Toomre (1969), a lopsided impulse vertical force

$$F_{imp} = f(r) \cdot \cos \theta \cdot \delta(t), \tag{1}$$

where r and θ are the inertial polar coordinates in the plane of the disk and f(r) is any smooth function. As a result, the particles will populate new orbits, inclined to the old ones, and the resulting response can be written as

$$\begin{aligned} h(r,\theta,t) &= \frac{f(r)}{v_{rot}(r)} \cdot r \sin \Omega t \cdot \cos(\theta-\Omega t) \\ &= g(r)\{\sin\theta - \sin[\theta-2\Omega(r)t]\}, \end{aligned} \tag{2}$$

where h is the height of the distorted disk above the old one and $\Omega = \Omega(r)$ is the angular velocity. The frequencies of oscillation form a continuous spectrum,

$$\omega \varepsilon [\Omega(r_i), \Omega(R)], \tag{3}$$

where $r_i < R$ is the limiting radius of our approximation and R is the outer edge of the disk. This means that the initial impulse excites a definite number of modes. As the different modes have different evolution rates, the response as in eq. (2) does not approach a stationary limit with time: all the shapes of the disk are transitory. Because of the differential rotation the initial shape becomes more and more corrugated, the number of modes, N, in any radial direction growing with time. Assuming a constant linear velocity, v_{rot}, this number increases by one every interval

$$\tau = \frac{dN}{dt}^{-1} \approx \frac{\pi r_i}{v_{rot}} \approx 3 \cdot 10^8 \text{ years}, \tag{4}$$

where the numbers are from the recent model of our Galaxy with a corona (Einasto, Joêveer, and Kaasik, 1976). Thus, during the orbital period of the Magellanic Clouds the initially smooth plane has developed about ten maxima and minima and is better described as a thickened plane.

The gaseous component of the disk is not self-gravitating, but feels the potential caused by the massive flat disk of stars. So, maybe it is the stellar disk that is bent, and maybe its self-gravitation can stabilize differential rotation? This problem was considered in detail by Hunter and Toomre (1969), and by Hunter (1969). They studied the limiting cases of totally self-gravitating disks and found that the edges of disks will almost always spoil the picture. There, the gravitational influence of the other parts of the disk is minimal, and, in order to possess only a discrete spectrum of oscillations, the surface mass-density, μ, near the edge, at R, has to behave as

$$\mu(r) \sim (R-r)^{1/2}. \tag{5}$$

For more realistic edges, where locally $\mu(r) \sim \delta r = R-r$, there always exists a range of continuous spectra, caused by differential rotation, which cannot be neutralized near the edge (Hunter and Toomre, 1969; Polyachenko and Fridman, 1976).

As noted by Hunter and Toomre (1969), factors tending to diminish the self-gravitation of the disk only worsen the situation. Evidently a massive spherical corona works in the same sense. And, indeed, supposing the halo effects to be small, it can be shown that the region of continuous spectrum widens with the growing role of halo. Consequently, this mechanism does not function in the case of heavy halos.

2.2. Intergalactic Wind

The failure of vertical oscillations as a bending mechanism led Kahn and Woltjer (1959) to propose a gas dynamical explanation. They supposed the Galaxy to be surrounded by a gaseous halo and estimated the pressure differences at its boundary due to intergalactic gas flow around the halo. This mechanism leads to a qualitatively correct picture, if one supposes a rather "rigid" halo. It is clear that potential wells generated by massive halos can serve as containers for the gaseous halos and can provide high temperatures for the gas, leading thus to high sound velocities.

Let us disregard all the difficulties listed by Hunter and Toomre (1969) and by Binney (1977) and estimate the resulting displacement of the Galactic gas layer.

To obtain the upper limit on the displacement we shall retain the value of $\approx$ 10% for the pressure differences at the boundary (Kahn and Woltjer, 1959) and take for the pressure the maximum value for the interstellar medium, $p_m = 2 \cdot 10^4$ k, suggested by Shapiro and Field (1976). Kahn and Woltjer supposed that the total pressure difference $\Delta p = 2 \cdot 10^3$ k is applied directly to the gaseous disk of the Galaxy. This, evidently, cannot be done, because the pressure differences have to cause a pressure gradient throughout the whole corona. So the pressure difference acting on the galactic gas (we suppose it for the moment to be homogeneous) is seriously reduced and the equilibrium of pressure forces and the gravitation of the stellar component leads to the amplitude of bending

$$h = \frac{\Delta p}{R_{cor}\ \alpha\ \rho_G}, \tag{6}$$

where

$$\alpha = K_z/z\big|_{z\,=\,0}. \tag{7}$$

Here R_{cor} is the outer radius of the gaseous corona, ρ_G is the gas density in the galactic disk and K_z the vertical acceleration. For R_{cor} = 300 kpc and our Galaxy, this height works out to $h \approx 1.3$ pc at $r = 15$ kpc, which is 10^{-3} times the observed value.

2.3. Tidal Forcing

Thus there remains, as the last hope to produce warped layers, tidal forces during the close passages of the companion galaxies. It turned out to be rather difficult to model the warp in our Galaxy: we saw the first results only in 1970 (Toomre). Here I can show you the most recent, and excellent, theoretical model (Fig. 1) of a warp (Spight and Grayzeck, 1977). They used an enormous number of test particles (12,000), a retrograde elliptic orbit ($\varepsilon = 0.5^5$) of the Large Magellanic Cloud, and a perigalactic distance of 20 kpc. The closest passage had to occur 4 10^8 years ago. So, all seems to be natural?

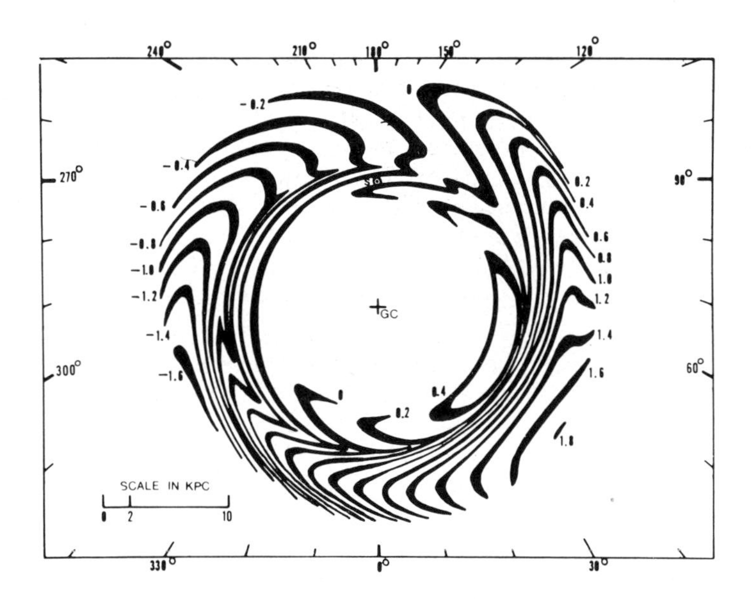

Figure 1. Relief map of the distorted disk after Spight and Grayzeck (1976). The z-distances are given in kpc with respect to the galactic plane.

There are only some minor points to worry about, some dark clouds on the horizon.

First, the gravitational field used by Spight and Grayzeck is spherical; there is no flat component. The degree of isotropy of a force field can be described by the ratio γ of the frequency of z-oscillations Ω_z to the angular velocity of rotation:

$$\gamma = \frac{\Omega_z}{\Omega} = \left(\frac{K_z \cdot r}{K_r \cdot z}\right)^{1/2} . \tag{8}$$

It approaches unity in the case of isotropy. The model of the Galaxy by Einasto, Joêveer and Kaasik (1976) gives in the region shown in Figure 1, $\gamma \varepsilon [2.4,1.2]$ for $z = 0$ kpc and $\gamma \varepsilon [2.0,1.2]$ for $z = 1$ kpc.

Second, the mass of the LMC was taken to be $3 \cdot 10^{10}$ $M_\odot$--about two times larger than the observational limit. Third, for so close a passage there arises the problem of disruption of the LMC-SMC system.

Fujimoto and Sofue (1976, 1977) tried to model the Galactic warp, considering all the points made above, and their pictures are not half as nice as Figure 1. So it seems that the model warp shown here belongs rather to some other Galaxy with a more centrally condensed corona than ours.

Summing up, the only mechanism that benefits by the introduction of spherical coronas is that of tidal forcing. The required amplitudes are simpler to produce and the condition $\gamma \to 1$ assures that there remains only one oscillation frequency at every radius instead of two different ones. But we have not answered the problem posed by Sancisi (1976)--how are warps produced in galaxies without companions?

3. NONSPHERICAL HALO

Considering the formation of massive halos we see that there is no reason for them to be spherical. Both the dissipative gravitational collapse picture (Doroshkevich, Sunyaev, and Zeldovich 1974) and the non-dissipative one (Aarseth and Binney 1978) give rise to triaxial mass distributions. Such a halo can lead to pleasant surprises when considering the problem of bending. This was realized recently by Binney (1977).

3.1. Free Oscillations

Binney represented his triaxial ellipsoid, for simplicity, as a superposition of two spheroids with perpendicular rotation axes. To be more specific, let us consider one of them oblate--this would be the conventional massive galaxy, bulge and all. The other one could be an oblate cake or prolate cucumber with the major axis in the plane of the first spheroid.

Representing now the gaseous disk by test particles in circular orbit, we find that they are moving in a periodic potential; there appears a kind of periodic driving force. For free vertical oscillations Binney obtained an equation

$$\frac{d^2h}{dt^2} + \Omega^2(r)\ \{1+b(r) - q(r)\ \cos\ [2\Omega(r)t]\}\ h = 0 \tag{9}$$

and demonstrated that it has regions of resonance, which leads to growing modes. This occurs whenever the two spheroids are both of the same shape and of comparable eccentricity. If those conditions are not satisfied, vertical oscillations cannot be coordinated, and we obtain once more a thickened disk. By assuming the "perturbing" spheroid to be rotating, the condition for resonance can be satisfied only in a specific range of radii.

So you can see yourself what can be done with two spheroids only!

One has not to possess second sights to predict that this work will open a wide field for people who love to derive dispersion relations, calculate eigenfunctions, and so on. As we have got our instability at last, there rises the nonlinear problem of final amplitudes, and, as we enter into the nonlinear domain, the problems of interaction of planar and vertical perturbations (bending and spiral structure), pumping of energy and so on, become important. It may really be a major breakthrough in the problem.

3.2. Hydrodynamical Models

The picture of halos proposed by Binney is extremely fruitful. Namely, it allows me to introduce the notion of accretion layers (Jaaniste and Saar 1976) and to divide the responsibility with Binney. Let us see: if we consider it natural that the plane of symmetry of one spheroid is populated with gas, then what about the other spheroid? We have listed a number of examples where such perpendicular gaseous planes can be suspected (Jaaniste and Saar 1976); additional examples are given by Gunn (1977). The gas in both layers has a cloudy structure, which seems to be best described by the recent model of McKee and Ostriker (1976), but, of course, the "extragalactic" gas is more rarefied, and the number of warm clouds is smaller. Bending can be produced now by the crude interaction of the two planes.

3.2a. Perpendicular rotation. It is difficult to imagine a mechanism causing rotation in perpendicular planes, but if we suppose it possible we come to the models of bending constructed by Haud (1977). He used the simple "snowplow" approximation for calculating the impulse changes of intergalactic clouds falling through the galactic plane. His models are numerical ones, as he used a realistic model of the gravitational potential and followed the clouds through many collisions. But to obtain an estimate of the amplitude of bending we can discard all the finessess and write

$$h = \frac{v_{cloud}\cdot r}{v_{rot}\cdot\phi(r)}\ \sin\left[\theta_o - \frac{\Omega_z(r)}{\Omega(r)}\ \theta\right], \tag{10}$$

where θ_o defines the intersection region of two planes and

$$\phi_{(r)} = \frac{\sigma_{gal}}{\sigma_{cloud}} \tag{11}$$

is the ratio of the surface densities of the (diffuse) gas in the galactic plane and of the infalling clouds. This ratio defines, in fact, the radial dependence of the amplitude of the warp. As it approaches zero in the outer regions of the disk, any bending amplitudes can be explained, in principle. In order to get a correct angular dependence one has to work near the halo region $\Omega_z \approx \Omega$, but this condition is not so strong as in the case of tidal forcing.

3.2b. Asymmetrical thickening. This variant of hydrodynamical forcing was proposed by Jaaniste (1977). He dropped the assumption of rotating accretion planes, but assumed inclined accretion layers. Now the infalling clouds will oscillate at both sides of the galactic gas, producing the thickening of the plane. But the clouds moving in retrograde orbits will lose more of their momentum than those in direct orbits. This leads to an asymmetry of the thickening. If the angle between the two planes α does not differ much from $\pi/2$ and $v_{rot} \approx v_{cloud}$, the thickness of the gas layer is

$$\Delta h = \frac{r \sin \alpha}{\sqrt{2}\ \phi} \sin (\theta_o - \theta) \tag{12}$$

and the amplitude of the warp

$$\bar{h} = \frac{1}{2} \Delta h \cot \alpha . \tag{13}$$

A typical result of density distributions produced in this way is shown in Figure 2. We note that it is the only mechanism predicting both bending and thickening to occur simultaneously.

Both hydrodynamical mechanisms predict a cloudy, irregular picture of the outer disk. But they need more input information than the case of vertical oscillations. A crucial, but unknown, factor is the degree of energy dissipation in collisions, for example.

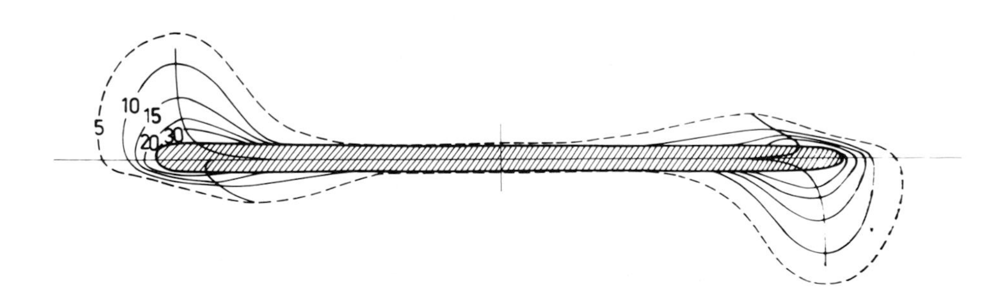

Figure 2. Total hydrogen density map for NGC 5907 after Jaaniste (1977). The densities given are in units of 10^{19} cm^{-2}, the tilt angle α = 45°, infall velocity v_{cloud} = 160 km sec^{-1}.

4. CONCLUSION

We have now once more three (or four) different possibilities to explain bending:

(1) tidal forcing has remained,
(2) instabilities of vertical oscillations in triaxial ellipsoids has emerged, as has
(3) gas-dynamical interactions with
  (a) perpendicular rotation and with
  (b) inclinded disks (or layers).

All of them can be functional given appropriate initial conditions. It is not simple to discard a priori any of them. The final words remain to be said by observers.

And, when fitting our models with observations, we have to choose with care the right things to fit. So, we do not know really the $\rho(h,\theta,r)$ dependence of hydrogen in the Galaxy to give observational plots like the theoretical one shown in Figure 1. From observations we get the functions $\rho(b,\theta,v)$, where the $r = r(v)$ dependence is fixed more by the mechanism of bending than by the undisturbed circular velocity law. The latter could be used only in case of a theory of small vertical oscillations in an axisymmetric galaxy. Both the tidal and hydrodynamic mechanisms lead to large perturbations of radial velocity, and radial oscillations are usual in disks immersed in the axial ellipsoids (Binney 1977). Thus only observed and model $\rho(b,\theta,v)$ profiles should be compared, as has been done by Fujimoto and Sofue (1976, 1977). It is worth displaying one of their figures to demonstrate the complexity of such diagrams, the velocities typical for tidal pictures, and the need for additional observations.

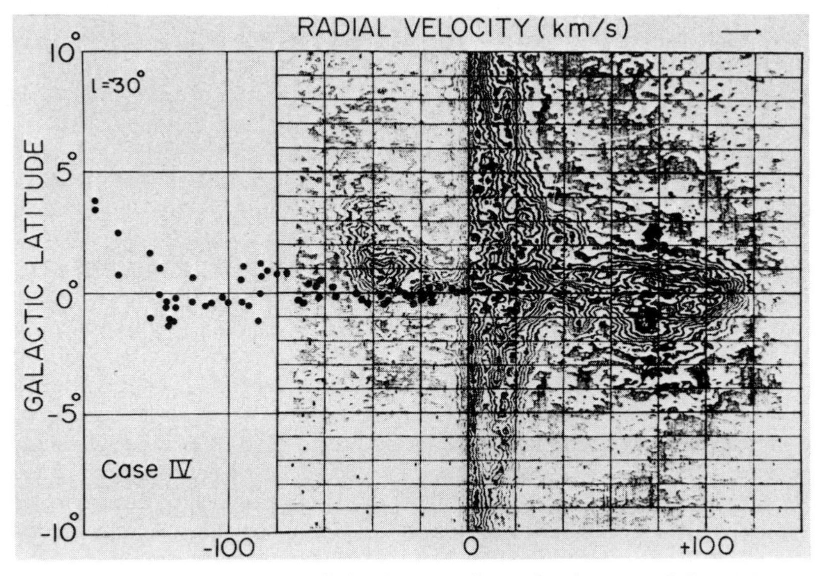

Figure 3. The computed (black dots) and observed bending of the galactic disk on the latitude-radial velocity diagram at L = 30°, after Fujimoto and Sofue (1977).

ACKNOWLEDGMENTS

My ideas on the subject have formed as a result of close collaboration with Drs. J. Jaaniste, J. Einasto, A. Kaasik and U. Haud. I am very grateful to Dr. J. Binney for sending his preprint just in time, and to Dr. A. Toomre for illuminating discussions.

REFERENCES

Aarseth, S. J. and Binney, J: 1978, preprint.
Avner, E. S., and King, I. R.: 1967, Astron. J. 72, 650.
Binney, J.: 1977, preprint.
Burke, B. G.: 1957, Astron. J. 62, 90.
Doroshkevich, A. G., Sunyaev, R. A., and Zeldovich, Ya.B.: 1974, in M. S. Longair (ed.) Confrontation of Cosmological Theories with Observational Data, IAU Symp. 63, 213.
Einasto, J., Joêveer, M., and Kaasik, A.: 1976, Tartu Astrophys. Obs. Teated No. 54, 3.
Elwert, C., and Hablick, D.: 1965, Z. Astrophys. 61, 273.
Fujimoto, M., and Sofue, Y.: 1976, Astron. Astrophys. 47, 263.
Fujimoto, M., and Sofue, Y.: 1977, Astron. Astrophys. 61, 199.
Gunn, J. E.: 1977, in B. M. Tinsley and R. B. Larson (eds.). The Evolution of Galaxies and Stellar Populations, Yale University Observatory, 445.
Haud, U.: 1977, Thesis.
Hunter, C.: 1969, Stud. Appl. Math. 48, 55.
Hunter, C., and Toomre, A.: 1969, Astrophys. J. 155, 747.
Jaaniste, J.: 1977, Thesis.
Jaaniste, J., and Saar, E.: 1976, Tartu Astrophys. Obs. Teated No. 54, 93.
Kahn, F. D., and Woltjer, L.: 1959, Astrophys. J. 130, 705.
Kerr, F. J.: 1957, Astron. J. 62, 93.
Lynden-Bell, D.: Monthly Notices Roy. Astron. Soc. 129, 299.
McKee, C. F., and Ostriker, J. P.: 1977, preprint.
Polyachenko, V. L., and Fridman, A. M.: 1976, The Equilibrium and Stability of Gravitating Systems, Moscow.
Sancisi, R.: 1976, Astron. Astrophys. 53, 159.
Shapiro, P. R., and Field, G. B.: 1976, Astrophys. J. 205, 762.
Spight, L., and Grayzeck, E.: 1977, Astrophys. J. 213, 374.
Toomre, A.: 1970, in the Spiral Structure of our Galaxy.

DISCUSSION

Sanders: What is the halo-mass to disk-mass ratio in your model which accounts for the persistence of a galactic warp.

Saar: This warp is generated in a spherically symmetric potential.

van Woerden: A washboard effect is seen in the deep photograph of the SO galaxy NGC 4762, the only warped edge-on galaxy in the Hubble Atlas.

Kerr: Your point that radial displacements would occur is an important one. Have you computed the size of such displacements?

Saar: Yes. In case of the model 3A (inclined accretion layer), the systematic deviations of radial velocities were about 40-60 km s^{-1}. This would lead to displacements in radii of 4-5 kpc, depending slightly on the type of smooth rotation curve adopted.

Giovanelli: What sort of accretion rates do you need for your models 1978.3 and 1978.4?

Saar: About two or three solar masses per year. Chemical evolution theories with infall need a little more.

Basu: At large distances from the galactic center, modes of density waves become unstable. Instability of density waves is likely to thicken the disk, which may in turn initiate the warping of the disk. So warping may be associated with the propagation of density waves.

Saar: It seems likely that nonlinear planar oscillations will generate vertical oscillations. But in order to get a warp you must tune up oscillations at different radii. It cannot be done, using the total self-gravitation of the disk, and it seems that density waves also cannot manage it.

Lynden-Bell: It is a flaw in your model that the halo has been taken exactly spherically symmetrical in a problem where departure from that symmetry is a crucial matter. Do you know what would happen to a self-gravitating disk placed at an angle to a somewhat non-spherical halo? If most perturbations eventually lead to a thickening of the disk edge, do we see more galaxies with thick edges than with warped edges?

Miller: F. Hohl has constructed some computer models such as you described: a self-consistent disk and bulge combination, that remained reasonably stable. The disk was not as thin as we would like for disk galaxies (axis ratios around 1:5), but they were reasonably thin for his grid. The disk thickened a bit from the starting condition. However Hohl did not do any experiments with a warp in the disk.

Cesarsky: I wonder if primordial tilts are a likely occurrence.

X. HIGH-VELOCITY CLOUDS AND THE MAGELLANIC STREAMS

INTRODUCTION TO THE SESSION

G. L. Verschuur, University of Colorado

The generic label "high-velocity cloud" is a very unspecific term that tells us only that there are clouds of hydrogen gas moving with velocities which are anomalous for their position on the sky, if we assume that we live in a flat, uniformly rotating galaxy. If one attempts to explain them all away in one fell swoop we might draw an analogy with the situation extant in the 19th century, when optical astronomers were aware of patches of emission that were called nebulae. We now know that there are extragalactic as well as galactic nebulae, all quite different from one another. They were recognized to be so different as soon as the quality of data on these objects improved sufficiently. We may be in a similar, somewhat frustrating, position with regard to the high-velocity clouds. Soon after their discovery in 1963 it was believed that the clouds were mostly at negative velocities and that they were very local. Now we know of many more clouds at both positive and negative velocities. The models proposed for their existence cover a range of distances. We still have local (infall and explosive event) models, while other models place the clouds on relatively nearby spiral arms, in the high-z extension or warp areas of distant arms, in the outskirts and immediate neighborhood of the Galaxy, at the distance of the Magellanic Clouds, and even beyond. There may well be some truth in all of these models.

Even as one strives for the simplest explanation for all the clouds, one should always remain aware that we can already identify several distinctly different phenomena in the data. The clouds near the galactic center have one explanation, those near the south galactic pole appear to be associated with the Magellanic Clouds, while those near the galactic anti-center may manifest yet another phenomenon. While we should strive for the simplest model we should bear in mind that Nature does not necessarily work that way. We should attempt to recognize systematic

W. B. Burton (ed.), The Large-Scale Characteristics of the Galaxy, 523–524.

differences between clouds in different areas of the sky. This might help in their identification. Hopefully by the end of this century we will have labels for several classes of the high-velocity clouds, in analogy with the progress made in the recognition of optically visible nebulae.

HIGH-VELOCITY CLOUDS: REVIEW OF OBSERVATIONAL PROPERTIES

Aad N.M. Hulsbosch
Astronomical Institute of the University of Nijmegen
The Netherlands

Abstract - Earlier extended surveys with moderate resolution and sensitivity had shown that high-velocity gas can appear as small clouds ($\sim 2^{\circ}$) either isolated, or in elongated complexes with lengths of up to 30°. The profiles were rather broad (halfwidths of 25 km/s). While the brightest HVC's were found in the northern galactic hemisphere, the highest negative velocities were found in the southern latitudes. The only known high-positive velocity objects were a few at low latitudes between $l = 245^{\circ}$ and 330°.

Recent high-resolution observations of a number of individual clouds reveal much small-scale structure. The most conspicuous property is the existence of small (6'), bright (up to 65 K), and narrow (5 km/s) cores, embedded in smooth envelopes (2°, 4 K, 25 km/s). Also sharp gradients, ringlike structures and large velocity gradients (up to 10 km/s/arc-degr.) are reported.

New highly sensitive observations (ΔT_b(rms) = 0.02 K) in the region $l = 100^{\circ}$ to 200° and $b = 0^{\circ}$ to -45° (partly -60°) with a velocity range from -1000 to +1000 km/s show the existence of many small (1°), faint (up to 0.3 K), isolated clouds with very high negative velocities, with the extreme value of -465 km/s ($l = 110^{\circ}.5$, $b = -7^{\circ}$, $T_b \approx 0.3$ K). A second important result is the existence of high-velocity gas with velocities around -250 km/s, continuing from HVC 160-46-330 to $l,b \approx 188^{\circ},-24^{\circ}$, near the Anticentre complex. No high-positive velocities are found.

CHARACTERISTICS OF HIGH-VELOCITY CLOUDS.

The objects, widely known as high-velocity clouds (HVC's), are neutral hydrogen features exhibiting a velocity not readily explainable by a simple model of galactic structure. In this paper we will confine ourselves to velocities in excess of the largely arbitrary value of 100 km/s. As at lower latitudes the high-velocity gas gets more and more intermingled with the spiral-arm gas we will concentrate ourselves largely on the latitudes above 20° (except when the gas is clearly different from spiral arm gas). An important limitation is that, until now, HVC's can be studied practically only by means of the 21-cm emission line of HI.

W. B. Burton (ed.), The Large-Scale Characteristics of the Galaxy, 525–533.

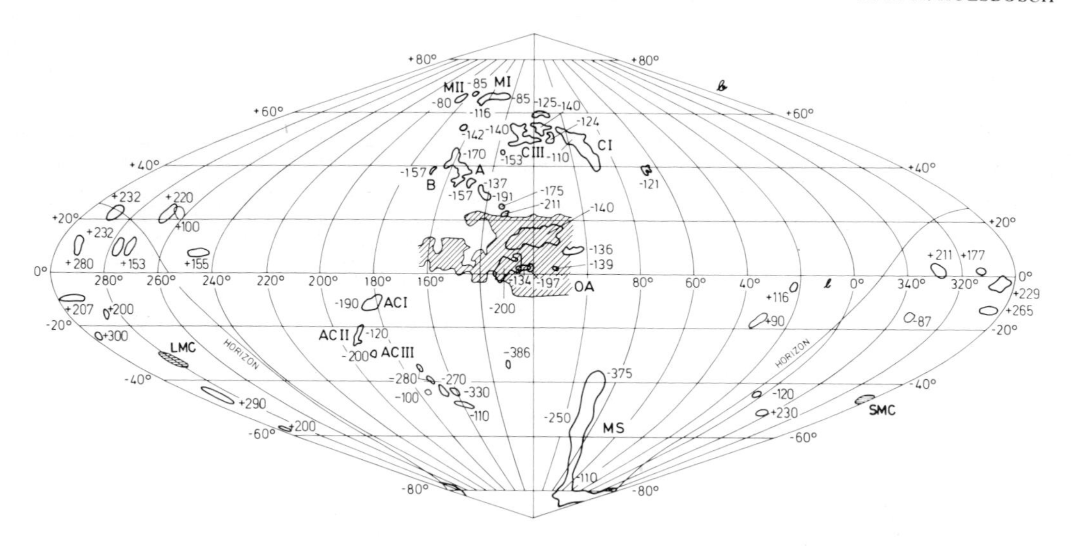

Figure 1 The distribution of features with $|V_{lsr}| > 100$ km/s. The figure is believed to be roughly complete for clouds larger than 2° or 3° over the whole sky above $\pm 20^{\circ}$ galactic latitude, for positive as well as negative velocities. The contour level is approximately 2×10^{19} at/cm^2. For references see text.

In the past a number of papers have reviewed the properties of HVC's. In this paper we shall give an overview and add recent new results. Figure 1 shows the distribution of all the more intense HI features with radial velocities $|V_{lsr}|$ higher than about 100 km/s relative to the local standard of rest (lsr). It is thought to be complete in both hemispheres down to galactic latitudes of 20° for clouds larger than 2° or 3° with column densities greater than 2×10^{19} HI atoms per cm^2. This map was taken from Hulsbosch (1975), supplemented for positive velocities and for the southern region (from Wannier _et al_ 1972, Mathewson _et al_ 1974, and Hulsbosch 1978b).

1. The distribution appears to be very uneven.

In the northern galactic hemisphere practically all the known HVC's are confined to a region between roughly 85° and 170° longitude and latitudes below $+70^{\circ}$, exept for the small HVC 66+39-121 (the numbers denote respectively longitude, latitude and radial velocity). In the southern latitudes the brighter ones occur between $l = 178^{\circ}$ and 195°, $b = -10^{\circ}$ and -32° (the Anticentre complex), in a narrow strip between $b = -30^{\circ}$ and -85° around $l = 80^{\circ}$ (the Magellanic Stream), and in a complex around $l,b = 165^{\circ},-45^{\circ}$. See, however, section 5.

2. There is a preponderance of negative velocities, especially in the northern galactic hemisphere.

Although the search for high-negative velocities has been made in greater detail, figure 1 should be practically complete also for the positive velocities, at least for the more extended objects. It appears that most high-positive velocity gas is found within 20° of the galactic equator in the longitude interval between 245° and 330°; at $b > +25^\circ$ no high-positive velocity gas has been found at all. For $b \geq +15^\circ$ we can make a crude estimate of the masses involved and find $\Sigma Mr^{-2} = 10 \times 10^4\ M_\odot kpc^{-2}$ for the negative velocities and $0.5 \times 10^4\ M_\odot kpc^{-2}$ for the positive velocities. For $b < -15^\circ$ an estimate is more difficult but we can roughly say that for $100^\circ < l < 200^\circ$ (thus excluding the Magellanic Stream) there is exclusively negative velocity gas, about $0.7 \times 10^4\ M_\odot kpc^{-2}$; the mass of the high-positive velocity gas, found near $l = 280^\circ$, amounts to about $0.5 \times 10^4\ M_\odot kpc^{-2}$.

It has been suggested that the velocity distribution is affected by the rotation velocity of the lsr. However, it is now clear from figure 1 that galactic rotation, though it may play a role, is not the main determining factor in the distribution, especially of the negative velocity gas. There are more dense HVC's in the Anticentre quadrant than in the quadrants around $l = 90^\circ$ and 270° and the distribution in latitude does not indicate any relation with the rotation.

3. Several high-velocity complexes form long "strings".

Examples of complexes forming long strings are the Magellanic Stream (Mathewson 1974) and complex A (Hulsbosch 1975). The latter extends from $l,b = 160^\circ,+43^\circ$ to $l,b = 132^\circ,+23^\circ$. It is extremely thin, the mean distance of the components to the chain axis being less than 1°. It may extend to still lower latitudes, possibly even $b = 0^\circ$, but there it cannot be unambiguously identified. At the higher latitudes it might be connected with the MI complex via the small cloud HVC 166+56-142 but this is rather uncertain.

Though not nearly so pronounced as string A the CI-CIII complex has likewise a longish shape. Davies (1973) believes that it can be connected with high-velocity hydrogen between $b = +10^\circ$ and $+20^\circ$ around $l = 50^\circ$, but the connection is not convincing. Large velocity jumps are observed in the Anticentre complex, while a considerable velocity gradient is seen in the elongated structure between $l,b = 157^\circ,-46^\circ$ and $188^\circ,-23^\circ$ (see figure 4).

Although in its thinness string A resembles the Magellanic Stream, in many other aspects the latter differs from the other high-velocity features and it was clear from the beginning that the Magellanic Stream is a quite different object (Hulsbosch, 1975). For an elaborate discussion see the papers of Mathewson and of Fujimoto.

4. There appears to be much small scale structure.

Most large scale surveys were performed with medium sized telescopes like that of Dwingeloo, with beamwidths of about $0\overset{\circ}{.}5$, bandwidths of 5 to 10 km/s, detection limits of 0.5 K and observation grids of $1^\circ \times 1^\circ$ or coarser. From these it was inferred that a typical HVC measures 2° to 4°,

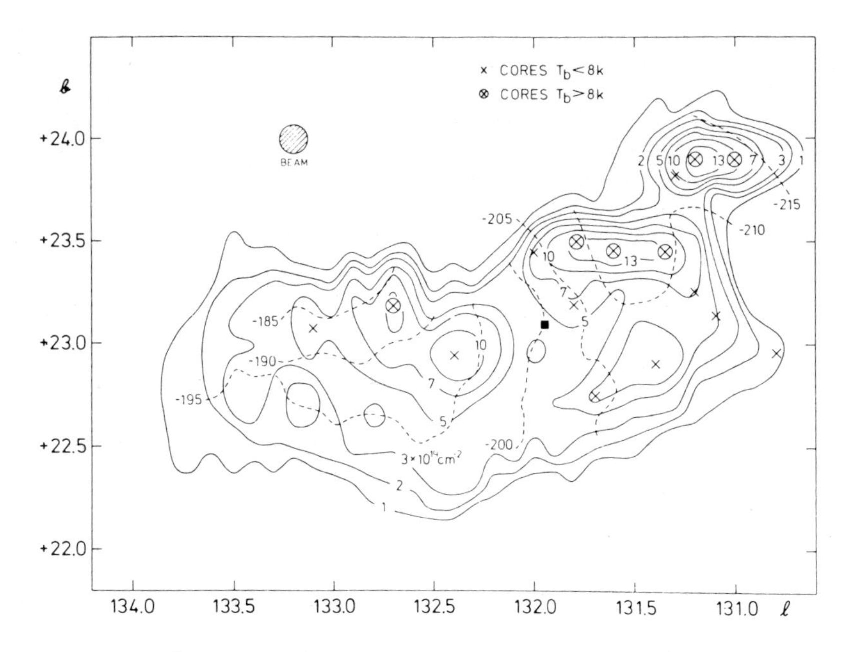

Figure 2 The high-velocity cloud HVC 132+23-211 (from Hulsbosch 1978a). Crosses indicate condensations having diameters of 5' or less.

has a T_b of up to 5 K and a profile halfwidth (W) in general between 20 and 30 km/s (e.g. Hulsbosch 1975).

Observations with higher resolution have shown a wealth of detailed structures. Chain A appears to be extremely clumpy and contains some ringlike structures, while jumps in velocity up to 50 km/s occur between various concentrations (Giovanelli et al 1973, their figure 2).

An important characteristic of several HVC's are their sometimes extremely steep edges, which make the impression of being shockfronts (Giovanelli and Haynes 1977, their figure 1). Several edges appear to be unresolved by 10' beams. Most HVC's contain small condensations, often unresolved, and having velocity halfwidths between 5 and 10 km/s, in contrast to the halfwidths of typically 25 km/s of the clouds in which they lie embedded. The intensities of these condensations may be as high as 65 K at full resolution. (Cram and Giovanelli 1976, Davies et al 1976, Giovanelli and Haynes 1976, 1977, Greisen and Cram 1976, Hulsbosch 1975, 1978a, Verschuur et al 1975). The cores are often numerous. An example is given by the Effelsberg 100-m telescope observations of HVC 132+23-211 (figures 2 and 3) showing the existence of several cores of which at least 6 are brighter than 8 K, have diameters hardly resolved by the 9' beam and profile halfwidths of 5 km/s (Hulsbosch 1978a). Their data, corrected for the beamsmoothing, are collected in table 1. With the spectral line interferometer of NRAO Greisen and Cram (1976) got striking similar results (T_b = 56 K, W = 5 km/s and diameter = 5!3).

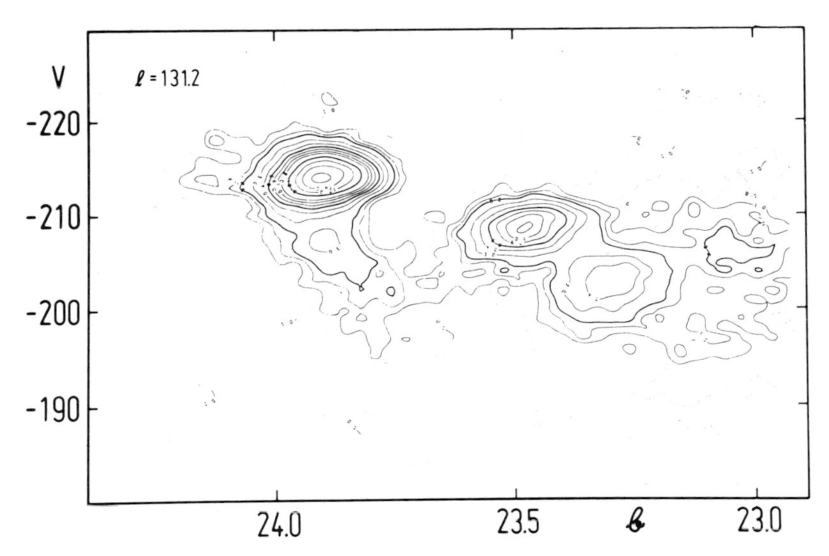

Figure 3 Contourmap of HVC 132+23-211 (from Hulsbosch 1978a). Shown are the T_b contours in the $(b,V)_l$ plane at l = 131°.2.

It should be noted that the small-scale structure of HVC's, notably the core-envelope structure and the existence of two velocity-dispersion components might not be typical for HVC's only but a common feature for hydrogen clouds in general, commonly, hidden because they cannot be studied individually.

5. Surveys at greater sensitivity show velocities as extreme as -465 km/s.

Cooled receivers which have become available in recent years have made it possible to observe features an order of magnitude fainter than the limiting contours in figure 1; at the same time searches were extended to higher velocities. Encrenaz et al (1971) had thereby found a "bridge" of weak emission at velocities around -150 km/s extending between CIII and the upper region of A. Faint high-positive velocity hydrogen was found between 252° and 322° longitude and +10° and +30° latitude by Wannier

Table 1 The brightest cores of HVC 132+23-211

l	b	V_r (km/s)	W (km/s)	T_b (K)	$\emptyset_{\frac{1}{2}}$	N_H (10^{19}cm^{-2})	rn_H (cm^{-3}.kpc)	Mr^{-2} ($M_\odot$ kpc^{-2})
131°.00	23°.90	-213.2	5.7	18	8!5	20	25	11
131.20	23.90	-214.3	4.5	62	5.4	55	110	12
131.35	23.45	-210.6	4.6	36	7.0	32	50	12
131.61	23.46	-211.4	4.8	58	5.6	54	100	13
131.79	23.50	-210.6	5.0	65	4.5	63	147	10
132.70	23.19	-186.6	5.0	18	9.0	18	21	11

et al. (1972). A large and relatively bright cloud was discovered by Wright (1974) around l,b = 128°,-33°, V = -380 km/s, while Davies (1975) found a small cloud near M31 at a velocity of -447 km/s. Two small clouds, HVC 39+5-353 and HVC 69+4-250 are reported by Shostak (1977, see also Cohen and Mirabel 1978).

A systematic survey of a large part of the southern galactic hemisphere down to a detection limit of $T_b \approx 0.05$ K (corresponding to ~ 3 x 10^{18} at/cm^2) covering a velocity range from -1000 to +1000 km/s has recently been made by Hulsbosch (1978b). The survey will be extended over the whole sky observable from Dwingeloo. Up to the present only the area l = 100° to 200° and b = -45° and -60° to 0° has been covered. Figure 4 shows the results. It is apparent that this part of the southern galactic hemisphere is full of high-velocity features, extending to quite high negative values. Not a single high-positive velocity relative to the lsr has been found. While at the northern galactic hemisphere no velocities lower than -210 km/s are yet known, in the southern latitudes velocities down to -465 km/s are found (HVC 110.5-7-465, see also the detailed observations of this cloud by Cohen and Mirabel 1978). In the region l = 100° to 140° we found 8 objects with $-465 < V < -364$ km/s, including those of Davies (1975) and Wright (1974). The next lowest velocity in this region is "only" -204 km/s (HVC 124-12-204) indicating that the highest velocities may depict the existence of a separate (intergalactic?) stream with a mean velocity of -227 km/s with respect to the galactic centre, possibly associated with the Magellanic Stream (the rotation of the lsr being 250 km/s). At longitudes greater than 140° the picture is dominated by two large complexes, one at l,b = 165°,-45°, V ≈ -330 km/s, discovered by Cohen and Davies (1976), and the Anticentre complex (ACI-ACII-ACIII) with velocities between -100 and -210 km/s. It now appears that the former continues through HVC 168-42-240 (Meng and Kraus 1970) towards l,b = 188°,-23° where it meets the AC complex at roughly a right angle. The same feature runs nearly parallel to a string with velocities around -110 km/s, first reported by Van Kuilenburg (1972). The weakest of the many small clouds found at the southern latitudes is HVC 158-32-318 with a T_b of only 0.06 K. It is confirmed by four independent observations.

REMARKS

An attempt to observe absoption of HVC 131+1-200 against the galactic radiosource 3C58 proved to be unsuccessful, indicating that this cloud either has a spin temperature larger than 200 K or lies beyond 3C58, i.e. its distance is larger than 8 kpc (Westerbork observations of Schwarz and Wesselius 1978).

HVC 132+23-211 has cores with densities of 100 cm^{-3}.kpc. A search for CO by F.P. Israel (private communication) was unsuccessful. This should, however, not be conclusive because, even at a distance as low as a few hundreds pc the HI density is much lower than in an average molecular cloud.

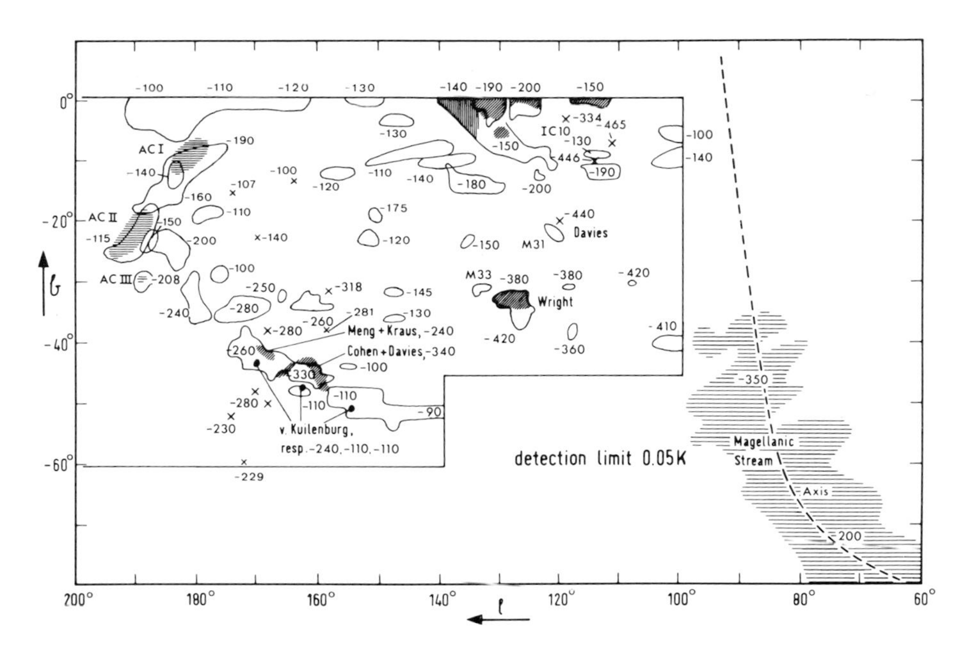

Figure 4 High-velocity features observed in a survey of high sensitivity extending from -1000 to +1000 km/s relative to the lsr (Hulsbosch 1978b, and unpublished data). Shown are the lowest observed contours (~ 0.05 K) of all the objects with $|V_{lsr}|$ > 100 km/s, labeled by their radial velocity in km/s with respect to the local standard of rest (for the extended objects rounded to the nearest multiple of 10). Crosses are single positions, shaded areas denote objects already known from previous investigations; for those discovered by others the name of the discoverer is indicated.

In this report we have paid no attention to the distances of HVC's and their nature. Something, however, should be said about this. Let us concentrate on complex A. Its thin structure together with the erratic velocity jumps suggests that the present appearance has been set up only recently (within the last ten million years), most likely as a consequence of its penetration into the Galactic System. The velocity differences would then have been caused by collision of the clumps in the stream with clouds or cloud complexes in the galactic halo. It is possible that steep edges of the clumps are shockfronts associated with such collisions. The distance would be no more than let we say 3 kpc. Following a discussion of Hulsbosch (1975) in connection with the observations of interstellar absorption lines, a minimum distance of roughly 2 kpc can be inferred so that, at least for chain A, we have some idea of the distance. Another case in which a distance can be inferred (a few tens kpc) is the Magellanic Stream. Other nearby intergalactic features might be the group of very-high-velocity clouds on the southern galactic hemisphere between longitude 100° and 140°, and the string connecting HVC 160-46-330 to l,b = 188°,-23°.

The HVC's have been enigmatical for many years. At this moment we realize that different kinds of high-velocity gas exist with each its own explanation (e.g. Magellanic Stream, distant-arm extensions to high z, infalling intergalactic gas?). To get a better knowledge further observations are urgently needed. In the next years the search for weak objects in the Dwingeloo sky in the velocity range -1000 to +1000 km/s will be continued. A few bright cores are now being investigated with the new Westerbork line receiver (results are not yet available). The investigations of the small-scale structures with 10'-beam telescopes should also be continued. One of the most important (and difficult!) optical programs is the search for interstellar absorption lines in the spectra of distant stars in order to get some more idea of distances of HVC's (Hulsbosch 1975). Such an investigation should be strongly stimulated.

Acknowledgements - The cooperation of Oort in preparing this paper is gratefully acknowledged. Many of the observations described in this paper have been carried out with the Dwingeloo radio telescope which is operated by the Netherlands Foundation for Radio Astronomy with the support of the Netherlands Organisation for the Advancement of Pure Research (ZWO).

REFERENCES

Cohen, R.J., Davies, R.D.: 1975, Mon. Not. R. astr. Soc. 170, 23P
Cohen, R.J., Mirabel, I.F.: 1978, Mon. Not. R. astr. Soc. 182, 395
Cohen, R.J., Mirabel, I.F.: 1978 (preprint)
Cram, T.R., Giovanelli, R.: 1976, Astron. Astrophys. 48, 39
Davies, R.D.: 1973, Mon. Not. R. astr. Soc. 160, 381
Davies, R.D.: 1974, in F.J. Kerr and S.C. Simonson III (eds.), 'Galactic Radio Astronomy', IAU Symp. 60, 599
Davies, R.D.: 1975, Mon. Not. R. astr. Soc. 170, 45P
Davies, R.D., Buhl, D., Jafolla, J.: 1976, Astron.Astrophys.Suppl. 23, 181
Encrenaz, P.J., Penzias, A.A., Gott, R. III, Wilson, R.W., Wrixon, G.T.: 1971, Astron. Astrophys. 12, 16
Giovanelli, R., Verschuur, G.L., Cram, T.R.: 1973, Astron. Astrophys. Suppl., 12, 209
Giovanelli, R., Haynes, M.P.: 1976, Mon. Not. R. astr. Soc. 177, 525
Giovanelli, R., Haynes, M.P.: 1977, Astron. Astrophys. 54, 909
Greisen, E.W., Cram, T.R.: 1976, Astrophys. J., 203, L119
Hulsbosch, Aad N.M.: 1975, Astron. Astrophys. 40, 1
Hulsbosch, Aad N.M.: 1978a, Astron. Astrophys. Suppl. (in press)
Hulsbosch, Aad N.M.: 1978b, Astron. Astrophys. (in press)
Mathewson, D.S., Cleary, M.N., Murray, J.D.: 1974, Astrophys. J. 190, 291
Meng, S.Y., Kraus, J.D.: 1970, Astron. J. 75, 535
Schwarz, U.J., Sullivan, W.T. III, Hulsbosch, Aad N.M.: 1976, Astron. Astrophys. 52, 133
Schwarz, U.J., Wesselius, P.R.: 1978, Astron. Astrophys. 64, 97
Shostak, G.S.: 1977, Astron. Astrophys. 54, 919
Van Kuilenburg, J.: 1972, Astron. Astrophys. Suppl. 5, 1
Verschuur, G.L., Cram, T., Giovanelli, R.: 1972, Astrophys.Letters 11, 57
Verschuur, G.L.: 1975, Ann. Rev. Astron. Astrophys. 13, 257
Wannier, P., Wrixon, G.T., Wilson, R.W.: 1972, Astron. Astrophys. 18, 224
Wright, M.C.H.: 1974, Astron. Astrophys. 31, 317

DISCUSSION

Wollman: Can you estimate the net flow rate of the gas?

Hulsbosch: It depends on a number of unknown factors, including the cloud distances and their thicknesses along the line of sight. If we assume for the gas at $b > +15°$ an average distance of 2 kpc, a layer thickness of 250 pc and a mean velocity of -150 km s^{-1}, the observed total mass of 10^5 $M_\odot$ kpc^{-2} corresponds to an inflow of 0.3 solar masses per year.

Giovanelli: I would like to report negative results of molecular searches in the cores of HVC's, performed with Martha Haynes and Tom Guiffrida. We searched for OH and for CO at, respectively, the 140-ft and the 36-ft telescopes of the NRAO. One of the positions sampled in the OH species, in the denser part of complex A, was observed for a total integration time of 35 hours. Leo Blitz at Columbia has also searched for CO with negative results.

van Woerden: Do your results imply an anomalous abundance of OH in HVC's?

Giovanelli: Not knowing anything about distance and therefore about cloud density and environment, it is hard to set a limit on OH column density.

Felten: You said that the total mass of the clouds is of order $2x10^5$ $M_\odot$ if their typical distance is of order 1 kpc. Does the inferred mass scale as the square of the assumed distance?

Hulsbosch: Yes.

van den Bergh: Did your survey turn up any dwarf galaxies?

Hulsbosch: Yes, IC 10 (HVC 119-3-334), but the observed profile was beforehand suspected to belong to such a feature because it showed the exceptionally large width of 65 km s^{-1}.

THE ANTICENTER HIGH-VELOCITY-CLOUD STREAMS AS A GALACTIC PHENOMENON

R. L. Moore
University of Arizona, Tucson, AZ, U.S.A.

and

W. B. Burton†
National Radio Astronomy Observatory*, Green Bank, WVa, U.S.A.

The galactic anticenter region contains several streams of high-velocity clouds of HI. The principal stream is associated with a localized HI feature at forbidden negative velocities near ℓ, b = 197°, 2° (Weaver 1970, 1974). Here we show that the forbidden-velocity feature, in addition to being the culmination of three streams of high-velocity clouds, is correlated with a disturbance in the permitted-velocity gas. Evidence suggests that this disturbance is located within the Galaxy, implying by association that the anticenter high-velocity clouds and the culmination feature are at distances interior to the Galaxy.

The high-velocity cloud streams in the anticenter have been mapped (from data of Weaver and Williams 1973) by Weaver (1974), Simonson (1975), and Burton and Moore (1978). There are three streams of negative-velocity HI which converge to a focus near ℓ, b = 197°, 2°. The principal stream extends continuously from ℓ, b = 200°, 2° to 160°, 8°. Two less intense secondary streams also converge to the focus. These extend from the focus towards ℓ, b = 191°, -7° and towards 184°, -5°.

In addition to the positional convergence, it has been shown that these streams converge in velocity space to the negative-velocity feature (the focus). A latitude-velocity cut at ℓ = 197°.3 (Burton and Moore 1978) reveals a constant db/dv gradient of the primary stream from b, v = 8°, -80 km s^{-1} to join the focus near 2°, -50 km s^{-1}. Longitude-velocity maps of the primary stream show continuity of the stream from ℓ, v = 197°, -50 km s^{-1} to 160, -130 km s^{-1}. The gradient dℓ/dv is similar to that of differential galactic rotation; however, the pattern

†Current address: Department of Astronomy, University of Minnesota, 116 Church Street, S.E., Minneapolis, Minnesota 55455

*Operated by Associated Universities, Inc., under contract with the National Science Foundation.

W. B. Burton (ed.), The Large-Scale Characteristics of the Galaxy, 535–540.

is offset by -90 km s^{-1} from circular velocities. There is no significant emission at anomalous positive velocities throughout the region considered.

From the convergence in both velocity and spatial coordinates of three streams of high-velocity material towards the region near ℓ, b = 197°, 2°, it is apparent that this focus is the center of activity for the anomalous-velocity HI. To examine the small-scale structure of this feature, we have obtained high-sensitivity HI spectra on a densely-sampled grid of the focus region. These observations were made with the NRAO 140-foot telescope (HPBW = 20'), at a velocity resolution of Δv = 1.4 km s^{-1}. Antenna temperature spectra are mapped; the conversion to brightness temperature is $T_B = 1.44\ T_A$.

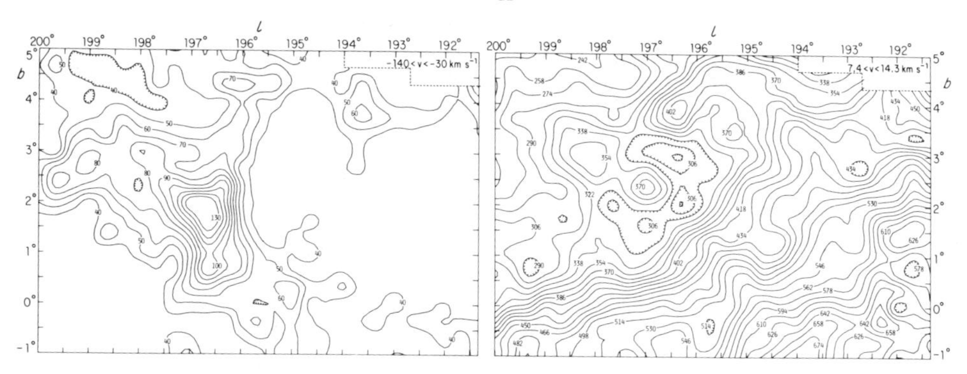

Fig. 1. Intensities integrated over the indicated velocity ranges. The left panel (a) shows the three streams converging to the focus. The right panel (b) shows the permitted-velocity minimum at the focus position.

Figure 1 shows the spatial arrangement of intensities integrated over two velocity ranges. The total extent of the anomalous negative velocities, $-140 < v < -30$ km s^{-1}, is represented in Figure 1a and shows the streams and focus as discussed above. Figure 1b shows the spatial behavior of emission integrated over the velocity range $7.4 < v < 14.3$ km s^{-1}. Perturbing the general pattern of emission in this velocity range is an isolated region of relatively low intensity.

It is crucial to the interpretation of the focus that it coincides in position with this disturbance in the permitted-velocity material. Figure 2 shows orthogonal position, velocity cuts through this region. Centered near ℓ, b, v = 197°, 2°, 10 km s^{-1} is a localized deep minimum in the permitted-velocity material. The rareness of such features in this area of the sky implies that the coincidence in position of the minimum with the stream-focus is not fortuitous. The possibility of hydrogen absorption can be ruled out because (1) the velocity width of the intensity minimum is larger than would be expected from cold-cloud absorption and (2) there is no extragalactic continuum source or concentration of the galactic continuum in the direction of the feature.

For the above reasons, we conclude that the permitted-velocity intensity minimum represents a true absence of material, and that this absence is associated with the forbidden-velocity focus of the high-velocity streams. These conclusions are supported by additional arguments. The total column density across the extent of the streams is approximately equal to the "missing" density in the minimum, suggesting that the focus region is the origin of the stream material. Comparing the spatial structure of the permitted-velocity depression (Fig. 1b) with that of the forbidden-velocity focus (Fig. 1a), we note the repetition of a characteristic "boomerang" shape in both features, suggesting association.

Three arguments support the conclusion that the intensity minimum is local. First, it occurs at a permitted velocity. Second, no plausible extragalactic phenomenon could cause such an intensity minimum. It does

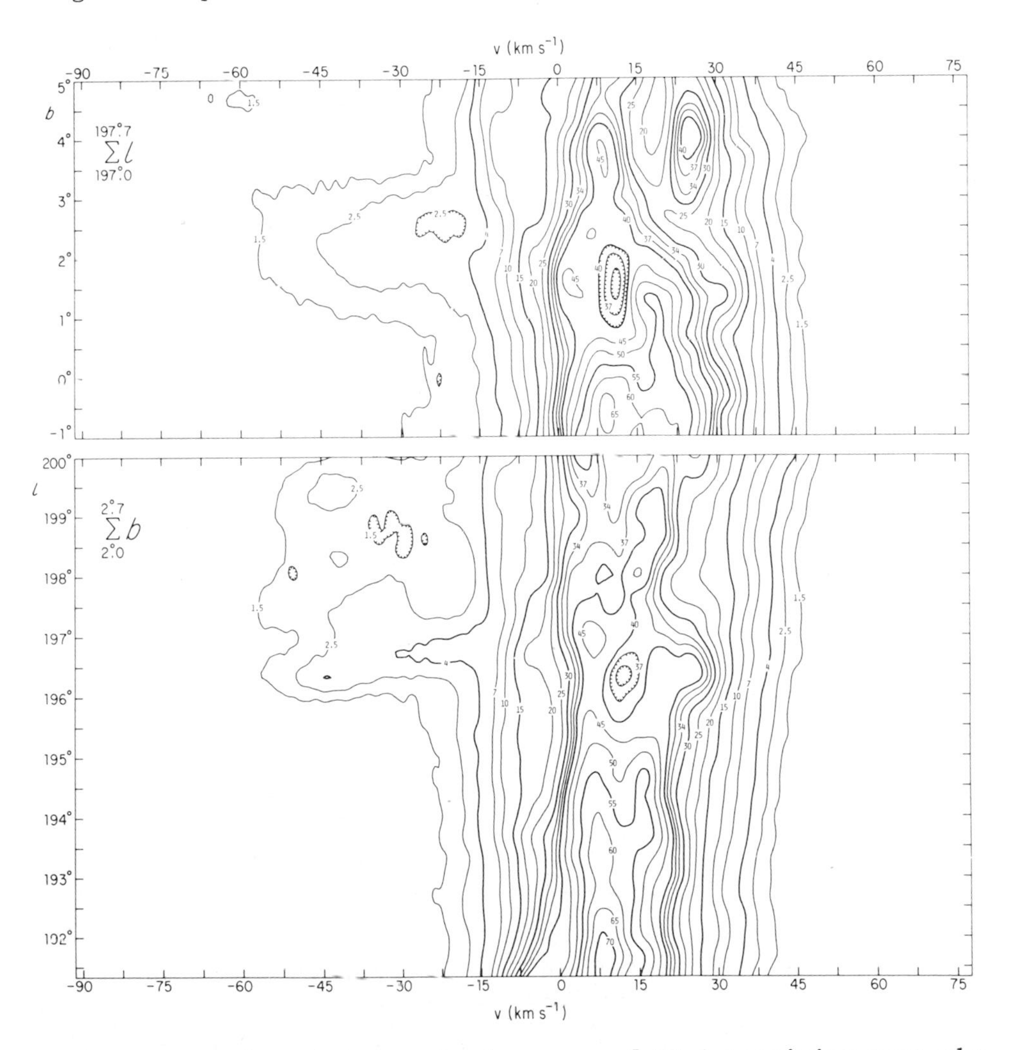

Fig. 2, Averaged position, velocity maps of HI intensities near the focus, showing the coincidence of the stream-focus with a disturbance in the permitted-velocity material.

not coincide with an extragalactic source of continuum radiation and neither does it have the attributes of an absorption feature. Finally, the general galactic-layer emission near the intensity minimum has a disrupted appearance (see Fig. 2).

Because of the convergence of the high-velocity streams to a focus which coincides in position with a depression in the permitted-velocity gas, we consider these features to be different aspects of a single phenomenon. We believe that the intensity minimum is located within the Galaxy, and that because of their intimate association with this feature, the anticenter high-velocity streams and their focus are similarly of a galactic nature.

Several models have been proposed for this complex. Simonson (1975) proposed that the forbidden-velocity feature at the focus represents a dwarf galaxy at a distance of 17 $\pm$ 4 kpc. The primary stream is postulated to be debris from the dwarf galaxy tidally removed by the Milky Way. The observations discussed above, in particular the permitted-velocity minimum, argue against this interpretation. The culmination of the streams and their association with the localized disturbance within the Galaxy weigh against interpreting the high-velocity clouds as independent entities falling in towards the Galaxy, or as a spiral arm. While a supernova remnant model is attractive as a disruptive galactic phenomenon, it is difficult to explain the marked asymmetry of the complex with this type of model. Impingement on the galactic disk of a stream originating outside the Galaxy is, however, a situation compatible with a localized disturbance and with the observed kinematic structure.

REFERENCES

Burton, W. B., and Moore, R. L.: 1978, Astron. J., submitted.
Simonson, S. C.: 1975, Astrophys. J. (Letters) 201, pp. L103-108.
Weaver, H. F.: 1970, in H. Habing (ed.), "Interstellar Gas Dynamics", IAU Symp. 39, pp. 22-50.
Weaver, H. F.: 1974, in F. J. Kerr and S. C. Simonson III (eds.), "Galactic Radio Astronomy", IAU Symp. 60, pp. 573-586.
Weaver, H. F., and Williams, D.R.W.: 1973, Astron. Astrophys. Suppl. 8, pp. 1-503.

DISCUSSION

Verschuur: Why don't you see positive velocity features associated with the presumed removal of HI from the region of deficiency at permitted velocities.

Moore: The velocity asymmetry, as well as the spatial elongation of the streams, pose considerable difficulties for an explosive disruption such as would be given by a supernova. We have not constructed a specific model for the complex. A model in which a stream of gas impinges on the disk would provide agreement in a number of respects; most importantly, one can introduce a preferred direction of momentum.

Heiles: I believe that the weak emission you reported is part of a much larger loop, 10° to 30° in diameter, which is visible on my photos of the Weaver and Williams survey.

Moore: It is possible that these features are part of an even larger complex.

Burton: But why the association with the intensity minimum? We satisfied ourselves that such minima are extremely rare in the anticenter region, and that the association is therefore unlikely to be fortuitous.

Heiles: Your emission is, I believe, just a part of a huge loop...yet your hole is only 1° in diameter (?).

Simonson: The low-velocity gas has nothing to do with the small galaxy, but it may be correlated with the obscuration in front of it. Consequently, Moore and Burton's mass should be of some value to that optical astronomer who wants to establish the definitive value for the mass of the Milky Way. It may provide a guide to the nearest areas, where the 20 or so RR Lyraes we expect in the small galaxy may be observed.

Let us hope that Moore and Burton will also publish their line profiles, as Weaver and Williams did. This will allow somewhat greater accuracy than is now available in defining the central position, central velocity, and maximum velocity of the small galaxy. These quantities are used, together with the distance, in deriving the mass of the Milky Way. Moore and Burton have brought forth nothing of a quantitative nature to verify or support their contention. The model I computed is a simple dynamical model of the familiar kind of tidal interaction. It accounts for all the observed features in both space and velocity and leaves nothing unexplained.

Moore: We have approached the observations phenomenologically; our conclusions follow directly from the data, and require no modelling.

Giovanelli: What is an upper limit to the energy liberated in the event, for the upper limit of the distance that you quoted?

Moore: For a distance of 1 kpc, the mass present in the high-velocity streams is about 10^4 $M_\odot$, with a kinetic energy of 10^{50} - 10^{51} ergs.

Giovanelli: A feature similar (though of smaller angular extent) to Weaver's jet, which you have described, coincides with IC 443, a supernova remnant. In relation to your worry that an asymmetry in velocity is present in Weaver's jet, I would like to report that the feature IC 443 has been mapped at Arecibo by Haynes and myself and found to constitute an HI shell that closely matches the optical and radio continuum emission. It also shows mainly negative velocities; this may be explained by the presence of denser gas in the foreground ISM, which is being encountered by the blast wave.

Heiles: I want to point out that it is very rare to see both the approaching and receding halves of an expanding shell. This fact has two possible interpretations; one, that these structures are not, in fact, expanding shells; or two, that we shouldn't worry when our models of shells predict the existence of the "other half" of the shell which isn't there. Personally, I subscribe to the latter viewpoint.

Dickey: I should like to point out that the identification of the hole in the allowed-velocity gas on the basis of its line width is dangerous because even cold clouds may often show broad emission and absorption lines. The Arecibo data in particular show little correlation between line width and spin temperature. I find the spatial correlation very good evidence for your interpretation, but the line width is not.

Burton: Liszt and I searched the region of the permitted-velocity hole for CO emission, and found none. This also seems to rule out HI self-absorption because of the demonstrated general correlation of CO emission with HI self-absorption (Ap. J. 1978, 219, L67).

Baker: Dr. Burton and I have obtained high-resolution data from Arecibo for the putative galaxy. The maps confirm the features mentioned by Moore. We had hoped that if the object were a galaxy we might resolve some cloud structure recognizable by its small linewidths. The 3 arcmin resolution did indeed pick out small substructures but all showed broad, often asymmetric, profiles that are not typical of normal galactic gas. The gas looks like highly perturbed material within our system. The data will appear in A.&A. Suppl. (1978).

A SENSITIVE HIGH-VELOCITY HI SURVEY

Riccardo Giovanelli
National Astronomy and Ionosphere Center*, Arecibo, Puerto Rico

The observational studies of the neutral hydrogen high velocity clouds (HVC's) in the last decade have for the most part concentrated on the detailed study of the prominent complexes discovered by the surveys of the late 1960's. More recently, several HVC's were discovered, often by accident, with very high velocities, which are typically outside of the velocity range and below the sensitivity limit of the old surveys. The tendency to associate these clouds with the closest galaxy at hand became a rather common practice. The idea that the Local Group, or the Local Supercluster, is populated with such "intergalactic clouds" was fed by these discoveries, and generalized to the whole field of the HVC's. I have repeatedly argued against the latter generalization (Giovanelli 1977, 1978). In this paper I shall report on the progress of a new survey that intends to constitute a test of the intergalactic approach itself.

For the past year and a half I have been conducting a large scale, high sensitivity survey in the 21-cm line, using the 300-foot (92.6 m) telescope of the N.R.A.O.† Earlier observations of more limited scope (concentrating on the region between the northern tip of the Magellanic Stream and the northern HVC complexes A and C) were conducted with the 140-foot (42.3 m) telescope. The surveyed region is bounded by declination +50° and -10°, with the grid spacing about 1° in R.A. and varying between 2°.5 and 5° in declination. Each of about 3500 sampled positions consists of two independent total-power spectra of orthogonal polarization with total integration time varying between 4 and 7 minutes. The bandwidth used was 10 MHz, yielding a 2100 km/s coverage, centered on zero km/s with respect to the LSR, and a channel separation of 11 km/s. The observational mode and the stability of the receiving equipment allowed the combination, for each profile, of an "off" spectrum of effectively indefinite integration time, yielding typical rms noise

* Operated by Cornell University under contract with the National Science Foundation.

† Operated by Associated Universities, Inc., under contract with the National Science Foundation

W. B. Burton (ed.), The Large-Scale Characteristics of the Galaxy, 541–544.

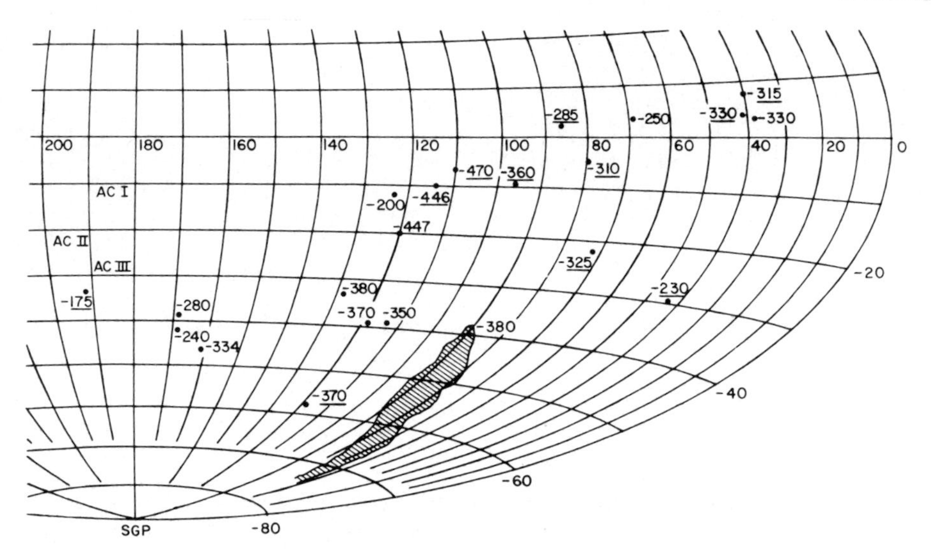

Fig. 1. Distribution of very HVC's. The shaded area is the Magellanic Stream.

fluctuations on the order of 0.01 K of antenna temperature. More than 400 sampled points yielded detections of high velocity gas; more than half of these refer to newly discovered features. The main results may be summarized as follows.

(a) A considerable number of clouds with LSR velocities more negative than -200 km/s are found, up to a maximum of -460 km/s; they all lie in one quadrant of the sky: $b < 0°$, $\ell < 180°$. Figure 1 shows their distribution superimposed on a grid of galactic coordinates. Clouds with underscored labels were found in the course of this survey; the remainder by other authors. This distribution is most interesting, since it is not counterbalanced by anything similar in the northern hemisphere, or, with comparable positive velocities, in the third or fourth quadrants of galactic longitude. There is no correlation with the general distribution of galaxies in the Local Group or with the distribution of their velocities. The centroid of the cloud distribution falls close to the tip of the Magellanic Stream and reinforces the argument that I brought forth in the discussions at IAU Symp. 77 last year, that they may be associated with the Magellanic Stream phenomenon.

(b) An extended complex exists at intermediate positive latitudes, between $10° < \ell < 60°$; it is broken up into several clouds with a fairly disordered velocity field, showing both negative and positive velocities which range between -180 km/s and +140 km/s. At lower latitudes confusion occurs with the high-z extension of galactic disk material.

(c) A new stream is found in the anticenter region. The distribution of new detections in the area is illustrated in the lower panel of Figure 2; the stream is outlined by a roughly drawn contour, and seems to merge with the ACII high velocity complex. Calling this feature a stream is encouraged by an inspection of the velocity field, shown in the upper panel of Figure 2. It is surprisingly smooth and almost constant with respect to the LSR.

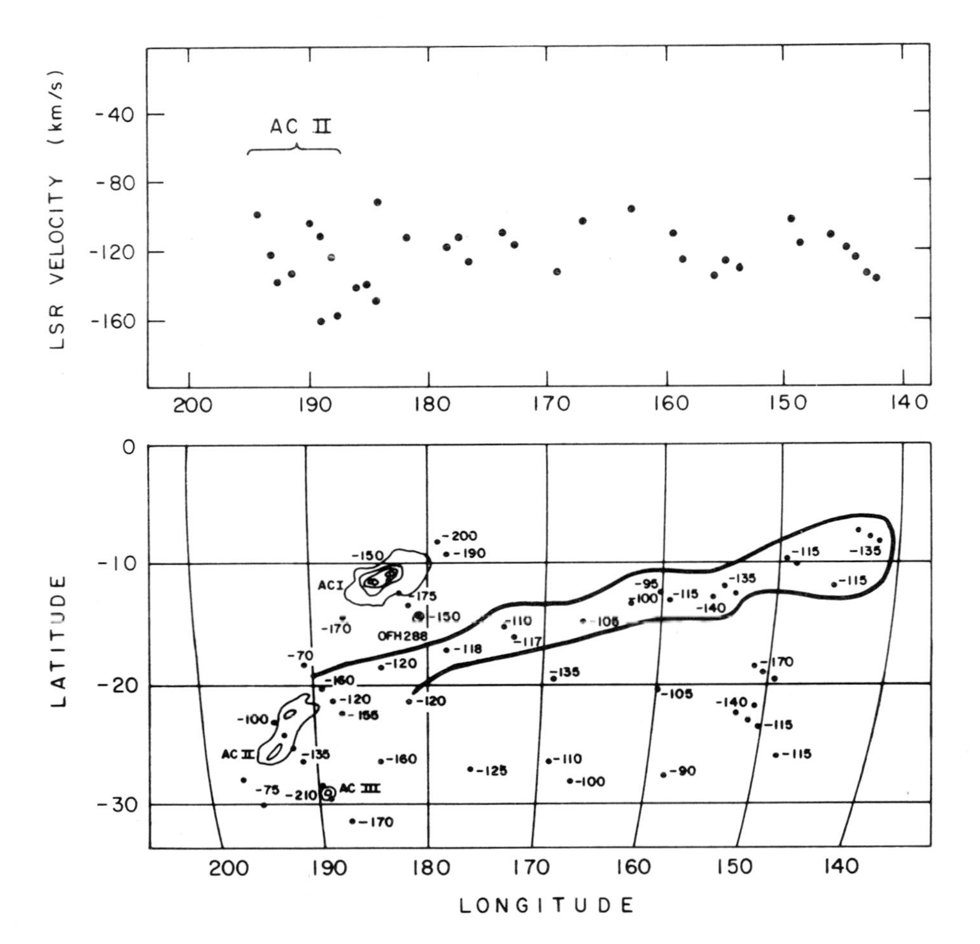

Fig. 2. Stream in the galactic anticenter region. Diagram above represents the velocity field.

(d) In contrast to the statement made by Hulsbosch in this symposium, numerous high positive velocity clouds exist, in addition to the low latitude complexes near $\ell = 260°$. They are all found at $\ell > 180°$, with velocities up to +195 km/s. A number of them overlap the region where the Local Group galaxies Sex A, B, C and Leo I lie, but their velocities are consistently 100-200 km/s more positive than the velocities of those galaxies. The similarity of these clouds with ones found far from this region further discourages an association with the galaxies.

(e) To the little surprise of most, the previously known complexes which are within the region sampled by this survey, such as the anticenter complexes, are revealed to be much more extended than previously thought.

A bonus of this survey was that I stumbled onto a very nearby galaxy, unaccounted for, as far as I know, in any of the galaxy catalogs. A profile with the characteristic signature of rotating HI disks was detected in a region dominated by foreground galactic dust, in Orion. Line profiles taken with the Arecibo telescope, along the North-South axis of the feature, assumed to be centered at R.A. (1950) = 5^h 42^m 24^s, Dec. (1950) = 5° 03', are shown in Figure 3. A red photographic plate kindly taken by Quintana and Melnick at the 4-m telescope at Cerro Tololo, centered on the peak of HI emission reveals a low surface brightness elongated extragalactic object.

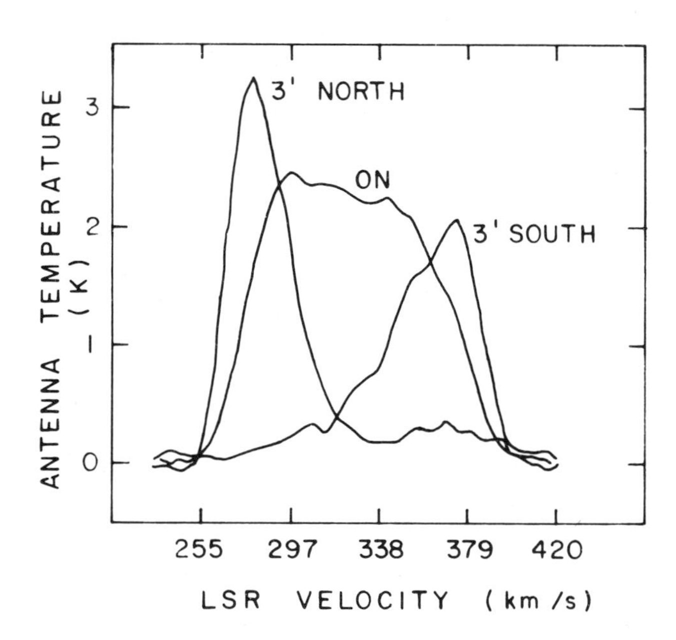

Fig. 3. 21-cm Arecibo profiles of features in Orion.

REFERENCES

Giovanelli, R.: 1977, Astron. and Astrophys. 55, pp. 395-400.
Giovanelli, R.: 1977, in "Structure and Properties of Nearby Galaxies", eds. E. M. Berkhuijsen and R. Wielebinski.

21-CM OBSERVATIONS OF HIGH-VELOCITY CLOUDS

I. F. Mirabel* and R. J. Cohen
University of Manchester, Nuffield Radio Astronomy Laboratories, Jodrell Bank, Macclesfield, Cheshire SK11 9DL England.

(a) Four high-velocity clouds (HVCs) with large negative velocities have been mapped at high sensitivity, with an angular resolution of 12 arcmin and a velocity resolution of 1.8 km s^{-1} (Cohen & Mirabel, in press). Although these are small isolated clouds they show the same two-component velocity structure that is observed in the large Northern HVC complexes at lower velocity. In addition the cool dense core of one of the clouds is found to be rotating. This rotation, and also the regular velocity patterns within the four clouds, indicates that the clouds are unlikely to be interacting with galactic material. It is interesting that similar observations of the Magellanic Stream do not show narrow velocity components, suggesting that there is a real physical difference between the Stream and other HVCs (Mirabel, Cohen & Davies, in press).

(b) A systematic search has been made for intergalactic HVCs associated with nearby dwarf irregular galaxies. Four regions containing eight such galaxies were searched to a detection limit of $\sim$0.1 K, the observations covering a velocity range of 2000 km s^{-1}. New HVCs were detected in three of the regions, and several lay within small projected distances of the galaxies. However their velocities differ by more than 100 km s^{-1} from those of the galaxies, so the HVCs cannot be gravitationally bound to the galaxies unless the latter contain much more mass than has so far been observed. One of the HVCs shows unusual properties which suggest that it may be a proto-dwarf irregular galaxy in the Local Group.

*Supported by a fellowship from Consejo Nacional de Investigaciones Cientificas y Tecnicas, Argentina.

W. B. Burton (ed.), The Large-Scale Characteristics of the Galaxy, 545–546.

DISCUSSION (After Dr. Giovanelli's talk)

Heiles: I got the impression from the spectra of your new galaxy that the rotation velocity was smaller at the points which are located furthest from the center. This makes it one of the few galaxies known which does not have a flat rotation curve at the last measured point.

Giovanelli: The profiles on the outskirts of the galaxy are fairly noisy, although there does seem to be such an effect as you point out. However, I want to underscore that these are still raw profiles, uncorrected for the effect of the sidelobes which may be picking out radiation near the center of the galaxy.

de Vaucouleurs: The HI clouds that you have observed in the general direction of M31 share the large negative velocities (about 400 km s^{-1}) characteristic of the IC 10-M31-M33 and associated galaxies. This is suggestive of distances of order 0.5 Mpc. As Dr. Einasto showed yesterday, objects closely associated with the Galaxy (distances $<$ 0.25 Mpc) have lower characteristic velocities.

Heiles: An upper limit to the distance can be obtained from the consideration that as you move the cloud further away the density drops and, at some point, the density is too low to collisionally excite the 21-cm transition, in which case it will have an excitation temperature of 3 K and thus be invisible. I suspect that at 1 Mpc distance the density would be less than 10^{-3}.

DISCUSSION (After Dr. Mirabel's talk)

Giovanelli: What is your velocity resolution in the cloud near NGC 6822, where you report rotation on the order of 5-10 km s^{-1}?

Mirabel: The rotation in the cloud adjacent to the Sagittarius galaxy and NGC 6822 is $\sim$ 15 km s^{-1} per degree. Our velocity resolution is 1.8 km s^{-1}.

Mathewson: If the clouds found near galaxies are contrails due to the passage of the galaxies through a hot intergalactic medium, then no association would be expected between the velocity of the galaxies and the velocity of the HI clouds.

Haynes: The region occupied by the very-high-negative-velocity clouds is also the region where the tip of the Magellanic Stream shows velocities of -380 km s^{-1} itself.

Mirabel: Relative to your comment on the possibility that the very-high-velocity clouds recently discovered may be associated with the stream: High-resolution observations of the stream and of four of these new, very-high-velocity clouds show different spectral properties, suggesting that they are different kinds of objects.

THE MAGELLANIC STREAM: OBSERVATIONAL CONSIDERATIONS

D. S. Mathewson, V. L. Ford, M. P. Schwarz
Mt Stromlo and Siding Spring Observatories
The Australian National University
and
J. D. Murray
Division of Radiophysics, CSIRO, Australia

GLOBAL CHARACTERISTICS

The Magellanic Stream is an arc of neutral hydrogen which nearly follows a great circle and which contains the Magellanic Clouds - hence its name (Mathewson, Cleary and Murray 1974). This great circle passes within a few degrees of the south galactic pole and lies close to the supergalactic plane. Mathewson and Schwarz (1976) argued that this indicates that the Magellanic Stream and Magellanic Clouds are not bound to the Galaxy. To reinforce this argument, they pointed out that around the supergalactic plane there is a similar systematic variation in the velocities of the Local Group and those of the Stream which may be due to the reflection of the motion of the galactic center if the velocity of rotation of the Sun is 225 km s^{-1}; if it is 290 km s^{-1} then the grounds for this argument would disappear.

THE SECTION BETWEEN THE MAGELLANIC CLOUDS AND THE GALACTIC PLANE AT l=306°

The Magellanic Stream (as first defined) has two sections; the main section is on the south galactic pole side of the Magellanic Clouds and the velocity varies in a systematic manner along its length whilst the other section is comprised of some small clouds scattered between the Magellanic Clouds and the galactic plane which show no systematic velocity variation (Wannier and Wrixon 1972 and Mathewson et al. 1974). Figure 1 shows the results of an HI survey of four of these clouds with the 64-m reflector at Parkes. The clouds 312+1+180 and 306-2+230 have regions with narrow velocity half-widths of about 8 km s^{-1} whilst other parts have much broader half-widths of about 30 km s^{-1}. This two component structure is common in the northern high velocity complexes (c.f. Hulsbosch 1975). It is not found in the main section of the Magellanic Stream where velocity half-widths mostly lie between 20-40 km s^{-1}.

It is interesting that the reflections of 312+1+180 and 306-2+230 in the galactic plane almost coincide with the northern high velocity complexes 131+1-200 and 122+1-197. This may mean that the Sun is moving

W. B. Burton (ed.), The Large-Scale Characteristics of the Galaxy, 547–556.

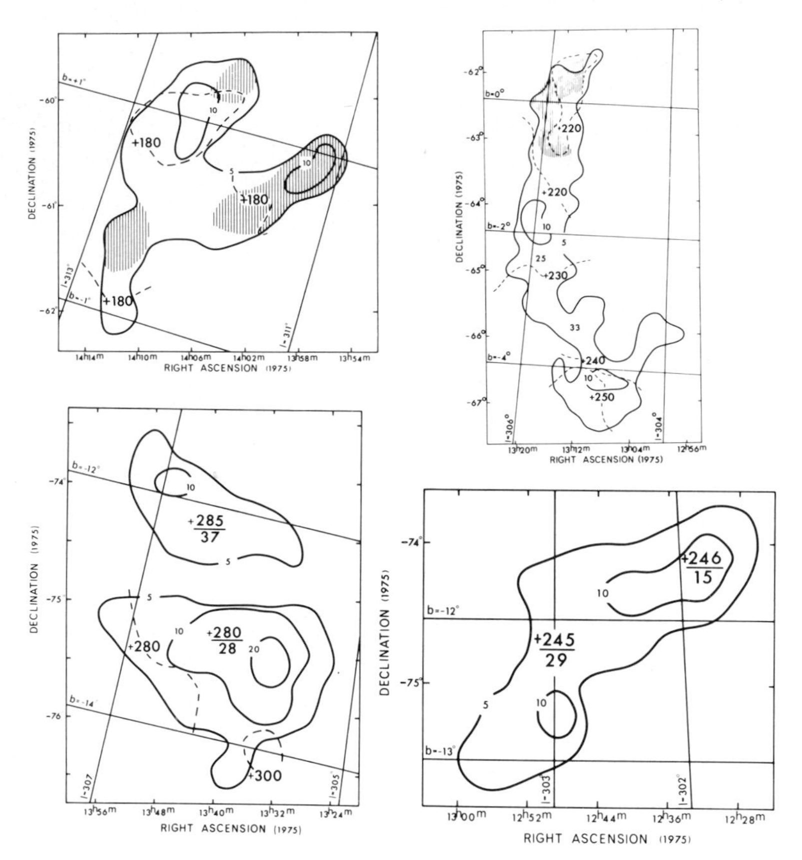

Figure 1. HI surface densities (full lines) of four high velocity clouds between the Magellanic Clouds and the galactic plane observed with the 64-m reflector at Parkes. The contour unit is 10^{19} atoms cm^{-1}. Radial velocity contours are dashed and the contour numbers are V_{LSR} (km s^{-1}). In the bottom two diagrams, the numbers above the lines are regionally representative V_{LSR} whilst those below the lines are velocity half-widths (km s^{-1}). In the top two diagrams, the shaded areas are regions where the velocity half-widths are very narrow, about 10 km s^{-1}.

through an elongated HI complex which is not partaking in galactic rotation. This, together with the two component velocity structure and lack of systematic velocity variation, suggest that this section does not belong to the Magellanic Stream.

A SEARCH FOR OTHER COMPONENTS

Deep plates of the region of the Magellanic Stream reveal nebulosities in some areas. Spectroscopy shows that these must be reflection nebulosities similar to those found by Sandage (1976) at high latitudes and illuminated by the galactic plane. Hα plates taken by Dr. K. H. Elliot with the SRC Schmidt telescope confirm that there are no emission nebulosities associated with the Stream.

Star counts of SRC Schmidt blue plates show no stellar component of the Stream which confirms the negative result from photographic photometry of UBV plates taken with the Uppsala Schmidt telescope. It should be mentioned here that Philip (1976a) claimed to have discovered a blue stellar component of the Magellanic Stream but has since withdrawn his claim as further work showed that the stars were either subdwarfs F-type or white dwarfs (Philip 1976b).

As part of the search program, the region between the LMC and SMC was observed and the stellar wing of the SMC (Westerlund and Glaspey 1971) was found to extend to R.A.03^h15^m. Radial velocity measurements of these stars showed that several were born in regions of HI surface density of only 3×10^{20} atoms cm^{-2} whilst some other regions of higher density had no stellar component. The maximum HI surface density in the Magellanic Stream is 5×10^{20} atoms cm^{-2} at R.A.00^h54^m, Dec.$-49°30'$ (1975.0).

Galaxy counts were made on the SRC Schmidt plates to about 250 galaxies per square degree and an upper limit of 0.2 mag was placed on A_V in the directions of the denser parts of the Stream. However a real decrease in galaxy counts was found in the wing of the SMC as far out as R.A.02^h36^m (HI surface density of 5×10^{20} atoms cm^{-2}) where the absorption was estimated to be 0.3 mag. It is worth pointing out that the dust ridge line lies 40 min of arc to the south of the HI ridge line. This displacement may be produced by radiation pressure of the light from the LMC and SMC on the dust particles (Chiao and Wickramasinghe 1973).

Lynden-Bell (1976) hypothesised on the association between some outlying satellites of the Galaxy and the Magellanic Stream and some northern high velocity HI complexes. Hartwick and Sargent (1978) recently measured radial velocities of these objects and in all cases except Ursa Minor, there were large differences between their velocities and that of the HI. They deduced that no association exists.

Radio continuum observations of the Magellanic Stream at 408 MHz and 1420 MHz using the 64-m reflector show no evidence for emission although the background is irregular which makes an accurate measurement difficult. HI absorption measurements were made using the background continuum radio source at R.A.$00^h39^m48^s$, Dec.$-44°29'16''$ (1950) which allowed a lower limit of 30°K to be placed on the spin temperature of the HI of the Stream in that direction.

To sum up, the Magellanic Stream has only been detected by HI emission measurements. However it would be worthwhile to make UV observations with the IUE around R.A.00^h54^m, Dec.$-49°30'$ (1975.0), the region of maximum HI emission in the Magellanic Stream.

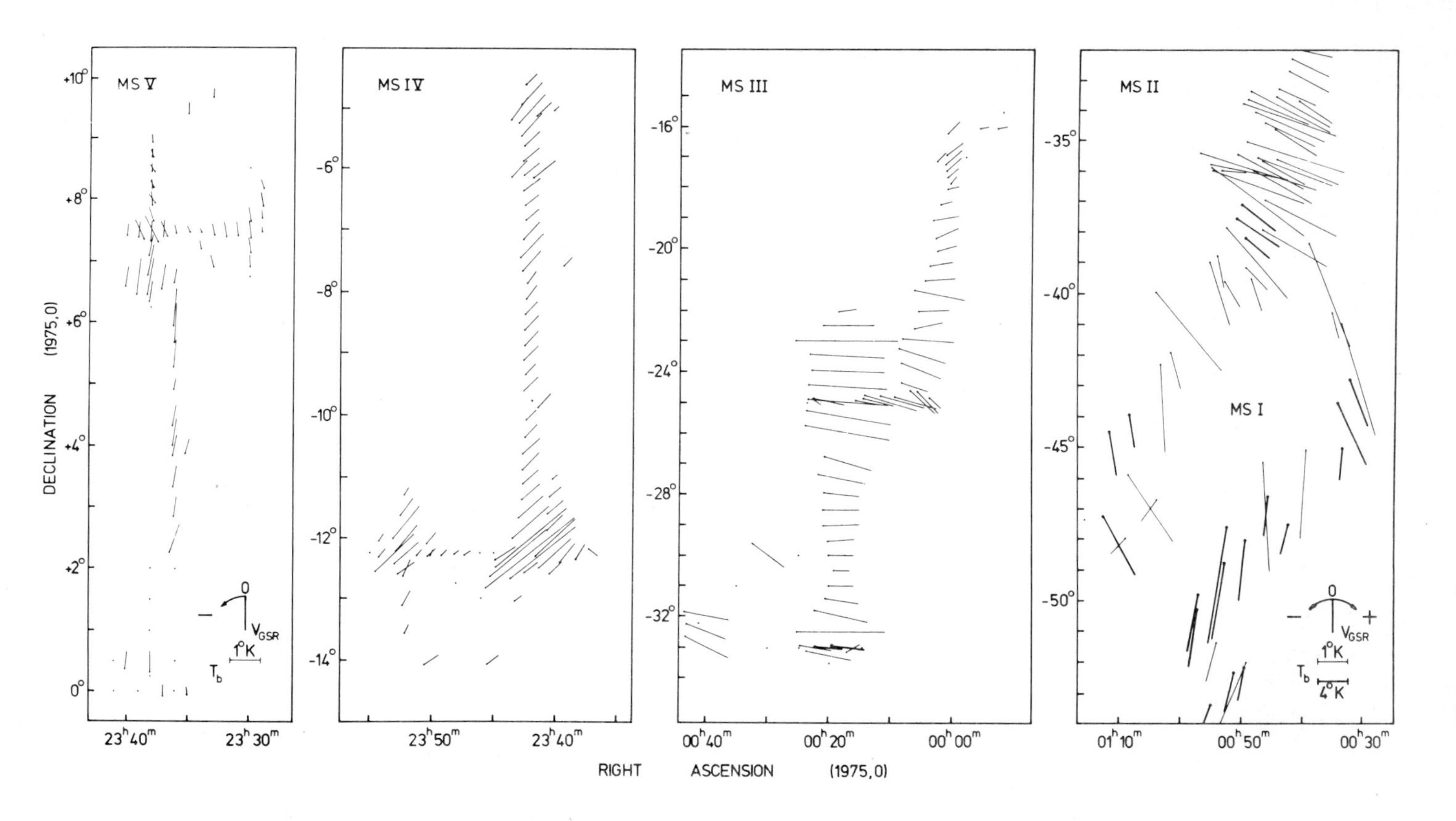

Figure 2. The velocity field along the ridge lines of the clouds of the Magellanic Stream observed with the 64-m reflector. The angle the lines make with the vertical represent V_{GSR}, the radial velocity in km s^{-1} corrected for a galactic rotation at the Sun of 225 km s^{-1}, of the peaks in the HI profiles. Dots indicate the head of each line e.g., MSIV and MSV have an average V_{GSR} of -135 km s^{-1} and -175 km s^{-1}, respectively. Anticlockwise rotation indicates a negative velocity. The length of the lines represent T_b (°K) of the peaks (see scales in bottom RH corner).

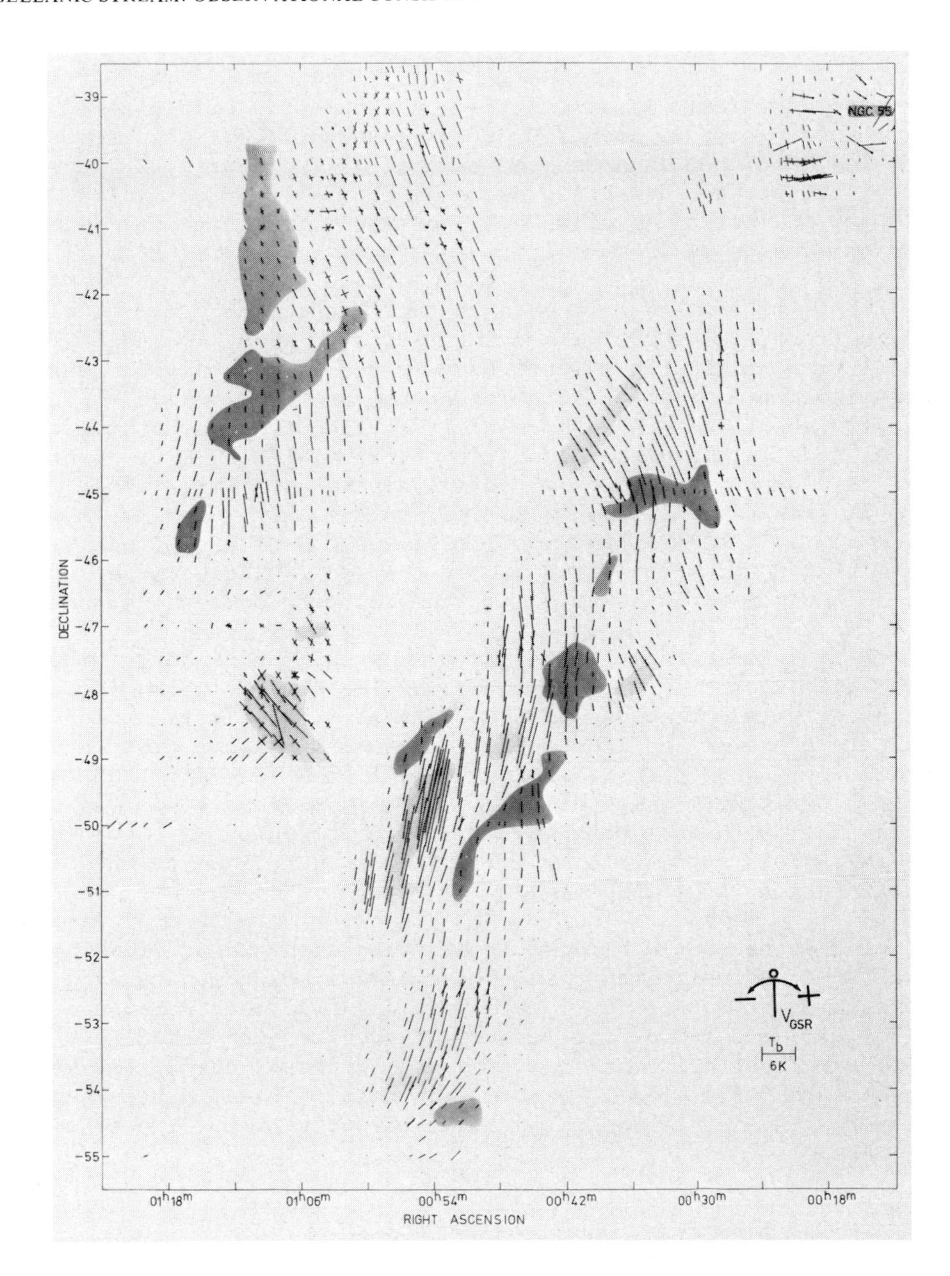

Figure 3. The velocity field of MSI obtained with the 64-m reflector. The angle the lines make with the vertical represent the radial velocity V_{GSR} (km s^{-1}) of the peaks in the HI profiles. Where velocities exceed 90 km s^{-1}, a dot marks the head of each line. Anticlockwise rotation indicates negative velocity. The length of the lines represent T_b (°K) of the peaks (see scale in bottom RH corner). The light and dark shaded regions are areas where velocity half-widths are about 20 km s^{-1} and >40 km s^{-1}, respectively. The co-ordinates are for 1975.0.

THE MAINSTREAM

The Magellanic Stream on the south galactic pole side of the Magellanic Clouds is composed of six large discrete gas clouds MSI-VI with looped or horseshoe-shaped structures (Mathewson, Schwarz and Murray 1977). HI surface densities decrease from an average of 26×10^{19} atoms cm^{-2} for MSI to 2×10^{19} atoms cm^{-2} for MSVI. Velocity half-widths increase from an average of 25 km s^{-1} for MSI to 40 km s^{-1} for MSV.

There is a systematic velocity variation from cloud to cloud which ranges from V_{GSR} = -200 km s^{-1} for MSVI to 0 km s^{-1} for MSI (V_{GSR} are radial velocities with respect to a nonrotating galaxy and they have been corrected for a galactic rotation at the Sun of 225 km s^{-1}). However Figure 2 shows that in each cloud, the velocity is amazingly constant which places constraints upon the degree of translational motion and suggests that they have predominantly radial velocities. This rather featureless velocity structure is seen also in Figure 3 which displays the entire velocity field of MSI, the most massive cloud of the Stream (mass approximately $10^8 M_\odot$ if at the distance of the SMC). There may be some evidence for a rotation of 10-20 km s^{-1} in the velocity pattern (Mathewson 1976).

A line drawn through the six clouds and the center of mass of the Magellanic Clouds is a small circle of latitude 7° which is parallel to the great circle that passes through the south galactic pole and cuts the galactic plane at $l = 280^\circ$. If the reasonable assumption is made that the Stream would form part of a great circle when seen from the galactic center, this 7° parallax gives a distance to the Stream of 70kpc (taking the Sun's distance from the galactic center as 8.5kpc) which is about the distance of the SMC. This suggests that the Stream is not close to the Galaxy, certainly not closer than 30kpc.

The Stream would be exactly overhead when viewed from the galactic center. If this implies that the Magellanic Clouds have an overhead orbit, it is interesting to ask if this is a coincidence. Perhaps not, as Lynden-Bell (1976) and Kunkel and Demers (1976) have both noted that there is a preferred plane for other outlying satellites of the Galaxy which is close to the Magellanic plane. Is the fact that overhead orbits seem to be preferred telling us something about the gravitational potential field of the Galaxy?

THE REGION BETWEEN THE LMC AND SMC

Figures 4, 5, 6 and 7 show some results from an HI survey of the inter-Cloud region using the 18-m and 64-m reflectors at Parkes. Briefly the main features of these new observations germane to this discussion are:
a) HI spurs are found to run out almost at right angles to the strong bridge of gas connecting the LMC and SMC and point in the direction of the Magellanic Stream. In particular, the strongest spur at R.A.01^h50^m, Dec. -68° (Fig. 6) points almost directly at MSI. These spurs make this side

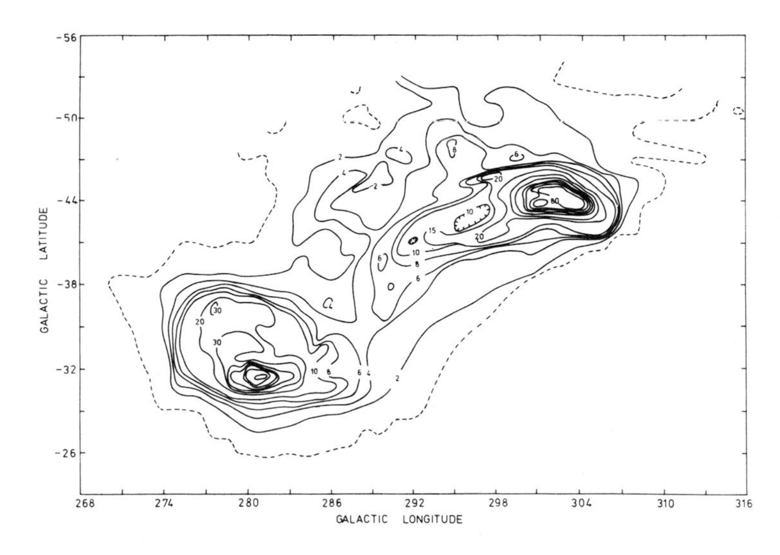

Figure 4. Contours of the HI surface densities of the region of the Magellanic Clouds obtained using the 18-m reflector at Parkes. The contour unit is 5×10^{19} atoms cm^{-2}.

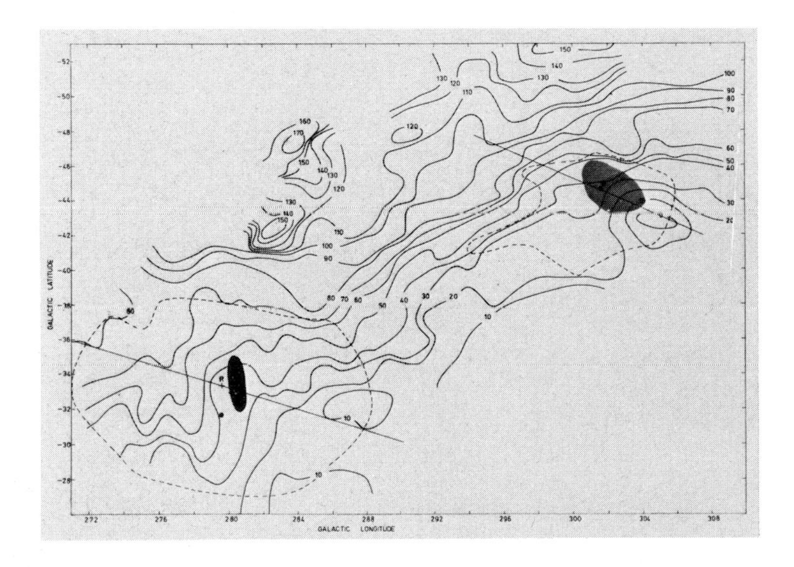

Figure 5. Contours of the center radial velocity V_{GSR} (km s^{-1}) of the HI profiles obtained using the 18-m reflector in a survey of the Magellanic Clouds. The effects of a translational motion of 250 km s^{-1} of the Magellanic Clouds along the great circle of the Stream in the direction of the galactic plane has been removed from the observed velocities. This correction does not appreciably alter the velocity field. The dashed contours give the outer limits of the clusters in the LMC and SMC, the straight lines indicate the major axes of the two galaxies and their centers of rotation are labelled R.

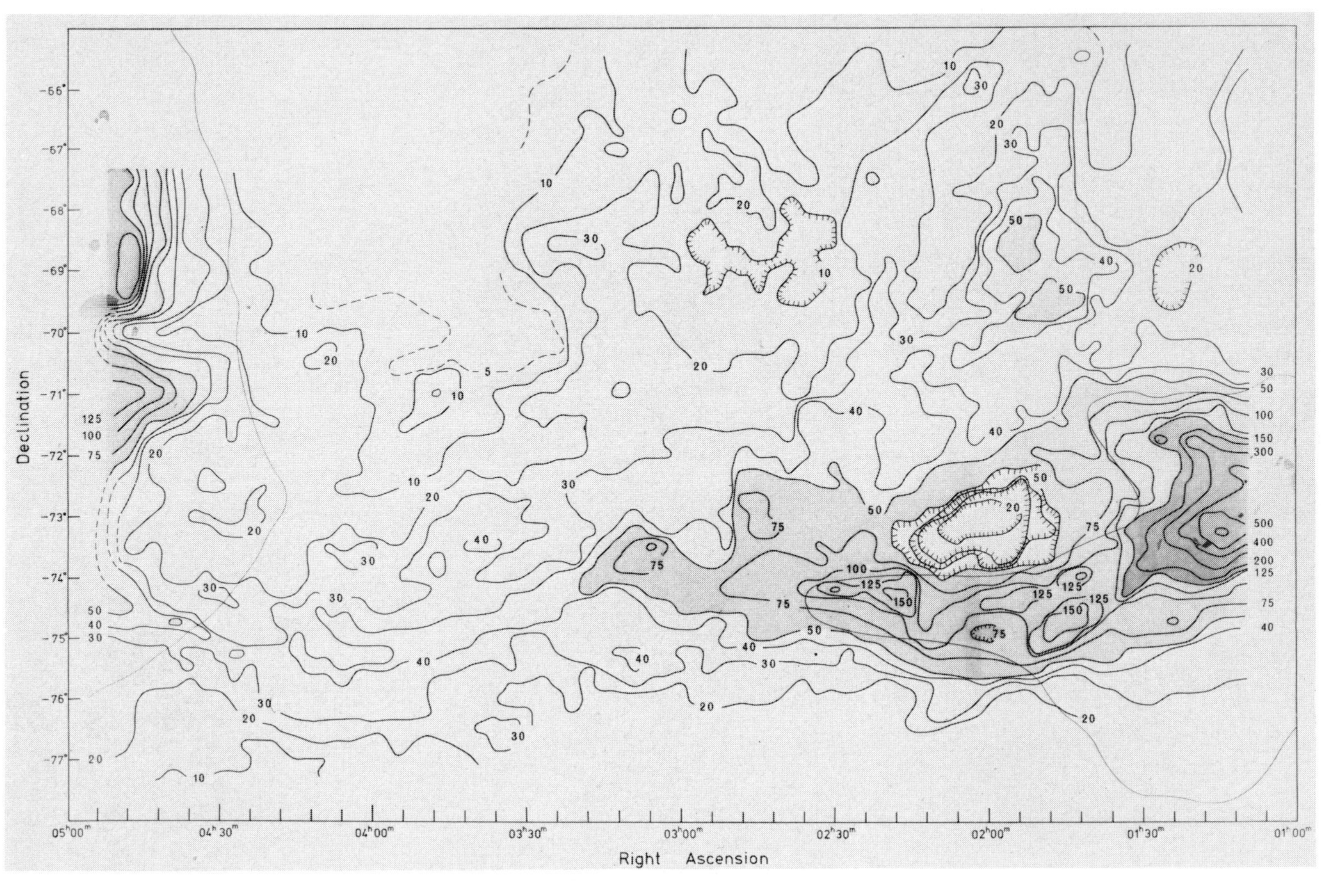

Figure 6. Contours of HI surface densities for the region between the Large and Small Magellanic Clouds obtained using the 64-m reflector at Parkes. The contour unit is 10^{19} atoms cm^{-2}. The shading indicates regions of different contour interval. Co-ordinates are for 1975.0.

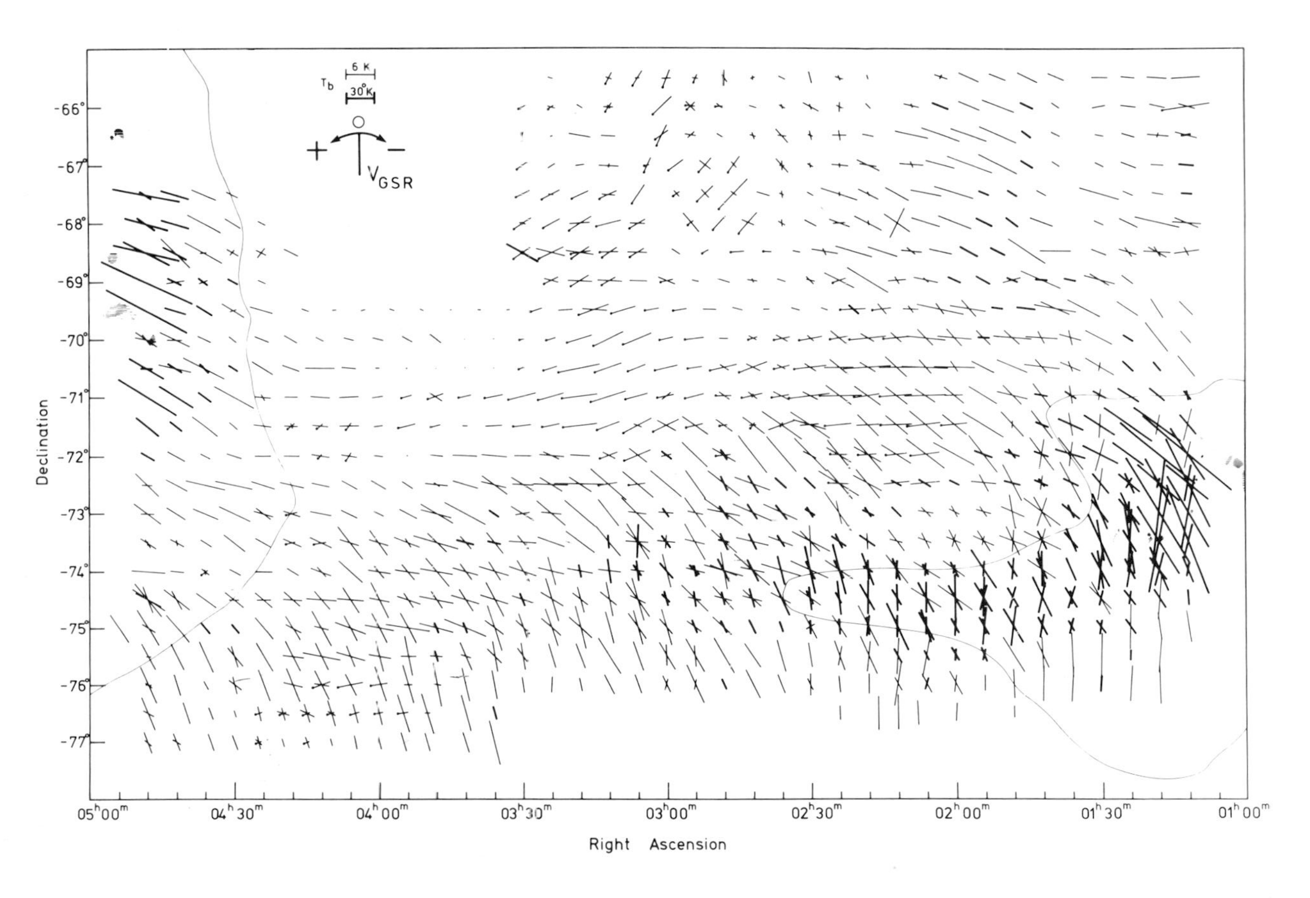

Figure 7. The velocity field of the region between the LMC and SMC obtained using the 64-m reflector. The angle the lines make with the vertical represent the radial velocity V_{GSR} (km s^{-1}) of the peaks in the HI profiles. Anticlockwise rotation indicates + ve velocity. Most velocities are + ve and always <180 km s^{-1}. Negative velocities are confined to a small area in the bottom RH corner. The length of the lines represent T_b ($^{\circ}$K) of the peaks (see scales in top LH corner).

of the inter-Cloud region extremely irregular which is in marked contrast to the straight, steep edge of the HI on the opposite side.
b) The radial velocity contours of the inter-Cloud region connect smoothly with the radial velocity contours of the HI in the LMC and SMC. This, plus the continuity of the general velocity gradient across the whole Magellanic System strongly suggests that the two galaxies form a bound system. This large velocity gradient of 100 km s^{-1} across the inter-Cloud region leads to a large velocity discontinuity of about 80 km s^{-1} between the end of this gas envelope around the Magellanic Clouds and the start of MSI. Indeed in the region around R.A.03^h15^m, Dec.-68°, the radial velocity of the HI is greater than the escape velocity.

REFERENCES

Chiao, R.Y., and Wickramasinghe, N.C.: 1973, Astrophys. Lett. 14, 19.
Hartwick, F.D.A., and Sargent, W.L.W.: 1978, Ap.J. 221, 512.
Hulsbosch, A.N.M.: 1975, Astron. & Astrophys. 40, 1.
Kunkel, W.E., and Demers, S.: 1976, R.Greenwich Obs. Bull. No. 182, 241.
Lynden-Bell, D.: 1976, M.N.R.A.S. 174, 695.
Mathewson, D.S., Cleary, M.N., and Murray, J.D.: 1974, Ap.J. 190, 291.
Mathewson, D.S.: 1976, R. Greenwich Obs. Bull. No. 182, 217.
Mathewson, D.S., and Schwarz, M.P.: 1976, M.N.R.A.S. 176, 47p.
Mathewson, D.S., Schwarz, M.P., and Murray, J.D.: 1977, Ap.J. 217, L5.
Philip, A.G.C.: 1976a, Bull. A.A.S. 8, 352.
Philip, A.G.C.: 1976b, Bull. A.A.S. 8, 532.
Sandage, A.R.: 1976, A.J. 81, 954.
Wannier, P., and Wrixon, G.T.: 1972, Ap.J. 173, L119.
Westerlund, B.E., and Glaspey, J. 1971, Astron. & Astrophys. 10, 1.

DISCUSSION

Basu: You showed that among the Local Group of galaxies the spirals are the most massive ones. In general, spiral galaxies have high masses. This may suggest that only those galaxies which are capable of forming a superdense body at the center, which subsequently undergoes explosions, can appear as spirals. This would not be the case if a galaxy has a mass much less than 10^{10} $M_\odot$. This also suggests that spiral phenomena are associated with explosions at the center.

Felten: In my ignorance of the subject, the following point escaped me: You deduced a distance of 50-60 kpc on the basis that if this distance is correct then the clouds lie on a great circle as viewed from the galactic center. But why should they lie on such a circle?

Matthewson: If they are in orbit about the galactic center, then they will lie on a great circle as viewed from the center.

Verschuur: The uniform velocity patterns within clouds suggest that they are falling in radially.

THE MAGELLANIC STREAM: THEORETICAL CONSIDERATIONS

M. Fujimoto
Department of Physics
Nagoya University, Nagoya

ABSTRACT

The tidal and the primordial theories for the Magellanic Stream are examined in a frame of test-particle simulation for the interacting triple system of the Galaxy, the Large and Small Magellanic Clouds (LMC and SMC). Difficulties of the radial velocity of the Stream still beset these two theories. Several new models for the Stream and the Clouds are briefly discussed in relation to the bending of the galactic disk, the past binary orbits of the LMC and SMC and also the Local Group and the Local Supercluster of galaxies.

1. INTRODUCTION

The "tidal" theory considers that the Magellanic Stream (van Kuilenberg 1972, Wannier and Wrixon 1972, Mathewson et al. 1974) is hydrogen gas which has been pulled out of the Large and Small Magellanic Clouds (LMC and SMC) during their close approach to the Galaxy. This theory is based on a number of test-particle computations which successfully reproduced well-known filamentary bridges and tails in gravitationally interacting galaxies (see Toomre and Toomre 1972 and references therein).

The "primordial" theory has been proposed by Mathewson et al. (1974) and Mathewson and Schwarz (1976) to explain the high negative velocity at the tip of the Magellanic Stream which the tidal model could not reproduce. The Stream is considered as a band of primordial gaseous debris left from the formation of the Clouds and moving along a hyperbolic orbit unbound to the Galaxy.

In the present paper, we examine these two models for the Magellanic Stream on the basis of the test-particle computations, and compare our results with the observed geometry and motion of the Stream.

W. B. Burton (ed.), The Large-Scale Characteristics of the Galaxy, 557–566.

2. TIDAL THEORY OF THE MAGELLANIC STREAM

2.1. Gravitationally-interacting triple system of the Galaxy, LMC and SMC

As preliminaries to the test-particle simulation, Fujimoto and Sofue (1976, 1977) obtained several series of orbits for the LMC and SMC passing the center of the Galaxy at 20, 30, 40 and 50 kpc, and along which the two Clouds were in a binary state for at least the past 5×10^9 years. The binary-orbit condition is based on the presence of the common envelope of diffuse gas and the systematic distribution of optical polarization planes of star-light and HII regions in and around the LMC and SMC (Hindman et al. 1963; Mathewson and Ford 1970; Schmidt 1970). Thereby the binary orbits of the LMC and SMC must satisfy at the present epoch the observed kinematical quantities listed in Table 1.

Table 1. Observed kinematical quantities of the LMC and SMC

	LMC	SMC
(l,b)	(280°,-33°)	(303°,-45°)
Distance from the Sun	52 kpc	63 kpc
Observed radial velocity (corrected for the galactic rotation and the motion of the Sun in the LSR)*	51 km s^{-1}	0
Mass (assumed in the text)	1-$2\times10^{10}M_\odot$	$2\times10^9M_\odot$

* V_θ=250 km s^{-1} and R_0=10 kpc are assumed: V_θ is the rotation velocity of the Sun at R_0 of the galactic center.

As shown in the next subsection, the orbit of the LMC is approximately on a plane perpendicular to the line joining the present position of the Sun and the galactic center. For such overhead orbits of $D\geq20$ kpc, the binary state of the LMC and SMC is guaranteed, at least, for the combinations of the masses of the Galaxy, LMC and SMC given in Table 2 where D is the perigalactocentric distance.

Table 2. Combinations of the masses of the Galaxy, LMC and SMC adopted in the present paper

m_G (Galaxy)	m_L (LMC)	m_S (SMC)
$1.2\times10^{11}M_\odot$	$2\times10^{10}M_\odot$	$2\times10^9M_\odot$
1.4	2	2
2.0	2	2
2.7	2	2
2.7	1	2

Three representative overhead orbits of the LMC are given in Fig.1, seen from the direction of l=180°, b=0°: the Sun is located on the galactic plane at 10 kpc on this side of the figure page. When the SMC's binary motion is taken into account, they are slightly waved. The bars at the present positions of the LMC and SMC, L and S, indicate the size of the tidal limit δr_L and δr_S, outside of which material would be pulled out easily.

When the LMC is at the perigalacticon, δr_L reduces to half of the present value, and the strong tidal disruption would be expected. If $m_G=2\times10^{11}M_\odot$ and D=20 kpc, the scale of the tidal limit becomes as small as $\delta r_L=4.6$ kpc.

We have $\delta r_L>\delta r_S$ for the present separation of the LMC and SMC, and δr_S is determined mostly by the LMC; the contribution from the Galaxy is only 20 percent or less. This is one of the reasons why Toomre (1973) predicted the presence of a long streak of gas emerging from the SMC or the vicinity of the Clouds region (see Mirabel and Turner 1973).

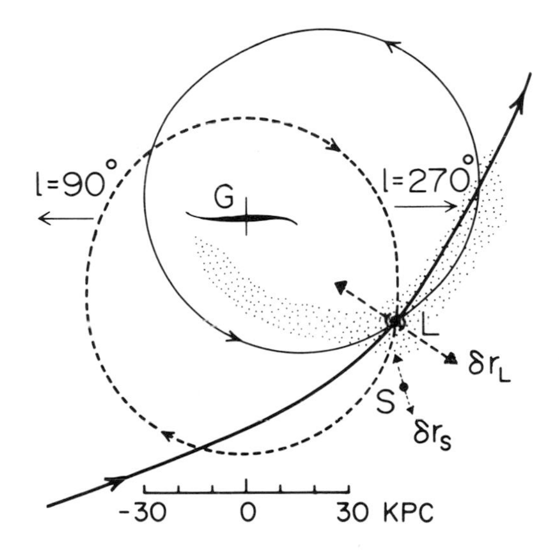

Fig.1. Three typical orbits of the LMC for D=30 kpc and 50 kpc. The direction of motion is indicated. The model Stream by Davies and Wright is given by numerous dots. $m_G=1.2\times10^{11}M_\odot$ and $m_L=2\times10^{10}M_\odot$ are assumed.

2.2. Tidal models for the Magellanic Clouds

About four hundred test-particles are distributed around the centers of the LMC and SMC within the radii of δr_L and δr_S, respectively. Numerical integrations of the motion of the test-particles began when the LMC was at the apogalacticon, and were performed toward the present. Fig.2 shows the post-interaction configuration of four hundred test particles, projected onto the plane of the sky. The binary orbits of the LMC and SMC are overhead and in a counterclockwise sense in Fig.1.

Two streaks—head and tail—are found to emerge from the Clouds region. When D=30 kpc, a good geometrical reproduction is obtained of the high-velocity-cloud (HVC) in the northern hemisphere and some HI gas complexes at l=260° to 330°, b=-20° to 30° on a great circle extrapolated from the Clouds to l=90°, b=-30°, passing near the south galactic pole (Fig.3). The radial velocities averaged over nearby particles are given (in km s^{-1}) relative to the Sun of the non-rotating Galaxy.

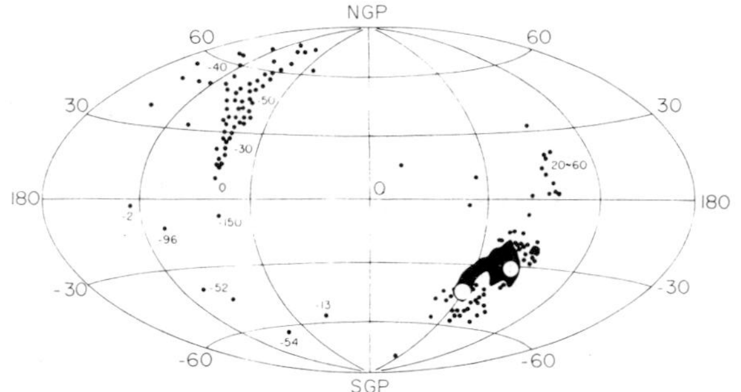

Fig.2. Present distribution of the test particles on the (l,b) plane in the overhead, D=30 kpc orbit of the LMC. Numbers show the radial velocities (km s^{-1}) to be observed at the Sun of the non-rotating Galaxy.

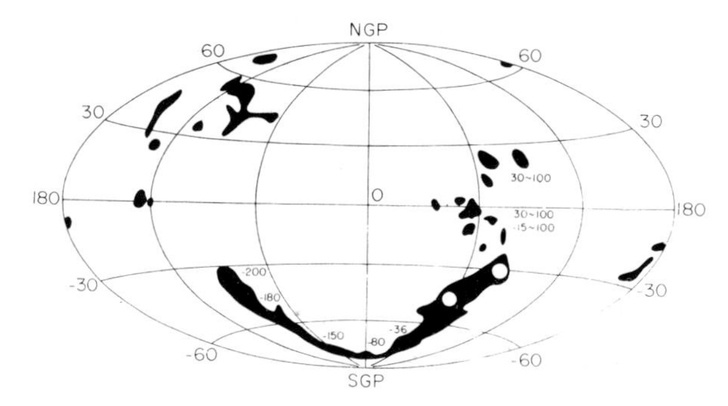

Fig.3. The Magellanic Stream of hydrogen gas, high-velocity HI clouds and some HI gas complexes at l=260° to 330°, b=-20° to 30°, after Mathewson et al. (1974). Numbers refer to the observed radial velocities relative to the Sun of the non-rotating Galaxy (in km s^{-1}).

When D=30 kpc the radial velocities of the particles in the HVC region are sufficiently high to regard the particles as the high -velocity gas clouds. The radial velocities of the test-particles corresponding to the Magellanic Stream are, although the sample number is small, much smaller than the observed values by about 100 km s^{-1}.

This discrepancy cannot be removed only by changing the parameters such as D, m_G and m_L, and it is exactly what Mathewson et al. (1974) stressed in their "primordial" theory of the Magellanic Stream.

If the orbital plane of the LMC is inclined to the galactic plane, the pulled-out particles do not lie on a great circle defined by the Magellanic Stream. This is the reason why the orbital plane of the LMC is considered as approximately perpendicular to the line joining the Sun and the galactic center.

Davies and Wright (1977) have made similar test-particle simulations for the LMC orbits circling in a clockwise and overhead sense seen from the Sun (see Fig.1). The leading and trailing streaks of the particles emerge from the LMC when it reaches the perigalacticon (Lynden-Bell 1976). The leading part, which is on the side towards the Galaxy, becomes a bridge between two galaxies and its tip is captured to the Galaxy when the LMC approaches the present position. Davies and Wright claimed that the high-negative velocities can be produced at the northern tip of the Stream.

As shown schematically in Fig.1, the narrow bridge of the test -particles seems to be clear and realistic seen from outside of the Galaxy-LMC system. However, when we plot the positions of these particles on the plane of the sky, they cover a considerable area of the sky because of the very large parallaxes involved for the particles which fall down so close as to hit the galactic plane in the solar neighborhood. If we choose a narrower and more likely Stream from their results, the radial velocity is much smaller than the observed values by 100 km s^{-1}.

3. PRIMORDIAL-GAS MODELS FOR THE MAGELLANIC STREAM

We examine in our scheme of test-particle simulation the behavior of primordial gaseous debris left from the condensation to the Clouds and moving along a hyperbolic orbit unbound to the Galaxy, with a special regard to the high-negative-velocity difficulties besetting the tidal theory.

3.1. A line of test-particles

Following the primordial models of the Mathewson et al. (1974), we placed two hundred particles in a line on the same hyperbolic orbit of the LMC, with the LMC and SMC at the extreme end near the Galaxy. The LMC, SMC and each particle were given velocities sufficient to move along the orbit whose perigalactocentric distance is 50 kpc and whose plane is perpendicular to the line joining the Sun and the galactic center.

The integration commenced when the LMC was at 550 kpc distant from the Galaxy. The particle distribution at the present epoch is shown projected onto the plane of the sky in Fig.4. From a comparison with Fig.3, we find a good reproduction of the linear distribution of gas and a good fit to the high negative velocity at the tip of the Stream. Moreover, if such initial linear distribution of gas is realistic, the hydrogen gas clouds at 270° to 330°, b=-20° to 30° could be regarded as primordial gas; they overtook and passed the Magellanic Clouds from behind, and now in a disordered state. The radial-velocity distribution of these particles is rather dispersed, 70 to 100 km s^{-1}, which is not inconsistent with observations.

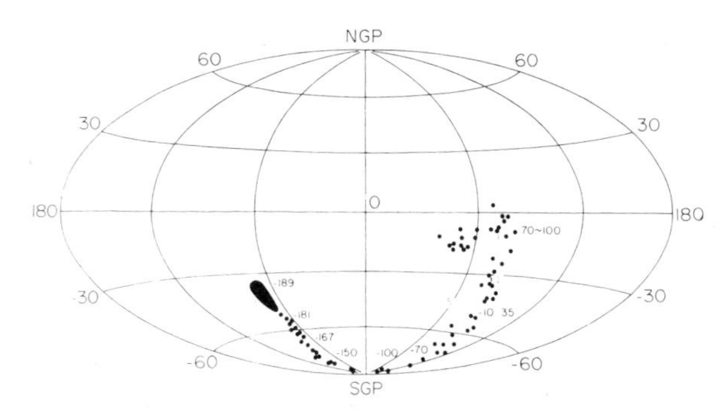

Fig.4 Present distribution of test particles for the initial linear case.

3.2. A spherical cloud of test-particles

Since the above initial configuration is too specific, Fujimoto and Sofue (1976, 1977) considered an elongation of a spherical gas cloud due to a drag force of intergalactic gas: the primordial gas cloud would be decelerated relative to the LMC and SMC and be stretched to become a narrow band of gas.

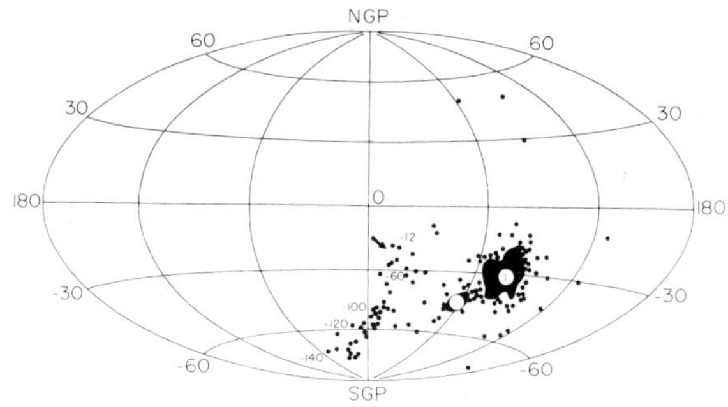

Fig.5. Present distribution of the test particles on the sky. $1/k=2\times10^{10}$ years is assumed. Note a hump in the particle distribution (arrow). The particles do not fall on the Magellanic Stream region adjacent to the south galactic pole.

We made similar computations for a sphere of two hundred particles, having the LMC and SMC at its center and moving on the hyperbolic orbit in Fig.1. A drag force is assumed on each particle in the form $-k\mathbf{V}$. Figs.5 and 6 show the post-interaction configurations at the present epoch for $1/k=2\times10^{10}$ and 2×10^{9} years, or 10^{-5} and 10^{-4} hydrogen atoms per cm^3 of intergalactic gas, respectively. A trailing structure is naturally produced.

From a comparison with the Stream in Fig.3, however, we find some unsatisfactory results such as a gap and hump in the particle distribution and its large deviation from a great circle defined by the Stream. Since such behavior of the particles is intrinsically associated with our scheme of computation of particles in the gravitational potential and the drag force $-k\underset{\sim}{V}$, we could not remove them from the primordial model of the Stream.

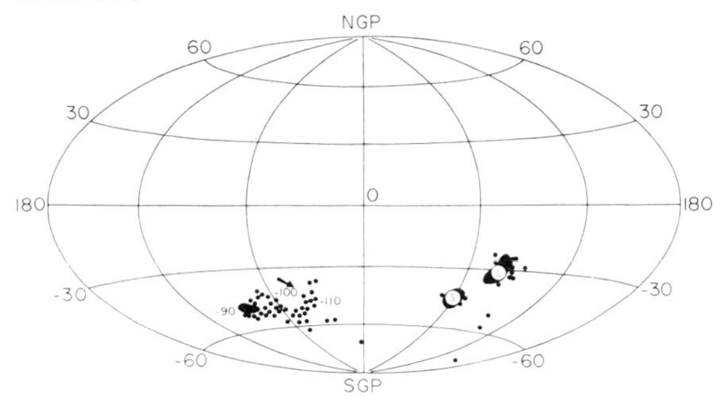

Fig.6 Same as Fig.5, but $1/k=2\times10^{9}$ years

4. PROBLEMS RELATED TO THE MAGELLANIC STREAM

The tidal and primordial models for the Magellanic Stream have been examined in sections 2 and 3 for the masses of the Galaxy, LMC and SMC assumed in Table 1. Parallel with this computation, Fujimoto and Sofue (1976, 1977) obtained the bending of our galactic disk for various binary orbits of the Clouds, and concluded that the bending is reproduced when D=20 to 30 kpc, $m_G\lesssim2.0\times10^{11}M_\odot$ and $m_L=2$ to $3\times10^{10}M_\odot$. For the theoretical analysis of the bending, see Hunter and Toomre (1969) and references therein, and Spight and Crayzeck (1977). In other words, if the bending of our galactic plane is due to the tidal force, the total mass of the Galaxy within the radius of 50 kpc is not so large as suggested by Ostriker and Peebles (1973). We stress that the current controversy concerning the mass of our Galaxy is but part of a much larger problem involving the mass of the LMC, the mechanism of the bending of the galactic disk and also dynamical state of the Magellanic Clouds —whether they are a steady binary circling round the Galaxy or are accidental passers-by on hyperbolic orbits.

Mathewson et al. (1977) considered that the primordial models in section 3 is not enough to explain the geometry and high negative velocities of the Stream, and proposed a new theory for the origin of the Stream in which the observed six discrete gas clouds were formed from thermal instability arising behind the Magellanic Clouds passing through the hot halo of our Galaxy (see Mathewson at the present symposium). The high negative velocities are considered to be achieved from the work done by the gravitational potential of the Galaxy. When $m_G=2.8\times10^{11}M_\odot$, they are reproduced for only 44 kpc of the closest distance of approach to the galactic center. The past binary state of the Clouds are guaranteed, but the origin of the bending is left undiscussed.

Tremaine (1976) has shown that the present proximity of the Clouds can be explained by the decay of the orbit due to dynamical friction in an extended massive halo of our Galaxy. Since he could take 40 and 190 kpc as the peri- and apogalactocentric distance of 10^{10} years ago, the initial state of the triple system of the Galaxy, LMC and SMC seems to be related to the Local Group and perhaps to the Local Supercluster of galaxies (de Vaucouleurs and Corwin 1975). The total mass of the Galaxy is assumed as 5 to $7\times10^{11}M_\odot$, within 50 kpc of the galactic center, which is much greater than that we have adopted so far, and which makes it difficult to maintain the Clouds in a gravitationally bound state for the last 5×10^9 years.

In various attempts to understand the Magellanic Stream, many models of the Magellanic Clouds have been introduced. It becomes more difficult to explain the Stream in detail in a simple way such as our test-particle computations. We consider that these difficulties must be resolved not by examining only one or two aspects separately, but by synthesizing the whole situations that would be associated with the Galaxy-LMC-SMC system. At the same time we must search for other possible mechanism which may produce the galactic bending structure.

REFERENCES

Davies, R.D. and Wright, A.E.: 1977, Mon. Not. R. astr. Soc. 180, pp.71-88.
de Vaucouleurs, G. and Corwin, H.G.: 1975, Astrophys. J. 202, pp.327-334.
Hindman, J.V., Kerr, F.J., and McGee, R.X.: 1963, Australian J. Phys. 16, pp.570-883.
Hunter, C. and Toomre, A.: 1969, Astrophys. J. 155, pp.747-776.
Fujimoto, M. and Sofue, Y.: 1966, Astron. Astrophys. 47, pp.263-291.
Fujimoto, M. and Sofue, Y.: 1977, Astron. Astrophys. 61, pp.199-215.
Lynden-Bell, D.: 1976, Mon. Not. R. astr. Soc. 174, pp.695-710.
Mathewson, D.S. and Ford, V.L.: 1970, Astrophys. J. (Letters) 160, pp.43-46.
Mathewson, D.S., Cleary, U.N., and Murray, J.C.: 1974, Astrophys. J. 190, pp.291-296.
Mathewson, D.S. and Schwarz, M.P.: 1976, Mon. Not. R. astr. Soc. 176, 47P-51P.
Mathewson, D.S., Schwarz, M.P., and Murray, J.D.: 1977, Astrophys. J. (Letters) 217, pp.5-8.
Mirabel, I.E. and Turner, K.C.: 1973, Astron. Astrophys. 22, pp.437-440.
Ostriker, J.P. and Peebles, P.J.E.: 1973, Astrophys. J. 186, pp.467-480.
Schmidt, Th.: 1970, Astron. Astrophys. 6, pp.294-308.
Spight, L. and Crayzeck, E.: 1977, Astrophys. J. 213, pp.374-378.
Toomre, A. and Toomre, J.: 1972, Astrophys. J. 178, pp.623-666.
Toomre, A.: 1973, in Mirabel and Turner (1973) given above.
Tremaine, S.D.: 1976, Astrophys. J. 203, pp.72-74.
van Kuilenberg, J.: 1972, Astron. Astrophys. 16, pp.276-281.
Wannier, P. and Wrixon, G.T.: 1972, Astrophys. J. (Letter) 173, pp.119-123.

DISCUSSION

Giovanelli: What happens with your models if you take a mass on the order of 10^{12} $M_\odot$ for the Galaxy instead of $1\text{-}2\times10^{11}$ $M_\odot$ as you used?

van Woerden: Are stability of the Magellanic Cloud pair and the formation of a warp in our Galaxy also excluded if 10^{12} $M_\odot$ would be distributed over a volume 60 kpc in radius?

Fujimoto: Yes. More exactly, if the mass within 30 to 50 kpc of the galactic center exceeds 3×10^{11} $M_\odot$, the overhead binary orbits of the Clouds are easily disrupted.

Mathewson: I believe that you have touched too lightly on the Turbulent Wake Theory of the Magellanic Stream. It explains rather nicely: (1) the near great circle of the Stream which suggests that it is at least at a distance of 50 kpc. There is then no difficulty in keeping the Magellanic Clouds bound to each other; (2) the looped structure of the six clouds which make up the Stream; (3) the extremely constant velocity field of each Stream cloud which places rigid constraints on the translational velocity of the clouds and suggests that their motion is purely radial; (4) the fact that the Stream clouds have all blue shifts; (5) the ease with which one can explain the high negative velocity of MS VI at the tip of the Stream; (6) the direction of motion of the Magellanic Clouds implied by the steep HI gradients on one side of the gas envelope containing the Magellanic Clouds; (7) the large velocity discontinuity between the nearest part of the gas envelope containing the Magellanic Clouds to the start of the Stream and MSI.

The main difficulties with the Turbulent Wake Theory are: (1) that it needs a hot gaseous halo around our Galaxy which we are not sure exists and (2) the question of whether or not the passage of the Magellanic Clouds will produce the thermal instabilities in the halo necessary to condense out the cold clouds.

I would like to point out that if this is the correct explanation of the Magellanic Stream then HI clouds of $10^7\text{-}10^8$ $M_\odot$ will fall into the galactic center at intervals separated by about 10^8 years.

Felten: If I understood Dr. Fujimoto's presentation, the great-circle geometry of the Magellanic Stream arises provided the clouds in the stream all originate in the orbital plane of the Magellanic Clouds. But I believe this geometry would be destroyed if the clouds were to have appreciable random velocities. Dr. Toomre's skepticism about the tidal models moves me to ask whether this planar geometry is telling us which models are correct. Would we necessarily get this geometry in non-tidal models?

Haynes: This geometry is obtained, for example, in the non-tidal, galactic-wake model of Mathewson.

Lynden-Bell: It is possible to compute away and to fit the lie of the Magellanic Stream on the sky exactly. All tidal models produce streams roughly in a plane and the parallax due to the Sun's offset is not too large. Thus rough great circles arise from all theories in which the clouds move in a planar orbit that does not come closer to the Galaxy than say 20 kpc. The fact that the stream is on a large small circle is not really very surprising.

Kerr: It is important for people who speak of a Galactic Standard of Rest (GSR) or of "velocities corrected to the center" to define quite carefully what they mean. Several systems are in use now, and others are possibly in the future.

van Woerden: Magellanic-Stream and intergalactic-hydrogen workers may be using various formulae to correct velocities to the galactic standard of rest. Most extragalactic observers use a velocity vector of 300 km s^{-1} towards $\ell = 90°$, $\ell = 0°$ (even if this is not the correct motion of the Sun with respect to the galactic center) to refer their observations to an extragalactic standard of rest, intended to be (the center of gravity of?) "the Local Group".

van Woerden: Do I remember correctly that three years ago there were reports about stars in the Magellanic Stream clouds?

Humphreys: I am curious to know the magnitudes, colors, and spectral type of the stellar objects you have identified in the gas between the clouds. Finding charts would be very useful.

Mathewson: I would like to point out how tricky it is to define an intergalactic gas cloud. For example, Murray and I found HI emission in the Centaurus Group a few degrees to the north of NGC 5236 with a velocity characteristic of the Group. The Palomar Sky Atlas showed no optical emission at this position. However, a deep SRC Schmidt plate showed a faint smudge of emission which Louise Webster, using the AAT, showed to be an HII region. It then becomes a matter of definition whether you call this an intergalactic gas cloud or a low surface brightness galaxy.

While Toomre's tidal theories have had outstanding success in explaining many astrophysical phenomena, one should be careful in using the tidal theory to explain all phenomena. Rather than going into a long-winded discussion as to why I do not believe Magellanic Stream has been produced by tidal forces, I would just like to point out that Alar Toomre has never attempted to model the Stream.

INTERGALACTIC HI AND TIDAL DEBRIS WITHIN GROUPS OF GALAXIES

M. P. Haynes
National Radio Astronomy Observatory* and Indiana University

Abstract: A population of isolated intergalactic HI clouds has yet to be recognized in the intergalactic medium. On-going high sensitivity studies of loose groups of galaxies, however, show that tidal debris in the form of HI streams is commonly found in such aggregates.

The possible existence of neutral atomic hydrogen clouds within the intergalactic medium has been the subject of a variety of experiments in recent years. An impetus for investigating the occurrence of such clouds stems from a desire to understand the intergalactic environment and its bearing on the large-scale structure of galaxies. Intergalactic neutral hydrogen clouds, as remnants of an inefficient galaxy formation process, would serve as probes of a more pervasive intergalactic medium in which they are immersed (Cowie and McKee 1975). They could also be responsible for the disruption of neighboring galaxies (Freeman and de Vaucouleurs 1974). But the very existence of such HI clouds is questionable.

The inability to detect a homogeneously distributed neutral hydrogen medium has set stringent upper limits on the contribution of HI to the critical density (Penzias and Scott 1968). Likewise, searches for HI gas within rich clusters of galaxies have failed (Haynes, Brown and Roberts 1978). Shostak (1977) and Roberts and Steigerwald (1977) have shown that the density of HI clouds is low. Despite all of these negative results, three hypotheses have stimulated the search for intergalactic clouds:

a) The explanation of the high velocity clouds as extragalactic clouds within the Local Group (Verschuur 1969; Kerr and Sullivan 1969).

b) The suggestion by Freeman and de Vaucouleurs (1974) that ring galaxies have been formed by the collision of a disk galaxy with an intergalactic cloud.

c) The interpretation of HI clouds in the vicinity of NGC 55 and NGC 300 as HI companions within the Sculptor Group (Mathewson, Cleary, and Murray 1975).

*Operated by Associated Universities, Inc. under contract with the National Science Foundation.

W. B. Burton (ed.), The Large-Scale Characteristics of the Galaxy, 567–573.

New high sensitivity searches have failed to uncover a population of isolated, intergalactic HI clouds but have demonstrated a frequent presence of HI tidal debris.

ISOLATED INTERGALACTIC CLOUDS ?

HI companions have been attributed to the Sculptor Group which itself is unique in both being the closest to the Local Group and in possessing the highest population index of all the de Vaucouleurs' (1975) groups because of its composition of late-type galaxies. It is reasonable to expect that similar intergalactic HI clouds might be found near late-type galaxies in other groups. Searches for such clouds in other groups have been made using various instruments. Specifically, Haynes and Roberts (1978) conducted a study of numerous nearby groups using the 140-foot telescope of the N.R.A.O. for intergalactic clouds of masses $M_H \gtrsim 10^8$ $M_\odot$ and sizes of 30-50 kpc, the parameters of the Sculptor clouds; no such isolated clouds were found. In light of the negative results of the several searches undertaken by different observers, suspicion was raised that the HI clouds in the Sculptor area might not, in fact, be associated with the Sculptor Group.

The principal clues in the intergalactic interpretation of clouds in the Sculptor region are (1) their proximity to the two galaxies, (2) the similarity of their velocities with those of the two galaxies, and (3) their separation in velocity from the clouds of the Magellanic Stream which runs through the region. The similarity in velocity of features associated with the Galaxy, the Magellanic Stream, and the Sculptor Group prevents any meaningful differentiation by velocity in this area. Haynes and Roberts (1978) obtained 1200 line profiles in the region of Sculptor and to the south as far as $\delta = -46^o$. The major result of this survey is the discovery of a large number of additional clouds at both positive and negative velocities with respect to the local emission and that traditionally associated with the main ridge of the Magellanic Stream. This larger population of HI clouds in the Sculptor region raises considerable doubt from both a spatial and a spectral standpoint as to their interpretation as bona fide members of the Sculptor Group. The locus of HI clouds is confined to the southern sector of the Group and extends far to the east and south to a radius of 0.7 Mpc from the nearest Sculptor galaxy. In contrast, no clouds are found near other principal members of the Group, NGC 45, 247 or 253. The velocity distribution of the clouds does not mimic that of Sculptor galaxies for any likely membership, while the negative velocity clouds in the region could certainly not be interpreted to lie within the Sculptor Group. The velocity distribution could only be reconciled if the positive velocity clouds were contained within a subgroup defined only by NGC 55 and 300 which still had a radius of 0.7 Mpc; we find this interpretation unattractive. The lack of fine structure observed within these clouds either spatially or spectrally distinguishes them from typical high velocity clouds in the northern galactic hemisphere, including those studied with the same instument (Giovanelli and Haynes 1976).

Although Sculptor is distinctive in possessing only late-type galaxies, it is unique in being the only nearby group which is seen in projection with the Magellanic Stream. Indeed, in this region near the south galactic pole, the Stream emits strongly in competition with the zero velocity gas. Because of the difficulties in associating the HI clouds with Sculptor and because the Stream seems to possess a substantial amount of spatial structure not revealed by previous surveys, we see no a priori reason why the HI clouds in the region, including the ones near NGC 55 and 300, cannot by included in the Magellanic Stream phenomenon. In this picture, the Stream is defined both by the bright sharp ridge as presented by Mathewson, Schwartz and Murray (1977), but also by a collection of smaller clouds, separated in velocity as well as sometimes in space, from the main body (Haynes 1978). The HI clouds then are still extragalactic, but are contained within the Local Group, and hence are much smaller, less massive, and not isolated.

HYDROGEN STREAMS IN GROUPS ?

Within the Local Group, we see two instances of hydrogen streams which may be interpreted as the tidal debris of encounters between the neighboring galaxies, namely, the bridge between the Magellanic Clouds (Hindman, Kerr and McGee 1963) and the Magellanic Stream (Mathewson, Cleary and Murray 1974). Historically, several other systems have been recognized to typify tidal interactions delineated by wide hydrogen distributions: the M81/M82/NGC3077 system (Roberts 1972) and the NGC4631/56 system (Roberts 1968).

The presence of these hydrogen streams in the Local Group and in other nearby groups stimulated an investigation of other pairs and groups for similar hydrogen appendages. These searches are being carried out with the 300-foot telescope of the N.R.A.O. and the 1000-foot telescope of the Arecibo Observatory*. The high sensitivity of the Arecibo instrument makes it ideal for the study of low emissivity hydrogen in relatively small groups, while the 300-foot was used for larger groups and those outside the range of Arecibo. Table I summarizes the presently available data on known hydrogen streams in all nearby groups which have been surveyed by ourselves and by others. Columns (1) and (2) give the group identification either from de Vaucouleurs (1975) or after the brightest galaxy in the group. Our surveys both at Green Bank and Arecibo have concentrated on small aggregates of galaxies selected from the Second Reference Catalog of Bright Galaxies. The sample is restricted to galaxies which lie within 250 kpc of one another and where the velocity difference is less than 500 km/s. Many of these smaller aggregates appear as subconcentrations within the more general de Vaucouleurs' groups; hence the entire field of a group may not have yet been sampled. Column (3) indicates the status of present observations: does the aggregate contain a <u>known</u> hydrogen stream which suggests past tidal interaction?

* Part of the National Astronomy and Ionosphere Center, operated by Cornell University under contract with the National Science Foundation.

TABLE I

Summary of Present HI Evidence for Tidal Interaction

Group # (1)	Common Name (2)	Known Tidal Appearing Stream ? (3)	Types of Galaxies in Interaction (4)	N_H^* ($\times 10^{20} cm^{-2}$) (5)	Reference (6)
	Local Group	Yes A	Ir, Sm	4	a
		B	Sbc, Ir, Sm	3	b
G1	Sculptor	No			
G2	M81	Yes	Sab, I0, I0	5	c
G3	CVn I	Maybe	Im, Scd	0.4	d
G5	M101	Yes+	Sbc, I0	1.5	e
G6	N2841	No			
G7	N1023	Maybe+	Scd, Sm	0.3	f
G9	M66	Yes	Sb, Sb	0.9	g
G10	CVn II	Yes	Sd, Sm	1	h
G11	M96	No			
G13	Coma I	Yes	Sab, Scd	0.6	f
G29	Virgo III	No			
G30	N5866	No			
G40	N488	No			
G42	N2964	No			
G47	N3190	No			
G50	N5846	No			
	N697	Yes	Sbc, Sc	0.3	f
	N2775	Yes	Sab, Sab	0.2	f
	N3166	Yes	Sa, S0/a	0.6	f
	N4038/9	Yes	Sm, Sm	2.3	c
	N5044	No			
	N5363	Maybe	Sbc, I0	0.3	f
	N5576	Maybe	E0, Sbc	0.7	f
	N5953	No			
	N7332	No			
	N7448	Yes	Sbc, ?	0.5	f
	N7541	Maybe	Sbc, Sbc	0.5	f

* beam-averaged peak column density within the limits of current data.
\+ galaxies involved are only "possible" members of group.

a Mathewson, D.S., Cleary, M.N., Murray, J.D. (1974).
b Mathewson, D.S., Schwartz, M.P., Murray, J.D. (1977).
c van der Hulst, J.M. (1977).
d Haynes, M.P., Roberts, M.S., Green Bank survey.
e Haynes, M.P., Giovanelli, R., Burkhead, M.S. (1978).
f Haynes, M.P., Roberts, M.S., Arecibo survey.
g Haynes, M.P., Giovanelli, R., Roberts, M.S. (1978).
h Weliachew, L., Sancisi, R., Guelin, M. (1978).

This list is not meant to be complete. Since not all of the galaxies within an aggregate have been studied thus far, the possibility of additional streams within these groupings is not ruled out. However, the frequent occurrence of tidal debris in the form of HI is clearly demonstrated. Column (4) lists the morphological classification of the galaxies apparently involved in the interaction. Not unexpectedly, we find that the gas has been drawn out from spiral galaxies. To date, we have preferentially studied late-type galaxies, and hence our sample is biased. We intend to extend this investigation to earlier systems. Column (5) lists the beam-averaged peak column density of the hydrogen appendages derived from the observations referenced in column (6); in some cases, the observations are incomplete, so that the peak column densities may be greater than that given.

We have already implied that tidal interactions are the likely cause of the extended hydrogen distributions. In their classic 1972 paper on the formation of galactic bridges and tails, Toomre and Toomre have shown the power and elegance of computer simulations of galaxy encounters. With the availibility of 21 cm data, their techniques have been applied to model additional interactions with impressive success in reproducing the overall characteristics of the HI distribution, although in detail the model parameters must be strained to fit the observations, as in the Leo Triplet NGC3623/7/8 (Haynes, Giovanelli and Roberts 1978). The computer simulations in Leo, in the NGC4038/9 system (van der Hulst 1977) and in the NGC4631/56 system (Combes 1978) strongly reinforce the hypothesis that tidal encounters can draw material far out from the disks of the galaxies involved. In contrast, the hydrogen distribution within the M 51 system defies a clear-cut interpretation (Haynes, Giovanelli and Burkhead 1978), although the optical features are understood in terms of the eccentric passage of NGC 5194 by NGC 5195 as presented by Toomre and Toomre (1972).

SUMMARY

We see in the Magellanic Stream the best example of intergalactic hydrogen, and while some of its characteristics elude simulation by tidal modelling (Lin and Lynden-Bell 1977), it is perhaps best understood as being the tidal relic of the recent passage of the Galaxy by the Magellanic Clouds. At least 10 other examples of hydrogen streams within aggregates and groups are recognized at present. The existence of extensive hydrogen distributions in groups may contribute to the evolution of galaxies within such aggregates, either in the accretion of such gas by a passing galaxy, or conversely, by the removal of gas from spirals involved in the interactions (Binney and Silk 1978). It should be noted that the amount of gas involved may be significant; the combined hydrogen mass of the appendages of NGC 3628 amounts to almost 20% of that still observed within the galaxy itself. While isolated intergalactic HI clouds of significant mass and size are unlikely to be found in the intergalactic medium, hydrogen streams such as we find in our own Galaxy's backyard tell a common tale of tidal interaction in nearby loose groups.

It is a pleasure to thank Morton S. Roberts for his continuing insight and assistance during the course of these investigations. Partial financial support has been provided by the Joseph Swain Fellowship of Indiana University.

REFERENCES

Binney, J., Silk, J.: 1978, Comments on Astrophys. 7, pp.139-49.
Combes, F.: 1978, Astron. Astrophys. 65, pp.47-55.
Cowie, L.L., McKee, C.J.: 1975, Astrophys. J. Lett. 209, pp.L105-9.
de Vaucouleurs, G.: 1975, in "Galaxies and the Universe", eds. A. Sandage, M. Sandage, J. Kristian (Univ. of Chicago Press), pp.557-600.
Freeman, K.C., de Vaucouleurs, G.: 1974, Astrophys. J. 194, pp.569-85.
Giovanelli, R., Haynes, M.P.: 1976, Mon. Not. R. astr. Soc. 177, pp.525-30.
Haynes, M.P.: 1978, preprint.
Haynes, M.P., Brown, R.L., Roberts, M.S.:1978, Astrophys. J. 221,pp.414-21.
Haynes, M.P., Roberts, M.S.:1978, preprint.
Haynes, M.P., Giovanelli, R., Burkhead, M.S.:1978, Astron. J.(to appear).
Haynes, M.P., Giovanelli, R., Roberts, M.S.:1978, preprint.
Hindman, J.V., Kerr, F.J., McGee, R.X.:1963, Aust. J. Phys.16, pp.570-83.
Kerr, F.J., Sullivan, W.T.:1969, Astrophys. J. 158, pp.115-22.
Lin, D.N.C.,Lynden-Bell, D.:1977, Mon. Not. R. astr. Soc. 181, pp.59-81.
Mathewson, D.S.,Cleary, M.N.,Murray, J.D.:1974, Astrophys. J. 190,pp.291-6.
Mathewson, D.S.,Cleary, M.N.,Murray, J.D.:1975, Astrophys. J. Lett. 195, pp.L97-100.
Mathewson, D.S.,Schwartz, M.P.,Murray, J.D.:1977, Astrophys. J. Lett. 217, pp.L5-8.
Penzias, A.A., Scott, E.H.:1968, Astrophys. J. Lett. 153, pp.L7-9.
Roberts, M.S.:1968, Astrophys. J. 151, pp.117-31.
Roberts, M.S.:1972, in "External Galaxies and Quasi-Stellar Objects", ed. D.S. Evans, I.A.U. Symp. #38, pp.12-36.
Roberts, M.S., Steigerwald, D.G.:1977, Astrophys. J. 217, pp.883-91.
Shostak, G.S.:1977, Astron. Astrophys. 54, pp.919-24.
Toomre, A., Toomre, J.:1972, Astrophys. J. 178, pp.623-66.
van der Hulst, J.M.:1977, thesis, Groningen.
Verschuur, G.L.:1969, Astrophys. J. 156, pp.771-7.
Weliachew, L., Sancisi, R., Guelin, M.:1978, Astron. Astrophys. 65, pp. 37-45.

DISCUSSION

Mathewson: The only criteria by which we may associate an HI cloud with a galaxy is spatial and velocity coincidence, and the clouds near NGC 300 and 55 satisfy these. I also notice that the clouds near these galaxies are more intense than the clouds you find further away which strengthens my original belief that they are associated with the galaxies. The weaker and lower velocity clouds which you find may be lying between the Local Group and the Sculptor Group. I do not agree with your conclusion that the clouds near NGC 300 and NGC 55 are associated with the Magellanic Stream. You disregard one of the main criteria (the coincidence in velocity), as the Stream in the direction of NGC 300 and 55 has a velocity of over 100 km s^{-1} different from that of HI clouds.

Haynes: With regard to the region near NGC 55; we do not identify some of the features observed very close to the galaxy as separate clouds as pictured by Mathewson, Cleary, and Murray. They have a smaller beam but we have higher sensitivity and the features seem to be spread at weak emission over a much broader velocity range. On the other hand, some of the clouds more distant from the galaxy with velocities $V_{GSR} \sim +80$ to +100 km s^{-1} coincide with anomalous negative velocity clouds, $V_{GSR} \sim -80$ km s^{-1}, whereas the Magellanic Stream in this region has a velocity $V_{GSR} \sim 0$ km s^{-1}. I do not believe anyone would want to associate the negative velocity clouds with NGC 55. Our arguments that the spatial and velocity distributions of the clouds and galaxies do not match are based on the total anomalous-velocity cloud population. One cannot consider only some of the clouds while simultaneously ignoring others.

van Woerden: The case for tides between galaxies seems to be weakening this morning. I wish to record that a convincing case for tidal effects has been made by van der Hulst (Thesis, Groningen, 1977), who showed that Westerbork observations of both density distribution and velocity field in the "Antennae" (NGC 4038/39) are fit well by a Toomre-Toomre model.

Thonnard: Was Toomre's tidal-interaction model for the NGC 3623, 27 and 28 group able to reproduce the very regular HI velocity field that you see in the plume of NGC 3628?

Haynes: Toomre's model is able to give a good fit but not a great fit. Along the plume of N3628, we see that the radial velocity does not vary by more than about 10 km s^{-1} over 50 kpc; at the far tip, a separate component arises abruptly with a velocity difference of 35 km s^{-1} from the other component. Such structure is not seen in Toomre's model. The overall characteristics of the narrow tail and patchy bridge are well simulated by the model, but the model parameters must be strained somewhat to fit the small-scale features of the observations.

Tinsley: Toomre remarked that maybe a lot of intergalactic gas may be not of tidal origin but rather leftover gas from inefficient galaxy formation. Perhaps it is relevant that the amount of gas inferred from x-ray models of the Coma cluster is several times more than can be plausibly accounted for by loss from galaxies by any one of the several processes that have been proposed in the literature (winds from ellipticals, later loss from stars, sweeping spirals to make SO's, etc.). Although the result is not firm because of model-dependent uncertainties in the amount of hot gas outside the observed core of the cluster (Ostriker, private communication), the discrepancy suggests to me that most of the intergalactic gas is left over from galaxy formation.

THE MAGELLANIC STREAM AS A PROBE OF THE GALACTIC HALO

R. J. Cohen, R. D. Davies and I. F. Mirabel*
University of Manchester, Nuffield Radio Astronomy
Laboratories, Jodrell Bank, Macclesfield, Cheshire SK11 9DL
England

Recent observations of the Magellanic Stream can be used to set limits on a possible hot halo surrounding the Galaxy. The observations are described in detail elsewhere (Mirabel, Cohen & Davies, submitted to Mon.Not.R.astr.Soc.). Briefly, the neutral hydrogen in the northern end of the Magellanic Stream is concentrated in narrow filaments which contain small elongated clouds of typical size $0\overset{\circ}{.}4 \times 0\overset{\circ}{.}6$. These clouds have a large velocity halfpower width (25 km s^{-1}) and are gravitationally unstable, unless there is a massive low luminosity stellar component. If we consider only the observed gas the expansion age of a typical cloud is 6×10^5 D years, where D is the distance in kpc from the Sun, and this falls at least a factor of ten short of the age of the Stream predicted by current models. This strongly suggests that some containment mechanism is operating.

We have considered the possibility that containment is provided by a hot gas. A gas pressure of $nT \sim 2 \times 10^4 D^{-1}$ K cm^{-3} is required, where n is the halo density and T the temperature. For the halo to support itself against the gravitational field of the Galaxy a temperature of $\sim 10^6$K is necessary, so for a distance D = 50 kpc a halo density of $n \sim 4 \times 10^{-4}$ cm^{-3} would be required to contain the clouds. If the Magellanic Stream were moving through such a halo at a relative velocity of ~ 100 km s^{-1} then ram pressure effects should lead to momentum transfer, and hence produce differential motion between the dense and the faint parts of the Stream. However no such differential motion has been observed in any of the regions surveyed, suggesting a very low drag on the Stream. Our observations set a limit of a few percent on the efficiency of momentum transfer in the case of the halo just mentioned ($T \sim 10^6$K, $n \sim 4 \times 10^{-4}$ cm^{-3}). Alternatively the containment could come from a very hot intergalactic medium ($T \sim 10^8$K, $n \sim 4 \times 10^{-6}$ cm^{-3}), but in this case a small magnetic field would also be necessary to prevent this very hot gas from entering the clouds and evaporating them.

*Supported by a fellowship from Consejo Nacional de Investigaciones Cientificas y Tecnicas, Argentina.

W. B. Burton (ed.), The Large-Scale Characteristics of the Galaxy, 574.

THE GALACTIC WAKE MODEL OF THE MAGELLANIC STREAM

Joel N. Bregman
Columbia University, New York

Tidal interaction models for the origin of the Magellanic Stream have been fairly successful in reproducing the radial velocities of the Stream (Lin and Lynden-Bell 1977, Davies and Wright 1977). However, no investigator has yet attained a self consistent treatment in which (1) the LMC and SMC are bound for at least 5×10^9 yr, (2) the passing Magellanic Clouds warp the galactic plane, and (3) the Stream velocities are reproduced (Fujimoto and Sofue 1977). Also, Mathewson, Schwarz, and Murray (1977) argue that their 21 cm observations are evidence against the tidal model. To avoid these problems, they suggest that the Magellanic Clouds pass through a hot coronal gas and produce vortices in their wake which radiatively cool to form the HI clouds comprising the Stream.

Here we examine the possibility of forming HI clouds in this manner. The steady-state fluid equations for a non-expanding, viscous, heat-conducting-vortex were solved for a hot, fully-ionized gas in which vortices would appear in the wake (i.e. Reynolds number between 20 and 2500 and no magnetic field). These solutions show that the pressure decreases toward the center of the vortex and that the force generated by this pressure gradient balances the centrifugal force of the spinning gas. Because electron conduction is fairly efficient (Prandtl number is 0.05), the temperature gradient is much smaller than the density gradient so the cooling time increases toward the center of the vortex. Thus vortices can never be nucleation centers for HI clouds since they will remain hotter than the more rapidly cooling coronal gas. Furthermore, radiative cooling causes vortices to expand into the coronal gas and lose their integrity (for $Pr \leq 0.78$). So, the suggestion of Mathewson _et al_. (1977) appears to be an inappropriate way of forming the Stream.

A more detailed discussion of this topic and a critical examination of other non-tidal theories will be presented elsewhere.

REFERENCES

Davies, R. D., and Wright, A.E.:1977,Mon.Not.Roy.Astron.Soc. 180, 71.
Fujimoto, M., and Sofue, Y.: 1977, Astron. and Astrophys. 61, 199.
Lin, D.N.C. and Lynden-Bell, D.:1977, Mon.Not.Roy.Astron.Soc. 181, 59.
Mathewson,D.S., Schwarz,M.P.,Murray, J.D.:1977, Astrophys. J. 217, L5.

W. B. Burton (ed.), The Large-Scale Characteristics of the Galaxy, 575–576.

DISCUSSION

Giovanelli: Are any of your calculations particularly sensitive to the chemical composition of the gas?

Bregman: No. My conclusions are extremely insensitive to chemical composition.

Mirabel: Do you expect to find vortex motions on a scale of one degree? Our high-resolution observations of the Magellanic Stream made at Jodrell Bank do not show such motions for scale sizes of up to two degrees.

Bregman: The size of a vortex is comparable to the width of the wake behind the Magellanic Clouds. The angular extent in the sky depends on the distance to the Stream.

Mathewson: It is hazardous to extrapolate from the case of incompressible subsonic flow to that of diffuse material flowing supersonically past the Magellanic Clouds. The viscosity responsible for the formation of these eddies cannot be normal molecular viscosity but must be some form of turbulent viscosity associated with supersonic turbulent cells. A good illustration that this is a controversial point is Dr. S. Ikeuchi's (Hokkardo University) recently submitted paper (Astrophysics and Space Science) in which he supports the Turbulent Wake Theory of the Magellanic Stream and concludes that condensations can form in the wake of the Magellanic Clouds.

Basu: What is the time-scale for the survival of the vortex motion of which you have proposed?

Bregman: The vortex motion will last no longer than the cooling time of the gas. If vortex motion exists, it will linger for at least L/v, where L is the size of the Magellanic Cloud complex and v is their velocity through the hot coronal gas.

OUR GALAXY AS A MEMBER OF THE LOCAL GROUP*

Sidney van den Bergh
Dominion Astrophysical Observatory
Herzberg Institute of Astrophysics
National Research Council Canada
Victoria, B.C.

1. INTRODUCTION

Galaxies are like people. When you get to know one well it always turns out to be peculiar in some way or other. In many cases such peculiarities appear to be inherent whereas in others they seem to result from the environment in which a galaxy has evolved. Very few galaxies live in total isolation; most are members of cluster families. The purpose of the the present paper is to introduce the known members of the Local Group to you in the hope that a closer acquaintance with our closest relatives in space will ultimately lead to a deeper understanding of the structure and evolution of our own Milky Way system.

By definition the Local Group is a dynamical unit (Yahil, Tammann and Sandage 1977) which does not expand with the Hubble flow. The diameter of the Local Group is ~ 3 Mpc. An up to date census of Local Group members is given in Table 1. References to new or probable new members of the Local Group, which have been added since my review ten years ago (van den Bergh 1968), are given below:

IC 10: de Vaucouleurs and Ables (1965), Shostak (1974)
Leo A: Fisher and Tully (1975), Yahil et al. (1977)
Wolf-Lundmark-Melotte: Ables and Ables (1977)
IC 5152: Baade (1963)
Pegasus: Yahil et al. 1977
DDO 210: Fisher and Tully (1975), Yahil et al. (1977)
And I, And II, and And III: van den Bergh (1972ab, 1974)
Sagittarius: Cesarsky et al. (1977), Hawarden et al. (1977)
Carina: Canon, Hawarden and Tritton (1977)

The status of the Phoenix dwarf galaxy (Schuster and West 1976, Canterna and Flower 1977, Laustsen et al. 1977) is not yet clear. It

*Dominion Astrophysical Observatory Contribution No. 369=NRC. No. 16669

W. B. Burton (ed.), The Large-Scale Characteristics of the Galaxy, 577–582.

TABLE 1

DATA ON PROBABLE LOCAL GROUP MEMBERS

Name	α 1950	δ	Type	M_V
M31=NGC 224	00 40.0	+41 00	SbI-II	-21.1
Galaxy	17 42.5	-28 59	Sbc	-20.5:
M33=NGC 598	01 31.1	+30 24	ScII-III	-18.9
LMC	05 24	-69 50	IrIII-IV	-18.5
IC 10	00 17.6	+59 02	IrIV?	-17.6
SMC	00 51	-73 10	IrIV/IV-V	-16.8
M32=NGC 221	00 40.0	+40 36	E2	-16.4
NGC 205	00 37.6	+41 25	E6p	-16.4
NGC 6822	19 42.1	-14 53	IrIV-V	-15.7
NGC 185	00 36.1	+48 04	dE0	-15.2
NGC 147	00 30.4	+48 14	dE4	-14.9
IC 1613	01 02.3	+01 51	IrV	-14.8
WLM=DDO 221	23 59.4	-15 44	IrIV-V	-14.7
Fornax	02 37.5	-34 44	D Sph	-13.6
Leo A=DDO 69	09 56.5	+30 59	IrV	-13.6
IC 5152	21 59.6	-51 32	IrIV/IV-V	-13.5:
Pegasus=DDO 216	23 26.1	+14 28	IrV	-13.4
Sculptor	00 57.5	-33 58	D Sph	-11.7
And I	00 42.8	+37 46	D Sph	-11:
And II	01 13.6	+33 11	D Sph	-11:
And III	00 32.7	+36 14	D Sph	-11:
DDO 210	20 44.1	-13 02	Ir	-11:
Leo I=DDO 74	10 05.8	+12 33	D Sph	-11.0
Sagittarius	19 27.1	-17 47	Ir	-10:
Leo II=DDO 93	11 10.8	+22 26	D Sph	-9.4
Ursa Minor=DDO 199	15 08.2	+67 18	D Sph	-8.8
Draco=DDO 208	17 19.4	+57 58	D Sph	-8.6:
Carina	06 40.4	-50 55	D Sph	...

has therefore not been included in the present listing of probable numbers of the Local Group.

The total number of probable Local Group galaxies listed in Table 1 is 28, i.e. ~ 2 galaxies per Mpc^3. It should be emphasized that the present census of Local Group members is probably quite incomplete below M_V ~ −10. Furthermore some Local Group members might be hidden at low galactic latitudes.

Table 1 shows that the three brightest Local Group galaxies are all spirals. Of the remaining objects 11 are irregulars and 14 are dwarf elliptical/spheroidal galaxies. The luminosity function of known Local Group members, which is shown in Fig. 1, gives no sign of a turndown at the low-luminosity end.

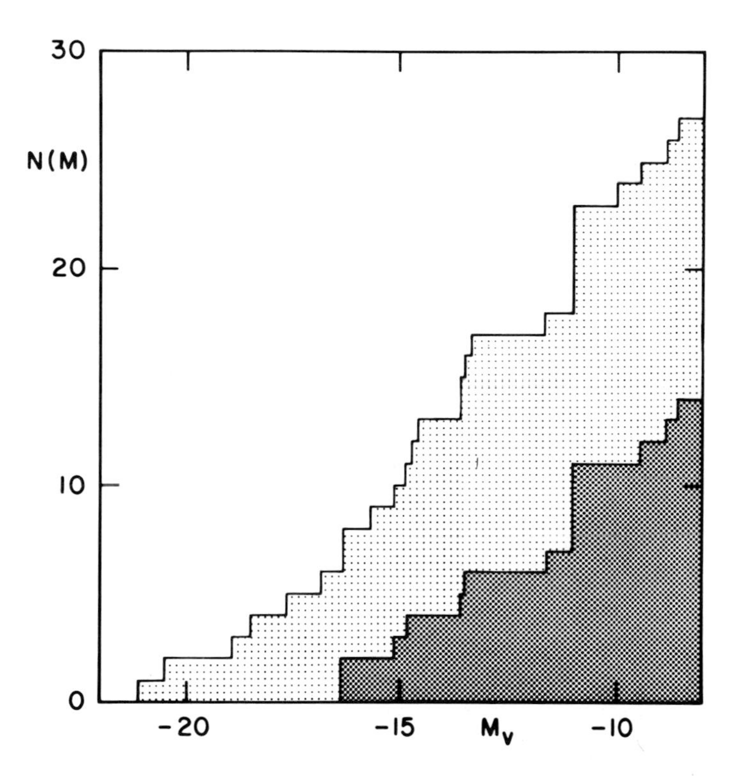

Figure 1. Integral luminosity function of Local Group galaxies. The upper histogram shows the total luminosity function for all types of galaxies. The lower histogram refers to elliptical and dwarf spheroidal galaxies only.

Table 2 shows that there is a hierarchy of subclustering within the Local Group. Two major subgroups are centered on M31 and on the Galaxy. It is interesting to note that the majority of early-type (dE + D sph) galaxies are located in subgroups, whereas 9 out of 11 of the irregulars occur outside them. This may indicate that irregular galaxies form preferentially in a low-density environment. Due to their low mean density dwarf spheroidal galaxies are extremely fragile and are easily disrupted by tidal forces. As a result the present number of dwarf spheroidals is probably much smaller than it was originally.

TABLE 2

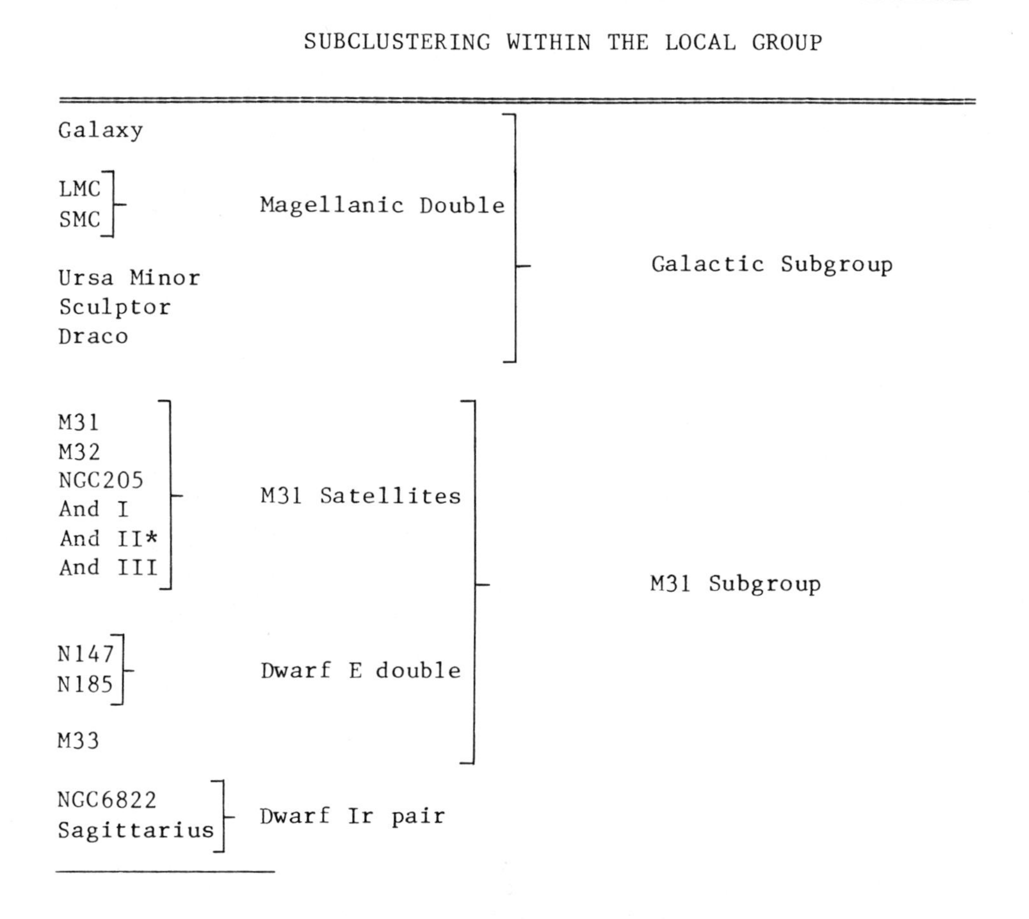

SUBCLUSTERING WITHIN THE LOCAL GROUP

Galaxy	Subclustering	Subgroup
Galaxy		Galactic Subgroup
LMC, SMC	Magellanic Double	Galactic Subgroup
Ursa Minor		Galactic Subgroup
Sculptor		Galactic Subgroup
Draco		Galactic Subgroup
M31, M32, NGC205, And I, And II*, And III	M31 Satellites	M31 Subgroup
N147, N185	Dwarf E double	M31 Subgroup
M33		M31 Subgroup
NGC6822, Sagittarius	Dwarf Ir pair	

*And II is actually located closer to M33 than it is to M31

The Local Group may be regarded as a gigantic laboratory in which the effects of differing initial conditions and environmental factors on chemical evolution and morphology may be studied. Probably the most striking conclusion that can be drawn from presently available chemical abundance studies is that heavy element abundance is strongly correlated with galactic mass. In first approximation $Z/Z_{\odot} \propto \mathcal{M}^{\frac{1}{2}}$. Within the rather low accuracy of presently available data Elliptical, Spiral and Irregular galaxies of comparable mass have similar metallicity. Within the Local Group the only obvious exception to this conclusion is M32 which has a metallicity (cf Faber 1973ab) which is appropriate to its mass before it was stripped by tidal encounters with M31.

Within individual galaxies mean metallicity appears to correlate with density in the sense that the high density cores of galaxies have above-average metallicity whereas low density halos are generally metal-poor. The mean metallicity of globular cluster families is found to correlate with the masses of their parent galaxies (van den Bergh 1975).

REFERENCES

Ables, H.D. & Abels, P.G. 1977, Astrophys. J. Suppl. 34, 245.
Baade, W. 1963 "Evolution of Stars and Galaxies", Harvard University Press, Cambridge, p.24.
Canterna, R. & Flower, P.J. 1977, Astrophys. J. Letters 212, L57.
Cannon, R.D., Hawarden, T.G. & Tritton, S.B. 1977, Monthly Notices Roy. Astron. Soc. 180, 81p.
Cesarsky, D.A., Laustsen, S., Lequeux, J., Schuster, H.E. and West, R.M. 1977, Astron. Astrophys. 61, L31.
de Vaucouleurs, G. & Ables, H. 1965, Publ. Astron. Soc. Pacific 77, 272.
Faber, S.M. 1973a, Astrophys. J. 179, 423.
______.1973b, Astrophys. J. 179, 731.
Fisher, J.R. Tully, R.B. 1975, Astron. Astrophys. 44, 151.
Hawarden, T., Goss, M., Longmore, A., Mebold, U., Webster, L. 1977, preprint.
Laustsen, S., Richter, W., van der Lans, J., West, R.M. & Wilson, R.N. 1977, Astron. Astrophys. 54, 639.
Schuster, H.E. & West, R.M. 1976, Astron. Astrophys. 49, 129.
Shostak, G.S. 1974, Astron. Astrophys. 31, 97.
van den Bergh, S. 1968, "The Galaxies of the Local Group", David Dunlap Obs. Commun. No. 195.
______.1972a, Astrophys. J. Letters 171, L31.
______.1972b, Astrophys. J. Letters 178, L99.
______.1974, Astrophys. J. 191, 271.
______.1975, Ann. Rev. Astr. Astrophys. 13, 217.
Yahil, A., Tammann, G.A., Sandage, A. 1977, Astrophys. J. 217, 903.

DISCUSSION

Peimbert: I would like to make two comments with respect to the carbon-to M-stars ratio: 1. The distribution of WC9 stars goes in the opposite direction, i.e., most of the WC9 stars in the Galaxy are located closer to the galactic center than the Sun. 2. It is possible that carbon stars in the direction of the galactic center are embedded in dust clouds that prevent their detection; there are some infrared observations by Grasdalen and Joyce that seem to indicate that this is the case.

van den Bergh: 1. The abundances observed in WR stars might well be due to stellar evolution. 2. Many of the globular clusters in the Magellanic Clouds contain carbon stars with B-V > 2.0. Not a single star has been observed in any galactic globular cluster. It seems rather artificial to assume that all the carbon stars in galactic globules are hiding in dust clouds!

Mezger: You mentioned the very interesting result that $z \propto (\text{mass of galaxies})^{1/2}$. What is known about the He-abundances in the local galaxies?

van den Bergh: Peimbert's results seem to indicate that the helium abundance ranges from ~ 0.07 (by number) in the SMC to ~ 0.10 in the Galaxy.

Lequeux: I have recently studied (Astron. Astrophys., in press) the rate of star formation in several galaxies of the Local Group, using a direct comparison between the populations of identical portions of the upper HR diagrams in these galaxies and in the solar neighborhood. The main result is that the rate of star formation per unit mass of gas is largest in the solar neighborhood, three times smaller in the LMC, and eight times smaller in the SMC. Thus the often-quoted vague statement that "the rate of star formation is extremely large in the MC's" has no basis. There is no evidence from my data for important bursts of star formation in the studied galaxies, and these data are consistent with a similar age for all these galaxies. There is also no evidence for strong variations in the Initial Mass Function.

Felten: You remarked that the luminosity function of the Local Group is roughly flat at the faint end. The statistical significance of this is low (with only 28 objects). Nevertheless, the Local Group is one of the few handles we have on the luminosity function for galaxies of low luminosity. This prompts me to ask whether you could say a few words about the accuracy of distance determinations in the Local Group, particularly for the smaller members.

van den Bergh: For dwarf spheroidals the distances are pretty good, because most of the dwarf spheroidals are evidently companions of larger galaxies. For dwarf irregulars, the distances are much more uncertain.

Tinsley: Of course you are right that the Coma cluster is very different from small nearby groups. However, Coma is relevant because I expect that if galaxy formation were inefficient in the Coma cluster, it must have been even less efficient in sparse groups.

XI. SUMMARIES

AFTERTHOUGHTS FROM A THEORETICAL POINT OF VIEW

Virginia Trimble
Physics Dept. U. California, Irvine CA 92717, & Astronomy Program, U. Maryland, College Park MD 20742 USA

APOLOGIA: The edifice we call science is built of bricks contributed by many, many workers, laid according to a design too large to be comprehended by any one reviewer. My apologies, therefore, to all those whose bricks I have, through ignorance, prejudice, or lack of space, put in the wrong place, attributed to someone else, or ignored completely. Names cited below are generally those of the speakers who mentioned particular points. Proper references to who did what can thus be found in their papers and discussion remarks.

I. A LIST OF DISPUTED ITEMS

It is customary to begin and end conferences with lists of topics on which progress has either been made since the last meeting or is expected before the next one. Mine (ordered roughly from large to small scales) is, less ambitiously, merely a list of topics about which I learned something in the past week. (1) What is the total mass of the galaxy, and how is it distributed among components? (2) What is our galaxy's type, and is it normal for that type? (3) What processes have entered into its formation and evolution? (4) How does chemical composition depend on position in the disc and halo? (5) Where and how numerous are our spiral arms, which things are confined to them, and what makes them? (6) Do we have a bar? (7) A ring? (8) A central hole in the disc? (9) Where are the warps, wiggles, and twists in the disc, and what makes them? (10) How old is the disc? (11) What are the dominant phases of the interstellar medium, and what keeps them the way they are? (12) What processes can trigger star formation, and how do their effects differ? (13) What is the position of the sun and its motion relative to nearby objects and the galactic center? (14) What is the nature of the non-circular motions in the inner part of the galaxy? (15) Where and why is the high velocity gas? (16) What, if anything, is at the very center? The discussion of these that follows is divided into issues concerning the galaxy-as-a-whole, the disc, the interstellar medium and star formation, dynamics, and the nucleus.

W. B. Burton (ed.), The Large-Scale Characteristics of the Galaxy, 585–590.

II. THE GALAXY AS A WHOLE

If we knew the galactic rotation curve out to ~100 kpc, we would know the total mass and its distribution with radius, though not whether halo or disc dominated at various places. What we have are an HI and CO rotation curve that remains flat, implying mass still increasing linearly with radius, out to 18 kpc (Gordon & Burton, Jackson), curves for 17 external Sa-c galaxies (Rubin) which are flat out as far as they can be measured, implying minimum total masses of 3-4 X 10^{11} $M_\odot$, and several indirect data. The escape velocity near the sun, determined from the few high-velocity stars with reliable motions, is at least 400-450 km sec^{-1} and probably 560 km sec^{-1} (Toomre, Ostriker). This requires a mass out to ~70 kpc of at least 6X10^{11} $M_\odot$ (Ostriker) and is consistent with masses still larger by factors of 2-4 (Einasto). There is a 7% chance that the Local Group is merely a chance encounter of two large spirals and their satellites (van den Bergh), but if we require it to be gravitationally bound, then total masses of at least 10^{12} $M_\odot$ are required, the situation becoming more extreme if θ_o is only 220 km sec^{-1} (Knapp, Einasto), implying a radial velocity for Andromeda of -115 km sec^{-1} relative to the galactic center and a tangential velocity of about 80 km sec^{-1} (since θ_o is 220 relative to our satellite galaxies but 300 or more relative to the Andromeda satellites). The chief objection to masses ~10^{12} $M_\odot$ is that they make it impossible for the Magellanic Clouds to have formed a bound system for 10^{10} yr (Fugimoto). The alternative, given that we see a bridge of star, gas, and dust between the Clouds (Mathewson), is to say that they were once a single entity, torn apart by a recent close encounter with the galaxy, which may also have given rise to the Magellanic Stream (Fujimoto) and a burst of star formation in the Clouds (Ostriker).

De Vaucouleurs assigns our galaxy to type SAB(rs)bc II. We return to the B, r, and s in Sect. III. The Milky Way appears disgustingly normal for its type, having a bulge luminosity of 2X10^{10} $L_\odot$ (Okuda) and M/L = 7.6 (Maihara), much the same as M 31. Its nuclear magnitude is close to the average of Virgo spirals (most like NGC 4421, which has a bar); its infrared properties are bracketed by those of normal S's; and its Hβ and X-ray luminosities are much less than those of Seyferts and qso's (Weedman). Our M/L = 5.5 inside R_o, M(HI)/L = 0.15, HI radius exceeding optical radius, and central hole in HI are all typical for Sbc's (Rots). The same can be said for central deficiencies in HII (Berkhuijsen), the existence of disc warps that are more conspicuous in HI than in visible light and not necessarily dependent on a companion (Sancisi), and radio emission from at least several kpc outside the plane (Ginzberg, Wielebinski), but we cannot say whether the more limited extent of the gamma-ray producing halo (Stecker) is also usual.

It is impossible to cover galaxy formation and evolution in one paragraph, but there are interesting connections visible among the ideas concerning mergers and accretion of intergalactic gas (Tinsley), the radial-infall model of the Magellanic Stream (Mathewson), and Burton's tilted disc, which _may_ be what we would see if a 10^{7-8} $M_\odot$ Stream-type cloud had arrived at the galactic center 10^{7-8} yr ago. There are assorted extragalactic precedents. A wary eye should, however, be kept on the possible differences between E galaxies and spheroidal parts of S's (Strom) and large-scale environmental effects on astration, galaxy types, and stellar initial mass function (van den Bergh).

Composition gradients in the inner part of the halo and the 8-14 kpc part of the disc are well known (Kraft, Peimbert). Outside R_0, the upper envelope of Z(R) stays flat, but because the standard indicators of metal abundance in RR Lyr stars and globular clusters lose sensitivity below [Fe/H] = -2, this need not imply that the average value of Z(R) stays constant (Kraft), though theorists will probably be happier if it does. In the disc, the local gradient cannot continue to the center, because intensities of NeII and H recombination lines and the detection of low-ionization states of Ar and S imply a metal abundance 2-3 times solar (Wollman). Z(R) may peak in the 5 kpc ring (Peimbert).

III. THE DISC

Observers all agree that our galaxy has spiral arms, but not on precisely how many or where, or on whether the arms are of the tidy, ordered variety or the feathery incoherent variety. Since optical and HI arms in other galaxies are only partially correlated (M. Roberts), this may not be surprising. The arms show up in HI density and velocity fields, though not with the relative phasing expected from density waves (Wielen), in HII recombination line velocities (Lockman), and perhaps in CO velocities (Scoville). Though other galaxies have dust lanes on arm inner edges, this is not obvious in the Milky Way (Lyngå), but the dense H_2CO probably is in such lanes (Davies). We see portions of three arms and a spur (containing the sun) locally by optical techniques (Humphreys), the innermost of these being double in radio continuum surveys (Kerr). On a larger scale, both HI and radio HII arms are probably best fit by a four-armed pattern, but with a good many irregularities, especially where the data are good (Henderson, Georgelin). Things that are confined to or correlated with arms in our own or other galaxies include the peaks of gamma ray production (Paul), giant HII regions (Mezger), the more conspicuous OB associations (Humphreys), and a magnetic field running along the arms (Kronberg). Things apparently not confined to the arms are CO (Solomon, Rickard), pulsars (Taylor), small HII regions (Mezger), Cepheids and other bright red stars (Humphreys).

Density wave theory is obviously a good thing and should be encouraged. Although there are other possible explanations for spiral arms (Pişmiş, Schmidt-Kaler), spiral density waves have the merit of explaining many things simultaneously. A nice case is the determination of the radial part of the solar motion relative to objects with different velocity dispersions (i.e. ages). U drops from 6 to 1 km/sec with increasing σ_v. This both agrees well with density wave theory and resolves a long-standing discrepancy between HI rotation curves determined from the northern and southern hemispheres. It is, however, necessary to be rather careful about comparing predictions with observations because of major differences in velocity fields implied by linear and assorted non-linear forms of the theory (Wielen). Since our galaxy seems to have 4 arms, it is lucky that there are unstable modes beside m = 2! On the "con" side, some galaxies have spiral arms extending over too large a radius to be covered by a single pattern, and "feathery" arms are more readily made by stochastic star formation in a differentially rotating disc (Rubin).

Bar instabilities occur and can facilitate arm formation (Mark). The problem of their stabilization by massive halos may be overcome when the bar is caused by an "inner-inner" Lindblad resonance (Lynden-Bell). The chief theoretical difficulty in giving our galaxy a bar is in locating the corotation radius to make both dust lanes and outer spiral arms (Sanders). Other SB's have managed to solve this problem (Hubble Atlas). There are several advantages to having a bar. We can blame the 3 kpc arm and other non-circular features on it (W. Roberts, Oort, de Vaucouleurs). It may be able to clear gas out of the inner galaxy, as observed (Sanders). And excess star formation near 5kpc in the Milky Way and, eg, NGC 4449 can be attributed to gas piling up at the ends of the bar (van Woerden, Lynden-Bell). It may be relevant that 60-70% of external galaxies having central "hot spots" in star formation, as has the Milky Way, are SB's (van den Bergh).

Many galaxies show an optically bright ring a few kpc in diameter, which can coexist with both bars and arms and need not be concentric with the nucleus (eg NGC 5728, Rubin). The enhancements of many galactic components in the region R = 4-8 kpc strongly suggest such a ring in our galaxy. Things commoner in this region than elsewhere include most molecules (Solomon, Downes, Johannson), type II OH/IR masers (Oort) HII regions and supernova remnants (Rohlfs), infrared and gamma ray production (Puget, Stecker), and perhaps pulsars (Taylor). There may also be an enhanced star formation rate with initial mass function favoring A stars (Puget, Lequeux). Along with all these excesses go deficiencies of the same things and HI inside the ring. Some deficiencies would probably disappear if the high nuclear densities could be smoothed out (Mezger, Solomon). An interesting question is whether the total mass density of the galactic disc also falls in the inner few kpc. We would expect it to if the gas has been swept out (as suggested by the large non-circular velocities in the region; Rohlfs) but not if the gas has all been turned into stars in the past. A genuine hole is favored by models of the galactic mass distribution, the hole and ring giving the observed inner peak in the rotation curve (Ostriker, Einasto).

The galactic "plane" is far from flat, the outer HI disc having a warp (Henderson) reaching about 0.8 kpc at $R = 1.5\ R_o$, which may or may not be shared by the young stars (Kerr, Lequeux). Inside R_o, there are residual large-scale waves above and below the mean plane in both HI (Henderson) and HII (Lockman). All the gas inside $\sim$1.5 kpc may be tilted relative to $b = 0^o$ (Oort), and the dust responsible for absorption features against the central IR source seems to be similarly tilted (Okuda).

A rotating, expanding 1.5 kpc diameter disc, tilted at about 22° to the galactic plane could be responsible for many non-circular features (Burton). No tilt appears in the central distributions of thermal or non-thermal radio continuum emission, in H166α , or in IR emission (Sanders, Terzian, Okuda), but the radiation may be mostly line-of-sight accumulation, not close to the center. HI profiles in many velocity intervals are tilted (Davies, Kerr). Physical models have been attempted only for the large-scale warping. There are several possibilites not requiring a nearby companion, some of which can co-exist with a massive halo (Saar).

If the maximum age of disc stars is nearer 5 than 10 billion years, this has important implications for theories of galactic formation and evolution (Tinsley) and for stability of spiral structure (Rubin).

IV. THE INTERSTELLAR MEDIUM AND STAR FORMATION

The observations require very small amounts of H_2 at T $\lesssim$ 8 K (Stark, Zuckerman), large amounts at higher temperatures, HI over a wide range of T, HII at Strömgren sphere temperatures and some hot enough to make OVI and soft X-rays. There is now no concensus on whether H_2 makes up 80-90% (Solomon) or much less (R. Cohen, Davies) of the mass of the ISM or on whether HII at SNR temperatures occupies most of the volume of the disc, including the region around the sun (Heiles) or less volume than the hot HI (Baker). Several other phases are also required (Turner, Salpeter), but the only one in which it seems to be hard to maintain the observed density-temperature-ionization conditions is the H_2 in giant molecular clouds. CO in clouds must last much longer than the 10^{6-7} yr free-fall times inferred for them (Scoville, Solomon), even though observable CO tends to disappear from OB associations in $\sim$3X10^{7} yr (Bash). Measured turbulent and rotation velocities do not provide adequate stabilization, and magnetic fields probably leak out (Turner, Baker). It may be possible to tell a coherent story by saying that GMC's exist only in spiral arms, where collapse and star formation occur on a short time scale, but that the clouds are rather quickly torn apart by shocks, SNe, or whatever, leaving the interarm clouds smaller, thus both longer-lived and less readily observable (Scoville, Elmegreen).

Gas compression that can lead to star formation has been blamed on shocks caused by density waves, expanding SNR's and HII regions, and infalling intergalactic gas (especially in mergers; Tinsley). That spiral shocks hitting GMC's is not the only possibility is demonstrated by the widths of spiral arms (Kaufman), the presence of two Herbig-Haro objects at the edge of a 25 $M_{\odot}$ Barnard globule (Bok), young stars at the edges of HII shells (Sivan), the lovely outer arms made by stochastic star formation (Rubin), and the environments in which SO's and anemic spirals are most common (Tinsley). Star formation is inefficient by any process, only 1-3% of the gas passing through spiral arms turning into stars per passage (Mezger). The nucleocosmochronological data suggest that two processes contributed to forming the solar system. Mg^{26}

anomalies imply that the collapse of the particular cloud that became the solar system was triggered by a nearby SN, while the 10^8 yr latency period inferred from I^{129} and Pu^{244} suggest that the SN was part of a wave of star formation caused by spiral arm passage and that no important star formation had occurred in the region since the last passage 10^8 yr ago.

V. DYNAMICS AND THE NUCLEUS

A fast-moving bandwagon appears headed toward 8.5 kpc and 225 km sec^{-1} for the solar galactocentric distance and rotation velocity (Knapp, Einasto, Feast, Graham). The mass interior to R_o is thereby reduced to about 10^{11} $M_\odot$. Since extragalactic observers often use 300 km sec^{-1}, the galactic standard of rest is not at present very well defined! The problems in establishing the LSR (Clube, Upgren) are probably not very serious. Although several speakers suggested improved values for the Oort constants A and B, evidence from rotation curves for velocity ripples in spiral arms means that the local values may not have much large scale significance (Rubin).

Gas velocities that do not fit on a smooth rotation curve are found from the nucleus to the Magellanic Clouds. Time scales in the inner regions (3 kpc arm, 200 pc molecular ring, 1 pc molecular cloud, etc) are so short that the gas motions must include sloshing about, streaming along bars, tilted discs or whatever and not just coherent expulsions or infall (Wollman, Sanders). If the time scale for star formation in the 5 kpc ring is really $\ll 10^{10}$ yr, then the gas there must be replenished by systematic inward flow from disc, halo, or intergalactic medium. The disc has many expanding HI and HII shells (Heiles, Sivan, Weaver), probably attributable to expanding SN and OB star shells. The situation is more complex further out. If all the non-circular HI belongs to our galaxy, it is an awful mess (Verschuur). Some of it must belong to tidal features in our own and other small groups (Haynes). The total absence of primordial, left-over hydrogen would be extremely unlikely on theoretical grounds (Toomre, Tinsley). The problem is which features to attribute to which mechanisms. My own prejudice is to blame those features that show bridges to "permitted" gas (Moore) on distant and out-of-plane spiral arms; to give the Sculptor group credit for its own primordial gas; and to hope that the Magellanic Stream is an example of genuine infall, probably of fresh, intergalactic gas (Mathewson), or perhaps of gas torn out of a previously-single Magellanic Cloud (Fujimoto).

If the galactic nucleus has to act as the power source for many of these non-circular phenomena as well as for compact infrared and radio sources (Paczyński), we can only sympathize with its having to do so without the assistance of a massive black hole (Ozernoy).

THE GALAXY IN PERSPECTIVE - AN OBSERVER'S SYNOPSIS

Hugo van Woerden
Kapteyn Astronomical Institute
Groningen, The Netherlands

As mentioned already in Morton Roberts' fine introduction, two features were striking in this Symposium's programme: the broad spectral range of observational data, and the strong emphasis on comparisons with external galaxies. I shall return to the latter item towards the end of my summary. Let us first review the spectral panorama, then scrutinize what it has taught us about our Galaxy.

THE SPECTRAL PANORAMA

For the first time in this series of symposia about our Galaxy, gamma-ray studies made significant contributions: Paul discussed the γ-ray disk, thin as that of the gas; and Stecker the γ-ray halo, more properly perhaps called a thick disk. X-rays and ultraviolet were less prominent now than five years ago, at IAU Symposium 60.

Optical observations contributed some of the highlights of this Symposium. First and foremost, Vera Rubin's beautiful rotation curves of external galaxies gave strong evidence for extended mass distributions and demonstrated density-wave effects in the differential motions across spiral arms. The fine photographic Hα surveys of Sivan and the Georgelins testify to the wide spread of ionized hydrogen, and provide ring diameters - of key importance for determinations of the extragalactic distance scale. Spectral studies of HII regions (Peimbert) show that abundances of heavier elements decrease with increasing galactocentric distance. Abundances also come from stellar spectra (Kraft). Indispensable, too, are stars as distance indicators in studies of spiral structure (Roberta Humphreys) and size of the Galaxy (Graham). Our only estimate of the Galaxy's morphological type (by De Vaucouleurs) was based on photometric work.

In the infrared, Fabry-Pérot studies of the [NeII]-line have given new, deep insights into the galactic nucleus (Wollman). The far infrared (Mezger, Puget) is our best probe of the dust clouds in the inner regions, and their role in star formation.

W. B. Burton (ed.), The Large-Scale Characteristics of the Galaxy, 591–598.

Observationally, the CO-surveys at 2.6 mm (Burton, Liszt, Bania, Solomon, Scoville) may represent the most massive advance since 1973. Our views of the whole "inner half" of the Galaxy have been changed drastically by this new tool. Thus, H_2 molecules are now thought to rival or surpass HI atoms as the major constituent of the interstellar gas. The high angular resolution in this millimeter line has caused severe undersampling in the major surveys, but this situation is now improving considerably (R.S. Cohen). CO is also being developed as a tracer of the molecular gas in other galaxies (Rickard). Potentially, the CO-lines could become as important as the 21-cm HI-line for studies of distribution, kinematics and dynamics of interstellar gas in external galaxies.

A recent addition to the spectrum is the CH-line at 9 cm. Johansson presented results of the first survey of this widespread molecule. At centimeter wavelengths, we must further mention the massive surveys of H_2O, H_2CO, recombination lines and continuum carried out at Effelsberg (Downes).

The hydrogen line at 21 cm has been a major galactic research tool since 1951. At this Symposium, it brought new views on the Galactic Centre (Burton, Liszt, R.J. Cohen), on the physics of the interstellar medium (Salpeter, Baker, Crovisier), on the rotation, mass and size of the Galaxy (Gill Knapp, Jackson), and on its environment (Hulsbosch, Giovanelli, Mathewson). In particular also, hydrogen-line studies of external galaxies at resolutions $\lesssim$ 1 kpc have a strong impact on our thoughts about properties of our Galaxy.

Long radio waves play a minor role in galactic radio astronomy nowadays. It is ironic that the galactic halo, discussed at IAU Symposium 5 in 1955 on the basis of meter wave observations, was now the subject of gamma-ray studies.

Strong improvements in the panorama are required at gamma- and IR-wavelengths, and in the CO- and CH-lines. And, of course, optically we still know our Galaxy very little, though there is good hope for more: O-stars at 50 kpc distance have been reported.

THE GALAXY UNDER SCRUTINY

What then have all these photons, large and small, told us about the large-scale characteristics of our Galaxy?

Nucleus and Central Region (R < 4 kpc)

Let us start at the smallest structural unit, the nucleus. As noted by Oort, VLBI results (not reported at this Symposium) give evidence of an 0".001-diameter core source. Virginia Trimble has summarized the views of Paczyński and Ozernoj on what might reside or be missing within those nuclear light-hours, and supplemented them with her own thoughts.

Wollman's 12.8-μm NeII observations have revealed a $10^{6.5}$ $M_\odot$ concentration within 1 pc diameter; the dynamical centre coincides with IRS 16, the peak of the infrared nuclear bulge. This galactic core contains, within the Sgr A molecular complex, a disk of HII, rotating and expanding at 200 km/s, and tilted 45° to the galactic plane. Whether this dust-filled HII-region (Mezger) produces young stars, is under debate.

Within 150 pc of the Centre, 6-cm continuum observations (Mezger) show 10^6 $M_\odot$ of extended, low-density (10 cm^{-3}) HII, plus several giant-HII-regions caused by massive star formation. The molecular ring at R = 200 pc (Scoville, Sanders et al.) now is viewed as part of a tilted nuclear disk. Burton and Liszt have shown that a thin disk, of 1.5 kpc radius, ~25° tilt and with rotating and expanding motions, can represent all sofar isolated features in one unified model; the disk contains 10^7 $M_\odot$ of HI, 10^{9-10} $M_\odot$ of H_2 gas, and $10^{10.4}$ $M_\odot$ of stars.

Sanders has derived a bulge rotation curve from the 2-μm IR brightness distribution. This curve nicely fits to the 21-cm disk rotation curve; also, the bulge may be viewed as the core of a 5×10^{10} $M_\odot$ halo. The double-peaked rotation curve allows an explanation of the gas deficit at $1 < R < 4$ kpc in terms of angular-momentum transfer through turbulent-shear viscosity. This explanation differs fundamentally from that proposed by Oort (IAU Symposium 58), who ascribed the gas deficit to exhaustive star formation induced by frequent density-wave passages. Sanders further represents the noncircular motions of the 3-kpc arm ("Rougoor's arm", I prefer to call it) by motion along elliptical streamlines around a slight oval distortion in the mass distribution.

As wishes for further work in these regions, I note: multi-λ IR maps of the inner core, and the dynamics of the various tilted-disk features.

The 5-kpc Ring: Molecules and Star Formation

To an outside observer, the region $3 < R < 8$ kpc (assuming still R_0 = 10 kpc for the Sun-Centre distance) would be the dominant "Population-I" feature in our Galaxy. Here we find the peaks of the radial distributions of molecules, HII-regions, supernova remnants (SNR), pulsars, and gamma-ray sources. As noted by Burton and Gordon in the 1976 Annual Reviews, only HI deviates from this.

Extensive new surveys (Solomon) place the CO peak at R = 5.5 kpc. The vertical distribution peaks at $z_0 \sim -30$ pc, the layer thickness (FWHM) is ~120 pc. The gas is concentrated in "clouds" of 50 pc size, density 2 H_2 cm^{-3} and mass 10^5 $M_\odot$; in toto, our Galaxy would contain 4×10^9 $M_\odot$ of H_2. These densities and masses are based on an assumed density ratio $^{13}CO/^{12}CO/H_2 = 0.5/20/10^6$. As noted by Peimbert, the abundance ratio C/O varies with R, hence the ratio H_2/CO is probably overestimated. Can we find effective ways to observe H_2 directly?

A pressing theoretical problem is that of support of the molecular clouds against gravitational collapse (Solomon, Turner). The spatial CO-distribution suggests that clouds live 10^{8-9} yr rather than form stars at a rapid rate. Scoville speculates that they grow by collision.

The Onsala Group is to be congratulated on their first CH-survey (Johansson). With a longitude spacing of $2\overset{\circ}{.}5$ and 5°, this work is at the same stage as HI in 1953. Already, it is clear that the CH-lines trace the diffuse component of the molecular gas, including warm gas, while the CO-lines favour the dense, cool clouds. The z-distribution of CH resembles that of CO, but its radial range (4-10 kpc) is greater. Davies notes that H_2CO represents an intermediate-density regime between CO and CH. The physics of molecular clouds was reviewed in detail by Turner.

Lockman finds peaks at R = 5 and $7\frac{1}{2}$ kpc in the radial distribution of HII-regions. The vertical distribution is the thinnest known: $\sigma(z) \sim 30$ pc. Of great interest is the "ripple" in the run of z with R, a "corrugation" in the galactic plane, also found in HI, CO and SNR.

I mentioned earlier the beautiful Hα surveys of Sivan and Georgelin.[2] Another phantastic atlas is that of Downes et al., who mapped a $2^{\circ} \times 62^{\circ}$ region along the galactic equator at 6 cm with $2\overset{\prime}{.}6$ resolution, using the Effelsberg telescope. For ~1000 sources, kinematic distances were obtained from 6-cm line studies. This is a giant jump ahead in the mapping of HII regions.

Local Interstellar Physics

The region $R \sim 10$ kpc is where we know most about structure and physical conditions in the interstellar medium. Forty years ago, Strömgren distinguished HII-regions and HI-clouds. Ten years ago, another two-component model: clouds and intercloud medium (ICM), was invented by Pikel'ner, Spitzer, Field and associates. Recently, a hot ionized component (traced by OVI lines) was added to the three proposed earlier: warm ionized, warm neutral, and cold neutral gas. This week, Salpeter added molecular clouds, lukewarm clouds, interfaces, and coronal gas, bringing the total to 8 "phases". And Turner distinguishes 6 types of molecular clouds! At this pace, the next symposium on interstellar physics may well see a classification scheme for interstellar structures as detailed as current ones for galaxies and even stars. Nevertheless, a special meeting to review the observational evidence for, and physical processes in, these various phases might well be worthwhile. Salpeter already has given estimates of mass flux, energy balance etc. It struck me that he considers most of the electrons to be inside HII-regions rather than in the (late?) ICM.

Baker discussed the physical processes in cold, warm and hot neutral hydrogen. Important was his emphasis on the strong effects that sidelobe responses ("stray radiation") have on 21-cm line profiles at high galactic latitudes. These stray radiation contributions have strongly affected both ICM parameters and studies of correlation between gas and dust. I further recall his remarks about the transient nature of interstellar structures such as filaments and sheets: " interstellar weather". Gone are the days of spherical standard clouds ...

About the motions of these structures, much information is contained in the new Nançay Survey of 21-cm absorption in ~1000 source spectra (Crovisier). The numbers are similar to those of earlier surveys: $\sigma \sim 6$ km/s for external, 0-3 km/s for internal motions, but the data base is far broader.

Weaver discussed the relationship between continuum loops and supernova remnants. He ascribes the North Polar Spur to a SN explosion inside a giant bubble blown by the Sco-Cen cluster. Heiles showed us supershells of kiloparsec size, due to 10^{53} -erg explosions that no astronomer has yet seen or photographed, here or far out in extragalactic space.

Spiral Structure

We have seen maps showing the distributions of HI, HII-regions (optical: stellar distances, and radio: kinematical), giant-HII's, H_2CO and CO molecular clouds, young stars, clusters and Cepheids. There is no full agreement among these maps, but Georgelin's 4-armed pattern seems to have a strong case. Given the observational evidence and theoretical case for noncircular motions, the direct stellar distances of Roberta Humphreys are of great value. An important need is for a wider range of such distances. Can we find suitable (infra)red, young objects?

Wielen has given a lucid review of various aspects of density-wave theory. I recall his remark that only old objects such as stars and HI gas can show the waves (with shocks for the gas), while the distribution of young objects (including HII-regions) is determined by migration from their birth places near the shock-front. The migration depends on the initial velocities (pre- or post-shock?); predictions are fairly complex, and comparison with observations difficult. Comparison of the density-wave predictions with observations has been very successful for HI in M81 (Visser). A similar confrontation of shock theory and observations for HI in our Galaxy would be of major interest.

Another important point in Wielen's paper is his discussion of average motions and the local standard of rest. It seems that our Symposium Chairman, Kerr, was closer to truth in 1962 than many of us believed.

Woodward's calculations of star formation in galactic shocks suggest a possible process that starts at one end, and then propagates through a cloud. Bash described calculations of cluster evolution after formation in a density wave; he can represent disk colours and brightness profile.

Galactic Rotation and Integral Properties

Both Vera Rubin, from optical spectra, and Albert Bosma, from HI syntheses, conclude that rotation curves are flat in the great majority of galaxies; hence, there is much mass in the outer parts, the M/L ratio must increase outward, and masses approaching 10^{12} $M_\odot$ now appear likely in many giant galaxies. From HI observations in our Galaxy, Jackson also finds a flat rotation curve, with large-scale regional deviations; at $12 < R < 14$ kpc, $d\Theta/dR > 0$, suggesting a density-wave effect across the Perseus Arm. Note that, on a rippled rotation curve, the Oort constants A and B have only local meaning. We have been lucky for 50 years not to have been led astray by them!

Although Einasto and Gill Knapp propose a local rotation speed of 220 km/s, and $R_0 = 8.5$ kpc, the latter finds flat rotation to such great distances that our Galaxy, too, must be placed in the 100-kpc diameter, 10^{12} - $M_\odot$ class. An interesting related point is that the velocity of escape becomes much higher (550 km/s, see Ostriker) than we believed some time ago. Meanwhile, there is enough dispute about these rotation parameters, and their effects on the fate of Magellanic Clouds and Stream, that a workshop to study these problems in detail might be fruitful.

While the case for flat rotation curves and consequent high masses appears very strong, one must realize that small errors in the derived curves - such as may be caused by noncircular motions and by disk warps - may have strong effects in the masses. It is thus important to supplement such mass determinations with other methods, especially: measurement of relative motions in galaxy pairs.

De Vaucouleurs has presented us with an impressive list of parameters for our Galaxy. On many of these parameters the errors are so small that I have wondered whether a trip far outside the galactic plane would still pay.

Rim, Fringes, Loose Ends, etc.

Henderson has shown new data on the shape and thickness of the galactic gas layer in its outer parts. In fact, although Habing, Kepner, Davies, Verschuur and others had drawn attention to the large vertical extent of these outer parts, no complete, systematic mapping had been done since the fifties. Henderson finds a warp of $z_0 = 1.6$ kpc at $R \sim 1.7\ R_0$, and 1.5 kpc disk thickness along the outer edge of the Perseus Arm.

Warps stronger than ours are frequent among external galaxies, as discovered by Sancisi at Westerbork, and new theoretical work was overdue. Saar's review showed that massive haloes of various shapes can help in various ways: spherical haloes can facilitate tidal forcing, triaxial haloes can produce unstable oscillations, and in elliptical haloes accretion may lead to bending and thickening of the gas layer. Tubbs and Sanders demonstrated that both primordial and tidal warps can persist for 5×10^9 yr in galaxies where the halo mass exceeds that of the disk, within the latter's radius. A prediction that invites further observational test is that these warps should develop only in the tenuous gas outside the (truncated!) stellar disk. Further observation also should strengthen the statistics of warps, and reveal their shapes farther out.

The nature of high-velocity clouds has been the subject of fierce debate for many years. I was happy to hear Verschuur acknowledge that the name HVCS may stand for a variety of phenomena of different nature. Hulsbosch and Giovanelli have made new, extremely sensitive surveys at velocities out to ± 1000 km/s; both find clouds with velocities down to -465 km/s, and their results agree in detail. The distance and nature of these objects is unclear, though a suggestion prevails that they may be intergalactic hydrogen clouds in the Local Group. In fact, some may be related to the Magellanic Stream.

Mathewson and Fujimoto summarized observations and theory of the Magellanic Stream. So far, nothing but HI has been observed in the Stream, but its "magellanic" distance and "overhead" geometry appear secure. No satisfactory model has yet been developed, but it seems likely that we must interpret the Stream as a tidal effect, caused in the Magellanic Clouds by a high-mass ($\gtrsim 5 \times 10^{11}$ $M_\odot$) galaxy. The Cloud pair would then be in grave danger of disruption (a fact to be regretted by Bart Bok), and our warp have to be blamed on non-magellanic causes.

Martha Haynes presented a convincing case that isolated intergalactic hydrogen clouds outside the Local Group have sofar eluded detection. What intergalactic gas has been observed in other galaxy groups can be properly described as tidal debris. To me it seems, however, that such debris must dissipate; does it then become unobservable?

Formation and Evolution

Our collective scrutiny of the Galaxy has been strongly descriptive - both as regards observations and theoretical models. Few papers only discussed how its present structure developed. However, Beatrice Tinsley's impressive discussion of the formation and evolution of galaxies well made up for this. One item I mention: not only high-velocity clouds, or even the warp of the Galaxy's disk, may be due to accretion; the disk itself may have been accreted rather than be primordial.

THE GALAXY? - A GALAXY!

Throughout this Symposium, detailed information on external galaxies has served not only as a comparison for our Galaxy, but indeed to guide our understanding. I note Strom's massive review of the light and mass distribution, dynamics and evolution (chemical and morphological) of bulge and disk components. I recall again the optical and radio rotation curves and mass distribution, and the demonstration of density-wave theory. I further mention Rots' summary of the distributions of atomic and molecular hydrogen and of continuum radiation. Weedman has shown how minute our Galaxy's "nuclear activity" is by extragalactic standards. It appears obvious that our quest for understanding of the origin of spiral structure and of the Galaxy itself will ultimately be answered from the realm of the nebulae.

This emphasis on comparisons with other galaxies is a sign that galactic research has grown up. We, inhabitants of our Galaxy, have learned: "know thy neighbours in order to understand thyself". But this does not say that galactic research will die, now that we have proper tools for research into galaxies. In fact, only at home can we determine three-dimensional kinematics and distribution. And it is here that we can go ten magnitudes fainter (minus wavelength-dependent extinction) and achieve 100 times better linear resolution than even in Andromeda. "Γνωθι σεαυτον" remains as another brief for galactic research.

These two adagia may strike you as philosophical or anthropological rather than astronomical. However, our science - as any other - is and remains a human enterprise. One of the great joys of astronomy is its strongly international character. In this connection, it is sad that this Symposium's audience was geographically so poorly balanced. Asia and - alas - Africa were badly underrepresented. So were South America and - especially - Australia, whose powerful telescopes have such a unique job to do in the southern hemisphere. Also, we have sorely missed our colleagues from Eastern Europe, particularly the Soviet Union. Finally, I regret the small representation of my home country with its strong galactic tradition. Has Westerbork drawn us too far away from our Galaxy? May the next galactic symposium be more truly international!

Acknowledgements

I wish to thank the organizers, and especially Frank Kerr, for a most stimulating Symposium. Many speakers have contributed to this written version of my off-the-cuff summary by sending preprints of their papers.

A NON-ESTABLISHMENT VIEW OF THE SYMPOSIUM

R. P. Sinha
University of Maryland, College Park, Md 20742, USA

Many interesting new results from studies of external galaxies as well as of the Milky Way Galaxy have been presented at this meeting. I shall confine my remarks to but a few observations which appear to be crucial in deriving an acceptable picture of the large scale structure in the Galaxy.

Do we know the extent of the Galaxy? Flat rotation curves appear to be the rule in disk galaxies rather than the exception. Dr. Jackson has presented direct measurements of the rotation curve for the Galaxy. It appears to be flat up to 17 kpc from the center. A substantial portion of the mass must be hidden in a "dark" halo to account for such a rotation curve. If the new determinations of the rotational velocity at the distance of the Sun, of 220 km s^{-1}, reported by Dr. Einasto and Dr. Knapp were to be adopted, the problem of missing mass in the Local Group will become further aggravated. A direct check of the contents and mass of the halo is called for. It is necessary to adopt a uniform terminology in order to refer to various aspects of the z-extent of galaxies such as the radio halo, cosmic ray halo, spheroidal central bulge and the like.

The mass distribution in the spheroidal bulge and in the larger halo has significant repercussions for the dispersion relation of the density waves in the disk for different values of the stability parameter Q. Some work in this connection has been published by Terzides (1977). The questions of persistence and amplification of spiral density waves will have to be reexamined in the context of the halo. It is also conceivable that the halo mass will have a dynamical effect on the warps in the outer parts of galaxies.

How well are the parameters of the "grand design" spiral pattern in the Galaxy known? Does one such pattern exist? These questions were not explicitly raised at the symposium. The relevance of inquiries of the type described by Dr. Wielen in his talk, to actual confrontation of observations with the theory is unclear unless the grand design spiral is assumed to exist. Dr. Clube has shown that if

W. B. Burton (ed.), The Large-Scale Characteristics of the Galaxy, 599–601.

one starts with an assumption of the existence of a two-armed spiral in the Galaxy and if one then demands a bi-symmetry in the observed kinematics, one is forced to invoke an expansion velocity field. Such a velocity field contravenes observations in the solar neighbourhood as well as other observations spanning larger parts of the Galaxy. Further, it is well known that a two-armed tightly wound spiral pattern (e.g. Lin et al. 1969) which has been derived to fit the kinematics of HI in the first quadrant of galactic longitudes completely fails to account for the observations in the fourth quadrant. It is surprising to recognize that in spite of this anomaly tightly wound two-armed spirals are the basis of so many of the models discussed at the meeting. A four-armed 13° pitch angle spiral similar to one proposed by Dr. Y. P. Georgelin and presented at the meeting by Dr. Sivan has been shown by Dr. Henderson (1977) to fit some of the HI data both first and the fourth quadrants, as well as in the region outside the solar circle. The fit in the first quadrant alone is not any worse than for the two-armed tightly wound spiral model. It should be of interest to check the consistency of such a model in the context of the density wave theory both for its persistence and its expected amplitude.

Only a limited portion of the HI data, namely the extreme velocity edges of the profiles, has been fitted to a spiral structure in the Galaxy. The features of the data over the remaining part of the profiles are too confusing for unambiguous fitting with a model. It was hoped that, with the advent of new techniques and the detection of CO molecular lines from a large part of the Galaxy, it would be possible to resolve spiral features clearly. This hope has not materialized. Based on inadequately sampled maps, at one time it was claimed that CO longitude-velocity diagrams do not show the characteristic loops and arcs which have been conventionally identified with spiral arms in the study of HI observations. However, a more complete set of maps presented by Dr. Cohen of the Columbia University does show the existence of the expected spiral-like features. The problem of fitting a model spiral will critically depend upon the availability of data from the southern half of the Milky Way.

Much of the misunderstanding in the galactic structure stems from overinterpretation of a limited amount of observations. The HI distribution in the nuclear region of the Galaxy is one such example. Observations of HI mostly confined to the galactic plane have been interpreted earlier in terms of a disk with a one-third sector devoid of neutral hydrogen. A further examination of profiles covering latitudes on either side of the plane has led to a more likely model of HI distributed in an inclined tilted disk. CO data is consistent with such a distribution of gas in the nuclear disk. It is desirable to derive a dynamical model in which the apparent expansion velocities can be explained in terms of a coherent organized streamline structure in the field of force of the tilted disk.

Do we have a bar at the galactic center? A major limitation in arriving at an answer to this question arises from the lack of a specific model of a bar and from the relative ignorance regarding the kinematics of gas in and around the bar. There is a gap in the observations of the nuclear region between a region of radius a few parsecs studied with the [NeII] line emission and of radius a few hundred parsecs accessible for study with the molecular lines of CO. A new tracer appropriate for studies of this region is desirable. As of now there is no direct evidence for a bar at the center of the Galaxy. The tilted disk mentioned in the last paragraph is different in its appearance than the bars seen in a barred galaxy. An oval ring of radius between 3 and 4 kpc may be a manifestation of a bar-like potential at the center. An oval deformation of the central region can induce velocity asymmetries and apparent expansion velocities similar to what is observed in the Galaxy (Manabe and Miyamoto 1975, Sanders 1977). Such oval distortions are quite common in external galaxies. A feature like this can provide a driving force for a spiral pattern in the disk of the Galaxy. Is such a spiral pattern, together with the bar, self-consistent?

REFERENCES

Henderson, A. P.: 1977, Astron. Astrophys. 58, 189.
Lin, C. C., Yuan, C., Shu, F. H.: 1969, Astrophys. J. 155, 721.
Manabe, S., Miyamoto, M.: 1975, Pub. Astron. Soc. Japan, 27, 35.
Sanders, R. H.: 1977, IAU Colloq. No. 45, ed. Basinska-Grzesik, E., Mayor, M., 103.
Terzides, Ch. K.: 1977, ibid, 297.

INDEX

W. B. Burton (ed.), The Large-Scale Characteristics of the Galaxy, 603–611.